The Earth's Magnetic Interior

IAGA Special Sopron Book Series

Volume 1

Series Editor

Bengt Hultqvist
The Swedish Institute of Space Physics, Kiruna, Sweden

The International Association of Geomagnetism and Aeronomy is one of the eight Associations of the International Union of Geodesy and Geophysics (IUGG).

IAGA's Mission
The overall purpose of IAGA is set out in the first statute of the Association:

- to promote studies of magnetism and aeronomy of the Earth and other bodies of the solar system, and of the interplanetary medium and its interaction with these bodies, where such studies have international interest;
- to encourage research in these subjects by individual countries, institutions or persons and to facilitate its international coordination;
- to provide an opportunity on an international basis for discussion and publication of the results of the researches; and
- to promote appropriate standardizations of observational programs, data acquisition systems, data analysis and publication.

Volumes in this series:

The Earth's Magnetic Interior
Edited by E. Petrovský, E. Herrero-Bervera, T Harinarayana and D. Ivers

Aeronomy of the Earth's Atmosphere and Ionosphere
Edited by M.A. Abdu, D. Pancheva and A. Bhattacharyya

The Dynamic Magnetosphere
Edited by W. Liu and M. Fujimoto

The Sun, the Solar Wind, and the Heliosphere
Edited by M.P. Miralles and J. Sánchez Almeida

Geomagnetic Observations and Models
Edited by M. Mandea and M. Korte

For titles published in this series, go to
http://www.springer.com/series/8636

The Earth's Magnetic Interior

Editors

Eduard Petrovský

Institute of Geophysics AS CR, Prague, Czech Republic

Emilio Herrero-Bervera

SOEST-HIGP, University of Hawaii, Honolulu, HI, USA

T Harinarayana

National Geophysical Research Institute, Hyderabad, India

David Ivers

University of Sydney, Australia

Springer

planetearth Earth Sciences for Society

IAGA

Editors
Dr. Eduard Petrovský
Institute of Geophysics AS CR, v.v.i.
Bocni II/1401
141 31 Praha 4
Czech Republic
edp@ig.cas.cz

Dr. T Harinarayana
National Geophysical Research Institute
Uppal Road
Hyderabad 500007
India
thari54@yahoo.com

Dr. Emilio Herrero-Bervera
Paleomagnetics and Petrofabrics Laboratory
School of Ocean & Earth Science & Technology (SOEST)
Hawaii Institute of Geophysics and Planetology (HIGP)
1680 East West Road
Honolulu
Hawaii 96822
USA
herrero@soest.hawaii.edu

Dr. David Ivers
School of Mathematics & Statistics
University of Sydney
New South Wales 2006
Australia
david.ivers@sydney.edu.au

ISBN 978-94-017-7781-0 ISBN 978-94-007-0323-0 (eBook)
DOI 10.1007/978-94-007-0323-0
Springer Dordrecht Heidelberg London New York

Cover illustration: Background: 2-D resistivity model of Western Cordillera and the Southern Altiplano at 21S (Fig. 6 from Chapter 4 in this book); Front left: Archaeomagnetic directional secular variation curve for Italy plotted in an inclination-declination diagram (Bauer plot) (Fig. 9 from Chapter 15 in this book); Front right: Radial component of geomagnetic field as result of one of the geodynamo models based on rotating convection (unpublished figure, by Dr. Jan Simkanin).

Printed on acid-free paper

Springer is part of Springer Science+Business Media (www.springer.com)

Foreword by the Series Editor

The IAGA Executive Committee decided in 2008, at the invitation of Springer, to publish a series of books, which should present the status of the IAGA sciences at the time of the IAGA 2009 Scientific Assembly in Sopron, Hungary, the "IAGA Special Sopron Series". It consists of five books, one for each of the IAGA Divisions, which together cover the IAGA sciences:

Division I – Internal Magnetic Field
Division II – Aeronomic Phenomena
Division III – Magnetospheric Phenomena
Division IV – Solar Wind and Interplanetary Field
Division V – Geomagnetic Observatories, Surveys and Analyses.

The groups of Editors of the books contain members of the IAGA Executive Committee and of the leadership of the respective Division, with, for some of the books, one or a few additional leading scientists in the respective fields.

The IAGA Special Sopron Series of books are the first ever (or at least in many decades) with the ambition to present a full coverage of the present status of all the IAGA fields of the geophysical sciences. In order to achieve this goal each book contains a few "overview papers", which together summarize the knowledge of all parts of the respective field. These major review papers are complemented with invited reviews of special questions presented in Sopron. Finally, in some of the books a few short "contributed" papers of special interest are included. Thus, we hope the books will be of interest to both those who want a relatively concise presentation of the status of the sciences and to those who seek the most recent achievements.

I want to express my thanks to the editors and authors who have prepared the content of the books and to Petra van Steenbergen at Springer for good cooperation.

Kiruna, Sweden
October 2010

Bengt Hultqvist

Preface

Division I of the International Association of Geomagnetism and Aeronomy (IAGA) is probably the most variable among the five IAGA divisions. It consists of four working groups (WG), dealing with generation of the geomagnetic field and geodynamo modeling, electromagnetic studies of the Earth's interior, paleomagnetism and magnetic dating, and, finally, with rock and environmental magnetism. Many research teams, involved in the activities of these WGs, do not deal directly with the Earth's magnetic field. Their research has become interdisciplinary, combining knowledge of geology, physics, mathematics, etc., with applications aiming at geological as well as environmental problems.

In this volume, we present altogether 28 contributions coming from all four WGs of Division I. We aimed at demonstrating the variability and complexity of recent studies carried out within the IAGA community, combining several review papers and standard-like contributions. The first seven contributions represent studies of natural electromagnetic signals available near the Earth's surface. These can provide information on the distribution of electrical conductivity of rocks below the surface, from shallow to as deep as even sub-lithospheric depths. In the present volume, authors have addressed the applications of both natural as well as artificial electromagnetic signals to understand some important geological problems. The next set of contributions deals with geodynamo modeling and features of historical records of the geomagnetic field. The third set of contributions represents diverse rock-magnetic studies aimed at revealing the history of the geomagnetic field, magnetic dating of archeological artifacts or understanding present geomagnetic anomalies. These studies would not be possible without progress in terms of experimental facilities, data collection and processing. Three contributions provide reports on such developments, in particular in the field of measurements of anisotropy of magnetic susceptibility and remanent magnetization. Last but not least, five contributions represent environmentally oriented applications of rock magnetism, including paleoclimatic studies, as well as assessment of anthropogenic effects (pollution) on magneto-mineralogical population in various environments.

It was not our aim to compile a complete handbook on some specific subject of the Earth's magnetic field. Instead, we provide an overview of recent progress across all the variable and complex disciplines included in IAGA Division I.

Prague, Czech Republic	Eduard Petrovský
Honolulu, Hawaii, USA	Emilio Herrero-Bervera
Hyderabad, India	T Harinarayana
Sydney, Australia	David Ivers

Acknowledgements

The editors thank the contributing authors and reviewers who all helped greatly in the compilation of this volume. We also gratefully acknowledge the support given by our own institutions: the Institute of Geophysics ASCR, Prague, Czech Republic; the University of Hawaii at Manoa, USA; the National Geophysical Research Institute (CSIR), Hyderabad, India; and the School of Mathematics & Statistics at the University of Sydney, Australia. The work of E.P. within IAGA has been supported by the Ministry of Education, Youth and Sports of the Czech Republic through Project LA09015. E.H.-B. would like to thank the financial support from SOEST-HIGP of the University of Hawaii as well as the National Science Foundation grants JOI-T309A4, OCE-0727764, EAR-IF-0710571, EAR-1015329 and NSF EPSCoR Program.

Contents

Contributors

Gary Acton Department of Geology, University of California, Davis, CA 95616, USA, gdacton@ucdavis.edu

Harald Böhnel Centro de Geociencias, Universidad Nacional Autónoma de México, Querétaro 76230, México, hboehnel@geociencias.unam.mx

Pavel Bosák Institute of Geology AS CR, v.v.i., Rozvojova 269, 165 00 Praha 6, Czech Republic, bosak@gli.cas.cz

Heinrich Brasse Freie Universität Berlin, Fachrichtung Geophysik, 12249 Berlin, Germany, heinrich.brasse@fu-berlin.de

Laurie L. Brown Department of Geosciences, University of Massachusetts, Amherst, MA 01003, USA, lbrown@geo.umass.edu

Edgardo Cañón-Tapia Departamento de Geología, CICESE, Ensenada, BC 92143, Mexico; CICESE, Geology Department, PO Box 434843, San Diego, CA 92143, USA, ecanon@cicese.mx

Martin Chadima AGICO Inc., Ječná 29a, Brno, Czech Republic; Institute of Geology AS CR, v.v.i., Prague, Czech Republic, chadima@agico.cz

E. Chandrasekhar Department of Earth Sciences, Indian Institute of Technology Bombay, Powai, Mumbai 400076, India, esekhar@iitb.ac.in

Rosa Hilda Compagnucci Dpto. de Cs. Geológicas y Dpto. de Cs. de la Atmósfera, FCEyN-Universidad de Buenos Aires, Buenos Aires, Argentina; Consejo Nacional de Investigaciones Científicas CONICET, Buenos Aires, Argentina, rhc@at.fcen.uba.ar

Mark J. Dekkers Paleomagnetic Laboratory 'Fort Hoofddijk', Department of Earth Sciences, Utrecht University, Budapestlaan 17, 3584 CD Utrecht, The Netherlands, dekkers@geo.uu.nl

Miroslav Dušek AGICO Inc., Ječná 29a, Brno, Czech Republic, mdusek@agico.cz

Ramon Egli Department of Earth and Environmental Sciences, Ludwig-Maximilians University, 80333 Munich, Germany, egli@geophysik.uni-muenchen.de

Michael E. Evans Institute for Geophysical Research, University of Alberta, Edmonton, AB, Canada T6G 2G7, tedevans.evans403@gmail.com

Karl Fabian Norwegian Geological Survey, 7491 Trondheim, Norway, karl.fabian@ngu.no

Tomas Grygar Institute of Inorganic Chemistry AS CR, v.v.i., 250 68 Rez, Czech Republic, grygar@iic.cas.cz

T. Hanstein KMS Technologies, Houston, TX, USA, Tilman@kmstechnologies.com

T Harinarayana National Geophysical Research Institute (CSIR-NGRI), Hyderabad 500606, India, thari54@yahoo.com

Florian Heidelbach Bayerisches Geoinstitut, Universität Bayreuth, 95440 Bayreuth, Germany, florian.heidelbach@uni-bayreuth.de

Friedrich Heller Institut für Geophysik, ETH Zürich, CH-8092 Zürich, Switzerland, heller@mag.ig.erdw.ethz.ch

Emilio Herrero-Bervera Paleomagnetics and Petrofabrics Laboratory, School of Ocean & Earth Science & Technology (SOEST), Hawaii Institute of Geophysics and Planetology (HIGP), 1680 East West Road, Honolulu, Hawaii 96822, USA, herrero@soest.hawaii.edu

Ann M. Hirt Institute for Geophysics, ETH Zürich, 8092 Zürich, Switzerland, ann.hirt@erdw.ethz.ch

František Hrouda AGICO Inc., Ječná 29a, CZ-621 00 Brno, Czech Republic; Institute of Petrology and Structural Geology, Charles University, CZ-128 43 Praha, Czech Republic, fhrouda@agico.cz

Jozef Hus Centre de Physique du Globe, de l'Institut Royal Météorologique de Belgique, B-5670, Dourbes, Belgium, jhus@meteo.be

Diana Jordanova National Institute of Geophysics, Geodesy and Geography, BAS, 1113 Sofia, Bulgaria; Faculty of Physics, Sofia University "St. Kl. Ohridski", Sofia, Bulgaria, vanedi@geophys.bas.bg

Neli Jordanova National Institute of Geophysics, Geodesy and Geography, BAS, 1113 Sofia, Bulgaria; Institute of Geophysics BAS, 1112 Sofia, Bulgaria, neli_jordanova@hotmail.com

Masashi Kamogawa Department of Physics, Tokyo Gakugei University, Tokyo, Japan, kamogawa@u-gakugei.ac.jp

Aleš Kapička Institute of Geophysics AS CR, v.v.i., Bocni II/1401, 141 31 Prague 4, Czech Republic, kapicka@ig.cas.cz

Gunther Kletetschka NASA's Goddard Space Flight Center, Code 691, Greenbelt, MD, USA; Catholic University of America, Washington, DC, USA; Institute of Geology, Academy of Sciences of the Czech Republic, Prague, Czech Republic, gunther.kletetschka@gsfc.nasa.gov

David Krása European Research Council Executive Agency, B-1049 Brussels, Belgium, david.krasa@ec.europa.eu

A. Manglik National Geophysical Research Institute (CSIR-NGRI), Hyderabad 500606, India, ajay@ngri.res.in

Suzanne A. McEnroe Norwegian Geological Survey, N-7491 Trondheim, Norway, Suzanne.McEnroe@ngu.no

Toshiyasu Nagao Earthquake Prediction Research Center, Tokai University, Shizuoka, Japan, nagao@scc.u-tokai.ac.jp

Lars Petter Nilsson Norwegian Geological Survey, N-7491 Trondheim, Norway, Lars.Nilsson@ngu.no

María Julia Orgeira Dpto. de Cs. Geológicas, FCEyN-Universidad de Buenos Aires, Buenos Aires, Argentina; Consejo Nacional de Investigaciones Científicas CONICET, Buenos Aires, Argentina, orgeira@gl.fcen.uba.ar

William H. Peck Department of Geology, Colgate University, Hamilton, NY 13346, USA, wpeck@colgate.edu

Petar Petrov National Institute of Geophysics, Geodesy and Geography, BAS, 1113 Sofia, Bulgaria, petar.petrov76@abv.bg

Eduard Petrovský Institute of Geophysics AS CR, v.v.i., Bocni II/1401, 141 31 Prague 4, Czech Republic, edp@ig.cas.cz

Vilém Podrázský Czech University of Life Sciences, Kamýcká 129, 165 21 Prague 6, Czech Republic, podrazsky@fld.czu.cz

Jiří Pokorný AGICO Inc., Ječná 29a, CZ-621 00 Brno, Czech Republic, jpokorny@agico.cz

Petr Pokorný AGICO Inc., Ječná 29a, CZ-621 00 Brno, Czech Republic, petrpok@agico.cz

Petr Pruner Institute of Geology AS CR, v.v.i., Rozvojova 269, 165 00 Praha 6, Czech Republic, pruner@gli.cas.cz

V. Vijaya Rao National Geophysical Research Institute (CSIR-NGRI), Hyderabad 500606, India, vijayraov@yahoo.co.in

M. Irene B. Raposo Institute of Geosciences, São Paulo University, 05508-080, São Paulo, SP, Brazil, irene@usp.br

Paul H. Roberts Institute of Geophysics and Planetary Physics, University of California, Los Angeles, CA 90095, USA, roberts@math.ucla.edu

Peter Robinson Norwegian Geological Survey, 7491 Trondheim, Norway, peter.robinson@ngu.no

Sedelia Rodriguez Paleomagnetics and Petrofabrics Laboratory, SOEST-HIGP, University of Hawaii at Manoa, 1680 East West Road Honolulu, Hawaii 96822, USA, sedelia509@gmail.com

K. Sain National Geophysical Research Institute (CSIR-NGRI), Hyderabad 500606, India, kalachandsain@yahoo.com

Ataru Sakuraba Department of Earth and Planetary Science, School of Science, University of Tokyo, Tokyo 113–0033, Japan, sakuraba@eps.s.u-tokyo.ac.jp

Simo Spassov Section du Magnétisme Environnemental, Centre de Physique du Globe de l'Institut Royal Météorologique de Belgique, B-5670 Dourbes (Viroinval), Belgium, simo.spassov@meteo.be

Viacheslav V. Spichak Geoelectromagnetic Research Centre IPE RAS, Troitsk, Moscow Region, Russia, v.spichak@mail.ru

C.H. Stoyer KMS Technologies, Houston, TX, USA, charles@kmstechnologies.com

K.M. Strack KMS Technologies, Houston, TX, USA, kurt@kmstechnologies.com

Petr Suza AGICO Inc., Ječná 29a, CZ-621 00 Brno, Czech Republic, psuza@agico.cz

Evdokia Tema Dipartimento di Scienze della Terra, Università degli Studi di Torino, Via Valperga Caluso 35, 10125 Torino, Italy, evdokia.tema@unito.it

L.A. Thomsen KMS Technologies, Houston, TX, USA, Leon@kmstechnologies.com

Seiya Uyeda Tokai University, Earthquake Prediction Research Center, Shizuoka 4248610, Japan, suyeda@st.rim.or.jp

Jean-Pierre Valet Institut de Physique du Globe de Paris, 4 Place Jussieu, 75252 Paris Cedex 05, France, valet@ipgp.fr

S.K. Verma National Geophysical Research Institute (CSIR-NGRI), Hyderabad 500606, India, skvngri@gmail.com

Tilo von Dobeneck Fachbereich Geowissenschaften der Universität Bremen, D-28334 Bremen, Germany, dobeneck@uni-bremen.de

Leping Yue Department of Geology, Northwest University, 710 069 Xi'an, Shaanxi Province, China, yleping@nwu.edu.cn

Natural Signals to Map the Earth's Natural Resources

1

T Harinarayana

Abstract

Exploration and exploitation of natural resources is one of the main concerns of an earth scientist to benefit humanity. Any country's economic growth and development can be judged from the availability of natural resources. Electrical and electromagnetic methods are quite useful in the exploration of oil, mineral, water, etc., as they exhibit anomalous electrical conductivity compared to the surrounding media and thus can be identified with ease. In this chapter, the discussion is restricted to four themes – hydrocarbons, geothermal resources, water and mineral. Fair treatment is provided on the global scenario for each item and the use of the methodology with a case study is explained. Detection of hydrocarbons in a most difficult region – sub-basalt – is discussed by considering an example from Gujarat, India. Mapping of geothermal potential is discussed with an example from Korea. A case study of assessment of groundwater potential is provided considering the Parnaiba Basin area in Brazil. Application of 3D modelling is provided for McArthur Basin area, Canada, for exploration of uranium. In the treatment of all these problems, the chapter is mainly focused on the use of natural signals for deep electromagnetic method – magnetotellurics.

1.1 Introduction

The presence of natural resources in a country helps to increase its economy. Richness of a country can be judged indirectly from the study of its natural resources. Although there are various natural resources that include air, water, forest, animal resource, oil, mineral, gas and heat, in our present study we restrict ourselves to hydrocarbons, minerals, geothermal resources and groundwater, which are not easily available on the surface but need to be explored at the subsurface depths. Many of them are essential for our survival, while others are used to increase our comforts. Thus, identification and exploitation of the natural resources is of utmost importance. The level of the available technology, the type of economy and preferences of the culture in a given society are some of the important factors that demand its use. The presence of natural resource is much more important for the low- and medium-level developed countries. It

T Harinarayana (✉)
National Geophysical Research Institute (CSIR-NGRI),
Hyderabad 500606, India
e-mail: thari54@yahoo.com

E. Petrovský et al. (eds.), *The Earth's Magnetic Interior*, IAGA Special Sopron Book Series 1,
DOI 10.1007/978-94-007-0323-0_1, © Springer Science+Business Media B.V. 2011

is imperative to say that sustainable use of natural resources is necessary to reach higher levels of human development.

Among the various geophysical methods, electromagnetic method forms one of the important tools to explore natural resources (Meju 2002, Harinarayana 2008). The method is based on the electrical resistivity property of the earth. Most of the natural resources are known to be sources of low resistive regions compared to the surroundings and thus electromagnetic methods play a vital role in delineation of the resources. More details on the methodology follow.

1.2 Methodology

Application for finding the natural resources using shallow and deep electromagnetic methods is found to be effective in the majority of cases. This is mainly due to the strength of these methods in mapping the anomalous conductors. Natural resources like sedimentary regions, geothermal reservoirs, water-enriched aquifers and mineralized zones exhibit usually highly conductive compared to the surrounding media. This helps in identification of the resources with ease. Although hydrocarbons exhibit as thin resistors, they exist within a large conductive media in a sedimentary basin environment. Identification and mapping of large basins that possess conducting sediments helps to locate the hydrocarbons.

Among the deep electromagnetic methods magnetotellurics is known to be most effective method. The method, since its inception in the 1950s, had a phenomenal growth, both in development of the methodology and also in the capability of the method for solving different geological problems. In the following, brief details are provided on theoretical development and its application along with major results deduced to delineate the resource. Attempts have been made to describe the significant results.

1.2.1 Theoretical Developments of the Method

Magnetotelluric methods depend on electromagnetic (EM) theory originating from four fundamental equations proposed by James Clerk Maxwell. These equations are from the well-known principles of Ampere's and Faraday's laws related to electric and magnetic fields. They are further analysed by Tikhonov (1950) and Cagniard (1953) who proposed a relation between the resistivity of a medium and electromagnetic field variations.

These relations gave birth to a new geophysical method, namely 'magnetotellurics (MT)'. The fundamental relations in MT can be written as

$$\rho_a = 0.2\ \mathrm{T}\ |Z|^2 \tag{1.1}$$

$$\Phi_{.} = \tan^{-1}(Z) \tag{1.2}$$

where

$$Z = E/H$$

ρ_a = apparent resistivity in Ωm, E = electric field variation in mV/km, H = magnetic field variation in nT and $\Phi_{.}$ phase in degrees.

These scalar equations are modified with the introduction of tensor concepts (Rokityanski 1961, Wait 1962). This means that the induced (electric or telluric) field in x direction can be related to the variations of magnetic fields in both x and y directions. The relations can thus be expressed as

$$Ex = Zxx\,Hx + Zxy\,Hy \tag{1.3}$$

$$Ey = Zyx\,Hx + Zyy\,Hy \tag{1.4}$$

where Zxx, Zxy, Zyx and Zyy are called impedance elements relating electric and magnetic fields. They contain vital information on the resistivity parameter, the dimensionality (1D, 2D or 3D) and directionality (geoelectric strike direction) of the subsurface. Their estimation is therefore the main task in MT data processing.

These equations are valid on the assumption that the incoming natural EM field is a plane wave. The validity of plane wave assumption was questioned by Wait (1954), which was resolved by Price (1962) and Madden and Nelson (1964). Later the use of natural variations of EM field of micropulsations that was limited in the frequency range, from 1s to a few thousand seconds, had been extended to audio frequencies and higher, a few kilohertz. These changes in methodology and revolutionary concepts in digital instrumentation and microprocessor-based computers paved the way for both exploration problems and deep crust–mantle studies. The extension of the method to audio frequency magnetotelluric (AMT) technique was introduced during the mid-1970s (Strangway et al. 1973, Hoover et al. 1975). The AMT method uses the signals originating from worldwide thunderstorm activity. This means that in order to know the information from near-surface to deep crust–mantle depths, one needs to acquire the data in the range from a few kilohertz to a few thousands of seconds (i.e. 10^4 Hz–10^{-4} Hz covering eight decades on a log scale). Initially, major problems were faced by the data acquisition systems due to extremely low amplitudes of the variations of natural electric and magnetic fields which are in the range of a few tens of microvolts per kilometre and a few tens of picotesla, respectively. As mentioned before, with innovative approaches in amplifying the signals by instrumentation and also with the help of digital microprocessors these problems were solved. Since the method depends on the natural source fields, the cultural noise posed a major problem to get good quality data. Clarke et al. (1983) have introduced remote reference techniques, in which the noise at a field site can considerably be reduced by cross-correlation of the signals from the base site located at a relatively noise-free location, provided the data are recorded simultaneously. Decomposition techniques are proposed by Groom and Bailey (1991) to reduce local near-surface distortions on the data. Static shift effects can distort the data. Several procedures are suggested to eliminate the distortions (Jones 1988, Pellerin and Hohmann 1990, Ogawa and Uchida 1996, Harinarayana 1999).

The method has undergone phenomenal growth in different sectors such as data acquisition, i.e. developments in instrumentation and recording of the data, and processing procedures using robust methods (Egbert and Booker 1986), i.e. computation of earth response functions from time-varying electric and magnetic fields and modelling schemes in 1D (Marquardt 1963, Constable et al. 1987), 2D (Wannamaker et al. 1986, Rodi and Mackie 2001) and 3D (Smith and Booker 1991, Santos et al. 2002), to derive the subsurface parameters of variation of resistivity as a function of depth from earth response functions. In recent years, the data for a remote location can be viewed online using Internet protocol file transfer facility. This is another milestone in data acquisition procedures.

1.3 Natural Resources

In the following, brief details on the global natural resources, it's occurrences with a case study on the application of MT, and the exploration procedures to delineate the resources are explained.

1.3.1 Hydrocarbons

Hydrocarbons naturally occur at subsurface depths, usually at 1–5 km, due to the processes related to thermogenesis – thermal decomposition of fossils and other organic matter, and bacteriogenesis – microbial activity on the organic matter (Sherwood Lollar et al. 2002). In the energy sector, although various forms of generation of energy such as hydro, thermal, nuclear, and renewable are well established, energy usage from hydrocarbons is still dominant. From the estimated world total energy, approximately 63% consumption during 2008 came from hydrocarbons alone and thus formed a single dominant resource in the energy sector. Search for locating the oil and oil structures is a continuous process and it is estimated that over 1600 billion barrels oil equivalent still need to be located in different parts of the world. Figure 1.1 shows the estimation of the distribution of undiscovered oil and gas areas in different parts of the world. The U.S. Geological Survey periodically estimates the amount of oil and gas remaining to be found in the world (USGS report 2000). These results have important implications for energy prices, policy, security, exploration programmes and the global resource balance.

It is clear that there is still huge potential available in the former Soviet Union and not much known in the Asia Pacific and South Asia regions. This demands more focus in these regions to identify the hydrocarbon resource. Table 1.1 (USGS 2000) provides information on the estimated amount of oil and gas and their percentage in a global scenario.

The hydrocarbons after their generation flow as fluid in a porous media of sedimentary strata and get trapped at suitable geological structures, for example, faults, anticlinal and synclinal folds. From a geophysical point of view, mapping of these structures is of utmost importance to locate the occurrence of

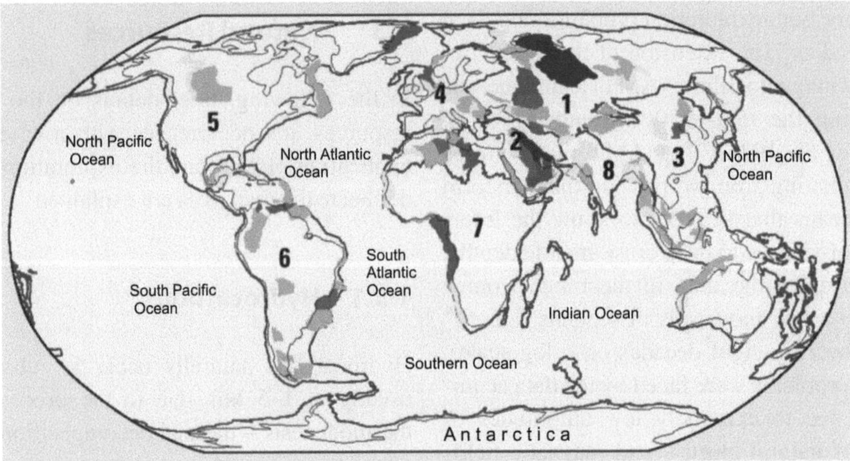

Fig. 1.1 Global assessment of the location of undiscovered oil potential in different parts of the world (modified from USGS 2000) (See more details in Table 1.1)

Table 1.1 Volumes of undiscovered oil and natural gas for different regions of the world, including percentages of world total (USGS 2000)

S. No.	Region	Oil (billion barrels)	Percent of world total	Natural gas (billion barrels of oil equivalent)	Natural gas (trillion cubic feet)	Percent of world total
1	Former Soviet Union	116	17.9	269	1611	34.5
2	Middle East and North Africa	230	35.4	228	1370	29.3
3	Asia Pacific	30	4.6	63	379	8.1
4	Europe	22	3.4	52	312	6.7
5	North America	70	10.9	26	155	3.3
6	Central and South America	105	16.2	81	487	10.4
7	Sub-Saharan Africa and Antarctica	72	11.0	39	235	5.0
8	South Asia	4	0.6	20	120	2.6
9	Grand total	649		778	4669	

Values are mean estimates and are exclusive of the United States

hydrocarbons in sedimentary formations. Its physical property – highly resistive – is distinct from its surroundings. Thus the first target for a geophysicist is to delineate the conductive sedimentary formation and map the structures within the sedimentary formation. As explained earlier, deep electromagnetic methods – magnetotellurics – are highly sensitive to map the conductive formations. Prior to MT study in Saurashtra, India, the region was considered as low priority from the point of view of oil occurrence. Its importance has been upgraded after the delineation of hidden sedimentary formation below the trap cover. In the following, the dependence of the oil industry on deep electromagnetic methods to delineate subtrappean sediments is presented considering a case study from Saurashtra region, Gujarat, India.

1.3.1.1 Oil Exploration – An Example from Saurashtra, Gujarat, India

The presence of hydrocarbons in the sediments (Mesozoics) buried below the volcanic rock covered area, limitations of seismic methods to delineate a low-velocity middle layer and effective mapping of such layer by MT are some of the factors that encouraged the Indian Oil Industry to support MT for exploration programmes. Due to large resistivity contrast between volcanic rock and the buried sediments, MT has proven to be superior compared to other geophysical methods as demonstrated in India and also at many locations around the world. Prior to MT survey in Saurashtra, the geological model from palaeo river channel study indicated gradual thickening of sediments towards the south from the exposed Dhrangadhra and Wadhwan sediments towards the north near Chotila. MT studies carried out by the National Geophysical Research Institute on an experimental basis were initiated along two profiles – one in NS and the other in EW direction (Sarma et al. 1992). Contrary to the earlier geological concepts, MT study has inferred thin sediments towards the south and thick sediments towards the northwestern part. This result has changed the whole concept of the basin structure. This new information encouraged the oil industry to verify the results and venture into a deep drilling exercise (~3.5 km) near Lodhika. The drilling result showed the existence of 2 km thick buried sediments below 1.5 km thick Deccan traps. Validation of MT interpretation with deep drilling (~3.5 km) at Lodhika, Saurashtra (Fig. 1.2), and also at Latur, Maharashtra (Gupta and Dwivedi, 1996), has proved the efficacy of the MT method in the exploration of subtrappean sediments for oil exploration. Due to this, Saurashtra region, which was considered as a low-priority region became important for oil exploration. Later on, with the aid from oil industry, the whole of Saurashtra region was covered with 600 stations along with other geophysical surveys as an integrated approach.

The results show thick sediments toward the northwestern part of Saurashtra below the trap cover as shown in Fig. 1.3. The success story of Saurashtra region has paved the way to investigate other major trap-covered regions like Kutch, Narmada-Cambay and Narmada-Tapti rift regions by integrated geophysical studies along with electromagnetic methods involving gravity, seismics, deep resistivity and MT surveys (Sarma et al. 1998, Harinarayana et al. 2003).

1.3.2 Geothermal Resources

Among the various forms of renewable energy sources such as solar, wind, geothermal and biomass, the use of geothermal energy is not being exploited to its fullest potential in developing countries. This is mainly due to the lack of proper investigation to locate the resources and technology to use low-temperature fields. Due to recent awareness on global warming, this energy sector is becoming more important now. A few articles on the use of electrical and electromagnetic techniques for geothermal and volcanoes is provided in a special issue (Harinarayana and Zlotnicki 2006).

If we study the geothermal power plants in the world (Fig. 1.4), it can be observed that only a limited number of locations are being exploited. This is not due to the reasons that other regions are not suitable. In reality, it is mainly due to lack of information about the subsurface structure suitable for geothermal potential. For example, let us take the case of India. No geothermal power plant is found in the whole country. In the high Himalayan regions, the only source of uninterrupted power supply may be geothermal. Conventional power plants face many problems in this region such as transportation problem of the fuels for thermal power plants and freezing temperature problem for hydroelectric

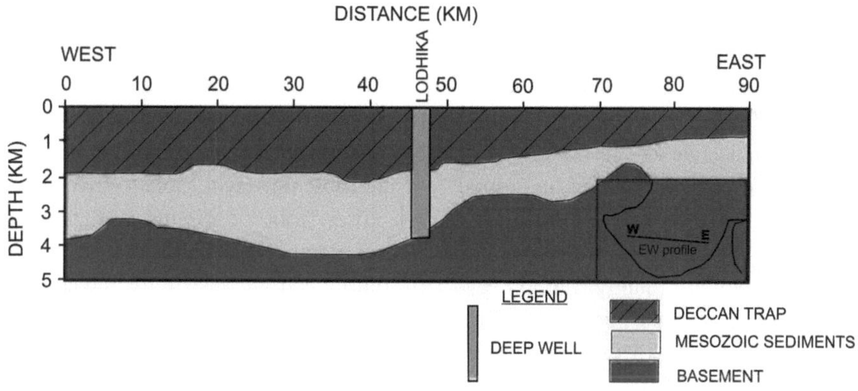

Fig. 1.2 Subsurface section showing thick sediments towards the western part of Saurashtra. This is a breakthrough result which has opened up a new scenario for the magnetotelluric studies in India for oil exploration

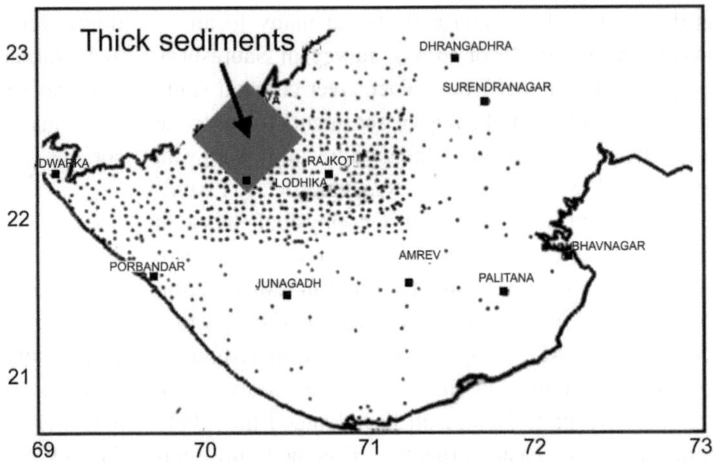

Fig. 1.3 Location map showing the distribution of 688 stations (*black dots*). Thick sediments buried below 1.5 km thick trap cover are indicated towards the NW quadrant of Saurashtra (indicated by a *square* in the figure) mainly from magnetotelluric study along with other geophysical methods as part of integrated studies

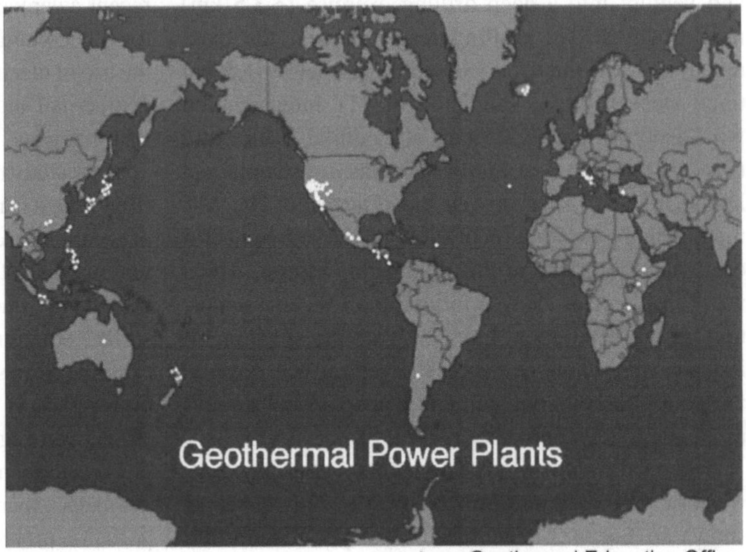

courtesy: Geothermal Education Office
Source : http://geothermal.marin.org/GFOpresentation/sld069.htm

Fig. 1.4 Distribution of power plants based on geothermal resource in different parts of the world

power plants. In applied geophysics scientific objectives, it is rather rare to get benefit of deep borehole data to get in situ information or to get enough budget to carry out scientific drillings. Therefore, exploration of either the hydrothermal state or the structure of geothermally active regions is mainly based on ground investigations.

In geothermally active regions as on volcanoes, the structure and the topography are mainly 3D and several co-existing factors such as faults, magma reservoir or groundwater fluids drastically disturb inhomogeneously the electrical resistivity with depth. These difficulties, which blurred the EM tomography, are now partly solved by the large development of new EM methods based on advanced equipments and supported by outstanding progress in modelling.

A multi-parametric approach combining several techniques of different penetration depths is now commonly applied, which makes the EM methods one of the most powerful tools to investigate complicated structures from the ground surface to several kilometres of depth. The shallow structure is well-imaged by multielectrode resistivity meter, up to some hundreds of metres in depth. Depending on the electrode spacing and the power of the current injection system into the ground, spatial details of the resistivity from several tens of centimetres to hundreds of metres can now be analysed with a depth resolution reaching 10 m or so. The flexibility of state-of-the-art devices allows to perform 3D surveys on the field. In geothermally active regions, the groundwater and more generally the fluid flow are the sources, which control the hydrothermal activity. Recent studies reveal that self-potential (SP) anomalies (up to some hundreds of millivolts) are also observed on volcanoes, active fissure zones and/or fumarolic areas, suggesting that the SP anomalies are closely related to the heat-triggered phenomena such as thermoelectric and electrokinetic effects due to hydrothermal circulations. Information on the hydrothermal state can be obtained by drilling the first several hundred meters below the surface. The preceding techniques become less powerful as the investigation on the structure requires a larger depth penetration. The well-known audiomagnetotelluric soundings fill the gap. It has made the most advanced progress in the last few years. The benefit of very low noise magnetic sensors associated with accurate data acquisition and filtering systems and the modelling of 3D structures is now achieved by several international EM groups. The frequency domain of the natural EM field in the high-frequency band (several kilohertz to 1 kHz) is enlarged in order to get information on the shallow structure and deep structure with low-frequency band (1 kHz to 0.001 Hz).

1.3.2.1 Geothermal Studies: A Case Study from Korea

As mentioned earlier, since MT depends on the earth's natural EM signals, it is not easy to acquire good data in all the countries. Korea is one such difficult country. This is mainly due to crisscrossing of the major power lines covering most of the regions. Additionally, the geothermal study region is covered by old sedimentary rocks and granite rocks. These rock types generally possess higher resistivities. This facilitates the EM noise from the power lines to travel large distances disturbing the MT measurement signals. In addition to these disturbances, the presence of DC-operated trains and HVDC (high-voltage direct current) power transmission distorts the signal with step-like noises, which have a wide frequency spectrum. However, Uchida et al. (2004) made concerted efforts to acquire fairly good-quality data in Pohang geothermal area (Fig. 1.5) using remote reference (Gamble et al. 1978) technique. Another magnetotelluric survey applied to geothermal exploration at Seokmo Island, Korea (Lee et al. 2010). In the following, the details of MT study of Pohang geothermal area are presented.

Uchida et al. (2004) have made detailed MT survey in 2002 and 2003 in Pohang low-enthalpy geothermal area (Fig. 1.5) of southeastern Korea. It is a joint

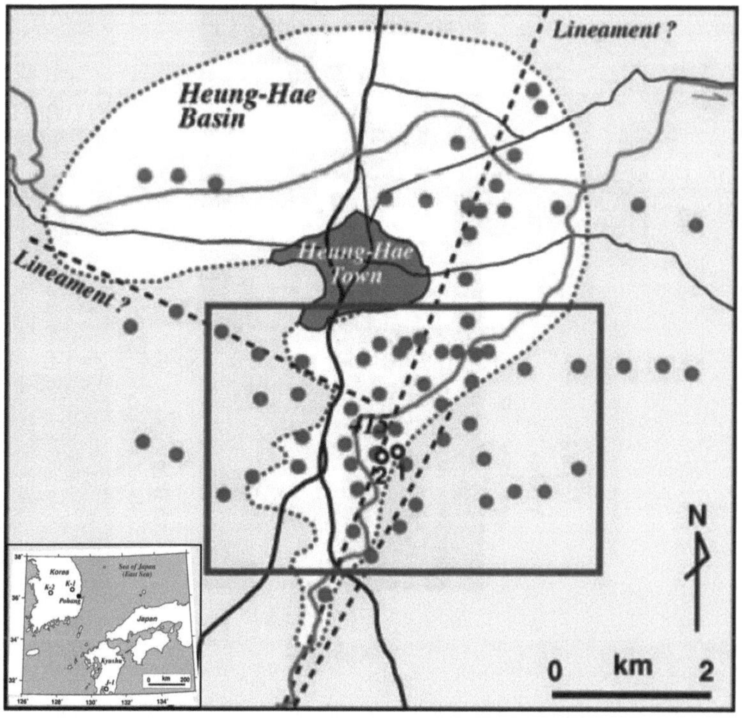

Fig. 1.5 Location of MT stations in the Pohang low-enthalpy geothermal area (*solid circle*), southeastern Korea. *Open circles* are remote reference stations in Korea (K-1 and K-2) and Japan (J-1) (from Uchida et al. 2004)

project between the Korea Institute of Geoscience and Mineral Resources (KIGAM) and the Institute of Geo-resources and Environment and Geological Survey of Japan (GSJ). Remote stations are deployed in Korea as well as in Kyushu, Japan. Use of these remote stations, which act like a base station, has helped to improve the data quality to a considerable extent. Interpretation of the data using 3D modelling has helped to identify the geothermal resources of Pohang (Lee et al. 2004).

The survey was initiated with a dense network of MT stations located in a near-grid fashion (Fig. 1.5) around the borehole. Interpretation of the data has been carried out with 44 MT sites and 11 data points between the 0.063 and 66 Hz frequency range using 3D inversion. A finite difference scheme is used for forward computation and the linearized least squares inversion with regularization for inversion algorithm (Sasaki 1989, Uchida and Sasaki 2003, Sasaki 2004). As the study area in Korea is close to the sea towards the eastern side, a 100 m thick shallow layer with 0.3 Ωm layer representing the sea water is incorporated in their model. Noise floor of 1% is assumed to tackle the static shift effects on the data. Figure 1.6 (modified from Uchida et al. 2004) shows the 3D geoelectric section in the form of depth slices. A thin (200–300 m) low resistive layer is seen towards the north and gradually thickens towards the south and southcentral portions. Below this near-surface layer, a

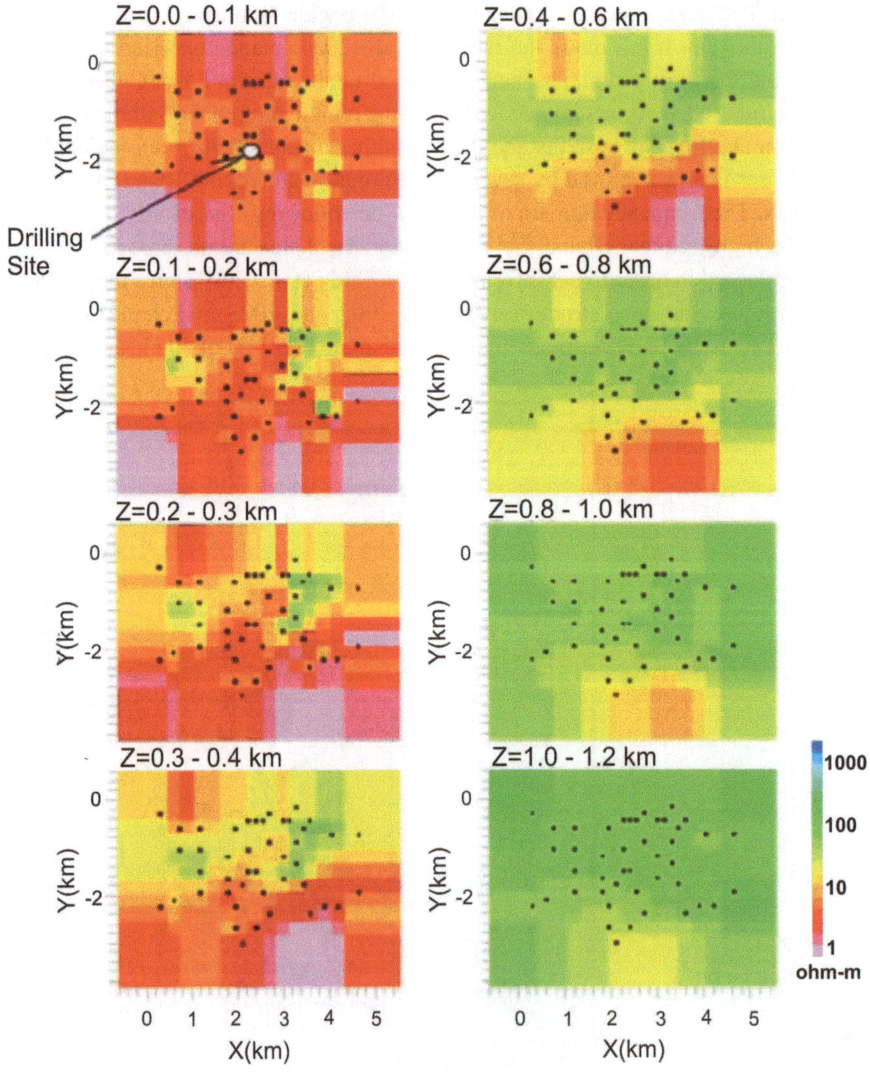

Fig. 1.6 Areal distribution of resistivity at different depths from 3D geoelectric section (from Uchida et al. 2004)

high resistive layer (100 Ωm) is observed to a depth of about 3 km, below which a low resistive layer is delineated towards the southern part.

There are two deep (1500 m) boreholes – BH-1 and BH-2 – in the study area with resistivity log information. From this information, geological formations are noted with 352 m of semi-consolidate mudstone; 173 m of basalt, tuff and Cretaceous mudstone; intruded rhyolite of Tertiary Period with a thickness of 475 m; followed by sandstone and mudstone of Cretaceous Period to the bottom of the hole.

The 2D modelling exercise has also been conducted for 5 EW traverses for the same data set. It is observed that 2D model resistivities derived for the deep high resistivity layer are lower than the observed borehole logging data. From these observations, modelling using 3D algorithm seems to be more reliable. From comparison of the borehole data it is seen that the high resistive rhyolite layer from the borehole data could not be seen clearly from the derived 3D model but clearly demarcates the lower boundary of the exposed low-resistive semi-consolidated mudstone. High-resistive rhyolite and deep sandstone/mudstone are interpreted as a single high-resistive layer (100–800 Ωm) in the 3D model. The bottom hole temperature observed for the borehole is approximately 70°C. Based on these results a production well of 2 km depth is planned and the expected temperature may go up to 80°C in the deep hole. Further steps include preparation of circulation system design, injection well design and pipelines to residential areas and nearby areas.

1.3.3 Water Resources

Groundwater forms one of the important resources of mankind. It is a renewable source as this is being replenished through annual precipitation from rainfall. Many developing countries like Bahrain, Taiwan, Mauritius, Malta, Pakistan, Bangladesh, India and Israel depend heavily on this resource. Large variations of groundwater are observed in terms of its occurrence, renewal rate and storage capacity. Storage of groundwater mainly depends on the geological setting of the surrounding region. Large sedimentary basins can store groundwater both at shallow depths in fractured rocks and also at deeper level as large aquifers.

Figure 1.7 (Vrba and Gun 2004) shows the assessment of groundwater resource distribution in the world. It shows major river basins; freshwater lakes; major, medium and low recharge areas, etc. The world is divided into 36 global groundwater regions and their predominant hydrogeological setting can be classified into four categories: basement, sedimentary basin, high-relief folded mountains and volcanic regions. The water stored beneath the earth surface is an order large in magnitude (96%) compared to all the earth's unfrozen freshwater. This demands for a concerted geophysical approach to map its resources. While moving in the underground it dissolves certain salts and becomes highly conductive compared to the surroundings and forms a good target to detect through electromagnetic and electrical methods.

In the following, a case study for identifying the deep aquifer zones from magnetotellurics along with aeromagnetics is provided.

1.3.3.1 Assessment of Groundwater Resource, a Case Study from Parnaiba Basin, NE Brazil

A large lineament that spans about 2700 km across Brazil in the NE direction connects the features from Paraguay and Argentina developed during the Brazilian Orogenic cycle. Later due to subsidence, three major intracratonic sedimentary basins have formed, namely the Amazon in the NW, the Parnaiba in the NE and the Parana in the South (de Brito Neves 1991). In the present case study part of the Parnaiba Basin is considered for groundwater exploration. The basin measures about 1000 km in NW and 800 km in SE direction with an oval shape. Figures 1.8 and 1.10 show the structural features of the basin (after de Sousa 1996). The study region consists of Palaeozoic sediments of Silurian and Devonian groups with lithology consisting of alternate sequences of thin layers of shales.

Towards the eastern part of the basin, Meju and Fontes (1996) and Meju et al. (1999) have described joint analyses of resistivity sounding, TEM and AMT for groundwater assessment.

Figure 1.9 shows the derived geoelectric section from MT/AMT data along a NW–SE profile (Fig. 1.10). The model is derived considering the data in two modes – TE and TM. From the 2D section presented in Fig. 1.9 it is seen that the average thickness of conductive sedimentary formation is about 1–2 km above a resistive basement (Chandrasekhar et al. 2009). The basement comprises of granite, granite

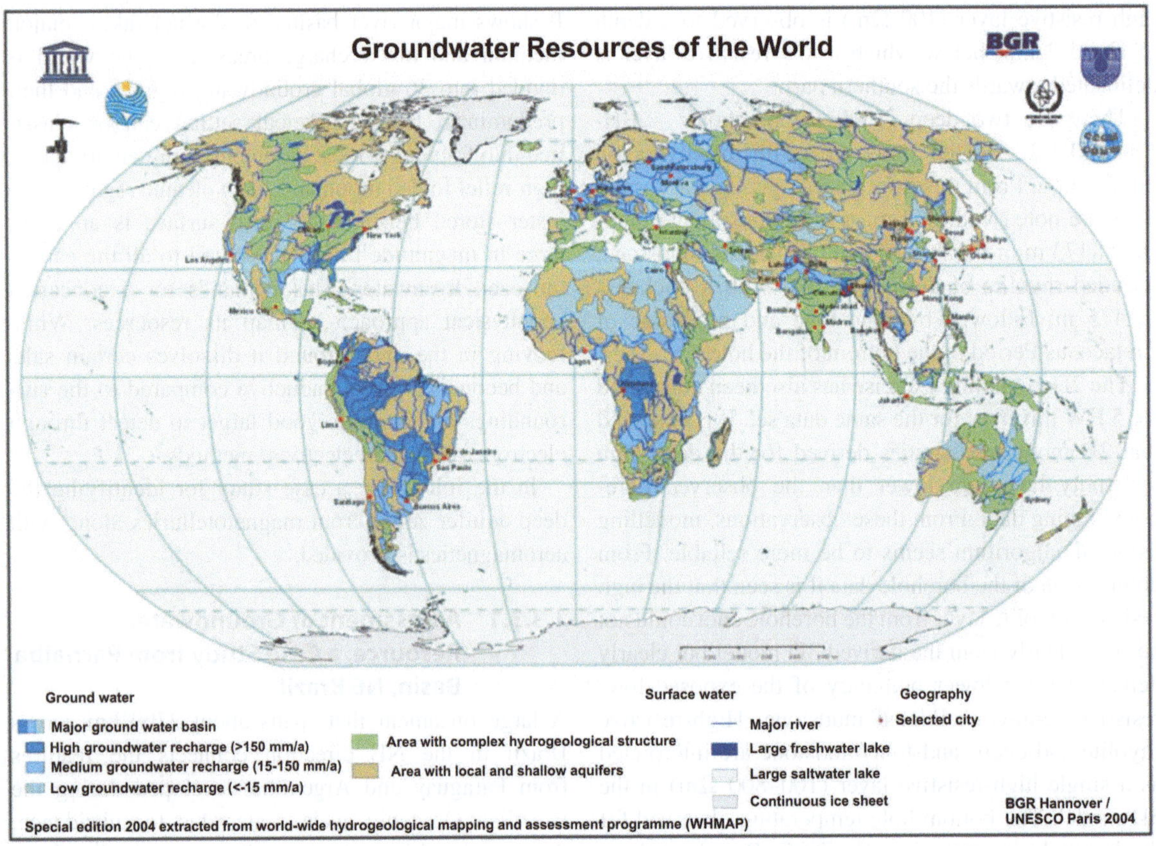

Fig. 1.7 Groundwater resource basin distribution in the world showing major, high, medium and low recharge areas (Source: BGR Hannover, Vrba and Gun 2004)

gneisses and schists. The lithology data along with well log resistivity data (after Meju et al. 1999) can also be seen for a location towards the NW part of the profile. Two anomalous conductive zones have been delineated at a depth of about 1 km below the surface. They probably represent highly porous but poor permeable shale formation. The lithology data, resistivity log from the borehole and also the anomalous conductors delineated from the present study strongly support and strengthen the conclusions as derived above.

From the inferences drawn using both magnetotelluric modelling and aeromagnetic data analysis deep drilling towards the NW end of the profile between the stations gb03 and gb13 is recommended. At this location, the sediments are exposed and it is far away from the high resistive Senador Pompeu Lineament (SPL). Additionally, it is located near the Vioeleta well, where large groundwater yield (900,000 L/h) is reported. Although another anomalous conductor is delineated towards the SE part of the profile, the location is not recommended for drilling because it is close to SPL where high resistive basement rocks are expected and may likely pose a problem during drilling.

Thus the natural source magnetotelluric results in combination with other geophysical methods, in the present example, airborne magnetics, can provide vital information for drilling sites for tapping the deep groundwater resource.

1.3.4 Mineral Resources

The world mineral map (Fig. 1.11) shows availability of minerals in the world (courtesy Maps of the world). Minerals marked on the mineral map of the world are uranium, silver, oil, lead and zinc, iron, diamond, bauxite, coal, copper and gold. As mentioned earlier, mineral reserves are important for the economy of every country as the products of the mineral mining industry are used as inputs for consumer goods,

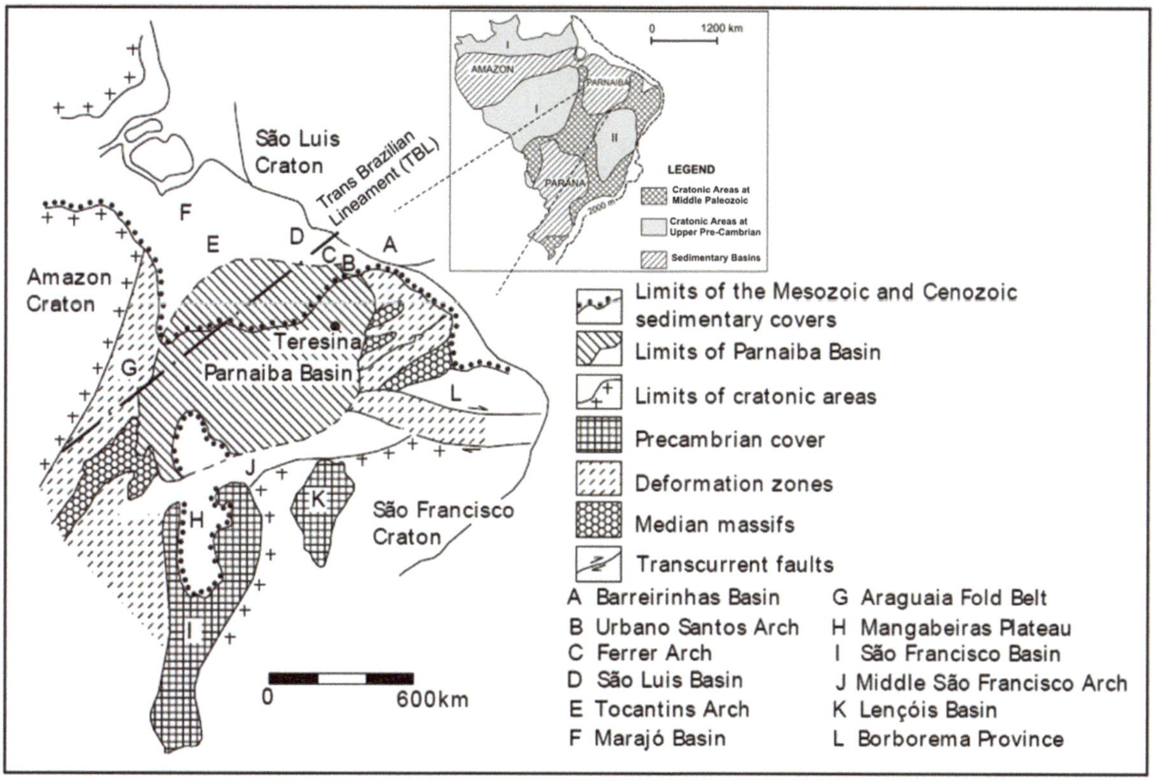

Fig. 1.8 Major geological units of Parnaiba Basin, Brazil (after De Sousa 1996)

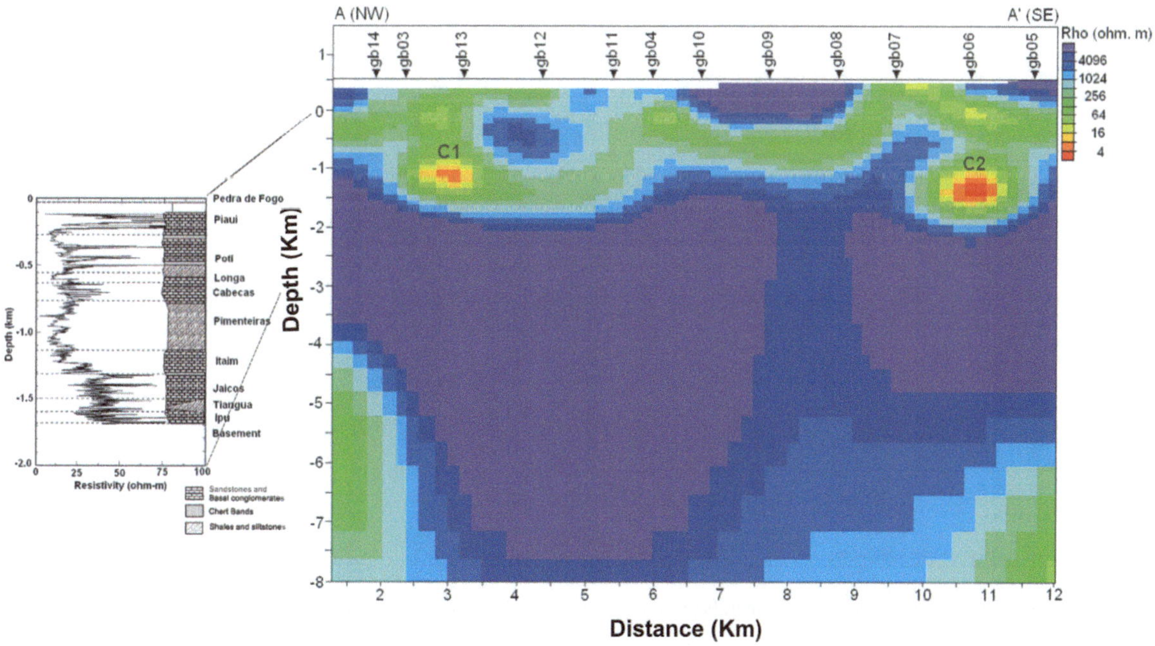

Fig. 1.9 Derived geoelectric section from MT/AMT data along a NW–SE profile (from Chandrasekhar et al. 2009)

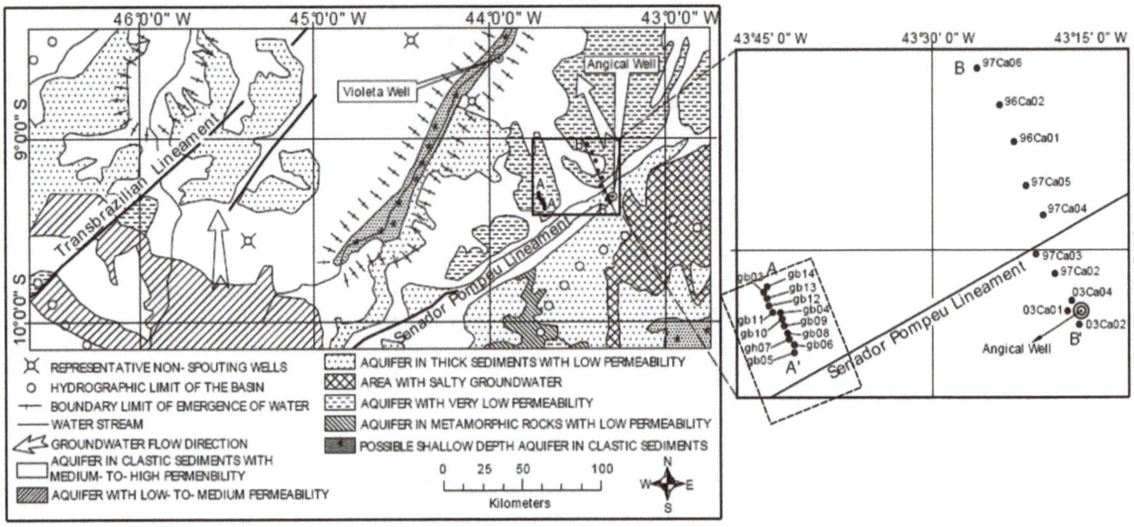

Fig. 1.10 Major geohydrological features in the study area and the location map of MT/AMT stations (from Chandrasekhar et al. 2009)

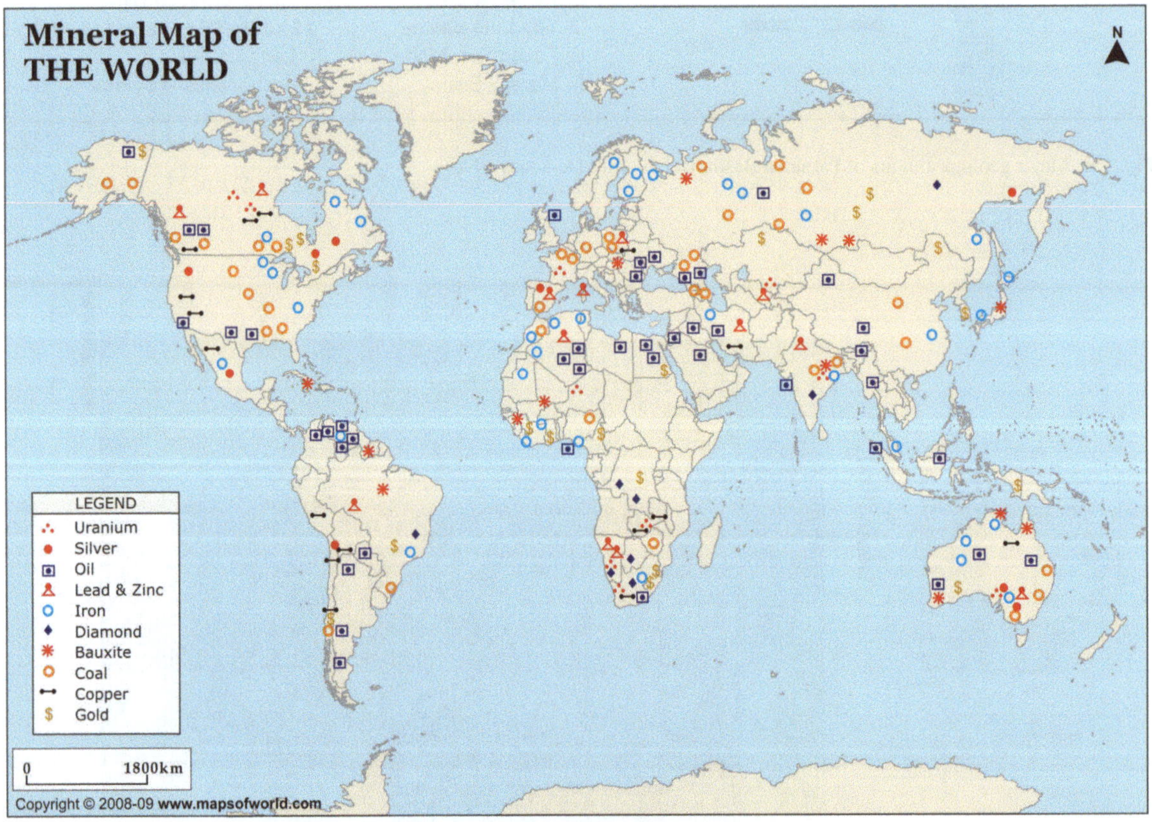

Fig. 1.11 Distribution of major economic minerals presently mapped in different parts of the world (Courtesy: www.mapsofworld.com)

services, etc. If we analyse the distribution of major minerals in different countries, Australia is the continent with the world's largest uranium reserves. Canada is the largest exporter of uranium ore. Mexico is the largest silver exporter. Earlier, diamonds were found only in alluvial deposits of southern India. At present diamond deposits are also found in Africa, South Africa, Namibia, Botswana, Tanzania and Congo. The most common element found on the earth is iron. The five largest producers of iron ore are China, Brazil, Australia, Russia and India. These five countries account for about 70% of the world's iron ore production. Huge quantity of gold comes from South Africa. Large deposits of copper ore can be found in Chile, Mexico, the United States, Indonesia, Australia, Peru, Russia, Canada, China, Poland and Kazakhstan.

Apart from other geophysical methods, electrical, electromagnetic, induced polarization and self-potential methods play a vital role to locate various mineral deposits. In the following, the use of magnetotellurics to locate one of the important minerals – uranium – is presented.

1.3.4.1 Exploration for Uranium in McArthur River Mine, Athabasca Basin, Canada

The recent trend in the mining industry is to explore deep targets. This is mainly due to the search for new sources for minerals as the easily detectable shallow deposits have become rather scarce. As a part of Extech IV project (Jefferson et al. 2003), different exploration methods are applied for delineating the deep uranium deposits of McArthur river mine of Athabasca Basin, Canada.

Uranium deposits the world over are often closely associated with highly conducting graphite deposits. At Athabasca Basin, the uranium deposits are typically found close to the locations of steeply dipping graphitic fault zones in the basement which in turn formed as unconformity between the sandstones and the basement. The depth range for the unconformity varies roughly from 100 m at known deposits and extends to about 1500 m near the centre of the basin. Delineation of the association of uranium deposits with graphitic zones – which are highly conductive relative to the high resistive metamorphic basement rocks –

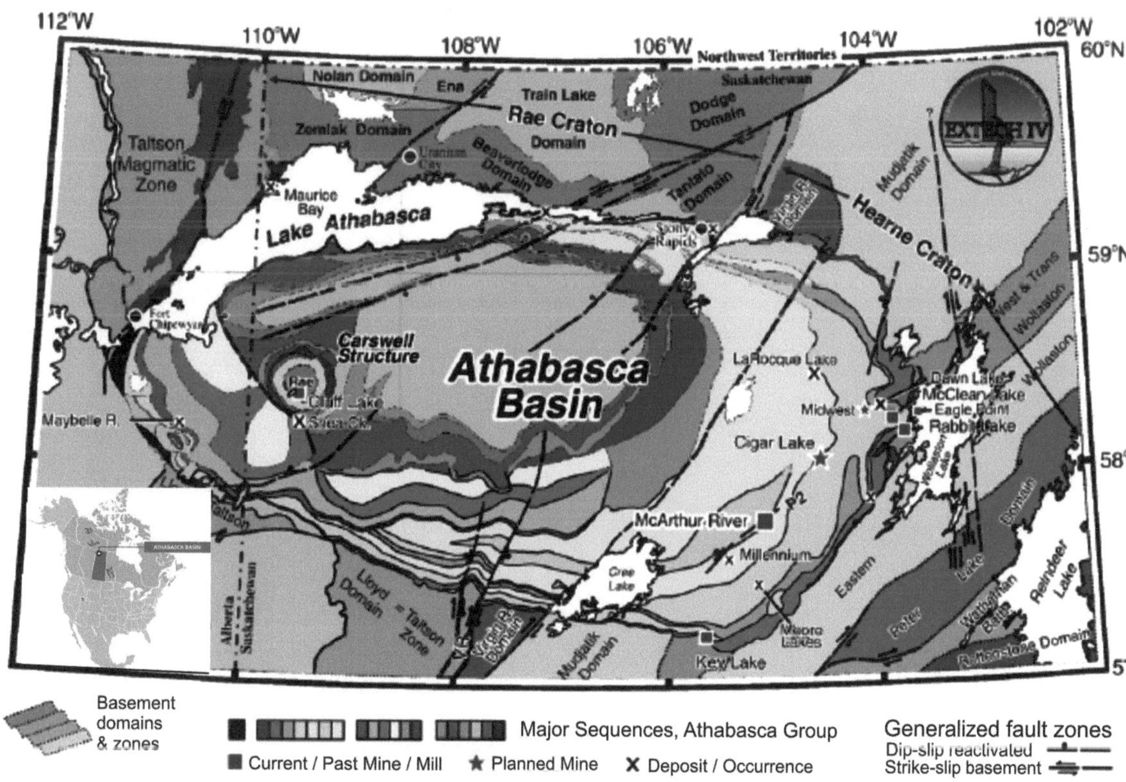

Fig. 1.12 Major geological units of Athabasca Basin (after Jefferson et al. 2003)

is a favourable zone to apply deep electromagnetic methods and also the newly developed time domain airborne electromagnetics to explore further deposits, especially around the edges of the basin (Cristall and Brisbin 2006, Craven et al. 2003, Leppin and Goldak 2005). Farquharson and Craven (2009) have made 3D modelling to locate anomalous conductive zones in Athabasca Basin. Details of their work are briefly presented here.

The study area (Fig. 1.12) is covered with 135 AMT stations along 11 lines (Craven et al. 2003). The survey is planned with a close interval of about 300 m with a separation of 800 m between the 11 lines (Fig. 1.13). The lines of orientation are in a NW–SE direction which is approximately perpendicular to the general strike of the graphitic basement fault zone. Remote reference processing was used to get high-quality response functions. For inversion, 11 frequencies are used for the data from 3.8 to 1280 Hz. Inversion process was carried with an initial model of uniform half-space of 10^{-4} S/m. Inversion results are presented (Fig. 1.6, Farquharson and Craven 2009) in the form of conductivity layers at different depths – 0, 232, 572, 1069, 1797 and 2864 m. Distribution of conductivity structure at a depth of 1797 m is reproduced in Fig. 1.14. Among the results derived, the most impressive is the delineation of an elongated conductive zone at 700–1500 m depth range. It is evident under the central and southern part of the study area and more prominently under the northern extremities. The anomalous conductive structure is related to the graphitic basement fault, at which McArthur river uranium deposit occurs. The 3D modelling results are also presented in the form of slices (Fig. 7, Farquharson and Craven 2009). The southward dipping signature of graphitic basement fault is clear. The anomalous conductive features at the northern part of the area represent another basement fault. In

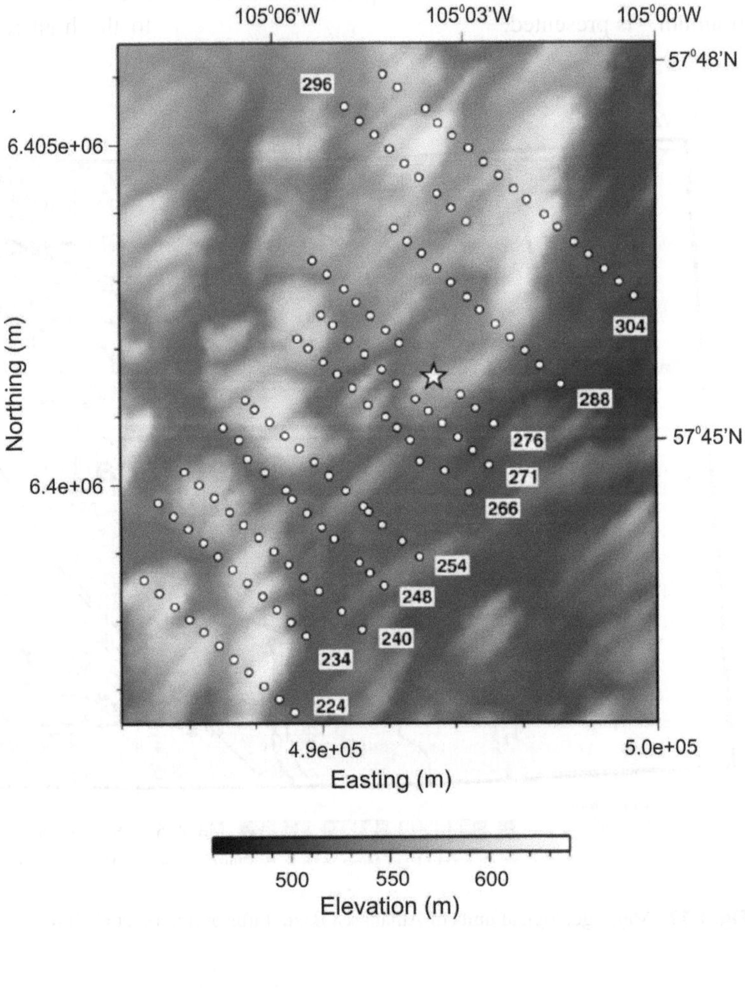

Fig. 1.13 Location of AMT stations along 11 lines projected on the surface elevation map of the Athabasca Basin (from Farquharson and Craven 2009)

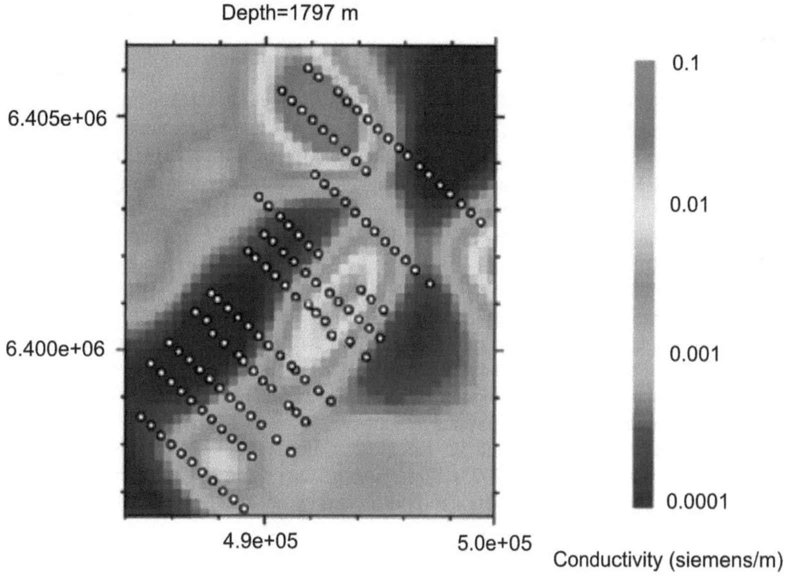

Fig. 1.14 Distribution of electrical conductivity at a depth of 1797 m in Athabasca Basin (from Farquharson and Craven 2009)

the same basin, recently another study using transient audiomagnetotelluric (TAMT) survey was carried out in the Athabasca Basin of northern Saskatchewan as part of a uranium exploration programme at Pasfield Lake (Goldak and Kostenik 2010) to locate deep anomalous conductors.

Thus from the present case study, it is clear that AMT is very effective to delineate the deep anomalous conductors related to the presence of graphitic zones which in turn are associated with the presence of rich uranium mineral.

1.4 Conclusions

Exploration and exploitation of natural resources for the benefit of humanity is one of the important goals for scientists in general. Due to interaction of solar wind with magnetosphere–ionosphere and also due to global thunderstorm activity natural electromagnetic signals are available on the surface of the earth over a wide band of frequencies from a few kilohertz to a few thousand seconds. This facilitates to probe the earth's electrical conductivity structure from shallow depths to even more than 100 km. From the measurement of these natural electromagnetic signals, earth's natural resources can be mapped for profitable use. Although great concern has been expressed in recent years on the conventional use of energy fuels – oil and gas – this sector will be dominating for at least the next two or three decades. The use of magnetotellurics in a volcanic rock – Deccan traps – covered region in Saurashtra, Gujarat, India, has paved the way for delineation of large thickness of sedimentary rock formation in the form of a basin – Jamnagar basin. This has greatly increased the potential of the region for exploitation with deep drilling and also detailed studies using 3D seismics. Some locations in Brazil suffer from acute shortage of water and finding groundwater is the only solution in some areas. By combining the aeromagnetic results with magnetotellurics, suitable location for drilling sites is identified in Parnaiba Basin, Brazil. Locating the mineralized zones, which are the sources of high conductivity, is relatively easier for natural source method as compared to resistive targets. Uranium-rich deposits are often present in association with graphite ore bodies. Using 3D modelling of the AMT data of McArthur Basin area in Canada, anomalous conductive features are delineated as a basement fault with the occurrence of graphite. Thus, while there are several techniques available for delineation of natural sources, the use of natural signals is demonstrated with specific examples in this chapter.

Acknowledgments I would like to thank the Director for permission to publish this work. I gratefully acknowledge Dr. Uchida, Dr. E Chandrasheker, Dr. Farquharson for permission to use figures from their published papers. I also thank Arvind Kumar Gupta for his help in preparation of this manuscript.

References

Cagniard L (1953) Basic theory of the magnetotelluric method of geophysical prospecting. Geophysics 18:605–635

Chandrasekhar E, Fontes SL, Flexr JM, Rajarm M, Anand SP (2009) Magnetotellurics and aeromagnetic investigation for assessment of ground water resources in Parnaiba basin in Piaui state of north-east Brazil. J Appl Geophys 68:269–281

Clarke J, Gamble TD, Goubau WM, Koch RH, Miracky RF (1983) Remote-reference magnetotellurics equipment and procedures. Geophys Prospecting 31:149–170

Constable SC, Parker RL, Constable CG (1987) Occam inversion: a practical algorithm for generating smooth models from EM sounding data. Geophysics 52:289–300

Craven JA, McNeice G, Powell B, Koch R, Annesley L, Wood G, Mwenifumbo J (2003) First look at data from a three-dimensional audio-magnetotelluric survey at the McArthur river mining camp, Northern Saskatchewan. Current Research 2003-C25. Geological Survey of Canada

Cristall J, Brisbin D (2006) Geological sources of VTEM responses along the collins bay fault, Athabasca basin. Giant uranium deposits: exploration guidelines, models and discovery techniques short course, prospectors and developers. Association of Canada International Convention, Toronto, Canada, 5–8 Mar 2006

De Brito Neves BB (1991) Os dois 'Brasis' geotectonicos in simp6sio de Geologia do Nordeste, 15, Recife-PE, SGB-NE, Boletim 12 da SBG-NE. 6

De Sousa MA (1996) Regional gravity modeling and geohistory of the Parnaiba basin. PhD thesis, Department of Physics, University of New Castle upon Tyne, UK

Egbert GD, Booker JR (1986) Robust estimation of geomagnetic transfer functions. Geophys J Roy Aust Soc 87:173–194

Farquharson CG, Craven JA (2009) Three-dimensional inversion of magnetotelluric data for mineral exploration: an example from the McArthur River uranium deposit, Saskatchewan, Canada. J Appl Geophys 68:450–458

Gamble TD, Goubau WM, Clarke J (1978) Magnetotellurics with a remote magnetic reference. Geophysics 44:53–68

Goldak D, Kosteniuk P (2010) 3D Inversion of transient magnetotelluric data: An example from Pasfield Lake, Saskatchewan, EGM International Workshop on adding new value to electromagnetic, gravity and magnetic methods for exploration, Capri, Italy, 11–14 Apr 2010

Groom RW, Bailey RC (1991) Analytical investigations of the effects of near surface three-dimensional galvanic scatterers on MT tensor decomposition. Geophysics 56:496–518.

Gupta HK, Dwivedi KK (1996) Drilling at Latur earthquake region exposes a peninsular gneiss basement, short communication. Geol Soc India 47:129–131

Harinarayana T (1999) Combination of EM and DC measurements for upper crustal studies. Surv Geophys 20(3–4): 257–278

Harinarayana T, Someswara Rao M, Veeraswamy K, Murthy DN, Sarma MVC, Sastry RS, Virupakshi G, Rao SPE, Patro BPK, Manoj C, Madhusudhan Rao, Sreenivasulu T, Abdul Azeez KK, Naganjaneyulu K, Begum SK, Francis Kumar B, Sudha Rani K, Sreenivas M, Prasanth V, Aruna P (2003) Exploration of sub-trappean mesozoic basins in the western part of Narmada-Tapti region of Deccan Syneclise. NGRI Technical Report No: NGRI-2003-Exp-404

Harinarayana T, Zlotnicki J (2006) Electrical and electromagnetic studies in geothermally active regions. J Appl Geophys 58:263–264

Harinarayana T (2008) Applications of Magnetotelluric Studies in India. Memoir. Geol Soc India 68:337–356

Hoover DB, Frischknecht FC, Tippens CL (1975) Audio-magnetotelluric sounding as a reconnaissance exploration technique in Long valley, California. JGR 81:801–809

Jefferson CW, Delaney G, Olson RA (2003) EXTECH IV Athabasca uranium multidisciplinary study of northern Saskatchewan and Alberta, Part 1: overview and impact. Current Research 2003-C18. Geological Survey of Canada

Jones AG (1988) Static shift of magnetotelluric data and its removal in a sedimentary basin environment. Geophysics 53:967–978

Lee TJ, Han N, Song Y (2010) Magnetotelluric survey applied to geothermal exploration: An example at Seokmo Island, Korea. Exploration Geophys 41:61–68

Lee T, Song Y, Uchida Y, Mitsuhata Y, Oh S, Graham GB (2004) Sea effect in three-dimensional magnetotelluric survey: an application to geothermal exploration in Pohang, Korea. In: Proceedings of the 7th SEGJ international symposium, Korea, pp 279–282

Leppin M, Goldak D (2005) Mapping deep sandstone alteration and basement conductors utilizing audio magnetotellurics: exploration for uranium in the Virgin River area, Athabasca basin, Saskatchewan, Canada. SEG Expanded Abstr 24: 591–594

Madden T, Nelson P (1964) A defense of Cagniard's magnetotelluric method. Project Report NR-371-401, Office of Naval Research, USA

Marquardt DW (1963) An algorithm for least-square estimation of non-linear parameters. J SIMA 11:431–441

Meju MA (2002) Geoelectromagnetic exploration for natural resources: models, case studies and challenges. Surv Geophys 23:133–205

Meju M, Fontes SL (1996) An investigation of dry water well near Floriano in Northeast Brazil using combined VES/TEM/EMAP techniques. In: Proceedings of the 8th EEGS symposium on application of Geophysics to Engineering and Environmental problems, pp 353–362

Meju MA, Fontes SL, Oliveira MFB, Lima JPR, Ulugergerli EU, Carrasquilla AA (1999) Regional aquifer mapping using combined VES-TEM-AMT/EMAP methods in the semiarid eastern margin of Parnaiba Basin, Brazil. Geophysics 64(2):337–356

Ogawa Y, Uchida T (1996) A two-dimensional magnetotelluric inversion assuming Gaussian static shift. Geophys J Int 126:69–76

Pellerin L, Hohmann GW (1990) Transient electromagnetic inversion: a remedy for magnetotelluric static shifts. Geophysics 55:1242–1250

Price AT (1962) The theory of magnetotelluric methods when the source field is considered. J Geophys Res 67:1907–1918

Rodi W, Mackie RL (2001) Nonlinear conjugate gradient algorithm for 2-D magnetotelluric inversion. Geophysics 66: 174–187

Rokityanski (1961) On the application of the magnetotelluric method to anisotropic and inhomogeneous masses. Bull (Izv) Acad Sci USSR Geophys Ser 11:1607–1613

Santos FAM, Matos L, Almedia E, Mateus A, Matias H, Mendes-Victor LA (2002) Three-dimensional magnetotelluric modeling of the Vialarica depression (NE Portugal). J Appl Geophys 49:59–74

Sarma SVS, Virupakshi G, Murthy DN, Harinarayana T, Sastry TS, Someswara Rao M, Nagarajan N, Veeraswamy K, Sarma MS, Rao SPE, Bhaskar Gupta KR (1992) Magnetotelluric studies for Oil Exploration over Deccan Traps, Saurashtra, Gujarat, India, NGRI Tech. Report No: NGRI-92-LITHOS-125

Sarma SVS, Virupakshi G, Harinarayana T, Murthy DN, Someswara Rao M, Sastry RS, Nandini Nagarajan, Sastry TS, Sarma MVC, Madhusudhan Rao, Veeraswamy K, Rao SPE, Gupta KRB, Lingaiah A, Sreenivasulu T, Raju AVSN, Patro BPK, Manoj C, Bansal A, Kumaraswamy VTC, Sannasi SR, Cyril Stephen, Naganjaneyulu K (1998) Integrated Geophysical Studies for Hydrocarbon Exploration Saurashtra, India. Integrated geophysical studies for hydrocarbon exploration Saurashtra, India. NGRI Tech. Report No: Ngri-98-Exp- 237

Sasaki Y (1989) Two dimensional joint inversion of magnetotelluric and dipole–dipole resistivity data. Geophysics 54:254–262

Sasaki Y (2004) Three-dimensional inversion of static-shifted magnetotelluric data. Earth, planets and space 56:239–248

Sherwood Lollar B, Westgate TD, Ward JA, Slater GF, Lacrampe-Couloume G (2002) Abiogenic formation of alkanes in the Earth's crust as a minor source for global hydrocarbon reservoirs. Nature 416:522–524

Smith JT, Booker JR (1991) Rapid inversion of two and three dimensional magnetotelluric data. J Geophys Res 96: 3905–3922

Strangway DW, Smft CM Jr, Holmer RC (1973) The application of audio frequency magnetotellurics (AMT) to mineral exploration. Geophysics 38(6):1159–1175

Tikhonov AN (1950) Determination of the electrical characteristics of the deep strata of the earth's crust. Doklady Akadamia Nauk 73:295–297

Uchida T, Sasaki Y (2003) Stable 3-D inversion of MT data and its application to geothermal exploration. In: Macnae J, Liu G (eds) Three dimensional electromagnetics III. ASEG, pp 12.1–12.10

Uchida T, Song Y, Lee TJ, Mitsuhata Y, Lee SK, Lim SK (2004) 3D magnetotelluric interpretation in Pohang low-enthalpy geothermal area, Korea. In: Proceedings of the 17th workshop on electromagnetic induction in the earth, Hyderabad, India, 18–23 Oct 2004

USGS World Petroleum Assessment (2000) New estimates of undiscovered oil and natural gas, natural gas liquids, including reserve growth, outside the United States. http://energy.cr.usgs.gov/oilgas/wep/. USGS Report 2000, June 2003

Vrba J, Gun JVD (2004) The world's ground water resources – contribution to chapter-4 of WWDR-2 Report No. IP-2004-1, Dec 2004, pp 1–10

Wait (1954) On the relation between telluric currents and the earth's magnetic field. Geophysics 19:281–289

Wait (1962) Theory of magnetotelluric fields. J Res Nat Bureau Stand Radio Propagation 66D:509–541

Wannamaker PE, Stodt JA, Rijo L (1986) A stable finite element solution for two-dimensional magnetotelluric modeling. Geophys J R Astron Soc 88:277–296

World Oil Resource Forecast Increases – USGS Sees 20 Percent Hike (2000) AAPG explorer. June issue. USGS report

Application of ANN-Based Techniques in EM Induction Studies

2

Viacheslav V. Spichak

Abstract

Recent advances in application of the artificial neural networks in EM induction studies are discussed. Special attention is paid to 3D reconstruction of the target macroparameters, initial resistivity model construction without prior information about 1D layering, inversion of inhomogeneous magnetotelluric (MT) data, compensation for lack of MT data by estimating the resistivity values using related proxy parameters, joint cluster analysis of the resistivity and other physical properties as well as their indirect estimation from surface EM data.

2.1 Introduction

Pattern recognition methods, in particular, the artificial neural network (ANN) technique (Haykin 1999), became popular during the last decade. The following properties of ANNs are helpful in their successful application in geophysics:

- ANNs are very effective for the solution of nonlinear problems;
- ANNs can conclude from incomplete and noisy data;
- ANNs admit the interpolation and extrapolation of the available database;
- ANNs provide a means for the synthesis of separate series of observations to obtain an integral response, which allows a joint interpretation of diverse data obtained by different geophysical methods;
- the time necessary for ANN recognition depends on the dimension of the unknown parameter space rather than on the physical dimension of the medium, which makes ANN particularly promising for interpretation in the class of 3D geoelectric structures.

A review of the ANN paradigms and the detailed analysis of their application to various geophysical problems are given in Raiche (1991), Poulton (2002), and references therein. The ANN methods were used in electromagnetic (EM) induction studies for data processing (Poulton 2001, Popova and Ogawa 2007) and 1D inversion (Hidalgo et al. 1994, Poulton and Birken 1998). The parameters of the 2D structure were estimated from synthetic and real time-domain electromagnetic data in Poulton et al. (1992a, b).

The scope of ANN applications to solve EM induction problems is not restricted by the directions mentioned above. The main idea of this chapter is to provide a review of new ANN-based algorithms and their implementation that was reported recently.

V.V. Spichak (✉)
Geoelectromagnetic Research Centre IPE RAS, Troitsk, Moscow Region, Russia
e-mail: v.spichak@mail.ru

E. Petrovský et al. (eds.), *The Earth's Magnetic Interior*, IAGA Special Sopron Book Series 1,
DOI 10.1007/978-94-007-0323-0_2, © Springer Science+Business Media B.V. 2011

2.2 Three-Dimensional Reconstruction of the Target Macroparameters

Three-dimensional inversion of EM data in terms of a "cell-by-cell" resistivity distribution is a challenging problem from both theoretical and computational points of view. In spite of some achievements in 3D inversion of EM data (see the review paper by Avdeev (2005) and references therein), it becomes evident that interpretation of real data requires a variety of tools to be used depending on the volume and quality of both the data and prior information available (Spichak 1999, Spichak et al. 1999).

Sometimes geophysicists have only an idea about the *type* of unknown resistivity distribution in the studied area (e.g., horst, graben, fault, magma chamber). In such a case none of the "regular" inversion techniques can transform EM data into a resistivity image that can relate to a structure. They are also inefficient for multiple inversion of data in the frames of the same model class (e.g., in the monitoring mode) since they do not "remember" the inversion way already found. Finally, interpretation of very noisy data by these methods may give results, which will be far from reality. Hence, it is important to basically use new approaches that would overcome or at least weaken the difficulties mentioned above.

An alternative inversion technique especially useful for reconstruction of the target macroparameters (say, blocks' resistivities) could be developed in the framework of the artificial intelligence paradigm (Spichak and Popova 2000). According to Spichak (2007), the process starts from creation of the training data base, followed by analysis, pre-processing (including finding "focusing" transforms of the data that help to increase the resolving ability, noise treatment, and data compression), and finally using the trained ANN for recognition of the target macroparameters (Fig. 2.1).

The best-fitting model reconstructed by ANN belongs to the guessed model class, on the one hand, and to the equivalence class formed by all models giving *rms* misfit less than the noise level in the data, on the other hand. The ANN model parameter reconstruction could be very effective when the geoelectrical model searched is among the model classes used for ANN training (see, for example, Spichak et al. 2002). In this case quick 3D interpretation of even incomplete and noisy data could be carried out in the field. The ability of the ANN to teach itself by real electromagnetic data measured at the same place during sufficiently long periods gives an impetus to use this approach for interpretation of the monitoring data in terms of the target macroparameters (see synthetic example in Shimelevich et al. 2007).

2.3 Initial Resistivity Model Construction

Solving the inverse problem requires the geophysicist to know in advance the 1D layering and supply with an initial guess (expressed in deterministic or probabilistic terms) on 3D resistivity distribution in the region of interest. Meanwhile, the "problem of starting model" could be solved efficiently by using ANN interpolation of initial apparent resistivity data in the spatial and/or frequency domains.

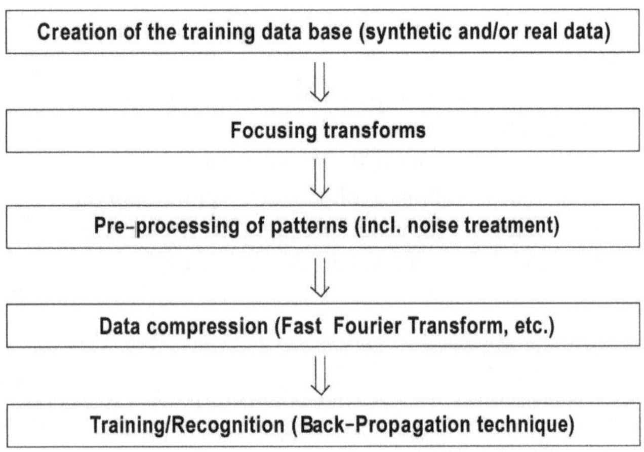

Fig. 2.1 The flowchart of data handling in the framework of the ANN approach using the back-propagation technique with a "teacher"

In order to find the geoelectric structure of the survey area, ANN imaging and full-range 3D inversion could be applied successively. First, 3D ANN synthesis of the apparent resistivity dependences on the apparent depth at each site yields a preliminary resistivity image of the studied area. Second, a regular inversion technique could be used for the refinement of the resistivity distribution in the domains of a special interest taking into account the supplementary information that comes from other methods, drilled boreholes or expert estimates. Below an example is given to demonstrate the use of ANN for constructing 3D resistivity model for the Minamikayabe geothermal area (Spichak 2002).

2.3.1 Preliminary 3D Resistivity Model of the Minamikayabe Geothermal Area

It is well known that a correct approach to 3D MT data interpretation should be based on using all the components of the MT impedance tensor. For instance, all its components can be taken into account by considering the "determinant" invariant determined using all elements of this tensor. Subsequent Bostick transformation of its frequency into a depth function in fact could yield the least deviated (though rather smoothed) image of the geoelectric structure. In turn, its subsequent 3D ANN interpolation into regular grid provides a good initial resistivity model. Figure 2.2a shows the volume resistivity model of the Minamikayabe geothermal area based on such an approach.

2.3.2 Refinement of the Resistivity Image by Means of the Bayesian Statistical Inversion

The resistivity image derived at the first stage was further refined using Bayesian statistical inversion technique. In the framework of this approach, the solution for the inverse problem is reduced to the search for the posterior resistivity distribution by means of successive solutions of the forward problem for the prior values of the resistivities in all domains of search. The effective algorithm developed based on this approach (Spichak et al. 1999) was used at the second stage for refinement of the initial resistivity model.

Figure 2.2b presents highly conductive areas with resistivity values not exceeding 6 Ωm, obtained on the basis of the Bayesian inversion taking into account the resistivity profiles from the wells MK-2 and MK-6. It is easy to see that, first, they cluster in the southern part of the zone in question and, second, their horizontal dimensions increase with depth, reaching a maximum depth range from about 200 to 800 m, and then decrease.

Thus, two-stage inversion of the invariant apparent resistivity data based on rough ANN imaging followed by refinement of the resistivity distribution by means of the Bayesian statistical inversion enabled to reconstruct a 3D geoelectric structure of the Minamikayabe area and to delineate a highly conductive zone that can be associated with geothermal reservoir.

2.4 Joint Inversion of Tensor and Scalar MT Data

Geophysical studies of the interior of the Earth are usually carried out along regional profiles. Despite the presence of well-developed tools for 2D data analysis and interpretation, the efficiency of the information used in the geophysical prospecting is noticeably reduced. This is due to several reasons. First, the geological medium is 3D and so its approximation by 2D models revealed from the profile data may result in unpredictable errors in determining the target location both vertically and horizontally. The assumption on two-dimensionality often leads to considerable over- or underestimation of the resistivity values (Spichak 1999) and sometimes results in false anomalies (for instance, caused by proximity of the MT profile to the edge of the anomalous zone). Second, in the model construction, only profile MT data are used, which a priori reduces the volume of information involved in its building. (The situation could be improved to some extent by 3D inversion of the profile MT data, especially if diagonal components of the impedance tensor are used in the inversion (Siripunvaraporn et al. 2005).)

Meanwhile, if some archive MT data (even scalar) are available in the vicinity of the MT profile the problem of building a volume resistivity model could be solved by means of the newly developed neural network-based technique (Spichak et al. 2010a).

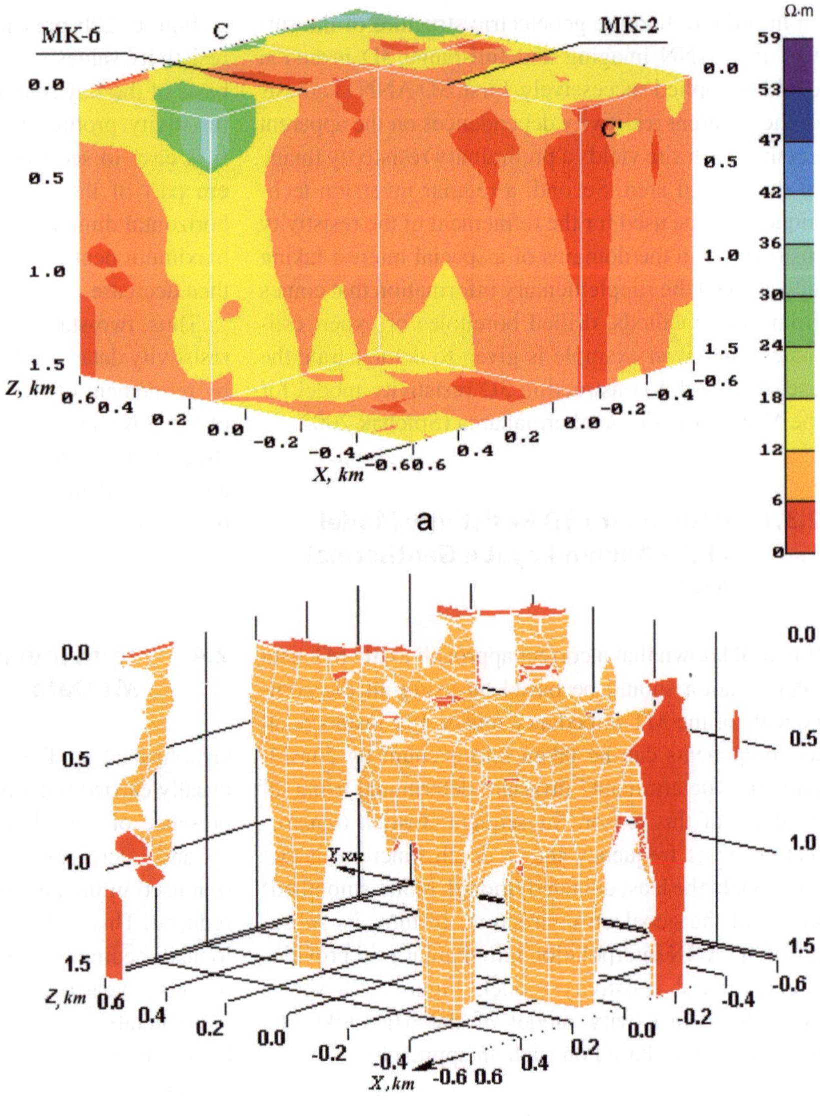

Fig. 2.2 Volume resistivity model built using ANN technique (**a**); highly conductive zone (resistivity is less than 6 Ωm) mapped by refinement of the initial model using Bayesian statistical inversion (**b**) (after Spichak 2002)

2.4.1 General Idea of the Algorithm

In principle, if the profile tensor MT data and a few scalar MT data in its vicinity are available, one could build a general 3D resistivity model of the region. This can be carried out with spatial synthesis of the resistivity cross-section determined from 2D inversion of tensor MT data and 1D models based on 1D inversion of scalar MT data in the rest part of the studied area. However, using the well-known similitude principle underlying the artificial neural network-based estimations one can increase the model accuracy over the *whole* region where the unknown parameters are to be sought for.

The approach developed in Spichak et al. (2010a) is based on the general idea of constructing locally 2D (rather than locally 1D) resistivity model of the region from a set of 1D models. The neural network is calibrated using the correspondence between the locally 1D and 2D resistivity functions determined in the same grid nodes of 2D section beneath the MT profile. The former function is estimated in 2D grid nodes in advance, using another ANN trained by the available 1D resistivity profiles in the whole region.

Thus, the resulting resistivity values estimated by ANN at some distance from the profile allow for both locally 1D values determined from scalar MT data and values in the nodes of 2D section obtained using

2D inversion of tensor MT data collected along the profile. Here, it is implied implicitly that the volume (2D+) model is closer to the real cross-section rather than a model that could be obtained by 3D synthesis of 2D resistivity cross-section and 1D resistivity profiles. This assumption is confirmed by the comparison between 1D and 2D MT data interpretation in 3D media, although it needs a further study using synthetic MT data.

2.4.2 Three-Dimensional Resistivity Model in the Vicinity of a 2D Profile

This algorithm was used for building the 3D resistivity model in the vicinity of the SB-1 profile in Eastern Siberia, Russia, situated within the southwest part of the Nepsko-Botuobinskaya anteclise. The tensor MT data were collected by Phoenix equipment at 26 periods within the range from 0.0028 to 1219 s in 64 sites with a site spacing of about 2 km extending from 105°E to 106°E (Fig. 2.3). Additional data were obtained by digitizing the archive MT master curves at 23 sites located in the vicinity of the profile. On the whole, these data were rather limited in their completeness since only the Bostick transforms of the apparent resistivity within a period ranging from 2 to 900 s were available.

The application of the algorithm has resulted in a 3D resistivity model as shown in Fig. 2.4. It is free of resistivity distortions typical of a 3D model based on archive scalar MT data only as well as of false anomalies in 2D resistivity cross-section obtained by bimodal inversion of profile tensor MT data collected close to the deep fault. It is worth noting that the proposed approach can be used not only for re-interpretation of archive scalar MT data but also for reasonable planning of future regional magnetotelluric soundings.

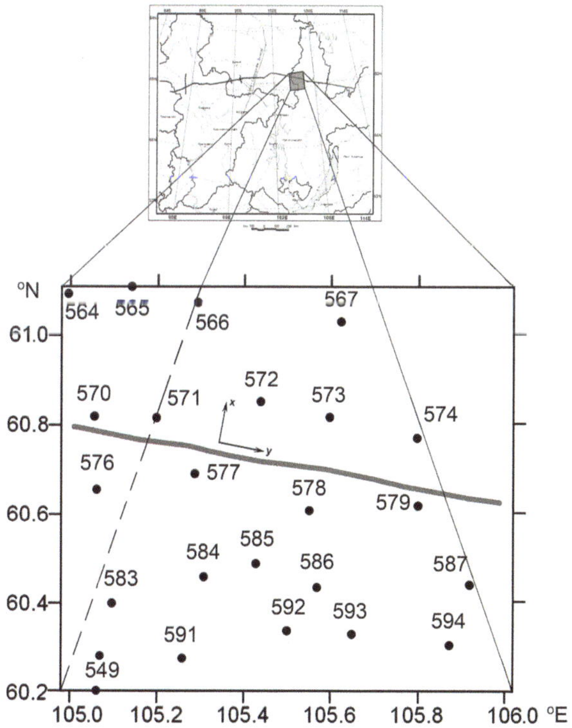

Fig. 2.3 Map of MT site location in the vicinity of the regional SB-1 profile (*solid line*) in Eastern Siberia (after Spichak et al. 2010a)

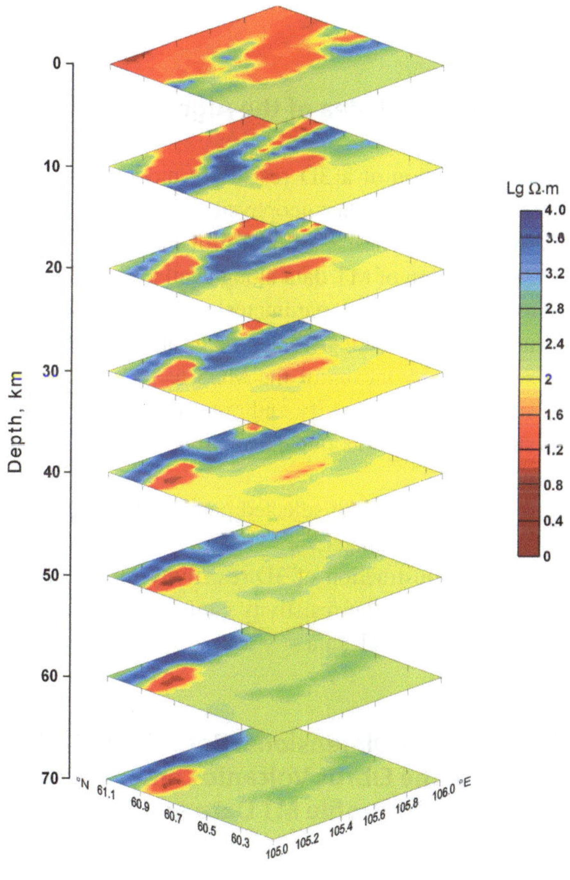

Fig. 2.4 The 2D+ resistivity model based on the profile (tensor) and array (scalar) MT data (after Spichak et al. 2010a)

2.5 Compensating for Lack of the Resistivity Data by Using "Proxy" Parameters

To build a 3D geoelectric model, one needs to have electromagnetic data measured over the entire region. Sometimes it is difficult to collect the required data due to a strong relief and/or inaccessibility of some parts of the studied area. One of the ways for solving this problem could be based on using an upward analytical continuation of MT data to the reference plane followed by their inversion (Spichak 2001). Alternatively, the problem can be solved directly in the resulting resistivity space using recently developed ANN-based maximal correlation similitude (MCS) technique (Spichak et al. 2006), which enables to fill the gaps in the resistivity values using related proxy parameters.

2.5.1 General Idea of the Algorithm

The construction of a 3D resistivity model using 2D MT data and also other geophysical data collected in the studied area consists of the following steps:

- 2D inversion of MT data along available profile;
- estimating another parameter values for the grid nodes, where the resistivity values are determined;
- selection (by means of the MCS technique) of the subset of the resistivity and other parameter values supporting the maximal correlation over the 2D grid nodes;
- ANN training using selected pairs of the resistivity and proxy parameter;
- ANN reconstruction of 3D resistivity model in the domain where the resistivity cannot be determined directly from MT data.

2.5.2 Three-Dimensional Resistivity Model of the Elbrus Volcanic Center Based on the Profile MT Data and Satellite Images

The approach mentioned above was applied to the construction of preliminary 3D resistivity model of the Elbrus volcanic center (EVC) from magnetotelluric data collected along one sub-meridional profile crossing the volcano summit and satellite images of the area (Spichak et al. 2007a). To fill the gaps in MT data, the maps of tectonic fractionation of the Earth's crust resulting from the lineament analysis of the satellite images were used. Tectonic fractionation coefficient (TFC) of the geological medium and resistivity values determined in the same grid nodes of 2D cross-section corresponding to sub-meridional MT profile were analyzed jointly in order to select a subset of pairs supporting the maximal correlation between these two parameters (Fig. 2.5). The artificial neural net was trained using selected samples and then used to reconstruct the 3D resistivity distribution within the entire area of interest.

Figure 2.6 indicates the volume resistivity model of the EVC basement. It contains two extended conductive bodies spaced in depth. One anomaly with resistivity of 25–40 Ωm lies within a depth interval of 0–10 km and has its most prominent manifestation at a depth of 5 km. Here the structure is quasi-isometric and, as outlined by 40–60 Ωm equiresistivity contours, has a radius of 10 km. Another structure with resistivity of 10–40 Ωm is located at a depth of ~45 km where its dimensions along the 40 Ωm contour are 35

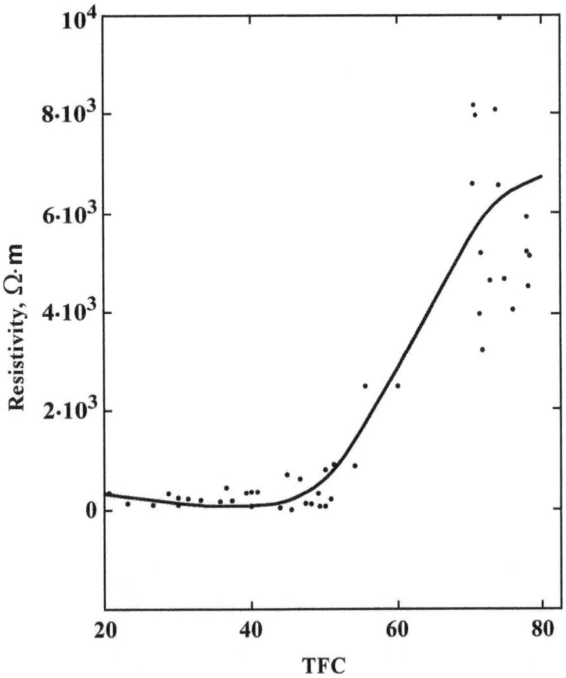

Fig. 2.5 Scatterplot of the resistivity vs. TFC. *Solid line* indicates the result of the ANN approximation based on maximally correlating values (correlation ratio = 0.71) (after Spichak et al. 2007a)

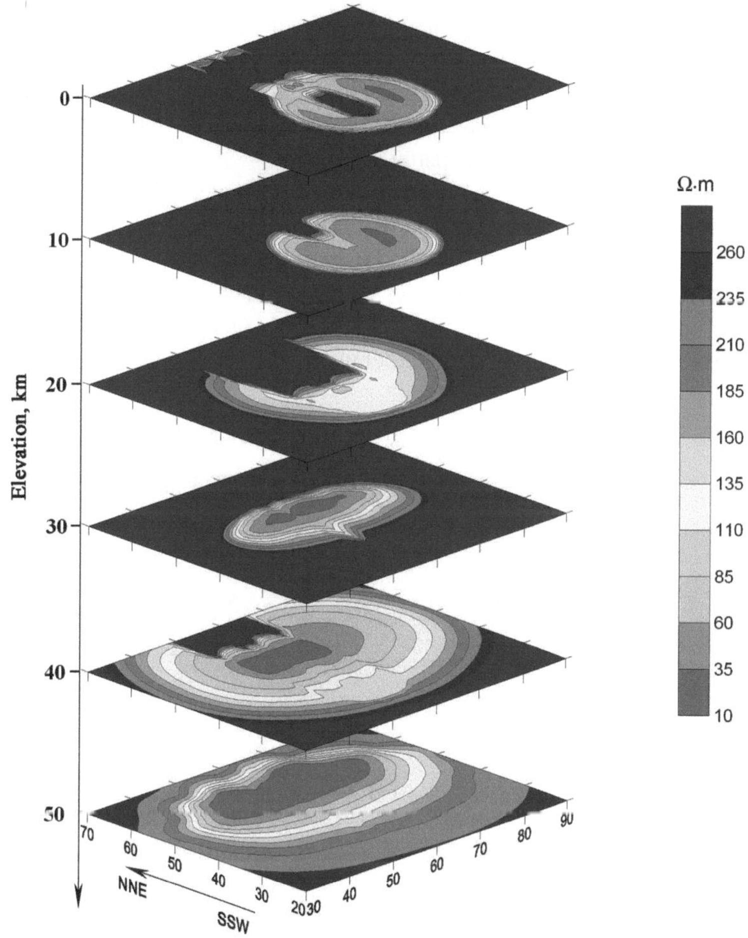

Fig. 2.6 Resistivity slices in the basement of the Elbrus Volcanic Center (after Spichak et al. 2007a)

and 15 km along latitude and longitude, respectively. The thickness of the conductive kernel of this object is about 20 km. The upper anomaly can be treated as a magma chamber of the volcano, while the lower one could be interpreted as a parent magma seat.

2.6 Joint Cluster Analysis of the Resistivity and Other Geophysical Parameters

The rocks composing the Earth's interior are characterized by different physical properties which are important in the construction of the geological model of the deep structure of the Earth. The key physical characteristics are the seismic velocity, electric resistivity, density, and magnetic permeability of rocks. In its turn, joint analysis of these parameters can be used for indirect assessment of several petrophysical properties such as the porosity and permeability, which play an important role, in particular, in the reservoirs' characterization.

2.6.1 Clustering with Self-Organizing Maps (SOM)

Different approaches are suggested for the identification of lithological structures from geophysical data measured on the ground (see review paper by Bedrosian (2007) and references therein). One group of methods is based on their simultaneous inversion taking into account a priori specified constraints on the resulting models, while another one invokes the posterior analysis of the results of independent inversions of the different geophysical data, in which the clusters of physical or lithological properties are determined in the space of parameters or coordinates.

In terms of the latter approach, the cluster cross-sections could be obtained by applying the technique based on the self-organizing maps (Kohonen 2001). The underlying idea of this method is to introduce prior information about the number of properties and to train Kohonen's neuronet in order to identify of areas with similar characteristics in the spatial domain.

In contrary to the back-propagation technique, Kohonen's net is trained without a "teacher." Once the training is completed, SOM classifies the input examples into groups (clusters) of similar elements. The entire set of neurons in the output layer exactly reproduces the distribution of the training samples in a multi-dimensional space. Thus, the uniqueness of the SOM technology is associated with the transformation of multi-dimensional parameter space into 2D space of clusters.

2.6.2 Assessment of the Regional Hydrocarbon Potential from the Cluster Petrophysical Cross-Section

SOM technique was applied in Spichak et al. (2008) for constructing the cluster petrophysical cross-section

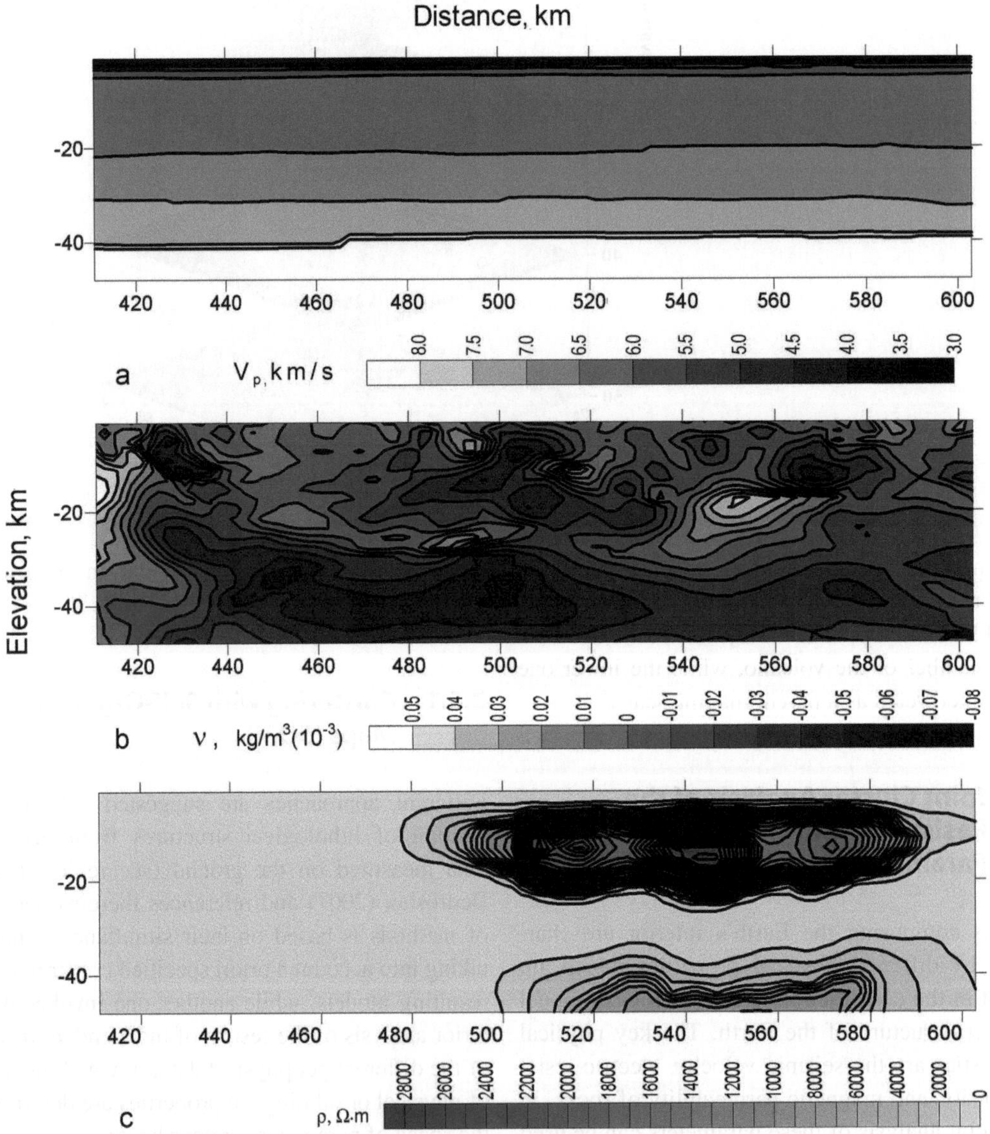

Fig. 2.7 Two-dimensional models of seismic velocity V_P, km/s (**a**), effective density ν, kg/m^3 $\times$ 10^{-3} (**b**), and electric resistivity, Ωm (**c**) (after Spichak et al. 2008)

from three sections of physical properties of the medium (seismic velocity (Fig. 2.7a), effective density (Fig. 2.7b), and electric resistivity (Fig. 2.7c)) inferred from geophysical data collected in Eastern Siberia along the SB-1 regional profile. Within 500–520 km, the profile intersects the Omorinskoe hydrocarbon (HC) field and within 540–580 km, the Yurubchen-Takhom HC field.

The appropriate cluster petrophysical cross-section, constructed by means of the SOM technique, is shown in Fig. 2.8. It is clearly seen (Fig. 2.8a) that the region of two HC deposits is encompassed by a deep-seated extended petrophysical "anomaly" (clusters 1–4), which at shallow depths is complicated by narrow vertical "channels" (probably associated with the upward fluid migration during the formation of HCs). The presence of such petrophysical anomalies in the vicinity of the HC deposit is a necessary, although insufficient, condition for HC identification. This approach can be a suitable tool for inferring petrophysical clusters based on joint posterior analysis of independently inverted geophysical data.

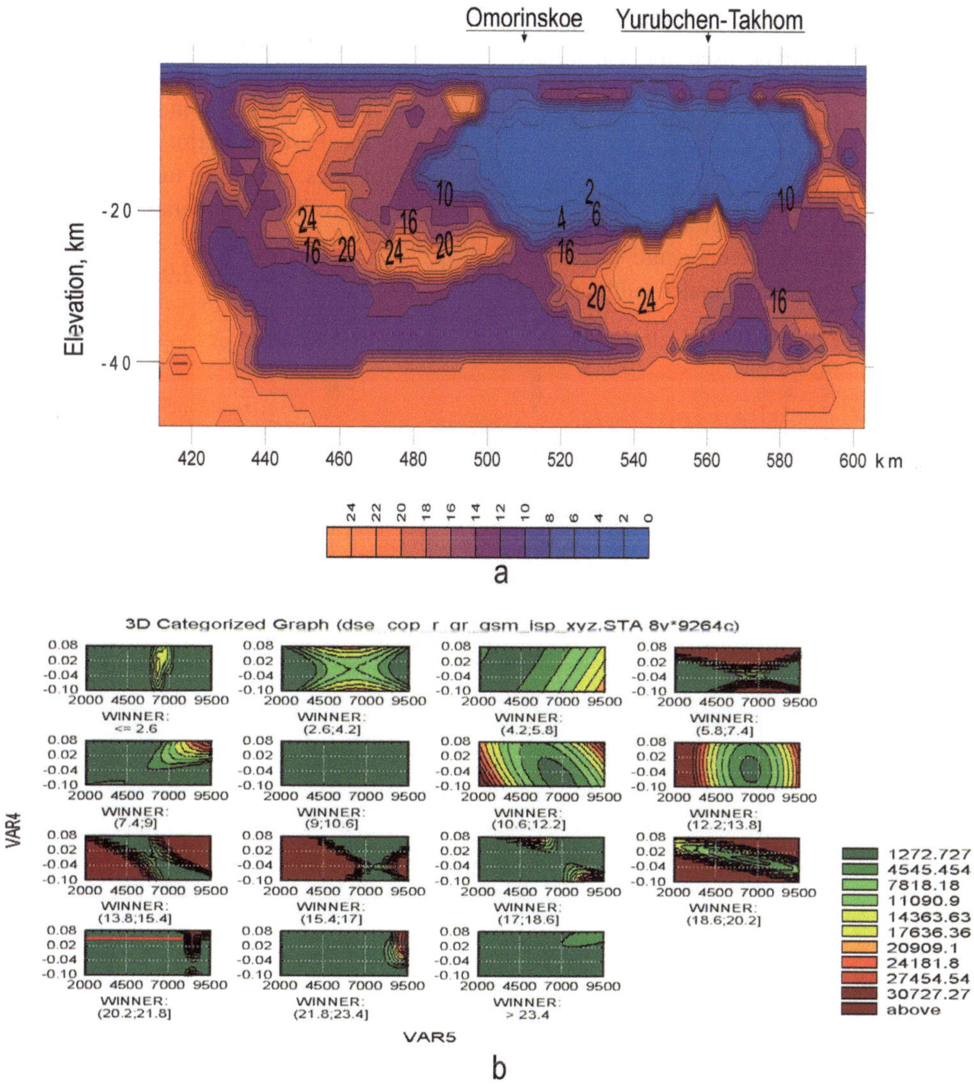

Fig. 2.8 Two-dimensional cluster cross-section of the petrophysical properties (**a**) and palettes corresponding to each cluster (horizontal axis-seismic velocity V_p, vertical axis-density ν, color indicates resistivity ρ) (after Spichak et al. 2008)

2.7 Indirect Estimation of the Rock Properties from Surface EM Data

Artificial neural networks are especially useful for indirect estimation of the rock physical properties from the surface electromagnetic data since the electric resistivity is closely related to such parameters like porosity, permeability, and temperature and could be used as an appropriate proxy parameter. In particular, Spichak et al. (2007b) have developed a new method for the indirect temperature estimation in the Earth based on the ground electromagnetic data (so-called indirect EM geothermometer). The approach used does not imply either the prior knowledge of the thermal gradient or initial guess of the electric conductivity mechanisms and could be used for sub-surface 3D temperature distribution forecast based on the ANN analysis of the implicit resistivity–temperature relations.

Optimal methodologies for interwell space interpolation are developed (Spichak et al. 2011). In particular, it is shown that the temperature estimation by means of the EM geothermometer calibrated by 6–8 temperature logs results in 12% average relative error (instead of 30% achieved by means of routine MATLAB-based interpolation of the temperature logs). Availability of prior geological information about the region under study and/or preliminary analysis of the local heterogeneities' indicators determined from the electromagnetic data may guide appropriate EM site locations that, in turn, may additionally decrease the average temperature estimation errors to only 1%.

Special attention is paid to the application of indirect EM geothermometer to the temperature extrapolation in depth (Spichak and Zakharova 2009). It is shown that in extrapolation to a depth twice as large as the well depth, the relative error is 5–6%, and in case of its threefold excess, the error is around 20%. This result makes it possible to increase significantly the deepness of indirect temperature estimation in the Earth's interior based on the available temperature logs, which, in turn opens up the opportunity to use available temperature logs for estimating the temperatures at depths of 3–10 km without extra drilling. For example, Fig. 2.9 shows the temperature

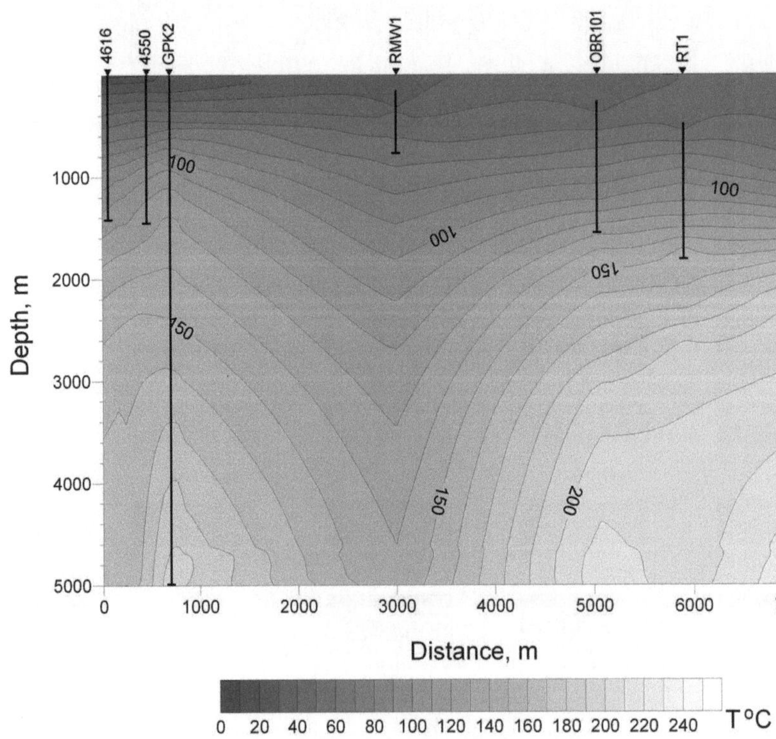

Fig. 2.9 The vertical temperature contour map (in °C) along the MT profile in Soultz-sous-Forêts geothermal area (after Spichak et al. 2010b). *Triangles* indicate the locations of boreholes

cross-section reconstructed in the Soultz-sous-Forêts geothermal area from the profile MT data and available temperature logs extrapolated to a large depth (Spichak et al. 2010b).

2.8 Conclusions

Thus, application of the artificial neural networks in EM induction studies provides very useful tools for data processing, analysis, and interpretation. The following lines illustrated in this review seem to be especially promising:

- 3D reconstruction of the target macroparameters (especially in the monitoring mode);
- initial resistivity model construction without prior information about 1D layering;
- joint inversion of inhomogeneous MT data (in particular, tensor and scalar);
- compensating for lack of EM data by estimating the resistivity values using proxy parameters;
- joint cluster analysis of the resistivity and other rock properties;
- indirect estimation of the rock physical properties from surface EM data.

Finally, it is worth mentioning that the area of potential ANN applications in EM induction studies is not restricted by the examples given in this chapter. The fundamental similitude principle underlying ANN implementation may assist in solving many other tasks, especially linked with processing and interpretation of inhomogeneous, noisy, or irregular data.

Acknowledgements The author acknowledges the Russian Basi Research Foundation (grants 11-05-00045, 11-05-12000) for support of this study.

References

Avdeev DB (2005) 3-D electromagnetic modeling and inversion: from theory to application. Surv Geophys 26:767–799

Bedrosian PA (2007) MT+ integrating magnetotellurics to determine earth structure, physical state and processes. Surv Geophys 28:121–167

Haykin S (1999) Neural networks: a comprehensive foundation, 2nd edn. Prentice Hall, N.J.

Hidalgo H, Gomez-Trevino E, Swinarski R (1994) Neural network approximation of an inverse functional. In: Proceedings IEEE world congress on computational intelligence, Orlando, pp 2715–3436

Kohonen T (2001) Self-organizing maps. Springer, Berlin

Popova I, Ogawa Y (2007) Processing of noisy magnetotelluric data using hopfield neural network. Izv. AN, Fizika Zemli 3:31–38

Poulton M (2002) Neural networks as an intelligence amplification tool: a review of applications. Geophysics 67(3): 979–993

Poulton M (Ed) (2001) Computational neural networks for geophysical data processing. Pergamon, London

Poulton M (2002) Neural networks as an intelligence application tool: a review of applications. Geophysics 67:979–993

Poulton M, Birken R (1998) Estimating one-dimensional models from frequency-domain electromagnetic data using modular neural networks. IEEE Trans Geosci Remote Sens 36: 547–559

Poulton M, Sternberg B, Glass C (1992a) Neural network pattern recognition of subsurface EM images. J Appl Geophys 29:21–36

Poulton M, Sternberg B, Glass C (1992b) Location of subsurface targets in geophysical data using neural networks. Geophysics 57:1534–1544

Raiche A (1991) A pattern recognition approach to geophysical inversion using neural networks. Geophys J Int 105: 629–648

Shimelevich MI, Obornev RA, Gavryushov S (2007) Rapid neuronet inversion of 2-D MT data for monitoring geoelectrical section parameters. Ann Geophys 50(1):105–109

Siripunvaraporn W, Egbert GD, Uyeshima M (2005) Interpretation of two-dimensional magnetotelluric profile data with three-dimensional inversion: synthetic examples. Geophys J Int 160:804–814

Spichak VV (1999) A general approach to the EM data interpretation using an expert system. In: Proceedings X workshop on EM induction in the earth, Ensenada, Mexico

Spichak VV (1999) Magnetotelluric fields in three-dimensional geoelectrical models. Scientific World, Moscow (in Russian)

Spichak VV (2001) Three-dimensional interpretation of MT data in volcanic environments (computer simulation). Ann Geofis 44(2):273–286

Spichak VV (2002) Advanced three – dimensional interpretation technologies applied to the MT data in the Minamikayabe thermal area (Hokkaido, Japan). In: Extended abstracts 64th EAGE conference, Florence, Italy, pp 243–246

Spichak VV (2007) Neural network reconstruction of macroparameters of 3-D geoelectric structures. In: Spichak VV (ed) Electromagnetic sounding of the earth's interior. Elsevier, Amsterdam, pp 223–260

Spichak VV, Bezruk IA, Goidina AG (2010a) Joint ANN inversion of tensor and scalar magnetotelluric data. In: Proceedings XX EM induction workshop, Giza, Egypt

Spichak VV, Bezruk I, Popova I (2008) Construction of deep claster petrophysical cross-sections based on geophysical data and forecasting the hydrocarbon potential of territories. Geofizika 5:43–45 (in Russian)

Spichak VV, Borisova VP, Fainberg EB et al (2007a) Three-dimensional electromagnetic imaging of the Elbrus volcanic centre from magnetotelluric and satellite data. J Volcanol Seismol 1:53–66

Spichak VV, Fukuoka K, Kobayashi T et al (2002) Artificial neural network reconstruction of geoelectrical parameters of the Minou fault zone by scalar CSAMT data. J Appl Geophys 49(1/2):75–90

Spichak VV, Geiermann J, Zakharova O et al (2010b) Deep temperature extrapolation in the Soultz-sous-Forêts geothermal area using magnetotelluric data. In: Expanded abstracts thirty-fifth workshop on geothermal reservoir engineering, Stanford University, Stanford, CA

Spichak VV, Menvielle M, Roussignol M (1999) Three-dimensional inversion of MT data using Bayesian statistics. In: Spies B, Oristaglio M (Eds) 3D electromagnetics. SEG Publications, GD7, Tulsa, OK

Spichak VV, Popova IV (2000) Artificial neural network inversion of MT – data in terms of 3D earth macro – parameters. Geophys J Int 42:15–26

Spichak VV, Rybin A, Batalev V et al (2006) Application of ANN techniques to combined analysis of magnetotelluric and other geophysical data in the northern Tien Shan crustal area. In: Expanded abstract 18th workshop of IAGA WG 1.2 on electromagnetic induction in the earth, El Vendrell, Spain

Spichak VV, Zakharova OK (2011) Methodology of the indirect temperature estimation basing on magnetotelluruc data: northern Tien Shan case study. J Appl Geophys 73:164–173

Spichak VV, Zakharova OK (2009) The application of an indirect electromagnetic geothermometer to temperature extrapolation in depth. Geophys Prospecting 57: 653–664

Spichak VV, Zakharova O, Rybin A (2007b) On the possibility of realization of contact-free electromagnetic geothermometer. Dokl Russ Acad Sci 417A (9):1370–1374

Regional Electromagnetic Induction Studies Using Long Period Geomagnetic Variations

3

E. Chandrasekhar

Abstract

Electromagnetic (EM) induction in the Earth by time-varying geomagnetic field variations at different frequencies facilitates to *look* into different layers of the Earth from surface up to upper mantle depths and beyond using electrical conductivity as a diagnostic parameter. Because of their ability to penetrate to greater depths, long period geomagnetic variations of periods of 1 day and above act as unique tools to probe the Earth's interior up to upper/lower mantle depths in the depth range of 200–1500 km. Thus, to provide a deeper insight in to the electrical behaviour of the upper mantle beneath Indian region, geomagnetic Sq, stormtime variations and 27-day variation and its harmonics data have been analyzed to obtain an integrated conductivity-depth profile up to upper mantle depths. The data were recorded at a dense network of 13 observatories situated along the narrow 150° geomagnetic longitude band extending from dip equator at the southern tip of India to the northern parts of Russia. Results of the regional deep mantle conductivity interpreted in the light of mantle mineral phase changes and compared with other geophysical parameters are discussed.

3.1 Introduction

3.1.1 Overview of the Electromagnetic Induction and Geophysical Significance of Electrical Conductivity

Interaction of the solar wind with the Earth's ionosphere and magnetosphere produces a variety of geomagnetic field variations with periods ranging from a few seconds to a few hundreds of years. Such naturally produced time-varying magnetic field recorded at the Earth's surface is a vector sum of its external and internal parts. The external (primary) part has its origin in the current systems generated in the ionosphere and the distant magnetosphere by the interaction of the solar wind with the Earth's permanent magnetic field. The internal (secondary) part arises due to the eddy currents induced in the conductive layers of the Earth by the diffusing external field within the Earth. Thus the geomagnetic field variations recorded at the magnetic observatories facilitate determination of not only the changes in the electrical conductivity within the Earth as a function of depth, but also the external source current systems associated with them (Schmucker 1970, Campbell 1987 and 2003).

E. Chandrasekhar (✉)
Department of Earth Sciences, Indian Institute of Technology Bombay, Powai, Mumbai 400076, India
e-mail: esekhar@iitb.ac.in

E. Petrovský et al. (eds.), *The Earth's Magnetic Interior*, IAGA Special Sopron Book Series 1,
DOI 10.1007/978-94-007-0323-0_3, © Springer Science+Business Media B.V. 2011

Electrical conductivity characterizes the composition and physical state of the Earth's interior. It is sensitive to temperature, pressure, partial melt, presence of volatiles, oxygen fugacity, iron content and mineral constitution of the rocks (Constable and Duba 1990, Karato 1990, Constable et al. 1992). Therefore, determination of electrical conductivity as a function of depth facilitates to provide knowledge of physical state as well as chemical composition at different depths. Anomalies of electrical conductivity are quite helpful in identifying the zones of melting and dehydration. Thus delineation of these zones is of acute importance in understanding the mobile areas of the Earth's crust and upper mantle, where tectonic movements and regional metamorphism lead to distinct patterns of subsurface conductivity. Therefore, studies of the physics of the Earth particularly in terms of the variation of the electrical conductivity with depth become important.

3.2 Implications of the Electrical Conductivity at Upper Mantle Depths

The important physical parameters that control the electrical conductivity at upper mantle depths are phase changes, temperature and water content. Olivine is generally considered to be the chief constituent of mantle rocks. It is not stable at high temperature and pressure conditions (Katsura and Ito 1989) implying that the mantle mineralogy changes with depth. The major phase transformation of olivine in the deeper mantle falls into two categories: The common olivine (α-phase) transforms under high pressure to a polymorph possessing the spinel structure with increase in density. The Mg-rich olivines transform into modified spinel (β-phase) and then to the post-spinel (γ-phase) with increasing depth. It is believed that the transformation of α-phase of olivine into β- and γ- phases is primarily responsible for the discontinuous changes in the electrical conductivity at greater depths and that the much reported seismic discontinuity at 410 km is attributed to be α-β transformation at high pressure and temperature (Ito and Katsura 1989). Both the β- and γ- phases have much higher elastic wave velocities than appropriate for mantle below 410 km (Anderson 1989).

The dissociation of γ-phase is believed to be related to the other well-known seismic discontinuity at 660 km (Ito and Takahashi 1988). The γ-phase dissociates into perovskite and magnesiowustite at this depth and beyond. The dissociation of the γ-phase is completed within a quite narrow depth interval (<4 km) (Ito and Takahashi 1988). This substantiates the *sharpness* of the 660 km seismic discontinuity. The extent to which the laboratory results simulate the in situ field conditions as well as field measurements, are still the issues of great debate (Lebedev et al. 2002). Recent advances on mantle phase transition studies can be found in the latest review by Ohtani and Sakai (2008).

Laboratory studies on olivine have shown that the electrical conductivity increases with increasing temperature. Intrinsic semiconduction is the most dominant mode of conduction throughout the mantle. The conductivity-temperature relation is given by (Xu et al. 1998)

$$\sigma = \sigma_0 \exp(-\Delta H/kT) \tag{3.1}$$

where σ_0 is the pre-exponential term, T is the absolute temperature, k is the Boltzman constant and $\Delta H (= \Delta U + P\Delta V)$ is the activation enthalpy (where ΔU and ΔV designate the activation energy and activation volume respectively and P is the pressure).

Water or presence of hydrogen ions in the mantle rocks also is responsible for variation of electrical conductivity in the upper mantle. The electrical conductivity of olivine is considerably enhanced by the presence of water (Karato 1990). A number of experimental observations have shown that the presence of water in the upper mantle considerably weakens olivine in its activation energy (Chen et al. 1998), thereby dominating the conduction processes in the olivine (Adam 1993). The correspondence between high conductivity and low-velocity seismic layers coupled with the laboratory results suggests that the hydrous phase of olivine is the dominant conductor at upper mantle depths (Arora et al. 1995). It is believed that the mechanism by which, water is transported to the upper mantle regions is through the subduction of cold oceanic lithosphere, the upper part of which is partially hydrated due to interaction with the overlying ocean (Irifune et al. 1998).

3.3 Different Types of Long Period Geomagnetic Variations and Electromagnetic Induction Response

Skin depth parameter (defined as $\delta = (2/\omega\mu\sigma)^{1/2}$ with σ(S/m) as the conductivity of the medium, $\omega (= 2\pi f)$ as the angular frequency of the inducing source and μ as the permeability of the free space) dictates the frequencies that permit the target depths of investigation. While the diurnal variation (Sq) and its harmonics can penetrate up to crust and uppermost mantle depths, the geomagnetic storm-time variations and the 27-day variation and its harmonics probe up to mid/lower mantle depths (Fig. 3.1).

The fundamental assumption in the applicability of long period geomagnetic variations for induction purpose is essentially based on the *plane wave* approximation of the inducing field (Cagniard 1953). The plane wave (expressed as $\exp(\pm i\omega t)$) implies that its spatial wavelength is large compared to its depth of penetration (Schmucker 1973). However, unlike in mid-latitude regions, the definition of plane wave in equatorial and polar regions differs in the sense that the inducing field is non-uniform and no longer *plane* in nature (see Chandrasekhar and Arora 1994).

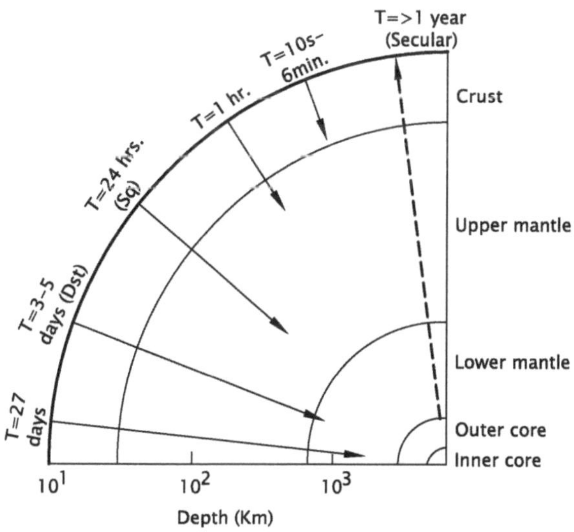

Fig. 3.1 Sketch of various types of geomagnetic variations and their corresponding depths of penetration. Note the depth axis is shown in logarithmic scale. While the solid arrows indicate external (solar) origin of respective geomagnetic variations, the reversed dashed arrow indicates the internal (core) origin of secular variations

Therefore, a correction term needs to be applied while solving the Maxwell's equations for determining the electromagnetic (EM) induction response (Wait 1954), in those regions.

EM induction response functions characterize the linear relation between the internal and external components of the magnetic field and between the electromagnetic field components themselves as a function of time or frequency. Three types of EM induction response functions are Q-, C- and, Z- responses, which respectively correspond to potential method, geomagnetic deep sounding (GDS) technique and magnetotelluric (MT) technique. The Q-response function, defined either in time domain or in frequency domain as

$$i(t) = Q(t) * e(t) \quad \text{or} \quad Q(\omega) = i(\omega)/e(\omega), \tag{3.2}$$

relates the internal (i) and external (e) components of the separated magnetic field variations. In the time domain representation, '*' denotes the convolution operation. The C-response designates the complex transfer function relating the vertical (Z) and horizontal (H) field variations and is defined as

$$Z(\omega) = C(\omega) \cdot H(\omega) + \delta Z(\omega) \tag{3.3}$$

with $\delta Z(\omega)$ being the error term. The real (in-phase) part of C(ω) describes the depth of penetration of the inducing field and the imaginary (quadrature) part of C(ω) is used to estimate the conductivity of the subsurface. The magnetotelluric Z(ω)-response signifies the surface impedance tensor that relates the spectral estimates of horizontal electric (E) and magnetic (H) field components and is defined by

$$\begin{bmatrix} E_x(\omega) \\ E_y(\omega) \end{bmatrix} = \begin{bmatrix} Z_{xx}(\omega) & Z_{xy}(\omega) \\ Z_{yx}(\omega) & Z_{yy}(\omega) \end{bmatrix} \cdot \begin{bmatrix} H_x(\omega) \\ H_y(\omega) \end{bmatrix} \tag{3.4}$$

The subscripts x and y respectively denote the north-south and east-west measurement directions of the E and H fields along the survey profiles. A fundamental difference between all these response functions exists. While the estimation of Q-response requires simultaneous measurements of magnetic field variations at many observatories, single site measurements are sufficient for estimation of C- and Z-responses. Accordingly, it is possible to distinguish the global, regional and local responses. Further details of these response functions and their mutual dependence on one another are given in excellent reviews by Schmucker (1985, 1987) and Olsen (1999). Here, we limit our discussion to the Q- and C- response functions and details about Z-response are beyond the scope of this chapter.

In the present work, we discuss an integrated conductivity-depth profile derived for the Indian region by calculating Q- and C- responses using Sq variations, geomagnetic storm-time (Dst) variations and 27-day variation and its harmonics data, recorded at a dense network of 13 magnetic observatories situated along the narrow 150° geomagnetic longitude band extending from dip equator at the southern tip of India to the northern parts of Russia, operated as a part of the International Magnetospheric Study campaign (Fig. 3.2).

3.4 Upper Mantle Conductivity Due to Sq

A simple two-layer model, in which, a resistive layer overlies a high conducting layer is defined as the substitute conductor model. Schmucker (1970) developed a method to calculate the conductivity and depth estimates of a substitute conductor model by estimating C(ω)-response. Campbell and Anderssen (1983) generalized the Schmucker's (1970) formulation and developed a method to estimate the complex C(ω)-response directly using the spherical harmonic gauss coefficients for Sq induction studies. If the EM induction responses over a substitute conductor model at a sequence of periods are computed for any region, then all such responses corresponding to all periods can be combined and modelled

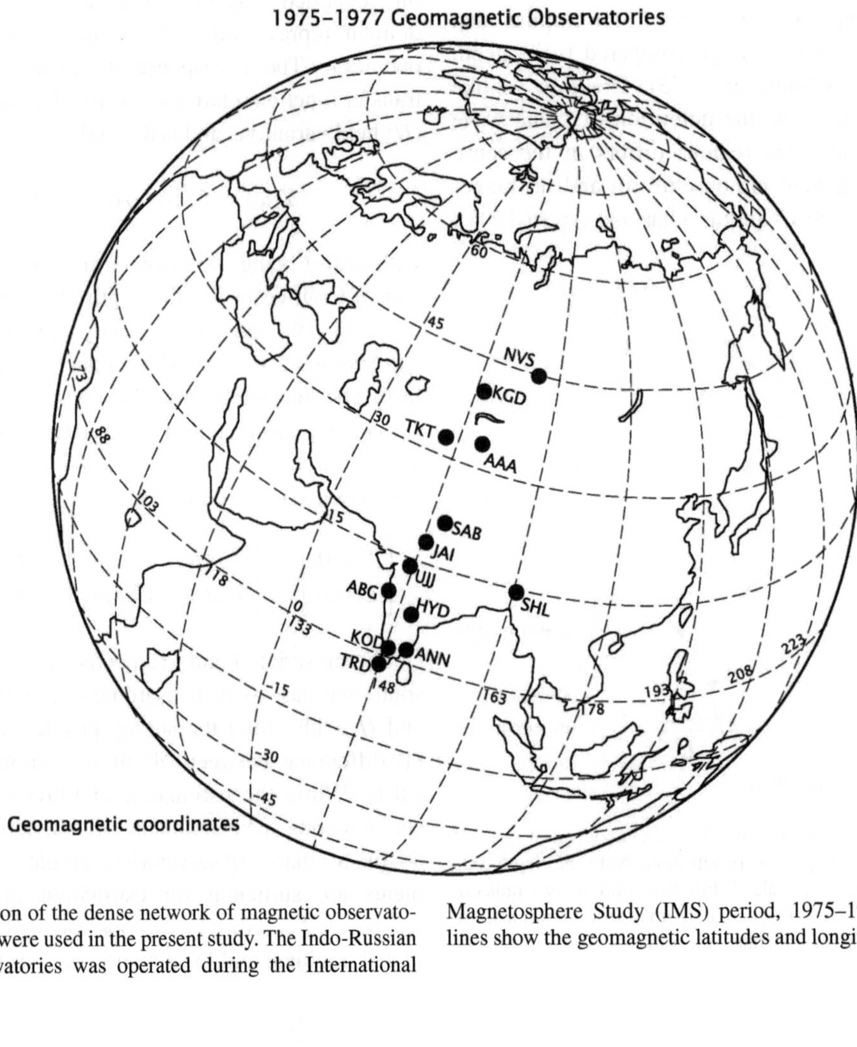

Fig. 3.2 Location of the dense network of magnetic observatories whose data were used in the present study. The Indo-Russian chain of observatories was operated during the International Magnetosphere Study (IMS) period, 1975–1977. The dashed lines show the geomagnetic latitudes and longitudes in 15° steps

together to determine a corresponding conductivity-depth profile. Highlighting this simple concept, Olsen (1998) determined upper mantle conductivity-depth estimates for European region. Adopting the procedure detailed by Campbell and Anderssen (1983), Arora et al. (1995) determined the conductivity-depth profile for Indian region by applying the spherical harmonic analysis technique up to degree ($n = 23$) and order ($m = 6$), to the uninterrupted Sq data of two years (1976–1977) from Indo-Russian network of observatories (Fig. 3.2), They estimated C(ω)-responses using 150 pairs of conductivity and depth values and delineated a robust conductivity-depth profile using locally-weighted regression technique (Fig. 3.3). Chandrasekhar and Alex (1996) inferred a strong localized conductivity anomaly at one of the Indian magnetic observatories by analyzing the anomalous nature of Sq and its harmonics at that observatory. Chandrasekhar et al. (2003) estimated the substitute conductor parameters for Pacific region due to Sq, besides quantifying the role of oceans on the induction response at periods of Sq and its harmonics.

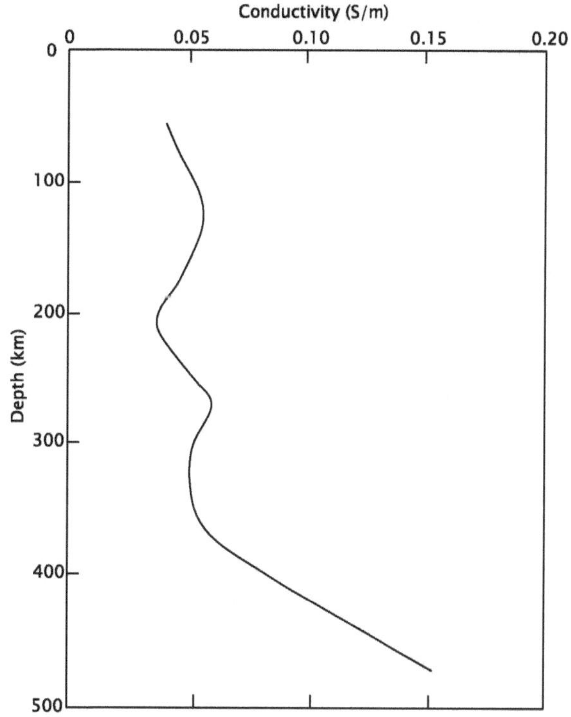

Fig. 3.3 Conductivity-depth profile obtained for the Indian region (see Fig. 3.2) by locally weighted regression of C-responses of Sq variation and its harmonics (after Arora et al. 1995)

3.5 Upper Mantle Conductivity Due to Magnetic Storms

Hourly mean values of five geomagnetic storms with well developed space-time characteristics, recorded during 1976–1977 at the observatories shown in Fig. 3.2 are utilized to compute the electromagnetic Q-response and determine the substitute conductor parameters. The duration and range of H field amplitudes of these storms are given in Table 3.1. Continuous records of five days (120 h) data for each storm are selected in a fashion that one day data preceding the actual commencement of the storm and remaining four days data following the storm onset time are chosen (Chandrasekhar and Arora 1992). This is usually done in order not to lose any important features of the storm under analysis.

The magnetic field perturbations associated with the geomagnetic storms can be expressed as a series of odd degree zonal harmonics. The magnetic scalar potential, ξ, defined for zonal harmonics as a function of geomagnetic co-latitude θ and

Table 3.1 Q-responses and depths to the substitute conductor model, obtained by using data of different geomagnetic storms. The values in the parentheses in the last two columns indicate respective errors. The average Q-response shown in the last row has been obtained by calculating the average Q-responses of each hour of all the storms. The thus obtained Q-value was used to calculate average h^* and error estimates

Storm date and duration	Range of H-amplitude (nT)	Q-response	h^* (km)
02-05-1976 (35 h)	175	0.275 (±0.05)	1153 (±325)
23-08-1976 (39 h)	140	0.130 (±0.04)	2305 (±255)
19-09-1977 (52 h)	170	0.291 (±0.05)	1052 (±313)
26-10-1977 (74 h)	225	0.312 (±0.05)	927 (±316)
10-12-1977 (44 h)	190	0.275 (±0.02)	1203 (±127)
Average		**0.269 (±0.03)**	**1190 (±190)**

the radial distance from the centre of Earth, r, is given by

$$\xi(r,\theta) = a\sum_n \left[e_n(\omega)\{r/a\}^n + i_n(\omega)\{a/r\}^{(n+1)}\right] \times P_n(\cos\theta) \quad (3.5)$$

where, $e_n(\omega)$ and $i_n(\omega)$ designate the external and internal parts of the field of degree n, as a function of frequency. a is the radius of the Earth (in km). P_n represents Schmidt normalized associated Legendre function of degree n.

The contributions of P_1, P_3 and P_5 terms to the observed field have been statistically tested and found that the P_3 and higher odd degree terms contribute less than 10% (Fig. 3.4). Hence only the terms corresponding to P_1 spherical harmonic are considered for further calculations. Accordingly, the external and internal parts of the magnetic field variations corresponding to P_1 spherical harmonic term are given by (Chapman and Price 1930)

$$H = -[e_1 + i_1]\sin\theta \quad \text{and} \quad Z = [e_1 - 2i_1]\cos\theta \quad (3.6)$$

Employing the statistical algorithm of Anderssen and Seneta (1969), the observed field has been separated into external and internal parts. Figure 3.5 shows the separated external and internal parts of

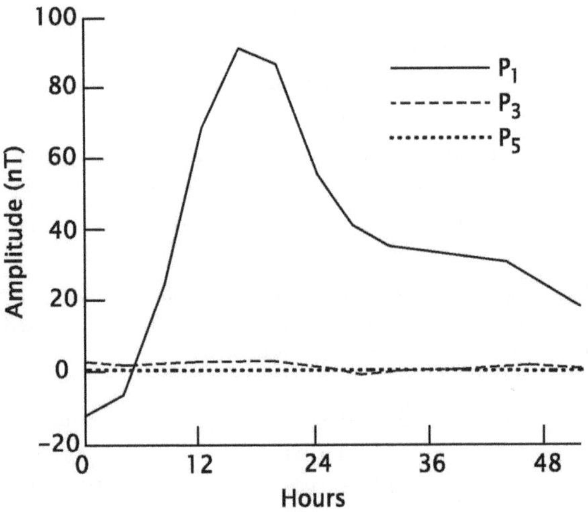

Fig. 3.4 An example of comparison of amplitudes of first three odd-degree zonal harmonics. Note the very low (<10%) amplitude of P_3 and P_5 terms compared to that of P_1 term. This signifies the dominant contribution of the latter, in the data considered for the present study

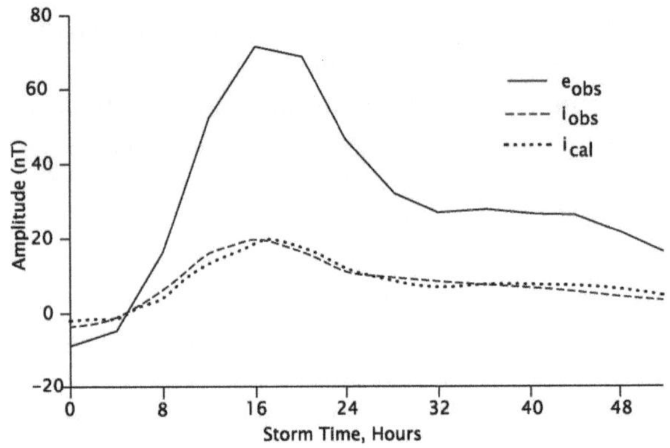

Fig. 3.5 Separated external and internal parts for the geomagnetic storm of 10-12-1977. The isolated external and internal parts were used to estimate the Q-response in least squares sense (see Chandrasekhar and Arora 1992). The estimated Q-response is substituted back in Eq. (3.2) and the calculated internal field was obtained. The dotted curve depicts these values. The good fit between the observed and calculated internal part validates the estimated robust Q-response

the field variations for one of the magnetic storms under study. The Q-response is evaluated in least-squares sense for all the hours of the storm event (see Chandrasekhar and Arora 1992).

The relation between the response function and the normalized radius, h_N of the substitute conductor model is given by (Chapman and Price 1930) $Q_n = \frac{n}{n+1}h_N^{(2n+1)}$. Here, $h_N = h/a$, h being the radius of the substitute conductor. For P_1 harmonic term, we have $Q_1 = 0.5h_N^3$. The depth to the substitute conductor, h^* is given by $h^* = a(1 - h_N)$. The errors associated with Q-responses and h^* are calculated following Schmucker (1999).

The obtained results are tabulated in Table 3.1 and their comparison with those of earlier analyses of geomagnetic storm-time variations is given in Table 3.2.

Table 3.2 Comparison of Q-responses and the depth estimates of substitute conductor model of the present study with those from earlier analyses

Storm-time variation analysis	Q-response	h^* (km)
Chapman and Price (1930)	0.37	610
Rikitake and Sato (1957)	0.35	715
Anderssen and Seneta (1969)	0.36	660
Anderssen et al. (1970)	0.28	1120
Langel (1990)	0.27	1138
Present study	**0.269**	**1190**

3.6 Determination of Conductivity Using Geomagnetic 27-Day Variation and Its Harmonics

3.6.1 Morphology of Geomagnetic 27-Day Variation and its Harmonics

The geomagnetic field variations with 27-day periodicity can be related to a source on the Sun through the solar rotation period, which varies between 25 days at the solar equator and 30 days at the poles. It has also been believed that the magnetic storms show a tendency to recur at 27-day intervals (Chapman and Bartels 1940). The source of recurrent magnetic storms at 27-day intervals is believed to lie in the high density, high velocity streams of plasma, originating in the regions of the Sun, where the magnetic field is weak, uniform and unidirectional (Banks 1969). The interplanetary plasma shows greatest order around solar cycle minimum and therefore, the 27-day variation is well defined with high signal to noise ratio at low levels of solar activity (Banks 1969). The presence or absence of harmonics associated with 27-day variation is controlled by the nature of the 27-day periodicity. It is because of this reason, while studying these variations, the data corresponding to the '*period bands*' centered at 27-day and its harmonics are generally considered, rather than the actual periods themselves.

3.6.2 Upper Mantle Conductivity Due to 27-Day Variation and Its Harmonics

Daily mean values of the geomagnetic *D*, *H* and *Z* components for 901 days, recorded at each station shown in Fig. 3.2 are considered for the analysis. Details of sequential data processing procedures and identification of period bands corresponding to 27-day (25–32 days) and its harmonics, i.e., 13.5-day (11–16 days) and 9-day (8–10 days) can be found in Chandrasekhar (2000).

Spectral estimates corresponding to each period band are subjected to complex demodulation technique due to Banks (1975). Complex demodulation technique facilitates examination of variation of a particular frequency in a given signal as a function of time. Accordingly, each computed demodulate represents an instantaneous estimate of amplitude and phase of a particular frequency, thereby allowing to treat each demodulate as an independent entity. This way, it is quite helpful to identify and use only those demodulates having high signal to noise ratio in any period band.

Exploiting the advantage that this technique offers, the zonal nature of 27-day and its harmonics was tested statistically, using demodulates corresponding to the above three period bands of all the stations (Fig. 3.6). Only those demodulates, which statistically fitted the P_1 dependence, were used for calculating C(ω)-responses corresponding to (i) a particular period band at each station (single-station responses) and (ii) all stations for each period band (band-averaged responses). While estimating the single-station and

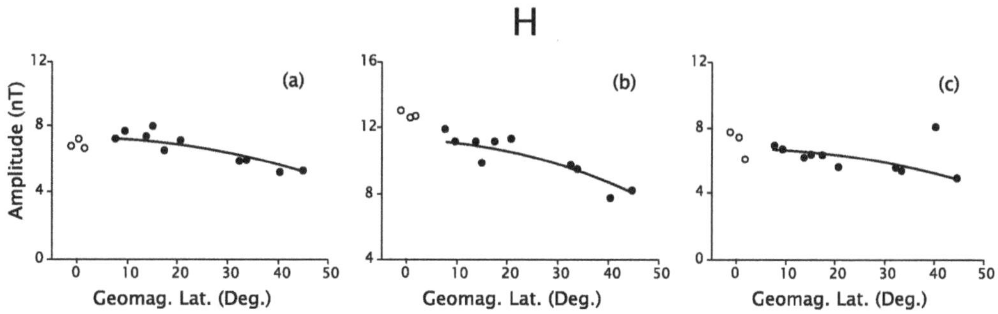

Fig. 3.6 Best fit P_1 curves (*solid lines*) for H component for a selected demodulate corresponding to (**a**) 25–32, (**b**) 11–16 and (**c**) 8–10 day period bands. The best fit is obtained in a least-squares sense, by considering only mid-latitude stations (*solid circles*) and by excluding the equatorial stations (*open circles*). Note the consistency in the P_1 representation of the inducing field at all period bands

band-averaged responses, the data corresponding to equatorial and polar regions have not been considered.

3.6.2.1 Determination of Single Station C-Responses

Details about the robust estimation of single-station C(ω)-response estimates for 27-day variation and its harmonics can be found in Chandrasekhar (2000). The errors associated with the C-responses are calculated following Schmucker (1999). As mentioned earlier (see Section 3.3), the real part of the C(ω)-response designates the depth (h^*) of the induced currents and the imaginary part is used to determine the conductivity (σ) of substitute conductor. The conductivity was estimated by (Lilley and Sloane 1976)

$$\sigma = [0.8\pi\,\omega\,C^2_{\text{imag}}(\omega)]^{-1} \quad (3.7)$$

Table 3.3 shows C(ω)-response estimates and their corresponding phases together with their respective error estimates obtained for individual sites corresponding to the 25–32 day period band. To check for the latitudinal dependence of the obtained depth estimates, a contour plot of the values given in Table 3.3 is made (Fig. 3.7).

3.6.2.2 Determination of Band-Averaged Response

To calculate the band-averaged C(ω)-response, first, the response for each individual *Z-H* demodulate-pair of all the stations corresponding to each band was calculated using equation 5 of Chandrasekhar (2000). In this exercise, *N* in equation 5 (of Chandrasekhar 2000) denotes the total number of stations. The rest of the calculations are repeated for each band, grouping the responses of all stations. Table 3.4 shows the band-averaged C(ω)-responses for the three period bands, together with their errors. The band-averaged responses corresponding to 27-day variation and its harmonics are augmented with the Sq analysis results (Fig. 3.3). The combined conductivity depth profile from surface to upper mantle depth is shown in Fig. 3.8a. The solid curve in Fig. 3.8a depicts the band-averaged responses. The derived upper mantle conductivity-depth profile for Indian region is compared with that of European region (Fig. 3.8b), Pannonian basin and other regions (Fig. 3.8c).

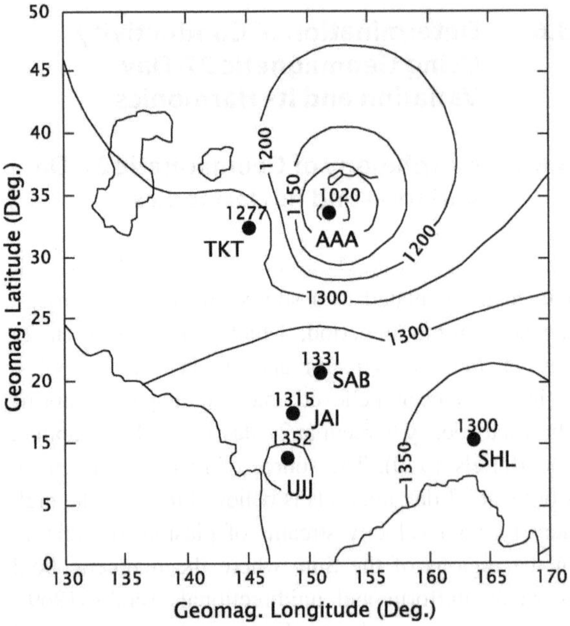

Fig. 3.7 Contour plot of the depths of substitute conductor model, h^* (in km) for the period band of 25–32 days for 6 mid-latitude stations (Table 3.3). The response is compatible both with the P_1 approximation of the inducing source as well as with a local 1-D internal structure of the Earth (after Chandrasekhar 2000)

Table 3.3 Response estimates (depth (h^*) and conductivity (σ)) of the substitute conductor and their corresponding phases, obtained by a robust method for 6 stations shown in Fig. 3.2 for the period band of 25–32 days, corresponding to 27-day period. The values given in the parentheses represent the errors associated with the respective estimates

Station Name	C-response ($C(\omega) = C_{\text{real}}(\omega) + iC_{\text{imag}}(\omega)$)	σ (S/m)	Phase
UJJ	1362 – 515*i* (±53)	0.60 (±0.12)	−20.7 (±2.1)
SHL	1400 – 523*i* (±62)	0.58 (±0.14)	−20.5 (±2.4)
JAI	1315 – 341*i* (±74)	1.36 (±0.59)	−14.5 (±3.1)
SAB	1331 – 1062*i* (±106)	0.14 (±0.03)	−38.6 (±3.6)
TKT	1275 – 520*i* (±51)	0.58 (±0.12)	−22.2 (±2.1)
AAA	1029 – 703*i* (±30)	0.32 (±0.03)	−34.3 (±1.4)

Table 3.4 Depth (h^*) and the conductivity (σ) of the substitute conductor for three different period bands, representing 27-day period and its harmonics. The values in the parentheses depict the corresponding error estimates

Period band (days)	Weighted mean h^* (km)	Weighted mean conductivity (S/m)
25–32	1177 (±191)	0.34 (±0.19)
11–16	900 (±101)	0.23 (±0.08)
8–10	825 (±296)	0.15 (±0.10)

3.7 Conclusions

The results derived from the interpretation of regional EM induction response for Indian region are briefly reviewed by this article. This article long period geomagnetic variations recorded at a dense network of magnetic observatories in the Indian region have been analyzed to delineate the depth and conductivity of the substitute conductor model. The spatial distribution of the observatories has aptly facilitated to statistically test the plane-wave approximation of the inducing field, which is essential for determining the EM induction response.

In agreement with the global conductivity models at upper mantle depths (for e.g., Kuvshinov and Olsen 2006, Kelbert et al. 2009), the conductivity-depth profile shown in Fig. 3.3 clearly suggests that the upper mantle beneath Indian region could be viewed as a sequence of inhomogeneous layers up to 400 km, with well-resolved conductivity highs at 125 km and 275 km. The concurrence of enhanced conductivity at 125–150 km beneath Indian region with similar observations made at Fennoscandian shield (Jones 1982), and Central Europe and Central Asian regions (Campbell and Schiffmacher 1988), suggests the lithosphere-asthenosphere boundary. The sharp increase in conductivity at 350–500 km depth matches well with the one observed beneath Central Asia and Africa regions (Campbell and Schiffmacher 1988). By comparing the conductivity-depth profile (Fig. 3.3) with the seismic velocity models for mantle, such as PREM (Dziewonski and Anderson 1981) and IASP91 (Kennet and Engdahl 1991), Arora et al. (1995) attribute the conductivity increase between 350 and 500 km depth to the well-known 410 km seismic discontinuity.

Katsura and Ito (1989) have estimated a temperature of 1400°C for the $\alpha \rightarrow \alpha + \beta$ transition of olivine at 410 km. Constable (1993) discussed that the estimated value of 1400°C for the $\alpha \rightarrow \alpha + \beta$ transition can be achieved by incorporating more conducting layers within the top 200 km. Thus the correspondence between the conductivity increases at 125 km and 275 km supports the hypothesis of Constable (1993). Possible mechanism for enhanced upper mantle conductivity can be found in Arora et al. (1995).

The averaged depth of the substitute conductor model beneath Indian region derived from robust estimation of Q-responses for 5 selected geomagnetic storms is about 1200 km (Table 3.1). The differences in the substitute conductor model parameters between the present study and those prior to 1970 (Table 3.2) may be attributed to (i) availability of data from limited number of observatories, which results in regional biases in the estimated values, (ii) simple *assumption* of source field characteristics rather than statistically testing for the same, (iii) use of less rigorous statistical techniques, to cite a few.

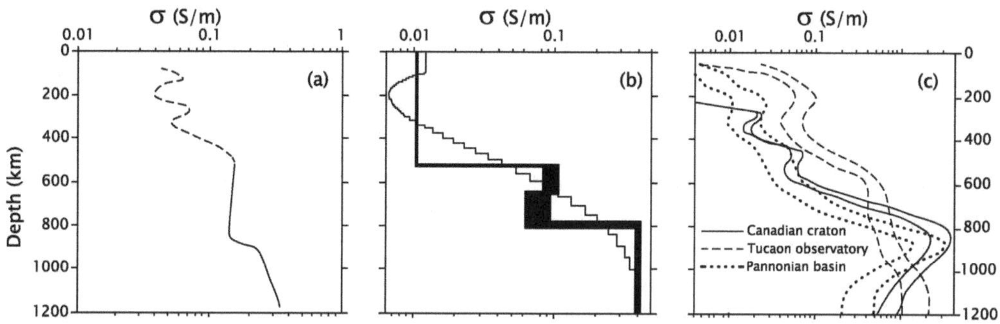

Fig. 3.8 Comparison of the geo-electrical structure of the mantle for (**a**) the Indian region, obtained by augmenting the Sq analysis results (*dashed line*) (Fig. 3.3) with those of the 27-day variation and its harmonics (*solid line*), (**b**) the European region (after Olsen 1998) and (**c**) the Pannonian basin (Semenov et al., 1997), the North American region (Egbert and Booker 1992) and the Canadian craton (Schultz et al. 1993). The conductivity increase at 850 km depth range, depicting the presence of a *mid-mantle conductor*, clearly seen in all the regions can be considered as a global phenomenon

Using the data of a single magnetic storm from as many as 51 global observatories, Anderssen et al. (1970) obtained a depth estimate of 1120 km for a similar model. Langel (1990) by analyzing MAGSAT data obtained a depth estimate of 1100 km. Thus the comparable values of the substitute conductor model parameters beneath the Indo-Russian sector suggests that the electrical character of the upper mantle in the depth range of 1200 km is not very different from the global model. This is further confirmed by comprehensive estimation of conductivity-depth estimates obtained for the same region using 27-day variation and its harmonics (Chandrasekhar 2000).

The robust approach implemented in determining the single station $C(\omega)$-responses has resulted in producing reliable response estimates for UJJ, SHL, JAI, SAB, TKT and AAA stations (Table 3.3), which have shown a consistent P_1 dependence of the source field (Fig. 3.6). These estimates are validated against the strict conditions prescribed by Weidelt (1972) and Whittal and Oldenburg (1992) for any model representative of a substitute conductor model. An observation of contour plot of the depth estimates (Fig. 3.7) reveals a more or less constant substitute conductor depth beneath Indian region (Chandrasekhar 2000), contrary to what has been observed by Petersons and Anderssen (1990) for North American region, where they report a decrease in depth of the substitute conductor with increase in latitude.

A comparison of the geo-electric profiles of the lower mantle for different regions clearly shows enhanced upper mantle conductivity at a depth of about 850 km. This feature is believed to be a *mid-mantle conductor* and could be considered as a global phenomenon, as it is also seen in the (i) combined MT and GDS response for Pannonian basin (Semenov et al. 1997) (ii) combined analysis of Sq and geomagnetic storm-time variations for European region (Olsen 1998) (iii) for Europe-Asia region (Semenov and Jozwiak 1999) and (iv) for Indian region (Chandrasekhar 2000). The main feature of such a mid-mantle conductor is, that it is situated much deeper than the well-known 660 km seismic discontinuity, where β–γ phase change is observed to produce a discontinuous conductivity change. However, the data set used in the present study could not resolve a conductivity jump at 660 km because of the restricted frequency range in the data. Another possibility would be to relate an electrical conductivity increase at 850 km with a 920 km seismic discontinuity, as reported by Kawakatsu and Niu (1994) for the Japanese islands. They argue that this discontinuity probably represents the bottom of the upper mantle transition zone at 660 km and thus be the top of lower mantle.

We conclude that the average depth of the substitute conductor model for Indian region is 1200 km (± 200 km) with an average conductivity of 0.6 (± 0.17) S/m. The results of the present study, where different EM induction responses have been computed for Indian region by using a variety of long period geomagnetic variations are concurrent with one another and also match well with the global estimates obtained by similar investigations.

Acknowledgments The author thanks the editor, T Harinarayana, for inviting him to write this article. An anonymous referee is thanked for his/her meticulous review. Figures prepared by Mr. P. Sawant and Mr. V.K. Vijesh are acknowledged with thanks. The authors thanks the staff of all the observatories for providing high quality data.

References

Adam A (1993) Physics of the upper mantle – a review. Act Geod Geophy Mont Hungary 28:151–195

Anderson DL (1989) Theory of the earth. Blackwell, London

Anderssen RS, Seneta E (1969) New analysis of geomagnetic Dst field of the magnetic storm of June 18–19, 1936. J Geophys Res 74:2768–2773

Anderssen RS, Doyle MA, Petersons HF, Seneta E (1970) On the smoothing and spherical harmonic analysis of the storm of September 25, 1958. J Geophys Res 75:2569–2577

Arora BR, Campbell WH, Schiffmacher ER (1995) Upper mantle electrical conductivity in the Himalayan region. J Geomagnetism Geoelectricity 47:653–665

Banks RJ (1969) Geomagnetic variations and the electrical conductivity of the upper mantle. Geophys J R Astr Soc 17:457–487

Banks RJ (1975) Complex demodulation of geomagnetic data and the estimation of transfer function. Geophys J R Astr Soc 43:87–101

Cagniard L (1953) Basic theory of magnetotelluric method of geophysical prospecting. Geophysics 18:605–635

Campbell WH (1987) The upper mantle conductivity analysis method using observatory records of geomagnetic field. Pure Appl Geophys 125:427–457

Campbell WH (2003) Introduction to geomagnetic fields. Cambridge University Press, Cambridge

Campbell WH, Anderssen RS (1983) Conductivity of the subcontinental upper mantle: an analysis using quiet-day geomagnetic records of North America. J Geomagnetism Geoelectricity 35:367–382

Campbell WH, Schiffmacher ER (1988) Upper mantle electrical conductivity for seven subcontinental regions of the Earth. J Geomagnetism Geoelectricity 40:1387–1406

Chandrasekhar E (2000) Geo-electrical structure of the mantle beneath the Indian region derived from the 27-day variation and its harmonics. Earth Planets Space 52:587–594

Chandrasekhar E, Alex S (1996) On the anomalous features of the geomagnetic quiet-day field variations at Nagpur, India. Geophys J Int 127:703–707

Chandrasekhar E, Arora BR (1992) Upper mantle electrical conductivity distribution beneath the Indian sub-continent using geomagnetic storm-time variations. Memoir Geol Soc Ind 24:149–157

Chandrasekhar E, Arora BR (1994) On the source field geometry and geomagnetic induction in southern India. J Geomagnetism Geoelectricity 46:815–825

Chandrasekhar E, Oshiman N, Yumoto K (2003) On the role of oceans in the geomagnetic induction by Sq along the 210° magnetic meridian region. Earth Planets Space 55: 315–326

Chapman S, Bartels J (1940) Geomagnetism. Oxford University Press, Oxford

Chapman S, Price AT (1930) The electrical and magnetic state of the interior of the earth as inferred from terrestrial magnetic variations. Phil Trans R Soc Lond A-229:427–460

Chen J, Inoue T, Weidner DJ, Wu Y, Vaughan MT (1998) Strength and water weakening of mantle minerals, olivine, wadsleyite and ringwoodite. Geophys Res Lett 25: 575–578

Constable SC (1993) Constraints on mantle electrical conductivity from field and laboratory measurements. J Geomagnetism Geoelectricity 45:1–22

Constable SC, Duba A (1990) The electrical conductivity of olivine, a dunite, and the mantle. J Geophys Res 95:6967–6978

Constable SC, Shankland TJ, Duba A (1992) The electrical conductivity of an isotropic olivine mantle. J Geophys Res 97:3397–3404

Dziewonski AM, Anderson DL (1981) Preliminary reference earth model. Phys Earth Planet Inter 25:297–356

Egbert GD, Booker JR (1992) Very long period magnetotellurics at Tucson observatory: implications for mantle conductivity. J Geophys Res 97:15099–15115

Irifune T, Kubo N, Ichiki M, Yamasaki Y (1998) Phase transformations in serpentine and transportation of water into the lower mantle. Geophys Res Lett 25:203–206

Ito E, Katsura T (1989) A temperature profile of the mantle transition zone. Geophys Res Lett 16:425–428

Ito E, Takahashi E (1988) Post-spinel transformations in the system MgSiO4-Fe2SiO4 and some geophysical implications. J Geophys Res 94:10637–10646

Jones AG (1982) Observations of the electrical asthenosphere beneath Scandinavia. Tectonophys 90:37–55

Karato S (1990) The role of hydrogen in the electrical conductivity of upper mantle. Nature 347:272–273

Katsura T, Ito E (1989) The system of Mg2SiO4-Fe2SiO4 at high pressures and temperatures: precise determination of stabilities of olivine, modified spinel and spinel. J Geophys Res 94:15663–15670

Kawakatsu H, Niu F (1994) Seismic evidence for a 920 km discontinuity in the mantle. Nature 371:301 305

Kelbert A, Schultz A, Egbert G (2009) Global electromagnetic induction constraints on transition-zone water content variations. Nature. doi:10.1038/nature08257

Kennet BLN, Engdahl ER (1991) Travel times for global earthquake location and phase identification. Geophys J Int 105:429–465

Kuvshinov A, Olsen N (2006) A global model of mantle conductivity derived from 5 years of CHAMP, Ørsted, and SAC-C magnetic data. Geophys Res Lett. doi:10.1029/2006GL027083

Langel RA (1990) Study of the crust and mantle using magnetic surveys by MAGSAT and other satellites. Proc Ind Acad Sci 99:581–618

Lebedev S, Chevrot S, van der Hilst RD (2002) Seismic evidence for olivine phase changes at the 410 and 660-kilometer discontinuities. Science 296:1300. doi:10.1126/science.1069407

Lilley FEM, Sloane MN (1976) On estimating electrical conductivity using gradient data from magnetometer arrays. J Geomagnetism Geoelectricity 28:321–328

Ohtani E, Sakai T (2008) Recent advances in the study of mantle phase transitions. Phys Earth Planet Int. doi:org/10.1016/j.pepi.2008.07.024

Olsen N (1998) The electrical conductivity of the mantle beneath Europe derived from C-responses from 3 to 720 hours. Geophys J Int 133:298–308

Olsen N (1999) Induction studies with satellite data. Surv Geophys 20:309–340

Petersons HF, Anderssen RS (1990) On the spherical symmetry of the electrical conductivity of the earth's mantle. J Geomagnetism Geoelectricity 42:1309–1324

Rikitake T, Sato S (1957) The geomagnetic Dst field of the magnetic storm on June 18–19, 1936. Bull Earthquake Res Inst 35:7–21

Schmucker U (1970) Anomalies of geomagnetic variations in south western United States. Bull Scripps Inst Oceanogr 13:1–165

Schmucker U (1973) Regional induction studies: a review of methods and results. Phys Earth Planet Inter 7:251–265

Schmucker U (1985) Magnetic and electric fields due to electromagnetic induction by external sources. In: Fuchs K, Soffel H (eds) Landolt Bornstein numerical data and functional relationships in science and technology. New series group V vol 2b. Springer, Berlin, pp 100–125

Schmucker U (1987) Substitute conductors for electromagnetic response estimates. Pure Appl Geophys 125: 341–367

Schmucker U (1999) A spherical harmonic analysis of solar daily variations in the years 1964–1965: response estimates and source fields for global induction—II Results. Geophys J Int 136:455–476

Schultz A, Kurtz RD, Chave AD, Jones AG (1993) Conductivity discontinuities in the upper mantle beneath a stable craton. Geophys Res Lett 20:2941–2944

Semenov VYu, Jozwiak, W (1999) Model of the geoelectrical structure of the mid- and lower mantle in the Europe–Asia region. Geophys J Int 138:549–552

Semenov VYu, Adam A, Hvozdara M, Wesztergom, V (1997) Geoelectrical structure of the earth's mantle in Pannonian basin. Acta Geod Geophys Hungary 32:151–168

Wait JR (1954) On the relation between telluric currents and the earth's magnetic field. Geophysics 19:281–289

Weidelt P (1972) The inverse problem of geomagnetic induction. J Geophys 38:257–289

Whittal KP, Oldenburg DW (1992) Inversion of magnetotelluric data for one dimensional conductivity. In: Fitterman DV (ed) Geophys monograph Series. Society of Exploration Geophysicists, vol 5. SEG Publication, pp 1–114

Xu Y, Poe B T, Shankland TJ, Rubie DC (1998) Electrical conductivity of olivine, wadsleyite, and ringwoodite under upper-mantle conditions. Science 280:1415. doi:10.1126/science.280.5368.1415

Electromagnetic Images of the South and Central American Subduction Zones

4

Heinrich Brasse

Abstract

Current and fossil plate margins offer some of the most rewarding targets for geophysical studies. Particularly, the fluid/melt cycle in subduction zones continues to be of major interest for seismological as well as deep electromagnetic (EM), specifically magnetotelluric investigations. In this contribution we describe a number of experiments which have been conducted in several ocean-continent convergence zones around the world, with a focus on the Andes and Central America, respectively. Zones of potentially high electrical conductivity range from bending-related faulting near the outer rise, the subduction channel at the tip of the continental plate, the dehydration-hydration cycles in and above the downgoing plate, the assumed melting of the asthenospheric wedge to the rise of melts toward the volcanic arc and the magma chambers beneath the volcano edifices. Further targets include fault zones in the forearc, accommodating tensional stress, as well as hydrothermal and mineral deposits, to mention a few. The following chapters emphasize on a variety of structures along continental margins and show the potential of deep EM in this geodynamic setting.

4.1 Introduction

It is one of the wonders of geoscience that volcanic ranges where hot, molten material reaches the surface, are often located at subduction zones in which cold oceanic lithosphere is subducted. This enigma has been resolved during the last decades by numerous geological, geophysical, mineralogical and petrological studies: As the downgoing plate reaches larger depth, i.e., higher temperatures and pressures, mineral-bound water is released in a succession of dehydration/hydration reactions. At certain depths, the released water leads to lowering of the solidus and partial melting of mantle material of the overriding plate. The melt then rises – perhaps with some intermediate storage at crustal density boundaries – to the surface, where it erupts at arc volcanoes. These processes are described, e.g., by Schmidt and Poli (1998), see also Fig. 4.1.

They should lead to distinct geophysical anomalies, particularly to low seismic velocity and high electrical conductivity (low resistivity) zones. Electrical conductivity σ, measured in S/m (Siemens per meter), of rocks is normally low at crustal or even upper

H. Brasse (✉)
Freie Universität Berlin, Fachrichtung Geophysik,
12249 Berlin, Germany
e-mail: heinrich.brasse@fu-berlin.de

E. Petrovský et al. (eds.), *The Earth's Magnetic Interior*, IAGA Special Sopron Book Series 1,
DOI 10.1007/978-94-007-0323-0_4, © Springer Science+Business Media B.V. 2011

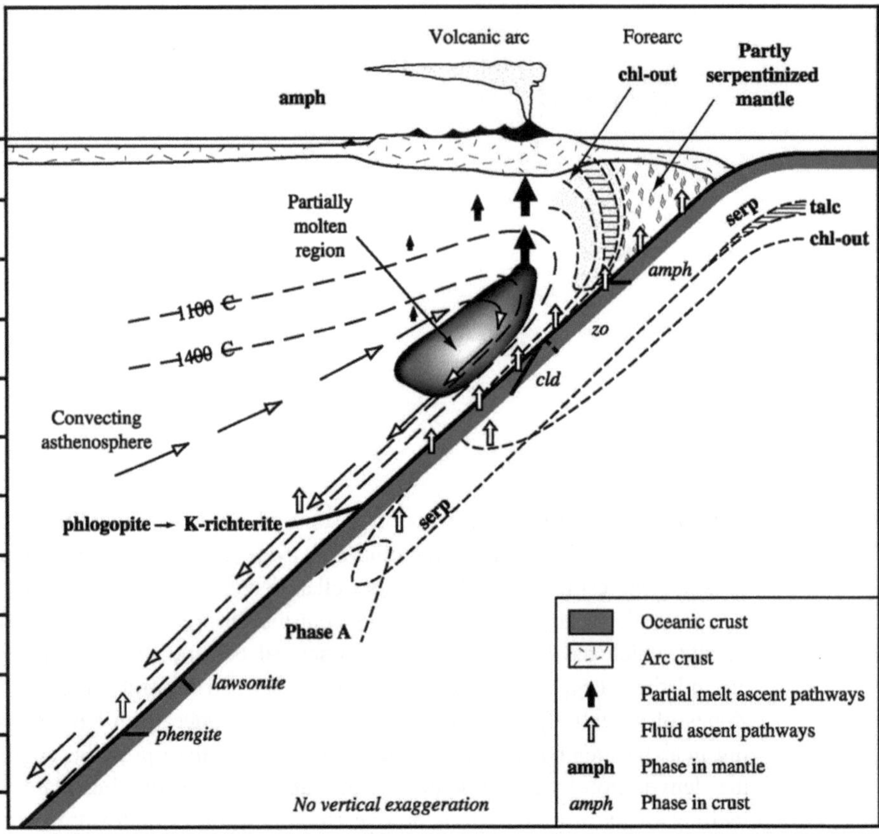

Fig. 4.1 The fluid-melt cycle in subduction zones (Schmidt and Poli 1998)

mantle temperatures. When metal or fluid phases come into play, a dramatic increase of conductivity is usually observed. Thus, investigation of the conductivity structure may be a valuable tool in the exploration of subduction zones, where fluids and melts play a significant role. And, indeed, numerous geophysical images in various parts of the world showing these effects have been presented in the past. Often, the resulting models do not image the complete fluid/melt cycle as evident from Fig. 4.1, but only aspects thereof, and reveal surprising aspects which where not known before.

The EMSLAB amphibious (onshore-offshore) electromagnetic experiment was intended to study the resistivity structure of the Cascadia subduction zone, where the Juan de Fuca plate subducts beneath the North American continent. Instruments were deployed along the Lincoln line in the state of Oregon with a seaward extension across the trench. Early 2-D forward and inversion models (Jiracek et al. 1989, Wannamaker et al. 1989) already showed low-resistivity zones (with resistivity $\rho = 1/\sigma$, measured in Ωm) on top of the subducting plate and beneath the volcanic range of the High Cascades. Later, various modeling and inversion attempts were conducted by the Russian EM community (Varentsov et al. 1996, Vanyan et al. 2002); from these studies the final model EMSLAB III of Vanyan et al. (2002) is shown in Fig. 4.2. The geometry of model domains was derived from seismological and gravity investigations, and resistivity was allowed to vary within these preset domains. Again, a low-resistivity zone is evident above the Juan de Fuca plate, extending to a depth of approx. 40 km. It is not clear yet, if this HCZ (high-conductivity zone) is related to the subduction channel or to the directly overlying zone where hydration (serpentinization) occurs again. A moderate, dike-shaped conductor is modeled beneath the High Cascades (HC in Fig. 4.2), which is rooted in the asthenosphere of the overriding plate. Interestingly, in the lower crust the conductor broadens

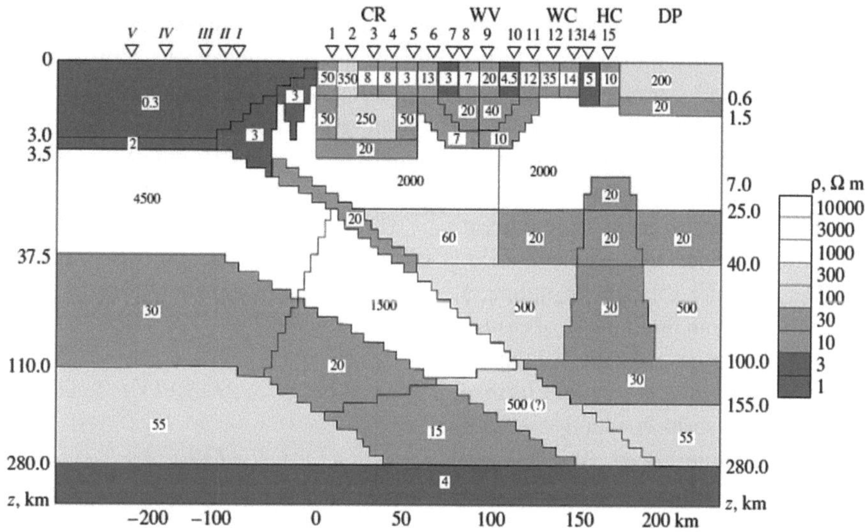

Fig. 4.2 Resistivity model of the Cascadia subduction zone (Vanyan et al. 2002). Note the non-linear depth scale

and extends into the forearc until the Willamette Valley and into the backarc beneath the Deschutes Basin (WV and DP in Fig. 4.2).

A HCZ beneath at least some of the High Cascades volcanoes was also imaged by Patro and Egbert (2008) with 3-D inversion of magnetotelluric impedance data from the USArray experiment, using broadly distributed stations from the northwestern United States. A very distinct anomaly is observed by Hill et al. (2009) below Mount St. Helens and an obvious connection towards Mount Adams. Although there is some debate about the cause of the enhanced conductivity, the most probable source seems to be partial melt.

Studies restricted to land measurements alone may often not image a HCZ above the slab or associated with the subduction channel, as was pointed out by Evans et al. (2002). And indeed, apart from the Cascadia models mentioned above, such a structure has never been observed on the basis of onshore data, neither in Canadian Cascadia (Soyer and Unsworth 2006) nor in Japan or New Zealand (Wannamaker et al. 2009). However, at the Mexican margin and farther inland, Jödicke et al. (2006) resolved a distinct anomaly above the subducting plate which they attribute to dewatering of the slab.

Apart from the investigation of Mount St. Helens mentioned above, arc volcanoes seem to be a difficult target for a detailed passive EM study, particularly due to often rugged topography, the shielding effect of hydrothermal systems, and the limited spatial extent of possible magma chambers and the feeder dike; see e.g. the 3-D model study at Merapi volcano, Indonesia (Müller and Haak 2004) or the combined MT/self-potential work at volcanoes in Japan (Aizawa et al. 2009). More promising targets seem to be large caldera systems, e.g., the Taupo volcanic zone in New Zealand (Heise et al. 2010).

4.2 Basic Principles of the Magnetotelluric Method

In this section an overview of the magnetotelluric method is given and the terms and concepts necessary to understand the following are briefly explained. For a comprehensive overview the reader is referred to the literature, e.g., Simpson and Bahr (2005) or Zhdanov (2009).

Electromagnetic (EM) exploration comprises a wide variety of sounding methods. While active methods, which involve the usage of transmitters, are primarily utilized in the exploration of near-surface structures, only passive methods may be employed for sounding of Earth's deeper crust and upper mantle.[1] For periods T larger than roughly 1 s, energy

[1] Hybrid cases are the so-called VLF (very low frequency) and RMT (radio magnetotelluric) methods which use the radiation

is provided by the time-varying solar wind, which carries a frozen magnetic field from the surface of the sun. Particles of the solar wind – primarily protons – interact with the Earth's magnetosphere and lead to variations of the magnetic field over a wide period range. Additionally, ionization of the ionosphere by ultraviolet solar radiation leads to large-scale current systems which in turn influence the magnetic field. For short periods $T < 1$ s, electromagnetic radiation ("atmosferics" or "sferics") from lightning strokes provides the source. A particular case are the Schumann resonances, standing waves in the waveguide formed by the ionosphere and Earth's surface, with frequencies $f \approx 7.8, 14.1, 20.3, \ldots$ Hz. Sprites, elves, blue jets and other phenomena associated with the upward discharge of thunderclouds are perhaps not frequent enough to act as a near-permanent source, as required for passive EM sounding; the same probably holds also for whistlers.

Altogether, a wide period range, generally from $T \approx 10^{-4}$ s to $T \approx 10^4$ s is available for EM sounding. The fields diffuse into the conducting earth, where they induce a current system which in turn induces a secondary magnetic field. A superposition of primary and secondary magnetic and electric fields is then measured at the surface. A scale length for the induction process in a homogeneous half space is the depth of penetration or skin depth

$$\delta = \sqrt{\frac{2}{\mu\omega\sigma}} \approx 0.503\sqrt{\rho T}\ [km] \qquad (4.1)$$

with μ denoting magnetic permeability of free space (the relative permeability of the Earth may be set to 1 in most cases), ω angular frequency, σ electrical conductivity, and electrical resistivity $\rho = 1/\sigma$; at depth δ the magnitude of the field has decayed to 1/e of its surface value. σ is measured in S/m, while Ωm is the unit of ρ.

In the magnetotelluric sounding method (MTS) horizontal electric and magnetic fields are measured at the surface of the earth. In geomagnetic depth sounding (GDS) the vertical magnetic field is related to the horizontal magnetic field, while in magnetovariational sounding (MVS) the magnetic fields at different locations are compared. The fields have to fulfill the quasi-homogeneous approach, i.e., their spatial structure may be approximated as a plane wave. This is certainly not the case for the ionospheric current systems, e.g., the Sq (solar quiet) current and the polar and equatorial electrojets (PEJ and EEJ).[2]

The magnetotelluric impedance[3] **Z** is defined as a complex transfer function between horizontal electric and horizontal magnetic fields in the frequency (or period) domain:

$$E_x(T) = Z_{xx}(T)B_x(T) + Z_{xy}(T)B_y(T) \qquad (4.2)$$

$$E_y(T) = Z_{yx}(T)B_x(T) + Z_{yy}(T)B_y(T), \qquad (4.3)$$

where x,y,z denote Cartesian, geomagnetic coordinates, E is the electric field, B geomagnetic induction and T period.

For an one-dimensional (1-D) subsoil the main diagonal of the impedance tensor vanishes and $Z_{xy} = -Z_{yx}$. In a 2-D environment a coordinate system may be found by rotation of **Z** with the strike angle α so that the main diagonal (in rotated coordinates) vanishes again, but now $Z_{xy} \neq -Z_{yx}$. In E-Polarization or TE mode (electric field $\parallel$ strike) a vertical magnetic field arises, while the vertical electric field in B-Polarization or TM mode (magnetic field $\parallel$ strike) is almost never measured. In 3-D, an assignment of TE and TM modes is no longer possible. Interpretation is often complicated by "static shift" or galvanic effects, the parallel shifting of apparent resistivity curves $\rho_a(T)$ due to small-scale, near-surface heterogeneities without affecting the phase $\phi(T)$. A distortion-free measure of the impedance is the phase tensor (Caldwell et al. 2004).

The tipper $\mathbf{W} = (W_x, W_y)^t$ (with t denoting transpose) is defined as a complex transfer function between vertical and horizontal magnetic fields (Schmucker 1970):

$$B_z(T) = W_x(T)B_x(T) + W_y(T)B_y(T). \qquad (4.4)$$

from distant transmitters for communication with submarines and LF radio stations, respectively.

[2] If the spatial structure of the field is known, an extension of the methods described above may be used to estimate conductivity of the Earth.

[3] The impedance is defined here via the magnetic induction B as is common in MT, yielding an unit of m/s. By replacing induction B with the magnetic field H the expected unit of Ω is obtained.

It is conveniently displayed as an "induction arrow" or "induction vector"[4] for both real and imaginary parts, calculated according to:

$$\mathbf{P}(T) = Re\{W_x(T)\}\mathbf{e}_x + Re\{W_y(T)\}\mathbf{e}_y \quad (4.5)$$

$$\mathbf{Q}(T) = Im\{W_x(T)\}\mathbf{e}_x + Im\{W_y(T)\}\mathbf{e}_y, \quad (4.6)$$

with $\mathbf{e}_x$ and $\mathbf{e}_y$ as unity vectors in x- and y-direction. If plotted on a map, real arrows point away from conductive zones in a 2-D environment, while imaginary arrows change signs at the period where the real arrow is maximal; this is the so-called *Wiese* convention. Sometimes the real arrow is reversed to make it point towards conductive anomalies; this style is referred to as *Parkinson* convention.

In magneto-variational sounding, magnetic fields at a site are referenced to a remote or base station (Schmucker 1970):

$$B_x(T) = M_{xx}(T)B_x^b(T) + M_{xy}(T)B_y^b(T) \quad (4.7)$$

$$B_y(T) = M_{yx}(T)B_x^b(T) + M_{yy}(T)B_y^b(T) \quad (4.8)$$

$$B_z(T) = M_{zx}(T)B_x^b(T) + M_{zy}(T)B_y^b(T) \quad (4.9)$$

with b denoting the base and **M** is the magnetic tensor. Often a 1-D base station is not available in the measuring area; then only "inter-station" transfer functions are calculated. Both **M** and the tipper **W** are practically not affected by static shift.

4.3 Electrical Resistivity Beneath the Andes

The Andes form a circa 8000 km long mountain range where the oceanic Nazca and Antarctic plates subduct beneath the South American continent; the oceanic plates are separated by the Chile Rise as a continuation of the East Pacific Rise. Convergence rate of Nazca-South American plates amounts to currently 6.5 cm/yr (Klotz et al. 2006). At the Chilean margin, convergence is oblique with a direction of ∼N80°E. Four volcanic provinces are identified along the margin: (1) The Northern Volcanic Zone (NVZ) of Colombia and Ecuador, (2) the Central Volcanic Zone (CVZ) in Peru, Bolivia, Argentina and Chile at the western rim of the Altiplano-Puna high plateau, (3) the Southern Volcanic Zone along the border of Argentina and Chile, and (4) the Austral Volcanic Zone (AVZ) south of the triple junction, where volcanism is predominantly occurring on Chilean territory. For the Central Andes volcanism spatially correlates with (relatively) steep subduction of the Nazca plate (Cahill and Isacks 1992), while large volcanic gaps occur in the adjacent regions of sub-horizontal (flat-slab) subduction in Peru and Chile/Argentina. Backarc volcanism occurs in various parts of Bolivia as well as Argentina.

The Peru-Chile Trench shows markedly different characteristics along the margin. In the Central Andes it reaches depths of 8000 m and is almost devoid of sediments due to the hyper-arid climate of the Atacama desert, while the southern trench is dominated by large sedimentary filling as a consequence of abundant rainfall; this can immediately be seen in a topographic/bathymetric map (Fig. 4.3). Depth of the trench slowly decreases from north to south.

In spite of the enormous dimensions of the mountain range and the abundance of geoscientific targets, the Andes remain until today largely undersampled with regard to deep EM sounding. The first observation of an extended high-conductivity zone beneath the Central Andes came from a magnetometer array established by the Carnegie Institution of Washington (Schmucker et al. 1966). Much later, in the eighties and early nineties of last century the Free University of Berlin carried out a magnetotelluric transect over the entire mountain range at approximately 22°S. This study revealed mid-crustal high-conductivity zones beneath the volcanic arc as well as beneath the Altiplano plateau (Schwarz and Krüger 1997). More recently, the Argentinian flat-slab segment was subject to a study by the universities of Washington and Buenos Aires. A spectacular electrical anomaly (see Fig. 4.4) was detected at large depths of $\geq$ 100 km and extending to at least 250 km (perhaps more than 400 km), which was attributed to image partial melting associated with the plunging of the Nazca plate (Booker et al. 2004).

[4] Although the term vector is often used (also in this text), note that, for coupled anomalies the "vectors" can't simply be added (Siemon 1997).

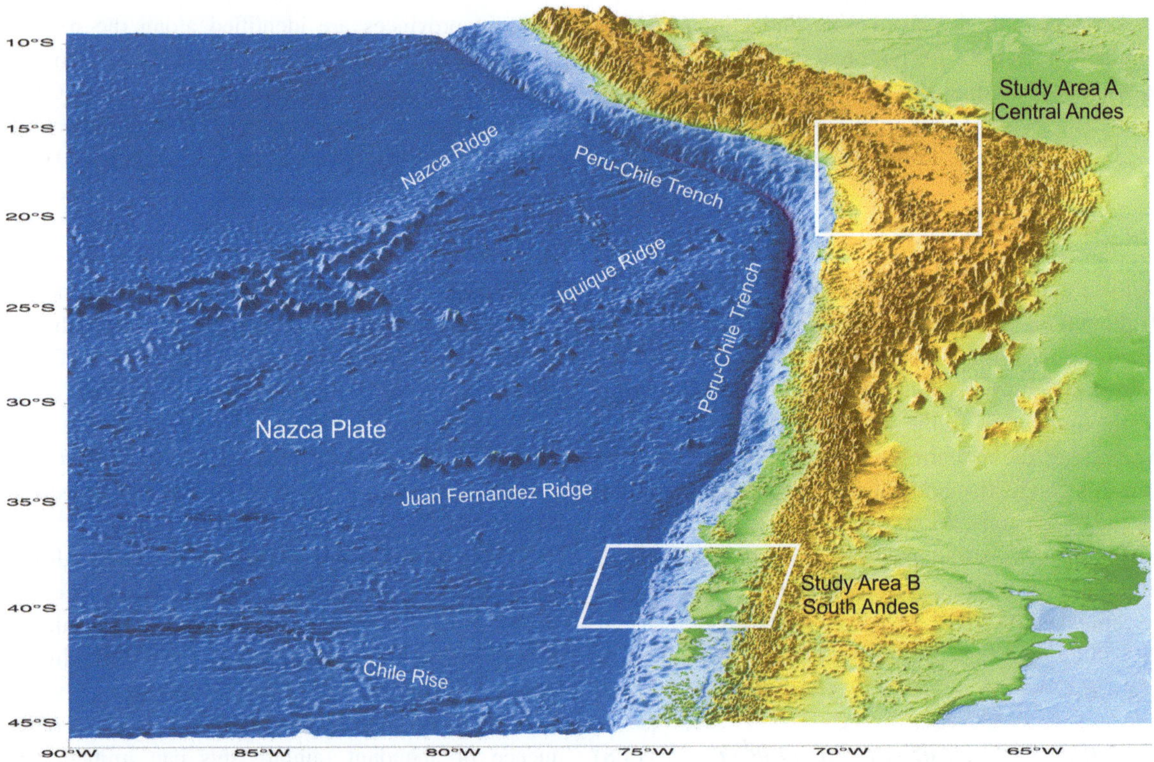

Fig. 4.3 Topographic-bathymetric map of the Central and Southern Andes and the Nazca plate, derived from ETOPO1 data. Note the change of morphology of the trench along the margin and the widening of the mountain range in the Central Andes (Altiplano-Puna Plateau). Study areas of this contribution are also shown

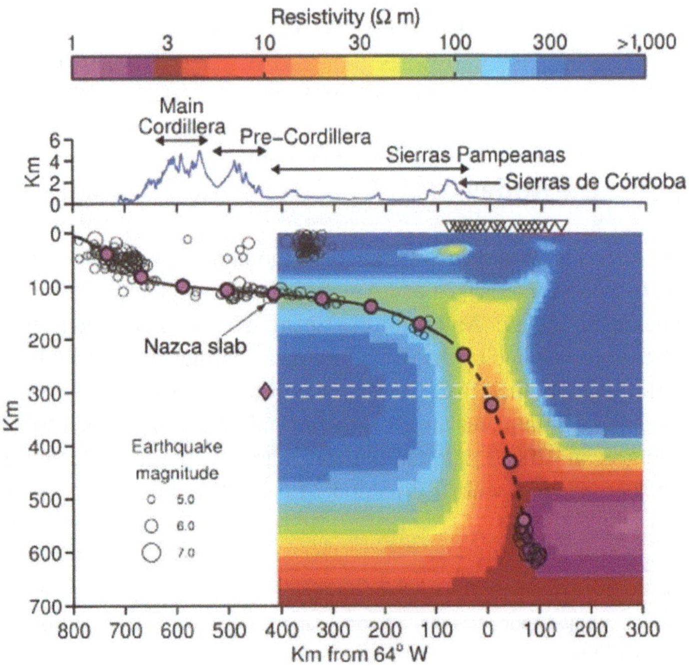

Fig. 4.4 Resistivity model of the Pampean flat-slab segment in Western Argentina, modified from (Booker 2004)

4.3.1 The Central Andes

In the Central Andes of Peru, Chile, Bolivia and Argentina the 1800 km long and up to 350 km wide Altiplano-Puna high plateau, with average elevations of 3800 m in its northern (Altiplano) and 4500 m in its southern part (Puna), has evolved since the late Oligocene; crustal thickness amounts to 70 km (James 1971, Wigger 1994, Yuan et al. 2002, ANCORP Working Group 2003). The causes for uplift are mainly seen in crustal shortening (e.g., Allmendinger et al. 1997, Elger et al. 2005) with perhaps minor contributions from magmatic addition beneath the volcanic arc (Dorbath and Masson 2000) or hydration of the lithospheric mantle (James and Sacks 1999). The uplift is thought to have been triggered by arid to hyperarid climatic conditions prevailing at the central western Andean margin since at least the Miocene (Lamb and Davis 2003), which led in turn to a trench nearly completely devoid of sediments (von Huene et al. 1999). At the Peru-Chile trench (reaching depths of up to 8 km in the region of the Central Andes) the Nazca plate currently subducts obliquely with an angle of N77°E and a velocity of 6.5 cm/a (Klotz et al. 2006); the angle of subduction is ~25–30° below the plateau (Cahill and Isacks 1992). The active volcanic arc – with individual edifices exceeding heights of 6000 m – developed at the western margin of the plateau.

A large number of long-period magnetotelluric measurements have been conducted along several profiles in the Central Andes. The current synthesis concentrates on transects A, B and C, which all cross the volcanic arc and reach into or traverse the high plateau (Fig. 4.5).

4.3.1.1 The Southern Plateau

At latitude 21°S the southern Altiplano crust in Bolivia is characterized by large, coincident geophysical anomalies, in particular a broad and deep-reaching zone of enhanced conductivity (Fig. 4.6, Soyer and

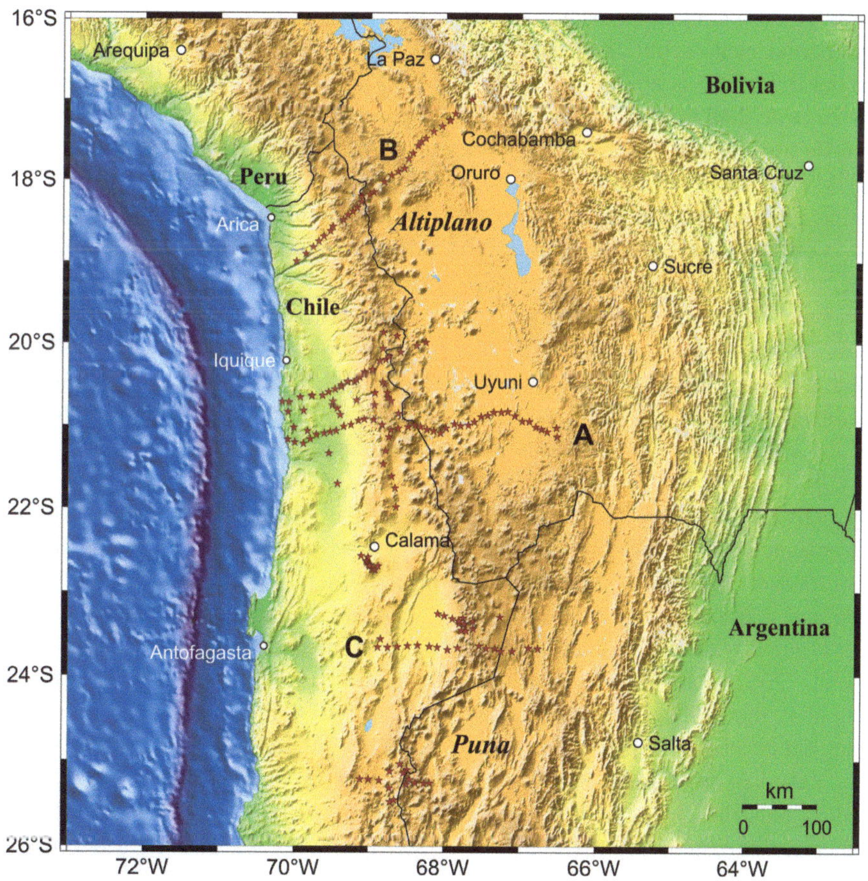

Fig. 4.5 Magnetotelluric sites in the Central Andes. Transects A, B and C are treated in the text

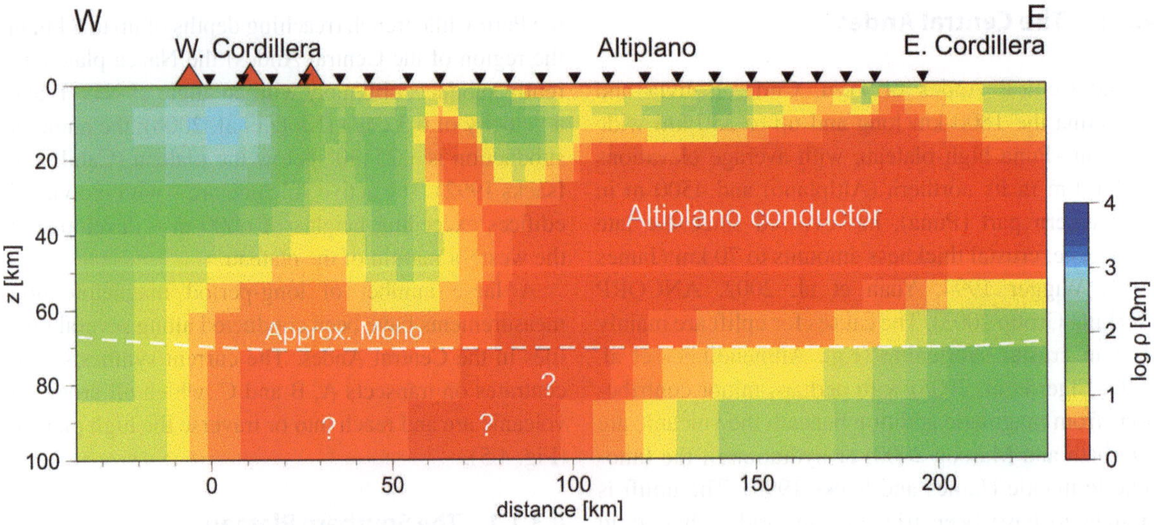

Fig. 4.6 2-D resistivity model of the Western Cordillera and the Southern Altiplano at 21°S

Brasse 2001, Brasse et al. 2002), high attenuation of seismic waves (Haberland et al. 2003) and low velocities (Heit 2005), all hinting at the occurrence of large volumes of partial melt (Schilling et al. 2006). Note that the model in Fig. 4.6 is a new compilation: Whereas in (Brasse et al. 2002) a Gauss-Newton version of the 2-D code of Rodi and Mackie (2001) was applied and only MT impedances were inverted, here tipper data were included and the non-linear conjugate gradient version was employed. The regularization parameter was set to $\tau = 10$ and the parameter β for balancing of horizontal to vertical smoothing was set to 1 (for details of the inversion algorithm see Section 4.3.1.3). The main difference to the original model is that the main anomaly is now clearly confined to the crust. Note that the volcanic arc is *not* underlain by high conductivity.

The high-conductivity zone leads to severe attenuation of electromagnetic fields which aggravates the resolution of deeper structures, particularly in the upper mantle and a possible root of the crustal anomaly. For the original model (Brasse et al. 2002) Schwalenberg et al. (2002) carried out a sensitivity analysis showing that the deeper parts of conductive structures are not resolved. In the forearc other crustal anomalies linked to arc-parallel but also arc-perpendicular faults blur the conductivity image at depth. It was thus a major aim to verify the lateral extent of the crustal conductor (i.e., may it be seen as a characteristic feature of the entire plateau?) and to prove if the subduction-related fluid/melt cycle may be resolved in an alternative location of the high Andes.

4.3.1.2 The Central Altiplano

This study area is situated in the Bolivian Orocline or "Arica Bend" (named after the coastal town of Arica) at about 18°S, where the coastline and the structural trends change from N-S to NW-SE. Morphological segments of the forearc parallel to the trench correspond to an ancient magmatic arc which shifted eastward through time, recordable since the Jurassic (Scheuber et al. 1994) in the Coastal Cordillera, afterwards in the Longitudinal Depression, the Precordillera and until recently in the Western Cordillera. The forearc is incised by deep, E-W running valleys which open to the ocean; some of them may be associated with active faults. A system of west-vergent, steeply dipping thrust faults in the Precordillera is thought to be the location where the plateau was uplifted (Muñoz and Charrier 1996). The smooth monoclinic western flank of the plateau locally collapsed to form the giant Oxaya block (Wörner et al. 2002). Plio-Pleistocene to recent andesitic to rhyodacitic stratovolcanoes of the Western Cordillera are built on thick layers of ignimbrites which are widespread across the forearc until the coast and are dated between 22 and 19 Ma (Oxaya Ignimbrites) and 2.7 Ma (Lauca-Pérez Ignimbrite) (Wörner et al. 2000). Volcanic activity is less than to the south, where the lithosphere is ~50 km thinner (Whitman et al. 1996); however, volcanoes are categorized from dormant (Parinacota) to solfataras state (Guallatiri) (Gonzáles-Ferrán 1994), with ongoing activity probably reflected by shallow seismicity beneath the arc (Comte et al. 1999).

The most prominent structure of the Altiplano is the 80 km wide Corque Syncline where Tertiary strata reach a thickness of up to 10 km (Hérail et al. 1997). To the west the basin is bordered by the San Andrés-Villa Flor Fault which corresponds to the eastern limit of the Arequipa block, part of the Arequipa-Antofalla Massif with up to 2 Ga old rocks which underpins the Western Cordillera probably along the whole plateau (James and Sacks 1999), whereas to the east the basin is controlled by the Chuquichambi thrust system. The Coniri-Laurani Fault System marks the transition to the Eastern Cordillera, the rough eastern flank of the plateau, where rocks up to Ordovician age are exposed. Farther to the east the Main Andean Thrust and the Subandean in the Andean foreland present its actual deformation front with thin-skinned folding and thrusting, and large shortening rates (Allmendinger et al. 1997).

The MT profile extends from the Longitudinal Depression over the Precordillera, the Western Cordillera and the central Altiplano basin to the Eastern Cordillera, following a general trend of N48°E and thus roughly perpendicular to the main structural units and the contours of the Wadati-Benioff zone. Data at 30 locations with a separation of ~10 km were collected during campaigns in late 2002 (Bolivia) and late 2004 (Chile, with a reference site in Bolivia occupied during both field studies). Several gaps in profile coverage were due to steep topographic gradients at the Western Cordillera Escarpment and in the Eastern Cordillera. Time series were processed by employing a robust, remote-reference code (Egbert 1997), yielding mostly stable and high-quality estimates of impedances and tippers in a period range from 10s–10000s. We may furthermore safely assume that the equatorial electrojet (EEJ) does not pose a problem as an inhomogeneous source in the investigation area according to an analysis of day/night time events (Friedel 1997) for data collected approx. 150 km farther south. The study presented here goes beyond a previous MT investigation on the central Altiplano (Ritz 1991) both by employing longer periods and by usage of magnetic field transfer functions.

Figure 4.7 exemplarily displays the real part of induction arrows along the profile for periods of 186 s and 1311 s. As mentioned above they should point away from good conductors in a two-dimensional setting. Remarkably, this is not the case in the forearc near the highly conductive Pacific Ocean (with a conductance of up to 25 000 Siemens at the trench); obviously conductive features running obliquely to the coastline (faults in the Coastal Cordillera?) deviate the direction of arrows to even coast-parallel. In the Precordillera,

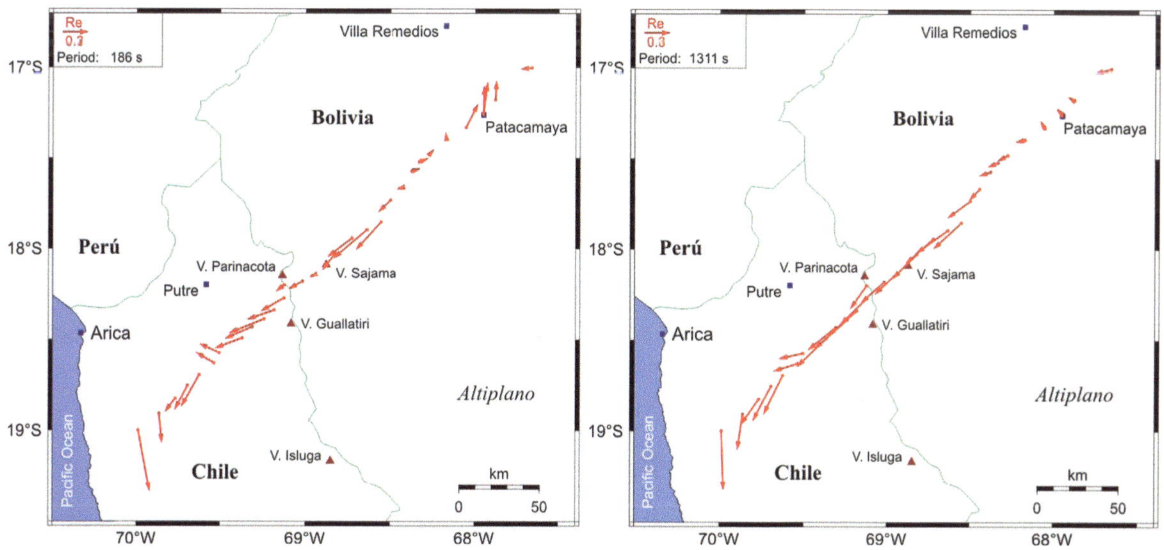

Fig. 4.7 Induction vectors for the Northern Altiplano at $T = 186$ s and 1311 s

the west-vergent thrust system and perpendicularly striking faults further complicate the situation. This is particularly obvious at the two sites near the Western Cordillera Escarpment and on the Oxaya Ignimbrites (*tic* and *oxa*). It shall be noted here that this deflection of arrow direction is observed in many other coastal and forearc parts of Chile as well (e.g., Brasse et al. 2009, Soyer 2002, Lezaeta 2001; see also below). In the volcanic arc and on the Altiplano proper, however, arrows at long periods point consistently parallel to the profile, indicating a strike direction of roughly NW-SE and thus parallel to main structural units. This is analogous to the situation farther south on the ANCORP profile (Brasse et al. 2002).

From the behavior of arrows several conclusions can be drawn immediately: (1) A 2-D interpretation is only meaningful for data on the plateau; the forearc data are clearly 3-D and thus discarded here for the time being. (2) Induction arrows at short periods show the margin of the well-conducting sediments of the central Altiplano basin. Unlike the south-western margin, the basement contact is not perpendicular to the profile near Patacamaya observatory. (3) The length of arrows at long periods is still large in the volcanic arc and decreases to almost zero only on the central plateau; this indicates that a deep high-conductivity anomaly is not situated below the volcanoes but rather beneath the central Altiplano. (4) At the northeastern end of the profile long-period induction vectors do not reverse direction as would be expected if they were caused by a single deep anomaly below the Altiplano. This hints at an additional anomaly below the Eastern Cordillera.

This crude image of electrical conductivity distribution is also supported by analysis of impedances. Figure 4.8 displays characteristic transfer functions for the Precordillera, volcanic arc, Altiplano and Eastern Cordillera. *Forearc*: The coast effect predominates (large splitting of app. resistivities ρ_a and phases ϕ for xy- and yx-components), but note that the induction arrows do not point W-E. *Volcanic arc*: The coast effect becomes smaller, induction arrows are large and point away from a conductive structure farther to the east. *Altiplano*: App. resistivities in the Corque basin are generally very small, resembling a three-layer case, and induction arrows are almost zero

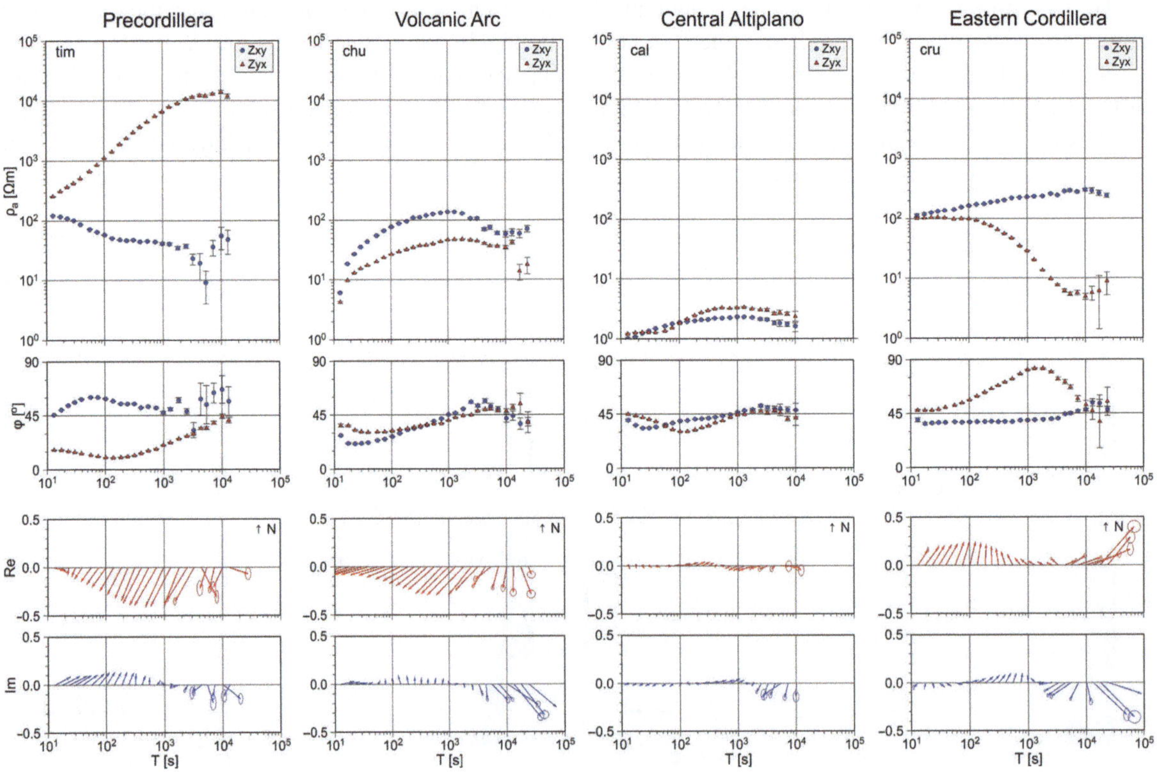

Fig. 4.8 Characteristic apparent resistivities, phases and induction vectors in the 4 morphological units crossed by profile A

indicating a conductive body at larger depth directly below. *Eastern Cordillera*: App. resistivities are higher again and strong splitting of ρ_a and ϕ is observed. Large, roughly northward pointing induction arrows at short periods indicate an additional conductivity contrast below the easternmost stations (*cru* and *lur*) or farther to the south.

A dimensionality analysis by calculating the skewness of impedance tensors reveals strong 3-D effects in the forearc while the rest of data may be regarded as 2-D: phase-sensitive skew values (being less prone to static effects) are below 0.3 – but note that this has to be regarded as a soft measure only; for definitions see, e.g., Bahr (1988). A notable exception is site *pat* (near Patacamaya geomagnetic observatory and the Coniri-Laurani Fault), where yx-phases leave the third quadrant reaching 45° at 20,000 s, an effect which may be explained by superposition of two anomalies with different strike directions at different depths and current channeling around a basement tip, extending into the highly-conducting sediments. Rotating these data into the coordinate system of regional strike (see below) yields normal phases in the first and third quadrant.

Decomposing the impedance tensor using a multi-site, multi-frequency scheme of McNeice and Jones (2001) provides estimates of regional strike for the whole profile as well as for single sites. According to the above mentioned 3-D influences, we first investigated the strike per station, which reflects again, besides the ocean effect at near-coastal sites, strong EW distorting influences at forearc stations. Thus excluding the forearc stations up to the western foothills of the high plateau provides quite a good fit for an unique strike direction of about N40°W, approx. perpendicular to the profile direction. This single-site strike (arithmetically averaged over profile) is more or less consistent with the results of the multi-site, multi-frequency routine, which yields N38°W strike direction for the plateau stations. Taking into account the average magnetic declination of approx. N4°W (MT fields are usually measured in geomagnetic coordinates), this strike is indeed almost exactly perpendicular to the profile direction (N48°E). Note, however, that this coincidence is fortunate: there is only a single accessible road along which sites could be deployed.

4.3.1.3 2-D Model Finding and Interpretation

With the arguments given above we assumed a common electrical strike of N38°W for all plateau sites and rotated all data by −38° into profile direction. Since strong 3-D effects do not permit 2-D modeling of forearc sites, they were discarded from further analysis as presented below. A qualitative explanation of coast-parallel induction arrows can be achieved by assuming structural anisotropy in the upper forearc crust, as was already shown for South Chilean data Brasse et al. (2009) by employing the anisotropic 2-D forward algorithm of Li (2002).

A basic model (Brasse and Eydam 2008) incorporates the Pacific Ocean with crude bathymetry, a highly-resistive forearc with an embedded anisotropic layer between $z = 5$ and 20 km and a strike direction of the conductive axis of N20°W, simulating a deeply fractured crust in accordance with the strike of the Precordillera Fault System in this area. At the margin of the anisotropic layer and without the ocean present, large induction arrows have a N110°W direction, perpendicular to the conductive axis. The coupling of the ocean and the anisotropic layer leads to a dramatic deflection of induction arrows from the expected W-E direction if only the coast effect would be present. For further explanation of the effects occurring at an anisotropic continental margin see the more systematic study of Brasse et al. (2009). A more quantitative investigation would have to take the true bathymetry into account as well as local structures in the forearc. This – or an equivalent full 3-D approach – goes far beyond the scope of this manuscript and is not followed for the time being.

Not taking into account the forearc stations has another positive consequence: the coast effect is much smaller for sites on the plateau (from the closest site *cat* distance to the coast is 90 km, to the trench 230 km) and important for very long periods only; it thus suffices to incorporate bathymetry in a crude manner. Nevertheless, the influence of the ocean is perhaps not purely 2-D due to bending of the trench near Arica; this has possibly to be considered in interpretation (Eydam 2008).

As in most of the modeling examples in this text, we employed the non-linear, conjugate gradient algorithm of Rodi and Mackie (2001) to carry out the 2-D inversions. It is based on a Tikhonov-type regularization by minimizing the objective function

$$S(\mathbf{m}) = \|W_d\,(\mathbf{d} - F(\mathbf{m}))\|^2 + \tau\,\|W_m\,(\mathbf{m} - \mathbf{m}_0)\|^2 \quad (4.10)$$

with **d** denoting the data vector, **m** the model vector, t the transpose, W_d a data weight matrix (usually the data variances) and τ the regularization parameter. F(**m**) is the model response, $\mathbf{m}_0$ the starting or an *a priori* model, ||...|| the norm, and W_m the regularization operator, often set as the Laplacian $(\nabla^2\,\mathbf{m})^2$. W_m incorporates a weighting function w(x,z) allowing to penalize horizontal or vertical exaggeration of model structures (R. Mackie, 2-D inversion manual):

$$\mathbf{m}^t\,W_m^t\,W_m\,\mathbf{m} \approx \int w(x,z)(\nabla^2\,\mathbf{m})^2 dA\,; \quad (4.11)$$

integration is over model area A. If an uniform grid is chosen for W_m, then

$$w(z) = (z(k)/z_0)^\beta \text{ for } z(k) > z_0 \text{ and} \quad (4.12)$$

$$w(z) = 1 \text{ for } z(k) \leq z_0 \quad (4.13)$$

where $z(k)$ is the thickness of the k-th row and z_0 the minimum block thickness, which has to be set manually.

The misfit between data and model response is calculated as a root mean square error according to:

$$rms = \sqrt{\frac{(\mathbf{d} - F(\mathbf{m}))^t\,W_d^{-1}\,(\mathbf{d} - F(\mathbf{m}))}{N}} \quad (4.14)$$

with N = number of data points. Under practical conditions, W_d contains only the main diagonal, i.e., the data variances. Note that this misfit measure depends on the data errors and care must be taken if analyzing a model fit by looking at the rms alone; this is particularly important if an error floor is set.

Numerous experiments have to be conducted to achieve a best-fitting and reliable final model. Although these tests should be a standard procedure in non-linear, regularized inversion, they are briefly mentioned here: (1) variation of the starting model and influence of the ocean; (2) exploration of model space by varying the regularization parameter yielding a trade-off curve (ideally L-shaped) between rms fit (root mean square) and model roughness (e.g., Asters et al. 2005); (3) assigning specific error floors to individual components, i.e., TE and TM mode (from tangential-electric and -magnetic, referring to the field parallel to conductivity contrasts) resistivities and phases, and tippers, as well as checking for static effects; (4) sensitivity tests and tests of (in)significant model features; (5) discretization; (6) convergence and number of iterations. Another important aspect is to check individual components separately: Due to different boundary conditions at interfaces, TE and TM mode impedances and tippers are sensitive to different subsurface structures.

Topographic effects play only a minor role along the profile with the exception of sites *par* (at the foot of Parinacota volcano) and *lur* in a river valley of the Eastern Cordillera. This may at first glance be surprising considering the high altitude of the study area, but the Altiplano is a vast plain with low intermittent ranges. Even in the volcanic arc, wide and only marginally undulating plains dominate the landscape between the individual volcanic edifices, making the terrain ideally suited for magnetotelluric work. Test modeling shows that at long periods >10 s the topographic effect reduces to a static shift problem (for 2-D only in TM mode) which is treated in the inversion process described above.

For the resulting model (Fig. 4.9), obtained by jointly inverting tippers, TE and TM mode apparent resistivities and phases, we chose a regularization parameter of $\tau = 10$ which lies in the corner of the trade-off curve (but note that the L-shape is not well expressed (Eydam 2008)). Error floors were set to 20% for apparent resistivities and 5 % (≈1.45° absolute) for phases, thus assigning a higher weight to phases in order to overcome the static shift problem. This procedure leads to app. resistivity curves shifted with respect to the data, while the phases are usually fitted well. Of course, the mentioned parameters are only for the data under consideration, other data sets may require different settings. The statics issue was further addressed by inverting for shift factors with the underlying (and perhaps not always unproblematic) assumption that the product of these factors should be one along the profile. Static shift in TM mode data is treated easily in a 2-D model by setting a small cell beneath a station to an adequate value; it may, however, be misleading in the case of TE mode. Here the shift results from inhomogeneities perpendicular to the profile (i.e., 3-D structures) and correcting this may require a change of structure at depth in the 2-D model (which is thus

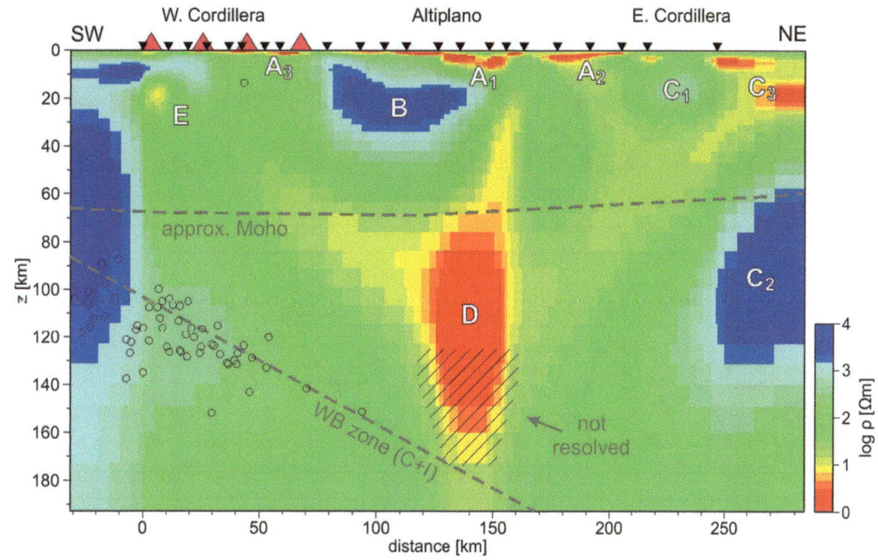

Fig. 4.9 Resistivity model for the central Altiplano in Bolivia and northernmost Chile (Brasse and Eydam 2008)

not a real static problem any more; shifts at neighboring sites may also be contradictory). Therefore some investigators prefer not to use TE mode at all or restrict themselves to TE phases only (see, e.g., Wannamaker et al. 1989).

Since tipper data are very consistent on the Altiplano, their weight was set very high with respect to impedance-derived data by assigning a small (absolute) error floor of 0.02. At site tak only tipper data could be used due to layout problems with the telluric lines. The general features of the model are not changed if the error floor for apparent resistivities is set higher, e.g., to 100 %. The resulting rms of the model displayed in Fig. 4.9 is 1.64, data fit is exemplarily shown for periods of 186 s and 1311 s along the profile in Fig. 4.10 and as pseudo sections in Brasse and Eydam (2008).

The "final" inversion model shows that the whole Altiplano crust and even the upper mantle are generally characterized by relatively low resistivities in the range of 100 Ωm, surrounded by high-resistivity zones at the SW and NE margins. Several structures which are consistently resolved throughout the inversion runs can additionally be identified (marked by letters in Fig. 4.9):

(A) The up to ~10 km thick Tertiary sedimentary layers of the central plateau, particularly in the Corque Syncline (A_1), are well resolved. The asymmetric geometry of the conductive layer fits well with the reconstruction of geological history for the region (Hérail et al. 1997, Sempere et al. 1990) and the prior study of Ritz et al. (1991). The eastern limit of the Corque Syncline is defined by the Chuquichambi thrust system (seen as an interruption of the low resistivities at sites *cal* to *rio*), marking a transition to another, less deep sedimentary basin to the east, a northern extension of the Poopó Basin (A_2) (Hérail et al. 1997). The very low resistivities of approx. 2 Ωm are most likely due to saline fluids which in turn may stem from buried salars (salt pans) – while salars are not exposed at the surface in the actual study area, they are a characteristic feature of the southern Altiplano. A_3 depicts the sediments between Parinacota and Sajama volcanoes; on the other hand the intra-arc Lauca Basin at the southwestern margin is either not very conductive or does perhaps not reach deep enough for the period range considered here to be characterized by low resistivities.

(B) An upper to middle crustal (up to 25 km depth) zone of high resistivities (>1000 Ωm) extends between the arc and the Corque Basin. Its northeastern margin coincides with the superficial trace of the San Andrés Fault which is also marked by a strong contrast in seismic velocities separating the relatively fast volcanic western Altiplano from the slow sedimentary basin to the east (Dorbath

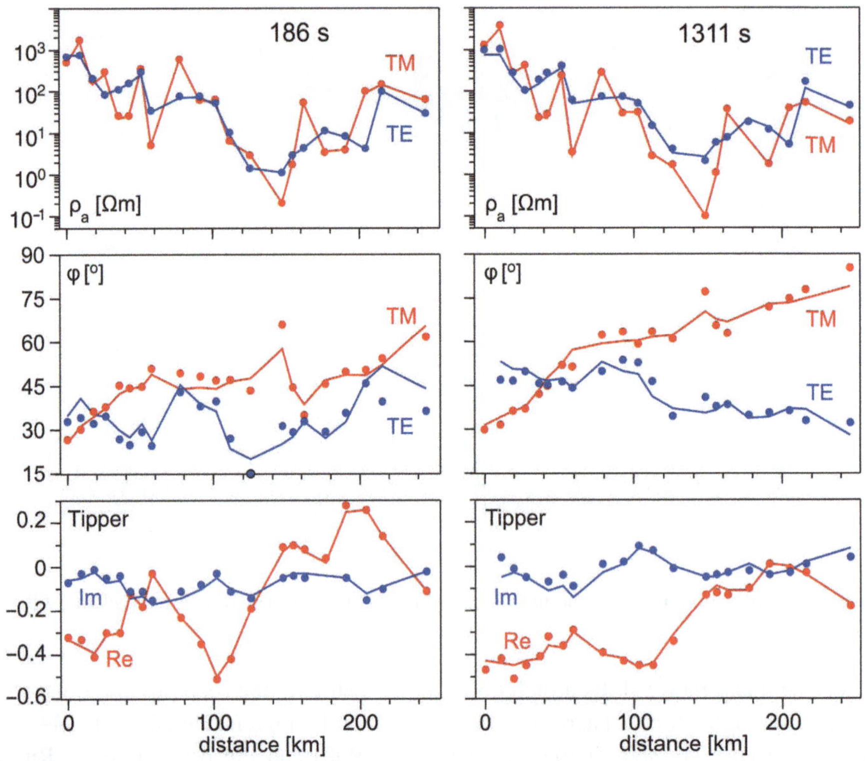

Fig. 4.10 Data and model response for 2 periods along the profile (Brasse and Eydam 2008)

and Granet 1996). This resistive, fast and thus rheologically strong structure may be correlated with old cratonic crust of the Arequipa block (see introduction).

(C) Passing the Coniri-Laurani Fault Zone to the east (sites *cru* and *lur*) higher resistivities (C_1) probably indicate the transition to the crystalline (Paleozoic) basement of the Eastern Cordillera. At larger depths the Eastern Cordillera is imaged as a very bad conductor (C_2), again in accordance with very high seismic velocities deduced by (Dorbath and Granet 1996). Interestingly, a good conductor (C_3) appears here at mid-crustal depths; this conductor was already mentioned in the section on data description and may well correlate with the conductivity anomaly running along the Eastern Cordillera, proposed by (Schmucker et al. 1966). Note, however, that this structure is at the margin of our study area and thus not well resolved.

(D) The most obvious anomaly is, however, located in the upper mantle and reaching into the lower crust below the central Altiplano, where resistivities are as low as 1 Ωm. This conductor would be in good accordance with the "standard model" of subduction (partially molten asthenospheric wedge) if its location were not 80-100 km northeast of the recent volcanic arc, i.e., already in the backarc (see discussion below).

(E) At the southwestern end of the profile, below the margin of the volcanic arc, a modest conductor is visible at mid-crustal depths. Although it may indicate a deep magma chamber, it should be noted that there is no recent volcanic activity observable at the surface; dormant Taapaca volcano is located 50 km farther to the NW. Due to its location at the edge of the profile and the increasing three-dimensionality in the adjacent forearc, this structure is poorly resolved.

Resolution analysis encompassed visualization of the sensitivity matrix and tests for significant structures by modifying their resistivity and/or geometrical borders. The highest sensitivity values are observed for the sedimentary basins and the mantle conductor. The more marginal conductors have higher sensitivities compared with their surroundings, too, but note that these features are not as consistently modeled in the different inversion runs as the basins and the mantle structure.

Several experiments to test resolution of the latter have been carried out, e.g., replacing the low resistivities with more normal values of 50 Ωm (leaving the other structures unchanged) worsens the rms from 1.64 to 4.94, an unacceptable increase of misfit. Then the inversion was started over, leading the conductor to appear again after a few iterations. The deeper parts (>120 km) on the other hand have only negligible influence on the model response and may thus be considered as unresolved (hatched area in Fig. 4.9). The principal features of the model in Fig. 4.9 did not change, if only a restricted period range (e.g., omitting all periods below 100 s) was used; the same statement holds for a variety of starting models. Another test concerned the location at depth: Can the measured data also be explained if the conductor is shifted entirely into the crust? As above, we replaced the low resistivities in the mantle with values of the surrounding mantle, placed a conductor in the mid-crust and started the inversion again. After 50 iterations the resulting rms is at an acceptable level of 1.85, but the conductor is moved deeper towards the Moho. The relatively good fit, however, is achieved at the cost of unrealistically low resistivities of 0.01 Ωm; this alternative model is therefore discarded.

The individual inversions of TE, TM and tipper data (which all achieved an rms in the range from 1.1 to 1.4) showed that the mantle conductor is best resolved by the tipper and (with slightly lower conductivities) in TE mode; both data subsets are in general agreement. On the other hand TM mode senses only a broad and diffuse high-conductivity zone at lower crustal/upper mantle levels, which is expected: TM data are less sensitive to deep, dyke-like structures. For upper crustal features all 3 subsets agree reasonably well in accordance with the model in Fig. 4.9.

4.3.1.4 The Asthenospheric Wedge

While most of the conductive and resistive features displayed in Fig. 4.9 are easily explained and agree well with other geological and geophysical background information, the deep high-conductivity zone in the upper mantle below the central Altiplano poses problems regarding its high core conductivity and its unexpected location with perhaps far-reaching consequences. Although such an anomaly as an image of melts in the asthenospheric wedge above the subducted plate is not surprising as such, it is not situated below the volcanic arc but rather shifted by 80–100 km toward the backarc. This does not seem to be unique for the Altiplano plateau: there are weak hints from profiles farther south at 21°S and 20.5°S that there exists a similar conductivity pattern off the volcanoes at large depths (e.g., Brasse et al. 2002, Schilling et al. 2006). On the southern Altiplano, however, this could not be resolved unambiguously due to lack of resolution below a huge crustal conductor. As Heit (2005) has shown the whole mid- and lower crust below the plateau at 21°S is also characterized by anomalously low seismic velocities, thus supporting the hypothesis of large volumes of partial melt in the mid- to lower crust. This large-scale conductor is missing below the central Altiplano (with the exception of near-surface conducting sediments), thus enabling insight into deeper segments of the subsoil.

An estimation of melt rates in the wedge may be achieved by assuming a simple two-phase system with near-perfect melt interconnectivity (Hashin-Shtrikman upper bound). The melt rate is critically dependent on conductivity of the melt phase (σ_m), which is in turn a function of temperature and composition. Melt composition for ultramafic material changes from nephelitic (above 2 GPa) through alkaline-basaltic to tholeiitic with increasing temperature or melt fraction, respectively (cf., Tyburczy and Fisler 1995, Roberts and Tyburczy 1999). In order to feed a volcanic arc, significant melting must occur in the wedge, which implies a minimum temperature of ~1 300°C (Schmidt and Poli 1998). Conductivity σ_s of the solid phase is comparatively unimportant; as function of temperature for dry, semi-conducting rocks it follows an Arrhenius law, and we may assume a σ_s in the order of 0.01–0.005 S/m at upper mantle depths.

The modeled bulk conductivity inside the wedge core is in the order of 1 S/m (Fig. 4.9), similar to the crustal conductor below the ANCORP profile and about an order of magnitude larger than the high-conductivity zones in South Chile (Brasse and Soyer 2001). For a basaltic to tholeiitic melt composition at 1 300°C, we may assume a melt conductivity of $\sigma_m = 5.5 - 6.5$ S/m, resulting in a melt fraction estimate of 21–25 vol.%.

This value is very high, but note that conductivities are for dry magmas which are less frequent in subduction zones. Subduction zone magmas contain at least 2–3 wt% water in the hot core of the mantle wedge Gaetani and Grove (2003); water content may even increase to 30 wt% for first magmas near the slab according to Grove et al. (2006). The supply of water enhances the melt conductivity considerably (Gaillard 2004); additionally higher temperatures or saline brines may further increase conductivity and thus reduce melt rate. For instance, a hotter wedge (1 400°C) implies a σ_m of 11.5 S/m for tholeiitic melt, yielding a value of 12.5 vol.%.

Conductivities of mantle fluids are difficult to assess. Deep crustal fluids are known to reach very high conductivities of up to 100 S/m (Nesbitt 1993), depending on their ionic composition. For slab-derived mantle fluids one may assume similar values based on high salinities found in fluid inclusions (Scambelluri and Philippot 2001). The input of saline fluids into the hot wedge may thus significantly reduce the bulk resistivities, yielding a melt rate in the order of 5 vol.%. In general the high melt rates estimated from electrical conductivities are principally in accordance with new geochemical results (Grove et al. 2006), revealing that the amount of hydrous melt may reach values of 10–15 wt.%.

Very recently, Gaillard et al. (2008) presented laboratory measurements of carbonatite melts with their conductivity values exceeding the numbers mentioned above by more than an order of magnitude. If such melts would be present in the Central Andes, they would reduce the necessary melt fraction significantly. However, such an occurrence is not documented to the author's knowledge.

To understand the location of the highly conductive part of the wedge, Fig. 4.9 also displays the projected locations of earthquakes (lying within a lateral distance of 50 km from the profile), extracted from the catalogue of Engdahl and Villaseñor (2002). As expected hypocenters are clustered below the volcanic arc at depths of 100–130 km (with an additional, single event in the crust below the volcanoes at the Chilean-Bolivian border). Major earthquake activity ceases at depths >150 km (leaving apart the very deep events below the eastern Andean foreland), still far from the anomaly, which is located above a slab depth of 160-200 km. The catalog only comprises larger events (M > 4.5), but there is no indication of smaller events according to the local tomography study of Dorbath and Granet (1996); this may, however, be a problem of resolution.

Attenuation data for the mantle below the central Altiplano are scarce. The images given by Baumont et al. (1999) and Myers et al. (1998) touch our study area only marginally and refer mainly to the crust; however, their conclusion about a heterogeneous and generally highly absorbing crust is in accordance with the conductivity image. The best correlation exists perhaps with an upper mantle high attenuation (low Q) zone (Myers et al. 1998) and low v_p velocities (Dorbath and Masson 2000) at 19.5°S. The well-conducting asthenospheric wedge is also in accordance with low densities in the uppermost mantle in the 3-D model of Tassara et al. (2006), explaining – together with the thickened crust – the large Bouguer anomaly of −400 mGal (locally −450 mGal) on the plateau.

There is an interesting correlation between our deep Altiplano model and the seismic attenuation model of Schurr et al. (2003) for the Puna plateau at about 24–25°S. It proposes vertical as well as non-vertical rise of fluids/melts in the backarc and fluid release from the downgoing slab at similar depths as the MT model presented here. The situation in the Puna, however, differs somewhat with respect to a zone of major seismicity in the backarc, directly below the anomaly. Due to the unfavorable distribution of earthquakes below the Andes at 18°S, a highly attenuating wedge corresponding to the resistivity model would be difficult to detect (i.e., only far to the east, perhaps beyond the Eastern Cordillera).

Intermediate-depth earthquakes are commonly understood as consequences of fluid release from the subducted slab (dehydration embrittlement, e.g., Kirby et al. 1996, Hacker et al. 2003). Due to propagation of hydraulic fractures rise of fluids into the overlying mantle is not vertical according to Davies et al. (1999). Corner flow may additionally transport the fluids farther away from the source region, including subsequent hydration and dehydration, perhaps in addition to a mechanism described by Mibe et al. (1999). Dependent on P-T conditions the released fluids may not form an interconnected network directly at the source (dihedral angle Θ above 60°); the hydrous peridotite (with fluids in isolated pockets) is dragged downward until an open network ($\Theta < 60°$) may form, e.g., at 1 000°C and 4 GPa.

A massive influx of fluids into the continental mantle may be facilitated by an increase of slab dip as was proposed for areas farther south (e.g., Kay and Mpodozis 2001); a larger subduction angle compared to the general Cahill & Isacks image of Wadati-Benioff contours has recently been deduced from local earthquake tomography by (David 2007) for the orocline, too. We can of course not exclude the scenario, but in this case one would assume the highly conductive wedge farther to the west.

From the asthenospheric wedge melt rises upwards, perhaps being stored in a MASH zone (melting, assimilation, storage, and homogenization) at the crust/mantle interface according to Hildreth and Moorbath (1988) before rising further into the crust, imaged by rather low resistivities in the MT model. The highly resistive Arequipa block seems to force a bifurcation; one branch leads subvertically to the central Altiplano. There are indications for another, non-vertical rise, too: From the upper part of the conductor a pathway with intermediate resistivities (20–60 Ωm) may be identified, passing obliquely through the crust until below the Western Cordillera volcanoes. Note that there is no indication of recent volcanic activity on the central plateau, perhaps with the exception of Cerro Colluma, a maar SE of the profile which may have been active in the Holocene (Smithsonian Global Volcanism Program, volcano no. 1505-024). But it should also be noted that heat flow on the plateau is widely enhanced – in the order of 100–120 mW/m^2 (e.g., Springer and Förster 1998, Hamza and Muñoz 1996) – and, although reliable data are sparse and scattered, likely higher than in the volcanic arc itself. One may perhaps speculate that we are observing the formation of a new volcanic arc and thus a momentary view of its eastward migration, as it has occurred already several times since the Jurassic.

The Altiplano plateau has been subject to extension during its initial build-up phases; if this tectonic pattern would be persisting until today, it would be difficult to explain why the partial melts would not reach the surface. Since about 18 Ma, however, shortening – i.e., compression – is observed on the entire plateau (Scheuber et al. 2006).

Individual volcanoes and possible magma chambers or conduits have not been the target of this study. To achieve a detailed conductivity image at upper crustal depths below the edifices a far denser data sampling including higher frequencies would have been necessary. Due to their enormous height (Parinacota 6348 m, Sajama 6542 m a.s.l.) a well-suited network of sites is difficult to construct.

Fluid release in and above the subducting slab further trench-ward has not been modeled along this profile, and including these as a-priori, well-conducting features did not prove to be successful; they were converted to highly resistive during the inversion process. As stated above, the conductivity image in the forearc is overprinted by 3-D crustal anomalies, and the part of the profile which is interpretable in a 2-D approach is too short to resolve a low-resistivity zone above the slab unambiguously.

The dominance of the backarc in the conductivity image is also emphasized in a further model more to the south at ≈ 24° S. Profile C extends from the Precordillera, crosses the Salar de Atacama and the Western Cordillera (S of highly-active Lascar volcano) and end for the time being in the western Puna plateau. Again the volcanic arc is not conductive (except near Lascar itself, where a HCZ is modeled south of the edifice Diaz (2010)), but beneath the Puna appears a huge conductor (Fig. 4.11). Its eastern and lower extensions can't be resolved due to the current lack of stations in Argentina, but there is striking correlation with seismic tomography (Schurr et al. 2003).

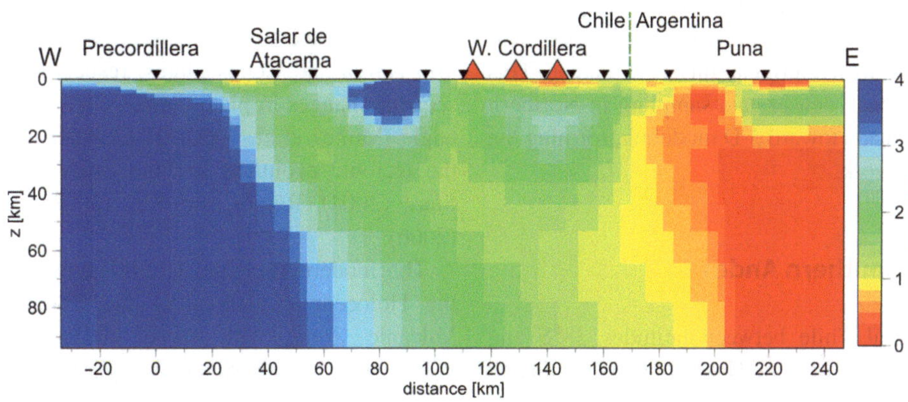

Fig. 4.11 Resistivity model for profile C

4.3.1.5 Conclusion for the Central Andes

New deep electromagnetic sounding data from the Bolivian orocline in the central Andes have been collected. They may be analyzed with a two-dimensional approach for the high plateau, but strong 3-D effects prevail in the Chilean forearc. Here data are tentatively explained by 2-D models incorporating structural anisotropy in the crust with a preference direction of electrical conductivity parallel to the strike of a prominent forearc thrust fault system. Classical 2-D modeling is thus restricted to the plateau, including the volcanic arc, the Altiplano basin and the margin of the Eastern cordillera. Strike directions deduced from impedance data agree well with those obtained from tipper data at long periods.

The 2-D inversion revealed several high-conductivity and high-resistivity zones in the crust and upper mantle of the central Andes. They are interpreted as images of sedimentary basins of the central Altiplano plateau, still uncertain (due to lack of resolution), partially molten regions below the Western Cordillera and the consolidated Precambrian Arequipa block, respectively. The main result is the image of the conductive asthenospheric wedge and a conductor rising into mid-crustal levels. This large anomaly is located already in the backarc, i.e., not beneath the current volcanoes. This may imply that we observe a snap-shot of the migration of the arc, as it has occurred frequently since formation of the subduction system.

On the other hand, no large-scale conductor in the crust as in the Southern Altiplano has been detected, which is thus not a characteristic feature of the entire plateau as was presumed earlier. To understand and incorporate transfer functions in the forearc, more complete approaches than the 2-D modeling investigations presented here are necessary, particularly near the coast and at the western margin of the plateau. This includes additional, denser data sampling in Northern Chile and extension toward 3-D modeling, forming the task for future work.

4.3.2 The Southern Andes

In South-Central Chile between latitudes 38°S and 41°S (Fig. 4.12) the oceanic Nazca plate is subducted beneath the South American continent and the great earthquake of 22 May 1960 (moment magnitude $M_W = 9.5$) initiated (Cifuentes 1989). Subduction is oblique with an angle of $\sim 25°$ (i.e., N77°E) with respect to the plate margin and with a current velocity of ~6.5 cm/a (Klotz et al. 2006). The study area is located in the northernmost Patagonian (Neuquén) Andes and can be subdivided into several main morphotectonic units (e.g., Folguera et al. 2006, Melnick et al. 2006): (1) a narrow Coastal Platform comprising uplifted Tertiary marine and coastal sequences; (2) the Coastal Cordillera, formed by a Permo-Triassic accretionary complex and a late Paleozoic magmatic arc; (3) the Longitudinal Valley, a basin filled with Oligocene-Miocene sedimentary and volcanic rocks, covered by Pliocene-Quaternary sediments; (4) the Main Cordillera, formed by the modern magmatic arc and intra-arc volcano-sedimentary basins; (5) the Loncopué Trough, already in Argentina, an extensional basin east of the Main Cordillera associated with abundant mafic volcanism; (6) the southern extension of the Agrio fold-and-thrust belt; and (7) the Mesozoic Neuquén Basin and the Cretaceous-Tertiary foreland basin to the east.

Subduction at the Chilean margin started already in the late Paleozoic, while Andean evolution began in the Jurassic, associated with the opening of the South Atlantic Ocean. In the Cretaceous widespread plutonism occurred in the Coastal Cordillera and in the area of the volcanic arc, where the Patagonian Batholith was formed. South of 38°S the position of the volcanic arc remained relatively constant through time with the exception of a significant broadening of the magmatic system (Muñoz et al. 2000) and an 80–100 km westward shift of the volcanic front in the late Oligocene-early Miocene with respect to its current position (Parada et al. 2007). This event was probably related to the breakup of the Farallon plate into Nazca and Cocos plates, respectively, and subsequent changes in plate convergence and subduction angle (Muñoz et al. 2000). For further description of the tectonic evolution see the overview articles by Stern (2004), Ramos and Kay (2006), and Glodny (2006).

The modern Principal Cordillera is dominated by the Holocene volcanoes of the Southern Andean Volcanic Zone, with some of the most active volcanoes in South America, e.g., Villarrica, Llaima and Lonquimay (Gonzáles-Ferrán 1994). The chain of

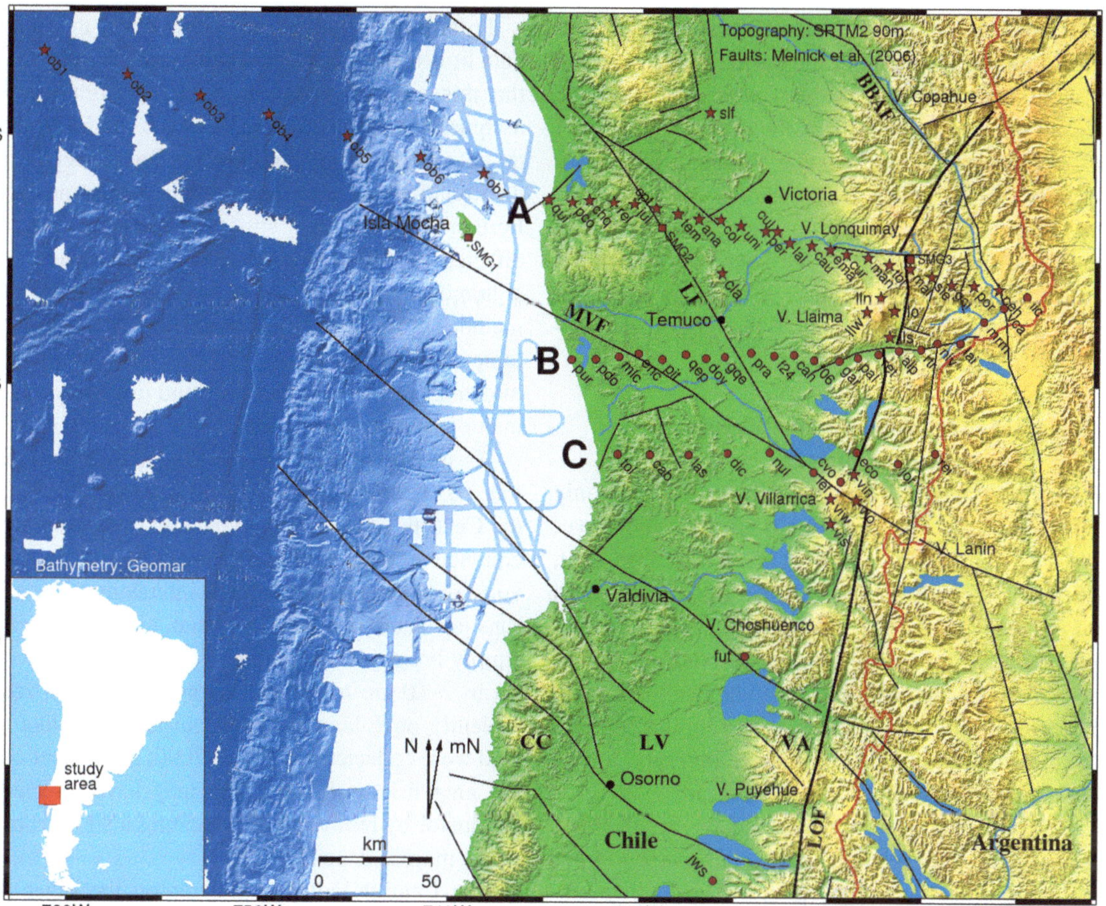

Fig. 4.12 Magnetotelluric sites in Southern Chile. Topography is based on SRTM data, while swath bathymetry was obtained aboard RV Sonne during various cruises (Scherwath et al. 2006)

stratovolcanoes is aligned parallel to the trench and along the Liquiñe-Ofqui Fault (LOF), a mega shear zone extending for over 1 000 km from the triple junction of Antarctic, South American and Nazca plates to ~ 38°S (Cembrano et al. 1996, 2007). A NW-SE – thus obliquely to the trench – oriented fault system crosses the arc and forearc, e.g., Melnick et al. (2006), which may have been of importance for a major eruption in the Cordon Caulle volcanic complex immediately after the $M_w = 9.5$ earthquake (Lara et al. 2004). The Lanalhue Fault, in particular, is regarded as an inherited, continuously reactivated, pre-Andean structure, which is associated with deep-reaching seismicity (Yuan et al. 2006).

Two long-period magnetotelluric campaigns were conducted, an earlier one in 2000 and an additional field experiment in austral summer 2004/2005, which also included an amphibious component employing sea-bottom instruments from Woods Hole Oceanographic Institution (WHOI). While the first experiment (Brasse and Soyer 2001, Soyer 2002) was carried out in the framework of the multi-disciplinary programme SFB 267 "Deformation Processes in the Andes", the second one (Kapinos 2011) was part of the TIPTEQ project ("From the Incoming Plate to Megathrust Earthquakes"), with other subprojects dealing with passive and active seismology, gravity and geology/tectonics. Structural information at the South Chilean margin in the study area concerning Moho depths and geometry of the downgoing plate may particularly be inferred from a large number of recent seismic experiments (Bohm et al. 2002, Lüth and Wigger 2003, Rietbrock et al. 2005, Haberland et al. 2006, Krawczyk et al. 2006).

During the two campaigns, a total of 72 stations were deployed between the Argentinian border and the Pacific Ocean, yielding electromagnetic (MT and GDS) transfer functions in the period range between 10 s – 20,000 s. On land, the network encompasses the areas of the Coastal Cordillera, the Central Depression or Longitudinal Valley and the Principal Cordillera (see Fig. 4.12). The seafloor stations (from Woods Hole Oceanographic Institution) were deployed across the Peru-Chile trench during RV Sonne cruise SO181. Since analysis of offshore sites is not completed yet, we restrict the following study to the onshore component of the experiment.

4.3.2.1 2-D Modeling – The Standard Isotropic Approach

An early modeling approach by inverting only impedance data was carried out (Brasse and Soyer 2001) for the central profile at 38.9°S in Fig. 4.12, which corresponds to the seismic ISSA line (Lüth and Wigger 2003). The major result was the detection of a good conductor beneath the Central Valley, probably associated with the Lanalhue Fault (formerly known as Gastre FZ, obliquely traversing the northern Patagonian Andes in a SE-NW direction), and a high conductivity zone beneath the volcanic arc.

We extended this modeling study by incorporating tipper transfer functions and known a-priori information like highly accurate swath bathymetry data, obtained during several cruises of RV Sonne (Scherwath et al. 2006); the result is shown in Fig. 4.14. Another feature included in the starting model is a highly-resistive slab of the subducted Nazca Plate, an assumption justified by EM measurements on the seafloor (Chave et al. 1991). As in the previous model a common strike direction of 0° was assumed, justified by multi-site, multi-frequency analysis of strike directions according to Smith (1997) and the phase tensor plot in Fig. 4.13. Regularized inversion was carried out with the non-linear conjugate gradient code of Rodi and Mackie (2001); the regularization parameter was set to $\tau = 10$. Error floors were set to 20% for apparent resistivities, 5% for phases, and 0.1 (absolute value) for real and imaginary parts of the tipper. Since tipper data are not consistent with impedance strike directions (see below), we used their projection on the y-(EW-)axis. Further details concerning inversion settings and sensitivity issues, which reach beyond the purpose of the study presented here, are described by Kapinos (2011).

In terms of a root mean square error, the obtained data fit is remarkably good with a rms = 1.1; but note that this is mainly due to the relatively high error floor assigned to the tipper data. However, model structures do not change significantly if a different weighting of data relative to each other is applied (Kapinos 2011). Furthermore, the main structures are only marginally affected, if different starting models are used (e.g., a homogeneous half space with the Pacific Ocean included). Since isotropic modeling is not the main topic of the study presented here, we skip discussion of data fit and sensitivities here and just investigate the major features of the model.

As in Brasse and Soyer (2001) the main conductors (B, C and C' in Fig. 4.14) are located beneath the Central Valley and the volcanic front. The highly-conductive zone C at mid-crustal depths beneath the volcanic arc seems particularly interesting: The profile runs just south of Llaima volcano, and the model may simply image a large magma deposit beneath, but offset by $\sim$10 km to the east. Note that Llaima erupted violently on 1 January 2008, but also note that these data were collected already in 2000, and it is of course not known how this eruption may have affected the conductor by removing or relocating a significant part of the magma deposit.

The Central Valley conductor on the other hand, beneath the trace of the Lanalhue Fault, is unlikely to originate from partial melts due to temperature considerations in the (relatively cold) forearc – fluids are seen here as the main cause of elevated conductivities. A consistent feature is conductor D east of the eastern margin of the profile (already in Argentina) – although not truly resolved with respect to location and resistivity due to the lack of stations, it appears in all inversion runs. A preliminary explanation may lie again in a root zone for the Holocene backarc volcanism in the Loncopué Trough. However, only a future extension of the profile into Argentina could unambiguously answer this question.

A very good conductor (A, not well visible on the southernmost profile) appears west of the transects already beneath the ocean, overlying the downgoing plate. This structure is not seen when only crude bathymetry is taken into account (Brasse and Soyer 2001) – this underlines the importance of exact bathymetry for near-coastal data. It may at first glance seem like an inversion artifact; however, it is also modeled at the northernmost TIPTEQ profile, where an

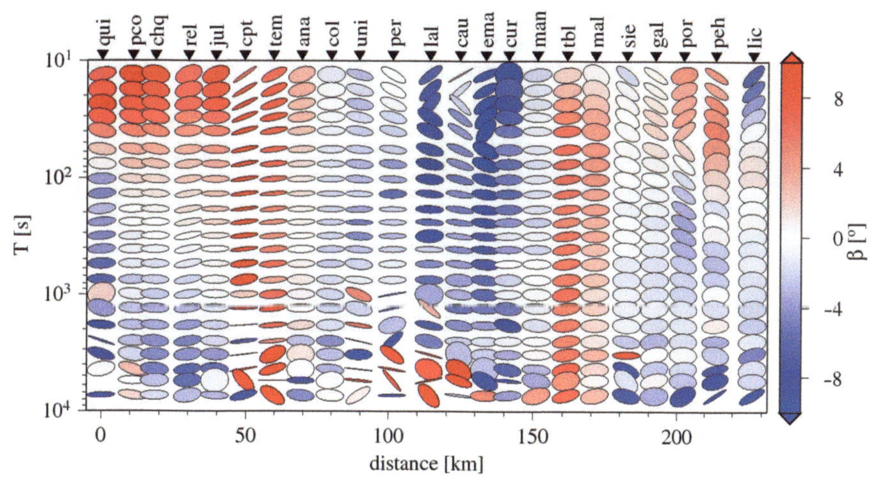

Fig. 4.13 Phase tensor ellipses for the northern profile A (Kapinos 2011)

additional offshore station was incorporated (Kapinos 2011). Furthermore, it correlates with a strong seismic reflector (Groß et al. 2007) beneath the TIPTEQ traverse. The origin of this structure remains enigmatic for the time being, but analysis of seismic tomography data suggests a possible low-velocity zone (Haberland, pers. comm.) and thus a fluid-rich accretionary wedge, perhaps fed by faults originating at the downgoing plate. Interestingly, the overall appearance of the model with its several conductors – particularly structure A off the coast – is quite similar to a model recently published by Soyer and Unsworth (2006) for the Cascadia subduction zone in SW Canada.

The subducted Nazca plate was modified in the course of the inversion process, leading to a much more heterogeneous image of the slab; additionally several poor conductors are now seen in the continental crust, rising from the plate interface. Apart from the main features in Fig. 4.14, minor structures also include the near-surface, but not-deep reaching sediments in the Central Valley in accordance with tectonic assumptions (H. Echtler, pers. comm.).

Although the 2-D inversion model appears like a plausible result particularly with respect to the resolved conductive features beneath the profile, it cannot represent a "true" model in an important aspect: its response only approximates the vertical magnetic field data, which are significantly and systematically distorted throughout the study area. Furthermore, the number of conductive "blobs" of the model, resembling the study of Heise and Pous (2001), suggest a different, anisotropic approach. This is treated in the following sections.

4.3.2.2 GDS Transfer Functions in South-Central Chile

The Chile trench reaches a depth of ~4 600 m in the study area, yielding a conductance of 14,000 S (Siemens); taking into account the – presumably well-conducting – sedimentary filling this value should even be higher. Tipper magnitude directly at the coast is $W \approx 0.8$ for long periods. This is not as large as would be the case near a deep ocean if the continental lithosphere would in total be highly-resistive (i.e., resistivity in the order of 1000 Ωm or more). Thus the continent must generally be less resistive (a few hundred Ωm maximum) or must at least contain anomalous high-conductivity zones.

Given the average N10°E trend of the trench and the similar overall course of the volcanic chain, it was expected that real parts of induction vectors would show a general W-E tendency. This is, however, not the case. Instead, at long periods, all real induction vectors point systematically NE for all sites in the measuring area, regardless on which geological unit data were collected (Fig. 4.15). Note that there is not a single site in the study area where this observation is opposed. Thus flow of anomalous current in the continent – itself caused by current concentration in the ocean parallel to the coast – is not NS as would be the case for a simple 2-D distribution of conductivity, but obliquely deflected on a large lateral scale.

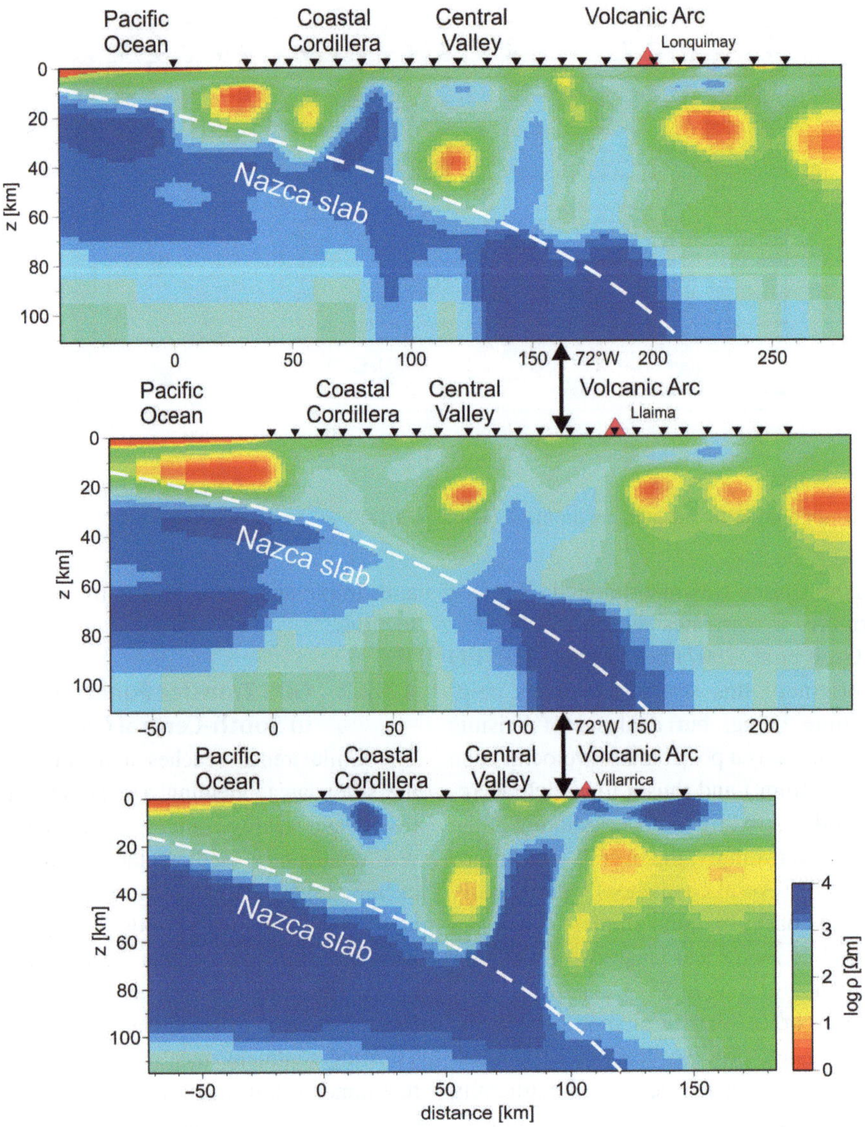

Fig. 4.14 Results of isotropic 2-D inversion of the northern, central and southern profiles in Southern Chile (Brasse et al. 2009, Montahei 2011, Kapinos 2011)

On the other hand, this effect is not observed at shorter periods: The coast effect is "normal" and vectors in the Coastal Cordillera point roughly perpendicularly away from the shoreline (not necessarily perpendicular to the trench, since local bathymetry is dominating at short periods). This also rules out an instrumental effect as we initially suspected (and which gave rise to a later extensive test program of stations near Niemegk geomagnetic observatory close to Berlin). Source effects should not play a significant role, too, because the study area is located in mid-latitudes and far from both polar and equatorial electrojets.

An interesting effect is visible around highly active volcanoes Villarrica (altitude 2847 m) and Llaima (3125 m, latest eruption in 2008), where a small network of sites was established: While induction vectors at 100-200 s point away from Villarrica at the closest sites, this is not seen at Llaima. Although all volcanic edifices lead to a topographic effect at their slopes, this cannot be the reason for the direction and magnitude of vectors at Villarrica. Since the slope of this mountain is

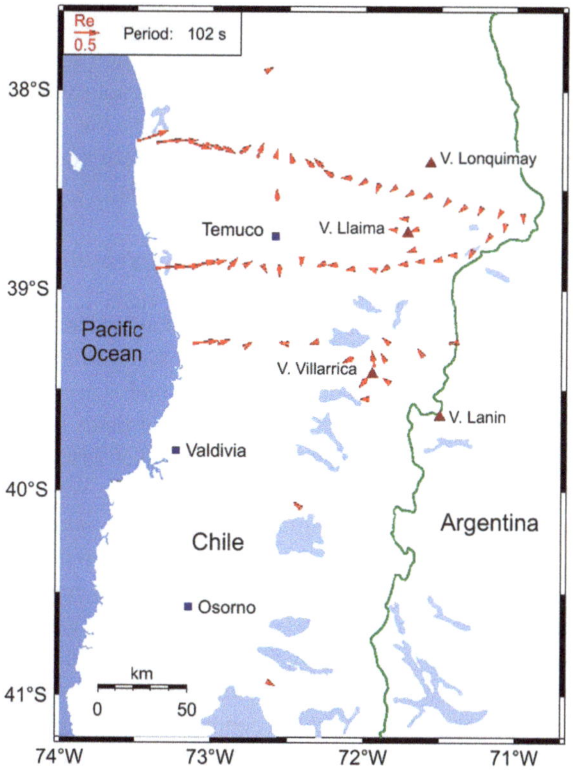

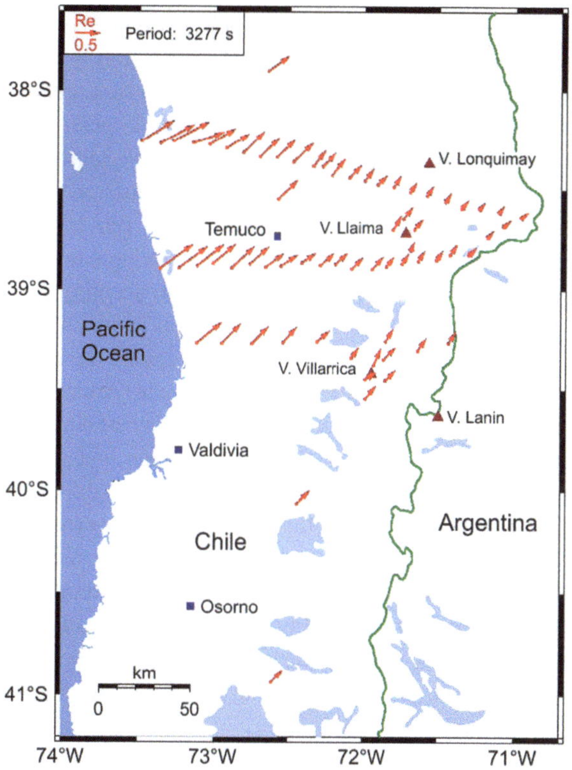

Fig. 4.15 Induction vectors in S. Chile at T = 102 s (*left*) and 3277 s (*right*)

only in the order of 25–30°, topography signals in MT transfer functions are restricted to short periods (<10 s) and only a static shift-like effect remains in apparent resistivities as was shown by 2-D and 3-D modeling of topographic effects in the central Andes (Bydam 2008, Brasse and Bydam 2008). We may thus assume that deeper in the crust beneath Villarrica a large-scale magma deposit might exist – note that the crater at the top of Villarrica is filled with a lava lake (Calder et al. 2004). A more detailed statement about conductivity distribution at depth below the volcano is not possible for the time being, since detailed 3-D modeling has not been carried out yet. Note that these long-period data only permit statements about the deeper crust; to assess a possible magma deposit just beneath the volcanic edifices would require measurements at shorter periods (AMT range) on a denser network.

4.3.2.3 3-D Modeling Attempts to Explain Induction Vectors

It is obvious that the anomalous deflection of induction vectors cannot be explained by pure 2-D models. We therefore tested several simple 3-D approaches to account for the deflection over a large area, in particular the N-S extent of at least 350 km (it is not known if transfer functions continue to be this anomalous to the north and south of the measuring area, but this may quite safely be assumed at least for some distance). Such models must incorporate the Pacific Ocean with an average depth of $\sim$4.5 km and an almost N-S running coastline, and some other structure of large, regional extent which accounts for the deflection. For the computations the algorithm of Mackie et al. (1994) was applied; seawater resistivity was fixed at $\rho = 0.3\ \Omega$m.

Test 1: It may be possible that a layer with increasing conductance (conductivity-thickness product) from north to south exists at some depth in the crust or even in the upper mantle. This may indeed explain the induction vectors but leads to inherently large conductances in the northern part (and unrealistic conductivities if layer thickness is not changed accordingly). Furthermore there exists no geological evidence whatsoever for such a layer and the idea is thus abandoned here.

Test 2: South of the southernmost site begins archipelagic Chile, i.e., the Central Depression is submerged and the Coastal Cordillera becomes a chain of islands, with Isla de Chiloé being the largest (not shown in Fig. 4.12). The distribution of seawater masses causes therefore a deflection of induction vectors near latitude 41°S, but according to our model results this effect does not reach far enough to the north, taking the known water depths into account.

Test 3: At 45–46°S, i.e., 450–550 km south of the study area the Chile Rise is subducted beneath the continent; this is the triple junction between Antarctic, Nazca and South American plates. It may be speculated that the location of the triple junction is associated with a deep seated plume structure of enhanced conductivity. Such a hypothetic good conductor (resistivity 1 Ωm) was incorporated into the 3-D model (which also takes the irregular coastline into account); the response is shown in Fig. 4.16. Again, the model response is only compatible with the data at the southernmost sites of the measuring area; in the north vectors point strictly W-E.

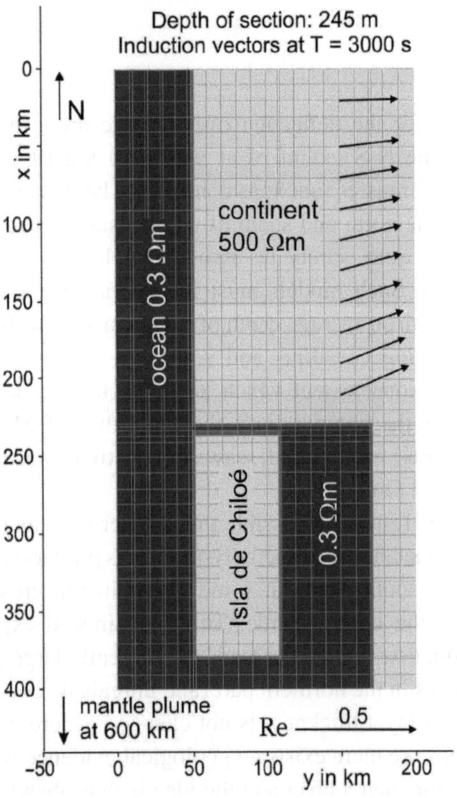

Fig. 4.16 One of the 3-D models calculated to explain deflection of induction vectors

Summarizing, simple and (geologically) realistic 3-D models explaining the observed induction vectors could not be found (which does of course not exclude more detailed classes of models, see later). We therefore tested another approach, the simulation of a deeply fractured crust with anisotropic 2-D models.

4.3.2.4 2-D Models with Anisotropy: Some Principal Considerations

Under anisotropy we may either understand microanisotropy as an inherent rock property or structural (pseudo- or macro-) anisotropy; in both cases the inductive scale length ("wave length") of fields in the earth is larger than the width of individual structural units which are thus not resolvable separately. We may assign either one or two directions to high conductivity: The first case may be interpreted as simulating a system of line currents while the second may be regarded as an image of a sequence of fault planes. A ratio of $\gg 100$ between directions of low and high conductivity seems reasonable if we assign the resistive part to the host rock and the conductive one to possible fault planes, assumed to be fluid-rich in the damage zone of the fault core.

Full anisotropy has to account for 6 variables, the 3 principal resistivities, strike, dip and slant (Pek and Verner 1997). Due to the resulting complexity we varied only the first 4 parameters leaving dip and slant constant at 90° and 0°, respectively. Dip and slant (if not too large) have a much smaller influence on transfer functions than the other parameters. Since ρ_z has only minor influence on induction vectors at least distant from the coast, this quantity was set equal to ρ_x for most model experiments.

If an anisotropic layer – extending to infinity in horizontal directions – is present in an otherwise isotropic and homogeneous or layered half space, a split of impedance phases and apparent resistivities is observed while the vertical magnetic field is zero; the vertical components of the secondary field from the parallel current lines – or planes – superpose destructively. A vertical field only arises if the anisotropic layer is bounded or if some other lateral inhomogeneity is present, in our case mainly from the coast effect at the Chilean margin. Under 2-D isotropic conditions only the W_y component of the tipper would exist due to the basically N-S trending coastline and trench. Here, however, W_x is of similar magnitude at long periods.

First we evaluate the responses of several principal models for periods of $T = 102.4$ and 3277 s, respectively, corresponding to period bands at which transfer functions are estimated by the time series analysis scheme. We thus analyze responses at periods corresponding to relatively short and large penetration depths. Calculations were carried out employing the algorithms of Pek and Verner (1997) and Li (2002); both yield comparable results.

(1) An anisotropic layer between $z = 5$ and 20 km, bounded at $y = 20$ km and $y = 200$ km is embedded in a homogeneous half space of resistivity 300 Ωm. Resistivities of the layer are set to $\rho_x = \rho_z = 1$ Ωm and $\rho_y = 300$ Ωm, anisotropy strike to $\alpha = 45°$. The conductive axis thus strikes SW-NE. We neglect any non-zero dip or slant and simply assume vertical conductive planes. Vertical resistivity has only minor influence on the vertical magnetic field at the surface (in contrast to impedance, which we don't further investigate here) and only near the layer boundaries. At both margins of the anisotropic block significant induction vectors are observed, pointing NW in the west and SE in the east (i.e., both perpendicular to strike); above the center of the block the combined effect of both margins results in a vanishing vertical field. According to depth of penetration vectors are largest for shorter periods – for longer periods penetration depth increases and the layer is too shallow to have a major effect.

(2) Since the observed induction vectors display a significant W_x component from the coast until the Argentinian border, the anisotropy has to persist eastward beyond the station network. This is taken into account by removing the layer bound in the east. Because no data exist in Argentina at this latitude, we have no information on the eastern margin of the layer and simply set it to infinity. Furthermore, a large W_x component is present also at coastal site PUR and even at sea-bottom site OB7 20 km offshore on the northern profile (Kapinos 2011) as well as on Isla Mocha (O. Ritter, pers. comm.). Therefore the western margin has to be extended below the ocean; here it is arbitrarily set to $y = -20$ km. As expected, oblique induction vectors are only observed at the western margin, but now the effect is largest for long periods due to the infinite extent towards east.

(3) The Pacific Ocean is simulated by a conductive half layer with a thickness of 5 km. For the time being we neglect the actual bathymetry and place the block at y = –80 km from the coast, i.e., 60 km away from the anisotropic layer. We also neglect dependence of seawater conductivity on salinity and temperature (i.e., water depth) and set it to an average value of $\sigma =$ 3.3 S/m ($\rho = 0.3$ Ωm).

The third model already grossly explains the observed induction vectors at long periods: They point NE over the whole profile, slowly decreasing in length with distance from the ocean. This behavior is not difficult to understand: If ocean and anisotropic layer are separated far enough from each other and thus not coupled, we may simply carry out a vector addition of contributions originating from the ocean and the layer, respectively. Accordingly the E-pointing "ocean vector" and the NW-directed "anisotropy vector" yield a combined vector pointing towards NE. Both real parts of W_x and W_y are positive in this case. At short periods the ocean effect is smaller due to the distance from the ocean with its crude geometry adopted here (note that this characteristic does not conform with observation and a more realistic bathymetry has to be incorporated; see below). Imaginary parts remain small in all cases, they are discussed later.

If, however, the anisotropic layer extends below the ocean, the anomalous fields of both structures will be coupled. Then a simple vector addition does not suffice; e.g., the secondary field of the anisotropic layer induces a "secondary secondary" field in the ocean which is no longer N-S but rather obliquely oriented with respect to the coast. This is taken into account by the modeling algorithm. For further discussion on coupled anomalies in the anisotropic case see, e.g., (Weidelt 1999) and (Soyer 2002).

From these fundamental considerations it is immediately evident, that the anisotropy strike does *not* reflect the NW-SE oriented fault pattern as displayed in Fig. 4.12. Highly conductive planes in that direction would produce SE-oriented induction vectors; thus information from tippers reveals other, less obvious structures (see discussion below).

(4) As already mentioned above, the uniform pattern of induction vector deflection is only observed at periods >1000 s, while at shorter periods local effects come into play. In a 4th test we thus shifted the anisotropic layer into the lower crust, i.e., it is located now between 20 and 38 km. This approach also reproduces the oblique vectors and differences to the responses of model (3) are only minor. It is thus difficult to discriminate between upper and lower crustal

anisotropy from one period alone, and the full period range has to be taken into account.

4.3.2.5 Anisotropic Models for the Chilean Margin

Due to the large number of parameters involved when carrying out 2-D anisotropic modeling, the search for a model that fits real and imaginary tippers at all sites for all frequencies, and that is preferably somehow geologically realistic is a time-consuming issue. Like in the preceding section we set slant, dip to constant values and varied only ρ_x, ρ_y (with $\rho_z = \rho_x$) and anisotropy strike α plus the isotropic background resistivities. Some of the features of the isotropic 2-D inversion results (see Fig. 4.14) were incorporated and adjusted where necessary.

The resulting model is displayed in Fig. 4.17 – calculated for the central profile of Fig. 4.12. It incorporates a homogeneous background with a resistivity of 200 Ωm, an anisotropic layer in the lower crust and the Pacific Ocean with detailed bathymetry, taken from ETOPO1 data and swath bathymetry obtained during several cruises of RV Sonne (Scherwath et al. 2006). The subducting oceanic Nazca slab is modeled as a poor conductor with a vertical extent of 150 km (lower limit not shown in Fig. 4.9). The dip angle of the slab is provided by seismological studies (Bohm et al. 2002, Yuan et al. 2006) and the TIPTEQ seismic transect along the northern MT profile (Groß et al. 2007).

The trench contains a sediment filling with a thickness of about 2 km (Völker et al. 2006) and with presumably low (but unknown) resistivity. The filling and its geometry – known from offshore reflection seismology, e.g., Sick et al. (2006) – is roughly taken into account by assigning an arbitrary low resistivity (5 Ωm), which suffices if only onshore (i.e., far away) stations are investigated. The same resistivity is set for the uppermost oceanic crust. We used also information from the isotropic inversion; particularly the near surface structures are motivated by isotropic models and allowed to fit at least in a crude manner the short-period induction vectors. On the other hand the compatibility between the isotropic and anisotropic models shows that they complement rather than oppose each other.

We carried out numerous tests to constrain upper and lower boundaries of the anomalous, anisotropic layer. Without going into further detail here it may be concluded that one cannot resolve (as might be expected) the thickness of the anisotropic layer; in our model it reaches until Moho depths. The crust-mantle boundary lies at 35–45 km depth in South Chile as inferred from seismological studies, e.g., Krawczyk et al. (2006) and Asch et al. (2006). Since the deflection of induction vectors persists for offshore stations until the trench (Kapinos 2011), the anisotropic layer may not be located at mantle depths, because the downgoing slab would cut this layer apart. It is additionally not easy to constrain the upper limit of the

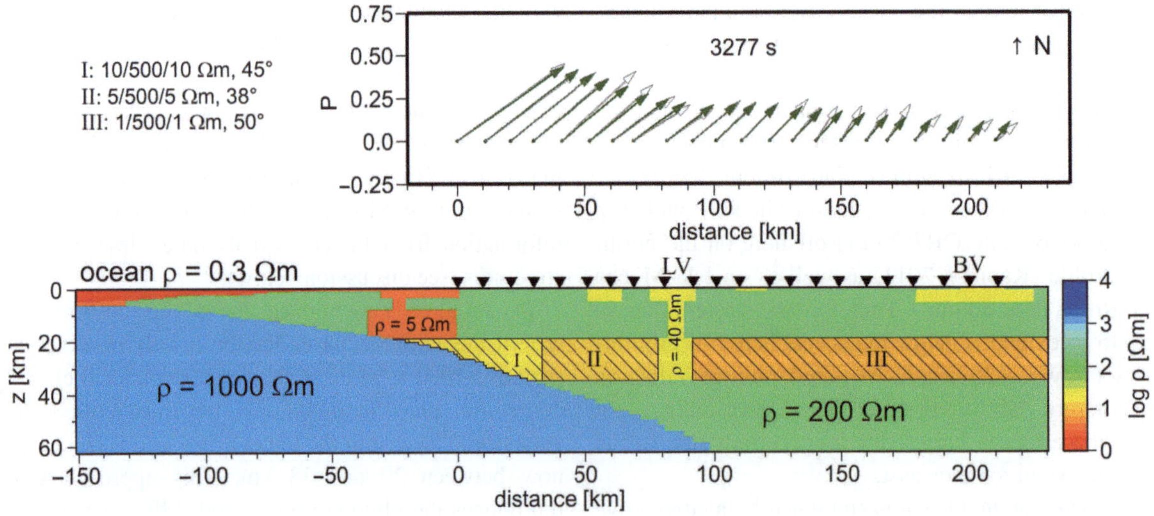

Fig. 4.17 *Bottom*: Anisotropic 2-D model for Southern Chile. *Top*: Measured (*open arrows*) induction vectors and model response (*filled arrows*)

layer since, at short periods, induction vectors show more local (also 3-D) features which are difficult to implement.

It would be unrealistic to assume that in a subduction zone setting (considering the movement of plates, resulting stress field and the oblique faults traversing the study area) the anisotropic layer will stay intact (unharmed) at the regional scale. It is rather to expect that features like resistivity or strike direction will change along the profile and which will also be reflected in the induction vectors. Indeed, the best fitting is obtained by dividing the anomalous layers in sections with minor variations in the anisotropic parameters. The values of resistivities in x-, y- and z-directions and the strikes are shown in Fig. 4.17. The ocean primarily accounts for the length of induction vectors near the coast while the strike of anisotropy is responsible for the deflection from W-E. The minimum of vector lengths at ~75 km may be accounted for by introducing a homogeneous and isotropic block at the location where the profile crosses the Lanalhue Fault, in accordance with the isotropic inversion results.

In the model of Fig. 4.17 this anisotropy strike – i.e., the conductive axis – is basically running NE-SW. This strike direction is also motivated by the horizontal stress field and lineaments of minor eruptive centers along the Liquiñe-Ofqui lineament (López-Escobar et al. 1995), which is discussed later.

Note that MT and GDS data are usually measured in geomagnetic coordinates and the coordinate system used here is set accordingly. Since the declination of the main geomagnetic field is in the order of N10°E (incidentally similar to the direction of the trench and the volcanic chain), it has to be taken into account when comparing geomagnetic results with geographic directions. A geomagnetic strike direction of 50° thus corresponds to geographic N60°E.

Also note that anisotropy persists along the whole profile and extends below the volcanic arc (where induction vectors are small but still deflected) and even into Argentina, east of the easternmost site location. Introducing an isotropic good conductor below or to the east of the profile levels the amount of induction vectors and debases the fitting considerably. The model in Fig. 4.17 also includes near-surface, well-conducting sediments ($\rho = 40\ \Omega$m with a very thin layer of resistivities of 10 Ωm on top, which is not visible in the plot) in the Central Depression and the Bío-Bío valley. The thickness of these sediments is not known and was arbitrarily set to 2 km. Finally, apart from responses at short periods (and at long periods for the imaginary part) the data fit is reasonable (Brasse et al. 2009).

We can of course not rule out that a certain degree of anisotropy exists in the uppermost continental mantle or in parts of the upper crust. In the 1st case this is not resolvable, in the 2nd case a possible difference between ρ_x and ρ_y could be so small that it has only negligible influence. We may also discard a possible anisotropy of the oceanic crust (which might intuitively be the case due to the numerous fracture zones entering the subduction system) as the cause for induction vector deflection – model studies showed that its influence would not reach far enough along the profile on the continent.

4.3.2.6 Discussion of Models in South Chile

The deduced overall preference direction of electrical conductivity (NE-SW) does not agree with the image of faults in the South-Central Andes, striking obliquely (NW-SE) to the continental margin as shown in Fig. 4.12 (Melnick et al. 2006), as could originally be suspected. Apart from the Lanalhue Fault – consistently modeled as a good conductor in both isotropic and anisotropic approaches – the strike direction in the lower crust is rather perpendicular to the overall forearc fault pattern. The (structural) deep-crustal anisotropy also crosses the most prominent mega shear zone in Chilean Northern Patagonia, the ~N10°E striking Liquiñe-Ofqui Fault Zone (LOF) which extends from the triple junction area at ~46°S until 38°S, where it terminates at the Bío-Bío-Aluminé Fault (BBAF). The LOF is assumed to largely control volcanism in S. Chile (e.g., Cembrano et al. 1996) because many of the Quaternary and active volcanoes are aligned along this lineament.

The structural anisotropy has to continue eastwards across the border into Argentina for at least several tens of kilometers, perhaps until the backarc volcanic centers of the Loncopué trough. The lateral extent to the east cannot be constrained, as there are no stations at this latitude in Argentina (the closest stations from the University of Washington and INGEIS Buenos Aires are still several hundred km to the north; J.R. Booker and C. Pomposiello, pers. comm.). On the other side of the Andean range, the anisotropy reaches most likely until the plate interface, at least significantly beneath the Pacific Ocean. The N-S extension of the anomalous zone is again not known due to missing data N of 38°S and S of 41°S.

A hint at the cause of the structural anisotropy comes from an early observation by (Nakamura 1977): Different from the ~N10°E alignment of the large stratovolcanoes along the LOF, minor eruptive centers, parasitic vents and flank craters in the Central Southern Volcanic Zone are predominantly aligned in a NE-SW direction. Nakamura (1977) related this preference direction to the maximum horizontal stress S_H in the arc region and assumed a system of dikes enabling the rise of molten or partially molten material to the surface. López-Escobar et al. (1995) and Muñoz et al. (2000) refined this study by analyzing the geochemistry of rocks and their source region. Indeed, most of the samples they analyzed are of mainly basaltic composition (in contrast to the more andesitic-basaltic composition of the large stratovolcanoes), indicating a short residence time of magmas in the crust. The generally NE-SW oriented stress in this part of the Chilean margin (Assumpcão 1992) was recently confirmed by Reuther and Moser (2007) for the uppermost crust until depth of ~500 m. The analysis of 2nd order structures (lineaments, dikes, drainage anomalies) by Rosenau et al. (2006) gives further evidence of the importance of the NE-SW direction.

According to Shaw (1980) dikes in the crust develop perpendicular to the direction of minimal effective stress (S_3). The maximum horizontal stress may then either be S_1 or S_2. In a strike-slip environment (like the Southern Volcanic Zone) S_1 and S_3 are horizontal while S_2 is vertical, thus allowing partial melts and fluids to rise in vertical dikes, parallel to S_H. Local features (e.g., gravitational load from the stratovolcanoes and the mountain range as such) may modify the overall stress pattern – this may in turn lead to local deviations of induction vector directions and to slightly different conclusions on anisotropy directions along the 3 profiles in the study area.

Our findings concerning structural anisotropy (if we regard it as a measure of a deeply fractured crust, but being unable to resolve individual dikes/faults due to wavelength considerations) strongly support the assumption of Nakamura (1977) and López-Escobar et al. (1995). The surprising result is, however, that the crust has to be deeply fractured in the forearc as well. Due to low temperatures in the forearc crust the conductive phase cannot be any partial melt here – instead we have to assume a relatively cold, but fluid-rich crust or even, at least partly, an occurrence of metallic phases. The extension of anisotropy beneath the Coastal Cordillera would be in accordance with the broadening of the mid-Tertiary volcanic arc until the Pacific coast, where volcanic outcrops occur south of our actual study area at 41°S (Muñoz et al. 2000). Our results suggest that this magmatic event may have reached even further to the west, beyond the coastline and perhaps until the continental slope.

Unfortunately the model in Fig. 4.17 explains the impedances only in a crude manner. For the time being we have been unable to construct a model which satisfactorily fits both magnetotelluric and vertical magnetic field observations; this constitutes the next task for the evaluation of the data set. However, several features of the model presented here correspond to the isotropic impedance model (Brasse and Soyer 2001), in particular the enhanced conductivity below the Central Depression and the generally higher conductivities beneath the volcanic arc. The characteristics of induction vectors outside the area depicted in Fig. 4.12 are not known and it would thus be a rewarding effort to establish further sites to the north of the Bío-Bío Fault and to the south in archipelagic Chile.

4.4 Magnetotelluric Studies in Central America

In the Central American subduction zone, the Cocos Plate subducts slightly obliquely in a northeastern direction beneath the Caribbean Plate with a convergence rate of ~8.5 cm/a (DeMets 2001). Depth of the Middle America Trench (MAT) gradually decreases from offshore Nicaragua ($z > 5000$ m) to less than 2000 m in southern Costa Rica, where the Cocos Ridge is subducted. Similarly, the subduction angle decreases from near vertical beneath Nicaragua to sub-horizontal beneath S Costa Rica (Protti et al. 1995, Husen et al. 2005), implying a significant diversity in the geothermal regime (Peacock et al. 2005). The varying slab dip is seen as a result of differences in the strength of the mantle wedge beneath Nicaragua and Costa Rica (Rychert et al. 2008). The position of active arc volcanoes above the downgoing slab changes abruptly near the border between the two countries: While volcanoes in the Cordillera de Guanacaste of NW Costa Rica are located roughly above the 100 km depth contour line, Nicaraguan volcanoes sit above a

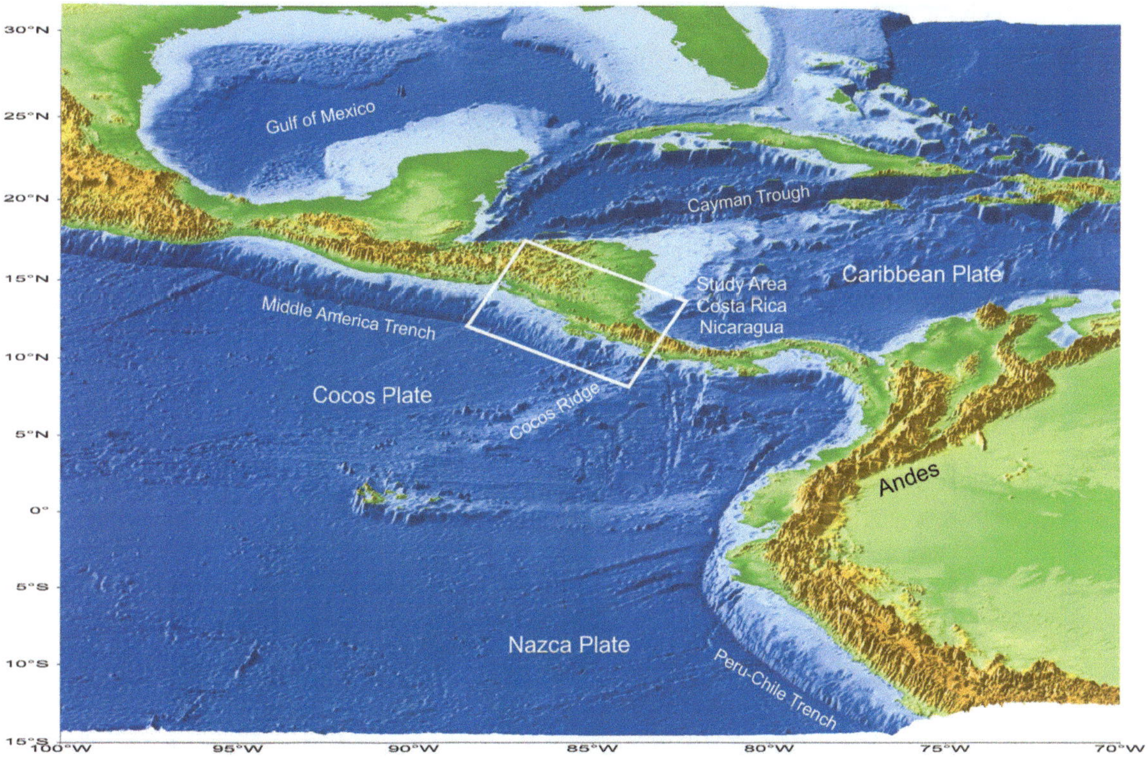

Fig. 4.18 Setting of Central America. Topography and bathymetry are plotted from ETOPO1 data

much deeper segment (>150 km) of the Cocos plate (Syracuse et al. 2008). A significant variation of lava chemistry is observed, with larger fluid and sediment signature beneath Nicaragua than beneath Costa Rica (Carr et al. 2003, Rüpke et al. 2002). Additionally, crustal thickness beneath the Costa Rican arc is significantly larger (~38 km) than beneath the Nicaraguan depression (~25 km)[79].

The Pacific margin of Central America has in the recent past been studied intensively, particularly with seismological (on- and offshore) and geochemical methods; it is a key location of the NSF Margins Program and also addressed by a number of projects funded by the German Science Foundation. We present here the first results of a complimentary project which aims to investigate the electrical resistivity distribution at the margin by employing long-period ($T = 10 - 10^4$ s) magnetotelluric and geomagnetic deep sounding (MTS and GDS). Data along a first profile comprising 18 sites with a spacing of ~10 km were collected in February/March 2008. It extends from the Pacific coast near Sámara, crosses the volcanic arc at Tenorio volcano and ends in the backarc near the Nicaraguan border at Los Chiles (see Fig. 4.18), and coincides with the line of an active seismic experiment (e.g., (Sallarès et al. 2001)) and the TUCAN seismological study (e.g., Mackenzie et al. 2008).

Further relevant geological structures which are traversed by the MT line (and which are expected to have a response on the transfer functions) are the mafic and ultramafic rocks (ophiolites) of the Nicoya Peninsula (Hauff et al. 2000, Hoernle and Hauff 2000) and the sedimentary basins in the fore- and backarc, i.e., the Tempisque and San Carlos Basins. Here a few drillings – mainly carried out for hydrocarbon exploration – provide some constraints on the structure of these basins (Barboza et al. 1997, Pizarro 1993). In the backarc the profile crosses the so-called "Santa Elena suture", extending from the peridotite outcrops of Santa Elena Peninsula in an easterly direction where it may connect with the Hess Escarpment as a major bathymetric step in the Caribbean plate and limit of the Caribbean Large Igneous Province. The Santa Elena suture is believed by some authors to mark the boundary between allochthonous Chortis and Chorotega blocks. Note, however, that this interpretation is

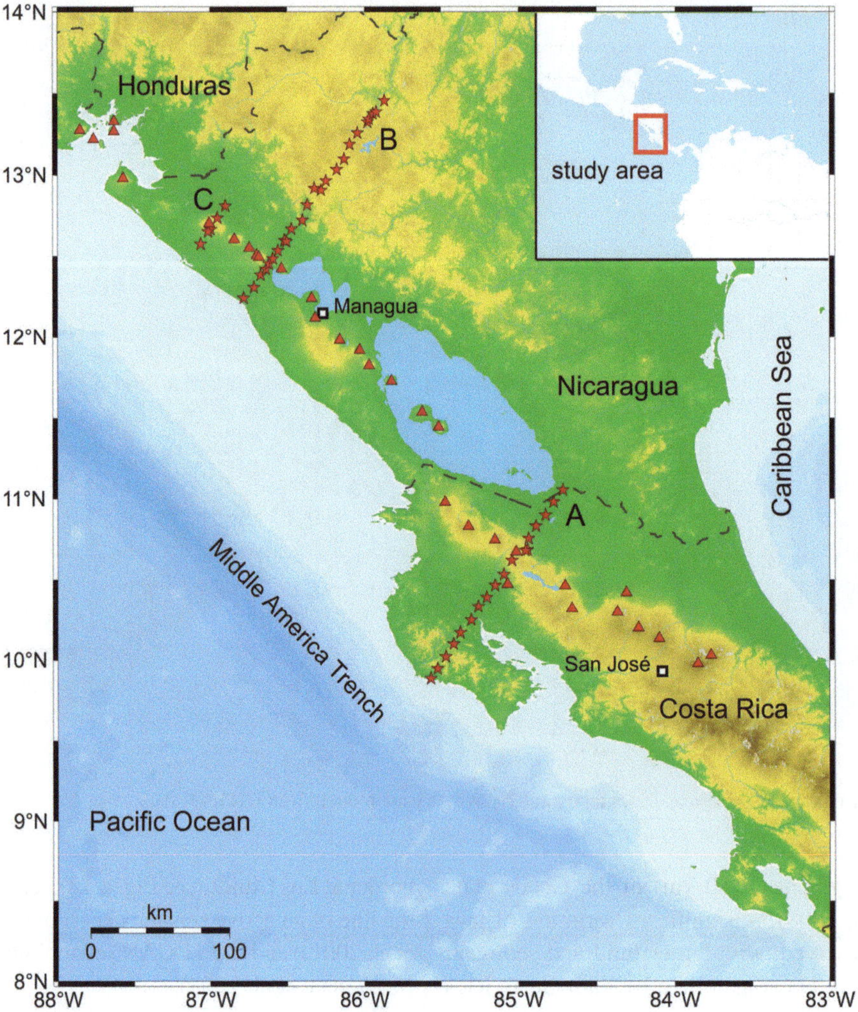

Fig. 4.19 MT sites in Central America (Costa Rica and Nicaragua)

contended (see discussion in Gazel et al. 2006, Mann et al. 2007). The Chortis block is apparently a continental fragment, while the Chorotega block is probably a continuation of the Caribbean LIP.

4.4.1 Data Characteristics

Data processing was carried out by employing the robust remote-reference scheme of Egbert (1997), which yielded mostly high-quality estimates of impedance and tipper in the period range 10–10^4 s. A crucial step for further investigation is the check if the transfer functions reflect a two-dimensional (2-D) subsurface or if three-dimensional (3-D) structures are required by the data. As can be seen from Fig. 4.20 (top), the assumption of two-dimensionality is roughly but not completely fulfilled. Electrical strike directions – calculated from the impedances after Smith (1995) – for the entire profile are shown in Fig. 4.20 (note the 90°-ambiguity inherent to impedances; the decision for the true strike may be based on geology or tipper information). Data are split into three period bands (short, intermediate and long) and show basically similar characteristics: The strike is between N45°W and N55°W with a slightly larger scatter at short periods (10–100 s). This scatter is not surprising, as the influence of near-surface anomalies is larger for short periods. Additionally, for near-coastal sites, the bending of the coastline (which is not parallel to the trench) becomes relevant. Summarizing, the impedance data suggest that a 2-D approach is suitable

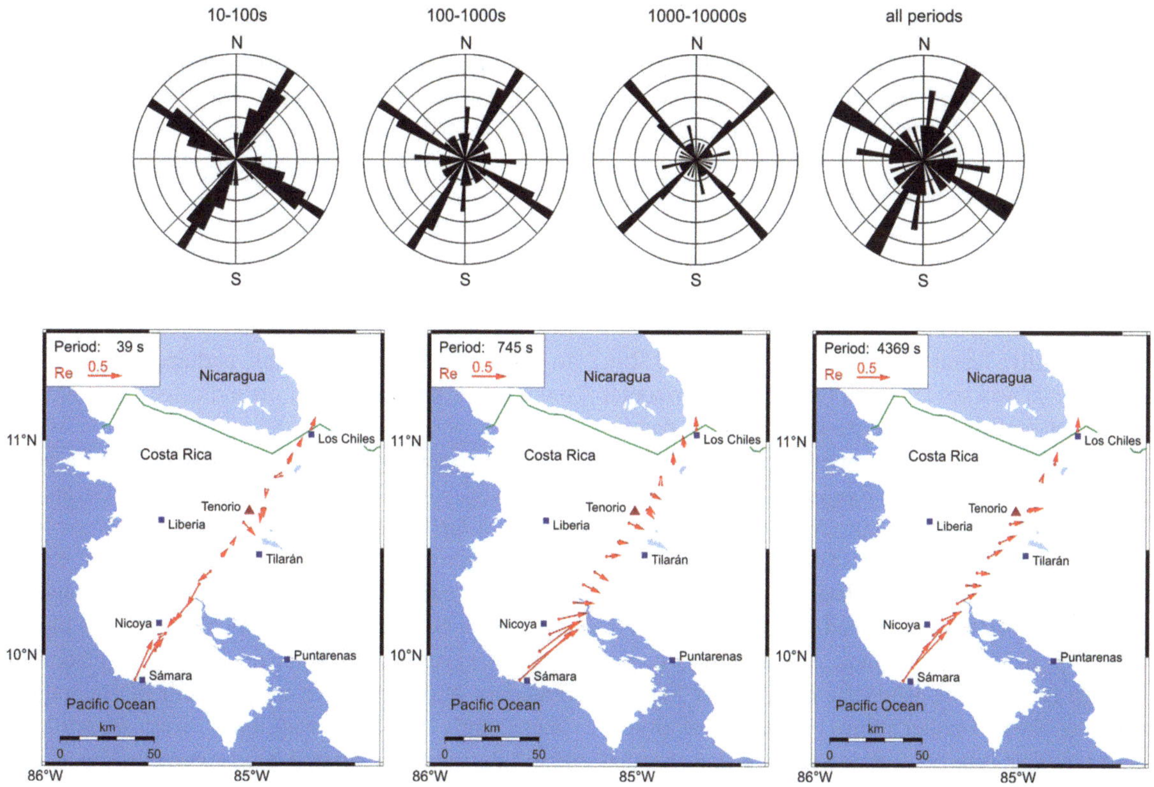

Fig. 4.20 Electrical strikes directions (*top*) and induction vectors (*bottom*) in Costa Rica

and the resulting strike is approximately perpendicular to the profile.

This clear statement is obscured somewhat by the induction arrows or vectors, as shown in Fig. 4.20 (bottom) for three periods. Induction arrows are calculated from the tipper; their real parts (**P**) point towards or away from high conductivity zones (at least in simple environments), depending on convention. We use the latter here, and consequently at the coast arrows point away from the well-conducting ocean. This coast effect reaches a large magnitude of over 1.1 at site *sam* for periods between 1000 s and 2000 s, which is due to the proximity of the 4 km-deep trench off Nicoya, an effect which is enhanced by the relatively resistive ultramafic rocks of the peninsula. Interestingly the coast effect does not reach far inland; it is obviously compensated by the sediments of the Tempisque Basin and perhaps another, deeper anomaly. Near Tenorio volcano (site *par*) arrows do not point away from the edifice at short periods, but rather hint at a conductive zone farther to the west, i.e., Miravalles volcano (where a large geothermal reservoir is encountered). Even more evident is the deviation of arrows in the Tempisque Basin itself at longer periods. This may either be caused by still deeper sediments in the NW or could even be attributed to the deepening of the trench near northernmost Costa Rica and off Nicaragua.

Thus induction vectors sense significant 3-D effects along the profile. One may argue that these arrows pointing perpendicularly to the profile in the center of the basin do not pose a problem for 2-D modeling; the projected arrows are almost zero and do not require a conductive structure directly beneath the line. Nevertheless, care must be taken when interpreting the following 2-D model.

4.4.2 Results of 2-D Modeling in NW Costa Rica

Taking the strike angle analysis and the direction of most induction arrows into account, we rotated all data by −53°, i.e., basically assuming a strike perpendicular to profile direction. As for the Andes data, the common non-linear conjugate gradient algorithm of Rodi and Mackie (2001) was used, which implements a

Tikhonov-type, regularized 2-D inversion scheme. The program allows for a multitude of settings with respect to regularization factor, weighting functions penalizing horizontal or vertical structure, and error floors, among others.

A number of tests were carried out to check dependence on starting models and inversion parameters. For the model shown in Fig. 4.21 crude bathymetry was included; seawater was given a fixed resistivity of 0.3 Ωm. Additionally, the oceanic lithosphere was set as a resistive feature with a resistivity of 1000 Ωm. The factor β was set to 1 (implying equal horizontal and vertical smoothing), and the regularization parameter was set to a value of 10 after a trade-off analysis between misfit and model roughness. To reduce the influence of static shift effects, error floors were set to higher values for apparent resistivities than for phases. All possible combinations of data sets were tested; the model of Fig. 4.21 was obtained by joint inversion of tipper, TE and TM mode data. The resulting model yields an rms of 2.09, while the inversion of individual modes shows smaller values (better fits) of around 1.3 to 1.5. This is to be expected and may be regarded as a sign for inherent three-dimensionality. However, the most relevant features of the model in Fig. 4.21 are recovered by all runs and a combination of TE, TM and tipper data. Data and model response are shown in the auxiliary material.

The crust beneath Nicoya Peninsula (structure A) is resistive (several hundred Ωm), but not as much as could be expected from its geological setting with predominantly mafic and ultramafic rocks, which should display resistivities in the order of 1 000 Ωm and more. This may hint at a certain fluid input from the downgoing slab (see discussion later). The fore- and backarc basins (B_1 and B_2) are well resolved; thickness of conductive sediments corresponds to results of drillings in the vicinity of the MT line. In the Tonjibe borehole (San Carlos Basin) ultramafic rocks – similar to those outcropping in Santa Elena peninsula – constitute the basement at a depth of ~2 km (Pizarro 1993). In the Tempisque Basin well-conducting sediments seem to reach slightly larger depths, perhaps with a maximum off-profile to the NW, as indicated by induction arrows.

The northeastern part of the section shows a poor conductor which rather abruptly terminates in the region of Caño Negro (D). This location agrees well with the proposed trace of the Santa Elena suture (compare, e.g., (Saltarès et al. 2001)), and may be regarded as the southern terminus of the Chortis block. Note, however, that (Mackenzie et al. 2008) place this boundary slightly further to the north in Nicaragua. Interestingly the mid- and lower crust in northern Costa Rica is underlain by a good conductor (E). It may have a connection (conductive path) to the upper mantle, slightly NE of the volcanic chain and thus already in the backarc (E'). An upper crustal conductor is visible just NW of Tenorio volcano beneath sites *ten* and *maq* (C). This may indicate a shallow magma deposit, but the lack of sites at the volcano edifice (where topographic gradients and dense rain forest limit accessibility) does not permit a more definite statement. In contrast, no enhanced conductivity is modeled at lower

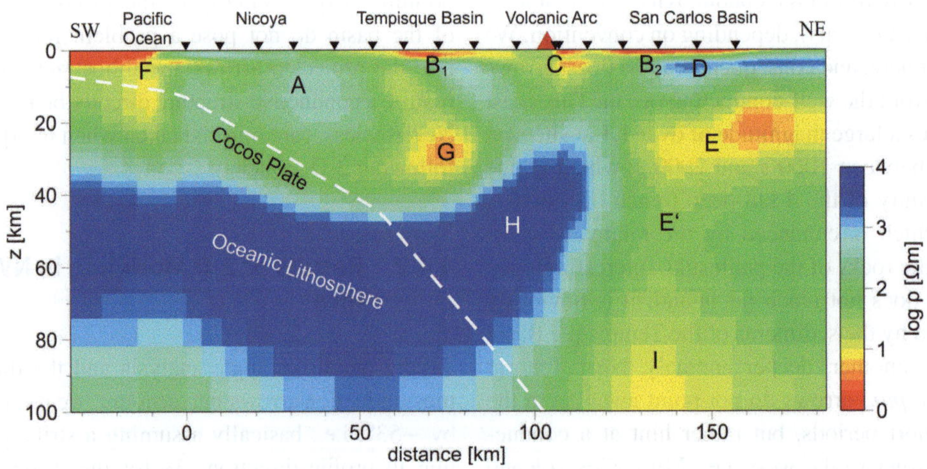

Fig. 4.21 2-D resistivity model for NW Costa Rica (Brasse et al. 2009)

crustal or upper mantle depth beneath the volcanic arc as was proposed by Elming and Rasmussen (1997) from the inversion of MT data in Nicaragua. Note, however, that Tenorio volcano is considered as dormant, unlike the highly active volcanoes in the Nicaraguan depression. This may be the consequence of the absence of a large magma deposit at depth.

Near-coastal conductive feature F in Fig. 4.21 poses a more severe problem for interpretation. It may be an inversion artifact (like the over-estimation of seafloor depth) as it is located seaward from and thus outside the profile (coastal site *sam* also has the worst fit, see electronic supplement). It may on the other hand signify a substantial fluid release from the slab and input into the crust, as was suggested for Central America from seismological observations due to bending-related faulting of the Cocos plate at the MAT (Ranero et al. 2003). Note that such a conductive feature seaward from the coastline appears in models of other subduction zones, too, for instance in South Chile (Brasse and Eydam 2008) and Cascadia (Soyer and Unsworth 2006). Without an offshore prolongation of the profile and ocean-bottom stations near the coast the resolution problems are difficult to overcome. Sea-bottom MT stations were deployed by IFM-Geomar (Kiel); these data are still under evaluation.

Note that the top of the slab (set resistive in the starting model) is changed to medium-resistive during the inversion process. The model in Fig. 4.20 is compatible with fluid release from the oceanic crust, serpentinization of the forearc mantle, and perhaps further dehydration further into the Caribbean plate (G). A serpentinized forearc mantle wedge beneath Nicoya has already been proposed by DeShon and Schwartz (2004) from seismological studies. Serpentinite itself is probably only a moderate conductor (Bruhn et al. 2004). Thus structure G cannot be explained by serpentinite alone; a free, interconnected fluid phase is necessary to explain the low resistivities in the order of 5–10 Ωm. This result concerning a fluid-rich forearc is in general agreement with seismological findings (Husen et al. 2003).

At depths between 40 and 90 km the plate interface and the Caribbean plate upper mantle have a similar high resistivity of several thousand Ωm (H in Fig. cr-model.jpg), comparable with the oceanic lithosphere and implying dry conditions throughout. Any occurrence of free fluids/melts is restricted to greater depth, as can be deduced from the low resistivities in the asthenospheric wedge (I). Note that a postulated rise of fluids/melts from the asthenospheric wedge towards the volcanic arc (from I via E' to E and perhaps C) cannot be vertical or direct – H is a robust structure in the inversion process. It is generally compatible with a forearc mantle sliver found by Walther et al. (2000) further NW beneath Nicaragua and a zone of high velocities above the slab as inferred by Syracuse et al. (2008) from data of the recent TUCAN seismological experiment in Costa Rica and Nicaragua.

4.4.3 Results of Preliminary 2-D Modeling in Nicaragua

Investigation of the Nicaragua profile (B in Fig. 4.19) was conducted in early 2009. It extends from the coast of the Pacific Ocean, crosses the Nicaraguan Depression (where the narrow volcanic arc is located) and end in the Nicaraguan Highlands. Processing, strike and dimensionality analysis were performed in a similar manner as for the Costa Rica data and is not detailed here. The profile runs NW of an earlier study by Elming and Rasmussen (1997), who deduced a mid-crustal conductor beneath the Masaya volcanic complex south of Managua.

A preliminary 2-D model of the new profile is displayed in Fig. 4.22. As in Elming and Rasmussen (1997) a HCZ (anomaly B) is modeled beneath the volcanic arc (the profile passes between Momotombo and El Hoyo volcanoes), although slightly displaced to the NE. A connection (conductive path) to the upper mantle seems possible, indicating rise of fluids/melts; however, this structure is not very well resolved. The basin of the Nicaraguan Depression is a shallow conductive feature with a thickness of 2–3 km (A in Fig. 4.22). Again, the oceanic lithosphere is poorly conducting, this is also the case for the continental wedge beneath the coast (C).

A disturbing feature is the series of conductive "blobs" beneath the Nicaraguan Highland (question marks in Fig. 4.22). This was first believed to be a result of inadequate inversion parameter, but they also appear if the factor β in the inversion program (Rodi and Mackie 2001) which controls the balance of horizontal vs. vertical smoothing, is set to a higher value (e.g., 2 instead of 1 as was usually done). As in South Chile, this may hint at an anisotropic crust of the

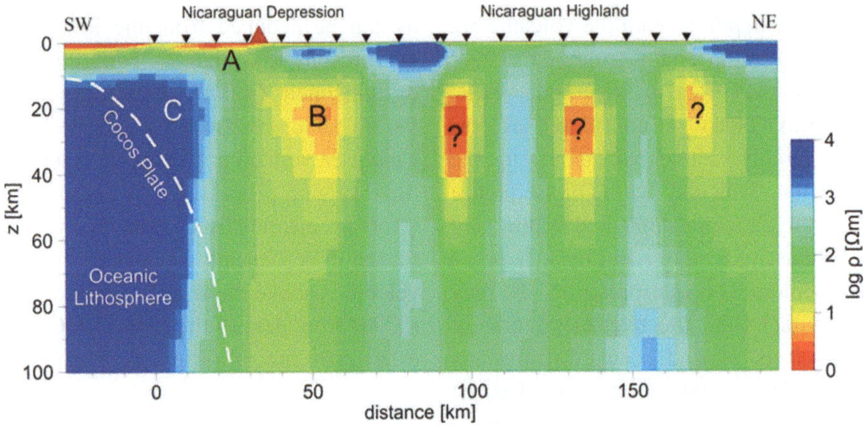

Fig. 4.22 Preliminary 2-D resistivity model for Nicaragua. Note the conductive "blobs", marked with question marks: they appear even if strong horizontal smoothing is applied

Chortis block – these model experiments still have to be carried out.

4.4.4 Conclusion for Central America

Long-period magnetotelluric investigations resulted in the first deep resistivity image at the Costa Rica margin and an enhanced model in Nicaragua. Their principal features and implications are (a) the depth and low resistivity of the fore- and backarc basins, (b) the apparent termination of the Chortis block at the Santa Elena suture in northernmost Costa Rica, (c) a highly conductive backarc mid-crust and upper mantle, (d) the image of fluid release from the downgoing slab in the forearc, and (e) a very resistive forearc mantle adjacent to the subducted plate. These statements also hold for Nicaragua, where in addition a complicated (anisotropic) structure of the Chortis block is obvious. The resistivity image of fluid release from the downgoing slab is generally consistent with seismological observations. We assume hydrous fluids as a cause for the modeled resistivities in the forearc regions, as the temperature in the slab is probably too low to allow for partial melting. Partial melts may, however, occur in the asthenospheric wedge of the overriding plate and explain the high conductivities in the backarc.

The obtained resistivity cross-sections should be seen as a first order approximation. More field data are necessary to constrain 3-D effects – their collection and additional, comparative profiles in Nicaragua are planned in the near future. Incorporating the sea-bottom MT data into the models will further contribute to the question of fluid content in the downgoing plate.

Acknowledgements The author wants to thank the partner institutions in Chile (Universidad Católica del Norte, Universidad de Concepción), Bolivia (Universidad Mayor de San Andrés), Argentina (Universidad Nacional de Salta, Universidad de Buenos Aires), Costa Rica (Instituto Costarricense de Electricidad) and Nicaragua (Instituto Nicaragüense de Estudios Territoriales); without their logistical support this work would not have been possible. The help of many members and students from these institutions and the Free University of Berlin is also gratefully acknowledged. Funding was provided by German Science Foundation (DFG) through numerous grants to the author.

References

Aizawa K, Ogawa Y, Ishido T (2009) Groundwater flow and hydrothermal systems within volcanic edifices: delineation by electric self-potential and magnetotellurics. J Geophys Res 114:B01208. doi:10.1029/2008JB005910

Allmendinger RW, Jordan TE, Kay SM, Isacks BL (1997) The evolution of the altiplano-puna plateau of the central andes. Ann Rev Earth Planet Sci 25:139–174

ANCORP Working Group (2003) Seismic imaging of an active continental margin and plateau in the central Andes (Andean Continental Research Project 1996 (ANCORP '96)). J Geophys Res 108 doi:10.1029/2002JB001771

Asch G, Schurr B, Bohm M, Yuan X, Haberland C, Heit B, Kind R, Woelbern I, Bataille K, Comte D, Pardo M, Viramonte J, Rietbrock A, Giese P (2006) Seismological studies of the central and southern andes. In: Oncken O et al. (eds) The Andes: active subduction orogeny, frontiers in earth sciences. Springer, Berlin, pp 443–458

Assumpcão M (1992) The regional intraplate stress field in South America. J Geophys Res 97:11889–11903
Asters RC, Borchers B, Thurber CH (2005) Parameter estimation and inverse problems. International Geophysics Series, 90. Elsevier, Amsterdam
Bahr K (1988) Interpretation of the magnetotelluric impedance tensor: regional induction and local telluric distortion. J Geophys 62:119–127
Barboza G, Fernández JA, Barrientos J, Bottazi G (1997) Costa rica: petroleum geology of the Caribbean margin. Lead Edge 16:1787–1794
Baumont D, Paul A, Beck S, Zandt G (1999) Strong crustal heterogeneity in the Bolivian Altiplano as suggested by attenuation of Lg waves. J Geophys Res 104: 20287–20305
Bohm M, Lüth S, Echtler H, Asch G, Bataille K, Bruhn C, Rietbrock A, and Wigger P (2002) The Southern Andes between 36° and 40°S latitude: seismicity and average seismic velocities. Tectonophysics 356:275–289
Booker JR, Favetto A, Pomposiello MC (2004) Low electrical resistivity associated with plunging of the Nazca flat slab beneath Argentina. Nature 429:399–403
Brasse H, Eydam D (2008) Electrical conductivity beneath the Bolivian Orocline and its relation to subduction processes at the South American continental margin. J Geophys Res 113:B07109. doi:10.1029/2007JB005142
Brasse H, Kapinos G, Li Y, Mütschard L, Eydam D (2009) Structural electrical anisotropy in the crust at the South-Central Chilean continental margin as inferred from geomagnetic transfer functions. Phys Earth Planet Inter. doi:10.1016/j.pepi.2008.10.017
Brasse H, Lezaeta P, Rath V, Schwalenberg K, Soyer W, Haak V (2002) The Bolivian Altiplano conductivity anomaly. J Geophys Res 107. doi:10.1029/2001JB000391
Brasse H, Soyer W (2001) A magnetotelluric study in the Southern Chilean Andes. Geophys Res Lett 28:3757–3760
Bruhn D, Siegfried R, Schilling F (2004) Electrical resistivity of dehydrating serpentinite. Eos Trans AGU 85(Fall Meet. Suppl):Abstract T41B-1176
Cahill TA, Isacks BL (1992) Seismicity and the shape of the subducted Nazca Plate. J Geophys Res 97:17503–17529
Calder ES, Harris AJL, Peña P, Pilger E, Flynn LP, Fuentealba G, Moreno H (2004) Combined thermal and seismic analysis of the Villarrica volcano lava lake, Chile. Rev geol Chile 31:259–272
Caldwell TG, Bibby HM, Brown C (2004) The magnetotelluric phase tensor. Geophys J Int 158:457–469
Carr MJ, Feigenson MD, Patino LC, Walker JA (2003) Volcanism and geochemistry in Central America: progress and problems. In: Eiler J (ed) Inside the subduction factory. Geophysical Monograph Series, vol 138. AGU, Washington, DC, pp 153–179
Cembrano J, Hervé F, Lavenu A (1996) The Liquiñe Ofqui fault zone: a long-lived intra-arc fault system in southern Chile. Tectonophysics 259:55–66
Cembrano J, Lavenu A, Yañez G, Riquelme R, García M, González G, Hérail G (2007) Neotectonics. In: Moreno T, Gibbons W (eds) The geology of chile. Geological Society London, pp 231–261
Chave AD, Constable SC, Edwards RN (1991) Electrical exploration methods for the seafloor. In: Nabighian MN (ed) Electromagnetic methods in applied geophysics, vol 2. Society of Exploration Geophysicists, Tulsa, pp 931–966
Cifuentes I (1989) The 1960 Chilean earthquakes. J Geophys Res 94:665–680
Comte D, Dorbath L, Pardo M, Monfret T, Haessler H, Rivera L, Frogneux M, Glass B, Meneses C (1999) A double-layered seismic zone in Arica, northern Chile. Geophys Res Lett 26. doi:10.1029/1999GL900447
David C (2007) Comportamiento actual del ante-arco y del arco del codo de Arica en la orogénesis de los Andes centrales. PhD thesis, Universidad de Chile, Santiago
Davies JH (1999) The role of hydraulic fractures and intermediate-depth earthquakes in generating subduction-zone magmatism. Nature 398:142–145
DeMets C (2001) A new estimate for present-day Cocos-Caribbean plate motion: implications for slip along the central American Volcanic Arc. Geophys Res Lett 28:4043-4046
DeShon HR, Schwartz SY (2004) Evidence for serpentinization of the forearc mantle wedge along the Nicoya Peninsula, Costa Rica. Geophys Res Lett 31. doi:10.1029/2004GL021179
Diaz D (2010) Magnetotelluric investigation of the volcanic arc in the Central Andes with special emphasis on Lascar volcano. PhD thesis, Free University of Berlin
Dorbath C, Granet M (1996) Local earthquake tomography of the Altiplano and the Eastern Cordillera of northern Bolivia. Tectonophysics 259:117–136
Dorbath C, Masson F (2000) Composition of the crust and upper-mantle in the central Andes (19°30′S) inferred from P wave velocity and Poisson's ratio. Tectonophysics 327: 213–223
Egbert GD (1997) Robust multiple-station magnetotelluric data processing. Geophys J Int 130:475–496
Elger K, Oncken O, Glodny J (2005) Plateaustyle accumulation of deformation: Southern Altiplano. Tectonics 24. doi:10.1029/2004TC001675
Elming SA, Rasmussen T (1997) Results of magnetotelluric and gravimetric measurements in western Nicaragua, central America. Geophys J Int 128:647–658
Engdahl ER, Villaseñor A (2002) Global seismicity: 1900–1999. In: Lee WHK, Kanamori H, Jennings PC, Kisslinger C (eds) International handbook of earthquake and engineering seismology, part A. Academic Press, Burlington, MA, pp 665–690
Evans RL, Chave AD, Booker JR (2002) On the importance of offshore data for magnetotelluric studies of ocean-continent subduction systems. Geophys Res Lett 29. doi: 10.1029/2001GL013960
Eydam D (2008) Magnetotellurisches Abbild von Fluid- und Schmelzprozessen in Kruste und Mantel der zentralen Anden. Diploma thesis, Fachrichtung Geophysik, FU Berlin
Folguera A, Zapata T, Ramos VA (2006) Late Cenozoic extension and the evolution of the Neuquén Andes. In: Kay SM, Ramos VA (eds) Evolution of an Andean margin: a tectonic and magmatic view from the Andes to the Neuquén Basin (35°– 39°S lat). Geol Soc Am Spec Paper 407. doi:10.1130/2006.2407(12)
Friedel S (1997) Elektromagnetische Tiefensondierungen in Nordchile unter Berücksichtigung der Sq-Variationen und des EEJ. Diploma thesis, Fachrichtung Geophysik, FU Berlin

Gaetani GA, Grove TL (2003) Experimental constraints on melt generation in the mantle wedge. In: Eiler J (ed) Inside the subduction factory. Geophysical Monograph vol 138. American Geophysical Union, Washington, DC pp 107–133

Gaillard F (2004) Laboratory measurements of electrical conductivity of hydrous and dry silicic melts under pressure. Earth Planet Sci Lett 218:215–228

Gaillard F, Malki M, Iacono-Marziano G, Pichavant M, Scaille B (2008) Carbonatite melts and electrical conductivity in the asthenosphere. Science 322. doi:10.1126/science.1164446

Gazel E, Denyer P, Baumgartner PO (2006) Magmatic and geotectonic significance of Santa Elena Peninsula, Costa Rica Geol Acta 4:193–202

Glodny J, Echtler H, Figueroa O, Franz G, Gräfe K, Kemnitz H, Kramer W, Krawczyk C, Lohrmann J, Lucassen F, Melnick D, Rosenau M, Seifert W (2006) Long-term geological evolution and mass-flow balance of the South-Central Andes. In: Oncken O et al. (eds) The Andes: active subduction orogeny, frontiers in earth sciences. Springer, Berlin, pp 401–428

Gonzáles-Ferrán O (1994) Volcanes de Chile. Instituto Geográfico Militar, Santiago de Chile, 640pp

Groß K, Micksch U, TIPTEQ Research Group, Seismics Team (2007) The reflection seismic survey of project TIPTEQ – the inventory of the Chilean subduction zone at 38.2°S. Geophys J Int. doi:10.1111/j.1365-246X.2007.03680.x

Grove TL, Chatterjee N, Parman SW, Médard E (2006) The influence of H_2O on mantle wedge melting. Earth Planet Sci Lett 249:74–89

Haberland C, Rietbrock A, Lange D, Bataille K, Hofmann S (2006) Interaction between forearc and oceanic plate at the south-central Chilean margin as seen in local seismic data. Geophys Res Lett 33. doi:10.1029/2006GL028189

Haberland C, Rietbrock A, Schurr B, Brasse H (2003) Coincident anomalies of seismic attenuation and electrical resistivity beneath the southern Bolivian Altiplano plateau. Geophys Res Lett 30. doi:10.1029/2003GL017492

Hacker BR, Peacock SM, Abers GA, Holloway SD (2003) Subduction factory, 2, Are intermediate-depth earthquakes in subducting slabs linked to metamorphic dehydration reactions? J Geophys Res 108. 10.1029/2001JB001129

Hamza VM, Muñoz M (1996) Heat flow map of South America. Geothermics 25:599–646

Hauff F, Hoernle K, van den Bogaard P, Alvarado G, Garbe-Schönberg D (2000) Age and geochemistry of basaltic complexes in western Costa Rica: contributions to the geotectonic evolution of Central America. Geochem Geophys Geosyst 1. doi:10.1029/1999GC000020

Heise W, Caldwell TG, Bibby HM, Bennie SL (2010) Three-dimensional electrical resistivity image of magma beneath an active continental rift, Taupo Volcanic Zone, New Zealand. Geophys Res Lett 37:L10301. doi:10.1029/2010GL043110

Heise W, Pous J (2001) Effects of anisotropy on the two-dimensional inversion procedure. Geophys J Int 147: 610–621

Heit BS (2005) Teleseismic tomographic images of the Central Andes at 21°S and 25.5°S: an inside look at the Altiplano and Puna plateaus. PhD thesis, FU Berlin

Hérail G, Rochat P, Baby P, Aranibar O, Lavenu A, Masclez G (1997) El Altiplano Norte de Bolivia, evolución geológica terciaria, El Altiplano: ciencia y conciencia en los Andes, Actas 2. In: Charrier R et al. (eds) Simposio Internacional Estudios Altiplánicos, Arica 1993. Universidad de Chile, Santiago, pp 33–44

Hildreth W, Moorbath S (1988) Crustal contributions to arc magmatism in the Andes of central Chile. Contrib Mineral Petrol 98:455–489

Hill GJ, Caldwell TG, Heise W, Chertkoff DG, Bibby HM, Burgess MK, Cull JP, Cas RAF (2009) Distribution of melt beneath Mount St Helens and Mount Adams inferred from magnetotelluric data. Nat. Geosci. doi:10.1038/NGEO661

Hoernle K, Hauff F (2000) Oceanic igneous provinces. In: Bundschuh J, Alvarado GE (eds) Central America: geology, resources, hazards, vol 1. Taylor & Francis, London, pp 523–548

Husen S, Quintero R, Kissling E, Hacker BR (2003) Subduction zone structure and magmatic processes beneath Costa Rica as constrained by local earthquake tomography and petrologic modeling. Geophys J Int 155:11–32

James DE (1971) Andean crustal and upper mantle structure. J Geophys Res 76:3246–3271

James DE, Sacks JW (1999) Cenozoic formation of the central Andes: a geophysical perspective. In: Skinner B (ed) Geology and ore deposits of the central Andes. Society of Economic Geologists Special Publication 7, Littleton, CO, pp 1–25

Jiracek G, Curtis J, Ramirez J, Martinez M, Romo J (1989) Two-dimensional magnetotelluric inversion of the EMSLAB lincoln line. J Geophys Res 94:14145–14151

Jödicke H, Jording A, Ferrari L, Arzate J, Mezger K, Rüpke L (2006) Fluid release from the subducted Cocos plate and partial melting of the crust deduced from magnetotelluric studies in southern Mexico: implications for the generation of volcanism and subduction dynamics. J Geophys Res 111. doi:10.1029/2005JB003739

Kapinos G (2011) Amphibious magnetotellurics at the South-Central Chilean continental margin. PhD thesis, Free University of Berlin

Kay SM, Mpodozis C (2001) Central Andean Ore deposits linked to evolving shallow subduction systems and thickening crust. GSA Today 11:4–9

Kirby SH, Engdahl ER, Denlinger R (1996) Intermediate-depth intraslab earthquakes and arc volcanism as physical expressions of crustal and uppermost mantle metamorphism in subducting slabs. In: Bebout GE et al. (eds) Subduction: top to bottom. Geophysical Monograph Series, vol 96. AGU, Washington, DC, pp 195–214

Klotz J, Abolghasem A, Khazaradze G, Heinze B, Vietor T, Hackney R, Bataille K, Maturana R, Viramonte J, Perdomo R (2006) Long-term signals in the present-day deformation field of the central and Southern Andes and constraints on the viscosity of the Earth's Upper Mantle. In: Oncken O et al. (eds) The Andes: active subduction orogeny, frontiers in earth sciences. Springer, Berlin, pp 65–89

Krawczyk CM, Mechie J, Tašárová Z, Lüth S, Stiller M, Brasse H, Echtler H, Bataille K, Wigger P, Araneda M (2006) Geophysical signatures and active tectonics at the South-Central Chilean Margin. In: Oncken O et al. (eds) The Andes: active subduction orogeny, frontiers in earth sciences. Springer, Berlin, pp 171–192

Lamb S, Davis P (2003) Cenozoic climate change as a possible cause for the rise of the Andes. Nature 425:792–797

Lara LE, Naranjo JA, Moreno H (2004) Rhyodacitic fissure eruption in Southern Andes (Cordón Caulle; 40.5°S) after the 1960 (Mw:9.5) Chilean earthquake: a structural interpretation. J Volc Geotherm Res 138:127–138
Lezaeta P (2001) Distortion analysis and 3-D modeling of magnetotelluric data in the Southern Central Andes. PhD thesis, FU Berlin
Lezaeta P, Brasse H (2001) Electrical conductivity beneath the volcanoes of the NW Argentinian Puna. Geophys Res Lett 28:4651–4654
Lezaeta P, Muñoz M, Brasse H (2000) Magnetotelluric image of the crust and upper mantle in the backarc of the NW Argentinean Andes. Geophys J Int 142:841–854
Li Y (2002) A finite element algorithm for electromagnetic induction in two-dimensional anisotropic conductivity structures. Geophys J Int 148:389–401
López-Escobar L, Cembrano J, Moreno H (1995) Geochemistry and tectonics of the Chilean Southern Andes basaltic Quaternary volcanism (37–46°S). Rev Geol Chile 22:219–234
Lüth S, Wigger P (2003) A crustal model along 39°S from a seismic refraction profile – ISSA 2000. Rev Geol Chile 30:83–101
MacKenzie L, Abers GA, Fischer KM, Syracuse EM, Protti JM, Gonzalez V, Strauch W (2008) Crustal structure along the southern Central American volcanic front. Geochem Geophys Geosyst 9:Q08S09. doi:10.1029/2008GC001991
Mackie RL, Smith JT, Madden TR (1994) Three-dimensional modeling using finite difference equations: the magnetotelluric example. Radio Sci 29:923–935
Mann P, Rogers RD, Gahagan L (2007) Overview of plate tectonic history and its unresolved tectonic problems. In: Bundschuh J, Alvarado GE (eds) Central America: geology, resources, hazards, vol 1. Taylor & Francis, London, pp 201–238
McNeice GW, Jones AG (2001) Multi-site, multi-frequency tensor decomposition of magnetotelluric data. Geophysics 66:158–173
Melnick D, Rosenau M, Folguera A, Echtler H (2006) Neogene tectonic evolution of the Neuquén Andes western flank (37–39°S). In: Kay SM, Ramos VA (eds) Evolution of an Andean margin: a tectonic and magmatic view from the Andes to the Neuquén Basin (35–39°S). Geol Soc Am Spec Paper 407. doi:10.1130/2006.2407(04)
Mibe K, Fujii T, Yasuda A (1999) Control of the location of the volcanic front in island arcs by aqueous fluid connectivity in the mantle wedge. Nature 401:259–262
Montahei M (2011) Investigation of electrically anisotropic structures employing magnetotelluric data. PhD thesis, University of Tehran
Müller A, Haak V (2004) 3-D modeling of the deep electrical conductivity of Merapi volcano (Central Java): integrating magnetotellurics, induction vectors and the effects of steep topography. J Volc Geotherm Res 138:205–222
Muñoz N, Charrier R (1996) Uplift of the western border of the Altiplano on a west-vergent thrust system, Northern Chile. J South Am Earth Sci 9:171–181
Muñoz J, Troncoso R, Duhart P, Crignola P, Farmer L, Stern CR (2000) The relation of the mid-Tertiary coastal magmatic belt in south-central Chile to the late Oligocene increase in plate convergence rate. Rev Geol Chile 27:177–203
Myers SC, Beck S, Zandt G, Wallace T (1998) Lithospheric-scale structure across the Bolivian Andes from tomographic images of velocity and attenuation for P and S waves. J Geophys Res 103:21233–21252
Nakamura K (1977) Volcanoes as possible indicators of tectonic stress orientation (principle and proposal). J Volcan Geotherm Res 2:1–16
Nesbitt BE (1993) Electrical resistivities of crustal fluids. J Geophys Res 98:4301–4310
Parada MA, López-Escobar L, Oliveros V, Fuentes F, Morata D, Calderón M, Aguirre L, Féraud G, Espinoza F, Moreno H, Figueroa O, Muñoz J, Troncoso Vásquez R, Stern CR (2007) Andean magmatism. In: Moreno T, Gibbons W (eds) The geology of Chile. Geological Society, London, pp 115–146
Pek J, Verner T (1997) Finite difference modelling of magnetotelluric fields in 2-D anisotropic media. Geophys J Int 128:505–521
Patro PK, Egbert GD (2008) Regional conductivity structure of Cascadia: preliminary results from 3D inversion of USArray transportable array magnetotelluric data. Geophys Res Lett 35:L20311. doi:10.1029/2008GL035326
Peacock SM, van Keken PE, Holloway SD, Hacker BR, Abers GA, Fergason RL (2005) Thermal structure of the Costa Rica-Nicaragua subduction zone. Phys Earth Planet Int 149:187–200
Pizarro D (1993) Los pozos profundos perforados en Costa Rica: aspectos litológicos y bioestratigráficos. Rev Geol Am Central 15:81–85
Protti M, Guendel F, McNally K (1995) Correlation between the age of the subducting Cocos plate and the geometry of the Wadati-Benioff zone under Nicaragua and Costa Rica. Geol Soc Am Spec Paper 295:309–326
Ramos VA, Kay SM (2006) Overview of the tectonic evolution of the southern Central Andes of Mendoza and Neuquén (35°–39°S latitude). In: Kay SM, Ramos VA (eds) Evolution of an Andean margin: a tectonic and magmatic view from the Andes to the Neuquén Basin (35°–39°S). Geological Society of America Special Paper, vol 407. doi:10.1130/2006.2407(01)
Ranero CR, Phipps Morgan J, McIntosh K, Reichert C (2003) Bending-related faulting and mantle serpentinization at the Middle America trench. Nature 425:367–373
Reuther CD, Moser E (2007) Orientation and nature of active crustal stresses determined by electromagnetic measurements in the Patagonian segment of the South America Plate. Int J Earth Sci (Geol. Rundschau). doi:10.1007/s00531-007-0273-0
Rietbrock A, Haberland C, Bataille K, Dahm T, Oncken O (2005) Studying the seismogenic coupling zone with a passive seismic array. EOS Trans AGU 86:293
Ritz M, Bondoux F, Hérail G, Sempere T (1991) A magnetotelluric survey in the northern Bolivian Altiplano. Geophys Res Lett 18:475–478
Roberts JJ, Tyburczy JA (1999) Partial-melt electrical conductivity: Influence of melt composition. J Geophys Res 104:7055–7065
Rodi W, Mackie RL (2001) Nonlinear conjugate gradients algorithm for 2-D magnetotelluric inversions. Geophysics 66:174–187
Rosenau M, Melnick D, Echtler H (2006) Kinematic constraints on intra-arc shear and strain partitioning in the

southern Andes between 38°S and 42°S latitude. Tectonics 25. doi:10.1029/2005TC001943

Rüpke LH, Morgan JP, Hort M, Connolly JAD (2002) Are the regional variations in Central American arc lavas due to differing basaltic versus peridotitic slab sources of fluids? Geology 30:1035–1038

Rychert CA, Fischer KM, Abers GA, Plank T, Syracuse EM, Protti JM, Gonzalez V, Strauch W (2008) Strong along-arc variations in attenuation in the mantle wedge beneath Costa Rica and Nicaragua. Geochem Geophys Geosyst 9:Q10S10. doi:10.1029/2008GC002040

Sallarès V, Dañobeitia JJ, Flueh ER (2001) Lithospheric structure of the Costa Rican Isthmus: effects of subduction zone magmatism on an oceanic plateau. J Geophys Res 106:621–643

Scambelluri M, Philippot P (2001) Deep fluids in subduction zones. Lithos 55:213–227

Scherwath M, Flueh E, Grevemeyer I, Tilmann F, Contreras-Reyes E, Weinrebe W (2006) Investigating subduction zone processes in Chile. EOS Trans AGU 87:265

Scheuber E, Bogdanic T, Jensen A, Reutter K-J (1994) Tectonic development fo the north Chilean Andes in relation to plate convergence and magmatism since the Jurassic. In: Reutter K-J, Scheuber E, Wigger P (eds) Tectonics of the Southern Central Andes. Springer, Berlin, pp 121–140

Scheuber E, Mertmann D, Ege H, Silva-González P, Heubeck C, Reutter K-J, Jacobshagen V (2006) Exhumation and basin development related to formation of the Central Andean Plateau, 21°S. In: Oncken O et al. (eds) The Andes: active subduction orogeny, frontiers in earth sciences. Springer, Berlin, pp 459–474

Schilling FR, Trumbull RB, Brasse H, Haberland C, Asch G, Bruhn D, Mai K, Haak V, Giese P, Muñoz M, Ramelow J, Rietbrock A, Ricaldi E, Vietor T (2006) Partial melting in the Central Andean crust: a review of geophysical, petrophysical, and petrologic evidence. In: Oncken O et al. (eds) The Andes: active subduction orogeny, frontiers in earth sciences. Springer, Berlin, pp 459–474

Schmidt MW, Poli S (1998) Experimentally based water budgets for dehydrating slabs and consequences for arc magma generation. Earth Planet Sci Lett 163:361–379

Schmucker U (1970) Anomalies of geomagnetic variations in the Southwestern United States. Bull Scripps Institution La Jolla, University of California Press, Los Angeles

Schmucker U, Forbush SE, Hartmann O, Giesecke AA, Casaverde M, Castillo J, Salgueiro R, del Pozo S (1966) Electrical conductivity anomaly under the Andes. Carnegie Inst Wash Yearb 65:11–28

Schurr B, Asch G, Rietbrock A, Trumbull R, Haberland C (2003) Complex patterns of fluid and melt transport in the central Andean subduction zone revealed by attenuation tomography. Earth Planet Sci Lett 215:105–119

Schwalenberg K, Haak V, Rath V (2002) The application of sensitivity studies on a two-dimensional resistivity model from the Central Andes. Geophys J Int 150:673–686

Schwarz G, Krüger D (1997) Resistivity cross section through the southern central Andes as inferred from magnetotelluric and geomagnetic deep soundings. J Geophys Res 102:11957–11978

Sempere T, Hérail G, Oller J, Bonhomme MG (1990) Late Oligocene-early Miocene major tectonic crisis and related basins in Bolivia. Geology 18:946–949

Shaw H (1980) Fracture mechanisms of magma transport from the mantle to the surface. In: Hargraves RB (ed) Physics of magmatic processes. Princeton University Press, Princeton, NJ

Sick C, Yoon M-K, Rauch K, Buske S, Lüth S, Araneda M, Bataille K, Chong G, Giese P, Krawczyk C, Mechie J, Meyer H, Oncken O, Reichert C, Schmitz M, Shapiro S, Stiller M, Wigger P (2006) Seismic images of accretive and erosive subduction zones from the Chilean margin. In: Oncken O et al. (eds) The Andes: active subduction orogeny, frontiers in earth sciences. Springer, Berlin, pp 147–169

Siemon B (1997) An interpretation technique for superimposed induction anomalies. Geophys J Int 130:73–88

Simpson F, Bahr K (2005) Practical magnetotellurics. Cambridge University Press, Cambridge

Smith JT (1995) Understanding telluric distortion matrices. Geophys J Int 122:219–226

Smith JT (1997) Estimating galvanic-distortion magnetic fields in magnetotellurics. Geophys J Int 130:65–72

Soyer W (2002) Analysis of geomagnetic variations in the Central and Southern Andes. PhD thesis, FU Berlin

Soyer W, Brasse H (2001) Investigation of anomalous magnetic field variations in the central Andes of N Chile and SW Bolivia. Geophys Res Lett 28:3023–3026

Soyer W, Unsworth M (2006) Deep electrical structure of the northern Cascadia (British Columbia, Canada) subduction zone: implications for the distribution of fluids. Geology 34. doi:10.1130/G21951.1

Springer M, Förster A (1998) Heat-flow density across the Central Andean subduction zone. Tectonophysics 291: 123–139

Stern CR (2004) Active Andean volcanism: its geologic and tectonic setting. Rev Geol Chile 31:161–206

Sylvester AG (1988) Strike-slip faults. Geol Soc Am Bull 100:1666–1703

Syracuse EM, Abers GA, Fischer K, MacKenzie L, Rychert C, Protti JM, González, V, Strauch W (2008) Seismic tomography and earthquake locations in the Nicaraguan and Costa Rican upper mantle. Geochem Geophys Geosyst 9:Q07S08. doi:10.1029/2008GC001963

Tassara A, Götze HJ, Schmidt S, Hackney R (2006) Three-dimensional density model of the Nazca plate and the Andean continental margin. J Geophys Res 111. doi:10.1029/2005JB003976

Tatsumi Y (2003) Some constraints on arc magma genesis. In: Eiler J (ed) Inside the subduction factory. Geophysical Monograph Series, vol 138. AGU, Washington, DC, pp 277–292

Tyburczy JA, Fisler DK (1995) Electrical properties of minerals and melts, mineral physics & crystallography. In: Ahrens TJ Handbook of physical constants, American Geophysical Union, Washington, DC, pp 185–208

Vanyan LL, Berdichevsky MN, Pushkarev PYu, Romanyuk TV (2002) A geoelectric model of the Cascadia subduction zone. Izvestiya Phys Solid Earth 38:816–845

Varentsov IvM, Golubev NG, Gordienko VV, Sokolova EYu (1996) Study of deep geoelectrical structure along EMSLAB Lincoln-Line. Izvestiya Phys Solid Earth 32:375–393

Völker D, Wiedicke M, Ladage S, Gaedicke C, Reichert C, Rauch K, Kramer W, Heubeck C (2006) Latitudinal variation in sedimentary processes in the peru-Chile trench off Central

Chile. In: Oncken O et al. (eds) The Andes: active subduction orogeny, frontiers in earth sciences. Springer, Berlin, pp 193–216

von Huene R, Weinrebe W, Heeren F (1999) Subduction erosion along the North Chile margin. Geodynamics 27:345–358

Walther CHE, Flueh ER, Ranero CR, von Huene R, Strauch W (2000) Crustal structure across the Pacific margin of Nicaragua: evidence for ophiolitic basement and a shallow mantle sliver. Geophys J Int 141:759–777

Wannamaker PE (1999) Affordable Magnetotellurics: interpretation in Natural Environments. In: Oristaglio M, Spies B (eds) Three-dimensional electromagnetics. Soc. Expl. Geophys., Tulsa, pp 349–374

Wannamaker PE, Booker JR, Jones AG, Chave AD, Filloux JH, Waff HS, Law LK (1989) Resistivity cross-section through the Juan de Fuca subduction system and its tectonic implications. J Geophys Res 94:14127–14144

Wannamaker PE, Caldwell TG, Jiracek GR, Maris V, Hill GJ, Ogawa Y, Bibby HM, Bennie SL, Heise W (2009) Fluid and deformation regime of an advancing subduction system at Marlborough. New Zealand. Nature 460. doi:10.1038/nature08204

Weidelt P (1999) 3-D Conductivity models: implications of electrical anisotropy. In: Oristaglio M, Spies B (eds) Three-dimensional electromagnetics. Society of Exploration Geophysicists, Tulsa, pp 119–137

Wessel P, Smith WHF (1998) New, improved version of the Generic Mapping Tools released. EOS Trans AGU 79:579

Whitman D, Isacks BL, Kay SM (1996) Lithospheric structure and along-strike segmentation of the Central Andean Plateau: seismic Q, magmatism, flexure, topography and tectonics. Tectonophysics 259:29–40

Wiese H (1962) Geomagnetische Tiefentellurik Teil II: Die Streichrichtung der Untergrundstrukturen des elektrischen Widerstandes, erschlossen aus geomagnetischen Variationen. Pageoph 52:83–103

Wigger P, Schmitz M, Araneda M, Asch G, Baldzuhn S, Giese P, Heinsohn W-D, Martínez E, Ricaldi E, Röwer P, Viramonte J (1994) Variation of the crustal structure of the southern Central Andes deduced from seismic refraction investigations. In: Reutter K-J, Scheuber E, Wigger P (eds) Tectonics of the Southern Central Andes. Springer, Berlin, pp 23–48

Wörner G, Hammerschmidt K, Henjes-Kunst F, Lezaun J, Wilke H (2000) Geochronology (^{40}Ar-^{39}Ar-, K-Ar-, and He-exposure-ages) of Cenozoic magmatic rocks from Northern Chile (18°-22°S): implications for magmatism and tectonic evolution of the central Andes. Rev Geol Chile 27: 205–240

Wörner G, Uhlig D, Kohler I, Seyfried H (2002) Evolution of the West Andean Escarpment at 18°S (N. Chile) during the last 25 Ma: uplift, erosion and collapse through time. Tectonophysics 345:183–198

Yuan X, Asch G, Bataille K, Bock G, Bohm M, Echtler H, Kind R, Oncken O, Wölbern I (2006) Deep seismic images of the southern Andes. In: Kay SM, Ramos VA (eds) Evolution of an Andean margin: a tectonic and magmatic view from the Andes to the Neuqúen Basin (35°–39°S lat), Geological Society of American, Special paper. doi:10.1130/2006.2407(03)

Yuan X, Sobolev SV, Kind R (2002) Moho topography in the central Andes and its geodynamic implications. Earth Planet Sci Lett 199:389–402

Zhdanov MS (2009) Geophysical Electromagnetic theory and methods. Methods in geochemistry and geophysics, 43. Elsevier, 868pp

Joint Inversion of Seismic and MT Data – An Example from Southern Granulite Terrain, India

5

A. Manglik, S.K. Verma, K. Sain, T Harinarayana, and V. Vijaya Rao

Abstract

Inherent ambiguity in the estimations of the subsurface structure made using single geophysical method is well recognized. Use of more than one method to map the same area greatly reduces the ambiguity. Recognizing this fact, there is an increasing trend for the application of multiple geophysical methods and integration of results to develop more confidence on the interpreted model. An earlier study based on integrated geophysical data along the Kuppam- Palani geo-transect in the Southern Granulite Terrain (SGT) revealed a four-layered crustal structure with a seismic low velocity, electrically conducting mid-crustal layer (LVCL). Reliable estimates of LVCL parameters are often difficult to achieve because of equivalence problem. We apply previously developed 1-D joint inversion algorithm to a segment of this transect to analyze the efficacy of the algorithm in reducing the ambiguity in the estimates of LVCL parameters. We take seismic refraction and reflection data corresponding to one shot point, and coincident magnetotelluric data at one location from the part of the profile where the crustal structure is interpreted as layered. Inversion with a large number of initial guesses and model appraisal indicates that LVCL parameters are better constrained with joint inversion than with individual inversions.

5.1 Introduction

It is well established that no single geophysical method provides unique solution. The finite data measured on the surface can be explained by a number of models through inversion process. The main difficulty arises in deriving a three dimensional distribution of the physical property of the subsurface from the finite number of data and also due to the insensitivity of the limited number of measurements made on the surface to the changes of the subsurface structure at greater depths. Additionally due to the well known 'principle of equivalence' in electrical and electromagnetic methods, it is difficult to derive unique solution. This is because different combinations of conductivity and thickness of subsurface layers for a 1 D earth can produce similar response. In order to resolve this issue several attempts were made earlier to develop joint inversion techniques utilizing electrical and electromagnetic data sets. For example, joint inversion of DC resistivity and magnetotellurics was developed (Vozoff and Jupp 1975) and applied to practical data (Harinarayana 1999). Several

A. Manglik (✉)
National Geophysical Research Institute (CSIR-NGRI), Hyderabad 500606, India
e-mail: ajay@ngri.res.in

E. Petrovský et al. (eds.), *The Earth's Magnetic Interior*, IAGA Special Sopron Book Series 1,
DOI 10.1007/978-94-007-0323-0_5,

other combinations have also been attempted e.g. DC electrical and electromagnetic data (Verma and Sharma 1993; Sharma and Verma 2011), transient electromagnetics and MT (Meju 1996), MT and dipole-dipole resistivity (Sasaki 1989), DC and AMT (Santos et al. 2007) etc. Additionally, joint inversion techniques have also been developed by several workers utilizing two different physical parameters like resisitivity and seismic velocity, gravity and magnetics, gravity and seismics etc. For example combinations of MT and seismic data (Manglik and Verma 1998), DC resistivity, MT and seismic data ((Manglik et al. 2009), DC resistivity and seismic data (Gallardo and Meju 2003), 3D gravity and magnetic data (Gallardo et al. 2003), structural approach (Haber and Oldenburg 1997), multiple geophysical methods (Kozlovskaya 2001), co-operative inversion (Lines et al. 1988), crosshole seismics and georadar (Musil et al. 2003), crosshole resistance and georadar (Linde et al. 2006), weighted inversion with different arrays (Athanasiou et al. 2007), gravity and electrical data (Santos et al. 2006), seismic and resisitivity data (Wisen and Christiansen 2005), etc.

The southern Indian shield is composed of various crustal blocks of Archaean to Neoproterozoic age separated by shear zones. The shield exhibits significant diversity in lithology, structural pattern, tectonics, and metamorphism. The oldest part of the shield, the Archaean Dharwar craton (DC), is separated from the Proterozoic Southern Granulite Terrain (SGT) in the south by two prominent shear zones, the Moyer-Bhavani and the Palghat-Cauvery shear zones. SGT mainly consists of high-grade granulites representing deep crustal origin in contrast to gneisses of DC. Thus, the region provides an assemblage of differentially exhumed crustal blocks which can provide insight into the evolution of the continental crust. An integrated geological and geophysical study was carried out along the Kuppam–Palani geo-transect (Fig. 5.1) with the objective to delineate the crustal structure of various blocks and to understand the nature of interaction between DC and SGT. The 300 km long transect starts from Kuppam in DC, traverses through the transition zone bounded between Moyar-Bhavani and Palghat-Cauvery shear zones, and ends at Palani in the SGT.

The integrated study delineated a four-layered crustal model with a mid-crustal low velocity layer (LVL) beneath the entire profile (Reddy et al. 2003) which is also found to be electrically conductive (Harinarayana et al. 2003) and having a low density of 2650 kg/m^3 (Singh et al. 2003). The thickness of the LVL is 10–15 km in the northern Kuppam-Kumarapalaiyam segment and it thins to less than 10 km in the southern Kolattur- Palani segment. Reddy et al. (2003) obtained the crustal structure by assuming a seismic velocity of 6.0 km/s for the LVL. However, presence of LVLs limits the interpretation of seismic

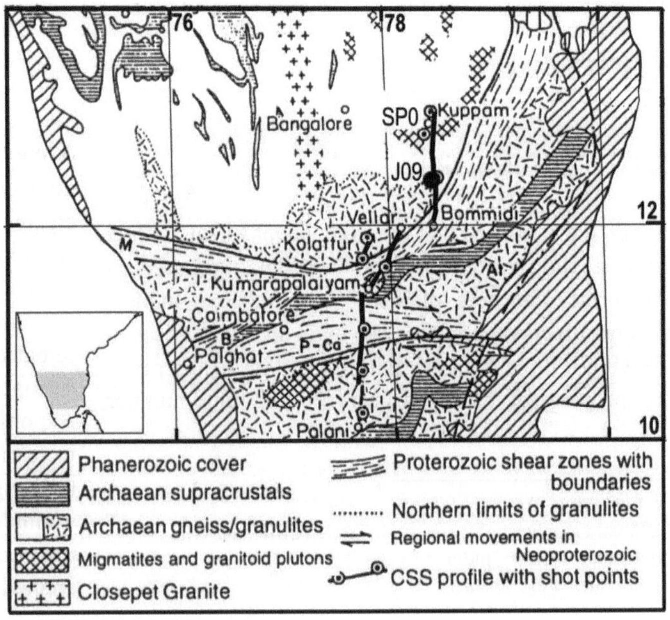

Fig. 5.1 Simplified geological map showing the location of the Kuppam-Palani geo-transect (Modified after Reddy et al. 2003)

data because of the absence of refraction from its top and reverberations due to the trapping of energy within the LVL. This problem has also been encountered in hydrocarbon exploration in Deccan trap covered areas and is expected to have similar effects in deep crustal seismic studies.

In hydrocarbon exploration, attempts have been made to integrate other geophysical data to delineate the structure of LVLs in complex geological settings e.g. integration of electromagnetic methods with seismics to delineate sediments buried beneath thick basalt cover (Morrison et al. 1996, Warren 1996, Withers et al. 1994). These studies are mainly based on cooperative interpretation. Some attempts have also been made to jointly invert two genetically different data sets under the assumption that both responses are produced by the same geometrical structure (Hering et al. 1995). Manglik and Verma (1998) showed that joint inversion (JI) of seismic (SI) and magnetotelluric (MT) data can help in the delineation of low velocity, electrically conducting thin sedimentary layer beneath a thick basalt cover. For crustal studies, Jones (1998) highlighted the significance of integrated interpretation in testing various hypotheses on the evolution of continental lithosphere.

In the present work, we jointly invert coincident seismic refraction and reflection travel time data corresponding to shot point SP0 and MT apparent resistivity and phase data of J09 (Fig. 5.1) to test if the 1-D JI (Manglik and Verma 1998) can provide better estimates of the parameters of electrically conductive LVL (hereafter called Low Velocity Conducting mid-crustal Layer or LVCL) in comparison to those obtained by individual inversions. Although the geological setting is very complex with the transitional nature of the contact between DC and SGT, we selected the data from the northern part of the profile where the crustal structure is interpreted as a simple layered model.

5.2 Seismic and MT Data

Seismic data along the Kuppam- Palani geo-transect were acquired in three segments with some offset between the segments due to logistic constraints. We take first arrival and reflection data of Reddy et al. (2003) corresponding to shot point SP0 (Fig. 5.1) of the northern segment, known as Kuppam- Bommidi (KB) profile. For this shot point, data were recorded up to a distance of 80 km and then between 120 and 135 km distance with a data gap of about 40 km. We use the data of only first 80 km in our computations as shown in Fig. 5.2a.

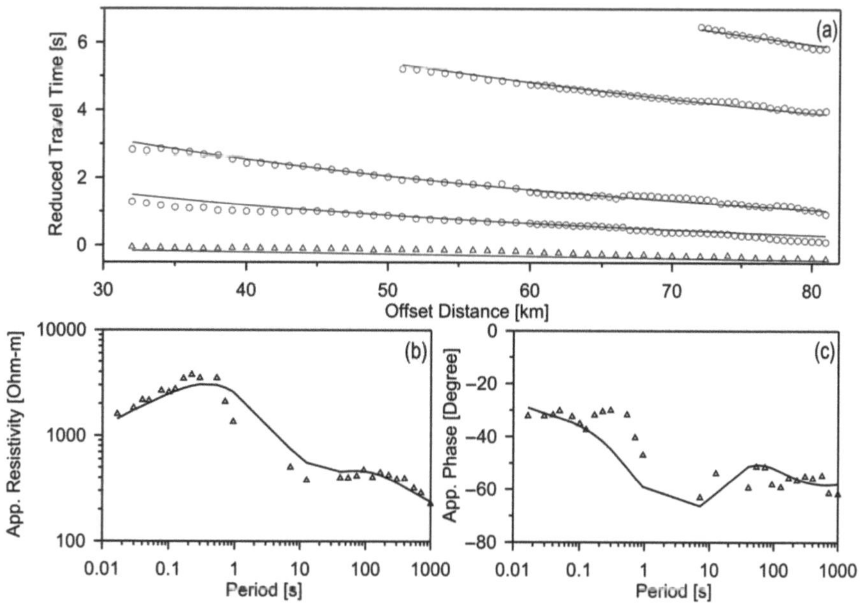

Fig. 5.2 Observed first arrival (*triangles*) and reflection (*circles*) travel times for shot point SP0, and (**b**) observed static shift corrected MT apparent resistivity (*triangles*) and (**c**) apparent phase (*triangles*) corresponding to location J09. Solid lines are responses computed for the final model (Table 5.1).

Table 5.1 Parameters of the final model obtained after model appraisal for LVCL

Final model			
Layer Nr.	Resistivity (Ω-m)	Thickness (km)	Velocity (km/s)
1	364	0.22	6.00
2	3467	13.68	6.20
3	852	6.84	6.36
4	28 ± 2.7	11.56 ± 0.36	5.44 ± 0.01
5	1010	9.75	6.24
6	527	61.07	8.05
	100	–	–

Similarly, five component broadband MT data were acquired at 13 stations along the KB profile (Harinarayana et al. 2003). Static shift correction was applied to the data on the basis of information available from deep resistivity sounding results obtained at some locations along the profile. The apparent resistivity curves indicated the presence of a conductive zone in the period of 1 to 20 s coinciding with the mid-crustal LVL. Harinarayana et al. (2003) inverted these data using Rapid Relaxation Inversion (RRI) to obtain a 2-D electrical conductivity distribution in the crust. We however, use a single station data (J09, Fig. 5.1) in our study because static shift corrected apparent resistivity curves for stations J07, J08, and J09 located approximately at the centre of the profile are similar in nature and have relatively less deviation between ρ_{xy} and ρ_{yx}, satisfying the assumption of 1-D structure. The apparent resistivity and phase curves corresponding to J09 are shown in Fig. 5.2b, c.

5.3 Inversion Results

We use singular value decomposition based joint inversion algorithm (Manglik and Verma 1998) for individual and joint inversion assigning equal weights to both the data sets and perform analysis in three steps. First, we perform joint inversion of seismic reflection and first arrival data using 4-layered crustal model (dashed line, Fig. 5.3a) of Reddy et al. (2003) as an initial guess. Inversion results are shown in Fig. 5.3a by solid line. Final model indicates no change in the velocity and thickness of the first layer alongwith a reduction of about 5 and 18% in the velocity and thickness of the second layer respectively, with respect to the initial guess. However, this model is significantly different in terms of velocity and thickness of the LVCL and thickness of the underlying low conducting layer (LCL). It is found that the velocity and thickness of LVCL are

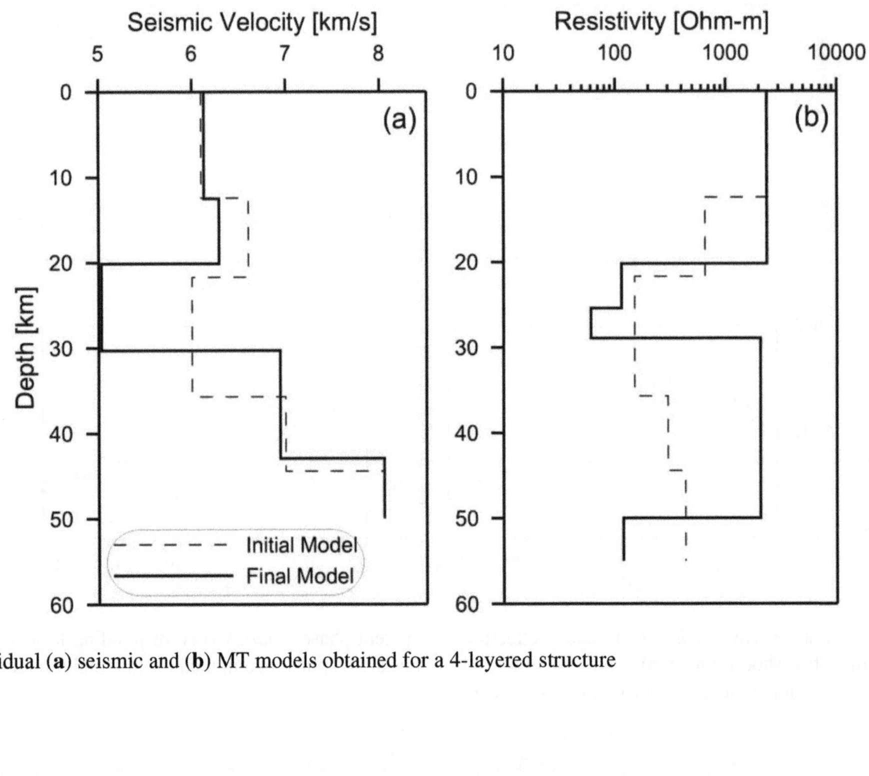

Fig. 5.3 Individual (**a**) seismic and (**b**) MT models obtained for a 4-layered structure

strongly coupled and a large number of models can satisfy the observations. For example, if we assume the LVCL velocity as 6.0 km/s we get a larger thickness of LVCL.

The above initial guess is used to invert MT data with the initial resistivity values of 2400, 650, 150, and 300 Ω-m for two upper crustal layers, LVCL, and LCL, respectively (dashed line, Fig. 5.3b). In the inverted model (solid line, Fig. 5.3b), top two upper crustal layers appear as a single resistive layer of 20 km thickness underlain by two mid-crustal conducting layers of 5.2 and 3.5 km thickness, respectively, representing LVCL. The lower crust and lithospheric mantle appear as a single resistive layer with a thickness of 20 km.

MT apparent resistivity curve in the period of 0.01 to 0.1 s (Fig. 5.2b) indicates the presence of a thin conducting near-surface layer. Similarly, seismic method detects the Moho which is not seen in MT for a 4–layered crustal structure although a drop in resistivity from 2065 to 118 Ω-m is observed at 50 km depth (Fig. 5.3b). Since these layers are detected by atleast one method, we modify the initial guess to a 6-layered model by including a thin near-surface conducting layer and a lithospheric mantle layer below the crust (dashed line, Fig. 5.4).

MT results (solid line, Fig. 5.4b) for the modified initial guess (IG) reveal a 260 m thick near-surface conducting layer (difficult to see in Fig. 5.4b due to linear depth scale) and a decrease in the resistivity from 1280 to 711 Ω-m at a depth of 45 km which may be correlated with the Moho. The sub-crustal layer is 50 km thick, below which the resistivity further drops from 711 to 109 Ω-m. Based on MT results we again looked into seismic records and picked near-offset first arrivals corresponding to the thin near-surface layer. However, it was not possible to get any reflection data from the sub-crustal layer and this layer was transparent in seismic inversion. The results (Solid line, Fig. 5.4a) show a thicker (13.7 km) and a higher velocity (5.98 km/s) LVCL compared to respective values in 4-layered model. However, the underlying LCL is considerably thin (9.3 km) and has low seismic velocity (6.17 km/s). A comparison of models obtained by individual inversions of seismic and MT data shows significant difference in the thicknesses of various crustal layers.

Now we perform joint inversion of seismic and MT data using the same 6-layered initial model as used above (dashed line in Fig. 5.4). The results (broken lines, Fig. 5.4) show slight modification in the thicknesses of the two upper crustal layers. Here, LVCL is about 5 km thicker in comparison to the MT model but it is approximately the same as obtained by seismic method. Similarly, LCL is thin compared to MT model.

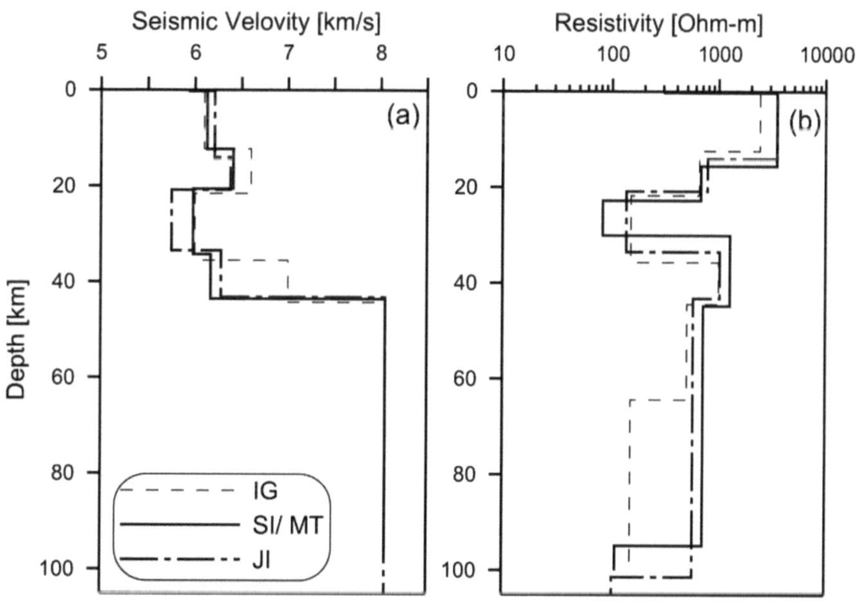

Fig. 5.4 Results of individual (**a**) seismic and (**b**) MT inversion considering 6-layered structure. Joint inversion model is also shown. IG: Initial guess; SI: Seismic inversion; MT: Magnetotelluric

The Moho depth, however, is approximately same for all the models. At sub-crustal depths only MT data provides information about the layering as no seismic data are available from these depths. The thickness of the sub-crustal layer is increased to 58 km.

The above results pertain to only one initial guess selected on the basis of 1-D structure given by Reddy et al. (2003). It is warranted to test the goodness of the obtained model and it's dependence on the choice of the initial guess because both LVCL and LCL suffer from the problem of equivalence, giving a range of acceptable models, all satisfying the minimum root mean square (RMS) error criterion. We also perform model appraisal for LVCL and LCL by inverting the data for a range of initial models to test if joint inversion can constrain the range of acceptable models.

5.4 Model Appraisal for LVCL and LCL

For model appraisal, we consider initial guesses for inversion by systematically choosing resistivity, thickness, and velocity of LVCL in the range of 50–200 Ω-m, 4–20 km, and 4–7 km/s, respectively (Open circles, Fig. 5.5a, b), keeping other parameters same as obtained in Fig. 5.4. We consider 15 models each for MT and SI methods and 45 models for JI. For all these models 20 iterations were used, although the convergence is achieved in less than 10 iterations, and there is no change in RMS error atleast up to 5th decimal place after 15th iteration. This exercise is also done for the LCL (Open circles, Fig. 5.5c–d).

Final models with resistivities, thicknesses, and velocities of LVCL and LCL are plotted in Fig. 5.5a–d, respectively. Figure 5.5a shows the relation between the resistivity and thickness of LVCL. All the models obtained either by MT (diamonds) or by JI (stars) fall in a narrow zone of least RMS error, suggesting that the parameters of LVCL suffer from the problem of equivalence. Similar pattern is observed for SI also (Fig. 5.5b). We don't get this pattern for the resistivity and thickness of LCL as it is a resistive layer (Fig. 5.5c). Here, the final model converges in the vicinity of the initial guess. However, seismic velocity and thickness show a narrow zone of convergence (Fig. 5.5d). These results indicate that the parameters of LCL obtained in the present analysis by individual/joint inversion are difficult to constrain and are not well resolved. Since the parameters of LVCL converge in a long narrow zone of minimum RMS error even with

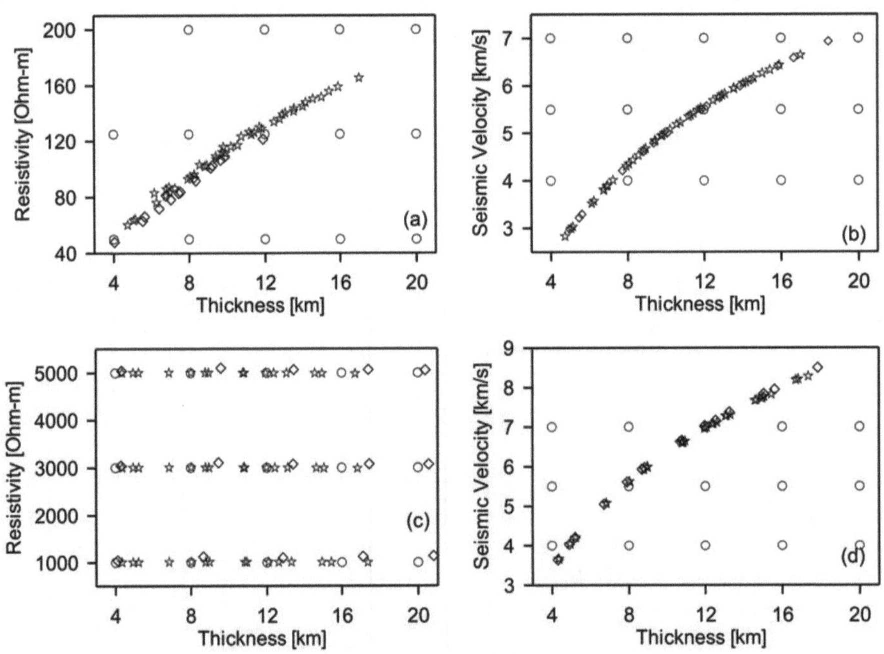

Fig. 5.5 Model appraisal for (**a, b**) low velocity conducting mid-crustal layer (LVCL) and (**c, d**) LCL. Circles, diamonds, and stars represent initial value, final value by individual method, and final value by JI, respectively

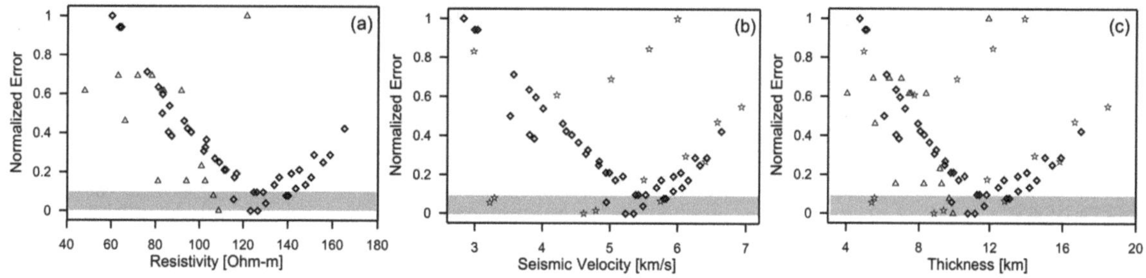

Fig. 5.6 Normalized RMS error vers. (**a**) resistivity, (**b**) seismic velocity, and (**c**) thickness for various models used in model appraisal of LVCL. Models with less than 10% error (shaded region) are considered as acceptable models. Triangle, stars, and diamonds represent MT, SI, and JI solutions, respectively

JI, a cursory look at Fig. 5.5a, b would suggest almost no improvement. We now analyze the RMS error of all the final models in some detail to identify if any pattern exists that can help in reducing the ambiguity of the model parameters.

We normalize RMS error of each final model in such a way that the model having largest RMS error has normalized error of 1 and that having least RMS error now has a normalized error of 0. This was done for MT, SI, and JI cases and the results are shown in Fig. 5.6 by triangles, stars, and diamonds, respectively. We consider the models falling within 10% of the least normalized RMS error as acceptable models (shaded region). Figure 5.6a shows the normalized error as a function of the resistivity of LVCL. It is seen that only two models (out of 15) obtained by MT inversion fall in this range and suggest a resistivity value of 106–110 Ω-m. On the other hand, JI gives 9 models (out of 45) in this range. These are confined in the range of 115–140 Ω-m.

The results of velocity of LVCL are more interesting. Here, we find that five seismic models (out of 15) fall in the range of acceptable models (Fig. 5.6b). These suggest three locations of acceptable models, one around 3.25 km/s, second around 4.7 km/s, and third at 5.75 km/s, indicating a large variation in the seismic velocity of LVCL. In contrast, JI models fall in the range of 5.0–5.8 km/s with the minimum at 5.4 km/s. The pattern is similar for the thickness (Fig. 5.6c). Here, SI gives three possible solutions (5.5, 9.1, and 12.8 km); MT gives 9.8 km; and JI gives a range of 10–13 km with the minimum at 11.5 km. These results indicate that JI gives a well defined pattern of normalized RMS error with a clear minimum unlike MT or SI cases. Corresponding values of LVCL parameters are also well constrained. Thus, JI reduces the ambiguity in parameters of LVCL. Final model with appraisal for LVCL is given in Table 5.1 and corresponding synthetic responses are shown in Fig. 5.2 by solid lines.

5.5 Conclusions

We perform 1-D joint inversion of seismic and MT data from the southern Indian shield to analyse whether such an approach can help in improving the estimates of the parameters of LVCL. The analysis suggests that JI can help in constraining the range of acceptable models.

Our best fit model suggests a velocity of 5.45 km/s for LVCL which is 0.55 km/s less than that proposed by Reddy et al. (2003) to obtain it's thickness based on geological considerations. In our analysis we allow the solution to evolve without any external bias. Here, our aim is not to discuss geological/tectonic implications of velocity of LVCL but to test the efficacy of JI in dealing with low velocity layer problem.

Our results also bring out the presence of a 60 km thick sub-crustal layer below which the resistivity decreases to about 100 Ω-m. This information, however, is based on long period MT data and no seismic data from this layer is available. Whether this can be correlated with the Hale's discontinuity or lithosphere-asthenosphere boundary needs to be explored with a larger data set.

In the above analysis we find that the electrical conductivity and low seismic velocity are produced by the same mid-crustal layer. However, this can not be generalized for every geological situation and JI should be applied with caution.

Acknowledgments Permission from the Director CSIR-NGRI to publish this work is gratefully acknowledged. The second author (SKV) would like to acknowledge the Raja Ramanna Fellowship awarded by DAE, Government of India.

References

Athanasiou EN, Tsourlos PI, Papazachos CB, Tsokas GN (2007) Combined weighted inversion of electrical resistivity data arising from different array types. J Appl Geophys 62:124–140

Gallardo LA, Meju MA (2003) Characterization of heterogeneous near-surface materials by joint 2D inversion of DC resistivity and seismic data. Geophys Res Lett 30(13):1658. doi:10.1029/2003GL017370

Gallardo LA, Perez-Flores MA, Gomez-Trevino E (2003) A versatile algorithm for joint 3-D inversion of gravity and magnetic data. Geophysics 68:949–959

Haber E, Oldenburg D (1997) Joint inversion: a structural approach. Inverse Probl 13:63–77

Harinarayana T (1999) Combination of EM and DC measurements for upper crustal studies. Surv Geophys 20(3–4): 257–278

Harinarayana T, Naganjaneyulu K, Manoj C, Patro BPK, Kareemunnisa Begum S, Murthy DN, Rao M, Kumaraswamy VTC, Virupakshi G (2003) Magnetotelluric investigations along Kuppam- Palani geotransect, South India – 2-D modeling results. In: Ramakrishnan M (ed) Tectonics of southern Granulite terrain: Kuppam-Palani geotransect. Memoir Nr 50, Geological Society of India, Bangalore, pp 107–124

Hering A, Misiek R, Gyulai A, Ormos T, Dobroka M, Dresen L (1995) A joint inversion algorithm to process geolctric and surface wave seismic data Part I: basic ideas. Geophys Prospecting 43:135–156

Jones AG (1998) Waves of the future: Superior inferences from collocated seismic and electromagnetic experiments. Tectonophysics 286:273–298

Kozlovskaya E (2001) Theory and application of joint interpretation of multimethod geophysical data. Ph.D. dissertation, University of Oulu, Oulu, Finland

Linde N, Binley A, Tryggvason A, Pedersen LB, Revil A (2006) Improved hydrogeophysical characterization using joint inversion of cross-hole electrical resistance and ground-penetrating radar traveltime data. Water Resour Res 42, W12404, doi:10.1029/2006WR005131

Lines LR, Schultz AK, Treitel S (1988) Cooperative inversion of geophysical data. Geophysics 53:8–20

Manglik A, Verma SK (1998) Delineation of sediments below flood basalts by joint inversion of seismic and magnetotelluric data. Gephys Res Lett 25:4015–4018

Manglik A, Verma SK, Singh KH (2009) Detection of sub-basaltic sediments by a multi parametric joint inversion approach. J Earth Syst Sci 118:551–562

Meju M (1996) Joint inversion of TEM and distorted MT soundings: Some effective practical considerations. Geophysics 61:56–65

Morrison HF, Shoham Y, Hoversten GM, Torres-verdin C (1996) Electromagnetic mapping of electrical conductivity beneath the Columbia basalts. Geophys Prospecting 44: 963–986

Musil M, Maurer HR, Green AG (2003) Discrete tomography and joint inversion for loosely connected or unconnected physical properties: application to crosshole seismic and georadar data sets. Geophys J Int 153:389–402

Reddy PR, Rajendra Prasad B, Vijaya Rao V, Sain K, Prasada Rao P, Khare P, Reddy MS (2003) Deep seismic reflection and refraction/ wide-angle reflection studies along Kuppam- Palani transect in the southern Granulite terrain of India. In: Ramakrishnan M (ed) Tectonics of Southern Granulite Terrain: Kuppam-Palani geotransect. Memoir Nr 50, Geological Society of India, Bangalore, pp 79–106

Santos FAM, Sultan SA, Represas P, Sorady ALE (2006) Joint inversion of gravity and geoelectrical data for groundwater and structural investigation: application to the northwestern part of Sinai, Egypt. Geophys J Int 165:705–718

Santos FAM, Afonso ARA, Dupis A (2007) 2D joint inversion of DC and scalar audio-magnetotelluric data in the evaluation of low enthalpy geothermal fields. J Geophys Eng 4:53, doi:10.1088/1742-2132/4/1/007

Sasaki Y (1989) Two-dimensional joint inversion of magnetotelluric and dipole-dipole resistivity data. Geophysics 54:254–262

Sharma SP, Verma SK (2011) Solutions of the inherent problem of the equivalence in direct current resistivity and electromagnetic methods through global optimization and joint inversion by successive refinement of model space. Geophys Prosp 59: (in press, July 2011)

Singh AP, Mishra DC, Vijaya Kumar V, Vyaghreswara Rao MBS (2003) Gravity- magnetic signatures and crustal architecture along Kuppam- Palani geotransect, south India. In: Ramakrishnan M (ed) Tectonics of Southern Granulite Terrain: Kuppam- Palani geotransect. Memoir Nr 50, Geological Society of India, Bangalore, pp 139–163

Verma SK, Sharma SP (1993) Resolution of thin layers using joint inversion of electromagnetic and direct current resistivity sounding data. J Electr waves and Appl 7:443–479

Vozoff K, Jupp DLB (1975) Joint inversion of geophysical data. Geophys J R Astr Soc 42:977–991

Warren RK (1996) A few case histories of subsurface imaging with EMAP as aid to seismic prospecting and interpretation. Geophys Prospecting 44:923–934

Wisen R, Christiansen AV (2005) Laterally and mutually constrained inversion of surface wave seismic data and resistivity data. J Environ Eng Geophys 10:251–262

Withers R, Eggers D, Fox T, Crebs T (1994) A case study of integrated hydrocarbon exploration through basalt. Geophysics 59:1666–1679

What We Can Do in Seismoelectromagnetics and Electromagnetic Precursors

6

Toshiyasu Nagao, Seiya Uyeda, and Masashi Kamogawa

Abstract

Earthquake (EQ) prediction, in particular short-term prediction, is one of the topmost challenges in modern science. However, the general view of the community is pessimistic. EQ prediction research has been rather heavily biased toward seismology for the last several decades. In addition to seismics, however, the importance of other methods is being recognized. We intend to evaluate the possible role of electromagnetic (EM) approach to this end by introducing examples of precursors needed for short-term prediction. Recent advances in the physics of critical phenomena to be applied to EQ generation mechanism and the possibility of EQ triggering effect of EM pulses will also be mentioned.

6.1 Introduction

EQ prediction is specifying the source location, magnitude *M*, and occurrence time of an EQ with certain accuracy before it occurs. It is often classified by the length of concerned time, i.e., into long-term (10^2–10^1 years), intermediate-term (10^1–1 years), and short-term (<1 year) predictions. These three categories are different in methodology, i.e., statistical/geological approach, crustal deformation and changes in seismicity, and immediate precursors, respectively. This chapter will focus on the precursors for short-term prediction and the possible role of electromagnetic (EM) research. We emphasize that more efforts are needed to scientifically confirm the existence of the EM precursors suggested so far.

6.2 History of Precursory Study

A number of precursory phenomena have been reported through human history (Tributsch 1982). Most of them were recognized without instrumental aides and are called macro-anomalies. They now include such items as radio and TV noise, e.g., at the 1995 M7.3 at Kobe, the 1999 M7.4 at Izmit, and the 1999 M7.7 at Chichi (Ikeya 2004). Most of the macro-anomalies, however, have not been studied well, although they played a significant role in the success of the 1975 M7.3 Haicheng EQ prediction (Raleigh et al. 1977).

From the 1960s to the 1970s, countries including U.S.S.R., Japan, China, and the USA had embarked on EQ prediction projects on a national scale. Most

T. Nagao (✉)
Earthquake Prediction Research Center, Tokai University, Shizuoka, Japan
e-mail: nagao@scc.u-tokai.ac.jp

E. Petrovský et al. (eds.), *The Earth's Magnetic Interior*, IAGA Special Sopron Book Series 1,
DOI 10.1007/978-94-007-0323-0_6, © Springer Science+Business Media B.V. 2011

projects aimed at long/intermediate-term prediction, but there was also deterministic approach to short-term prediction. In the early 1970s, the dilatancy–diffusion model (Sholtz et al. 1973) appeared based on various reported precursors, for example 10–20% V_P/V_S change in Garm, U.S.S.R. (Semyenov 1969), and the USA (Aggarwal et al. 1973); crustal uplift in Japan (Kato 1981); radon emission; and electrical resistivity (Sadovsky et al. 1972). The famous prediction of the 1975 M7.3 Haicheng EQ was also based on all kinds of precursors (Raleigh et al. 1977). These encouraging works in the 1960s–1970s made the whole prediction community highly optimistic (for example, see Press 1975).

However, the optimism did not last long for various reasons. First, the later studies (McEvilly and Johnson 1974) did not support the significant change of V_P/V_S, which was the backbone of the dilatancy/diffusion model. Second, no short-term prediction followed after the success of the Haicheng EQ. For example, the 1976 M7.8 Tangshan EQ was not short-term alerted, causing the loss of as much as 250,000 lives (Chen et al. 1988). In the case of Parkfield, where M6 class event was predicted in 1985 to come within 5 years, 72 h alert was issued in 1992 (Roeloffs 1994). However, it did not come until 2004, i.e., 12 years later (Bakun et al. 2005).

By the late 1970s, the general view of the community on EQ prediction became entirely pessimistic (e.g., Evernden 1982). This pessimism still prevails making the word prediction almost "forbidden." However, we note that many kinds of precursors for short-term prediction have been observed in non-seismic/geodetic fields, although only a few were convincingly demonstrated as in the cases of radon change before 1995 Kobe EQ (Fig. 6.1, Yasuoka et al. 2009) and anomalous behavior of mice before 2008 Wenchuan EQ (Fig. 6.2, Li et al. 2009). On top of these anomalies, electric, magnetic, and EM anomalous changes should not be overlooked.

6.3 Electric, Magnetic, and EM Precursors

Systematic research on EQ-related electric, magnetic, and EM phenomena, the seismoelectromagnetics, was initiated during the 1980s in two main streams, namely monitoring of possible EM emission from focal regions and anomalous transmission of man-made EM waves over focal regions.

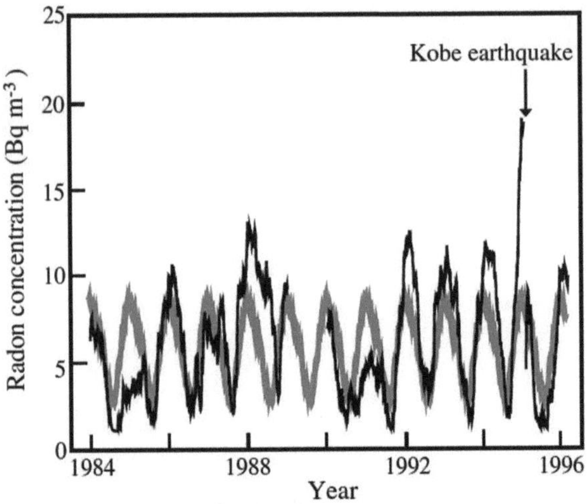

Fig. 6.1 Time series of the daily radon concentration observed from January 1984 to February 1996. The daily minimum variation is shown by *black line* and normal variation by *gray line* (Yasuoka et al. 2009). Clear enhancement of radon concentration can be seen before the M7.3 Kobe EQ

6.3.1 Telluric Current Precursors: The VAN Method

The VAN method, developed in Greece in the early 1980s, is named after the first letter of the founding scientists, P. **V**arotsos, K. **A**lexopoulos, and K. **N**omikos (Varotsos and Alexopoulos 1984a, b). This has been the only working system for real short-term prediction for almost three decades (Fig. 6.3). For $M \geq 5$ Greece EQs, their self-imposed criteria for successful prediction are less than a few weeks in time, less than 0.7 units in *M*, and less than 100 km in epicentral location. SES (seismic electric signals) are transient DC geopotential variations observed before EQs by dipoles of buried electrodes at separate sites at many stations (e.g., Varotsos 2005). At each station, several short (50–200 m) dipoles in both EW and NS directions and a few long dipoles (2–20 km) were installed. By adoption of the multiple dipole system they successfully rejected the noise. Single SES precedes single EQ, whereas the "SES activity," which consists of a number of SES in a short time, is followed by a series of EQs.

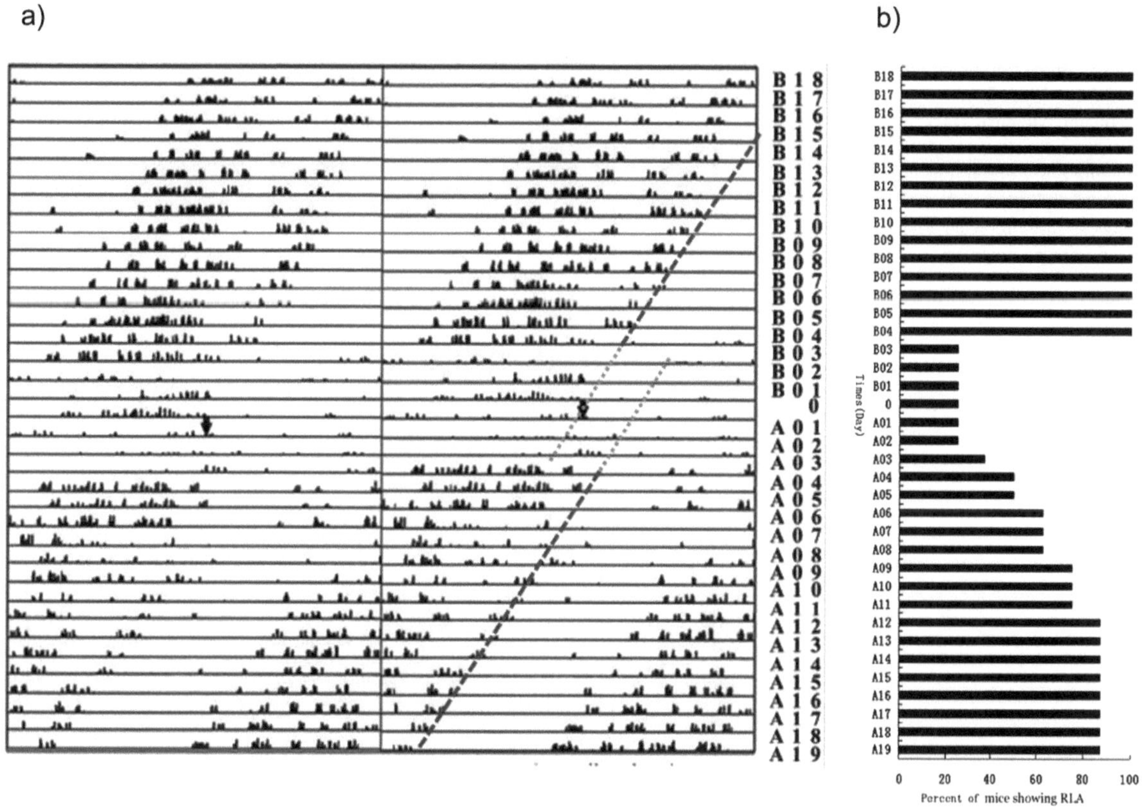

Fig. 6.2 Abnormal animal behavior observed just before the M8.0 Wenchuan EQ in 2008 (Li et al. 2009). (**a**) Under free-running conditions, mouse locomotor activity presents circadian rhythm from B18 (18 days before EQ) to B04 (4 days before) and no rhythm from B03 (3 days before EQ) to A02 (2 days after EQ) including the day of EQ. Horizontal axis is 48 h (2 days). *Diagonal solid lines* mean the circadian rhythm does not coincide with 24 h. (**b**) The percentage of mice showing RLA (circadian rhythm of mouse locomotor activity) among the eight mice indicates that some mice lost RLA around EQ

VAN group made two major discoveries. One is "selectivity," which means that only some sites are sensitive to SES, and a sensitive site is sensitive only to SES from some specific focal area(s). This information gives the probable location of expected EQs. The selectivity is considered to be originating from the inhomogeneity of the subsurface electrical structures, i.e., SES transmits selectively through underground conductive channels to sensitive sites. The other is "VAN relation," which means that the focal distance r, EQ magnitude M, and the potential difference ΔV for dipole of length L are related by $\log\left(\Delta V/L \times r\right) = aM + b$ where a is a constant varying between 0.34 and 0.37 and b is a site-dependent constant. Once the epicentral location is estimated from the selectivity data, M can be assessed from this relation since both $\Delta V/L$ and r are known.

In Japan, VAN-type monitoring has been tried since the late 1980s (Uyeda et al. 2000). Despite the high-level noise from DC-driven trains at many stations, the existence of the VAN-type SES has been confirmed under favorable conditions. For instance, in the year 2000, a 2-month long seismic swarm, with 7000 $M \geq 3$ shocks and 5 $M \geq 6$ shocks, occurred in Izu Island region. For this swarm activity, significant pre-seismic electric disturbances were observed (Fig. 6.4, Uyeda et al. 2002).

There have been a number of skeptic papers on VAN predictions. The VAN group, however, has refuted them (e.g., Geller 1996, Chouliaras and Stavrakakis 1999, Uyeda 2000, Uyeda and Kamogawa 2008, 2010, Papadopoulos 2010). According to these rebuttals, most of the skepticism was due to misunderstandings and misconceptions. Apparently, this situation was at

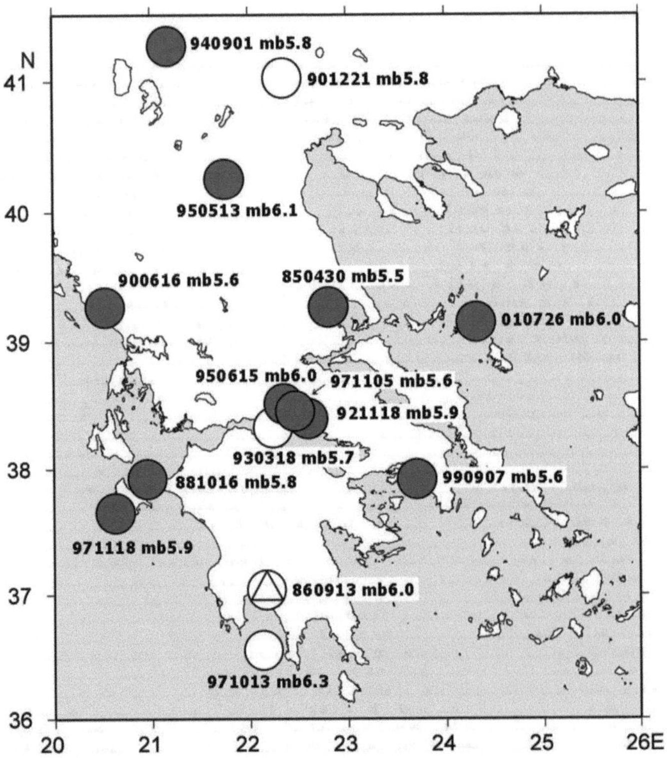

Fig. 6.3 Evaluation of VAN prediction. All EQs with USGS PDE magnitude larger than 5.5 for 1985–2003 (Uyeda and Meguro 2004). *Shaded*, *white* with *triangle*, and *white circles* represent "successfully" predicted, unsuccessfully predicted, and missed EQs, respectively

least partly caused by the fact that VAN publications are generally hard to read for multidisciplinary readers.

6.3.2 Ultra Low Frequency (ULF) Precursors

A well-known example is the case of the 1989 M7.1 Loma Prieta EQ (Fraser-Smith et al. 1990). The horizontal component of the geomagnetic field measured at a site 7 km from the epicenter started anomalous enhancement about 2 weeks prior to the EQ and culminated in a sharp increase a few hours before it. The disturbance lasted for about 3 months after the EQ. These have never been observed at any other time during the observation of more than 15 years. Reports of observing pre-seismic ULF geomagnetic anomalies have also been made for the 1988 M6.9 Spitak (Armenia) EQ (Kopytenko et al. 1993) and the 1993 M8.0 Guam (Marianas) EQ (Hayakawa et al. 1996). There are some discussions pointing out that these ULF signals might have been artifact of magnetotelluric origin (e.g., Thomas et al. 2009).

6.3.3 Higher Frequency EM Emission

Gokhberg et al. (1982) made a pioneering observation of pre-seismic EM wave emission in the LF range. Asada and his group in Japan monitored the waveforms of two horizontal magnetic components of VLF EM waves, through which apparent incoming directions of the VLF pulses were assessed (Asada et al. 2001). They found that, a few days before M5 class land EQs within 100 km of their stations, pulses with fixed incoming direction appeared and the EQs actually occurred in that direction, whereas the sources of overwhelmingly more numerous and stronger noises were moving along with lightning sources. This shows that the important key job of distinguishing precursory signals from lightning noises is possible. The same kind of VLF pulse investigation was also conducted by using a borehole sensor in Japan (Tsutsui 2002, 2005).

Warwick et al. (1982) reported on the observation of high-frequency radio waves possibly related to the precursory activity of the 1960 great M9.5 Chilean EQ. This was probably the first report of this kind. Enomoto et al. (2006) recorded two episodes of anomalous

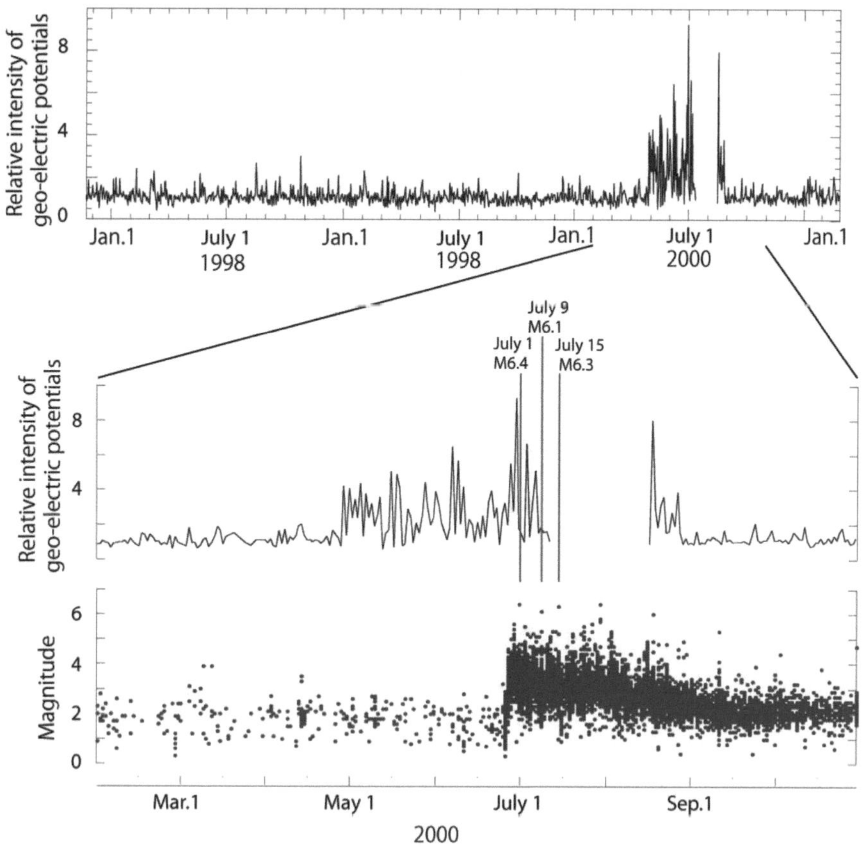

Fig. 6.4 Anomalous geoelectric potential changes in ULF (0.01 Hz) band prior to 2000 seismic swarm activity in Izu Island region, Japan. The bottom panel shows seismicity (Modified from Uyeda et al. 2002)

geoelectric current pulses (HF band) at Erimo station, Hokkaido, Japan (Fig. 6.5). One in 2000–2001 occurred before and during the volcanic activity of Mt. Usu (200 km away) and the second one started 1 month before the 2003 M8.0 Tokachi-Oki EQ (80 km away). These were the only geoelectric anomalies during the 10-year observation period.

6.4 Possible Mechanism of Pre-seismic EM Emissions

It is often questioned "why" EM signals, such as SES, appear only pre-seismically and not co-seismically. This point makes the scientific community dubious about EQ-related EM precursors in general because EQ is by far the most energetic event. Actually, co-seismic signals are routinely observed although they are not useful as precursors. However, all these so-called co-seismic EM signals were found to occur at the time of arrival of seismic waves. They are, therefore, "co-seismic waves" and not "true co-seismic." The fact that no "true co-seismic" signals are observed may be an important clue in exploring the physical mechanism of generation of both EM signals and EQs themselves. In fact, the slow pre-seismic stress buildup and the instantaneous stress release at EQ are physically very different processes and there is no reason why they should generate similar EM signals.

The generation mechanism is probably different for different frequencies. It may be the electrokinetic effects (e.g., Mizutani et al. 1976, Fitterman 1978) and/or pressure-stimulated current (PSC) effects for DC to low-frequency signals, and piezoelectric effects and exo-electron emission (Enomoto and Hashimoto

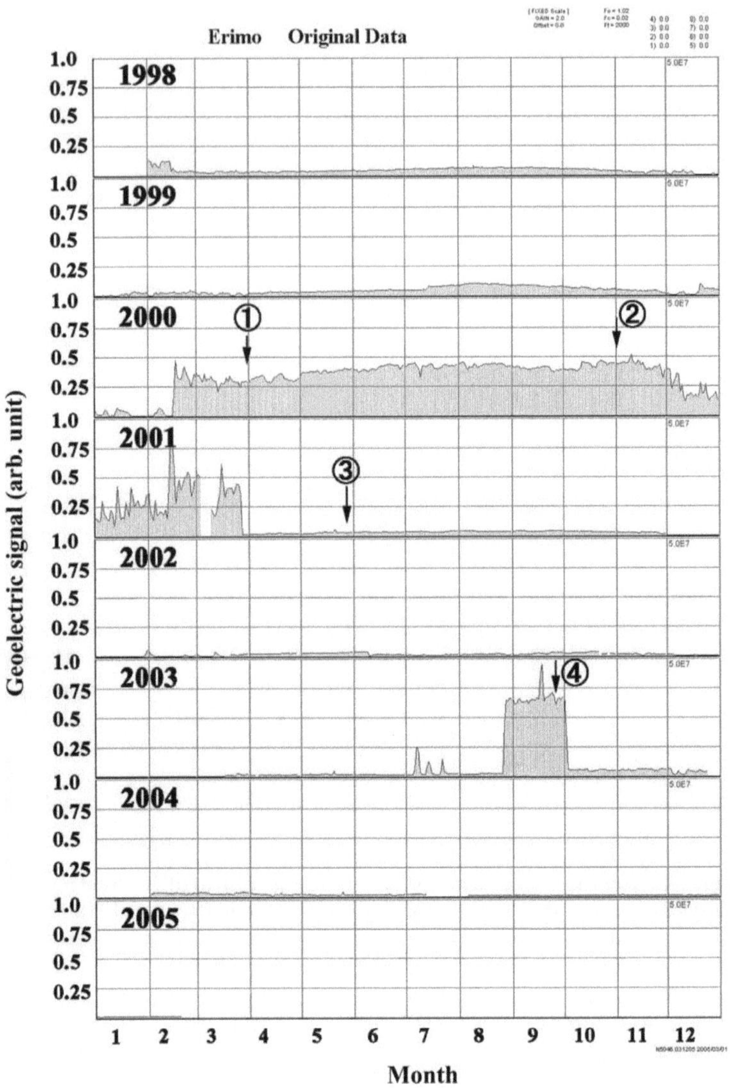

Fig. 6.5 Envelope of the geoelectric signal recorded at the Erimo site over 7 years since 1998. Japan Meteorological Agency (JMA) announcement of (1) the first eruption of Mt. Usu volcano, (2) decline in magma activity, and (3) complete cessation of the magma activity. (4) The main shock of M8.0 Tokachi-Oki EQ (Enomoto et al. 2006)

1990) for higher frequency signals. A mechanism involving the positive holes (p-holes) in rock-forming minerals under stress has also been proposed.

The PSC was proposed by Varotsos and Alexopoulos (1986). The impurities and vacancies in crystals form randomly oriented electric dipoles, which align their orientation in avalanche following an activation process to generate the electric current. The PSC model is unique because currents are generated during gradual stress buildup when the stress attains a critical level without requiring any sudden change of stress like micro-fracture. This explains the reasons why VAN signals are observed before EQs without any accompanying events and no co-seismic signals are necessarily expected. Freund and his colleagues propose a different mechanism for ULF electric signals (Freund et al. 2006). They discovered in the laboratory that a block of igneous rock under a locally applied stress turns into a battery without an external electric field. The charge carriers in this case are semiconductor-type positive holes (p-holes) which have oxygen in one valence state. They attribute the satellite image of pre-seismic IR anomalies over epicentral areas as a result of recombination of two p-hole charge carriers at the earth's surface (Freund 2009, 2010).

Although models related to micro-cracking have been proposed, based on laboratory experiments such as discharge of screening charge of piezoelectric polarization (Ikeya 2004, Yoshida et al. 1997), electrification of fresh crack surfaces (Yamada et al. 1989), and exo-electron, there have been no reliable field reports on the simultaneous occurrence of pre-seismic micro-cracking. Moreover, it might be pointed out that in all these models much stronger co-seismic signals would be expected, contrary to field observations.

6.4.1 Lithosphere–Atmosphere–Ionosphere Coupling

Sub-ionospheric anomalies before large EQs were reported (e.g., Molchanov and Hayakawa 1998) through monitoring of 10–20 kHz VLF ship navigation waves, although Clilverd et al. (1999) did not obtain similar results. Decrease in electron density in the ionosphere within 5 days before $M \geq 5$ EQs was found in Taiwan by the ionosonde measurements of critical plasma frequency, foF2 (Liu et al. 2006). Fujiwara et al. (2004) found pre-seismic atmospheric anomalies causing FM radio transmission anomalies within 5 days before $M \geq 4.8$ EQs. Similar phenomenon of anomalous reception of ULF waves was reported earlier by Kushida and Kushida (2002), and Moriya et al. (2010) further promoted this study to obtain statistical relations between the total time length of anomalous reception and EQ magnitude. The concept of pre-seismic lithosphere–atmosphere–ionosphere coupling (LAI coupling) has arisen from these investigations and is now an important issue in precursor studies (Pulinets and Boyarchuk 2005, Kamogawa 2006). There should be a causative factor in the lithosphere and some mechanism to transport its effect up to the high atmosphere. Recently, the French satellite DEMETER has observed anomalous depression of VLF EM wave intensity at ionospheric height 0–4 h before large EQs. This implies that the ionospheric disturbance may modulate the VLF propagation path (Němec et al. 2009). Possible LAI coupling processes have now been proposed by many researchers but they are all at the hypothetical stage.

6.4.2 Critical Phenomena

EQ may be a critical phenomenon defined in statistical physics (e.g., Bak and Tang 1989), which means that short-term EQ prediction is synonymous with identification of approach to criticality. Although this is difficult to achieve (see Main, 1995), it has recently been shown possible by analyzing the time series of seismicity in a newly introduced time domain called "natural time" (e.g., Varotsos 2005, Varotsos et al. 2002). They have shown that, by "natural time" analysis of signals from complex dynamic systems, it is possible to define a quantity by which critical approach can be identified. In a time series comprised of N events, the natural time χ_k serves as the index for the occurrence of the kth event. The time series as shown in Fig. 6.6a is expressed in natural time as in Fig. 6.6b. The quantity in question is the variance $\kappa_1 \left(\equiv \langle \chi^2 \rangle - \langle \chi \rangle^2\right)$ of natural time χ, where $\langle f(\chi) \rangle = \sum p_k f(\chi_k)$ and p_k is the normalized energy released by the kth event. It was shown that this quantity κ_1 converges to 0.070 at the critical state for a variety of dynamical systems, including both SES and pre-main shock seismicity. Originally, the condition $\kappa_1 = 0.07$ was theoretically derived for SES and later experimentally ascertained for actual SES activities preceding large Greek EQs. For the time series of pre-main shock seismicity, κ_1 is calculated each time a new small EQ occurs. The starting time of the κ_1 calculation of seismicity was set to the onset time of SES activity because it was considered close enough to seismic criticality. It was shown empirically that the condition $\kappa_1 = 0.07$ holds for Greek seismicity a few days before disastrous EQs.

This way, they seem to have succeeded in shortening the lead time of VAN prediction to only a few

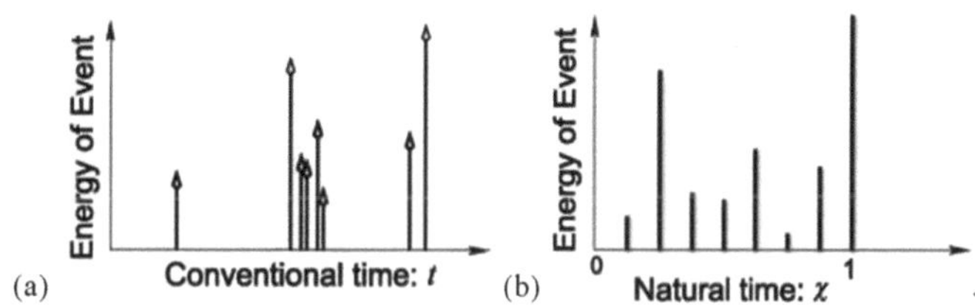

Fig. 6.6 Time series of events (**a**) in conventional time t and (**b**) in the natural time χ

days (Sarlis et al. 2008, Uyeda and Kamogawa 2008, 2010). This means that seismic data may play an amazing role of short-term precursor when combined with SES data.

6.4.3 EQ Triggering Effect of EM Pulses

Field experiments of injecting strong electric current have been conducted by the Research Station of Russian Academy of Sciences in Bishkek, Kyrgyz Republic. The idea of this experiment is to see if EM phenomena can induce seismic activity. It has been known that factors such as earth and ocean tides, earthquakes, volcanic eruptions, dam filling, water injection into the ground, underground nuclear test, and rock burst can trigger EQs. But whether high-energy EM pulses can trigger EQ has not been tested. Here we introduce the results obtained in Kyrgyz by injecting strong EM pulses from a magnetohydrodynamic generator (MHD) originally developed for EM sounding into the deep earth (Tarasov et al. 2001, Avagimov et al. 2004). The experiment was conducted from 1983 to 1990 in the eastern part of Kyrgyz. The MHD generator was installed at the Bishkek test site with a 4.5 km long dipole that has a contact resistance of 0.4 Ωm. The ranges of output current, power, and length of the pulses are 0.28–2.8 kA, 1.2–32.1 MJ, and 1.7–12.1 s, respectively. The total number of injections was 114. In superimposed epoch analysis with 20-day time windows before and after the injection, the total number of EQs 2 days after the injection was enhanced 1.7 times to those of normal days as shown in Fig. 6.7. After the enhancement, the number of EQs decayed with a 2–6 day relaxation time. The energy released by EQs within 20 days after the injection was larger by 2.03×10^{15} J than that within 20 days before. The energy injected from the MHD generator was 1.1×10^{9} J which was only 10^{-6} of the increased seismic energy. It is, therefore, concluded that the energy supply from MHD generator has activated the seismicity and released much larger energy stored in the crust, i.e., EM pulse did trigger the EQs. According to Avagimov et al. (2004), similar results were obtained also in Tajikistan. These results imply the possibility of lightning also activating seismicity.

6.5 Conclusions and Future Outlook

From all the above studies, it can be summarized that some macro-anomalous short-term precursors are real. Scientific research on them, however, has been inadequate so far. Conventional seismic and geodetic pre-seismic information is definitely useful for long/intermediate EQ prediction. But, it is unlikely that it can be useful for short-term purposes in the near future. Some of the non-mechanical precursors, such as geochemical, hydrological, and EM precursors, are plausible candidates for short-term EQ prediction. But so far, they were recognized mainly only after the main shock and they gave scarce constraints on the source location and magnitude *M*, with an exception of VAN's SES. Much more enhanced research is needed to make them practically useful tools. In particular, the electromagnetic ones, including the triggering effect of EM pulses, are in the territory where the IAGA members can effectively contribute to earthquake prediction. Close cooperative research with other disciplines on these aspects is also strongly recommended.

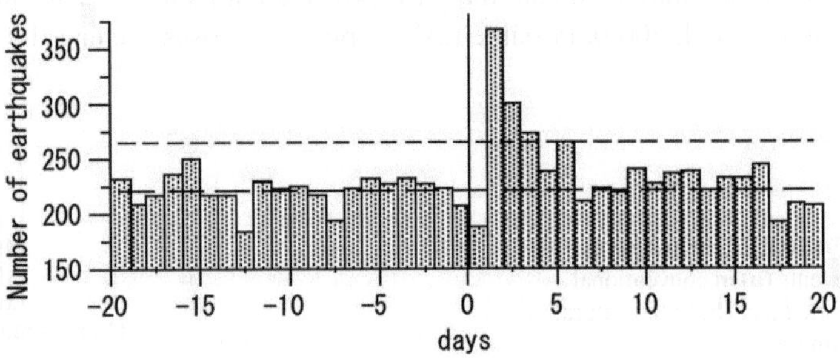

Fig. 6.7 Summary of the Bishkek MHD experiments. The horizontal axis shows the number of days. Day = 0 means the day of experiments

References

Aggarwal YL, Sykes J, Armbruster and Sbar M (1973) Premonitary changes in seismic velocities and prediction of earthquakes. Nature 241, 5385:101–104

Asada TH, Baba K, Kawazoe M, Sugiura (2001) An attempt to delineate very low frequency electromagnetic signals associated with earthquakes. Earth Planets Space 53: 55–62

Avagimov AL, Bogomolov T, Cheidze A, Ponomarev G, Sobolev N, Tarasov V, Zeigamik (2004) Induced seismicity by trigger stimulation from laboratory and field tests Proceedings of 1st international workshop on active monitoring in the solid earth geophysics, Mizunami, Japan, pp 56–59

Bak P, Tang C (1989) Earthquakes as a self-organized critical phenomenon. J Geophys Res 94:15635–15637

Bakun WH, Aagaard B, Dost WL, Ellsworth JL, Hardebeck RA, Harris C, Ji MJS, Johnston J, Langbein JJ, Lienkaemper AJ, Michael JR, Murray RM, Nadeau PA, Reasenberg MS, Reichle EA, Roeloffs A, Shakal RW, Simpson F, Waldhauser (2005) Implications for prediction and hazard assessment from the 2004 Parkfield earthquake. Nature 437: 969–974

Chouliaras G, Stavrakakis G (1999) Support for VAN's earthquake predictions is based on false statements. Eos Trans. AGU 80(19): 216

Chen Y, Tsoi K, Chen F, Gao Z, Zou Q, Chen Z (1988) The great Tangshan earthquake of 1976, Pergamon, Tarrytown NY

Clilverd MA, Rodger CJ, Thomson NR (1999) Investigating seismo-ionospheric effects on a long subionospheric path. J Geophys Res 104, A12:28171–28179

Enomoto Y, Hashimoto H (1990) Emission of charged particles from indentation fracture of rocks. Nature 346:641–643

Enomoto Y, Hashimoto H, Shirai N, Murakami Y, Mogi T, Takada M, Kasahara M (2006) Anomalous geoelectric signals possibly related to the 2000 Mt Usu eruption and 2003 Tokachi-Oki earthquake. Phys Chem Earth 31: 319–324

Evernden JF (1982) Earthquake prediction: what we have learned and what we should do now. Bull Seism Soc Am 72:343–349

Fitterman DV (1978) Electrokinetic and magnetic anomalies associated with dilatant regions in a layered earth. J Geophys Res 83:5923–5928

Fraser-Smith AC, Bernardi A, McGill PR, Ladd ME, Helliwell RA, Villard OG Jr (1990) Low-frequency magnetic field measurements near the epicenter of the Ms 7.1 Loma Prieta earthquake. Geophys Res Let 17:1465–1468

Freund (2009) Stress-activated positive hole charge carriers in rocks and the generation of pre-earthquake signals. In: Hayakawa M (ed) Electromagnetic phenomena associated with earthquakes. Research Signpost, India, pp 41–96. ISBN: 978-81-7895-297-0

Freund F (2010) Toward a unified solid state theory for pre-earthquake signals, Acta Geophys 58:719–766. doi: 10.2478/s11600-009-0066-x

Freund F, Takeuchi A, Lau BES (2006) Electric currents streaming out of stressed igneous rocks – A step towards understanding pre-earthquake low frequency EM emissions. Phys Chem Earth 31:389–396

Fujiwara H, Kamogawa M, Ikeda M, Liu JY, Sakata H, Chen YI, Ofuruton H, Muramatsu S, Chuo YJ, Ohtsuki YH (2004) Atmospheric anomalies observed during earthquake occurrences. Geophys Res Let 31:L17110. doi:10.1029/2004GL019865

Geller R (ed) (1996) Debate on VAN. Geophys Res Lett 23(Special Issue):11

Gokhberg MB, Morgounov VA, Yoshino T, Tomizawa I (1982) Experimental measurement of electromagnetic emissions possibly related to earthquakes in Japan. J Geophys Res 87 B9:7824–7828

Hayakawa M, Kawate R, Molchanov OA, Yumoto K (1996) Results of ultra-low frequency magnetic field measurements during Guam earthquake of 8 August 1993. Geophys Res Lett 23:241–244

Ikeya M (2004) Earthquakes and animals: from folk legends to science. World Scientific, Singapore. ISBN-13: 978-9812385918

Kamogawa M (2006) Preseismic lithosphere-atmosphere-ionosphere coupling. Eos 87:417, 424

Kato T (1981) Secular and earthquake-related vertical crustal movements in Japan as deduced from tidal records (1951–1981). Tectonophysics 97:183–200

Kopytenko YA, Matishvili TG, Voronov PM, Kopytenko EA, Molchanov OA (1993) Detection of ultra-low-frequency emissions connected with the Spitak earthquake and its aftershock activity, based on geomagnetic pulsation data at Dusheti and Vardzia observatories. Phys Earth Planetary Inter 77:85–95

Kushida Y, Kushida R (2002) Possibility of earthquake forecast by radio observations in the VHF band. J Atmos Electricity 22:239–255

Li Y, Liu Y, Jiang Z, Guan J, Yi G, Cheng S, Yang B, Fu T Wang Z (2009) Behavioral change related to Wenchuan devastating earthquake in mice. Bioelectromagnetics. doi 10.1002/bem20520

Liu JY, Chen YI, Chuo YJ (2006) A statistical investigation of pre-earthquake ionospheric anomaly. J Geophys Res 111:A05304, 10.1029/2005JA011333

Main I (1995) Statistical physics, seismogenesis, and seismic hazard. Rev Geophys 34:433–462

McEvilly TV, Johnson LR (1974) Stability of P and S velocities from central California quarry blasts. Bull Seismol Soc Am 64:342–353

Mizutani H, Ishido T, Yokokura, T, Ohnishi S (1976) Electrokinetic phenomena associated with earthquakes. Geophys Res Lett 3:365–368

Molchanov OA, Hayakawa M (1998) Subionospheric VLF signal perturbations possibly related to earthquakes. J Geophys Res 100:1691–1712

Moriya T, Mogi T, Takada M (2010) Anomalous pre-seismic transmission of VHF-band radio waves resulting from large earthquakes, and its statistical relationship to magnitude of impending earthquakes. Geophys J Int 180(2): 858–870

Němec F, Santolík O, Parrot M (2009) Decrease of intensity of ELF/VLF waves observed in the upper ionosphere close to earthquakes: a statistical study. J Geophys Res 114:A04303. doi:10.1029/2008JA013972

Papadopoulos GA (2010) Comment on "The prediction of two large earthquakes in Greece". Eos Trans AGU 91(18):162

Press F (1975) Earthquake prediction. Sci Am 232(5):14–23
Pulinets S, Boyarchuk K (2005) Ionospheric precursors of earthquakes. Springer, New York, NY, p 316
Raleigh B, Benett D, Craig H, Hanks T, Molnar P, Nur A, Savage J, Scholz C, Turner R, Wu F (1977) Prediction of Haisheng earthquake. EOS 58(5):2
Roeloffs E (1994) The earthquake prediction experiment at Parkfield, California. Rev Geophys 32(3):315–336
Sadovsky M, Nersesov I, Nigumatullaev S, Latynina L, Lukk A, Semenov A, Simbireva I, Ulmov V (1972) The processes preceding strong earthquakes in some regions of Middle Asia. Tectonophys 14:295–307
Sarlis NV, Skordas ES, Lazaridou MS, Varotsos PA (2008) Investigation of seismicity after the initiation of a Seismic Electric Signal activity until the main shock. Proc Jpn Acad Ser B 84:331–343
Semyenov AN (1969) Variation in the travel time of traverse and longitudinal waves before violent earthquakes. Izv Acad Sci USSR (Phys Solid Earth) 4:245–248 (English transl)
Sholtz C, Sykes LR, Aggarwal YP (1973) Earthquake prediction: a physical basis. Science 181:803–810
Tarasov NT, Tarasova NV, Avagimov AA, Zeigarnik VA (2001) The effect of electromagnetic implication on seismicity in the Bishkek geodynamic test ground. Russ Geol Geophys 42:1558–1566
Thomas JN, Love JJ, Johnston MJS, Yumoto K (2009) On the reported magnetic precursor of the 1993 Guam earthquake. Geophys Res Lett 36:L16301. doi:10.1029/2009GL039020
Tributsch H (1982) When the snakes awake: animals and earthquake prediction, MIT Press, Cambridge, MA
Tsutsui M (2002) Detection of earth-origin electric pulses. Geophys Res Lett 29:1194. doi:10.1029/2001GL013713
Tsutsui M (2005) Identification of earthquake epicentre from measurements of electromagnetic pulses in the earth. Geophys Res Lett 32:L20303. doi:10.1029/2005GL023691
Uyeda S (2000) In defense of VAN's earthquake predictions. Eos Trans AGU 81(1):3
Uyeda S, Hayakawa M, Nagao T, Molchanov O, Hattori K, Orihara Y, Gotoh K, Akinaga Y, Tanaka H (2002) Electric and magnetic phenomena observed before the volcano-seismic activity in 2000 in the Izu Island Region. Japan, Proc Natl Acad Sci USA 99:7352–7355
Uyeda S, Kamogawa M (2008) The prediction of two large earthquakes in Greece. Eos Trans AGU 89:39. doi:10.1029/2008EO390002
Uyeda S, Kamogawa M (2010) Reply to comment on "The prediction of two large earthquakes in Greece". Eos Trans AGU Eos Trans AGU 91(18)
Uyeda S, Meguro K (2004) Earthquake prediction, seismic hazard and vulnerability. In: Forecasting, prediction, and risk assessment. IUGG monograph, vol 19. pp 349–358
Uyeda S, Nagao T, Orihara Y, Yamaguchi Y, Takahashi I (2000) Geoelectric potential changes: possible precursors to earthquakes in Japan. Proc Natl Acad Sci USA 97: 4561–4566
Varotsos PA (2005) The physics of seismic electric signals. TerraPub Tokyo, Japan, p 358
Varotsos P, Alexopoulos K (1984a) Physical properties of the variations of the electric field of the earth preceding earthquakes I. Tectonophysics 110:73–98
Varotsos P, Alexopoulos K (1984b) Physical properties of the variations of the electric field of the earth preceding earthquakes II. Tectonophysics 110:99–125
Varotsos P, Alexopoulos K (1986) Stimulated current emission in the earth and related geophysical aspects. In: Amelinckx S, Gevers R, Nihoul J (eds) Thermodynamics of point defects and their relation with bulk properties. North Holland, Amsterdam
Varotsos P, Sarlis N, Skordas E (2002) Long range correlations in the electric signals that precede rupture. Phys Rev E 66(7):011902
Warwick JW, Stoker C, Meyer TR (1982) Radio emission associated with rock fracture: possible application to the great Chilean earthquake of May 22, 1960 J Geophys Res 87:2851–2859
Yamada IK, Masuda, Mizutani H (1989) Electromagnetic and acoustic emission associated with rock fracture. Phys Earth Planetary Inter 57:157–168
Yasuoka Y, Kawada Y, Nagahama H, Omori Y, Ishikawa T, Tokonami S, Shinogi M (2009) Preseismic changes in atmospheric radon concentration and crustal strain. Phys Chem Earth 34:431–434
Yoshida S, Uyeshima, M, Nakatani M (1997) Electric potential changes associated with slip failure of granite: preseismic and coseismic signals. J Geophys Res 102: 14883–14897

7 Time Domain Controlled Source Electromagnetics for Hydrocarbon Applications

K.M. Strack, T. Hanstein, C.H. Stoyer, and L.A. Thomsen

Abstract

During the past 10 years, marine electromagnetics has developed from infancy into a sizable geophysical industry. While this is feasible in the time and the frequency domains, most of the commercial marine hydrocarbon applications operate in frequency domain, i.e. Controlled-Source Electromagnetic (fCSEM). Until 5 years ago, it was generally assumed that time domain (tCSEM^{TM}) methods would not be of any use in oil exploration. Since then, however, many such measurements have been recorded with several independent tCSEM^{TM} systems that already exist or under development. Time domain measurements can be used everywhere and more suitable for shallow water. They have large anomalous responses than those in fCSEM. From the physical viewpoint, time domain measurements are complementary to frequency domain measurements, as they focus on different spatial regions. With recent advances in electronics time-domain CSEM data can reliably be acquired in an offshore environment. Multiple surveys using autonomous receiver nodes have successfully acquired marine time domain CSEM data. Our work takes the technology a step further, by developing a high-density marine cabled system with novel 5-component sensor package.

7.1 Introduction

For over 50 years, the seismic method has been the geophysical workhorse of the oil exploration industry. While it offers the best description of reservoir shape and stratigraphy, it falls short in describing the fluid properties of the pore space, since seismic waves respond to *both* rock matrix and fluid components of the rock, which are difficult to separate. Usually, fluid discrimination with seismic waves depends on amplitude analysis, which in turn depends on true-amplitude migration of high-quality seismic data that is not always possible.

In particular, many of the changes that take place during the production life of a reservoir do not exhibit a detectable acoustic property change. Since the CSEM response to thin resistors was understood, marine CSEM methods have found application for direct hydrocarbons detection (Eidesmo et al. 2002). While fCSEM has been studied for many years (Cox 1981, Cox et al. 1986, Sinha et al. 1990, Constable and Cox 1996, Constable 2006), little or no work has been done using time domain in an equivalent mode for hydrocarbon exploration. The use of marine CSEM has now gained momentum, and has become one of the most

K.M. Strack (✉)
KMS Technologies, Houston, TX, USA
e-mail: kurt@kmstechnologies.com

E. Petrovský et al. (eds.), *The Earth's Magnetic Interior*, IAGA Special Sopron Book Series 1,
DOI 10.1007/978-94-007-0323-0_7, © Springer Science+Business Media B.V. 2011

significant technology development in oil exploration, since the advent of 3D seismics. As mentioned earlier, most marine EM applications are in frequency domain, with limited ventures in time domain (Edwards 1987, 1997, Edwards and Yu 1993, Ziolkowski et al. 2006). Time domain measurements in marine environment have usually been restricted to near-surface applications with fixed length towed systems (Edwards et al. 1985, Chave et al. 1991, Cheesman et al. 1987, Edwards 1997), or for deeper hydrocarbon applications in shallow water and induced polarization (Strack and Petrov 2007, Strack et al. 2008, Veeken et al. 2009).

We have selected a time domain (tCSEM^{TM}) version, which employs time variant electromagnetic fields of either natural or artificial origin, causing eddy currents within the conductive sediment layers. Our choice is based on the success of time domain measurements on land and its response to thin resistive layers (Eadie 1979, Passalacqua 1983, Strack 1992, 1999, Strack et al. 1989). Unlike other time domain users (e.g. Holten et al. 2009a, b) that focus strongly on the induced polarization effect, we use a normal moving source system and can even use a standard frequency domain system with an additional source, tow over the receiver spread. The induced eddy currents are time variant as well, and they cause a secondary EM field that can be sensed with magnetic or electric sensors placed on the sea floor or in the wellbore. High resistivity lithologies and pore fluids are the resistors that alter the artificial electric field. Recent theoretical and practical evaluations on this aspect can be found in (Weiss 2007, Davydycheva and Rykhlinski 2009). Both favor time domain measurements although, most service providers of CSEM technology still transmit from a frequency-based source with a continuous sinusoid or a square wave. Our time domain signal uses much larger time (tens of seconds) between current switching, so that each switching constitutes a separate transient source. We record the Earth's transient response, while the current is off. We call this tCSEM^{TM}.

7.2 Background Physics

The theory of time-varying electromagnetic fields in a stratified Earth is described in a comprehensive fashion by Ward and Hohmann (1988) and for continuous and transient soundings by Kaufman and Keller (1983) with summaries for grounded dipoles given by Strack (1992, 1999). Responses for various 3D models are described by Hohmann (1988), Newman and Alumbaugh (2000), Druskin and Knizhnerman (1994), Davydycheva et al. (2003).

We follow the standard approach for marine application and fine-tune the digital filters that carry out the Hankel transform for high contrast boundaries. The measured voltages are corrected for frequency dependence of the sensors, processed for signal-to-noise improvement, and then input to the inversion.

With tCSEM^{TM}, one transmits current into the Earth, charging the zones where resistivity varies at the subsurface. The current is then switched "off" and the charge dissipates. Sensors that record the electric and magnetic components measure transient responses to this artificial electric field. Because the time domain method is only measuring the secondary field, it offers a solution to the shallow water limitations that confront frequency-targeted techniques (Weiss 2007 and Avdeeva et al. 2007). The duration of these "on" and "off" times of the source are optimized to each particular problem. Additionally, every current switching represents an initiation time, or time zero, for a given transient. Figure 7.1 shows a typical time domain source waveform and the resulting electric and magnetic fields.

To illustrate the behavior of the electromagnetic field, consider the survey configuration as shown in Fig. 7.2. A dipole source, usually about 300 m long, is

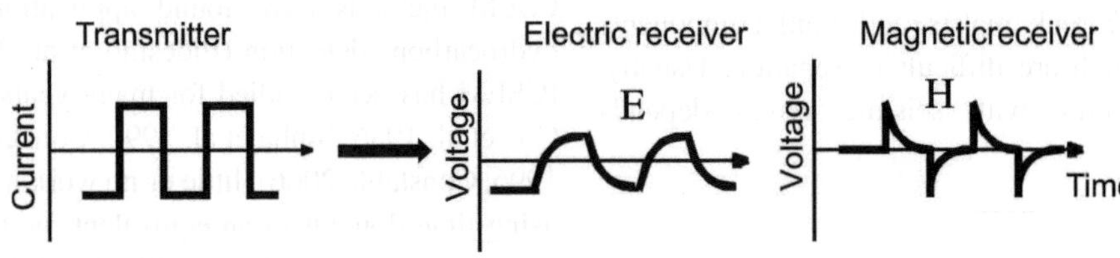

Fig. 7.1 Time domain source waveform and its resulting electric and time derivative of the magnetic fields

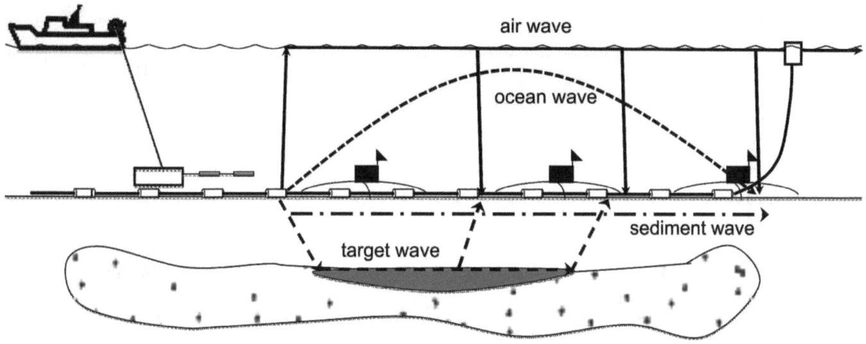

Fig. 7.2 Survey setup for a marine time domain electromagnetic system including nodes and cabled receivers

towed about 30 m above the sea floor. An electric current, as shown in Fig. 7.1, is injected between source electrodes, and the response of this current diffusing into the subsurface is measured with multi-component receivers that are either autonomous (nodes) or connected by a cable. The following wave model as illustrated in Fig. 7.2 can describe the received signal. (Note that more resistive is the path, faster is the propagation speed.)

- Part of the signal travels up to the surface from the transmitter below, then along the air-water interface with speed of light, arrives at the receiver; it is called the air wave.
- Part of the signal goes through the water column, and is called the ocean wave. Traveling through water, the ocean wave arrives later than the airwave and (depending on water depth) sometimes so much delayed that it arrives later than the subsurface response.
- Part of the signal diffuses through the subsurface sediments (usually more resistive than the reservoir) and is called the sediment wave.
- Finally, part of the signal diffuses deep into the subsurface, then refracts along the (resistive) reservoir where it travels faster than the sediment wave; this is called the target wave.
- When the electromagnetic energy diffuses into the medium, its energy (or Pointing vector) travels with similar speed as refracted seismic waves (Weidelt 2007), and we thus call it waves.

To illustrate the different "wave" components, we have calculated the responses and presented in Fig. 7.3. The Earth model is a half space that consists of ocean only; the transmitter and receiver are separated by 1000 m. We simulate the airwave and the ocean wave by placing a transmitter and receiver system close to the sea surface, so that the airwave dominates, and also far away from the air-water interface, so that the ocean wave dominates. The airwave response is shown as a black spike on the ordinate. Because it travels at the speed of light it appears at $t = 0$ on the time scale. The signal with only ocean wave response is the signal that has only one "hump" at approximately 1 s. The curves between these end-members are obtained as the transmitter-receiver and lowered together into the ocean. They exhibit first as airwaves and then ocean waves at later times. As the depth increases, the airwave spreads, due to dispersion in the water on the way up and down. At large depths, the two waves merge. In practice, an impulsive response is realized by time-differentiation of a step response.

We can now display the individual measurements similar to seismic-style traces, as suggested by Edwards (1997) for gas hydrates and by Wright et al. (2002) for land electromagnetic applications. Figures 7.4, 7.5, 7.6, and 7.7 show common-source gathers, where each vertical trace represents a measurement at an offset surface location. The gathers are 500 m apart, starting at 500 m from the transmitter to 10 km maximum offset. Each trace is normalized, to compensate for amplitude attenuation. This seismic-style operation serves much the same purpose as conversion to apparent resistivity. It removes the amplitude effects at far offsets. Of course the normalization factors are retained for later restoration of true amplitudes. On the vertical scale is the diffusion time after switching the transmitter; the scale goes from 0 to 25 s.

Figure 7.4 shows the gather for two of the models used in Fig. 7.3. The top of Fig. 7.4 shows the model

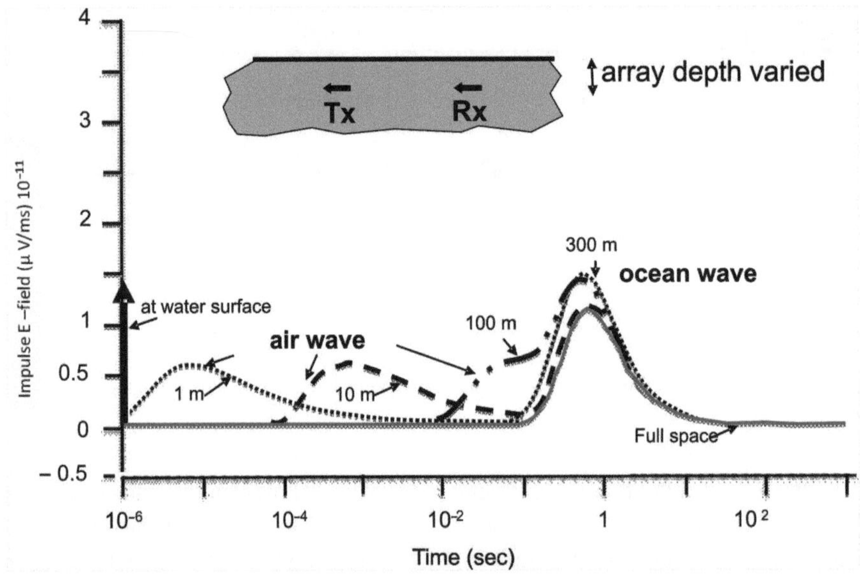

Fig. 7.3 Impulse response for an inline electric field marine CSEM setup for different transmitter-receiver depths below the water surface. The horizontal scale is time in seconds after the impulse

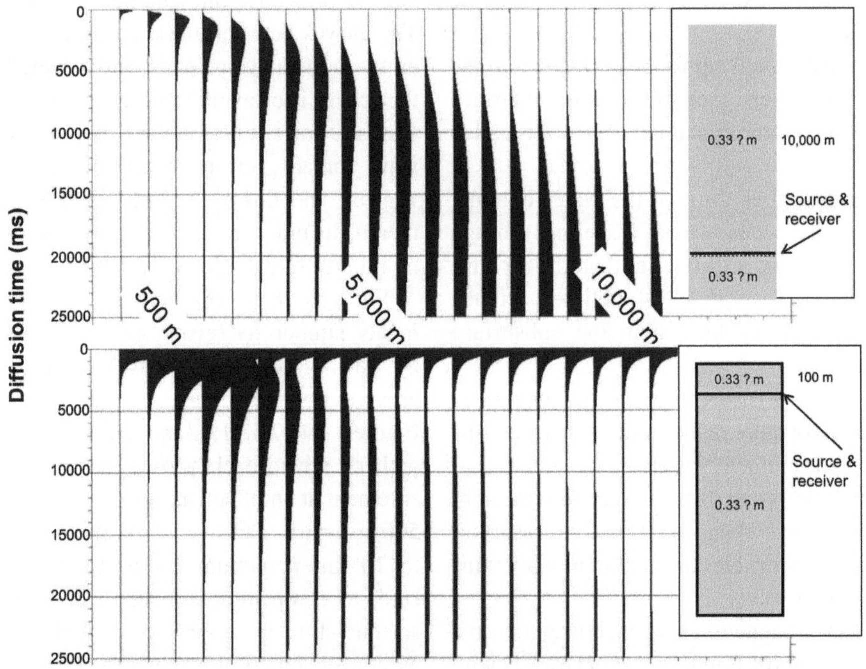

Fig. 7.4 Common-source gathers for impulse response of an inline electric field marine tCSEM setup. The traces represent different offsets between source and receiver; displayed are measured voltages. All traces are displayed trace-normalized. The Earth model is a half space with the resistivity of seawater. The *top* gather represents the case when the system is in deep water and far away from the sea-air interface. The *bottom* gather represents the case when the system is near the air-water interface and includes the airwave (After Allegar et al. 2008)

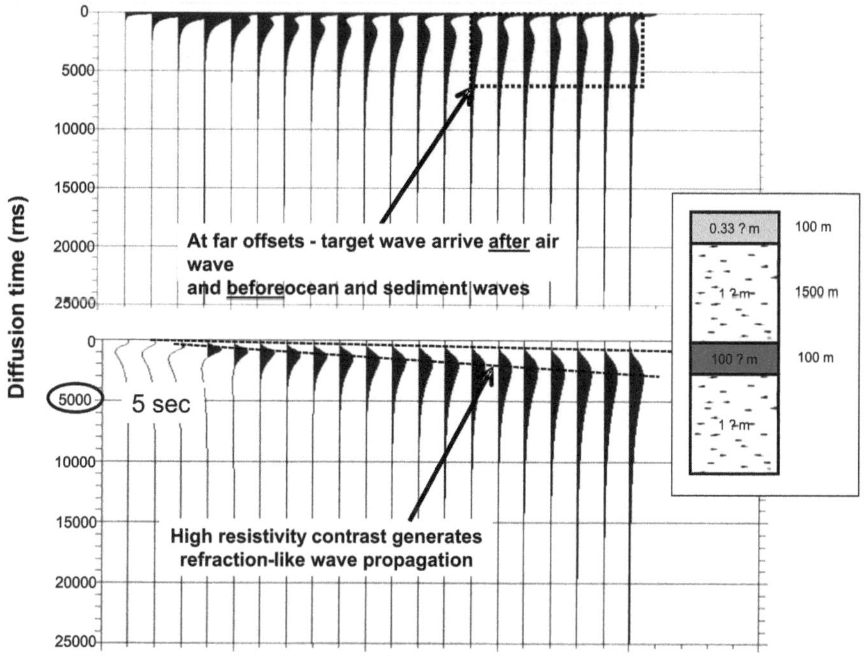

Fig. 7.5 Common-source gathers for the impulse response of an inline electric field marine tCSEM setup. The traces represent different offsets between source and receiver; displayed are measured voltages. All traces are displayed trace-normalized. The Earth model has an oil reservoir at 1500 m depth below the seafloor. The top gather contains all wave components (air wave, ocean wave, sediment wave and target wave). The bottom gather only contains the reservoir response after removal of all other components (After Allegar et al. 2008)

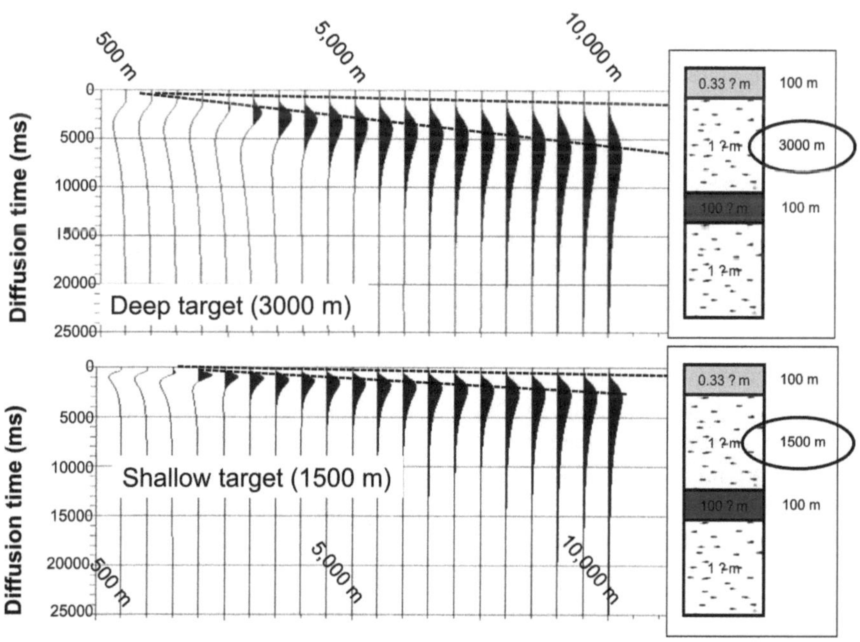

Fig. 7.6 Common-source gathers for the impulse response for an inline electric field marine CSEM setup for different reservoir depths. The traces represent different offsets between source and receiver; displayed are measured voltages. All traces are displayed trace normalized. The Earth model has an oil reservoir at 3000 m or 1500 m depth below the seafloor, respectively. Comparing both gathers shows that a response from a deep target arrives later (After Allegar et al. 2008)

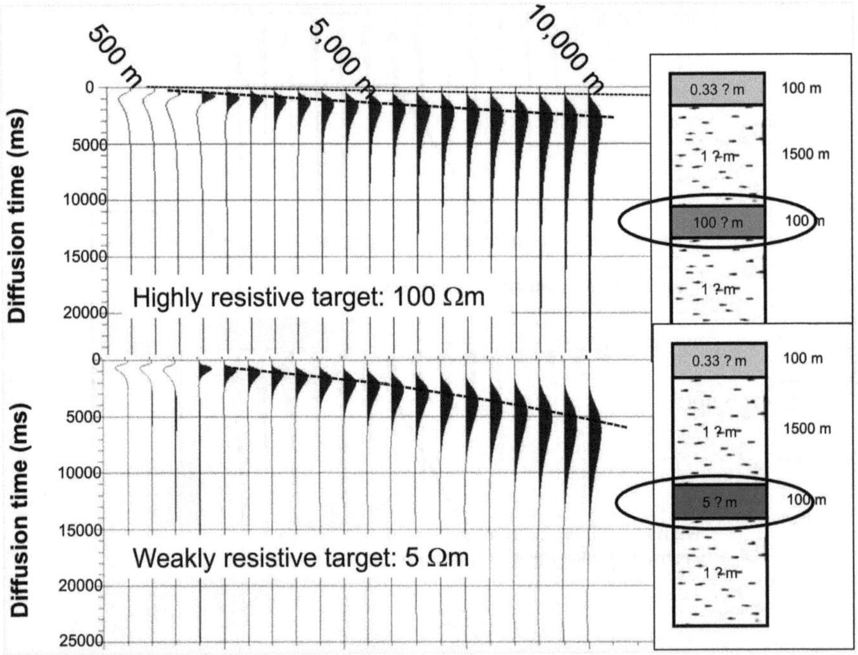

Fig. 7.7 Common-source gathers for the impulse response of an inline electric field marine tCSEM setup. The traces represent different offsets between source and receiver; displayed are measured voltages. All traces are displayed trace normalized. The Earth model has an oil reservoir at 1500 m depth below the sea floor with different resistivities. Comparing gathers shows that for a less resistive (brine saturated) reservoir the signals arrive later and are "smeared" more

with deep source and receivers, when the system only sees the ocean wave. The bottom of the figure shows the case when the system is close to the air-water interface and so also sees the airwave. Note that in this case the airwave gets less attenuated than the ocean wave.

Figure 7.5 shows the gather for the model shown with shallow water layer, with reservoir at 1500 m depth. At the top, all wave components are included, and at the bottom only the reservoir response is shown. This can be obtained by subtraction of the other components, based on the models run without a reservoir. This is easily done with modeled data. For field data, one usually requires a reference measurement over similar geology. This reference data set is then subtracted and the remainder contains the difference between reservoir and non-reservoir model. In the figure (top) we see that at far offsets, the reservoir response arrives after the airwave, but before the sediment and ocean wave, both of which are conductors and maintain induction currents flow longer.

Figure 7.6 displays offset gathers for varying depths to the reservoir, and illustrates that for greater depth the reservoir signal arrives later and gets dispersed more. These behaviors are expected from time domain electromagnetic signals.

Figure 7.7 shows the case of a strongly resistive versus a weakly resistive reservoir. In the case of a weakly resistive reservoir (bottom) the signals flow longer in the reservoir and consequently arrive later. Here, the dispersion at larger offset is even more visible, since the signal from less resistive targets arrives later.

It is often asked about the differences between time and frequency domain surveys. Both Weiss (2007) and Davydycheva (Davydycheva and Rykhlinski 2009) prefer time domain for shallow water applications. Of course for identical survey geometry and timing, data recorded in time domain can always be Fourier-transformed to the frequency domain for analysis. But, as the terms are used in the EM community, they imply *different* survey geometry and/or timing, so the data sets are NOT Fourier-equivalent. Perhaps better terms would be "continuous-source" for fCSEM, and "transient-source" for tCSEM^{TM}, but it is probably too late to change the conventional usage. In frequency domain, the data samples the entire volume between transmitter and receiver, as the secondary field is

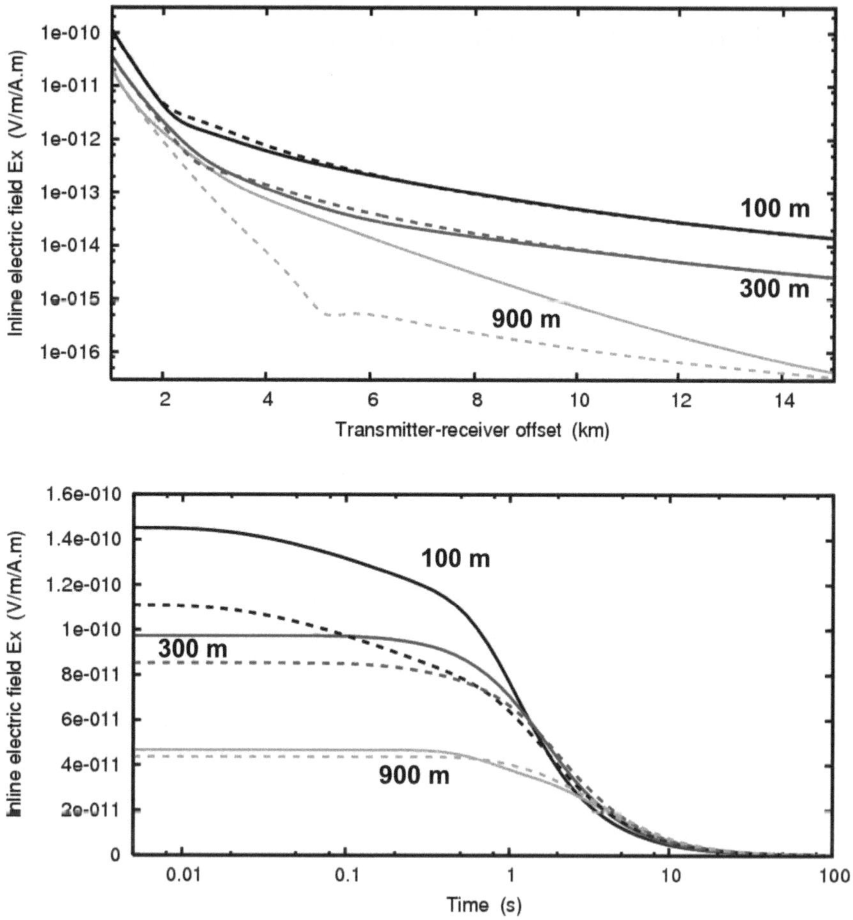

Fig. 7.8 Anomalous response for shallow and deep water for time and frequency domain, with (*solid lines*) and without (*dashed lines*) a resistive reservoir. The *top curves* show the frequency domain response averaged over all times, and the *bottom curves* show the time domain response for a 3 km receiver-to-transmitter offset

recorded in the presence of a continuous primary field. In time domain at each time step the energy is concentrated in a central diffusion volume and the corresponding data are thus more sensitive to the response from within that volume only. So looking at the underneath complex structure is obviously easier with focused EM than with the geometric averaging of fCSEM. While fCSEM has to measure a small secondary response in the presence of a large primary field, in the absence of a primary field, time domain CSEM can amplify the signal. But time domain CSEM is seriously hindered in the application of simple signal-to-noise ratio enhancement; e.g. when the boat is moving, vertical stacking can only be applied to a limited extent. *It seems that both should be used in complementary fashion: fCSEM to outline the resistors and time domain to focus on specific small volumes that are too detailed for fCSEM (like sub-salt targets).*

Figure 7.8 shows the comparison between deep and shallow water anomalies for time and frequency domain surveys. In both the cases, we plot normalized inline electric field magnitudes for 1-D models with 3 different water depths (100, 300, and 900 m), both with (solid lines) and without (dashed lines) a reservoir (100 m thick, 1000 m deep in a 0.7 ohm-m background). The top curves are the frequency domain curves and the bottom curves are the time domain curves. For the time domain plot, we selected offset of 3 km, which is not a significant parameter (Spies 1989). Comparing both plots it can be clearly seen that, for frequency domain measurements, the largest anomalous response can be obtained for deep water. For time domain, the anomalous response is largest for shallow water. The present state of instrumentation favors fCSEM but it will only be a matter of time before time domain will reach similar level for shallow water applications.

7.3 Data Processing and Interpretation

Marine time domain data is recorded using ocean bottom nodes or ocean bottom cables. Nodes are autonomous acquisition systems with multiple electric and magnetic field component sensors that are dropped from the surface onto the ocean floor. As the landing orientation is not known, the data needs to be rotated to a known orientation. Ocean bottom cables (OBC) are carefully laid on the sea floor and orientation is measured with various orientation devices. Nodes are commonly spaced at about 1 km (in special applications as closely as 500 m), while OBCs have 50 or 100 m sensor spacing. The data is usually recorded in a proprietary instrument format, and then converted to SEG D format or SEG Y format, which is the standard for data processing exchange.

The first step in data processing (Fig. 7.9) is the header completion with all survey parameters, navigation parameters, header check, and merging with the source records. The source-to-receiver timing synchronization is a key issue, in time domain as everything is based on the correct timing. While today's electronics drift very slowly, they still exhibit sudden clock jumps, which need to be corrected in data processing.

Next, the data is processed pre-stack to get the optimum signal-to-noise ratios. While today's computers are much faster, most of the filters were already invented decades ago and are described in (Strack et al. 1989, Strack 1992, Strack and Vozoff 1996). Today's improvements in digital processing techniques are mostly restricted to handling more data automatically or graphically.

After the pre-stack processing the stacking or vertical averaging follows. Consecutive shots from a moving boat are not really experimental replicates, but the limited subsurface resolution allows us to stack the data between 0.9 and 1.1 times each nominal offset from the receiver. As the boat moves at about 2 knots, this usually is not more than 4–8 shots in each stack. Thus it is very important to have a strong clean and repeatable source. However, high-frequency deviations from the ideal step source are not important, as they do not survive the propagation down to the target level and back; all the important data are of low frequency. After stacking we can process the data further and convert it into apparent resistivities for input in various interpretation schemes.

Figure 7.10 shows examples of raw data traces. In both the cases, a more conservative bi-polar source

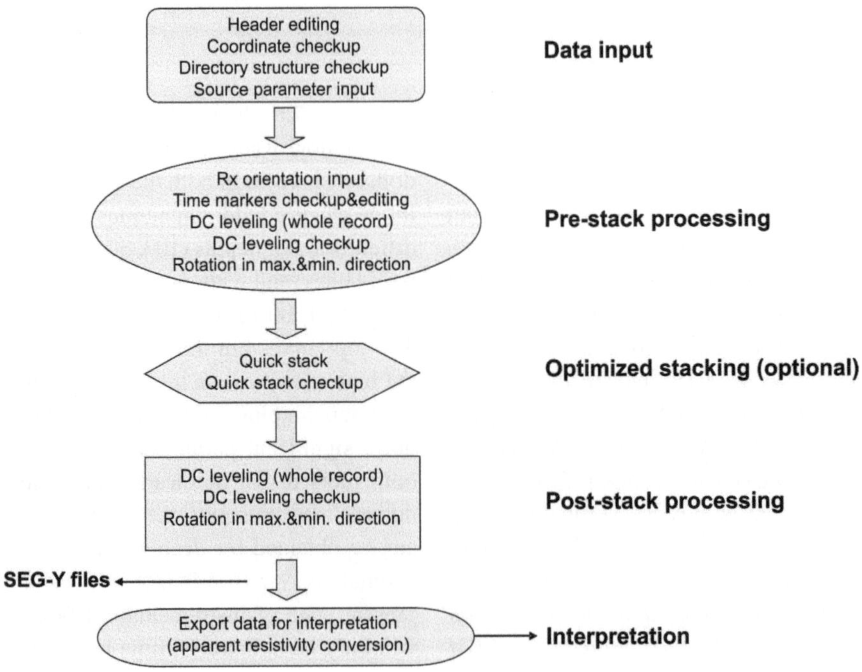

Fig. 7.9 Flow diagram of the various processing steps

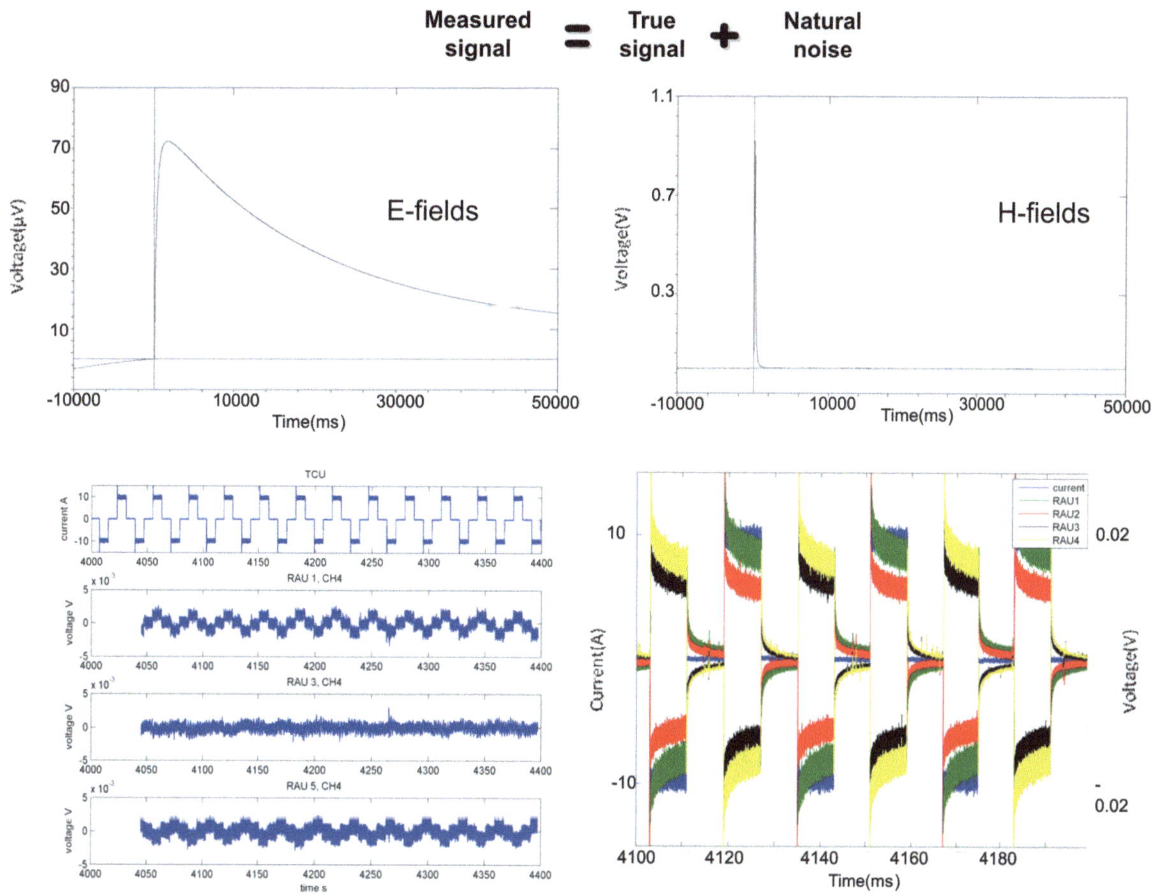

Fig. 7.10 Raw data example traces for a nodal and cabled time domain EM system

waveform (shown in the left-middle) was used. The top of the diagram shows electric and magnetic fields recorded with a nodal system. Both recordings are of good quality. The bottom two diagrams are recorded in very shallow water with a marine cable system under noise conditions. You can clearly see in both electric and magnetic field signals that the responses caused by the source switching. For the upper curves, the time between source switching was 50 s and for the lower curves it was 10 s. The lower curves show the data recorded with different remote acquisition units at different offsets (which accounts for the various amplitudes), in different colors.

Figure 7.11 shows two vector data sets, the top one is un-rotated in raw form (x,y), and the bottom one is rotated into the source-receiver direction (EI, HI) and the cross-line direction (EC, HC). This is the first step in processing to ensure that the amplitudes are treated in an undistorted fashion.

In the next step, one addresses various noise issues in the data. In particular, timing verification is important, because the skipped time marks can distort the processed data. After DC-leveling and filtering, the signal is smoothed with a time-variant smoothing filter as described in Strack et al. (1989) and Strack (1992). The results are shown in Fig. 7.12, whereas on the left is the original transient and on the right is the smoothed version. The smoothed data is then usually averaged (robust stacked) to further improve the signal-to-noise ratios. For normal marine acquisition, the processing stops here and the output data, filed either as EDI or as SEGY. This is usually the interface for interpretation.

Still there is one more still a step left to correct the data. This step may be necessary since, for the reasons of operational efficiency, the delay time between shots may not be sufficient to permit the charge set up from previous shots to completely dissipate. We call this incomplete relaxation as "run-on", and correct for

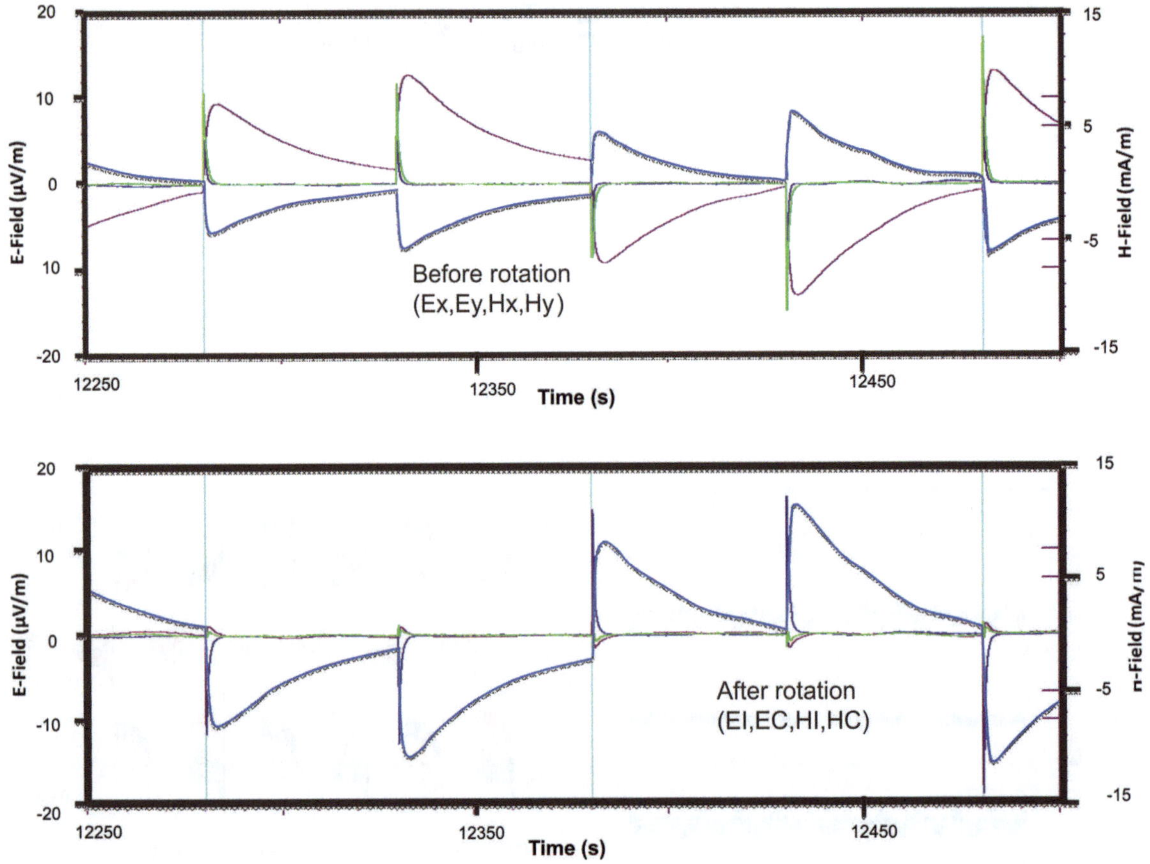

Fig. 7.11 Raw data sample traces (un-rotated) and rotated in the direction of the source dipole

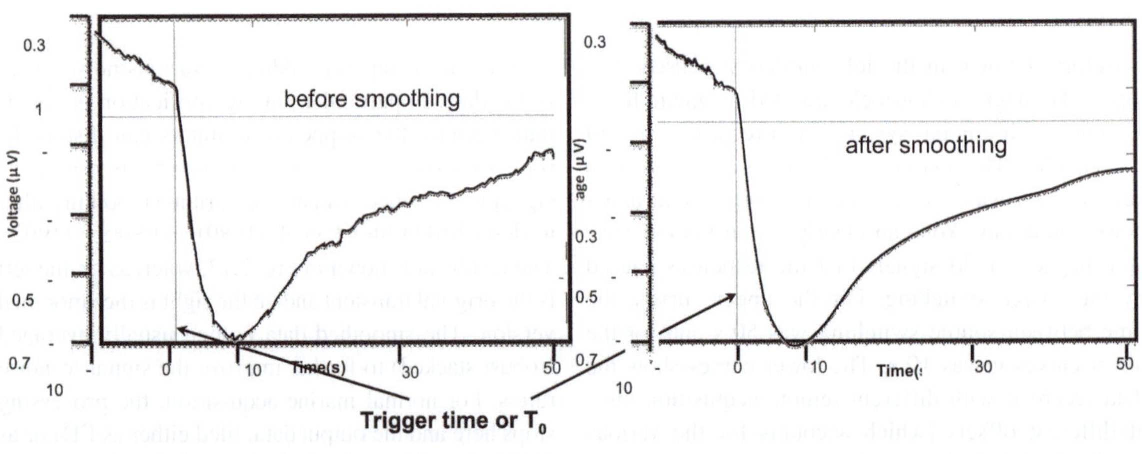

Fig. 7.12 Marine time domain EM sample data: On the left is a raw transient and on the right its smoothed equivalent

it iteratively so that each shot may be analyzed as if no other shots were interfering (Stoyer and Strack 2008).

Interpretation flow diagram using graphical representation is shown in Fig. 7.13. It shows that multiple components are the input into the inversion. Usually with two redundant electric field components, multi-components can be used in a joint or cooperative inversion mode.

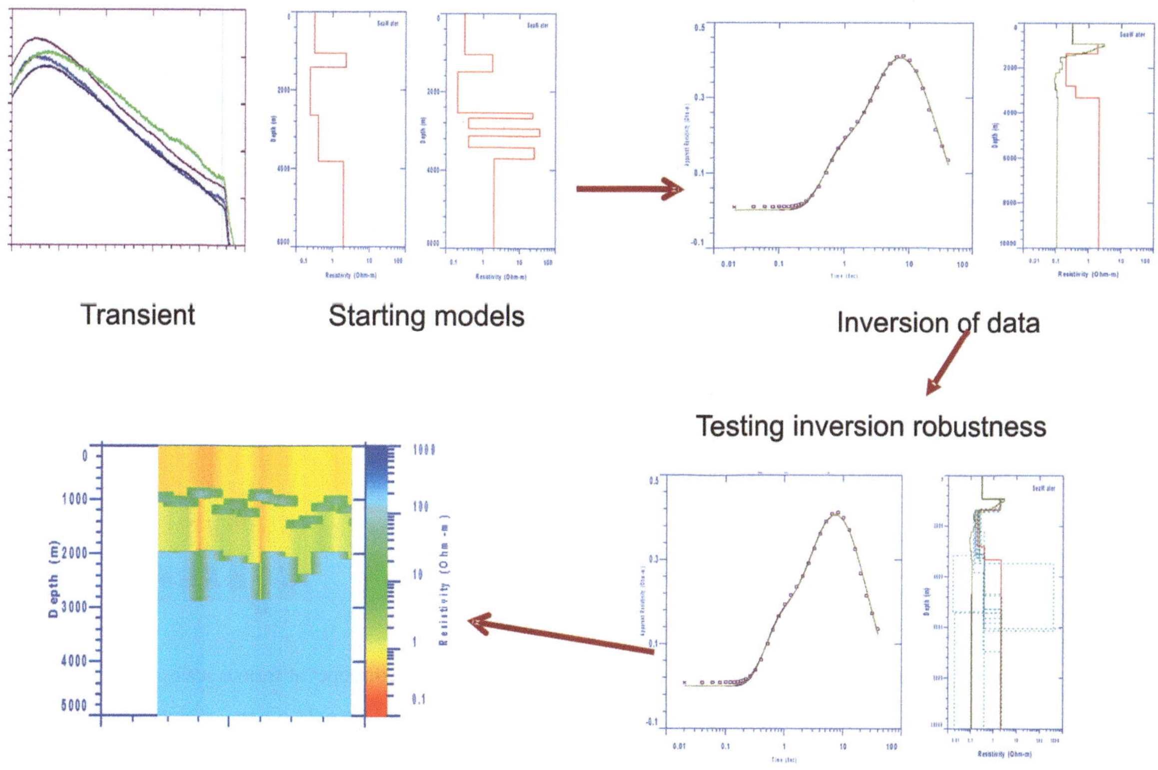

Fig. 7.13 Interpretation flow diagram for marine time domain data

When carrying out the interpretation, two starting models are often used: one for the case with reservoir and the other without. Both the cases are usually tested. When this reservoir/no-reservoir approach is not using, one would use an anisotropic model to ensure the best resistivity structure. When inverting the data, usually various inversion algorithms are compared. In the Fig. 7.13, on the upper right hand side, we compared a layered model with smooth model. When a satisfactory match has been found between the data and results, the final inversion model are compared as shown in this diagram.

When the match is satisfactory, the reliability of the result needs to be estimated to get the right input for the risk analysis. This estimation can be done in various ways. The most reliable method is to use the statistics of the inversion as described by Raiche et al. (1985) or Strack (1992). Other methods using equivalence analysis are also possible, though it is not always driven by data sensitivity. An example of an equivalence analysis is shown in Fig. 7.13 at the bottom right. Once the interpreter is satisfied with the inversion quality, color sections can be seen as shown on the bottom left side of the figure.

7.4 Anisotropy

The biggest factor influencing the surface resistivity measurements is electrical anisotropy. When using electric field measurements, anisotropy caused by layering is "transverse isotropy", i.e. with vertical resistivity different from horizontal resistivity (but the bulk resistivities being the same in horizontal/azimuthal directions). This effect was studied earlier also (Harthill 1968, Strack 1992). It is well known that when using an electric dipole source with primary horizontal diffusion paths, we measure the vertical resistivity with electric fields. Unfortunately, until the advent of tensor induction, this could not be calibrated with normal induction logs in vertical wells. Today i.e. more than 10 years after the 3D induction tool became available, we have enough calibration measurements to make reasonable estimates of the effect.

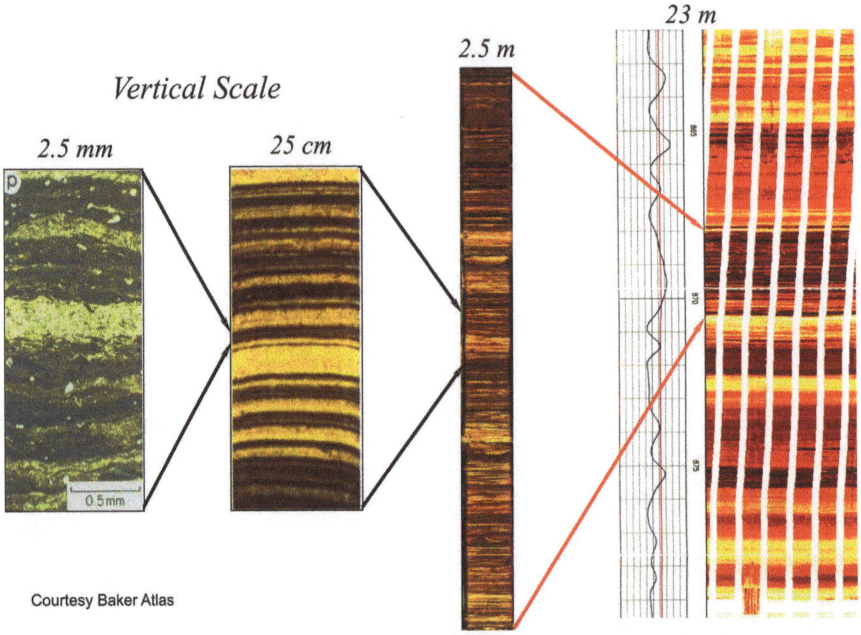

Fig. 7.14 Example of anisotropic images from image logs, core scans and electron microscopes at various scales

Figure 7.14 shows an example for the presence of anisotropy at all scales. On the right side of the figure, images from electrical logs are shown. On the left side of the figure, we see core images of various scales with electron-microscope images on the for left of the figure The light colors are the sand and the dark colors are the shales. We can clearly see thin laminations everywhere that result in electrical anisotropy. Normal anisotropy values (vertical resistivity/horizontal resistivity) are between 1.2 and 1.4 in sedimentary basin but they can go as high as 10.

A factor to be considered is azimuthal anisotropy and its interaction with dipole field from the transmitter. Hördt (1992) studied this for Long Offset Transient Electromagnetic (LOTEM) measurements and determined that there was an optimum azimuthal orientation of the source dipole, where one or the other measured component would be preferred (Fig. 7.15). Another interesting result from Hördt is that this optimum area is discerned only 2–3 times the transmitter dipole length away from the source, which confirms that TEM measurements can work with much closer offsets than frequency domain soundings. So far, from our experience we found that many elements from land measurements translate directly to the marine environment, and we assume that this is the case here also.

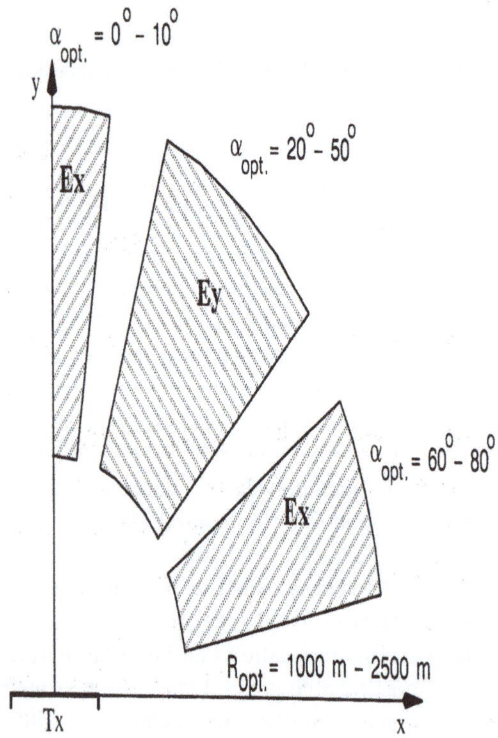

Fig. 7.15 Azimuthal distribution of the optimum component for surface measurements using electric fields and a Lotem setup (after Hördt 1992)

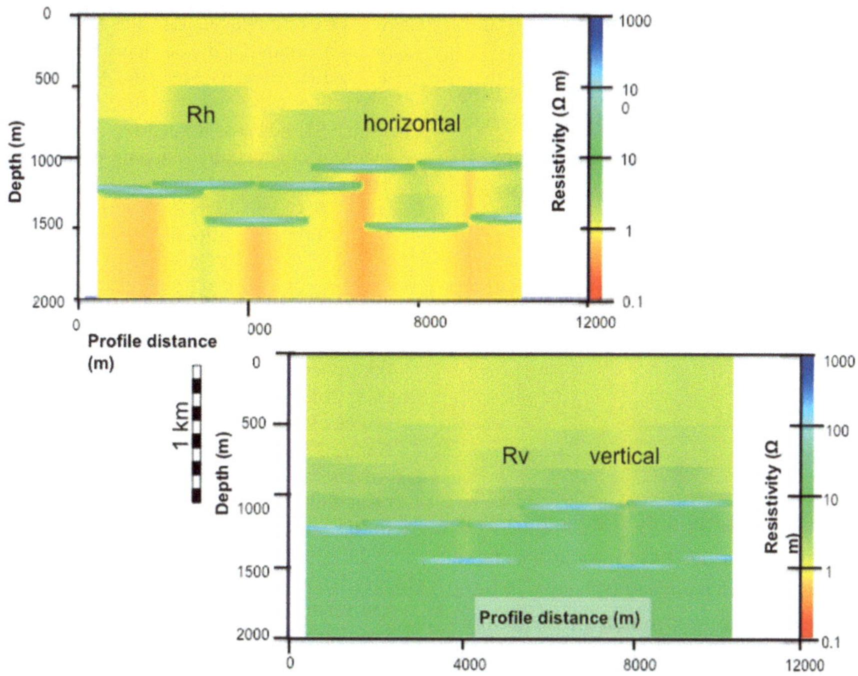

Fig. 7.16 Examples of two inverted stitched 1-D section using vertical and horizontal resistivities in a joint inversion mode

Figure 7.16 shows an example of an anisotropic inversion, with both vertical and horizontal resistivity sections. Note that horizontal resistivities often have more artifacts (vertical stripes), which are reduced in vertical resistivity inversion. Sub articificials may be a result of style of inversion. For example, in "stitched 1D" (i.e. independent 1D inversions repeated along a 2D profile, with varying results simply stitched together, despite existence of inconsistency. This procedure can work well with slow lateral variation and does not work well when the subsurface varies substantially on the scale of the source-receiver offset.

7.5 The Future and Pitfalls

The past twenty years have shown great progress in electrical geophysics especially in marine environment. With the advent of new logging tools, one can calibrate surface measurements and understand the anisotropy and production efficiently. Anisotropy is still the biggest pitfall, because its influence on resistivity measurements sometimes as high as 50%. Another factor influencing the EM signal is the induced Polarization. It is commonly observed and interpreted mainly in Russia and China. We have seen it in marine data but have been able to interpret the data using normal electromagnetic models without polarization effects.

A strong factor often ignored is the physical complexity of the subsurface resistivity distribution, not well imaged when using a system with several kilometers of source-to-receiver offset. It is well known that the Earth is rarely 1-dimensional. Thus shorter offsets are preferred to longer offsets, in order to resolve the subsurface complexity.

Predicting the future of marine EM is more complex. We have to make assumptions to do the right extrapolation. In 5 years period we assume:

- Marine Electromagnetics is in use (at least occasionally) by most oil companies, through their contractors.
- Reservoir monitoring will be the key focus of oil companies.
- Integration with seismic/geology will bring the value of EM to the forefront.
- Marine measurements will be done much denser with lower unit cost.
- Land measurements will still have limited applications unless integrated with seismic data acquisition.

- The integration value of EM will be better understood.

In summary, marine, electromagnetics will be fully complementary and integrated with other geophysical methods. The unit cost will have to be reduced by 10 fold in order to fit in a more routine scenario. The strongest market will be the monitoring market, which will require a change of business model for present service companies, as the oil companies will own the installations. Monitoring will include land and marine systems.

7.6 Conclusions

Time domain CSEM can reliably be acquired in a marine environment. Node based systems have acquired multiple data sets in a variety of basins, where the recorded transient responses match those of pre-survey models. We have adapted the methods from onshore environment to offshore and shown that the results are very similar. We also tested the next generation technology, a cabled version of the system and so far the signal behaves as anticipated. Time domain EM is optimum in shallow waters, but should really be used in a complementary fashion to frequency domain and marine magnetotellurics. Developments with a cabled system and initial functionality tests are very promising, in particular with respect to minimizing transmitter-to-receiver offset and resolving anisotropy.

Acknowledgements We thank KMS Technologies for permission to publish this material. Throughout the development of this work we received support from EMGS and BP and are grateful for that. Many colleagues have contributed to this work, a very special thanks to them. In particular, N. Allegar, I. Loehken, and Y. Martinez were very supportive in getting the task accomplished. A. A. Aziz and K. Vozoff were of great assistance in preparing the manuscript.

References

Avdeeva A, Commer M, Newman G (2007) Hydrocarbon reservoir detectability study for marine CSEM methods: time domain versus frequency domain: 78th international annual meeting. Soc Exploration Geophysicists, Expanded Abstr 26:628–632

Allegar NA, Strack KM, Mittet R, Petrov A (2008) Marine time domain CSEM – the first two years of experience. In: 70th conference & exhibition, European Association expanded abstract vol. Geoscientists & Engineers, Rome, Italy

Chave AD, Constable SC, Edwards RN (1991) Electrical exploration methods for the seafloor: investigation in geophysics no 3. Electromagn Methods Appl GeophysAppl Part B 2:931–966

Cheesman SJ, Edwards RN, Chave AD (1987) On the theory of sea-floor conductivity mapping using transient electromagnetic systems. Geophysics 52:204–217

Constable S (2006) Marine electromagnetic methods—a new tool for offshore exploration. The Leading Edge 25:438–444

Constable S, Cox CS (1996) Marine controlled source electromagnetic sounding 2: the PEGASUS experiment. J Geophys Res 101:5519–5530

Cox CS (1981) On the electrical conductivity of the oceanic lithosphere. Phys Earth Planetary Inter 25:289–300

Cox CS, Constable SC, Chave AD, Webb SC (1986) Controlled-source electromagnetic sounding of the oceanic lithosphere. Nature 320(6057):52–54

Davydycheva S, Druskin V, Habashy T (2003) An efficient finite-difference scheme for electromagnetic logging in 3D anisotropic inhomogeneous media. Geophysics 68: 1525–1536

Davydycheva S, Rykhlinski N (2009) Focused-source EM survey versus time-domain and frequency-domain CSEM. The Leading Edge 28:944–949

Druskin V, Knizhnerman L (1994) Spectral approach to solving three-dimensional Maxwell's diffusion equations in the time and frequency domains. Radio Sci 29(4):937–953

Eadie T (1979) Stratified earth interpretation using standard horizontal loop electromagnetic data. Research in applied geophysics, vol 9. Geophysics Laboratory, University of Toronto, Toronto

Edwards RN, Law LK, Wolfgram PA, Nobes DC, Bone MN, Trigg DF, DeLaurier JM (1985) First results of the MOSES experiment: sea sediment conductivity and thickness determination, Bute Inlet, British Columbia, by magnetometric offshore electrical sounding. Geophysics 50:153–160

Edwards RN (1987) Controlled source electromagnetic mapping of crust. In: James DE (ed) Encyclopedia of geophysics, geomagnetism and paleomagnetism volume. Van Nostrand Reinhold Co. Inc., Stroudsburg, Invited paper, pp 126–139

Edwards RN (1997) On the resource evaluation of marine gas hydrate deposits using the sea-floor transient electric dipole-dipole method. Geophysics 62:63–74

Edwards RN, Yu L (1993) First measurements from a deep-tow transient electromagnetic sounding system. Mar Geophys Res 15:13–26

Eidesmo T, Ellingsrud S, MacGregor LM, Constable S, Sinha MC, Johansen S, Kong FN, Westerdahl H (2002) Sea bed logging (SBL), a new method for remote and direct identification of hydrocarbon filled layers in deep water areas. First Break 20:144–152

Harthill N (1968) The CSM test area for electrical surveying methods. Geophysics 33:675–678

Hohmann GW (1988) Numerical modeling for electromagnetic methods of geophysics. In: Nabighian MN (ed) Electromagnetic methods in applied geophysics. Vol. 1, pp 313–363

Holten T, Flekkøy G, Måløy KJ, Singer B (2009a) Vertical source and receiver CSEM method in time domain. Soc Exploration Geophysicists, Expanded Abstr 28: 749–753

Holten T, Flekkoy G, Singer B, Blixt EM, Hanssen A, Maloy KJ (2009b) Vertical source, vertical receiver, electromagnetic technique for offshore hydrocarbon exploration. First Break 27:89–93

Hördt A (1992) Interpretation transient electromagnetischer Tiefensondierungen für anisotrop horizontal geschichtete und für dreidimensionale Leitfähigkeitsstrukturen: Ph D thesis, University of Cologne, Europe

Kaufman AA, Keller GV (1983) Frequency and transient soundings. Elsevier Science Publishers BV, Amsterdam

Newman GA, Alumbaugh DL (2000) Three-dimensional magnetotelluric inversion using non-linear conjugate gradients. Geophys J Int 140:410–424

Passalacqua H (1983) Electromagnetic fields due to a thin resistive layer. Geophys Prospecting 31:945–976

Raiche AP, Jupp DLB, Rutter H, Vozoff K (1985) The joint use of coincident loop transient electro-magnetic and Schlumberger sounding to resolve layered structures. Geophysics 50:1618–1627

Sinha MC, Patel PD, Unsworth MJ, Owen TRE, MacCormack MGR (1990) An active source electromagnetic sounding system for marine use. Mar Geophys Res 12:29–68

Spies BR (1989) Depth of investigation in electromagnetic sounding methods. Geophysics 54:872–888

Stoyer CH, Strack KM (2008) Method of acquiring and interpreting electromagnetic measurements. US Patent 07356411

Strack KM (1992 and 1999) Exploration with deep transient electromagnetics – introduction and indexes. Elsevier Science Publishers BV, Amsterdam

Strack KM, Petrov AA (2007) Marine time domain controlled source electromagnetics (tCSEM^{TM}): another way to illuminate marine, reservoirs. In: Proceedings of the 8th China international geo-electromagnetic workshop, Jingzhou, China, paper 3, pp 9–14

Strack KM, Vozoff K (1996) Integrating long-offset transient electromagnetics (LOTEM) with seismics in an exploration environment. Geophys Prospecting 44:99–101

Strack KM, Allegar N, Ellingsrud S (2008) Marine time domain CSEM: an emerging technology. Society exploration geophysicists. In: Annual meeting, extended abstracts, Las Vegas, pp 653–656

Strack KM, Hanstein TH, Eilenz HN (1989) LOTEM data processing for areas with high cultural noise levels. Phys Earth Planetary Inter 53:261–269

Strack KM, Hanstein T, Lebrocq K, Moss DC, Petry H, Vozoff K, Wolfgram PA (1989) Case histories of LOTEM surveys in hydrocarbon prospective areas. First Break 7:467–477

Veeken PCH, Legeydo PJ, Davidenko YA, Kudryavceva EO, Ivanov SA, Chuvaev A (2009) Benefits of the induced polarization geoelectric methods to hydrocarbon exploration. Geophysics 74(2):B47–B59

Ward SH, Hohmann G (1988) Electromagnetic theory for geophysical applications. Electromagn Methods Appl Geophys SEG 1:131–311

Weidelt P (2007) Guided waves in marine CSEM. Geophys J Int 171(1):153–176

Weiss CJ (2007) The fallacy of the "shallow-water problem" in marine CSEM exploration. Geophysics 72(6):A93–A97

Wright D, Ziolkowski A, Hobbs B (2002) Hydrocarbon detection and monitoring with a multicomponent transient electromagnetic (MTEM) survey. Leading Edge 60: 481–500

Ziolkowski A, Hall G, Wright D, Carson R, Peppe O, Tooth D, Mackay J, Chorley P (2006) Shallow marine test of MTEM method. 76th Annual International Meeting. Soc Exploration Geophysicists, Expanded Abstr 25:729–734

On Thermal Driving of the Geodynamo

8

Ataru Sakuraba and Paul H. Roberts

Abstract

It is widely believed that the main geomagnetic field is created by the dynamo action of motions in the Earth's fluid core that are driven by thermal and compositional buoyancy. Early numerical simulations of the geodynamo that succeeded in generating strong, Earth-like dipole magnetic fields had to assume, for computational reasons, an unrealistically high viscosity for the core fluid. Some recent high-resolution models have used more realistic, smaller viscosities, but have unexpectedly produced only non-dipolar or dipolar but comparatively weak magnetic fields, which are less Earth-like. We recently advanced a possible explanation for this paradoxical behavior: we argued that these models had used the geophysically unrealistic outer boundary condition of uniform temperature on the core-mantle interface. In support of this opinion, we integrated two otherwise identical models, in one of which we applied the uniform temperature condition and in the other the more realistic condition of horizontally uniform heat flux. In the latter model, we obtained large-scale convective flows and a comparatively strong dipole-type magnetic field; for the former, we found solutions resembling those obtained by other models that had assumed uniform temperature on the core-mantle boundary. Further explanations for the very different character of the solutions are given here.

8.1 Introduction

Convection in the Earth's mantle carries heat away from the core-mantle boundary (CMB) and allows convection to occur in the fluid outer core (FOC) necessary for the generation of the Earth's main magnetic field. As the core cools through this loss of heat, the solid inner core (SIC) grows, releasing latent heat at the inner core boundary (ICB) to drive thermal convection. Buoyancy is also created by the release at the ICB of light constituents that are incompatible with the solid phase (e.g., see Roberts and Glatzmaier 2000, Buffett 2000). Here we consider only the effects of thermal buoyancy.

We show in this paper that the boundary condition on the temperature at the CMB can be an important controlling factor on the flow in the FOC. In principle, the temperature, T, and the radial heat flux, q_r, must both be continuous across the CMB:

A. Sakuraba (✉)
Department of Earth and Planetary Science, School of Science, University of Tokyo, Tokyo 113–0033, Japan
e-mail: sakuraba@eps.s.u-tokyo.ac.jp

E. Petrovský et al. (eds.), *The Earth's Magnetic Interior*, IAGA Special Sopron Book Series 1,
DOI 10.1007/978-94-007-0323-0_8, © Springer Science+Business Media B.V. 2011

$$T^{(\mathrm{core})} = T^{(\mathrm{mantle})} \quad \text{and} \quad q_r^{(\mathrm{core})} = q_r^{(\mathrm{mantle})}. \tag{8.1}$$

In a medium convecting as strongly as the Earth's core, the temperature is everywhere, except in boundary layers, close to $T_a(r)$, the temperature of an adiabat, and therefore depends principally on radius, r. The associated adiabatic heat flux is $-K\mathrm{d}T_a/\mathrm{d}r$, where K is the thermal conductivity. This is supplemented by the outward convective heat flux, $\rho C_p\langle\Theta u_r\rangle_S$, where Θ is the temperature deviation from the adiabat, ρ is the density, C_p is the heat capacity per unit mass, and u_r is the radial velocity; the angle brackets, $\langle\cdot\rangle_S$, denote the average over spherical surfaces. The convective and adiabatic heat fluxes are both of the order of several TW. This implies that Θ is very small, of order 0.1 mK. The total outward heat flux is

$$\langle q_r\rangle_S = \rho C_p\langle\Theta u_r\rangle_S - K\frac{\mathrm{d}T_a}{\mathrm{d}r}. \tag{8.2}$$

This is continuous through the boundary layers at the top of the core and the bottom of the mantle even though u_r tends to zero as the CMB is approached from either side; increased thermal gradients within the boundary layers conduct the heat that convection can no longer carry.

In view of the enormous difference in the viscosities of the core and mantle, it is perhaps not surprising that the convective velocities in each differ greatly too. Typically, $|u_r^{(\mathrm{core})}| \approx 0.5$ mm/s compared with $|u_r^{(\mathrm{mantle})}| \approx 3$ cm/yr, so that $|u_r^{(\mathrm{core})}|/|u_r^{(\mathrm{mantle})}| \sim O(10^6)$. The difference in ρC_p is however not substantial. Since the convective heat fluxes are also comparable, the deviations, Θ, in mantle and core temperatures from their adiabats is very great: $|\Theta^{(\mathrm{mantle})}| \approx 100$ K but $|\Theta^{(\mathrm{core})}| \approx 0.1$ mK. Such small variations in core temperature over the CMB are utterly insignificant for mantle convection. This shows that the correct lower boundary condition for solving mantle convection is simply $T^{(\mathrm{mantle})} = T_a^{(\mathrm{core})}$. An end product of that calculation is the heat flux, $q_r^{(\mathrm{mantle})}$, on the CMB, and this provides the required top boundary condition for solving core convection. This completes the separation of the core-mantle system into core and mantle individually (see also Braginsky and Roberts 2007).

In most previous numerical simulations of core convection, the temperature is assumed to be fixed on the CMB, and is usually taken to be laterally uniform (e.g., Christensen and Aubert 2006, Takahashi et al. 2008, Kageyama et al. 2008). This defines our uniform surface temperature model (USTM), but it is not appropriate for the Earth's core for the reasons just given. Instead, the heat flux, q_r, on the CMB should be specified. Lacking precise knowledge of how q_r varies over the CMB, we make the simplest assumption that it is uniform. This defines our uniform heat flux model (UHFM), which apart from the different upper boundary condition is identical in all respects to our USTM. We have shown elsewhere that, when the core viscosity is sufficiently small, the USTM and UHFM solutions are dramatically different (Sakuraba and Roberts 2009). Here we give further information about the differences between the models, and some new interpretations.

8.2 Models and Equations

We use the Boussinesq approximation to model thermal convection in the FOC, $r_i \le r \le 1$, where r is the nondimensional radius scaled by the CMB radius, R_o. The time t is scaled by the magnetic diffusion time, R_o^2/η, where η is the magnetic diffusivity, so the velocity scale is η/R_o. The scales for the temperature and the magnetic field, $\mathbf{b}$, are respectively βR_o and $\sqrt{2\Omega\rho\eta\mu_0}$, where β is the temperature gradient at the CMB, Ω is the angular velocity and μ_0 is the magnetic permeability. The nondimensional governing equations for the convection in the FOC become

$$P_m^{-1}E\left[\frac{\partial\mathbf{u}}{\partial t} + (\mathbf{u}\cdot\nabla)\mathbf{u}\right] = -\nabla p + \mathbf{u}\times\mathbf{e}_z + \mathbf{j}\times\mathbf{b} + Ra\,\Theta\mathbf{r} + E\nabla^2\mathbf{u}, \tag{8.3}$$

$$\frac{\partial\mathbf{b}}{\partial t} = \nabla\times(\mathbf{u}\times\mathbf{b}) + \nabla^2\mathbf{b}, \tag{8.4}$$

$$\frac{\partial\Theta}{\partial t} + \mathbf{u}\cdot\nabla\left(\bar{T}+\Theta\right) = P_q\nabla^2\Theta, \tag{8.5}$$

$$\nabla\cdot\mathbf{u} = \nabla\cdot\mathbf{b} = 0, \tag{8.6}$$

where $\mathbf{j} = \nabla\times\mathbf{b}$ is the electric current density, $\bar{T}$ is the reference temperature that we take as a solution of the heat conduction equation, Θ is the temperature perturbation, p is the pressure perturbation, and $\mathbf{e}_z$ is the unit vector parallel to the polar (z) axis. We introduce four nondimensional parameters

$$\left\{\begin{array}{ll}\text{Ekman number}: & E = \dfrac{\nu}{2\Omega R_o^2},\\ \text{magnetic Prandtl number}: & P_m = \dfrac{\nu}{\eta},\\ \text{diffusivity ratio}: & P_q = \dfrac{\kappa}{\eta},\\ \text{Rayleigh number}: & Ra = \dfrac{\alpha g_o \beta R_o^2}{2\Omega\eta},\end{array}\right. \tag{8.7}$$

where ν is the kinematic viscosity, $\kappa = K/\rho C_p$ is the thermal diffusivity, g_o is the acceleration due to gravity at the CMB and α is the thermal expansivity.

We solve the magnetic diffusion equation inside the SIC which we suppose has the same density and electrical conductivity as the FOC and is free to rotate about the z-axis in response to the viscous and magnetic torques exerted by the FOC. Believing that the thermodynamics of the SIC does not significantly affect the results, we make the simple assumption that Θ inside the SIC is constant everywhere so that the surface temperature, which is laterally uniform, represents the whole inner core temperature. We assume that the temperature at the ICB obeys the heat conduction equation

$$\frac{4\pi r_i^3}{3}\frac{\partial\langle\Theta\rangle_S}{\partial t} = 4\pi P_q r_i^2 \frac{\partial\langle\Theta\rangle_S}{\partial r} \quad \text{at } r = r_i. \tag{8.8}$$

We adopt the no-slip condition for the velocity at the boundaries. We assume that the mantle is an electrical insulator, so that the magnetic field at the CMB is continuous with a source-free potential field in $r \geq 1$. For the temperature on the CMB, we consider two cases; in the USTM,

$$\Theta = 0 \quad \text{at } r = 1, \tag{8.9}$$

and in the UHFM,

$$\langle\Theta\rangle_S = 0 \quad \text{and} \quad \frac{\partial}{\partial r}(\Theta - \langle\Theta\rangle_S) = 0 \quad \text{at } r = 1, \tag{8.10}$$

so that averaged CMB temperatures are the same.

Thermal convection is driven by a uniform heat source distributed in the whole core and an additional heat source concentrated at the ICB. The reference temperature gradient in the FOC is

$$\frac{d\bar{T}}{dr} = -\frac{Q}{r^2} - (1 - Q)r, \tag{8.11}$$

where Q is the ratio of the heating at the ICB to the total heat flow out of the core. Once convection settles into a statistically steady state, the heat flow out of the CMB returns to its value before the onset of convection because the rate of heat generation in the core is constant in time and we did not include dissipation terms in (8.5).

The vector variables, **u** and **b**, are decomposed into toroidal and poloidal scalar fields that, together with Θ, are expanded in Chebyshev polynomials in r and in spherical harmonics in colatitude θ and longitude ϕ. The same spectral modes were used for all five scalars although the magnetic field is defined in the whole core while the others are defined only in the FOC. We typically use 256×256 spherical harmonic modes and 160 Chebyshev spectral modes for each variable.

8.3 Results of Numerical Experiments

In our previous paper (Sakuraba and Roberts 2009), we showed how differently the two models behaved when we reduced the viscosity as far as we could ($E = 5 \times 10^{-7}, P_m = P_q = 0.2, Ra = 3200, Q = 0.5, r_i = 0.353$). Here we summarize our results and provide some new information.

Both models reached statistically steady states in which the axial dipole dominates the magnetic field at the CMB. The UHFM has the larger dipole moment. The magnetic field intensity, averaged over time and over the volume of the FOC, is about 30 % greater in the UHFM than in the USTM (Fig. 8.1). Differences are even more evident when the internal structure of the two fields are compared. On the one hand, as shown by Figs. 8.2 and 8.3, the magnetic field of the UHFM is concentrated in two large-scale “flux tubes” outside the tangent cylinder, that spiral about the azimuthal direction in low latitudes, where $|z|$ is less than about 0.3. One flux tube is in each hemisphere, and is in the opposite direction to its partner in the other hemisphere. Because the flux tubes approach one another in some places, especially near the CMB, there are significant radial electric currents near the equatorial plane. Each flux tube meanders with an azimuthal wavenumber (m) of about six, and there is a general tendency for the magnetic helicity, $\mathbf{a} \cdot \mathbf{b}$, where $\mathbf{a}$ is the vector potential, to be positive (negative) in the northern (southern) hemisphere. This morphology is

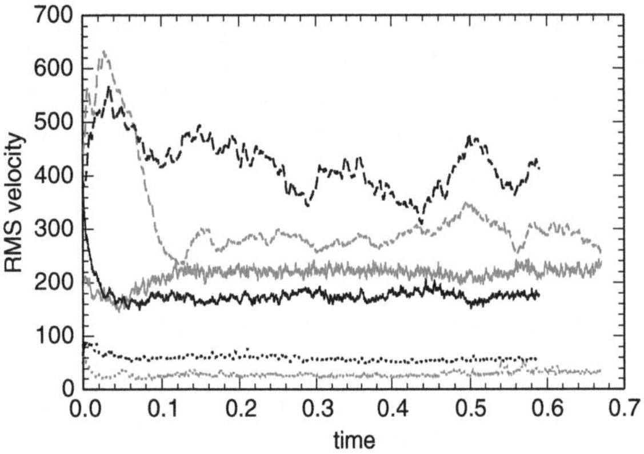

Fig. 8.1 The root-mean-square flow and Alfvén velocities in our UHFM (*black*) and USTM (*gray*) plotted versus time. The flow velocity (*solid line*) and the Alfvén velocity (*broken line*) are calculated from the volume-averaged kinetic and magnetic energy densities in the FOC, respectively. The axisymmetric toroidal flow velocities are also shown by dotted lines (the lowermost curve is for the USTM and the one above it for the UHFM). All velocities are scaled by η/R_o

consistent with the field generation process known as the α-effect, in which a flow that has negative kinetic helicity ($\mathbf{u} \cdot \nabla \times \mathbf{u}$), as is predominantly the case in the northern hemisphere, produces a magnetic field that has positive magnetic helicity and vice versa. This α-effect might be called a "macroscopic" because the flow and the generated magnetic field are both of large-scale. It resembles the macroscopic α-effect seen in earlier, higher-viscosity, dynamo models (Olson et al. 1999, Ishihara and Kida 2002). The flux tubes do not always encircle the FOC, but sometimes move to mid-latitudes or become too weak to be discernible. On the other hand, the magnetic field of the USTM is comparatively weak, small-scale, and predominantly radial outside the tangent cylinder; it has almost no net zonal component (see Figs. 8.4 and 8.5). Within the tangent cylinder, there is a large-scale, nearly axisymmetric magnetic field, whose intensity is as strong as, or rather stronger than, that of the UHFM in the same location.

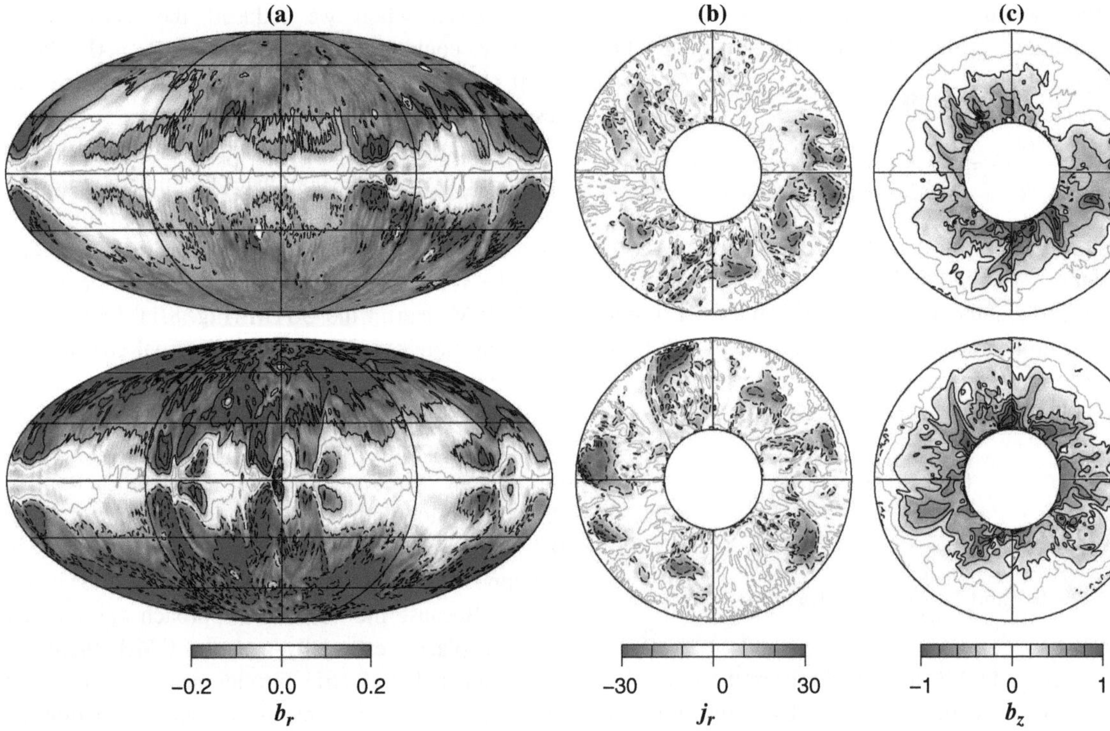

Fig. 8.2 The magnetic field structure of the UHFM at $t = 0.44$ (*top*) and $t = 0.48$ (*bottom*), the former representing respectively the times when the overall field intensity is weak and when it is strong. (**a**) The radial magnetic field at the CMB in the Mollweide projection. (**b**), (**c**) The radial electric current density and the axial magnetic field in the equatorial plane viewed from the north ($z > 0$). The *dotted contour lines* show negative values. The central meridian of the Mollweide projection is located on the *left-hand side* of the equatorial slices

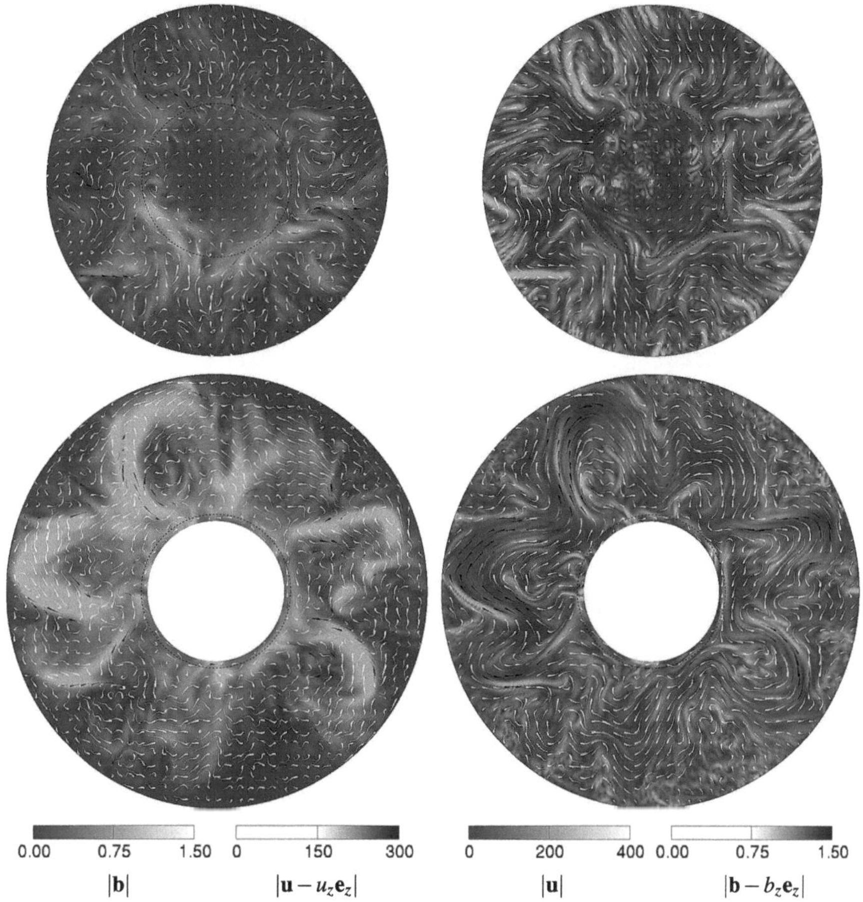

Fig. 8.3 The velocity and magnetic field structures in the planes $z = 0.6$ (*top*) and $z = 0.15$ (*bottom*) for our UHFM at $t = 0.48$. *Left*: The background shade shows the field intensity $|\mathbf{b}|$, while the streamlines show the velocity field parallel to the plane, $\mathbf{u} - u_z\mathbf{e}_z$. The length of the streamline is arbitrary but its brightness represents the vector amplitude, $|\mathbf{u} - u_z\mathbf{e}_z|$. The velocity field points in the same direction as the width of the streamline becomes narrower. *Right*: Same but the background shade showing the flow speed $|\mathbf{u}|$, and the lines showing the magnetic field parallel to the plane, $\mathbf{b} - b_z\mathbf{e}_z$. The tangent cylinder is represented by a *dotted circle*

To illustrate the differences in the magnetic field structures of the two models, we show in Fig. 8.6 the probability density functions $F(|\mathbf{b}|^2)$, calculated separately for the regions outside and inside of the tangent cylinder. For a volume V_0, $F(\xi)$ is defined by

$$F(\xi) = \frac{1}{V_0}\frac{\mathrm{d}V}{\mathrm{d}\xi}, \tag{8.12}$$

where $\mathrm{d}V$ is the subset of V_0 in which $\xi \leq |\mathbf{b}|^2 \leq \xi + \mathrm{d}\xi$. In our non-dimensional units, $|\mathbf{b}|^2$ is the local Elsasser number, Λ. Outside the tangent cylinder, $F(\Lambda)$ peaks at a smaller Λ for the USTM than for the UHFM and has a larger gradient for high-Λ. The volume fraction for which $\Lambda > 1$ is about 10% in the UHFM, indicating that the magnetic flux tube is well concentrated. The same fraction for the USTM is nearly an order of magnitude smaller. In contrast, $F(\Lambda)$ inside the tangent cylinder is not very different for the two models. The bump around $\Lambda = 0.6$ in our USTM represents the nearly axisymmetric field inside the tangent cylinder. Since the volume of the tangent cylinder is small, the averaged magnetic field intensity and the dipole moment are mostly controlled by the behaviors outside the tangent cylinder. This is why the UHFM maintains a stronger magnetic field than the USTM.

The flow structure of thermal convection at low-E usually has two main characteristics: it is almost independent of z, and it has high azimuthal wavenumbers. Both models show the first characteristic, but

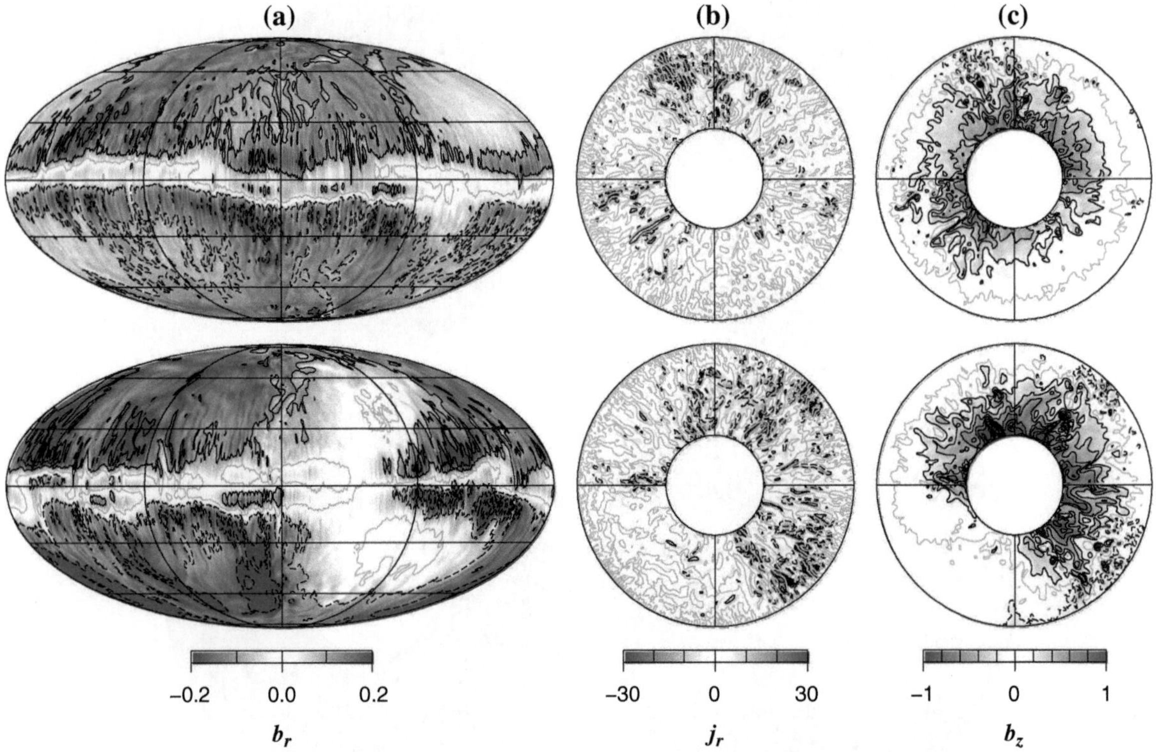

Fig. 8.4 The magnetic field structure of our USTM at $t = 0.3$ (*top*) and $t = 0.5$ (*bottom*), the former representing the time when the overall field intensity is weak and the latter when the solution is about to change to a strong-field state. **(a)** The radial magnetic field at the CMB in the Mollweide projection. **(b)**, **(c)** The radial electric current density and the axial magnetic field in the equatorial plane viewed from the north ($z > 0$)

the second is not everywhere seen in our UHFM; see Fig. 8.3. The flow structure shows small-scale turbulence in places where the magnetic field is weak, but inside the low-latitude flux tubes, the flow is slow and small-scale turbulence is suppressed. The flow direction inside the flux tube is primarily retrograde (westward) but there is a weak radial component too. A thin "sheet" of rapid flow exists that circumscribes the outer edges of the flux tubes. Because of the two-dimensionality of the flow structure, this sheet is clearly seen even in mid- and high-latitudes where the magnetic flux tube no longer exists. The direction of the sheet-like flow is mostly inward (toward the z-axis), and creates a sharp meandering of the flux tube. A similar sheet-like flow in a thin region where the magnetic field intensity changes abruptly was reported in low-E nonlinear magnetoconvection (Sakuraba 2007). We surmise that this jet-like flow is a fundamental feature of rapidly rotating magnetohydrodynamic systems. The low-latitude zonal flow advects the flux tubes westward. Therefore, if viewed from the frame co-rotating with the zonal flow, a weak, broad radial flow (upwelling) pushes the flux tubes against the CMB, and a thin sheet-like counter-flow creates a sharp edge to the spiraling flux tubes. In mid- and high-latitudes, the flow is small-scale and the zonal flow is not evident. The flow inside the tangent cylinder is relatively slow but has a fine-scale structure consisting of narrow tube-like helical flows parallel to the z-axis, as seen in thermal convection experiments and some other dynamo models (Aurnou et al. 2003, Aubert et al. 2008).

As we reported earlier (Sakuraba and Roberts 2009), the UHFM produces several low-latitude magnetic flux patches on the CMB (see Fig. 8.2) that move westward, like the observed geomagnetic westward drift that is also most prominent near the equator (Finlay and Jackson 2003). The westward drift of the UHFM is essentially a manifestation at the CMB of the low-latitude flux tubes that are advected by the

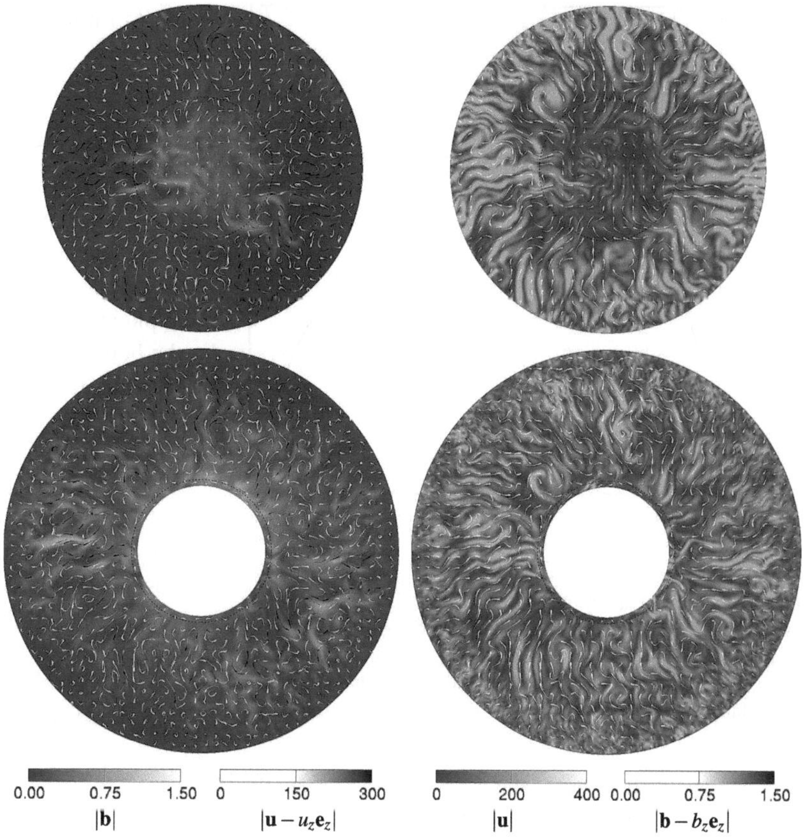

Fig. 8.5 The velocity and magnetic field structures at the planes $z = 0.6$ (*top*) and $z = 0.15$ (*bottom*) for our USTM at $t = 0.3$. *Left*: The background shade shows the field intensity $|\mathbf{b}|$, while the streamlines show the velocity field parallel to the plane, $\mathbf{u} - u_z\mathbf{e}_z$. The length of the streamline is arbitrary but its brightness represents the vector amplitude, $|\mathbf{u} - u_z\mathbf{e}_z|$. The velocity field points in the same direction as the width of the streamline becomes narrower. *Right*: Same but the background shade showing the flow speed $|\mathbf{u}|$, and the lines showing the magnetic field parallel to the plane, $\mathbf{b} - b_z\mathbf{e}_z$. The tangent cylinder is represented by a *dotted circle*

zonal flow. When the overall magnetic field intensity is strong, the interior flux tubes are pushed against the CMB and twin flux patches of opposite polarities are created near the equator. When dynamo activity is low, the interior flux tubes are comparatively weak and a single flux patch tends to be created. This low-latitude patch is probably maintained by the persistent thin sheet-like flows that sharpen the flux concentration.

Our USTM has a small-scale, high-wavenumber structure with a clear two-dimensionality that resembles other low-viscosity USTMs (Takahashi et al. 2008, Kageyama et al. 2008); see Fig. 8.5. We imposed the same heat sources in both models. Nevertheless, the volume-averaged flow speed is about 20% greater in the USTM than in the UHFM (Fig. 8.1). This probably shows that the small-scale convection in the USTM involves more thermal diffusion so that a more rapid flow is needed to transport heat. Similarly, the small scales of the USTM enhance magnetic diffusion so that dynamo action is inefficient despite the higher flow speed (larger magnetic Reynolds number). There is no net azimuthal flow outside the tangent cylinder (and no westward drift) in the USTM. The flow inside the tangent cylinder is relatively slow but has a larger-scale structure, including a net zonal component.

In our previous paper (Sakuraba and Roberts 2009), we showed that the kinetic and magnetic energy spectra at mid-depth of the FOC in the USTM have peaks around $m = 15$ and that these peaks have comparable magnitudes, implying equipartition. It was suggested that this may be a fundamental characteristic of USTMs. If so, USTMs would fail to maintain magnetic field intensities as strong as the geomagnetic field even if one could use Earth-like parameters in the simulation. We have now extended the time integration of the USTM and have found that the magnetic field gradually increases (see the curves when t is around 0.5 in Fig. 8.1). At that time, the magnetic field component of $m = 1$ is selectively intensified and reversed magnetic flux patches appear at the CMB (see Fig. 8.4). We anticipated that a transition to a convective state resembling that of the UHFM might

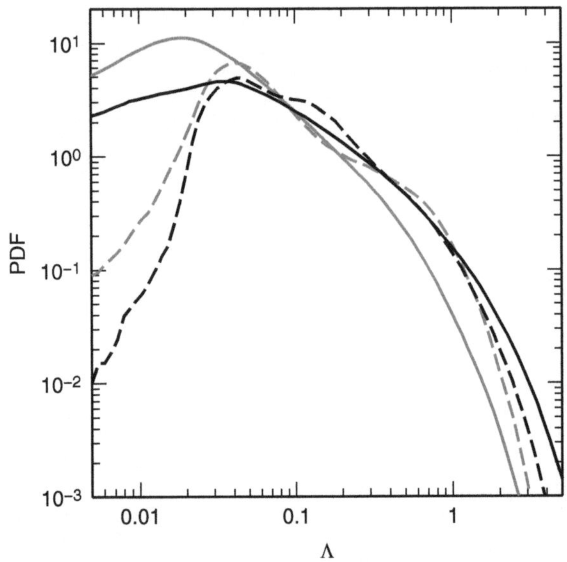

Fig. 8.6 The time-averaged probability density functions for our UHFM (*black*) and USTM (*gray*), with the Elsasser number, $\Lambda = |\mathbf{b}|^2$, taken as the independent variable. The solid and broken lines are those calculated for the outside and the inside of the tangent cylinder, respectively

become established, but instead the solution gradually reverted to the previous weak-field, equipartition state. Throughout this time integration, the flow structure remained small-scale and no net zonal flow appeared outside the tangent cylinder, which is illustrated by the fact that the axisymmetric toroidal (zonal) flow did not increase (Fig. 8.1). Although the time integration may not be long enough to capture all the intrinsic features of the solution, this numerical experiment suggests that energy equipartition at high-wavenumbers is the robust characteristic of the USTM and that it controls the overall magnetic field intensity.

8.4 Discussions on the Thermal Boundary Condition

In order to explain why the thermal boundary condition has such a profound impact on the dynamo solution, we average equation (8.5) over t and ϕ to obtain

$$P_q\left[\frac{\partial^2\langle\Theta\rangle}{\partial r^2}+\frac{2}{r}\frac{\partial\langle\Theta\rangle}{\partial r}+\frac{1}{r^2\sin\theta}\frac{\partial}{\partial\theta}\left(\sin\theta\frac{\partial\langle\Theta\rangle}{\partial\theta}\right)\right] = -C(r,\theta), \tag{8.13}$$

where $\langle\cdot\rangle$ now represents time and longitudinal averaging, and

$$C(r,\theta) = -\langle u_r\rangle\frac{d\bar{T}}{dr} - \langle\mathbf{u}\cdot\nabla\Theta\rangle \tag{8.14}$$

can be interpreted as a heat source created by the convergence of the convective heat flux. Equation (8.13) poses an elliptic, boundary-value problem on a meridional plane, and its solution is significantly influenced by the boundary condition applied at $r = 1$. We adopt the same simple $C(r,\theta)$ for both models:

$$C(r,\theta) = \frac{c(1-3\cos^2\theta)}{r} = -\frac{2cP_2(\cos\theta)}{r}, \tag{8.15}$$

where $P_2(\cos\theta)$ is the Legendre polynomial and c is a constant. We solve (8.13) using the same boundary conditions, (8.9) or (8.10), as in the numerical simulations, obtaining

$$\langle\Theta\rangle = \frac{c}{2P_q}\left[\frac{r^2}{r_i} - r + A\left(\frac{1}{r^3}-\frac{r^2}{r_i^5}\right)\right]P_2(\cos\theta), \tag{8.16}$$

where

$$A = \frac{r_i^4(2-r_i)}{2+3r_i^5} \quad \text{and} \quad \frac{r_i^4(1-r_i)}{1-r_i^5} \tag{8.17}$$

for the UHFM and the USTM, respectively. Figure 8.7 shows the resulting solution for $c/P_q = 1$ and $r_i = 0.35$.

Although (8.15) is too simple a model to represent dynamo solutions well, the greater magnitude of $\langle\Theta\rangle$ for the UHFM is apparent. Another important characteristic difference is that the maximal temperature perturbation is at the CMB in the UHFM, while it occurs near mid-depth in the USTM. In our simulations, both models produce an outward radial velocity on the equatorial plane (Fig. 8.8). This creates a positive $C(r,\theta)$, if the nonlinear term in (8.14) can be neglected, resulting in a positive temperature anomaly near the equatorial plane and outside the tangent cylinder. The effect of the nonlinearity is discussed later.

Another feature of $\langle\Theta\rangle$, also evident in Fig. 8.7, is the greater magnitude of $\partial\langle\Theta\rangle/\partial\theta$ for the UHFM. This means that this model drives a strong thermal wind.

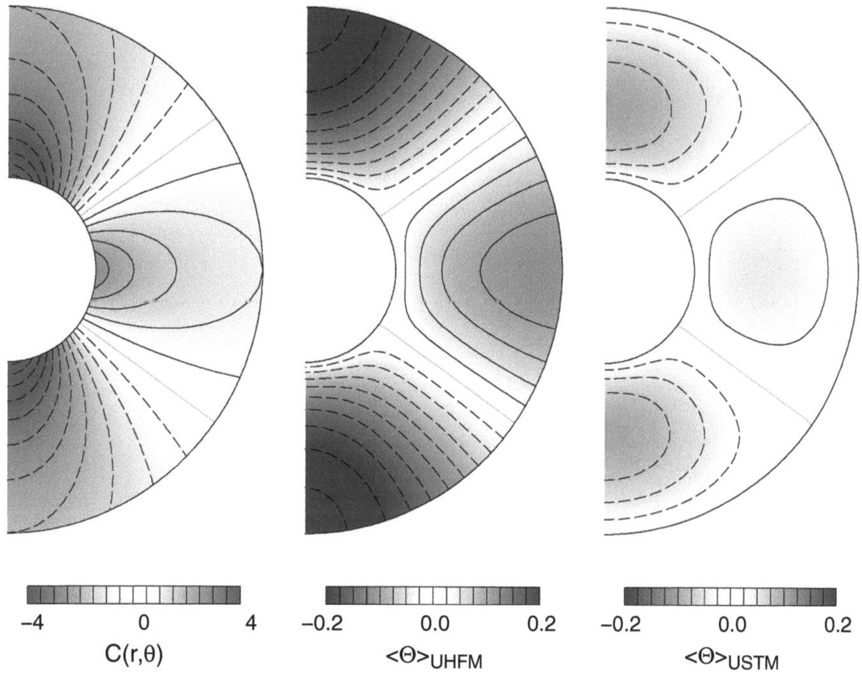

Fig. 8.7 Solutions of the equation for the time-averaged axisymmetric temperature perturbation $\langle\Theta\rangle$ when a source term $C(r,\theta)$, defined in (8.15), is assumed. Positive and negative values are represented by solid and broken contour lines, respectively. In this example, $\partial\langle\Theta\rangle/\partial r = 0$ at the CMB in the UHFM (middle), while $\langle\Theta\rangle = 0$ in the USTM (right). At the ICB, $\langle\Theta\rangle = 0$ in both models

As Aubert (Aubert 2005) demonstrated for his dynamo models, ours is consistent with the vorticity equation

$$\frac{\partial\langle u_\phi\rangle}{\partial z} \approx Ra\frac{\partial\langle\Theta\rangle}{\partial\theta} \tag{8.18}$$

because the Lorentz, viscous and inertial terms make only minor contributions. This explains why the UHFM creates a strong shear in zonal velocity; see Fig. 8.8. Such a shear is well known to be efficient in generating a toroidal magnetic field from a poloidal one (the ω-effect). As our models favor dipole dominated solutions, the thermal wind, which is largely concentrated in equatorial regions, creates zonal magnetic fields that have opposite directions in opposite hemispheres and are large near the equatorial plane; see Fig. 8.9.

These considerations lead us to the following conjecture on why the different thermal boundary conditions affect the dynamo solutions so strongly. Suppose for simplicity that our systems are in their linearly growing stages in which nonlinearities are negligible. If the total heat flux from the core is assumed to be the same for each model, the axisymmetric radial velocity $\langle u_r\rangle$ and the heat source function $C(r,\theta)$ are similar. The resulting temperature perturbations are, however, very different, that of the UHFM being greater than that of the USTM. The thermal wind and the resulting ω-effect are stronger too, so that the UHFM maintains a stronger toroidal field, which can be seen as the low-latitude flux tubes (Figs. 8.2 and 8.3).

Linear analyses of the onset of magnetoconvection in a rotating sphere have shown that the wavenumber of the convection pattern depends sensitively on the strength, B_0, of the applied magnetic field (Fearn 1979, Sakuraba 2002). For a sufficiently large rotation rate and for an Elsasser number, $\Lambda = B_0^2/2\Omega\rho\eta\mu_0$, smaller than some threshold value, which depends on the morphology of the imposed magnetic field but is around 0.1 to 1, the marginal wave number is large, of order $E^{-1/3}$. When a stronger, though not much stronger, magnetic field is applied, in the sense that Λ is still of order 1, the marginal wave number becomes of order 1, i.e., of the order of the scale of the system. It seems that the Elsasser number of the zonal magnetic field generated by the USTM never reaches the

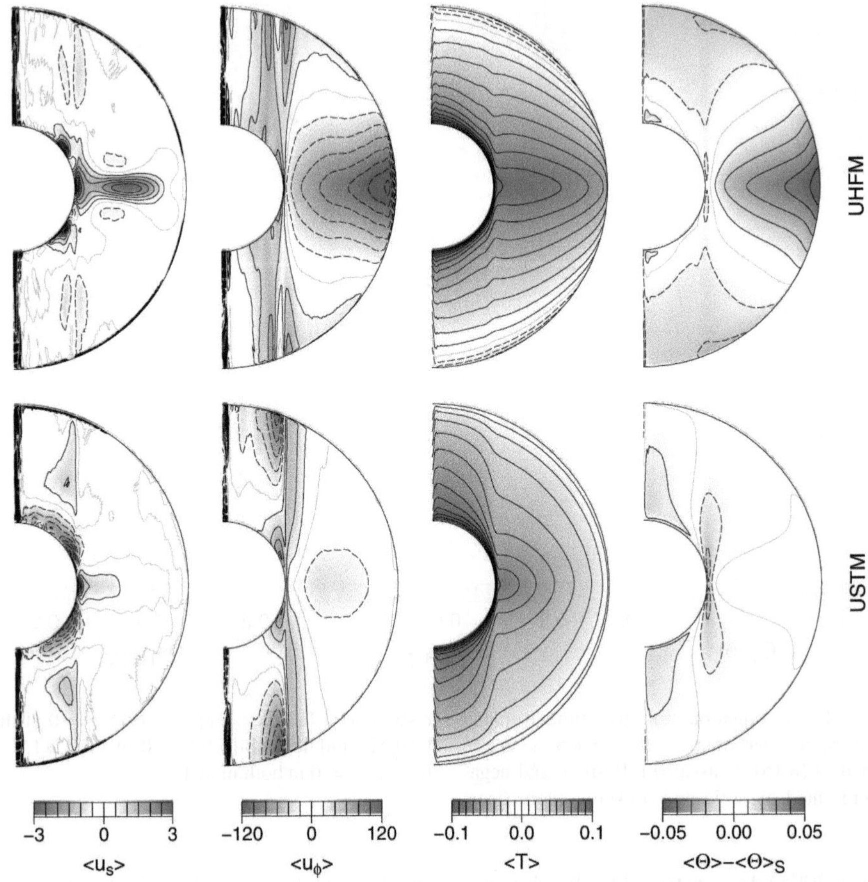

Fig. 8.8 The time-averaged axisymmetric velocity and temperature in our UHFM (*top*) and USTM (*bottom*). From *left* to *right*, the radial velocity $\langle u_s \rangle$ in the cylindrical coordinates, the zonal velocity $\langle u_\phi \rangle$, the temperature $\langle T \rangle$ including the reference field $\bar{T}(r)$, and the spherically asymmetric part, $\langle \Theta \rangle - \langle \Theta \rangle_S$, of the temperature perturbation are shown. Positive and negative values are represented by *solid* and *broken contour* lines, respectively

threshold for large-scale convection, and this explains why such models produce small-scale, sheet-like convection patterns.

Once a large-scale convection pattern becomes possible in the presence of a strong zonal magnetic field, as in our UHFM, dynamo action is efficient because of the smaller Joule losses, and this leads to the creation of the stronger magnetic field of this model. The presence of a strong magnetic field has a bearing on the axisymmetric radial velocity, $\langle u_r \rangle$, near the equatorial plane. Because viscous and inertial forces are unimportant in our solutions, the time-and longitudinally-averaged zonal momentum balance near the equatorial plane reduces to

$$\langle u_r \rangle \approx \langle j_r b_\theta \rangle - \langle j_\theta b_r \rangle, \tag{8.19}$$

the last term of which is unimportant in our dipole-dominated solutions. The first term on the right-hand side is well approximated by $\langle j_r \rangle \langle b_\theta \rangle$ (see Sakuraba and Roberts 2009). Then (8.19) becomes

$$\langle u_r \rangle \approx \langle j_r \rangle \langle b_\theta \rangle. \tag{8.20}$$

The thermal wind that is driven by the outward $\langle u_r \rangle$ near the equatorial plane induces, from the magnetic field crossing the equatorial plane, zonal fields that have opposite directions in the two hemispheres. This process also produces an axisymmetric radially directed electric current, $\langle j_r \rangle$, between these two oppositely-directed toroidal fields (Figs. 8.2 and 8.9). The resulting Lorentz force, $\langle j_r \rangle \langle b_\theta \rangle$, must be balanced by the azimuthal Coriolis force, as shown by

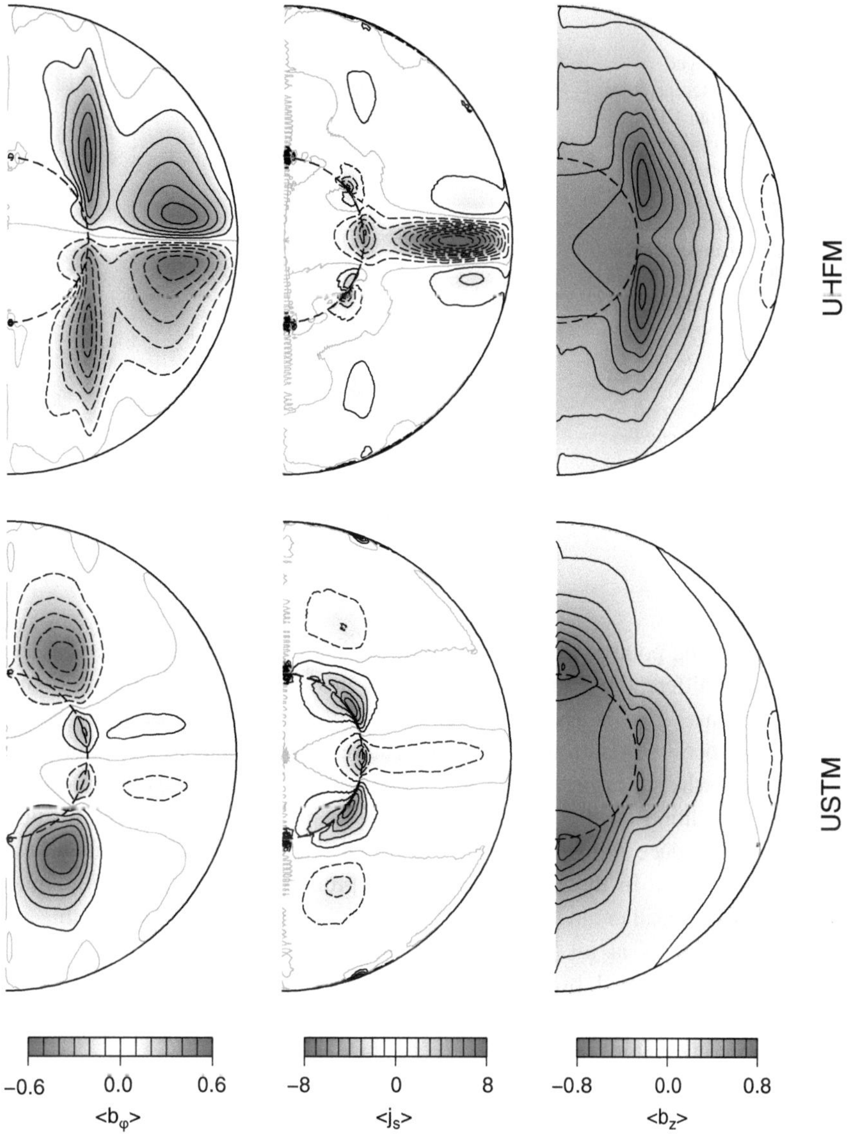

Fig. 8.9 The time-averaged axisymmetric magnetic field in our UHFM (*top*) and USTM (*bottom*). From left to right, the zonal (toroidal) magnetic field $\langle b_\phi \rangle$, the radial electric current density $\langle j_s \rangle$, and the axial magnetic field $\langle b_z \rangle$ are shown. Positive and negative values are represented by *solid* and *broken contour lines*, respectively

(8.20). The axial magnetic field therefore creates a linear relationship between $\langle u_r \rangle$, $\langle \Theta \rangle$, $\langle u_\phi \rangle$ and $\langle j_r \rangle$ near the equatorial plane. More importantly, the initial $\langle u_r \rangle$ creates a strong magnetic field in the UHFM and this magnetic field in turn promotes a stronger $\langle u_r \rangle$, as indicated by (8.20).

This positive feedback is eventually halted by nonlinear equilibration processes. For example, the strength of $\langle b_\theta \rangle$ (a poloidal magnetic field crossing the equatorial plane), which in our UHFM is primarily generated by a macroscopic α-effect, is limited by the back reaction of the Lorentz force. Another important process indicated by our numerical simulations is the saturation of the source term $C(r, \theta)$ in equation (8.13). As illustrated in Fig. 8.10, the first (linear) term of (8.14) makes a positive contribution to $C(r, \theta)$ near the equatorial plane because of the outward "jet" of $\langle u_r \rangle$, but the second (nonlinear) term tends to cancel this out, resulting in a reduction in the amplitude of $C(r, \theta)$. In the linear growth phase, the outward $\langle u_r \rangle$ in the equatorial plane increases $\langle \Theta \rangle$, enhances the thermal wind and creates a strong toroidal field. In the nonlinear phase, however, the effect of the equatorial outward jet on $\langle \Theta \rangle$ is significantly weakened by a compensating response from the nonlinear term.

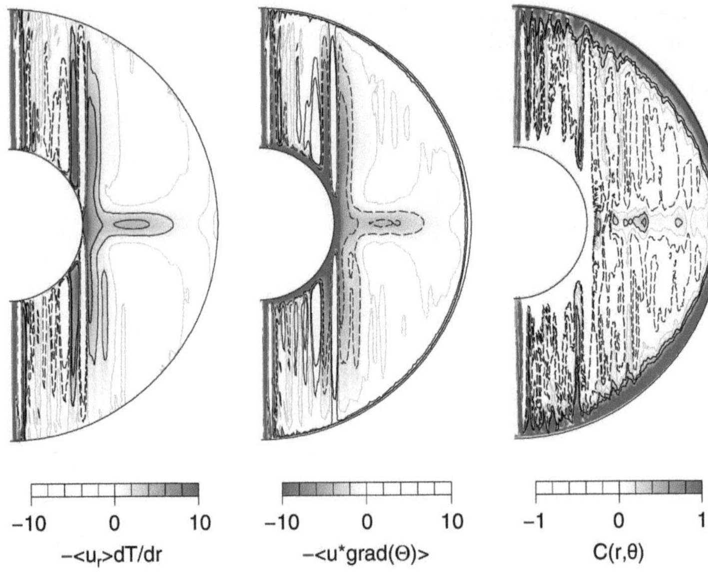

Fig. 8.10 The heat source term in the time- and azimuthally averaged equation for the temperature perturbation in our UHFM. The linear contribution (*left*) near the equatorial plane is positive because of an outward jet. The nonlinear term (*middle*) almost cancels out, resulting in small $C(r,\theta)$ (*right*). Positive and negative values are represented by *solid* and *broken contour lines*, respectively. Note that the gray scale is reversed for the nonlinear term (*middle*)

An interesting point is that the pattern of $\langle\Theta\rangle$ still satisfies the differential equation (8.13) even in the nonlinear phase. That is, thermal diffusion is still an important process in controlling the temperature pattern. This is why we argue for the importance of the boundary condition on the temperature and why we attempted to reduce the diffusivity ratio, P_q, to 0.2 in our simulations. In the Earth's core, P_q is probably very much smaller and thermal diffusion is not an efficient process at all. Nevertheless, the source term $C(r,\theta)$ is reduced by a subtle cancellation to balance the small diffusion term (*i.e.*, the convective heat flux becomes almost divergence-free). Figure 8.10 indicates that even though the contribution of the equatorial jet to $C(r,\theta)$ is mostly canceled by the nonlinear term, a remnant still remains and this translates into a major feature of $\langle\Theta\rangle$ in our UHFM.

The situation is totally different in the momentum equation. In an Earth-like condition, viscous diffusion is generally not an important process except in thin boundary shear layers and does not affect the main flow dynamics at all, because in the momentum equation there are three terms that are much greater (Coriolis, Lorentz and buoyancy forces), and these can control the velocity field independently of the viscous diffusion process. In the case of the heat equation, there is only one advection term to balance the thermal diffusion term, leaving room for the boundary condition to have a major impact on the dynamo. Conversely, we expect that the boundary condition on the velocity do not influence magnetostrophic states substantially, especially in the case of low-P_q dynamos where the Ekman-Hartmann boundary layer is essentially independent of viscosity. The opposite conclusion that boundary conditions on the velocity strongly influence Earth-like dynamo models (Kuang and Bloxham 1997) was drawn from results of simulations at much larger P_q and E, and may have been misleading. Numerical experiments using much smaller viscosities will be needed to decide this issue.

8.5 Conclusions

In our previous paper (Sakuraba and Roberts 2009), we reported on numerical experiments aimed at discovering how the boundary condition on the core surface temperature affects Earth-type dynamos. We focused on two extreme models, the USTM and the UHFM, which differed only in that boundary condition, the core surface temperature being uniform in the USTM but the heat flux from the core being uniform in the UHFM. We established that the magnetic fields these models produce are utterly different. The primary purpose of the present paper is to elucidate this surprising fact by analyzing the field regeneration mechanisms in each; a secondary aim is to provide more details about the solutions.

In order to maintain a zonal magnetic field strong enough to establish large-scale field and flow structures outside the tangent cylinder, a thermal wind is needed, driven by pole-equator temperature differences. We argued here that, no matter how tiny the thermal diffusivity, it nevertheless influences the axisymmetric part of the temperature perturbation. Even though the source term in the diffusion equation is the same for both models, the boundary condition on the USTM, which compels the surface temperature to be laterally uniform, also forces the latitudinal gradient of the temperature to be small outside the tangent cylinder. The UHFM lacks this strong constraint and can therefore drive a strong thermal wind. Once this creates a zonal magnetic field, the Lorentz force drives a radial flow near the equatorial plane. This enhances the thermal wind and accentuates the enormous differences between the UHFM and USTM solutions.

Acknowledgements The numerical simulations were performed in the Earth Simulator Center, Yokohama, Japan. A.S. was supported by a JSPS grant-in-aid for young scientists.

References

Aubert J (2005) Steady zonal flows in spherical shell dynamos. J Fluid Mech 542:53–67
Aubert J, Aurnou J, Wicht J (2008) The magnetic structure of convection-driven numerical dynamos. Geophys J Int 172:945–956
Aurnou J, Andreadis S, Zhu L, Olson P (2003) Experiments on convection in Earth's core tangent cylinder. Earth Planet Sci Lett 212:119–134
Braginsky SI, Roberts PH (2007) Anelastic and Boussinesq approximations. In: Gubbins D, Herrero-Bervera E (eds) Encyclopedia of geomagnetism and paleomagnetism. Springer, New York, NY, pp 11–19
Buffett BA (2000) Earth's core and the geodynamo. Science 288:2007–2012
Christensen UR, Aubert J (2006) Scaling properties of convection-driven dynamos in rotating spherical shells and application to planetary magnetic fields. Geophys J Int 166:97–114
Fearn DR (1979) Thermal and magnetic instabilities in a rapidly rotating sphere. Geophys Astrophys Fluid Dynam 14: 103–126
Finlay CC, Jackson A (2003) Equatorially dominated magnetic field change at the surface of Earth's core. Science 300: 2084–2086
Ishihara N, Kida S (2002) Dynamo mechanism in a rotating spherical shell: competition between magnetic field and convection vortices. J Fluid Mech 465:1–32
Kageyama A, Miyagoshi T, Sato T (2008) Formation of current coils in geodynamo simulations. Nature 454:1106–1109
Kuang W, Bloxham J (1997) An Earth-like numerical dynamo model. Nature 389:371–374
Olson P, Christensen UR, Glatzmaier GA (1999) Numerical modeling of the geodynamo: Mechanisms of field generation and equilibration. J Geophys Res 104:10383–10404
Roberts PH, Glatzmaier GA (2000) Geodynamo theory and simulations. Rev Mod Phys 72:1081–1123
Sakuraba A (2007) A jet-like structure revealed by a numerical simulation of rotating spherical-shell magnetoconvection. J Fluid Mech 573:89–104
Sakuraba A (2002) Linear magnetoconvection in rotating fluid spheres permeated by a uniform axial magnetic field. Geophys Astrophys Fluid Dynam 96:291–318
Sakuraba A, Roberts PH (2009) Generation of a strong magnetic field using uniform heat flux at the surface of the core. Nat Geosci 2:802–805
Takahashi F, Matsushima M, Honkura Y (2008) Scale variability in convection-driven MHD dynamos at low Ekman number. Phys Earth Planet Inter 167:168–178

Time-Averaged and Mean Axial Dipole Field

9

Jean-Pierre Valet and Emilio Herrero-Bervera

Abstract

Using the most recent global database of paleomagnetic directions for the past 4 Myr we have tested whether the far-sided effect of Wilson (1970, 1971) remains a stable feature of the time-averaged field. We found out that this characteristic persists for all sub-time intervals as well as for different sites distributions. The U-shaped pattern of the mean inclination anomaly (deviation from the inclination of the axial dipole) as a function of latitude is described by a small quadrupole contribution that amounts 5% of the dipole. There is no need for other terms which in any case cannot be properly described given the overall dispersion of the data. We have analyzed the evolution of the quadrupole/dipole ratio for periods characterized by different mean axial dipole strength using composite curves of relative paleointensity. We report that periods of weaker dipole field are effectively characterized by a larger mean inclination anomaly and thus by a larger quadrupole/dipole ratio. We infer that the mean value of the inclination anomaly could potentially be an indirect indicator of the mean dipole strength.

9.1 Introduction

It was originally thought that the paleomagnetic poles obtained for the past 5 Myr, i.e. roughly the period over which plate motions can be neglected, would plot around the north geographic pole (Hospers, 1954). The first compilations of paleomagnetic measurements showed that this was indeed the case (Cox and Doell, 1960, Irving 1964) and this crucial observation gave rise to the *Geocentric Axial Dipole* (GAD) hypothesis which is the underlying basis for plate reconstructions and the cornerstone of paleomagnetism. Early on Wilson (1970, 1971) noticed that the paleomagnetic poles from Europe and Asia tended to plot too far away from the observation sites along the great circle joining the site and the geographic pole. This bias was referred as the far-sided effect and is caused by paleomagnetic inclinations that are always too negative with respect to the expected inclination of an axial geocentric dipole (GAD). It was initially interpreted in terms of displacement of the dipole axis north of the equatorial plane along the Earth's rotation axis.

This problem received much attention from McElhinny and Merrill (1975) who re-analyzed data for the past 5 Myr by looking at the departure (ΔInc) of inclination (Inc) from the value expected with a

J-P. Valet (✉)
Institut de Physique du Globe de Paris, 4 Place Jussieu, 75252 Paris Cedex 05, France
e-mail: valet@ipgp.fr

E. Petrovský et al. (eds.), *The Earth's Magnetic Interior*, IAGA Special Sopron Book Series 1,
DOI 10.1007/978-94-007-0323-0_9, © Springer Science+Business Media B.V. 2011

GAD field as a function of latitude. They showed that the pattern of anomaly as a function of latitude can be modelled by a geocentric axial quadrupole and a geocentric octupole. More or less similar conclusions regarding the existence of a persistent quadrupole have been reached by models aimed at describing the harmonic content of the time-averaged field with the hope of extracting significant non zonal terms. The presence of such terms remains controversial, some authors (Schneider and Kent 1990, Gubbins and Kelly 1993, Johnson and Constable, 1995, 1997) claiming that they indeed are present and persist on the long-term while others (Quidelleur et al. 1994, McElhinny et al. 1996, Carlut and Courtillot, 1998, McElhinny 2004) conclude that terms greater than or equal to degree three are not significant, the only reliable geometry beyond the GAD being that of a persistent geocentric axial quadrupole. Among possible limiting factors could be the fact that many data have been acquired in the 1970s from poorly demagnetized rocks when demagnetization techniques were not as efficiently developed as now.

Since these early studies a large number of robust data have been obtained from various locations. In addition, dating techniques have considerably improved. Consequently, the present database may be more appropriate to discern long-term stationarity features of the time-avegared field.

9.2 Data Selection and Analysis of the Time-Averaged Field for the Past 4 Myr

The choice of the database has been discussed in all studies related to the time-averaged field. We found that lava flows are most suitable for this kind of analysis because they record the total internal geomagnetic field without any smearing or time-averaging processes as this is the case for sediments. It has been argued that such smoothing should not be a problem when dealing with very long-term variations. However we must keep in mind that all sediments are associated with a different resolution. Note that we will be considering here very small deviations so that any uncertainty regarding the fidelity of magnetization is a limitation. It is not appropriate to compare high resolution sedimentary records with ten to hundred times lower resolution records. In one case, we deal only with a unique axial dipole field while in the other case the non-dipole field variations have not been completely averaged out, which can potentially bias the results. A second limit inherent to the use of sediments is the possibility of flattening affecting inclinations. Although being limited to deviations which do not exceed a few degrees this effect can be large enough for generating a systematic bias which would then be misinterpreted in terms of field. For these reasons we have prefered to rely only on volcanic records. Cooling rates of lava flows do not exceed a few years so that their magnetization records the total geomagnetic field. The largest uncertainty may reside in dating but this should not affect the present analyses since we are essentially dealing with time-averaged field values.

Recent investigations have been conducted in the framework of the time-averaged geomagnetic field initiative (TAFI) to obtain high quality data from lava flows in the time interval 0–5 Ma. The results were summarized by Johnson et al. (2008) who focused on 17 locations, most selected to improve the geographical coverage of 0–5 Ma paleomagnetic directions at high latitudes and in the southern hempishere. These new data were supplemented with eight regional data sets based on compilations from the literature. The latitudinal structure of the time-averaged field was measured by ΔInc at each site. The results indicate that small negative inclination anomalies are present at most latitudes for the Brunhes. In contrast, the Matuyama inclination anomalies are dominantly negative in the Northern hemisphere and zero or positive in the southern hemisphere. The Brunhes inclination anomalies have been fitted by $g_2^0/g_1^0 = 2\%$ to which has been added a 1% contribution of g_3^0/g_1^0 while the Matuyama larger deviations are described by a g_2^0/g_1^0 ratio of 2% and a g_3^0/g_1^0 of 5%.

We have used directions summarized in the volcanic database initially produced by McElhinny and Lock (1995) and updated in 2004, but have restricted the analysis to the past 4 Myr because deviations by one or a few degrees in inclination could have been generated by rapid plate motions (e.g. the Pacific plate) prior to 4 Myr. We removed all directions characterized by a latitude of their geomagnetic virtual dipole (VGP) lower than 60° and we calculated a mean inclination for all sites that do not differ by more than 1° in latitude. As expected, the evolution of the inclinations as a function of site latitude (Fig. 9.1a) are in agreement with the corresponding values of the geocentric axial dipole which have been plotted in the same figure. However, little deviations from the GAD value

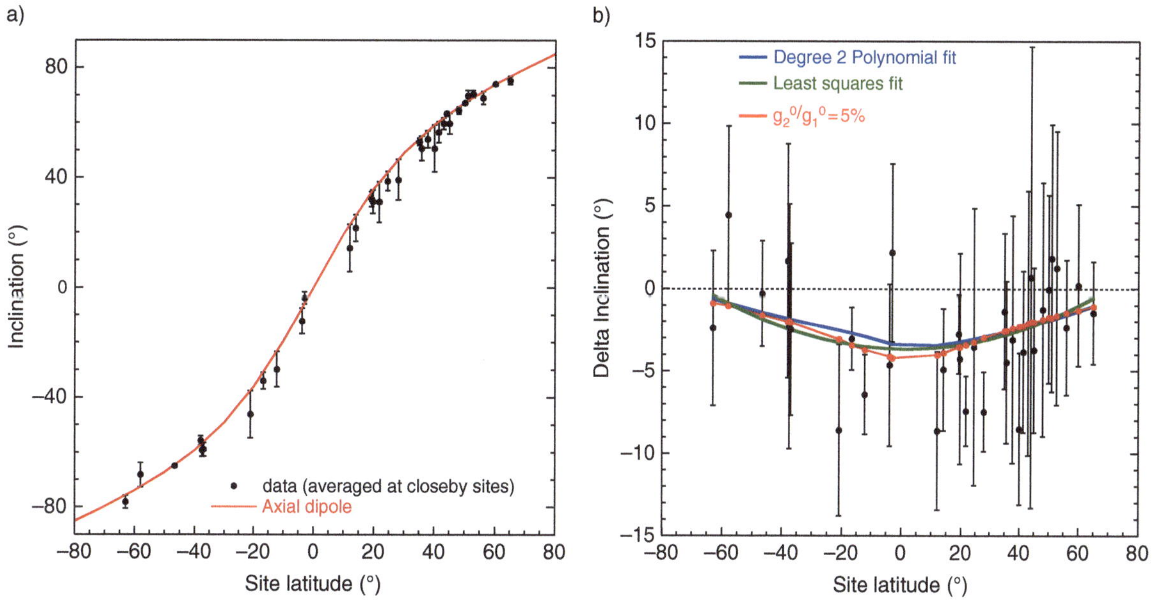

Fig. 9.1 (**a**) Inclination versus site latitude. Directions from sites distant by less than 1° of latitude have been integrated together. (**b**) deviation of the mean inclination with respect to the prediction at the site latitude versus site latitude. The fit by a degree 3 polynomial fgives similar results to the simulation obtained with a g_2^0/g_1^0 ratio of 5%

are clearly visible, particularly at mid-latitudes. These deviations have been reported in Fig. 9.1b as a function of site latitude. Similarly to the previous studies we note an increasing pattern in the amplitude of the deviation from high to low latitudes. This tendency can be simply fitted using a degree 2 polynomial which agrees also with the results obtained by least squares regression analysis. A few datapoints lie outside the fitting curve for sites between 20 and 40°N. They are heavily dominated by a unique location at the corresponding latitude (middle east at 40°N, Africa for 28°N, Hawaii at 21.7°N and the Indian ocean at 21°S) and thus may be slightly biased by local effects. The pattern of the curve remains basically unchanged if we remove these outliers. In the same figure is shown also the curve obtained by simulating an axial dipole to which has been added a small quadrupolar contribution that amounts to 5% of the dipole. This simulation fits the data as well as the polynomial or the least squares regression. Consequently we consider that it provides a first-order correct description of the far-sided effect.

We infer that the present volcanic database confirms the existence of a persistent small quadupole. These results do not strikingly differ from those derived from the TAFI project (Johnson et al. 2008) and we consider that they are robust at this stage. We have thus investigated whether this feature remains stable at all periods and if it is biased by specific locations with strong persisting anomalies. In order to investigate the effect of time we have compared the distributions of the inclination anomalies for the last million years, for the 1–2 Ma and 2–4 Ma time intervals, respectively. The plots shown in Fig. 9.2a–c display similar distributions with about the same number of datapoints within the bins characterized by an inclination anomaly from 2° to –6°. A few positive values are mostly linked to sites from the southern hemisphere. We thus conclude that the anomaly persists for periods characterized by different repartitions of normal and reversed polarities. If the inclination is lower than the dipole inclination at any site the VGP will lie farther away in latitude and thus in the hemisphere opposite to the site. A direct consequence is that the longitudinal distance between the VGP and the pole will exceed 90°. The distributions in Fig. 9.2d–f show that most VGPs lie effectively farther than 90° away in longitude from their site meridian.

As mentioned above, we cannot dismiss the possibility that strong anomalies could persist at certain locations. A mean inclination anomaly has been calculated for the past 4 million years at all sites with similar longitudes. Although different values might be

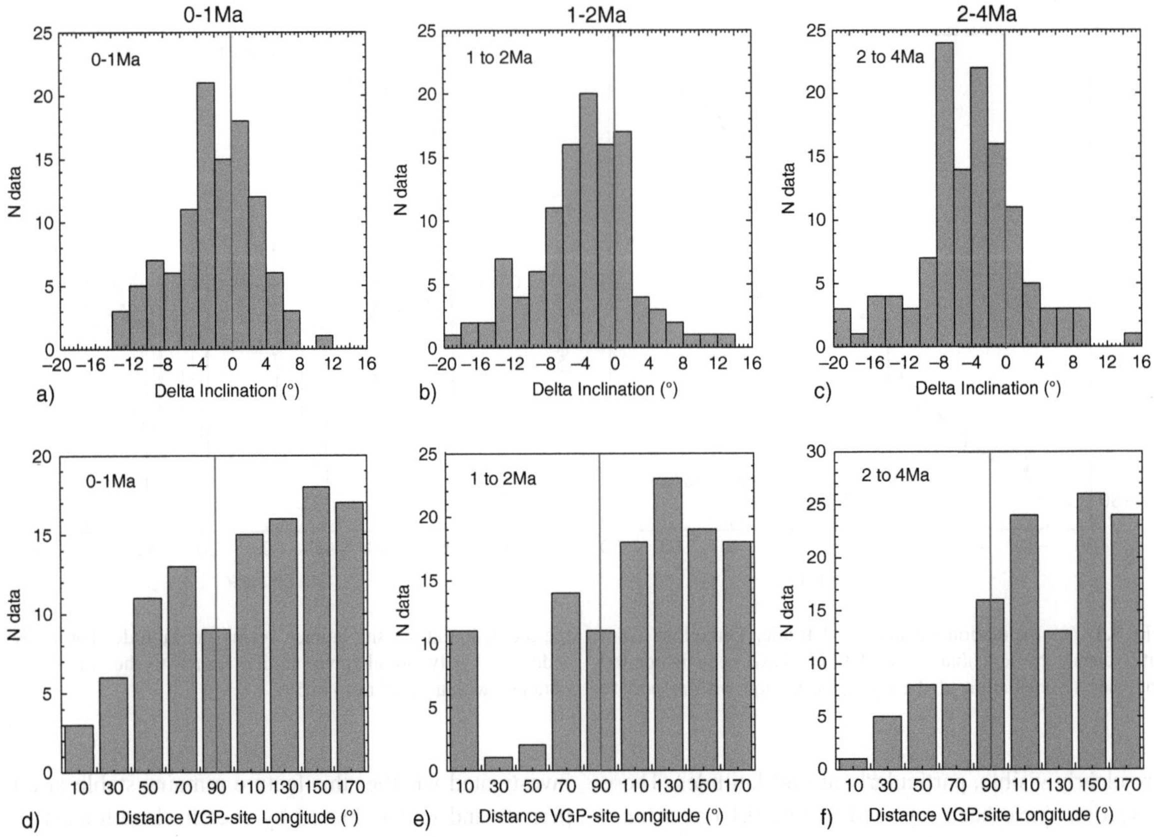

Fig. 9.2 **a**, **b**, **c**: Distribution of inclination anomaly with respect to the dipole for different time periods; (**d**, **e**, **f**) same for the distance of VGPs from their site longitude

expected for various latitudes along the same meridian and thus generate scatter, it is interesting that the deviation remains negative (Fig. 9.3a). Similarly the distance with respect to the site longitude (Fig. 9.3b) largely exceeds 90° in all but two cases for relatively high latitudes. It is interesting and more pertinent to investigate the distribution of the anomaly as a function of site latitude. In this case (Fig. 9.3c) there is a characteristic U-shaped distribution which is symmetric about the equator with a minimum anomaly for the equatorial band and zero deviation at high latitude sites. As expected a similar but opposite shape is reproduced by the distance of VGPs from their respective site longitude (Fig. 9.3d) which culminates at low latitude sites due to larger deviation in inclination and is negligible at high latitudes. These results confirm the analyses derived from the previous databases as well as the most recent study (McElhinny et al. 1996) looking at the departure of inclination with respect to the GAD. Note that the same pattern of Delta Inc with latitude has also been reported for shorter-term time-series averaged over 10^3 years (Lund and Banerjee 1985). We conclude that this small deviation from the GAD is a robust feature of the time-averaged field for the past 4 Myr and does not result from strong and local anomalies.

9.3 Time-Averaged Field and Mean Dipole Intensity

The question that comes to mind is the origin of the anomaly. Two distinct approaches have attempted to provide a first-order description of the field using spherical harmonics. The first one assumes that all non zonal terms of the magnetic potential will be averaged out over a long period leaving only the zonal terms. Similarly to the present study Merrill and McElhinny (1977), Merrill et al. (1996) have been looking at ΔInc as a function of latitude and used a least squares

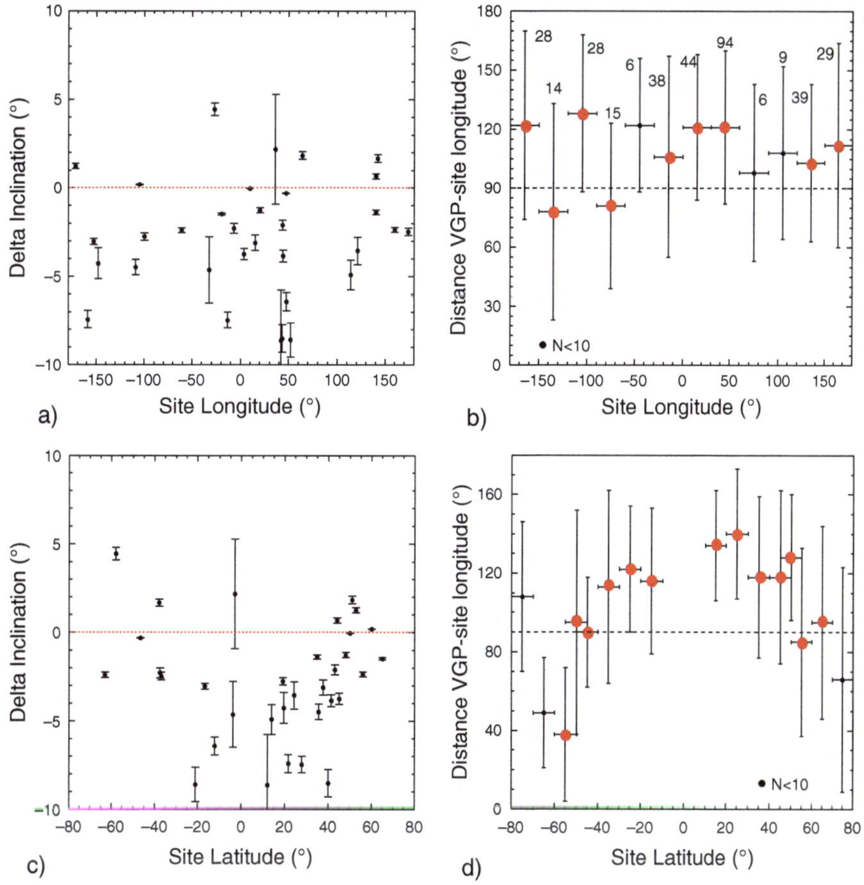

Fig. 9.3 Inclination anomaly and distance of VGPs from site longitude as a function of site longitude (**a**, **b**) and site latitude (**c**, **d**) for the 0–4 Ma period

analysis to find the best fitting g_2^0 term. Many other studies have attempted full spherical harmonic analyses of the field for the past 5 Myr. The results vary because different databases with either volcanics or sediments alone or both sediments and volcanics have been used. In many cases the authors attempted to detect some asymmetry between the normal and the reversed polarities. As a result values of the g_2^0/g_1^0 ratio found in the literature range from 2 to 5%. What is clear is that the uncertainties inherent to the data, the geographical site coverage and the time distribution of the directions do not allow to go beyond the dipolar and quadrupolar terms (Quidelleur et al. 1994, Carlut and Courtillot 1998).

One may wonder whether there is any reason to consider that the g_2^0/g_1^0 ratio has remained constant with time. In fact, the large standard deviations on the averaged Delta Inc values may result from noise but suggest also that the amplitude of the anomaly has changed with time. The detailed pattern of such variations cannot be investigated using the lava flow database due to poor time control. Schneider and Kent (1990) have approached this problem by studying equatorial deep-sea sediments i.e. at latitudes where the quadrupole contribution is larger. As for subsequent spherical harmonic analyses they found different quadrupolar contributions for reversed and normal periods. It is impossible to determine from the directional measurements alone whether the evolution of the ratio is linked to the dipole or to the quadrupole term. All studies performed so far evidently suffer from the absence of absolute paleointensity that would constrain the entire mean field vector. This aspect can now be envisaged differently. Indeed, a large number of coherent records of relative paleointensity with good time control has been acquired (Valet 2003,

Valet et al. 2005, Channell et al. 2009) over the past twenty years and integrated into global curves that depict the evolution of the axial dipole field intensity.

One of the characteristics of the Sint-2000 composite curve (Valet et al. 2005) is that the dipole field intensity becomes lower when the frequency of reversals gets higher (Valet et al. 2005). This pattern (especially visible between the Brunhes Chron and the upper Matuyama) is obviously independent from polarity and it would thus be pertinent to scrutinize how the g_2^0/g_1^0 ratio has evolved with respect to g_1^0. In order to meet this objective we compared the time-averaged directions of the volcanic database with the mean dipole intensity over successive time periods. We then calculated the mean value of Delta Inc over 500 kyr long successive time intervals. It is delicate to deal with shorter intervals because of large uncertainties in the ages of many volcanic data. Longer intervals would neither be relevant for comparison with the evolution of the dipole intensity. Similarly, we have considered the mean paleointensity value within the same periods for the Sint-2000 curve and also for the unique record of relative paleointensity (equatorial Pacific) in order to extend the comparison to the past 3 Ma. In Fig. 9.4 we show the mean value of Delta Inc as a function of the mean virtual dipole for the successive periods. The moment of the dipole has been obtained after calibrating the relative paleointensity with time-averaged values of absolute paleointensity derived from lava flows. Note that some uncertainties remain as this technique requires to average the absolute paleointensities over long enough intervals. We could restrain the calibration to the past 40 kyr which are very well documented by volcanics. However the upper part of the sedimentary records is frequently affected by coring disturbances so that the 10–20 last kyr may not be fully appropriate. It is evidently much better to deal with longer periods but we must keep in mind that the volcanic data remain relatively sparse. Despite this limit, it is striking that the pattern of the successive values is in good agreement with the evolution of the composite record of relative paleointensity (Valet et al. 2005). The most striking feature in Fig. 9.4 is that Delta Inc and g_1^0 are linearly correlated indicating that a lower dipolar field generates a larger inclination anomaly and therefore a larger g_2^0/g_1^0 ratio. Note that what is refered here as a quadrupole can be generated by other terms as well since this definition relies on symmetry considerations. For this reason we should probably consider this term

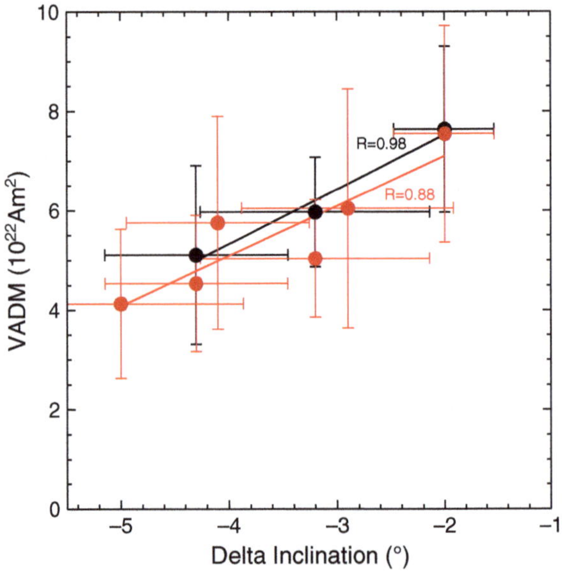

Fig. 9.4 Relation between the relative field intensity (shown here in terms of absolute dipole virtual moment after calibration) and the inclination anomaly for 500 yr long succesive time intervals

within the quadrupole family (McFadden et al. 1988). In any case, g_2^0 reverses its polarity with the dipole.

9.4 Conclusion

We conclude that, in the present state of the database, the time-averaged field for the past 4 Myr is correctly described by an axial dipole to which is superimposed a small quadrupolar component of the order of 5% of the dipole strength. We wonder whether the resolution and dispersion inherent to the data are consistent with higher terms and thus prefer to envisage only a first-order description. We report here that the evolution in amplitude of the time-averaged quadrupolar term seems to be mostly controled by the dipole strength. This observation is consistent with the TAFI data that are characterized by non-GAD contributions which are smaller during the Brunhes g_2^0 than during the Matuyama. If proven to be correct for longer time periods the inclination anomaly could then be used as an indirect indicator of the evolution of the long-term mean field intensity. A first application would obviously concern periods for which we have very poor knowledge such as for the superchrons which are crucial to our understanding of the processes generating instabilities in the outer core. However such applications are very limited by uncertainties in plate motions which are frequently of the same order of magnitude. In any case the association of this parameter with other indicators is pertinent.

Acknowledgements Financial support to J-P Valet and L. Meynadier was provided through the CNRS-INSU Interieur de la Terre Program, IPGP contribution # 3011. Financial support to E.H-B was provided by SOEST-HIGP and by the National Science Foundation grants EAR-0510061, EAR-0710571, EAR-1015329, and NSF EPSCoR Program. This is a SOEST 1145 and HIGP 1888 contribution.

References

Carlut J, Courtillot V (1998) How complex is the time-averaged field over the past 5Myr? Geophys J Int 134:527–544

Cox A, Doell RR (1960) Review of paleomagnetism. Geol Soc Am Bull 71:645–768

Channell JET, Xuan C, Hodell DA (2009), Stacking paleointensity and oxygen isotope data for the last 1.5 Myr (PISO-1500). Earth Planet Sci Lett 283:14–23

Gubbins D, Kelly P (1993) Persistent patterns in the geomagnetic field over the past 2.5 Myr. Nature 365:829–832

Hospers J (1954). Rock magnetism and polar wandering. Nature 173:1183

Irving E (1964) Paleomagnetism and its application to geological and geophysical problems. Wiley, New York, NY

Johnson CL, Constable CG (1995) The time-averaged geomagnetic field as recorded by lava flows over the past 5 Myr. Geophys J Int 122:489–519

Johnson CL, Constable CG (1997) The time-avergaed field: global and regional biases for 0–5 Ma. Geophys J Int 131:643–666

Johnson CL et al (2008) Recent investigations of the 0–5 Ma geomagnetic field recorded by lava flows. Geochem Geophys Geosyst 9:Q04032. doi:10.1029/2007GC001696

Lund SP, Banerjee SK (1985) Late quaternary paleomagnetic field secular variation from two Minnesota lakes. J Geophys Res 90:803–825

McElhinny MW (2004) The geocentric axial dipole hypothesis: a least squares perspective. In: Channell JET, Kent DV, Lowrie W (eds) Timescales of the internam geomagnetic filed. American geophysical union monograph, vol 145. pp 1–12

McElhinny MW, Lock J (1995) Four IAGA databases released in one package. Eos Trans AGU 76:266

McElhinny MW, Merrill RT (1975) Geomagnetic secular variation over the past 5 my. Rev Geophys 13:687–708

McElhinny MW, McFadden PL, Merrill RT (1996) The time averaged paleomagnetic field 0-5 Ma. J Geophys Res 101:25007–25027

McFadden PL, Merril RT, McElhinny MW (1988), Dipole/quadrupole family modelling of paleosecular variation. J Geophys Res 93:11,583–11,588

Merrill RT, McElhinny MW (1977) Anomalies in the time- averaged paleomagnetic field and their implications for the lower mantle. Rev Geophys Space Phys 15:309–323

Merrill RL, McElhinny MW, McFadden PL (1996) The magnetic field of the earth: paleomagnetism, the core and the deep mantle. Academic

Quidelleur X et al (1994), Long-term geometry of the geomagnetic field for the last 5 million years: an updated secular variation database from volcanic sequences. Geophys Res Lett 21:1639–1642

Schneider DA, Kent DV (1990) The time-averaged paleomagnetic field. Rev Geophys 28:71–96

Valet J-P (2003) Time variations in geomagnetic intensity. Rev Geophys 41:1/1004

Valet JP, Meynadier L, Guyodo Y (2005) Geomagnetic dipole strength and reversal rate over the past two million years. Nature 435:802–805

Wilson RL (1970), Permanent aspects of the Earth's non-dipole magnetic field over tertiary times. Geophys J R Astron Soc 19:417–439

Wilson RL (1971) Dipole offset – the time average paleomagnetic field over the past 25 million years. Geophys J R Astron Soc 22:491–504 instabilities in the outer core. However such applications are very limited by uncertainties in plate motions which are frequently of the same order of magnitude. In any case the association of this parameter with other indicators is pertinent.

Acknowledgments [illegible] supported by [illegible] NSF [illegible] and by the National Science Foundation grants, EAR-0510061, EAR 0711181, EAR 0115829, and NSF [illegible] Program. This is [illegible] and IPGP [illegible] contribution.

References

Carlut J, Courtillot V (1998) How complex is the time-averaged field over the past 5 Myr? Geophys J Int 134:527–544

Cox A, Doell RR (1960) Review of paleomagnetism. Geol Soc Am Bull 71:645–768

Channell JET, Xuan C, Hodell DA (2009) Stacking paleointensity and oxygen isotope data for the last 1.5 Myr (PISO-1500). Earth Planet Sci Lett 283:14–23

Gubbins D, Kelly P (1993) Persistent patterns in the geomagnetic field over the past 2.5 Myr. Nature 365:829–832

Hospers J (1954) Rock magnetism and polar wandering. Nature 173:1183

Irving E (1964) Paleomagnetism and its application to geological and geophysical problems. Wiley, New York, NY

Johnson CL, Constable CG (1995) The time-averaged geomagnetic field as recorded by lava flows over the past 5 Myr. Geophys J Int 122:489–519

Johnson CL, Constable CG (1997) The time-averaged geomagnetic field: global and regional biases for 0–5 Ma. Geophys J Int [illegible]

Johnson CL et al (2008) Recent investigations of the 0–5 Ma [illegible] Geochem Geophys Geosyst [illegible] doi:10.1029/2007GC001696

Lund SP, Banerjee SK (1985) Late quaternary paleomagnetic field secular variation from two Minnesota lakes. J Geophys Res 90:803–825

McElhinny MW (2004) The geocentric axial dipole hypothesis: a least squares perspective. In: Channell JET, Kent DV, Lowrie W (eds) Timescales of the paleomagnetic field. American Geophysical Union, Washington, DC, [illegible]

McElhinny MW, [illegible]

McElhinny MW, Merrill RT (1975) Geomagnetic secular variation over the past 5 my. Rev Geophys [illegible]

McElhinny MW, McFadden PL, Merrill RT (1996) The time averaged paleomagnetic field 0–5 Ma. J Geophys Res 101:25007–25027

McFadden PL, Merrill RT, McElhinny MW (1988) Dipole/quadrupole family modeling of paleosecular variation. J Geophys Res 93:11583–11588

Merrill RT, McElhinny MW (1977) Anomalies in the time-averaged paleomagnetic field and their implications for the lower mantle. Rev Geophys Space Phys 15:309–323

Merrill RT, McElhinny MW, McFadden PL (1996) The magnetic field of the earth: paleomagnetism, the core and the deep mantle. Academic

Quidelleur X et al (1994) Long-term geometry of the geomagnetic field for the last 5 million years: an updated secular variation database from volcanic sequences. Geophys Res Lett 21:1639–1642

Schneider DA, Kent DV (1990) The time-averaged paleomagnetic field. Rev Geophys 28:71–96

Valet J-P (2003) Time variations in geomagnetic intensity. Rev Geophys 41:1/1004

Valet JP, Meynadier L, Guyodo Y (2005) Geomagnetic dipole strength and reversal rate over the past two million years. Nature 435:802–805

Wilson RL (1970) Permanent aspects of the Earth's non-dipole [illegible] 19:417–439

Wilson RL (1971) Dipole offset – the time average palaeomagnetic field over the past 25 million years. Geophys J R Astron Soc 22:491–504

A Few Characteristic Features of the Geomagnetic Field During Reversals

10

Jean-Pierre Valet and Emilio Herrero-Bervera

Abstract

Volcanic records of reversals are mostly exempt of complications linked to their magnetization process and thus potentially tell us the most significant story about the field variations prevailing during these periods. We have found no convincing indication supporting the presence of long-term non-zonal features governing the transitional field. A few VGP paths seem to be controlled by flux patches of the present non-axial dipole field lying immediately below the sites, but the detailed reversal records are characterized by scattered VGPs that are not related to anomalies of the present non axial dipole field. Assuming that clusters of VGPs over Australia would be associated with an hypothetical time persistence of the present anomaly in this area, then the geometry of the transitional field would have to be controlled by the equatorial dipole, since it is responsible for the present Australian patch. This is difficult to reconcile with our present knowledge of the variability of the equatorial dipole as well as with the structure of most detailed VGP paths. In fact, the existence of complex directional changes with rebounds and precursors in the detailed volcanic records reflect the persistence and the amplification of secular variation following the collapse of the axial dipole. We have now learned much more about the evolution of the axial dipole from studies of relative paleointensity in sediments but also from the records of absolute paleointensity that have been obtained for a few volcanic records. The data converge to indicate asymmetrical pre- and post-reversal phases but also a systematic overshoot marking the end of the recovery phase. These features can be explained by a dynamical model assuming a coupling of the Earth's dipole with the quadrupolar mode during reversals.

10.1 Introduction

After being studied for almost fifty years, polarity reversals remain the most intriguing and fascinating feature of the geomagnetic field. Significant progress has been made to reveal important features which shed new light on the processes involved in the transitions between the two polarity states. The first reversal

J-P. Valet (✉)
Institut de Physique du Globe de Paris, 4 Place Jussieu, 75252 Paris Cedex 05, France
e-mail: valet@ipgp.fr

E. Petrovský et al. (eds.), *The Earth's Magnetic Interior*, IAGA Special Sopron Book Series 1,
DOI 10.1007/978-94-007-0323-0_10, © Springer Science+Business Media B.V. 2011

records were obtained from the remanent magnetization of basaltic lava flows that had cooled down during the transitional periods between the two polarity states. A few years later, sequences of marine sediments were commonly used, with the advantage of offering potentially continuous records of the field variations. The results indicated that most reversals are very short and are systematically associated with a huge drop in the field intensity at Earth's surface. Since the dipole represents about 90% of the total field measured at the surface it has to be responsible for the large field change, while the non-dipolar part could either be decoupled from the dipole or decrease with it. In the first case, the non-dipolar components would become dominant at some stage of the transition, whereas in the second case the hypothesis implies that the geometry of the field would always remain dipolar. If the field remains dipolar, then any vector at the Earth's surface points towards the same pole following the rotation of the unique dipole with its north pole passing to the south (or vice versa). In the opposite situation, the virtual geomagnetic poles (VGPs) are different at any site. Following earlier studies (Dagley and Lawley 1974, Hillhouse and Cox 1976, Herrero-Bervera and Theyer 1986, Herrero-Bervera and Runcorn 1997, Valet and Tauxe 1989), the accumulation of data has rapidly shown that each site records a different trajectory of the pole, which is incompatible with a dipolar field. Significant information regarding the first-order geometry of the field can certainly be derived from the pattern of the directional and intensity changes, but the exact nature of the dominant non-dipolar components remains a major challenge for reversals studies. Even in the most optimistic situation it is extremely delicate to inverse directions recorded during a specific reversal in order to determine the harmonic content of the field. This is due to the lack of several high-resolution records of the same reversal and to difficulties in correlating in detail, the field variations recorded at each site. For this reason, most studies have restrained their analysis to the respective role of zonal versus non-zonal components of the field during the transitional period and to the evolution of transition morphology across successive reversals. To our knowledge only two studies have attempted an inversion. The first one (Mazaud 1985) was concerned with the records of the Upper Olduvai and the second one dealt with the reversal that followed it (Leonhardt and Fabian 2007). We feel that in both cases, the site distribution, the resolution of the records, the absence of a full vector (most data are unit vectors) and the uncertainties in correlating the successive episodes generated large uncertainties.

10.2 Search for Axisymmetrical Components of the Transitional Field

Following the dynamo model of Parker (1969) and Levy (1972a, b) which assumed that reversals would be initiated by bursts of axisymmetrical cyclones at low or high latitudes, Hoffman (1977) proposed that the transitional field was dominated by axisymetrical components, namely an axial quadrupole or octupole. The role of axisymmetrical components has also been considered by McFadden et al. (1988) following a suggestion by Roberts and Stix (1972) who pointed out that the solutions of the dynamo could be separated between the dipole and the quadrupole families, the first one being symmetrical with respect to the equator and the second one being antisymmetrical. McFadden et al. (1991) have defended the idea that the latitudinal scatter of the VGPs with latitude during periods of normal polarity (and also at the present time) comes from the dipole family, while the quadrupole contribution is latitude-independent and thus produces all the scatter at the equator. They have constructed a reversal model that assumes that the dipole and quadrupole family members are decoupled during stable polarity intervals, excluding times of polarity transitions (McFadden et al. 1991, Merrill and McFadden 1999).

To envisage how broken symmetry can cause reversals, we refer to a very recent model describing the evolution of these two modes (Petrellis et al. 2009). The condition for the occurrence of reversals is that some instability increases the coupling between the two modes. Small fluctuations in the convective flow can push the system away from one pole to a position where it becomes attracted to the opposite pole. A reversal occurs in two phases: a slow phase during which the dynamics acts against the motion, and a fast phase during which the dynamics favors the motion. Sometimes at the end of the first phase, the system may simply return to the initial pole, completing an "excursion". However, if the system does reverse, the recovery happens much more abruptly than the decay phase and is characterized by an overshoot after reaching the opposite pole. These considerations are independently sustained by the dynamics of the magnetic field generated in the VKS laboratory experiment (Berhanu et al. 2007) involving a von Karman swirling flow of liquid sodium. This is the first and a unique experimental dynamo that has produced reversals. Although far from the earth's regime, the argument here is that a general mechanism can explain both magnetic fields, independent of the different symmetries and velocities of the two systems.

Because coupling can change with time, several interesting and testable predictions can be derived from the model. The first one is that the probability for a reversal increases when the axial dipole field is low and the quadrupole field family is high. Interestingly, the Sint-2000 composite curve of relative paleointensity (Valet et al. 2005) shows that a strong dipole is associated with a low reversal frequency (Fig. 10.1), as it has been also noticed for the Oligocene (Constable et al. 1998). Since there is no evidence that the non-dipole field decreases with the dipole (if that were the case the dipole would remain dominant), we infer that the reversal frequency is lower for a higher ratio between the axial dipole (AD) and the non-axial dipole (NAD), which is the part of the internal field remaining after the axial dipole has been removed. The model also predicts that the quadrupole family should increase in parallel with the decay phase of the dipole. However,

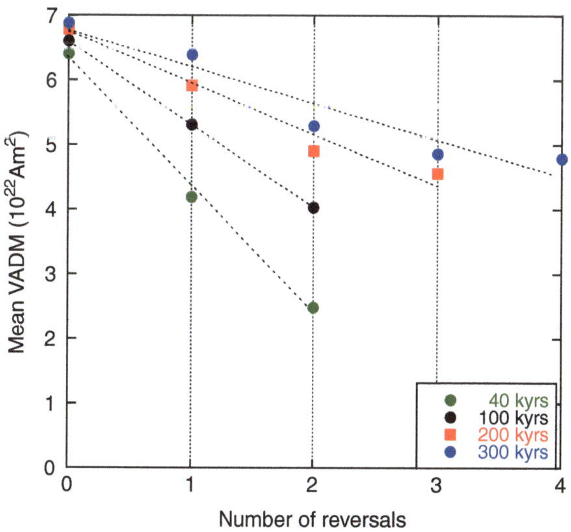

Fig. 10.1 Mean virtual dipole moment derived from the Sint-2000 composite curve as a function of the number of reversals (Valet et al. 2005)

this point is very delicate to test as long as the dipole remains dominant.

Testing the dominance of the quadrupolar mode is dependent on the condition that axisymmetrical terms become dominant with respect to other terms of the quadrupole family. As a consequence of the axisymmetry, the VGP trajectories during reversals are expected to be longitudinally confined along the site meridian (near sided) or the antipodal longitude (far sided). The first transitional VGPs appear to be rather consistent with the near or far-sided configurations (Fuller et al. 1979), but then, the accumulation of detailed sedimentary records (Clement 1991, Valet and Tauxe 1989, Herrero-Bervera and Runcorn 1997) have revealed VGP paths most frequently lying 90° away from their site meridian, thus far from an axisymmetrical field. At this stage, the question arises as to whether sediments are indeed appropriate to study reversals. If we assume that the directional changes between the two polarities did not last longer than a few thousand years, then they could not have been properly recorded by sediments that were accumulated at rate lower than 4 or 5 cm/kyr. Indeed, this resolution does not allow us to obtain records of rapidly-varying (a few hundred years at most) non-dipolar components without smoothing the signal, although such components certainly play a major role during the transitions. In addition, the magnetization of sediments is a progressive process which can also smear out the signal, ultimately change the directions and artificially introduce non-zonal components (Valet et al. 1992, Langereis et al. 1992, Quidelleur and Valet 1994, Barton and McFadden 1996). Magnetization of lava is exempt from such problems and provides almost instantaneous records which are thus sensitive to the non-dipolar components, provided that the stratigraphy is clearly established, which requires large sequences of overlying flows. In Fig. 10.2 we show the histogram of the distance of the individual VGPs from their site longitude for the seven most-detailed volcanic reversal records (Chauvin et al. 1990, Mankinen et al. 1985, Riisager and Abrahamsen 2000, Herrero-Bervera et al. 1999, Herrero-Bervera and Coe 1999, Herrero-Bervera and Valet 1999, 2005). These records have been selected because they incorporate a large number of transitional directions that were determined after stepwise demagnetization, as well as records of the periods preceding and following the reversal. This last point is crucial to assess the existence of the reversal and to test the presence and the antipodality of

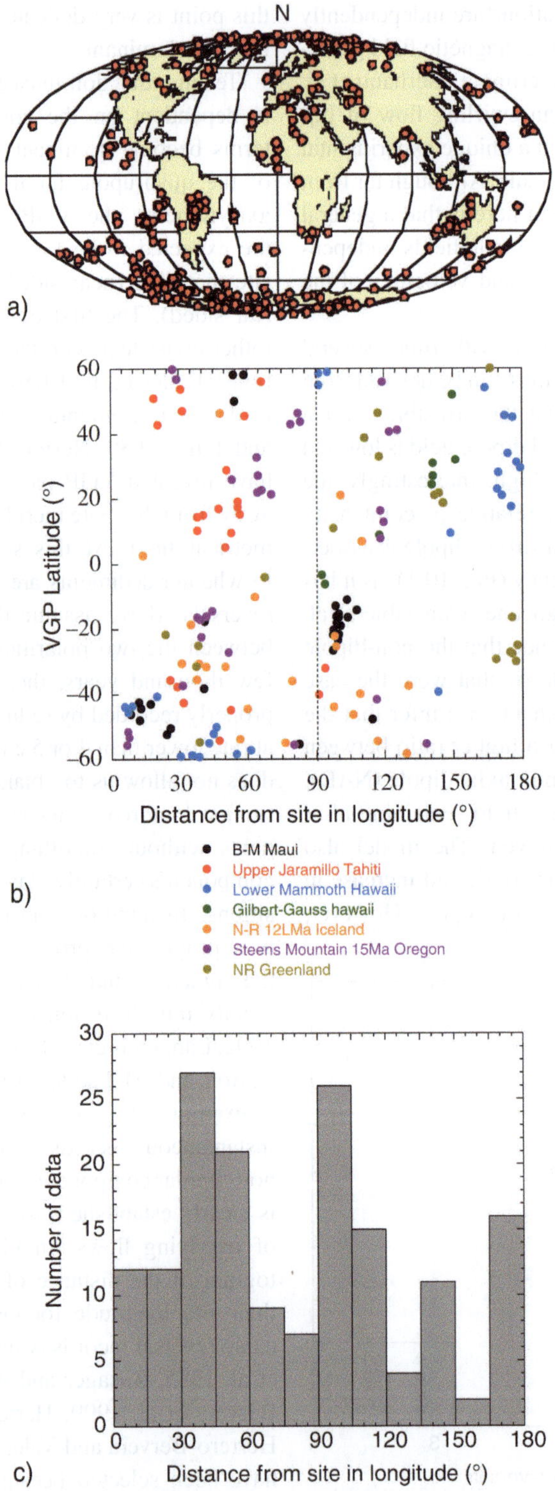

Fig. 10.2 (**a**) VGP positions of the seven most detailed volcanic records of reversals obtained so far. (**b**) Latitude of VGPs as a function of their distance in longitude from their respective site. (**c**) Distribution of the distance of VGPs with respect to their site longitude for the same volcanic records as in (**a**)

the normal and reverse directions. The distribution in Fig. 10.2 is not perfectly uniform, but there are no significant peaks over the site meridian or its antipode. Rather, they appear at about 45°, thus borderline in terms of field symmetry. A similar picture can be reported from a compilation of transitional directions (Love 1998) obtained from a very large number of volcanic units which were either obtained from isolated sites or within stratigraphic sequences. However, in this case, the authors mention a statistical preference of the VGP positions to occur in the longitudinal band over the Americas and its antipodal band over eastern Asia, which leads us to consider the importance of non-zonal components.

10.3 Does the Transitonal Field Reflect Long-Term Non-zonal Components?

An alternative model to a "flooding" process is one in which long-term persistent features of the field play a significant role in the geometry of the transition. Apparent motionless patches of vertical flux have been mentioned for the past 400 years. It is still debated as to whether such zones persist over a much longer period and thus would be indicative of long-term control over shallow core fluid by the lowermost mantle. This question is crucial because such flux concentrations would strongly influence the pattern of the directional changes during reversals. Thus, there are two issues. The first one is to determine whether stationary flux lobes are indeed persistent over several millions of years, and particularly over the past 5 Myr, while the second one concerns the existence of recurring features in the paleomagnetic records of the reversals. We will be dealing first, with the characteristics of reversal records.

10.3.1 Is There Evidence for Recurring Features Dominating the Field During Reversals?

The first significant step in favor of recurring field orientations during reversals was derived from a selection of sedimentary reversal records (Laj et al. 1991) characterized by VGP paths constrained either within two longitudinal bands over the Americas and eastern Asia. The coincidence of these two bands with the cold circum-Pacific regions in the lower mantle outlined by seismic tomography, was interpreted as evidence that the lower mantle plays a role in the dynamo processes prevailing during the transition. This coincidence remains controversial in the case of sedimentary records (Valet et al. 1992, McFadden et al. 1993), particularly if one considers the possible limits imposed by their magnetization processes (Rochette 1992, Langereis et al. 1992, Quidelleur and Valet 1994, Quidelleur et al. 1995, Barton and McFadden 1996). It is thus reasonable to turn towards the volcanic records of reversals. Prévot and Camps (1993) reported a uniform distribution of the volcanic VGPs at all longitudes, whereas Love (1998) found a statistical preference of the volcanic VGPs within the same bands over Americas and eastern Asia.

Following a similar approach, but relying solely on a selection of volcanic records from sites in the vicinity of Australia and Americas, Hoffman (1991, 1992, 1996) noticed a bimodal distribution of the VGPs above southern Australia and in the southwest Atlantic which corresponded to two locations of the present non-dipolar field at Earth's surface with the largest vertical field component. This relationship implies first, that the clusters of VGPs testify to strong non-axial field structures, and second, that these flux concentrations would persist over a long enough a period to be observed across successive reversals. The first hypothesis can be tested with reference to the present field. The north VGPs derived from the present non-axial dipole field (the part of the internal geomagnetic field remaining after the axial dipole has been removed) are dominant along the north American coast, and consequently the south VGPs are located along the eastern Asian longitudes (Valet and Plenier 2008). Interestingly, the VGPs located over Australia (thus in coincidence with the present flux patch) are due to the Pacific sites, while the south Atlantic anomaly is indicated by VGPs from sites lying immediately above it. However, these features disappear after removing the equatorial dipole from the residual non-dipolar field, indicating that they are constrained by the equatorial dipole.

Note that similar conclusions have been reached (Kutzner and Christensen 2002) from simulations using a 3D-convection-driven-numerical dynamo model. Assuming an imposed heat flux pattern derived from seismic tomography, the authors show that the maxima of the mean VGP density are located over the American and the east Asian longitudes. The low-latitude regions of high heat flow generate intense magnetic flux bundles which contribute to the equatorial dipole and are responsible for the accumulation of VGPs. Unless the configuration of the present field is purely accidental, we infer that these characteristics as well as the results of numerical simulations suggest a direct link between the tomographic features and the equatorial dipole that is responsible for the concentrations of VGPs. If the concentrations of VGPs in paleomagnetic records reflect a transitional state of the field, we must thus consider that they are primarily related to the equatorial dipole. Since the equatorial dipole appears to be constrained by tomographic features that are stable over several millions of years, it is thus expected to see the tilt of the equatorial dipole against the rotation axis along the preferred longitudes also during periods of stable polarity when the axial dipole dominates. Thus, two questions must be addressed. The first one evidently deals with the existence of preferred locations of the VGPs during reversals. The second one is related to a persistent orientation of the equatorial dipole during stable polarity intervals.

In a recent paper Hoffman and Singer (2008) mentioned that VGP clustering during reversals is heavily constrained by the position of the standing strong features of radial flux. They focused on volcanic records of reversals and excursions from Tahiti and from Eifel province in Germany for the past 780 Myr, that indicate that the VGPs are found west of Australia and across Eurasia, respectively. According to the authors, these clusters are constrained by persistent concentrations of the NAD flux lying in the vicinity of the sites, and the VGPs of the present NAD field are effectively in good coincidence with the paleomagnetic ones. Assuming that this model is correct, similar observations should be validated for other sites as well. Unfortunately, very few detailed high-resolution volcanic records of reversals have been obtained so far. However, we can select a few locations that can be adequately used for testing the model. Interestingly, the mid-Pacific islands can be seen as the best documented area with reversal records, given the large number of long volcanic sequences present there. Two detailed VGP paths document the last reversal, one recorded from Tahiti and mentioned by Hoffman and Singer (2008), and another one recorded from Hawaii. Both exhibit very different VGP paths, and we note that there are no VGP positions in the vicinity of Australia for the Hawaiian record. Despite being 40° away in latitude, both sites are roughly along the same longitude, so that the Australian patch should have a similar influence in both cases (see below). Other data from the Hawaiian area are not compatible with the suggestion of persistent strong flux patches driving the geometry of the transitional field. Indeed, persistent features should produce similar VGP characteristics for the three successive reversals (Matuyama-Brunhes, 0.78 Ma; Lower Mammoth, 3.33 Ma and Gilbert-Gauss, 3.58 Ma) recorded at Hawaii, but they are different (Fig. 10.3a). If we consider the European sites, the model predicts VGPs spreading over Europe for the Laschamp event (40 kyr old), which has been recorded in the center of France. It is striking that the VGP positions obtained so far, (Fig. 10.3b) are found almost everywhere except over Europe.

In fact, in accordance with what has been mentioned above (Valet and Plenier 2008), the apparent preference of VGPs for Australia appears to be controlled by the orientation of the equatorial dipole. This is explicit in Fig. 10.3c, which shows that the southern VGPs derived from the present NAD field for the Pacific sites are effectively constrained over western Australia in accordance with the finding of Hoffman and Singer (2008). However the VGPs from the same sites spread out over the southern hemisphere after removing the equatorial dipole. Unless the orientation of the equatorial dipole remains stable, which is in full contradiction with our present knowledge of secular variation, it seems difficult to associate the present flux concentration over Australia with a long-lived feature of the NAD.

10.3.2 The Question of Persistent Non-zonal Features in the Time-Averaged Field

Important information about the historical field has been gained from time-dependent models of the geomagnetic field. Stationary flux lobes at high northern and southern latitudes similar to the present ones are well identified in the geomagnetic field models averaged over the past 400 years (Jackson et al. 2000). If we extend the time interval to the past 7 kyr, we then refer to the archeomagnetic database (Genevey et al.

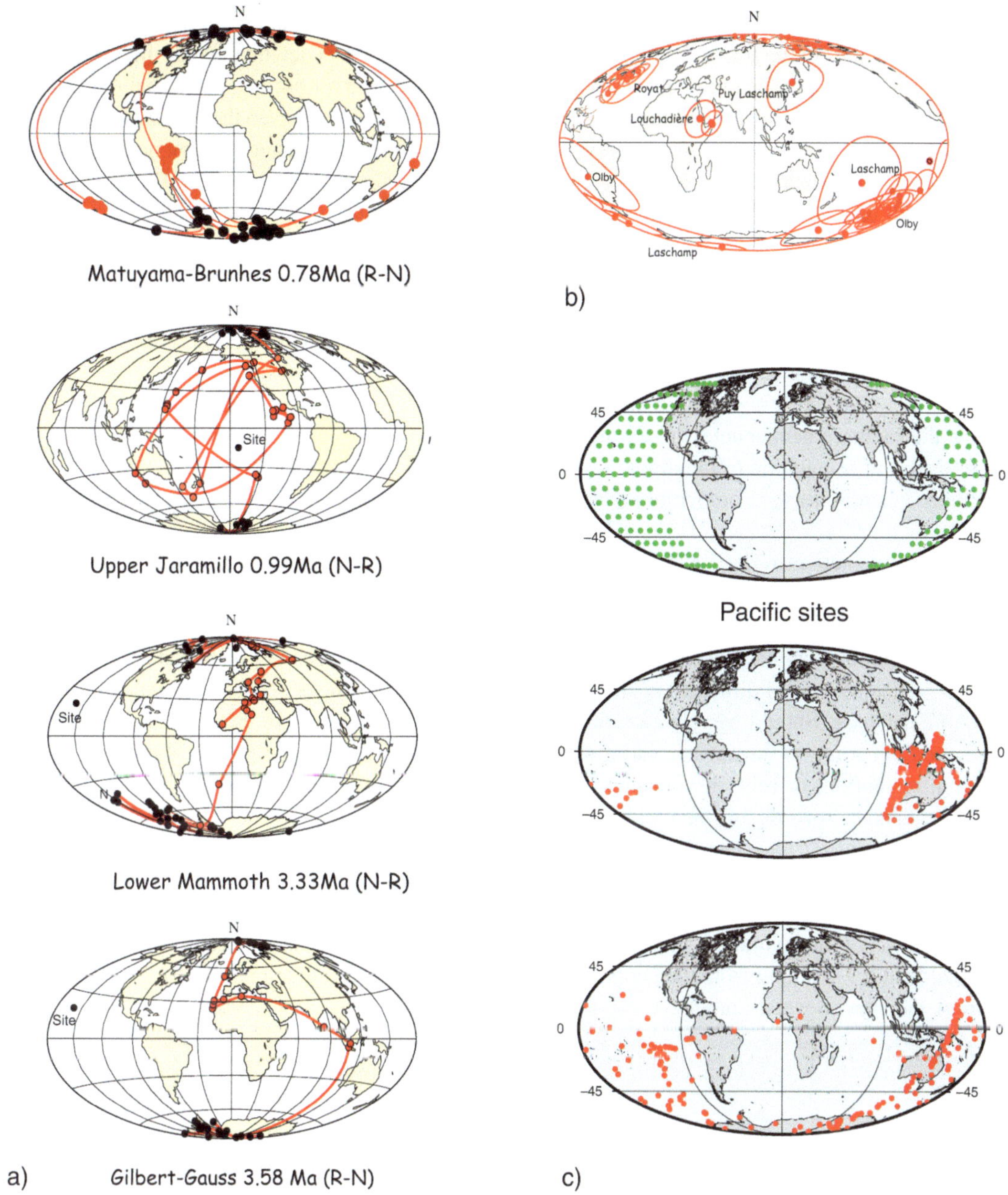

Fig. 10.3 (**a**) VGP paths of the four most detailed reversal records from central Pacific. The Matuyama-Brunhes, Lower Mammoth and Gilbert-Gauss reversals have been recorded from Hawaii while the Upper Jaramillo is from Tahiti. (**b**) VGP positions for the Laschamp event recorded in the Massif Central (France). (**c**) Virtual geomagnetic poles (medium planisphere) derived from the present non-axial dipole field for the Pacific sites shown in the upper plot. The lower plot shows the positions of the VGPs after removing the equatorial dipole for the same sites

2008) which incorporates more than 3000 data sets, but the most reliable period does not exceed the past 2 kyr. The successive models produced for the past 3 and 7 kyrs (Korte and Constable 2005) indicate small non-axial dipolar field contributions, but they fail to show the persistence of the same flux lobes as for the historical field. A restriction could be that the site distribution is far from being uniform, even for the past 2 kyrs. It may thus be premature to go beyond degree 2 in the description of the evolution of the harmonic content of the field (Valet et al. 2008), and even at this level, the absence of sites in the southern hemisphere is a limiting factor. Despite poor geographical coverage, the field variations are well constrained for the past 2 kyrs at a few locations with little age uncertainty. The best records from sites distributed around the globe have been used to calculate a dipole over successive time intervals (Valet et al. 2008). The stereoplot and the histogram in Fig. 10.4 show that the north pole is not preferably found within the two preferred bands mentioned for reversals, but instead shows large concentrations within the longitudinal zones centered at about 0° and 180°. Thus, keeping in mind the absence of data from the southern hemisphere, there is no indication of a persistent orientation of the dipole towards the longitudes inherent to the tomographic features described above.

Because paleomagnetism relies on the hypothesis of a time-averaged geocentric axial dipole, successive databases gathering paleomagnetic directions younger than the past 5 million years have been analyzed with controversial results. A typical difficulty is inherent in the choice of a database. Should we rely on sediments which averaged out some of the secular variation but with different resolution, or rely on volcanics which provide instantaneous records of the field but frequently are poorly dated? One can wonder whether it is pertinent to mix up records with resolutions that vary as much as a few weeks in the case of lavas, and several thousand years for sediments, without generating non-zonal features resulting from the geographical repartition of the sites. Several authors (Gubbins and Kelly 1993, Johnson and Constable 1995, 1997, Kelly and Gubbins 1997) defend the idea that the present flux patches are clearly observed and remain visible at least, for the last 1 Myr and even the past 5 Myr. Others argue for the lack of significance of any non-zonal variation due to poor site distribution (McElhinny and Merrill 1975, McElhinny et al. 1996, Quidelleur and Courtillot 1996, Carlut and Courtillot 1998), and sometimes poor data quality, and place very little, or no confidence at all in terms other than the axial dipole and quadrupole. This view is consistent with analyses derived from the archeomagnetic database.

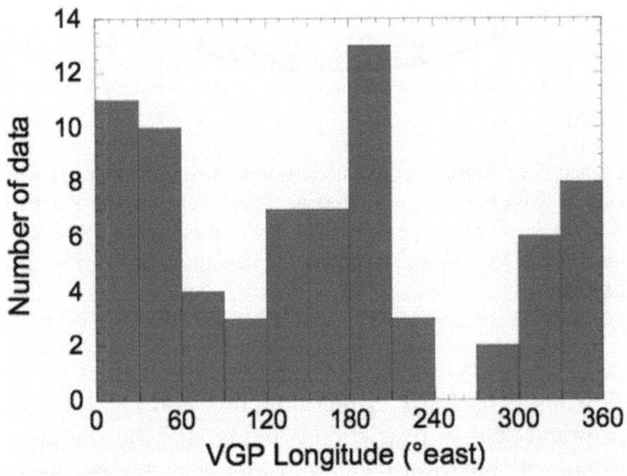

Fig. 10.4 Histogram of the VGP longitudes derived from a dipole model (Valet et al. 2008) for the past 2 kyr

10.4 The Importance of the Time-Varying Dipole

The records of relative paleointensity have revealed that the evolution of the dipole is governed by very large changes in amplitude which are punctuated by periods of low intensity associated with excursions or reversals. The longest records of relative paleointensity and the composite curve, Sint-2000, which depicts the field evolution over the past 2 Myr, show that reversals are preceded by a 60–80 kyr long dipole decay while its recovery only takes a few thousand years. This asymmetrical pattern has been controversial but we note that a similar evolution is shown by a record of ^{10}Be changes (Carcaillet et al. 2003), and in the most detailed volcanic records of reversals (Fig. 10.5) which incorporate determinations of absolute paleointensity (Bogue and Paul 1993, Valet et al. 1999, Herrero-Bervera and Valet 2005, Prévot et al. 1985, Riisager and Abrahamsen 2000). We can reasonably consider that the long dipole decay is associated with field diffusion while its recovery is driven by advection. It is also noticeable that a similar asymmetrical evolution of the dipole intensity is observed in the field evolution depicted by the VKS experimental dynamo.

Before going further on these aspects it is important to remind one that we are dealing with the evolution of a vector and therefore we may not disregard the relation with the directional changes. For the past 5 Myr the departure (delta-Inc) of inclination (Inc) from the value expected with a GAD field as a function of latitude is best modeled by a geocentric axial quadrupole. The existence of this small but persistent geocentric axial quadrupole contribution has been

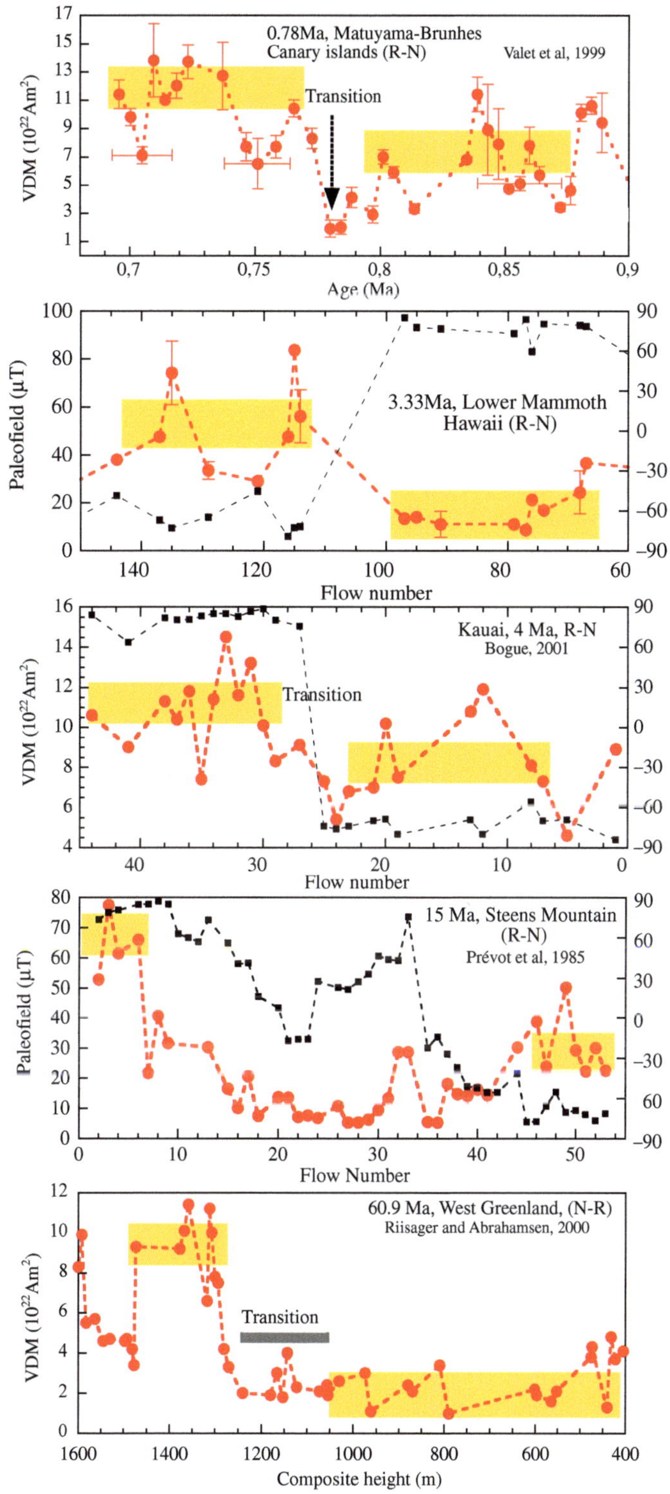

Fig. 10.5 Evolution of the field intensity across the 5 detailed volcanic records of reversals that incorporate absolute paleointensity studies. The *yellow boxes* indicate the mean field value prior and after the transitions. Note the asymmetry between the two phases in each case

confirmed by all models aimed at describing the harmonic content of the time-averaged field. Some believe that this axial quadrupole is likely to be overemphasized due to poor geographical coverage, but this is a stable feature. Interestingly, the amplitude of the long-term quadrupole is modulated by the strength of the dipole (Valet, Herrero-Bervera and Meynadier 2010, Chapter 9 this issue). A strong dipole increases the stability of the field, and one thus expects less or no excursions and reversals. The paleointensity data effectively show a correlation between dipole intensity and reversal frequency (Constable et al. 1998, Valet et al. 2005). It is logical to expect that periods of low dipole field are characterized by a larger contribution of the non-dipole components so that the time-averaged field is also much dependent on the strength of the dipole. For this reason it is not surprising to see a larger contribution of the time-averaged quadrupole during periods of weaker dipole, which are also periods of higher reversal frequency.

Finally, another significant characteristic in the evolution of the field intensity across reversals has not been mentioned until recently (Petrellis et al. 2009). This is the existence of a remarkable overshoot which concludes the recovery phase after the transition. Again, this feature appears clearly in sedimentary and in volcanic records of paleointensity (Fig. 10.6) as well as in the VKS experiment. We infer that it must be seen as inherent to the reversal process, and has been interpreted within the framework of the model described above (Petrellis et al. 2009).

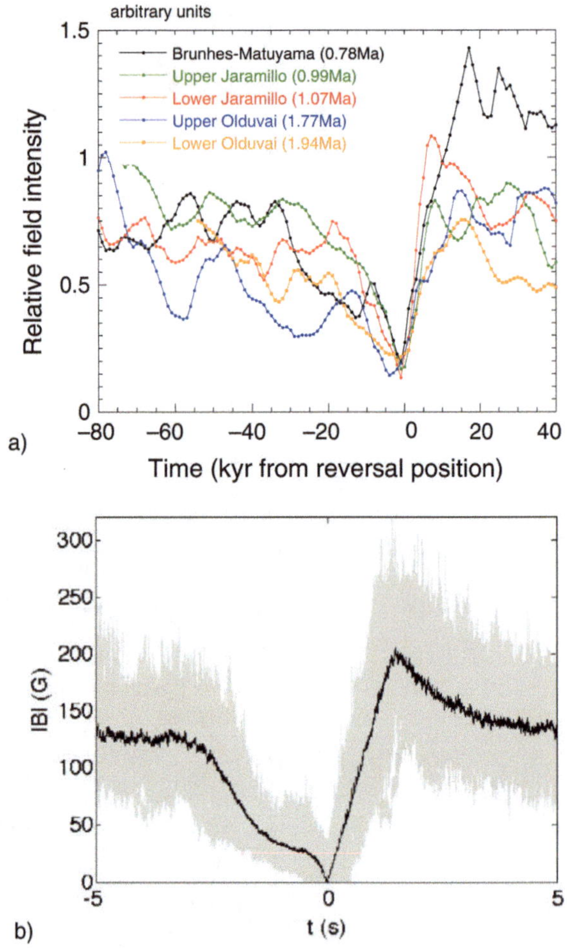

Fig. 10.6 (**a**) Relative field intensity across 5 successive reversals showing the overshoot that culminates at the end of the field recovery following the transitions. (**b**) Similarly an ovsershoot is also present in the reversals measured in the VKS experiment

10.5 Secular Variation, Precursors and Rebounds

The persistence of secular variation during reversals seems to be evident when considering the sedimentary records obtained with a very high resolution (Valet et al. 1992, Channell and Lehman, 1997, Herrero-Bervera et al. 1994). Large loopings of VGPs with time constants similar to the secular variation that prevails during periods of full polarity are common to all detailed records and likely reflect an amplification of the secular variation of the NAD in the presence of a very weak or zero dipole.

We can gain important information from the most detailed volcanic records by plotting the angular changes as a function of the successive lava flows in the sequence. This allows us to analyze the dynamical pattern of each transition without reference to the VGPs. It is striking that none of the reversals shown in Fig. 10.7 has a simple structure. In all cases the transit between north and south is either preceded or followed by a longer phase. If this phase happens prior to the transition, the vector attempts to reverse, and then comes back to the initial polarity then defining what is usually referred as a precursor. If it happens after the transition the vector attempts to return to the initial polarity but cannot reach it, it then comes back to the new one, thus completing a rebound. We cannot disregard the hypothesis that the vector may have reached the opposite polarity but the timing of the eruptions may miss this phase, especially if it was very

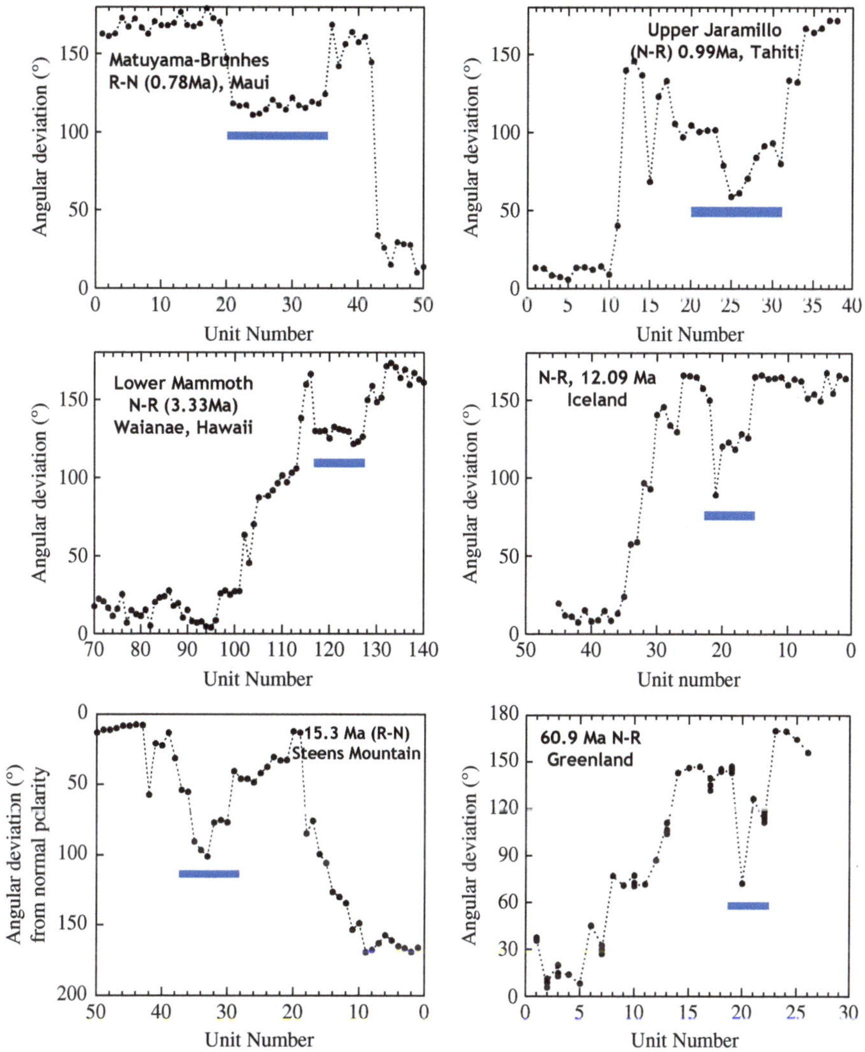

Fig. 10.7 Angular deviation of the successive directions recorded as a function of the unit number in stratigraphy for six detailed reversal records. The blue lines underscore the positions of "precursors" or "rebounds" which are part of the reversal process

short. This possibility is supported by the fact that the amplitude of the variations appears to be different in all records, suggesting partial recordings of the field changes. These episodes resemble episodes of "normal" secular variation but with a much larger amplitude caused by the weakness of the dipole. For this reason, we consider that these phases, which resemble excursions, must be considered in their entirety as being part of the reversal process. We infer that the secular variation with characteristics similar to those of the present field, indeed persist during the transitions. This observation must be accounted for by all models and be consistent with the concept of a decaying and vanishing dipole component leaving a complex non-dipolar field (Courtillot et al. 1992), which, in the present state of knowledge, appears to have a similar signature as the present one.

Conclusion

Most controversies about specific features of the reversing field are linked to the choice of the database. Despite a large number of sedimentary records having been published, it is not clear as to whether sediments can provide us with suitable information regarding the transitional field. In addition to other factors, they only partially record the field, and are thus mostly sensitive to the long-term dipolar component, which does not control the transitional period anymore and is thus not relevant. If we restrict the data to the volcanic records, we are thus forced to deal with a very limited database unless we use directions that are not integrated within the long sequences of overlying flows and are thus not suitable for proper consideration. Interesting features emerge from this volcanic database. The first is that there is no convincing indication that non-zonal features dominate during the transitions. Neither is there any striking evidence for recurring patterns between successive reversals despite sparse indications that do not seem to be valid over a wider range. However, all records display similar complex directional changes with rebounds and precursors which reflect the persistence and even the amplification of the secular variation in the presence of a very weak dipole. Turning towards the modulus of the vector, there is also a strong indication for asymmetrical pre- and post-reversal phases as well as a systematic overshoot marking the end of the recovery phase. Note that these features can be explained by a dynamical model assuming a coupling of the Earth's dipole with the quadrupolar mode, although this could work as well with an octupolar axisymmetrical mode. So far, we see that there is no obvious indication for dominant axisymmetrical components, but they may be partly masked by low-degree non-zonal terms.

Acknowledgements Financial support to J-P Valet and L. Meynadier was provided through the CNRS-INSU Interieur de la Terre Program, IPGP contribution # 3010. Financial support to E.H-B was provided by SOEST-HIGP and by the National Science Foundation grants EAR-0510061, EAR-0710571, EAR-1015329, and NSF EPSCoR Program. This is a SOEST 8145 and HIGP 1888 contribution.

References

Barton CE, McFadden PL (1996) Inclination shallowing and preferred transitional VGP paths. Earth Planet Sci Lett 140:147–157

Berhanu M, Monchaux R, Fauve S, Mordant N, Pétrélis F, Chiffaudel A, Daviaud F, Dubrulle B, Marié L, Ravelet F, Bourgoin M, Odier P, Pinton JF, Volk R (2007) Magnetic field reversals in an experimental turbulent dynamo. Europhys Lett 77:59001

Bogue SW, Paul HA (1993) Distinctive field behaviour following geomagnetic reversals. Geophys Res Lett 20:2399–2402

Carcaillet JT, Thouveny N, Bourlès DL (2003) Geomagnetic moment instability between 0.6 and 1.3 Ma from cosmonuclide evidence. Geophys Res Lett 30:1792–1795

Carlut J, Courtillot V (1998) How complex is the time-averaged geomagnetic field over the last 5 million years? Geophys J Int 134:527–544

Channell JET, Lehman B (1997) The last two geomagnetic polarity reversals recorded in high deposition-rate sediments drifts. Nature 389:712–715

Chauvin A, Roperch P, Duncan RA (1990) Records of geomagnetic reversals from volcanic islands of French Polynesia. J Geophys Res 95:2727–2752

Clement BM (1991) Geographical distribution of transitional VGPs: evidence for non zonal equatorial symmetry during the Matuyama-Brunhes geomagnetic reversal. Earth Planet Sci Lett 104:48–58

Constable CG, Tauxe L, Parker RL (1998) Analysis of 11 Myr of geomagnetic intensity variation. J Geophys Res 103:17735–17748

Courtillot V, Valet J-P, Hulot G, Le Mouël J-L (1992) The earth's magnetic field: which geometry? Eos Trans AGU 73:32337

Dagley P, Lawley E (1974) Paleomagnetic evidence for the transitional behaviour of the geomagnetic field. Geophys J R Astron Soc 36:577–598

Fuller MD, Williams I, Hoffman KA (1979) Paleomagnetic records of geomagnetic field reversals and the morphology of the transitional fields. Rev Geophys 17:179–203

Genevey A, Gallet Y, Constable CG, Korte M, Hulot G (2008) Archeoint: an upgraded compilation of geomagnetic field intensity data for the past ten millenia and its application to the recovery of the past dipole moment. Geochem Geophys Geosyst 9:Q04038. doi:10.1029/2007GC001881

Gubbins D, Kelly P (1993) Persistent patterns in the geomagnetic field over the past 2.5 Myr. Nature 365:829–832

Herrero-Bervera E, Coe RS (1999). Transitional field behavior during the Gilbert-Gauss and Lower Mammoth reversals recorded in lavas from the Wai'anae Volcano, O'ahu, Hawaii. J Geophys Res 104:29157–29173

Herrero-Bervera E, Runcorn SK (1997) Transition fields during geomagnetic reversals and their geodynamic significance. Philos Trans R Soc London Ser A 453:1–30

Herrero-Bervera E, Theyer F (1986) Non-axisymmetric behaviour of Olduvai and Jaramillo polarity transitions recorded in north-central Pacific deep-sea sediments. Nature 322:159–162

Herrero-Bervera E, Valet J-P (1999) Paleosecular variation during sequential geomagnetic reversals from Hawaii. Earth Planet Sci Lett 171:139–148

Herrero-Bervera E, Valet J-P (2005) Absolute paleointensity from the Waianae volcanics (Oahu, Hawaii) between the Gilbert-Gauss and the upper Mammoth reversals. Earth Planet Sci Lett 234:279–296

Herrero-Bervera E, Helsley CE, Sarna-Wojcicki AM, Lajoie KR, Meyer CE, McWilliams MO, Negrini RM, Turrin BD, Donelly-Nolan JM, Liddicoat JC (1994) Age and correlation of a paleomagnetic episode in the western United States by $^{40}Ar/^{39}AR$ dating and tephrochronology: the Jamaica, Blake, or a new polarity episode. J Geophys Res 99:24, 091–24,103

Herrero-Bervera E, Walker GPL, Harrison CGA, Guerrero-Garcia JC, Kristjansson L (1999) Detailed paleomagnetic study of two polarity transitions recorded in Eastern Iceland. Phys Earth Planetary Inter 147:171–182

Hillhouse J, Cox A (1976) Brunhes-Matuyama polarity transition. Earth Planet Sci Lett 29:51–64

Hoffman KA (1977) Polarity transition records and the geomagnetic dynamo. Science 196:1329–1332

Hoffman KA (1991) Long-Lived transitional states of the geomagnetic field and the two dynamo families. Nature 354:273–277

Hoffman KA (1992) Dipolar reversal states of the geomagnetic field and core-mantle dynamics. Nature 359:789–794

Hoffman KA (1996) Transitional paleomagnetic field behavior: preferred paths or patches? Surv Geophys 17:207–211

Hoffman KA, Singer BS (2008) Magnetic separation in Earth's outer core. Science 321:1800

Jackson A, Jonkers A, Walker M (2000) Four centuries of geomagnetic secular variation from historical records. Philos Trans R Soc London 358:957–990. doi:10.1098/ rsta. 2000.0569

Johnson CL, Constable CG (1995) The time-averaged geomagnetic field as recorded by lava flows over the past 5 Myr. Geophys J Int 122:489–519

Johnson CL, Constable CG (1997) The time-averaged field: Global and regional biases for 0–5 Ma. Geophys J Int 131:643–666

Kelly P, Gubbins D (1997) The geomagnetic field over the past 5 million years. Geophys J Int 128:315–330

Korte M, Constable CG (2005) Continuous geomagnetic models for the past 7 millennia: 2. CALS7K. Geochem Geophys Geosyst 6(2):Q02H16. doi:10.1029/2004GC000801

Kutzner C, Christensen UR (2002) From stable dipolar towards reversing numerical dynamos. Phys Earth Planetary Inter 131:29–45

Laj C, Mazaud A, Weeks R, Fuller M, Herrero-Bervera E (1991) Geomagnetic reversal paths. Nature 351:447

Langereis CG, van Hoof AAM, Rochette P (1992) Longitudinal confinement of geomagnetic reversal paths. Sedimentary artefact or true field behavior? Nature 358:226–230

Leonhardt R, Fabian K (2007) Paleomagnetic reconstruction of the global geomagnetic field evolution during the Matuyama-Brunhes transition: iterative bayesian inversion and independent verification. Earth Planet Sci Lett 253:172–195

Levy EH (1972a) On the state of the geomagnetic field and its reversals. Astrophys J 175:573–581

Levy EH (1972b) Kinematic reversal schemes for the geomagnetic dipole. Astrophys J 172:635–642

Love JJ (1998) Paleomagnetic volcanic data and geometric regularity of reversals and excursions. J Geophys Res 103(B6):12,435–12,452

Mankinen EA, Prévot M, Grommé CS, Coe RS (1985) The Steens Mountain (Oregon) geomagnetic polarity transition 1, Directional history, duration of episodes and rock magnetism. J Geophys Res 90:10,393–10,416

Mazaud A (1985) An attempt at reconstructing the geomagnetic field at the core-mantle boundary during the Upper Olduvai polarity transition (1.66 Myr). Phys Earth Planetary Inter 90:211–219

McElhinny MW, McFadden PL, Merrill RT (1996) The time-averaged paleomagnetic field 0–5 Ma. J Geophys Res 101:25,007–25,027

McElhinny MW, Merrill RT (1975) Geomagnetic secular variation over the past 5 my. Rev Geophys 13:687–708

McFadden PL, Barton CE, Merrill RT (1993) Do virtual geomagnetic poles follow preferred paths during geomagnetic reversals? Nature 361:342–344

McFadden PL, Merrill RT, McElhinny MW (1988) Dipole/quadrupole family modelling of paleosecular variation. J Geophys Res 93:11,583–11,588

McFadden PL, Merrill RT, McElhinny MW, Sunhee L (1991) Reversals of the Earth's magnetic field and temporal variations of the dynamo families. J Geophys Res 96(B3):3923–3933

Merrill RT, McFadden PL (1999) Geomagnetic polarity transitions. Rev Geophys 37:201–226

Parker EN (1969) The occasional reversal of the geomagnetic field. Astrophys J 158:815–827

Petrellis F, Fauve S, Dormy E, Valet J-P (2009) A simple mechanism for the reversals of Earth's magnetic field. Phys Rev Lett 102:144503

Prévot M, Camps P (1993) Absence of longitudinal confinement of poles in volcanic records of geomagnetic reversals. Nature 366:53–57

Prévot M, Mankinen E, Coe RS, Grommé CS (1985) The Steens Mountain (Oregon) geomagnetic polarity transition. Field intensity variations and discussion of reversal models. J Geophys Res 90:10,417–10,448

Quidelleur X, Courtillot V (1996) On low degree spherical harmonic models of paleosecular variation. Phys Earth Planetary Inter 95:55–77

Quidelleur X, Valet J-P (1994) Paleomagnetic records of excursions and reversals: possible biases caused by magnetization artefacts? Phys Earth Planetary Inter 82:27–48

Quidelleur X, Holt J, Valet J-P (1995) Confounding influence of magnetic fabric on sedimentary records of a field reversal. Nature 374:246–249

Riisager P, Abrahamsen N (2000) Paleointensity of west Greenland Paleocene basalts: asymmetric intensity around the C57n-C26r transition. Phys Earth Planetary Inter 118:53–64

Roberts PH, Stix M (1972) Alpha effect dynamos by the Bullard-Gellman formalism. Astron Astrophys 18:453–466

Rochette PE (1992) Rationale of geomagnetic reversals versus remanence recording processes in rocks. Earth Planet Sci Lett 98:33–38

Valet J-P, Plenier G (2008) Simulations of a time-varying non dipole field during geomagnetic reversals and excursions. Phys Earth Planetary Inter 169:178–193

Valet J-P, Tauxe L (1989) Clement BM, Equatorial and mid-latitudes records of the last geomagnetic reversal from the Atlantic Ocean. Earth Planet Sci Lett 94:371–384

Valet J-P, Brassart J, Quidelleur X, Soler V, Gillot P-Y, Hongre L (1999) Paleointensity variations across the last geomagnetic reversal at La Palma, Canary islands, Spain. J Geophys Res 104(B4):7577

Valet J-P, Herrero-Bervera E, LeMouël JL, Plenier G (2008) Secular variation of the geomagnetic dipole during the past 2 thousand years. Geochem Geophys Geosyst 9:Q01008. doi: 10.1029/2007GC001728

Valet J-P, Meynadier L, Guyodo Y (2005) Geomagnetic field strength and reversal rate over the past 2 Million years. Nature 435:802–805

Valet J-P, Tucholka P, Courtillot V, Meynadier L (1992) Paleomagnetic constraints on the geometry of the geomagnetic field during reversals. Nature 356:400–407

Rock Magnetic Characterization Through an Intact Sequence of Oceanic Crust, IODP Hole 1256D

11

Emilio Herrero-Bervera, Gary Acton, David Krása, Sedelia Rodriguez, and Mark J. Dekkers

Abstract

Coring at Site 1256 (6.736°N, 91.934°W, 3635 m water depth) during Ocean Drilling Program (ODP) Leg 206 and Integrated Ocean Drilling Program (IODP) Expeditions 309 and 312 successfully sampled a complete section of in situ oceanic crust, including sediments of Seismic Layer 1, lavas and dikes of Layer 2, and the uppermost gabbros of Layer 3. The crust at this site was generated by superfast seafloor spreading (>200 mm/yr full spreading rate) along the East Pacific Rise some 15 Ma ago. One goal of drilling a complete oceanic crust section is to determine the source of marine magnetic anomalies. For crust generated by fast seafloor spreading, is the signal dominated by the upper extrusive layer, do the sheeted dikes play any role, how significant is the magnetic signal from gabbros relative to that at slow spreading centers and what is the timing of acquisition of the magnetization? To address these questions, we have made a comprehensive set of rock magnetic and paleomagnetic measurements that extend through the igneous interval. Continuous downhole variations in magnetic grain size, coercivity, mass-normalized susceptibility, Curie temperatures, and composition have been mapped. Compositionally, we have found that the iron oxides vary from being titanium-rich titanomagnetite (TM60), which are commonly partially oxidized to titanomaghemites, to titanium-poor magnetite as determined semi-quantitatively from Curie temperature analyses and microscopy studies. Skeletal titanomagnetite with varying degrees of alteration is the most common magnetic mineral throughout the section and is often bordered by large iron sulfide grains. The low-Ti magnetite or stoichiometric magnetite is present mainly in the dikes and gabbros and is associated with higher Curie temperatures (550°C to near 580°C) and higher coercivities than in the extrusive section. Magnetic grain sizes predominantly fall in the pseudo single domain (PSD) grain size region on Day diagrams, with only a small numbers of samples falling within the single domain

E. Herrero-Bervera (✉)
Paleomagnetics and Petrofabrics Laboratory, School of Ocean & Earth Science & Technology (SOEST), Hawaii Institute of Geophysics and Planetology (HIGP), 1680 East West Road, Honolulu, Hawaii 96822, USA
e-mail: herrero@soest.hawaii.edu

E. Petrovský et al. (eds.), *The Earth's Magnetic Interior*, IAGA Special Sopron Book Series 1,
DOI 10.1007/978-94-007-0323-0_11,

(SD) or multi-domain (MD) regions. Overall the magnetic properties of this hole are strongly influenced by post-emplacement alteration, particularly the lower part of the section from the gabbros up into the transition zone. Some of the more prominent features of the rock magnetic data are the gradual increase in Curie temperatures with depth from about 200–350°C at the top of the extrusives to about 425°C just above the transition zone, the more variable Curie temperatures and less variable susceptibility and coercivity of remanence in the upper half of the extrusives relative to the lower half the near constant composition ($x = 0.6$) and oxidation ($z = 0.6$) of the iron oxide grains ($> 5\nu m$) in the extrusives (Chapter 12 this volume), the highly irreversible nature of thermomagnetic curves in the extrusives, in which the cooling curve has Curie temperatures higher (generally > 500°C) than indicated by the heating curve, the abrupt change in rock magnetic properties across the transition zone, particularly the Curie temperature., a somewhat finer grain size and increased intensity in the sheeted dike zone relative to the extrusives and gabbros, and the nearly constant Curie temperatures (530 and 585°C) for the dikes and gabbros.

11.1 Introduction

Oceanic crust formed along seafloor spreading centers contains magnetic minerals that record processes ranging from large-scale geodynamo and mantle convection to much more local processes like hydrothermal circulation and alteration. After 45 years of research into the formation and evolution of the oceanic crust, first-order principles are well understood but many key questions still remain unanswered. This is mainly caused by the extreme difficulty in sampling the middle and lower oceanic crust rocks, which recent drilling has sought to overcome (e.g., Wilson et al. 2003, 2006, Teagle et al. 2004, 2006, Expedition-309-Scientists 2005, Expedition 309/312 Scientists 2006).

Drilling a complete in situ section of oceanic crust had been an unfulfilled goal of Earth scientists since the inception of ocean drilling starting with Project MoHole, which sought to sample the first-order seismic boundary named the Mohorovičić discontinuity or 'Moho' that separates the crust from the mantle (Bascom 1961, Greenberg 1974). The principal goal of Project MoHole was to understand the nature of the oceanic crust and the underlying uppermost mantle, which required drilling roughly 4–6 km deep (Alt et al. 2007). Drilling that deep presented technical challenges that have yet to be fully overcome. Instead subsequent drilling targeted tectonic windows where intermediate and deep parts of the ocean crust could be sampled. Examples include Hess Deep (ODP Leg 147 eastern flank of the equatorial East Pacific Rise, Mevel et al. 1996), Southwest Indian Ridge (ODP Leg 176, Natland et al. 2002), and the Mid Atlantic Ridge (IODP Expeditions 304–305, Ildefonse et al. 2006).

Although technological limitations still hamper efforts to reach the Moho, scientists have persisted in their efforts to sample a complete section of intact oceanic crust away from fracture zones, tectonic windows, and other forms of disturbance. Such sections represent the best means to understand the formation and evolution of oceanic crust. To fulfill this task Hole 504B (1° 13.62′N 83° 43.81′W) located in 6.9 Ma crust on the southern flank of the intermediate spreading rate Costa Rica Rift (40–80 mm/yr) was drilled through the entire extrusive rock pile into the sheeted dike layer below it (Alt et al. 1996). Unfortunately, the sheeted dike-gabbro transition was not reached owing to hole instability.

A renewed effort was made at Site 1256 during ODP Leg 206 (Wilson et al. 2003) and IODP Expeditions 309 and 312 (IODP Expedition 309 Scientists 2005, IODP Expedition 309/312 Scientists 2006, Teagle et al. 2006, Wilson et al. 2006). Ocean crust at this site formed along a superfast spreading (>200 mm/yr full rate) segment of the East Pacific Rise. Selecting a site where spreading was fast takes advantage of the apparent inverse relationship between the depth of axial low velocity zone, hypothesized to be the magma chamber, and spreading rate (Alt et al. 2007, cf. Fig. 2). Gabbros of Seismic Layer 3 form from the cooling magma chamber as the crust migrates away from the ridge axis, and hence Layer 3 is at shallower depths for crust formed by superfast seafloor spreading crust. The strategy proved effective, with Holes 1256A, B, and C sampling sediments and the uppermost extrusives and Hole 1256D penetrating through the rest of the extrusive section, the entire sheeted dike layer, and into the uppermost gabbros (e.g., Wilson et al. 2006, Alt et al. 2007). Thus, with the drilling of Hole 1256D, a major unachieved goal of ocean drilling was accomplished.

In this chapter, we analyze the rock magnetic characteristics of a collection of samples spanning the entire intact section of upper oceanic basement through the sheeted dikes into the gabbro layer. Additional microscopy analyses are provided in a companion chapter (Chapter 12 this volume).

11.2 Importance of the Rock Magnetic Characterization

The magnetic minerals of the oceanic crust have proven to be high-fidelity recorders of the geomagnetic field. Geomagnetic reversals recorded by the crust and apparent in marine magnetic anomaly profiles were fundamental in the development of the theory of plate tectonics (Vine and Mathews 1963, Morley and Larochelle 1964) and in the construction of geomagnetic polarity time scales (e.g., Heirtzler et al. 1968, Cande and Kent 1992, 1995, Huestis and Acton 1997). Subsequently, marine magnetic anomalies have been used in estimating modern plate motions and intraplate deformation (e.g., DeMets et al. 1990, Royer and Gordon 1997), testing radiometric decay constants and orbitally tuned timescales (Shackleton, et al. 1990), computing paleomagnetic poles from oceanic plates (e.g., Acton and Gordon 1991, Acton and Petronotis 1994, Petronotis et al. 1994), determining the size of non-dipole components of the geomagnetic field (e.g., Schneider 1988, Acton et al. 1996), and monitoring relative changes in the intensity of the geomagnetic field (e.g., Valet 2003). Although the value of marine magnetic anomalies to geoscience research is profound, advances in fully understanding their origin have been slow in coming because samples representative of complete sections of oceanic crust are sparse. As mentioned above, until the most recent IODP expeditions, no in situ section of oceanic crust had been sampled through the extrusive basalts, the sheeted dikes, and into the gabbro layer.

Clearly, the basic premise of the Vine-Mathews hypothesis which states that lineated magnetic anomalies were derived from igneous oceanic crust that was magnetized in the ambient field as it cooled following extrusion at ridge axes, describes a significant part of the process of oceanic crust formation. Subsequent alteration must also play a role (e.g., Irving et al. 1970a, b, Pariso and Johnson 1991, Phipps Morgan and Chen 1993, Worm and Bach 1996, Matzka et al. 2003, Krása et al. 2005, Carlut and Horen 2007, Krása and Matzka 2007, Chapter 12), but the significance of this role and the timing of the alteration are still debated (e.g., Tivey and Johnson 1987, Beske-Diehl 1990, Bowles and Johnson 1999, Worm 2001). Other factors remain uncertain. In particular, tectonism may modify the anomaly signal, as would be the case for proposed tilting of the crust yielding NMR inclinations deviating from the Geocentric Axial Dipole (GAD) values (Verosub and Moores 1981, Schouten 2002). Similarly, uncertainty exists as to how much each part of the oceanic crust contributes to the total anomaly signal (e.g., Harrison 1976, Pariso and Johnson 1991, 1993a, b). If deeper portions of the crust carry a significant proportion of the signal, then the curvature of isotherms with depth due to seafloor spreading and the associated delay in the acquisition of magnetization become important and may explain features such as anomalous skewness (e.g., Arkani-Hamed 1991).

The uncertainty in the origin of the magnetic signal results mainly from a lack of direct observation of the composition and grain size of the remanent magnetic carriers throughout the oceanic crust. Dredges, tectonic slivers of lower crust exposed along fracture zones, and ophiolites all yield constraints, but the processes responsible for their exposure may have perturbed their magnetic state. Only through sampling complete or nearly complete sections of in situ oceanic crust can questions concerning the origin of marine magnetic anomalies be resolved. Such sampling was part of the original motivation for ocean drilling, but achieving the objective of sampling complete in situ sections of ocean crust has proven to be technologically challenging and expensive (Teagle et al. 2004). Only rarely have crustal sections with greater than 100 m basement penetration been cored.

Broadly speaking, the magnetic mineralogy of ocean crust is a complex function of magmatic, hydrothermal and tectonic processes. The composition of the original magnetic minerals varies with magma chamber composition and size; position of the chamber relative to transform faults, propagating rifts, and hotspots; magma temperature and cooling rate; seafloor spreading rate; oxygen fugacity, etc. Subsequent high and/or low temperature alteration may affect the (magnetic) minerals. This includes thermal metamorphism due to nearby heat sources and hydrothermal alteration due to interaction with (hot) seawater that penetrates via fractures and porous parts in the rock pile. Even with all these influencing factors, the magnetic minerals of the oceanic crust have shown to be reliable recorders of the main features of the geomagnetic field including polarity zones (e.g. Irving 1970, Bleil and Petersen 1983, Johnson and Pariso 1993, Gee and Kent 2007).

11.3 Background of IODP Site 1256

Site 1256 lies in 3635 m of water in the Guatemala Basin on Cocos plate crust. The crust formed at ~15 Ma on the eastern flank of the East Pacific Rise (EPR) (Fig. 11.1). The Guatemala Basin has relatively subdued bathymetry, and the immediate surroundings of Site 1256 (~300 km) are relatively unblemished by major seamount chains or large tectonic features high enough to penetrate the sediment cover (~200–300 m). The site sits ~5 km east of the magnetic Anomaly 5Bn–5Br polarity transition on the reversely magnetized crust of Chron 5Br age. This crust accreted at a superfast spreading rate (~200–220 mm/yr full rate) and lies ~1150 km east of the present crest of the EPR and ~530 km north of the Cocos-Nazca spreading center.

Four holes were cored at Site 1256, with the first two holes (1256A and B) focused on recovery of the 250.7-m thich section of sediment overlying the basement. Hole 1256C then penetrated 88.5 msb (meters sub-basement) with very good recovery (61%) (Shipboard Scientific Party 2003, ODP Leg 206). Following the installation of a reentry cone with a 16-in diameter casing string that extended 20 m into basement in Hole 1256D, the hole was cored during Leg 206 to a total depth of 752 mbsf (502 msb) with moderate to high recovery (48%). Hole 1256D was deepened to 1255 mbsf on IODP Expedition 309 and

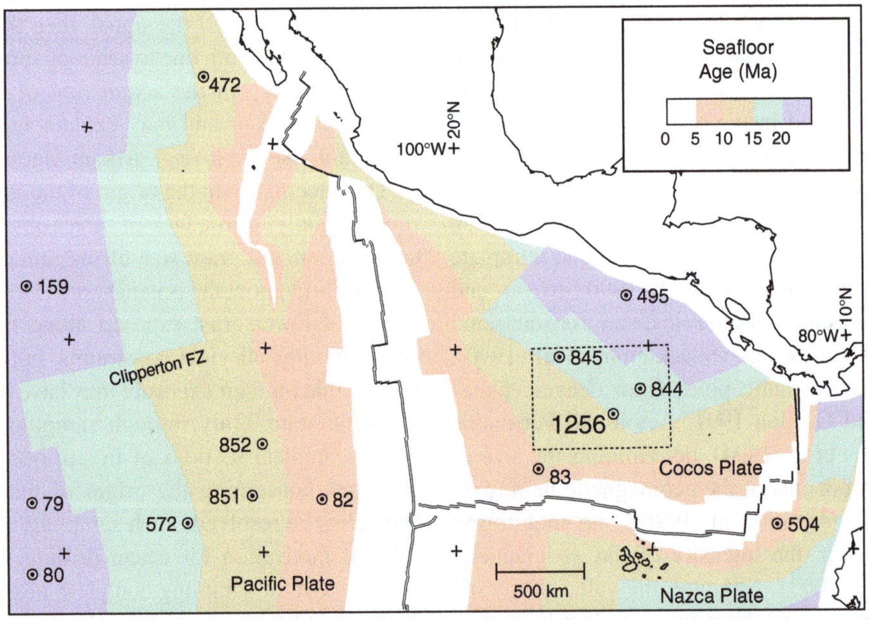

Fig. 11.1 Age map of the Cocos plate and East Pacific Rise with isochrones at 5-Ma intervals, converted from magnetic anomaly identifications according to timescale of Cande and Kent (1995). The wide spacing of 10–20 Ma isochrons to the south reflects the extremely fast (200–220 mm/y) full spreading rate. The locations of deep drill holes into the oceanic crust at Sites 1256 and 504 are shown. From Alt et al. (2007)

then to its current total depth of 1507.1 mbsf (1257.1 msb) during IODP Expedition 312, with recovery remaining moderate to high (~50%).

The basement sampled during Leg 206 is composed of axial (i.e. they are formed at the spreading ridge) sheet flows with subordinate pillow lavas, hyaloclastites, and rare dikes, all of which are capped by a more evolved off-axis massive ponded flow of >75 m thick and other sheet flows (Shipboard Scientific Party 2003, Wilson et al. 2006). The extrusive section is about 754 m thick, with the upper 284 m being comprised of off-axis massive and sheet flows and the lower 470 m being comprised of flows erupted at the ridge axis. The lavas have normal mid-ocean-ridge basalt (MORB) chemistries and display moderate fractionation up-section as well as heterogeneous incompatible element ratios (Wilson et al. 2003). The lavas are only slightly affected by low-temperature hydrothermal alteration, and very little interaction with oxidizing seawater is apparent (Shipboard Scientific Party 2003). Overall, the extrusives at Site 1256 are much less oxidized than those from Holes 504B (6 Ma) and 896A (6.9 Ma), but have similar alteration to the extrusives in Hole 801C (180 Ma), albeit with very little carbonate at Site 1256 (Shipboard Scientific Party 2003).

A lithologic transition occurs from 1004 to 1061 mbsf, where sub-vertical intrusions and mineralized breccias occur along with extrusives. Below this narrow "Transition Zone", sub-vertical intrusions are numerous, marking the top of the 346-m-thick sheeted dike complex. Alteration increases in stepwise fashion across the Transition Zone, with low-temperature (<150°C) alteration mineral phases in the extrusives and higher temperature (>250°C) phases at the top of the sheeted dike complex. This complex can be divided into an upper dike complex of massive basalts (1061–1348 mbsf) overlying a thin dike complex of recrystallized basalts with granoblastic texture (referred to as the granoblastic dikes, 1348–1407 mbsf) (Wilson et al. 2006). This recrystallization is attributed to contact metamorphism from the underlying gabbros (Wilson et al. 2006). In general, alteration graded increases with depth through the sheeted dike complex, with alteration temperatures approaching ~400° by 1300 mbsf (Wilson et al. 2006, Alt et al. 2007).

The top of the plutonic complex occurs at a depth of 1407 mbsf. The complex consists of an upper 52-m thick gabbroic body and a lower 24-m-thick gabbroic body separated by a 24-m-thick layer of granoblastic dikes (Alt et al. 2006). The gabbroic complex is highly altered, with alteration increasing with grain size near intrusive contacts (Alt et al. 2006).

11.4 Rock-Magnetic Methods

We have conducted a large number of paleomagnetic and rock magnetic measurements (Expedition-309-Scientists 2005, Shipboard Scientific Party 2003, Acton and Wilson 2005, Herrero-Bervera and Acton 2005 and 2011; Teagle et al. 2006, Alt et al. 2007) on samples acquired during Leg 206 and IODP Expeditions 309/312. Here, we focus on the rock magnetic results. We have analyzed the field-dependent and temperature-dependent magnetic behavior of over 350 discrete rock chips (~200 mg) from the working half sampled from the top of the core (i.e. just below the sediment cap) down to the end of the plutonic complex (i.e. well into the gabbroic layer). All analyses were conducted at the SOEST-HIGP Paleomagnetics and Petrofabrics Laboratory of the University of Hawaii at Manoa and at the University of California, Davis (UCD). The field-dependent measurements included determination of the mass-specific low-field susceptibility and hysteresis loops both carried out at room temperature. Temperature dependent measurements involved determination of the Curie temperature. Susceptibility measurements provide insight into the relative contributions of parts of the rock pile to the expression of the magnetic anomaly in the geomagnetic field. Hysteresis loops allow determination of the magnetic grain size that provides insight into magmatic and subsequent alteration processes. This together with the determination of the Curie temperature which constrains the magnetic mineralogy, alteration, oxidation, and (oxy)exsolution processes.

In order to determine the susceptibility of the samples we used an AGICO KLF-3 Minikappa bridge instrument that has been devised for laboratory measurements of magnetic susceptibility of cubic, cylindrical and fragment rock specimens, of nominal volume of 10 cc of cubic specimens (i.e. 20×20×20 mm) or cylindrical specimens of diameter of 25.4 mm and length of 22 mm, with a range of magnetic field intensity of 30 A/m or 300 A/m, and a measuring frequency of 2000 cycles/s, with a precision of measurement of within the whole range of 2% and a precision of absolute calibration of 5%. The sensitivity of the instrument is $1\times E^{-06}$ u.SI and a resolution of $1\times E^{-06}$ u.SI. Magnetic hysteresis measurements were performed on small chips of rocks (~200 mg) using a Petersen variable field translation balance (VFTB) capable of generating fields of up to 1.2 T located at the SOEST-HIGP Paleomagnetics and Petrofabrics Laboratory of the University of Hawaii at Manoa (for samples from sites 309 and 312) and a Princeton alternating gradient force magnetometer, MicroMag model 2900 at the Paleomagnetics laboratory of the University of California, Davis for samples from Site 206. For the MicroMag measurements samples of 4–10 mg were mounted on the sample probe, a maximum field of 2 Tesla was utilized and the field increment was 4 mT (averaging time 0.1s); samples appeared to be saturated below 1 Tesla (pole shoe saturation correction between 1 and 1.5 Tesla). Backfield curves of the SIRM were acquired to 0.1 Tesla to determine the remanent coercive force B_{cr} (waiting time 1s, averaging time 0.1s). The instrumental noise level is $\sim 2\times 10^{-10}$ Am^2, typical sample intensities were at least four orders of magnitude higher. Hysteresis parameters were determined after various thermal treatments and are based on at least three chips per treatment. For the VFTB measurements larger samples of up to 200 mg were used, in a maximum field of 1 Tesla. First, the hysteresis curve and the backfield demagnetization of SIRM were determined, followed by the thermomagnetic heating and cooling curves in a field of ~720 mT. The VFTB has a measurement range of 10^{-8}–10^{-2} Am^2. Saturation remanent magnetization (Mr), saturation magnetization (Ms), and coercive force (Hc) were calculated after removing the paramagnetic contribution.

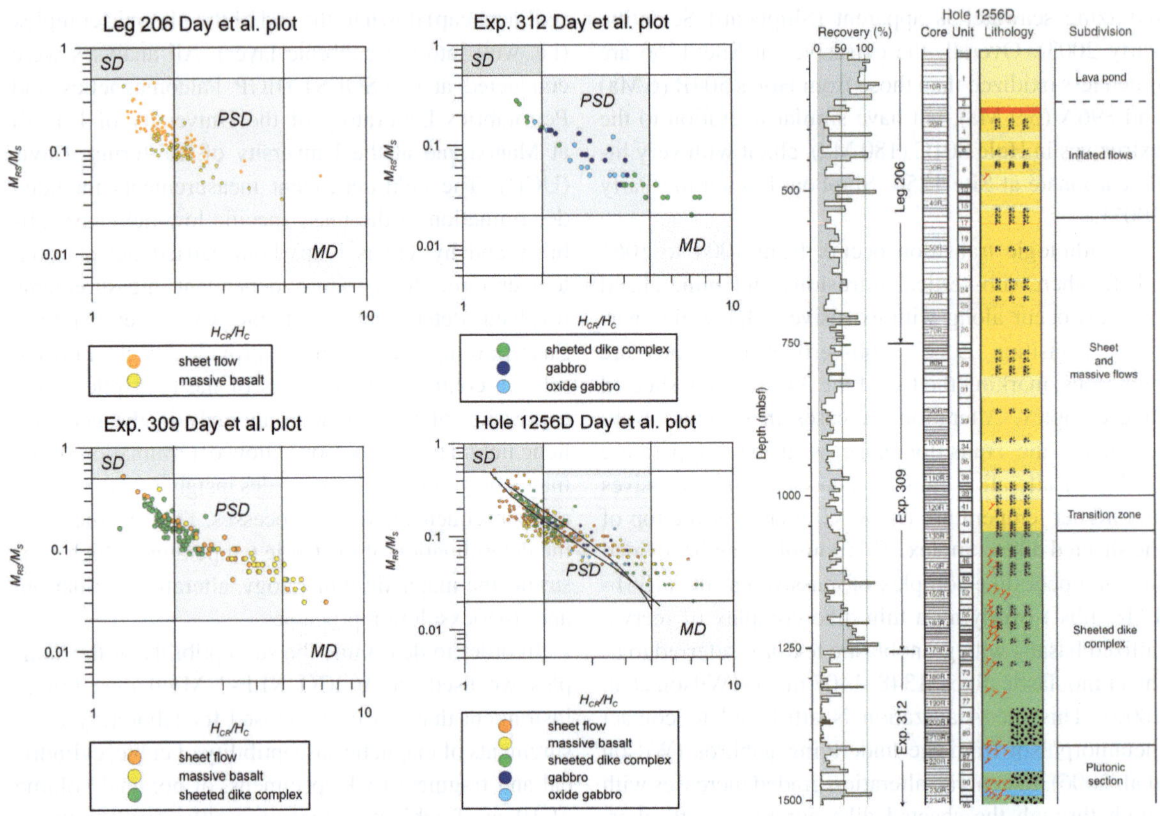

Fig. 11.2 Day plot (modified according to Dunlop 2002a, b) showing magnetic grain size variations of Leg 206 and IODP Expedition 309 and 312. Magnetic Grain Size analyses of the entire set of samples covering 1250 m of drilled core are color coded with respect to the lithology of Hole 1256D. As is depicted in this figure we have found that the great majority of the specimens studied cover a wide range of magnetic sizes but most of them are located in the Pseudo Single Domain (PSD) areas of the Day plot whereas some specimens are characterized by Multi Domain sizes (MD) and very few in the Single Domain (SD) areas of the plot

After measurement of the hysteresis loop and the back-field demagnetization curve of the SIRM, we carried out thermomagnetic experiments in air in a strong field of 718 mT from room temperature up to 700°C using the same Variable Field Translation Balance (VFTB) instrument (see for example Herrero-Bervera and Valet 2005, 2009, Valet et al. 2010).

11.5 Results

Rock magnetic results are plotted in stratigraphic order and correlated with respect to the lithostratigraphic column in, which is shown in Figs. 11.2, 11.3, and 11.4. The low-field susceptibility ranges between ~1 × 10^{-5} and ~2 × 10^{-5} m^3kg^{-1} in the upper 500 m of extrusive basalts with an increasing trend with depth. The lower portion of the basalts is notably more variable in susceptibility than the upper portion, which may result from more variably alteration in the lower segment. Susceptibility in the top half of the sheeted dikes roughly varies between the same values but increases up to ~5 × 10^{-5} m^3kg^{-1} in the lower half of the dikes. The gabbroic section has a lower and more variable susceptibility ranging from ~3 × 10^{-6} to ~1–2 × 10^{-5} m^3kg^1.

NRM intensities provide constraints on the relative contribution that each part of the oceanic crust contributes to the marine magnetic anomaly signal. The data have some caveats: The NRM depends on both the ability of the magnetic minerals to retain a stable remanence and on the magnitude of the magnetizing field. Because Site 1256 is located near a transition zone between normal and reversed polarity crust, some units may have acquired their magnetizations while the geomagnetic field was reversing. At such times, the ambient field has very low magnitude. Thus, some units could give low NRM values even though they are capable of carrying a stable remanence. Also, drilling overprints affect the NRMs of all samples to some degree. The overprints are not homogeneous, but instead depend on lithology (particularly the coercivity) and on the different drilling equipment used in collecting each core. The drilling overprint acts somewhat like a low-field isothermal remanent magnetization (IRM). Generally, this overprint can be removed by AF demagnetization in peak fields >5 mT and <25 mT. Hence, the low coercivity part of the NRM is altered by drilling and cannot be recovered from NRM measurements, but the medium and high coercivity is typically unaffected.

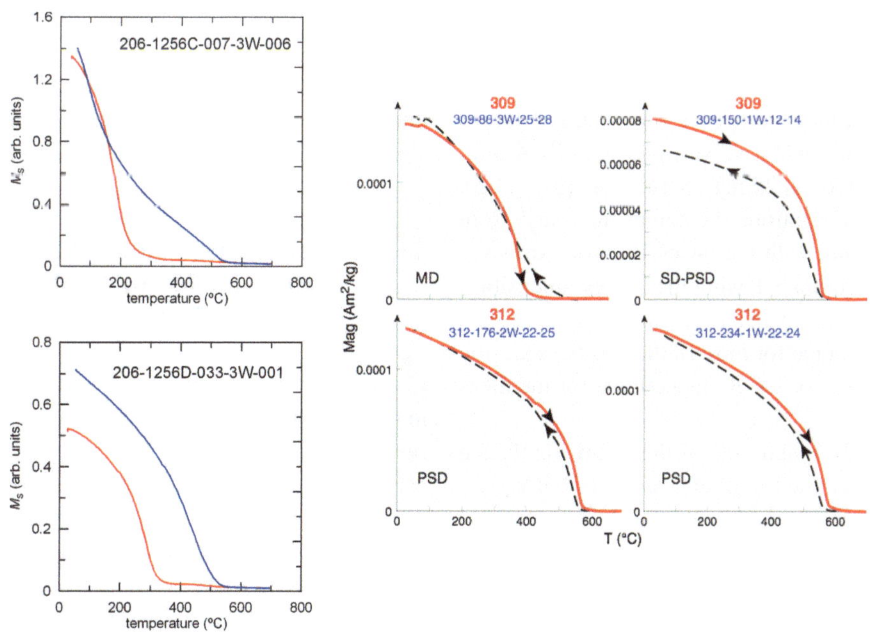

Fig. 11.3 Typical thermomagnetic curves of representative samples from ODP Leg 206 (two curves on the *left* of diagram), and IODP Expedition 309 and 312 (four curves on the *right side* of diagram). The *red* curve indicates the heating curve part of the experiment

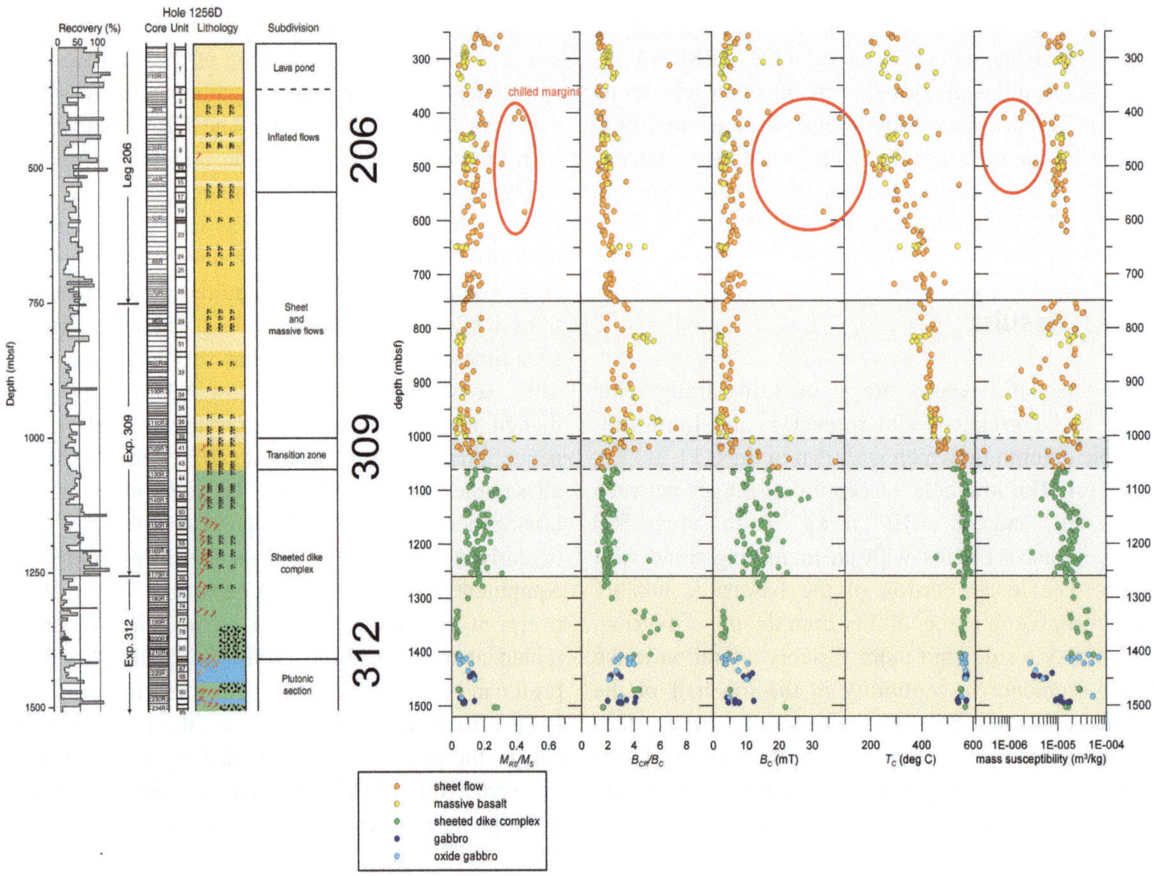

Fig. 11.4 Downhole variation of the rock magnetic results (ratio of saturation remanence over saturation magnetization, coercivity, Curie points and magnetic susceptibility)

With the caveats in mind, we show the results for both the original NRM (NRM_0) and the NRM after AF demagnetization at 25 mT (NRM_{25mT}) (Fig. 11.5). Prior to demagnetization, the intensities vary from about 2 to 50 A/m, with the lower 70 m of extrusives and upper dike complex having the highest intensities. Demagnetization at 25 mT reduces the by more than an order of magnitude for most of the extrusive section and by less than an order of magnetitude for the dikes and gabbros.

To quantify the magnitude of the NRM unaffected by drilling, we use the logrithmic values of $NRM_{25\,mT}$ to calculate mean intensities for the different crustal intervals (Fig. 11.6). The mean NRMs are 0.10 A/m for the Lava Pond (250–350 mbsf), 0.37 A/m for the extrusives from 350 to 650 mbsf, 0.14 A/m for extrusives from 650 to 1004 mbsf, 0.57 A/m for the transition zone, 1.03 A/m for the sheeted dikes, and 0.67 A/m for the gabbros. At least based on the NRMs, the sheeted dikes and gabbros both have the potential to contribute significantly to the marine magnetic anomaly signal.

11.5.1 Hysteresis Results: Coercivity, Saturation Magnetization, and Grain Sizes

Coercivity values are low as expected for titanomagnetite that is variably maghemitized. B_c values range between 1.5 and 26.8 mT for the extrusive rocks, between 1.5 and 29.0 mT for the sheeted dikes and between 2.8 and 21.9 mT for the gabbros (cf. Fig. 11.4). B_{cr}/B_c values are between 1.3 and 6.8 for the extrusive rocks (with an isolated point characterized by a value of 17.4), 1.3 and 7.6 for the sheeted dikes, and 1.6 and 4.8 for the gabbros. Most notably are the increase of B_{cr}/B_c values with depth in the extrusives and the large B_{cr}/B_c values within the

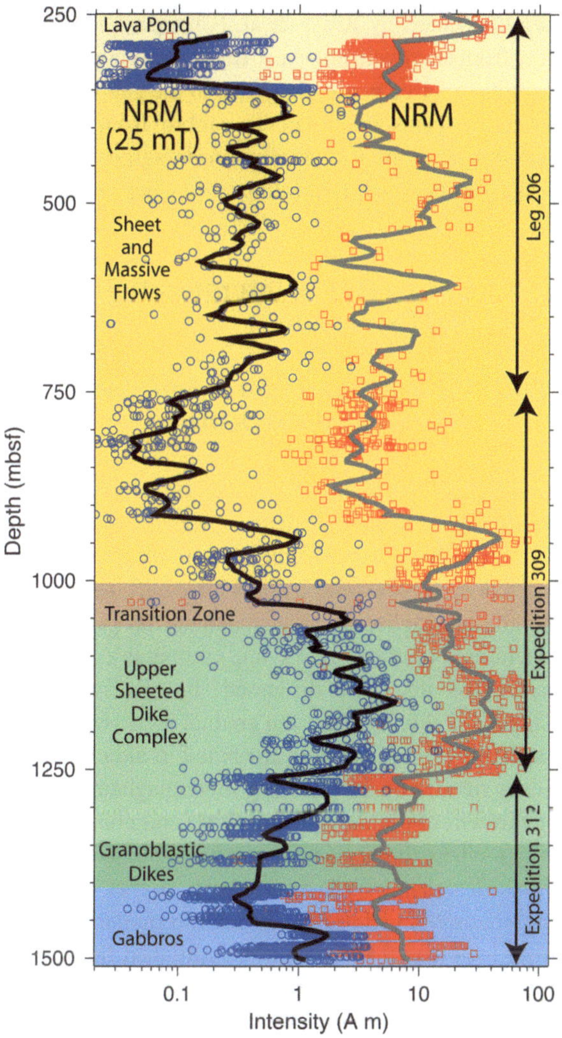

Fig. 11.5 Downhole variation of the NRM intensity before demagnetization (*red squares* with *thick gray curve*) and after 25 mT demagnetization (*blue circles* with *thick black curve*). The data are from both discrete samples and archive half split-core sections

granoblastic dikes that makeup the lowermost sheeted dike complex (1348–1407 mbsf). M_{rs}/M_s is ~0.15 for most of the extrusive rock pile with a very slight decrease with depth. In the transition zone between the extrusive rocks and the sheeted dikes it notably increases to values up to ~0.4. The uppermost half of the sheeted dikes is characterized by values of ~0.2 whereas M_{rs}/M_s drops to 0.05–0.1 in the dikes below 1300 mbsf. M_{rs}/M_s increases gradually downhole within the gabbros, with values averaging ~0.15 at the very bottom of Hole 1256D.

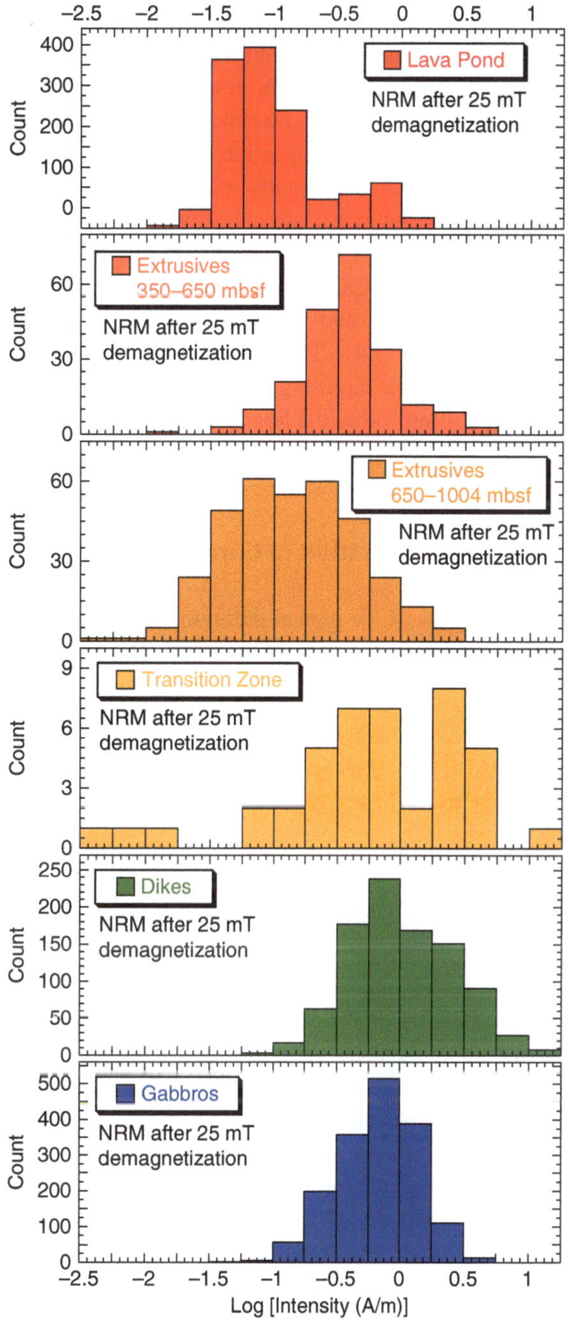

Fig. 11.6 Histograms of the logarithmic values of intensities (in A/m) for the natural remanent magnetization (NRM) demagnetized at 25 mT for different layers of the crust

Magnetic grain size is commonly deduced by plotting B_{cr}/B_c vs. M_{rs}/M_s in a diagram referred to as a Day plot (Day et al. 1977), which contains single domain (SD), pseudo-single domain (PSD), and multi-domain (MD) grain-size regions that have been defined experimentally for magnetite and titanomagnetite (with recent modifications by Dunlop 2002a, b). For each cruise – Leg 206, Expedition 309, and Expedition 312 – separate Day plots are shown, with the combined data depicted in the right-handed lower panel of Fig. 11.2. The results span a wide range of magnetic sizes, but with a vast majority of them located in the PSD region of the Day plot. Magnetically the most stable samples are those with small grain sizes (SD to PSD sizes) and large coercivities (compare Figs. 11.2 and 11.4). The granoblastic dikes consistently have the largest grain sizes of any of the intervals, falling within the large PSD and MD ranges, and hence are the least stable magnetically.

11.5.2 Determination of Curie Temperatures

In Fig. 11.3 we show typical thermomagnetic heating and cooling curves for a selection of representative samples. The curves are typically irreversible in the extrusives down to about 1029 mbsf and relatively reversible below that depth. The irreversibility results from inversion of titanomaghemite to nearly pure magnetite during heating, as can be noted by the steep increase in magnetization of the cooling curves near the Curie temperature of magnetite.

Curie temperatures are determined from the main decrease in magnetization in the thermomagnetic curves following the graphical method of Gromme et al. (1969). The uncertainty in the determinations is about ±10°C. Besides curves that have a single drop in magnetization, such as those in Fig. 11.6, some curves have a secondary smaller drop, indicating the existence of two magnetic minerals. The secondary Curie point could result from alteration during the laboratory thermomagnetic measurements (e.g., Krása and Matzka 2007).

Curie points range from about 200°C to 585°C. Most notably, they increase progressively downhole in the upper 750 m of the extrusives from about 200°–350°C at the top (250 mbsf) to about 425°C at 1004 mbsf. They then increase abruptly to values >500°C across the narrow transition zone at the very base of the extrusives, corresponding to the change in alteration temperatures from <150° to >250° (see Section 11.3). The Curie temperatures are between 530°C and 585°C for the dikes and gabbros, indicating the presence of almost pure magnetite (Figs. 11.3 and 11.4), which has been confirmed by Krása et al. (Chapter 12 this volume) from energy dispersive x-ray spectroscopy (EDS) analyses on large magnetic grains in the gabbros. This magnetite forms in the Fe-Ti gabbro from primary titanomagnetite by oxy-exsolution, in agreement with opaque mineralogy observations (see Chapter 12)

11.6 Observations, Interpretation and Discussion

11.6.1 Curie Temperature as Function of Lithology and Depth

Overall, the thermomagnetic analyses indicate that the dominant magnetic carrier for the extrusives, from the lava pond unit down to the transition zone, is low-temperature oxidized titanomagnetite/titanomaghemite. The magnetic properties and mineralogy of these rocks are comparable to other marine basalts. As discussed by Krása et al. (Chapter 12 this volume), the gradual increase in Curie temperatures with depth is attributed to alteration that causes submicron inversion of titanomaghemite to an intergrowth of titanomagnetite (TM) and nonmagnetic phases, where the Ti-content of the TM phase is continuously decreasing with depth due to higher inversion temperatures. In addition, sub-micron titanomagnetite (from low-Ti to TM60) that has sustained little or no alteration may occur within interstitial glass as has been documented in prior studies of ocean basalts (e.g., Smith 1986, Zhou et al. 2000).

The transition zone is essentially a mixture of extrusives and dikes and has been hydrothermally altered to greenschist facies. The Curie temperatures probably define this step in alteration more clearly than petrographic studies as there is a distinct increase in Curie temperatures at 1029 mbsf from a mean of 486°C just above this depth to a mean of 559°C below it. Below this depth, two processes lead to the drastic change in magnetic mineralogy: The slower cooling rate after initial emplacement below this depth causes oxy-exsolution of primary titanomagnetite, the higher degree of subsequent alteration in this part of the section leads to partial or complete replacement of the primary magnetominerals by secondary minerals. Alteration occurs in the manner originally described by Ade-Hall et al. (1968) as "granulation". In this process, a titanium-bearing phase, such as sphene or anatase is formed from the homogeneous primary titanomagnetite, and iron either remains in the magnetite host or is leached from the spinel lattice (e.g Pariso and Johnson 1991). In the 504B transition zone, the extremely high degree of alteration is reflected by original titanomagnetite grains which are partially and often completely, replaced by sphene (Alt et al. 1986, Pariso and Johnson 1991).

The dominant magnetic mineral is the Fe-rich spinel component of exsolved titanomagnetite, with composition approaching pure magnetite at the base of the transition zone where Curie temperatures are >500°C. The sheeted dike complex has a complex thermal and alteration history, with greenschist facies alteration continuing downward through it, resulting in the replacement of primary titanomagnetite by secondary minerals but with much less silicate replacement. The dominant magnetic mineral is the Fe-rich spinel component of exsolved primary titanomagnetite coupled with an uncertain amount of secondary magnetite produced as a silicate alteration product, similar to that observed in other ocean crust studies (e.g. Petersen et al. 1979, Smith and Banerjee 1986, Pariso and Johnson 1991, Carlut and Horen 2007). The primary oxide minerals within the dike complex are observed to have suffered two subsolidus processes, namely, a high-temperature deuteric oxidation and second, hydrothermal alteration. The observed hydrothermal alteration shows titanomagnetites that have been visibly altered by hydrothermal fluids and some of the grains are characterized by the presence of fine anatase or rutile granules within the host titanomagnetite or by the incipient replacement of ilmenite lamellae by sphene (see Fig. 12.2). The most striking feature of the oxide minerals is the large variation, both with depth in the crustal section and locally within the polished section, in the degree of both deuteric and hydrothermal alteration (Krása and Herrero-Bervera 2005). Overall, the increasing degree of high-temperature alteration with depth suggests the geologically reasonable hypothesis that the lower portion of the sheeted dike complex cooled at a substantially slower rate than the upper portion. These findings have been also observed in other oceanic sequences such as Hole 504B (e.g., Furuta and Levi 1983, Smith and Banerjee 1986, Pariso and Johnson 1991). These hydrothermal alteration changes can also be observed in the rock magnetic characteristics such as the high variability of B_C from about 10 to 30 mT (from 1060 to ~1300 mbsf), the striking uniformity of the Curie temperatures (540–585°C) and the variability of the susceptibility record. From about 1300 to 1400 mbsf the rock magnetic parameters change significantly because, even though the rocks are still part of the sheeted dike complex, the texture changes to a granoblastic texture and alteration is more severe (Koepke et al. 2008). In this granoblastic dike zone there is a higher uniformity of B_C dropping from 30 mT to only 10 mT, an increase in grain size, an increase of the dispersion of the susceptibility, and a decrease in NRM intensity. Curie temperatures, however, remain fairly constant (i.e. 562°C) across the granoblastic dikes and down through the gabbros.

The chemical and structural changes in the magnetic minerals that occur in the lower transition zone and sheeted dike complex suggest that the NRM carried by these units would not be a primary thermal remanent magnetization (TRM). Instead, these rocks probably carry a chemical remanent magnetization (CRM) or perhaps more accurately, a thermo-chemical remanent magnetization (TCRM), as the portion of the grains with blocking temperatures below the alteration temperature will actually carry a partial TRM. As high-temperature hydrothermal alteration appears to take place soon after emplacement, perhaps within a few hundred meters of the ridge axis, alteration in the lower transition zone and dike complex may have been either happening with cooling, resulting in a primary CRM, or it may have followed cooling, yielding an initial TRM followed by a secondary CRM (e.g. Smith and Banerjee 1986, Chapter 12).

Low-titanium magnetite is ubiquitous in the gabbros and is the end result of both igneous and alteration processes. It is known that the initial titanomagnetite solid solution [x[Fe2TiO4][1-x][Fe3O4]] which crystallizes from a tholeiitic melt initially contains 50 and 85% ulvospinel [Fe2TiO4] (Carmichael and Nicholls 1967). After crystallization, continued slow cooling of the gabbros results in high-temperature, or deuteric oxidation of titanomagnetite forming both ilmenite and low-titanium magnetite (Buddington and Lindsley 1964). Furthermore, secondary titanomagnetite, formed from the alteration of olivine and pyroxene, tends to form very close to the magnetite end-member of the solid solution. Even though it is known that magnetite forms as a result of hydrous alteration of oceanic gabbros, it has been difficult to elucidate which is the most important or pervasive type of alteration process, either at low or at high-temperature which influenced oceanic deep crustal rocks (e.g. Pariso and Johnson 1993a).

11.6.2 Discussion

Our results show that the rock magnetic properties of oceanic crust at Site 1256 are highly influenced by composition, alteration, and cooling history. Hence, rock magnetic changes are notable across lithologic contacts and alteration fronts. Some of the more prominent features of the rock magnetic data are:

(a) The gradual increase in Curie temperatures with depth from about 200°–350°C at the top of the extrusives to about 425°C just above the transition zone
(b) The more variable Curie temperatures and less variable susceptibility and coercivity of remanence in the upper half of the extrusives relative to the lower half.
(c) The near constant composition ($x = 0.6$) and oxidation ($z = 0.6$) of the iron oxide grains (>5 μm) in the extrusives (Chapter 12 this volume).
(d) The highly irreversible nature of thermomagnetic curves in the extrusives, in which the cooling curve has Curie temperatures higher (generally >500°C) than indicated by the heating curve.
(e) The abrupt change in rock magnetic properties across the transition zone, particularly the Curie temperature.
(f) A somewhat finer magnetic grain size and higher magnetization (NRM25 mT) in the upper sheeted dike zone relative to the extrusives and gabbros.
(g) The large grain size (large PSD to MD range) in the lower sheeted dikes (the granoblastic dikes) and the significantly reduced NRM25 mT intensity relative to the upper dikes.
(h) The nearly constant Curie temperatures (530 and 585°C) for the dikes and gabbros.

Conclusions

The rock magnetic experiments described here indicate that throughout Hole 1256D post-emplacement alteration has had a significant and often substantial effect on the magnetic properties of these oceanic rocks. Variation in the degree of alteration with depth are strongly influenced by the thermal structure and cooling rate.

The magnetic structure of the crustal section of Hole 1256D can be summarized into these four essential units.

1. The upper extrusive units (lava pond, inflated flows, and sheet and massive flows, from 250 mbsf down to ~1004 mbsf) have a magnetic mineralogy characterized primarily by low temperature oxidized titanomagnetites (=titanomaghemite). The degree of alteration increases with depth as evidenced by the increase in Curie temperature. Because both the composition and the oxidation state are relatively constant throughout the section, at least for the >5 μm grains analyzed in the companion chapter (Chapter 12), much or all of this increase is caused by inversion of titanomaghemite to titanomagnetite. Mean NRM_{25mT} intensities are 0.10 for the Lava Pond (250–350 mbsf), 0.38 for the extrusives from 350 to 650 mbsf, and 0.14 A/m for extrusives from 650 to 1004 mbsf (Figs. 11.5 and 11.6).
2. The transition zone (from 1004 to 1061 mbsf) is a relatively thin unit characterized by a mixture of extrusives and dikes that has suffered hydrothermal alteration to greenschist facies minerals throughout most of the zone. Curie temperatures display a step-like change within the transition zone at 1029 mbsf, with a mean of 486°C above this depth and a mean of 559°C below, indicating that very low-Ti magnetite is the primary carry of the magnetization below this depth. The NRM_{25mT} intensities within this interval are relatively high (0.58 A/m) but have high scatter (0.58 A/m).
3. The sheeted dike complex (1061–1407 mbsf) is a unit characterized by greenschist facies minerals. The upper dike complex (above 1348 mbsf) has been affected by similar alteration processes as the extrusives but to a higher degree. The lowermost part of the complex (1348–1407 mbsf) has a granoblastic texture resulting from contact metamorphism from underlying gabbros (Wilson et al. 2006, Koepke et al. 2008). Alteration is present but less intense overall relative to the transition zone for example, and the end product of the greenschist alteration is oxidation-exsolution of primary titanomagnetite but much less silicate replacement. Despite or perhaps because of this alteration and thermal history, the intensity of the natural remanent magnetization is larger in the dikes than any other interval, with a mean NRM_{25mT} intensity of 1.03 ± 1.11 A/m (see Figs. 11.5 and 11.6). The

stable and sufficiently large magnetization of the upper dikes provide evidence that it can contribute significantly to the marine magnetic anomaly signal, whereas the large grain sizes and reduced magnetization indicates the lowermost dikes have a less stable magnetization and probably contribute less per volume than of the other crustal intervals.

4. The Plutonic Complex composed mainly of gabbros, oxide gabbros, quartz-rich oxide diorites and small trondhjemite dikelets is characterized by a magnetic mineralogy that results from changes in the concentration of magnetite that is directly related to crystal fractionation of a tholeiitic melt. Iron-titanium rich gabbros are dominated basically by primary, igneous magnetite with small amounts of secondary magnetite. Magnetite in cumulate olivine gabbros is formed during high-temperature hydrous alteration of olivine and pyroxene, with secondary magnetite being the most important magnetic mineral phase, formed as small amounts of seawater penetrated into lower crustal rocks at relatively high temperatures. The most important aspect of these observations is that both primary and secondary forms of magnetite in this part of the section formed at temperatures near or above the Curie temperature. There are marked and dramatic differences in the physical grain size of magnetite in Fe-Ti and olivine gabbros. However, the magnetic behavior of these two lithologies is very similar, hysteresis loop parameters and NRM intensities (median NRM_{25mT} = 0.72 A/m, Figs. 11.5 and 11.6) indicate that all gabbros of this section are capable of also carrying significant and stable remanent magnetization and therefore also capable of recording reversals of the Earth's magnetic field as it has been reported e.g. for gabbros from ODP Hole 735B.

Based on the rock magnetic data from this study, virtually all the igneous section cored has the potential to retain a long-term remanent magnetization and thus has the potential to play a role in the total magnetization that produces marine magnetic anomalies. Perhaps the only exception to this is the relatively thin granoblastic dike unit at the base of the sheeted dike complex, where magnetic grain sizes are relatively large (the large end of the PSD range and into MD range) and the magnetization somewhat reduced relative to the upper dike section. Even though virtually all the section cored at Site 1256 has the potential to carry a significant remanence, the actual contribution of the different layers depends on the timing and duration of magnetization acquisition. The extrusive section and upper dikes cool and record the ambient geomagnetic field soon after their formation at or very near the ridge axis. Subsequent alteration (oxidation and inversion) as documented here and in the companion chapter (Chapter 12) is similar to what has been documented in laboratory experiments (Krása and Matzka 2007), in which the new magnetic minerals formed by alteration inherit the paleomagnetic direction of the parent minerals although with reduced magnetization intensity. Hence, the extrusives and upper dikes would carry some form of a primary and/or an inherited primary magnetization. This magnetization undoubtedly contributes signficantly to the marine magnetic anomaly signal for crust generated by fast seafloor spreading. The timing of acquisition of the thermal-chemical remanent magnetization carried by the gabbros is more difficult to establish based on rock magnetic data alone, but can be investigated by combining rock magnetic constraints from studies like this one with marine magnetic anomaly studies that seek to estimate and model the origin of anomalous skewness (e.g., Arkani-Hamed 1988, 1991, Petronotis et al. 1992, Dyment et al. 1997).

Acknowledgements We are grateful to Mr. James Lau for his laboratory assistance and help with the laboratory measurements. We thank the referees for their very constructive criticisms that made us improve greatly our manuscript. We also give special thanks to the participating scientists and crew members of JOIDES Resolution for their help and support during the scientific cruises. This research used samples and data provided by the Ocean Drilling Program (ODP) and the Integrated Ocean Drilling Program (IODP). Funding for this research was provided by the National Science Foundation (NSF) through its support of ODP, IODP, and the United States Science Support Program (USSSP) and through NSF grants JOI-T309A4, OCE-0727764, and EAR-IF-0710571 to Herrero-Bervera, and a USSSP Post-Expedition Activity Award and NSF grant OCE-0727576 to Acton. Additional financial support to Herrero-Bervera was provided by SOEST-HIGP. Krása received funding through a Royal Society of Edinburgh BP Trust Research Fellowship. The views expressed are purely those of the authors and may not in any circumstances be regarded as stating an official position of the European Research Council Executive Agency. This is an HIGP and SOEST contribution 1889, 8146 respectively.

References

Acton GD, Gordon RG (1991) A 65 Ma palaeomagnetic pole for the Pacific plate from the skewness of magnetic anomalies 27r-31. Geophys J Int 106(2):407–420

Acton GD, Petronotis KE (1994) Marine magnetic anomaly skewness data and oceanic plate motions. Eos 75:49–52

Acton G, Wilson D (2005) Paleomagnetic and rock magnetic signature of upper oceanic crust generated by superfast seafloor spreading: results from ODP Leg 206. Eos Trans AGU 85(47). Fall Meeting Suppl., Abstract GP11D–0868

Acton GD, Petronotis KE, Cape CD, Ilg SR, Gordon RG, Bryan PC (1996) A test of the geocentric axial dipole hypothesis from an analysis of the skewness of the central marine magnetic anomaly. Earth Planet Sci Lett 144(3–4):10

Ade-Hall JM, Khan MA, Dagley P, Wilson RL (1968) A detailed opaque petrological and magnetic investigation of a single Tertiary lava flow from Skye, Scotland, I, Iron-titanium oxide petrology. Geophys J R Astron Soc 16:375–388

Alt JC, Honnorez J, Laverne C, Emmermann R (1986) Hydrothermal alteration of a 1-km section through the upper oceanic crust, Deep Sea Drilling Project hole 504B;The mineralogy, chemistry and evolution of basalt-seawater interactions. J Geophys Res 91:10,309–10,335

Alt JC, Teagle DAH, Umino S, Miyashita S (2007) The IODP expeditions 309 and 312 scientists and the ODP leg 206 scientific party. Sci Drilling 4:4–10. doi:10,2204/iodp.sd.4.01

Alt JC, Laverne C, Vanko DA, Tartarotti P, Teagle DAH, Bach W, Zuleger E, Erzinger J, Honnorez J, Pezard PA, Becker K, Salisbury MH, Wilkens RH (1996) Hydrothermal alteration of a section of upper oceanic crust in the eastern equatorial Pacific: a synthesis of results from Site 504 (DSDP Legs 69:70, and 83, and ODP Legs 111, 137,140, and 148.). In: Alt JC, Kinoshita H, Stokking LB, Michael P (eds) Proceedings ODP, science results: (College Station, Texas), Ocean Drilling Program, pp 417–434

Arkani-Hamed J (1988) Remanent magnetization of the oceanic upper mantle. Geophys Res Lett 15:48–51

Arkani-Hamed J (1991) Thermoremanent magnetization of oceanic lithosphere inferred from a thermal evolution model: implications for the source of marine magnetic anomalies. Tectonophysics 192:81–96

Bascom W (1961) A hole in the bottom of the sea. Doubleday and Company, New York, NY, 352p

Beske-Diehl SJ (1990) Magnetization during low-temperature oxidation of seafloor basalts: no large scale chemical remagnetization. J Geophys Res 95:21413–21432

Bleil U, Petersen N (1983) Variation in magnetization intensity and low-temperature titanomagnetite oxidation of ocean floor basalts. Nature 301:384–388

Bowles JA, Johnson PH (1999) Behavior of oceanic crustal magnetization at high temperatures: viscous magnetization and the marine magnetic anomaly source layer. Geophys Res Lett 26:2279–2282

Buddington AF, Lindsley DH (1964) Iron titanium oxide minerals and synthetic equivalents. J Petrol 5:310–357

Cande SC, Kent DV (1992) A new geomagnetic polarity time scale for the Late Cretaceous and Cenozoic. J Geophys Res 97:13917–13951

Cande SC, Kent DV (1995) Revised calibration of the geomagnetic polarity timescale for the Late Cretaceous and Cenozoic. J Geophys Res 100:6093–6095

Carlut J, Horen H (2007) Oceanic crust magnetization. In: Gubbins D, Herrero-Bervera E (eds) Encyclopedia of Geomagnetism and Paleomagnetism. Springer, Germany, pp 596–599

Carmichael ISE, Nicholls J (1967) Iron-titanium oxides and oxygen fugacities in volcanic rocks. J Geophys Res 72:4665–4687

Day R, Fuller MD, Schmidt VA (1977) Hysteresis properties of titanomagnetites: grain size and composition dependence. Phys Earth Planet Int 13:260–267

DeMets C, Gordon RG, Argus DF, Stein S (1990) Current plate motions. Geophys J Int 101:425–478

Dunlop DJ (2002a) Theory and application of the Day plot (M_{rs}/M_s vs. H_{cr}/H_c) 1. Theoretical curves and tests using titanomagnetite data. J Geophys Res 107(EM 4-1-EPM):4–22. doi:10.1029/2001JB000486

Dunlop DJ (2002b) Theory and application of the Day plot (Mrs/Ms vs. Hcr/Hc) 2. Application to data for rocks, sediments, and soils. J Geophys Res 107:EPM5-1–EPM5-15. doi:10.1029/2001JB000487

Dyment J, Arkani-Hamed J, Ghods A (1997) Contribution of serpentinized ultramafics to marine magnetic anomalies at slow and intermediate spreading centres: insights from the shape of the anomalies. Geophys J R Astron Soc 129:691–701

Expedition-309-Scientists (2005) Superfast spreading rate crust 2: a complete in situ section of upper oceanic crust formed at a superfast spreading rate. IODP Prel Rep 309. doi:10:2204/iodp.pr.309

Expedition 309/312 Scientists (2006) Superfast spreading rate crust 2 and 3: a complete *in situ* section of upper oceanic crust formed at a superfast spreading rate. IODP Prel Rep 312. doi:10:2204/iodp.pr.309312.2006

Furuta T, Levi S (1983) Basement paleomagnetism of Hole 504B. Initial Rep Deep Sea Drilling Project 69:711–720

Gee JS, Kent DV (2007) Source of oceanic magnetic anomalies and the geomagnetic polarity timescale. In: Kono M (ed) Treatise on geophysics, vol 5, Geomagnetism. Elsevier, Amsterdam, pp 455–507

Greenberg DS (1974) MoHole: geopolitical fiasco. In: Gass IG, Smith PJ, Wilson RCL (eds) Understanding the earth. Open University Press, Maidenhead, pp 343–349. Cambridge, MA, MIT Press [1971]

Gromme CS, Wright TL, Peck DL (1969). Magnetic properties and oxidation of iron-titanium oxide minerals in Alae and Makaopuhi Lava Lakes, Hawaii. J Geophys Res 74:5277–5293

Harrison CGA (1976) Magnetization of the oceanic crust. Geophys J R Astron Soc 47:257–283

Heirtzler JR, Dickson GO, Herron EM, Pitman WC III, Le Pichon X (1968) Marine magnetic anomalies, geomagnetic reversals and motions of the ocean floors and continents. J Geophys Res 73:2119–2136

Herrero-Bervera E, Acton G (2005) Magnetic properties and absolute paleointensity of upper oceanic crust generated by superfast seafloor spreading, ODP Leg 206. Eos Trans AGU 86(52). Fall Meeting Suppl., Abstract GP23A-0028, 2005

Herrero-Bervera E, Acton G (2011) Absolute paleointensities from and intact section of oceanic crust cored at ODP/IODP Site 1256 in the Equatorial Pacific, The Earth's Magnetic Interior. IAGA Special Sopron Book Series, Springer, Germany

Herrero-Bervera E, Valet J-P (2005) Absolute paleointensity and reversal records. From the Waianae sequence (Oahu, Hawaii, USA). Earth Planet Sci Lett 287:420–433

Herrero-Bervera E, Valet JP (2009) Testing determinations of absolute paleointensity from the 1955 and 1960 Hawaiian flows. Earth Planet Sci Lett 287:420–433

Huestis SP, Acton GD (1997) On the construction of geomagnetic timescales from non-prejudicial treatment of magnetic anomaly data from multiple ridges. Geophys J Int 129:176

Ildefonse B, Blackman D, John BE, Ohara Y, Miller DJ, MacLeod CJ (2006) IODP Expeditions 304–305 scientific party (2006) IODP expeditions 304 & 305 characterize the lithology, structure, and alteration of an oceanic core complex. Sci Drilling 3. doi: 10.2204/iodp.sd.3.01.2006

Irving E (1970) The Mid-Atlantic Ridge at 45o N, XIV, Oxidation and magnetic properties of basalts: review and discussion. Can J Earth Sci 7:1528–1538

Irving E, Park JK, Haggerty SE, Aumento F, Loncarevic B (1970a) Magnetism and opaque Mineralogy of basalts from the Mid-Atlantic Ridge ridge at 45°N. Nature 228:974

Irving E, Robertson WA, Aumento F (1970b) The Mid-Atlantic Ridge near 45° N, VI. Remanent intensity, susceptibility and iron content of dredge samples. Can J Earth Sci 7:226–238

Johnson HP, Pariso JE (1993) Variations in oceanic crustal magnetization: systematic changes in the last 160 million yaears. J Geophys Res 98:435–445

Koepke J, Christie DM, Dziony W, Lattard D, Meclennan J, Park S, Scheibner B, Yamasaki T, Yamasaki S (2008) Petrography of the dike-gabbro transition at IODP Site 1256 (equatorial Pacific): the evolution of the granoblastic dikes. Geochem Geophys Geosyst 9:Q07O09. doi:10/1029/2008GC001939

Krása D, Matzka J (2007) Inversion of titanomaghemite in oceanic basalt during heating. Phys Earth Planet Inter 160:169–179

Krása D, Herrero-Bervera E (2005) Alteration induced changes of magnetic fabric as exemplified by dykes of the Koolau volcanic range. Earth Planet Sci Lett 240:445–453

Krása D, Shcherbakov VP, Kunzmann T, Petersen N (2005) Self-reversal of remanent magnetization in basalts due to partially oxidized titanomagnetites. Geophys J Int 162(1):115–136

Matzka J, Krása D, Kunzmann T, Schult A, Petersen N (2003) Magnetic state of 10–40 Ma old ocean basalts and its implications for natural remanent magnetization. Earth Planet Sci Lett 206:541–553

Mevel C, Gillis KM, Allan JF, Meyer PS (eds) (1996) Proceedings of ODP, scientific results, vol 147. Ocean Drilling Program, College Station, TX

Morley LS, Larochelle A (1964) Paleomagnetism as a means of dating geological events. R Soc Can Spec Publ 8:39–50

Natland JH, Dick HJB., Miller DJ, Von Herzen RP (eds) (2002) Proceedings of ODP, scientific results, vol 176 [Online]. Available from World Wide Web. http://www-odp.tamu.. edu/publications/176_SR/176sr.htm

Pariso JE, Johnson HP (1991) Alteration processes at Deep Sea Drilling Project/Ocean Drilling Program Hole 504B at the Costa Rica Rift: implications for magnetization of oceanic crust. J Geophys Res 96:11703–11722

Pariso JE, Johnson HP (1993a) Do lower crustal rocks record reversals of the Earth's magnetic field? Magnetic petrology of gabbros from Ocean Drilling Program Hole 735B. J Geophys Res 98:16013–16032

Pariso JE, Johnson HP (1993b) Do layer 3 rocks make a significant contribution to marine magnetic anomalies? In situ magnetization of gabbros at Ocean Drilling Program Hole 735B. J Geophys Res 98:16033–16052

Petersen N, Eisenach P, Bleil U (1979) Low temperature alteration of the magnetic minerals in ocean floor basalts. In: Talwani M, Harrison CGA, Hayes D (eds) Deep drilling results in the Atlantic Ocean: ocean crust. American Geophysical Union, Washington, DC, pp 169–209

Petronotis KE, Gordon RG, Acton GD (1992) Determining paleomagnetic poles and anomalous skewness from marine magnetic anomaly skewness data from a single plate. Geophys J Int 109:209–224

Petronotis KE, Gordon RG, Acton GD (1994) A 57-Ma Pacific plate paleomagnetic pole determined from a skewness analysis of crossings of marine magnetic anomaly 25r. Geophys J Int 118:529–554

Phipps Morgan J, Chen YJ (1993) The genesis of oceanic crust: Magma injection, hydrothermal circulation, and crustal flow. J Geophys Res 98:6283–6297

Royer J-Y, Gordon RG (1997) The motion and boundary between the Capricorn and Australian plates. Science 277:1268–1274

Schneider DA (1988) An estimate of the long-term non-dipole field from marine magnetic anomalies. Geophys Res Lett 15:1,105–1,108

Schouten H (2002) Paleomagnetic inclinations in ODP Hole 417D reconsidered: variable tilting or secular `variation? Geophys Res Lett 29. doi:10.1029/2001GL013581

Shackleton NJ, Berger A, Peltier WR (1990) An alternative astronomical calibration of the lower pleistocene timescale based on ODP Site 677. Trans R Soc Edinb 81: 251–261

Shipboard Scientific Party, Site 1256 (2003) In: Wilson DS, Teagle DAH, Acton GD et al (eds) Proceedings of ODP, initial Reports, vol 206 [CD-ROM]. Available from: Ocean Drilling Program, Texas A&M University, College Station, TX, pp 1–396

Smith GA (1986) Selective destructive demagnetization of breccias from DSDP Leg 83: a microconglomerate test. Earth Planet Sci Lett 78:315–321

Smith GM, Banerjee S (1986) Magnetic structure of the upper kilometer of the marine crust at Deep Sea Drilling Project Hole 504B, Eastern Pacific Ocean. J Geophys Res 91:10,337–10,354

Teagle DAH, Wilson DS, Acton GD, the ODP Leg 206 Shipboard Party (2004) The road to the MoHole, four decades on: deep drilling at Site 1256. Eos 49: 521–531

Teagle DAH, Alt JC, Umino S, Miyashita S, Banerjee NR, Wilson DS, the Expedition 309/312 Scientists (2006) Superfast Spreading Rate Crust 2 and 3. In: Proceedings of IODP, vol 309/312. Integrated Ocean Drilling Program Management International, Inc., Washington, DC, 2006 pp. doi:10.2204/iodp.proc.309312

Tivey MA, Johnson HP (1987) The central anomaly magnetic high: implications for ocean crust construction and evolution. J Geophys Res 92:12685–12694

Valet J-P (2003) Time variations in geomagnetic intensity. Rev Geophys 1–44. doi:10.1029/2001RG000104

Valet JP, Herrero-Bervera E, Carlut J, Kondopoulou D (2010) A selective procedure for absolute paleointensity in lava flows. Geophys Res Lett 37. doi:10.1029/2010GL044100

Verosub KL, Moores EM (1981) Tectonic rotations in extensional regimes and their paleomagnetic consequences for oceanic basalts. J Geophys Res 86:6335–6349

Vine FJ, Matthews DH (1963) Magnetic anomalies over oceanic ridges. Nature 199:947–949

Wilson DS, Teagle DAH, Acton GD, Firth JV (2003) An in situ section of upper oceanic crust created by superfast seafloor spreading. Proc ODP Init Res 206:1–125. (Texas A&M University, College Station, Texas) Ocean Drilling Program

Wilson DS, Teagle DAH, Alt JA, Banerjee NR, Umino S, Miyashita S, Acton GD, Anma R, Barr SR, Belghoul A, Carlut J, Christie DM, Coggon RM, Cooper KM, Cordier C, Crispini L, Durand SR, Einaudi F, Galli L, Gao Y, Geldmacher J, Gilbert LA, Hayman NW, Herrero-Bervera E, Hirano N, Holter S, Ingle S, Jiang S, Kalberkamp U, Kerneklian M, Koepke J, Laverne C, Lledo Vasquez HL, Maclennan J, Morgan S, Neo N, Nichols HJ, Park S-H, Reichow MK, Sakuyama T, Sano T, Sandwell R, Scheibner B, Smith-Duque CE, Swift SA, Tartarotti P, Tikku AA, Tominaga M, Veloso EA, Yamasaki T, Yamazaki S, Ziegler C (2006) Drilling to gabbro in intact ocean crust. Science 312:1016–1020. doi.org/10.1126/science.1126090

Worm H-U (2001) Magnetic stability of oceanic gabbros from ODP Hole 735B. Earth Planet Sci Lett 193:287–302

Worm H-U, Bach W (1996) Chemical remanent magnetization in oceanic sheeted dikes. Geophys Res Lett 223:1123–1126

Zhou W, Van der Voo R, Peacor DR, Zhang Y (2000) Variable Ti-content and grain size of titanomagnetite as a function of cooling rate in very young MORB. Earth Planet Sci Lett 179:9–20

Magnetic Mineralogy of a Complete Oceanic Crustal Section (IODP Hole 1256D)

12

David Krása, Emilio Herrero-Bervera, Gary Acton, and Sedelia Rodriguez

Abstract

Oceanic crust is the carrier of the marine magnetic anomalies and is therefore a valuable archive of geomagnetic information. ODP/IODP Hole 1256D was the first to sample an entire sequence of oceanic crust down to the gabbro. We studied the vertical variation of magnetic remanence carriers by means of scanning electron microscopy, microanalysis and rock magnetic measurements. The extrusive layer contains dendritic, low-temperature oxidized titanomagnetites (TMs), i.e. titanomaghemite, with initial compositions close to values previously reported for mid-ocean ridge basalts (MORB). The degree of low-temperature oxidation (maghemitisation) remains fairly constant across the extrusives. We explain the observed increase in Curie temperature with depth by submicron inversion of titanomaghemite to intergrowths of titanomagnetite and nonmagnetic phases, where the Ti-content of titanomagnetite is decreasing with depth. In the underlying sheeted dikes, TMs are again the primary magnetic mineral. Due to slower cooling, they are in most cases oxy-exsolved into lamellar intergrowths of Ti-poor TMs and ilmenite. The magnetominerals are altered to a much higher degree than in the extrusives. In the gabbroic part of the section, TMs reach sizes up to several mm, although the magnetic grain size remains consistently in the pseudo-single-domain range because of grain subdivision by exsolution lamellae. The extrusives carry a thermoremanent magnetisation (TRM), retaining the primary paleomagnetic direction but with a reduced remanence intensity. The sheeted dikes hold a thermo-chemical remanent magnetization (TCRM) or secondary TRM acquired during hydrothermal alteration, whereas the underlying gabbro acquired a TCRM significantly after emplacement due to slow cooling at this depth.

12.1 Introduction

Magnetic minerals in oceanic crust are the carriers of the marine magnetic anomalies – the feature that provided the first evidence for sea floor spreading and the paradigm of plate tectonics (Vine and Matthews 1963). More recently, it has been argued that the rocks making

D. Krása (✉)
European Research Council Executive Agency, B-1049 Brussels, Belgium
e-mail: david.krasa@ec.europa.eu

E. Petrovský et al. (eds.), *The Earth's Magnetic Interior*, IAGA Special Sopron Book Series 1,
DOI 10.1007/978-94-007-0323-0_12, © Springer Science+Business Media B.V. 2011

up oceanic crust are potentially a highly valuable, continuous archive of geomagnetic field variations, including directional information capable of constraining the size of non-dipole field components (Schneider 1988, Acton et al. 1996) and palaeointensity information that provides a measure of the strength of the geomagnetic field over the past 180 Ma (Gee et al. 1996, Juarez et al. 1998, Gee et al. 2000, Carlut and Kent 2002). However, these studies have also shown that the reliability of this geomagnetic record is highly dependent on emplacement conditions, magnetic mineral grain size, and alteration state.

Because of their value as carriers of directional and intensity information about the geomagnetic field, magnetic minerals in mid-ocean ridge basalts (MORB) have been studied in great detail (e.g. Petersen et al. 1979, Smith and Banerjee 1986, Johnson and Pariso 1993, Kent and Gee 1994, Xu et al. 1997, Zhou et al. 1999a, Zhou et al. 2000, Zhou et al. 2001, Doubrovine and Tarduno 2004, Doubrovine and Tarduno 2006, Wang et al. 2006, Gee and Kent 2007, Krása and Matzka 2007, Matzka and Krása 2007). MORBs have a tholeiitic composition, where the predominant primary ferrimagnetic mineral is titanomagnetite with an average composition of $Fe_{2.4}Ti_{0.6}O_4$, so-called TM60. In tholeiites, this titanomagnetite is often co-existing with primary hemoilmenite. However, the latter is paramagnetic at ambient temperatures, and therefore does not contribute to the magnetic remanence of MORBs. The age dependence of magnetic properties of MORBs, i.e. their lateral variation with regard to oceanic spreading centres, has received particular attention (Bleil and Petersen 1983, Xu et al. 1997, Tivey and Tucholke 1998, Zhou et al. 1999b, Zhou et al. 2001, Matzka et al. 2003). The variation of magnetic properties with depth below seafloor has been described with much less detail, also because of the technical difficulties to obtain an undisturbed and complete section of oceanic basement.

Therefore, sampling a complete crustal section down to the gabbro has been a longstanding goal of oceanic drilling. In situ samples are particularly needed to understand crustal accretion processes and the evolution of oceanic crust, but also to study the origin of the marine magnetic anomalies, and the magnetic recording properties of the different depth sections of oceanic crust.

ODP/IODP Hole 1256D, which was drilled during Leg 206 (Wilson et al. 2003), and Expeditions 309 and 312 (Teagle et al. 2006), has provided us with such a complete section through fast spreading oceanic crust and is therefore ideally suited to study the magnetic mineralogy of a complete vertical stack of oceanic basement rocks, covering the extrusive part, the sheeted dikes, and the underlying gabbros. A complete description of the drilling location, its geological and tectonic setting, and details on the lithology are contained in the companion chapter in this book (Chapter 11, this volume).

In the present chapter, our aim is to investigate the magnetic mineralogy of this crustal section by means of scanning electron microscopy and microanalysis, in order to constrain the mode and timing of magnetic remanence acquisition and the reliability of the palaeomagnetic record contained in these rocks.

12.2 Samples and Experimental Methods

The sampling of the complete section of fast spreading oceanic crust in Hole 1256D is reported in great detail in the companion chapter Herrero-Bervera et al. (Chapter 11). From the samples described there, we picked a subset of 43 representative samples throughout the section to study them by means of scanning electron microscopy (SEM) and energy dispersive x-ray spectroscopy (EDS) to observe trends in average magnetic grain size and grain morphology, magnetic mineral composition, and the change in magnetic mineral alteration state and mineral paragenesis depending on depth and lithology.

The analysis was carried out with a Jeol JSM-5900 SEM with EDS analysis capability. The EDS data were analysed with NORAN system six version 1.8 software using the quant fit method (filter without standards) and the PROZA Phi-Rho-Z matrix correction algorithm.

12.3 Electron-Microscopic Characterisation of Magnetic Minerals

In Fig. 12.1, some representative samples from the sheet flow and massive basalt part of the section are shown. The primary magnetic mineral is titanomagnetite, although as we will show below, for grains

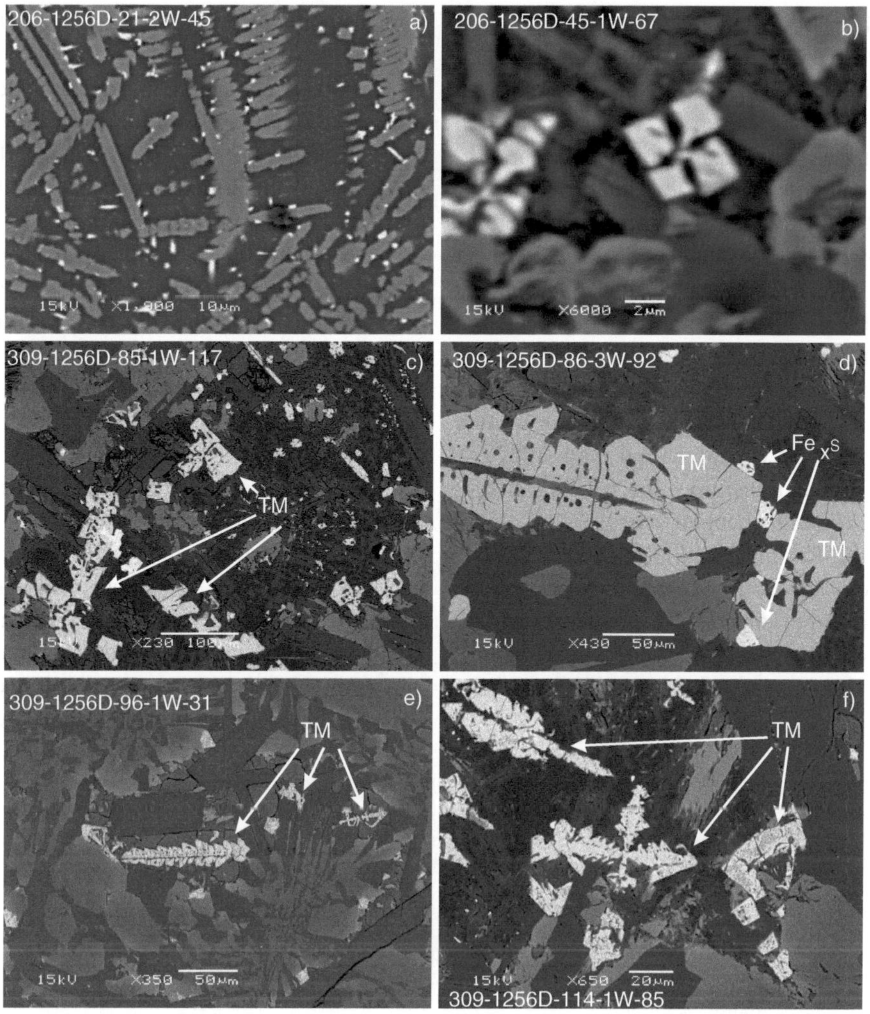

Fig. 12.1 Characteristic electron micrographs of samples from the sheet flow and massive basalt part of the section. Sample identifiers are shown according to IODP nomenclature. (**a**) Volcanic breccia consisting of aphyric cryptocrystalline basalt and glass with micron to sub-micron titanomaghemite grains, (**b**) aphyric cryptocrystalline basalt sheet flow with small cruciform titanomaghemite, (**c**) and (**d**) massive aphyric fine-grained basalt containing large (>100 μm) dendritic titanomaghemite and iron sulfides, (**e**) Aphyric cryptocrystalline basalt sheet flow, (**f**) aphyric fine-grained basalt sheet flow, both containing dendritic and cruciform titanomaghemite grains (all rock descriptions in this and the following figure captions after Wilson et al. 2003, Teagle et al. 2006)

that are several microns or larger, the titanomagnetite is always partially oxidised and so by definition is titanomaghemite. (In this chapter, we use the term "titanomaghemite" in a wide sense, i.e. for titanomagnetite of any composition that has undergone partial or full low-temperature oxidation ($0<x<1$, $z>0$)). As a very general trend, the physical grain size increases with depth, with magnetic mineral particles in the uppermost part of the section being in the range of below one to several tens of μm (Fig. 12.1a, b). In the deeper parts of the section, grain sizes increase to values from several tens to above 100 μm (Fig. 12.1c–f). However, the variability of grain size from flow to flow and even within flows is much more pronounced in magnitude than this general trend. In many cases, the grain size distribution is bimodal, such as in Fig. 12.1c, where larger particles dominate the micrographs but, in addition, smaller grains below 10 μm are present. We have not imaged nor analyzed sub-micron titanomagnetite in interstitial glass, which has been shown to exist in oceanic basalts and to be unoxidised or less oxidised than larger titanomagnetite grains and has been inferred to carry a significant part of the magnetic remanence (Smith 1979, Xu et al. 1997, Zhou et al. 2000).

Throughout the basaltic part, grains have the typical dendritic and cruciform shapes characteristic for titanomagnetites in MORBs. In terms of grain morphology and grain size, no distinction between sheet flows and massive basalts can be made. Again, the variation of these parameters within individual cooling units is much more pronounced.

Titanomagnetites in the basaltic part often contain shrinkage cracks, a typical result of low-temperature oxidation in MORBs (Petersen and Vali 1987), a gradual process that transforms the primary titanomagnetite into titanomaghemites with varying oxidation degree (e.g., Readman and O'Reilly 1972). The shrinkage cracks appear throughout this part of the section, and no depth dependence of oxidation state can be deduced from the microscopic observations. In addition to titanomagnetite/titanomaghemite, many samples contain Fe-sulphides as an additional opaque mineral phase (Fig. 12.1d).

A marked change in magnetic mineralogy takes place when moving from the sheet flow and massive basalt part of the section to the underlying sheeted dikes. While titanomagnetite was the dominant primary magnetic mineral formed during emplacement of the sheeted dikes, this phase appears completely altered in the electron micrographs (Fig. 12.2). The original titanomagnetites are only discernible by their grain shapes and only faintly visible, as they are partially to completely replaced by secondary minerals (Fig. 12.2a, b). The alteration products are mostly sphene and rutile. Similar to the basaltic part of the section, grain sizes are highly variable with an average around 100 μm. The particle morphology of the original titanomagnetites is more idiomorphic in comparison to the overlying sheet flows and massive basalts. In many cases, the remnants of a lamellar texture are visible within the altered titanomagnetite particles (Fig. 12.2c). This is the typical texture of high-temperature oxidized titanomagnetite, which appears

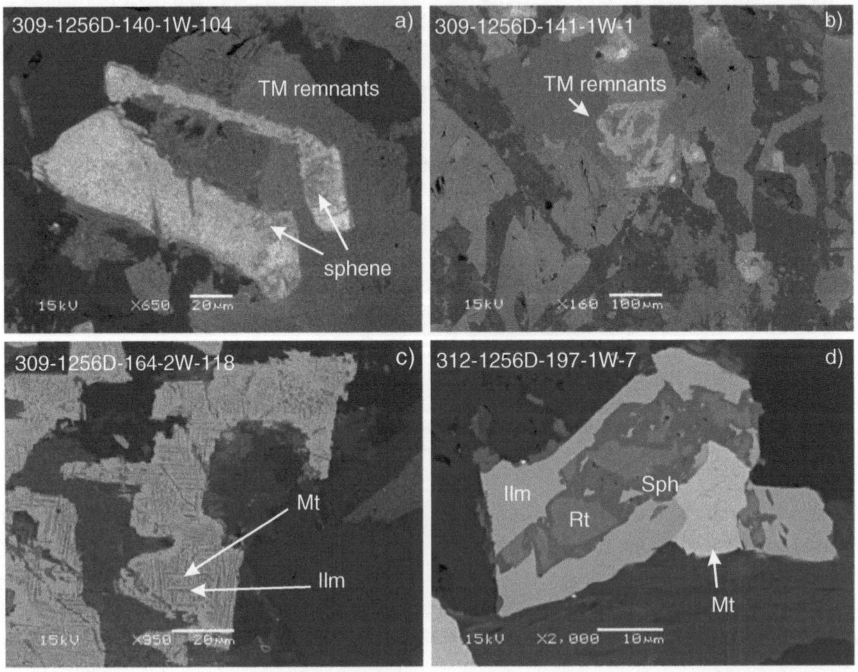

Fig. 12.2 Characteristic electron micrographs of samples from the sheeted dikes. Sample identifiers are shown according to IODP nomenclature. (**a**) Intrusive margin breccia consisting of aphyric cryptocrystalline-microcrystalline basalt, primary titanomagnetite partially replaced by secondary minerals; (**b**) massive aphyric fine-grained basalt, only outlines of the primary titanomagnetite are visible, which is partially to completely replaced by secondary minerals; (**c**) massive aphyric microcrystalline basalt containing high-temperature oxidised titanomagnetite, ilmenite lamellae are highly altered; (**d**) aphyric microcrystalline basalt containing secondary magnetite in close assemblage with rutile and sphene

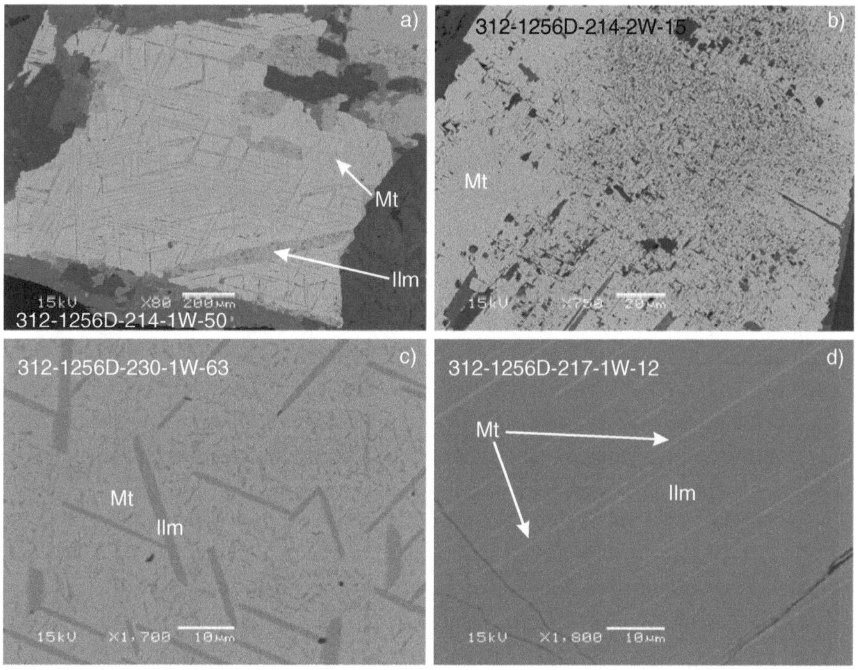

Fig. 12.3 Characteristic electron micrographs of samples from the gabbroic part of the section. Sample identifiers are shown according to IODP nomenclature. (**a**) Medium-grained quartz-rich oxide diorite with large (>1 mm) oxy-exsolved titanomagnetite grains; (**b**) Medium-grained disseminated oxide gabbro with large oxy-exsolved titanomagnetite grains, ilmenite lamellae are strongly altered; (**c**) Medium-grained oxide gabbro containing oxy-exsolved titanomagnetite with two generations of ilmenite lamellae; (**d**) Medium-grained disseminated oxide gabbro with ilmenite containing finely exsolved magnetite needles

in slowly cooled basaltic rocks. High-temperature oxidation causes oxy-exsolution of titanomagnetite, forming lamellae of magnetite and ilmenite. In the oxy-exsolved particles of the sheeted dikes of this section, magnetite seems to be more resistant to alteration and replacement by secondary minerals than the ilmenite fraction.

In the lowermost part of the sheeted dikes (1348–1407 mbsf), termed the "granoblastic dikes" (Koepke et al. 2008), where partial to complete recrystallization occurs, the magnetic mineralogy changes dramatically. Secondary magnetite appears as the dominant magnetic mineral phase. This magnetite has a very fresh appearance and often occurs in close assemblage with other secondary minerals such as rutile and sphene (Fig. 12.2d). Its morphology is idiomorphic, and it is discernible from the primary magnetic minerals by its freshness and its composition, as will be discussed in the next section.

Grain sizes in the gabbroic part of the section increase up to several millimeters, and magnetic minerals consist of lamellar or sandwich type oxy-exsolution intergrowths of magnetite and ilmenite (Fig. 12.3a). In some cases, two generations of exsolution lamellae can be found (e.g. sample 312-1256D-230-1w-63, Fig. 12.3c). Some ilmenites contain fine needles of magnetite, again representing oxy-exsolution (Fig. 12.3d). Generally, the magnetic minerals appear fairly fresh in the gabbroic part of the section. In cases where alteration features occur, the ilmenites are generally more altered than the magnetites and are frequently replaced by sphene and rutile as already described for the sheeted dikes (Fig. 12.3b). Regarding the magnetic mineralogy, no difference between gabbros and oxide gabbros could be observed.

12.4 Composition of Magnetic Minerals

Semi-quantitative, standardless, EDS analyses of magnetic minerals were carried out for the subset of 43 samples throughout the whole drilled section to determine the chemical composition of the ferrites and to determine the causes of the variations of rock magnetic parameters described in Herrero-Bervera et al.

(Chapter 11). For meaningful analyses, the grain size of magnetic mineral particles has to be above approx. 5–10 μm, as otherwise the surrounding silicate minerals contaminate the analysis. Therefore, analyses were only obtained for samples containing magnetic mineral particles above this grain size. The particles were analysed for O, Mg, Al, Si (to detect silicate contamination), S, Ca, Ti, V, Cr, Mn and Fe. The concentration of Fe and Ti was used to determine the titanomagnetite composition parameter x as defined by the general titanomagnetite formula $Fe_{3-x}Ti_xO_4$. Under the assumption of an inverse spinel structure X_3O_4 (X being the cations), the measurement software calculated the number of cations N_C per formula unit. The difference 3-N_C was used to calculate the titanomagnetite oxidation parameter z. For each studied sample, between 1 and 12 titanomagnetite particles were analysed (on average 5 particles). Successful analyses were obtained for a total of 34 samples spanning the full length of the cored section.

Figure 12.4 shows side by side the down-core evolution of the Curie temperature, the titanomagnetite composition parameter x and the titanomagnetite oxidation parameter z. Throughout the extrusive part of the section, both x as well as z stay remarkably constant, despite the rather large variation in Curie temperature. Even in the sheeted dykes, where the Curie temperature reaches values of around 570°C,

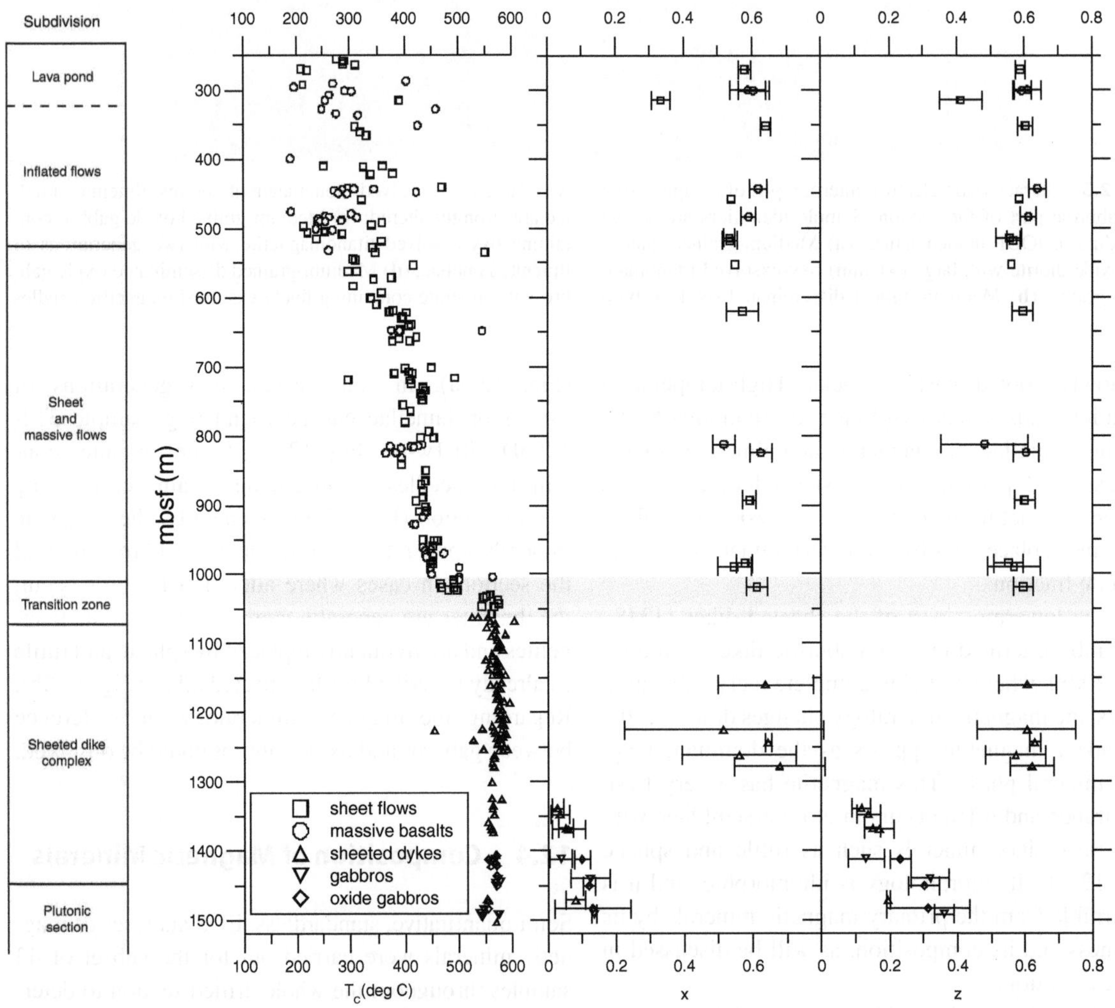

Fig. 12.4 Downcore plots of T_C (Chapter 11, this volume), composition parameter x and oxidation parameter z vs. depth of the titanomagnetite fraction. The error bars for x and z show ± one standard deviation. In cases where they are missing, the analysis is based on only one particle measured. The shaded zone marks the transition from extrusives to intrusives (see simplified lithological column on the *left-hand* side)

close to magnetite values, x and z both remain at values of around 0.6 for both parameters. The variance between individual grains in the sheeted dikes increases significantly, however, as expressed by the error bars. In the gabbroic part, x ranges from 0 to 0.15 and z has values between 0.1 and 0.4.

12.5 Discussion

A gradual increase of the Curie temperature of MORBs away from oceanic spreading centres is generally attributed to an increase in the titanomagnetite oxidation parameter z (e.g. Petersen et al. 1979, Zhou et al. 2001, Matzka et al. 2003). This increase in z is well established by microprobe determinations of titanomagnetite compositions and lattice parameters. In the present case, we observed an increase in T_C in a vertical section through coeval oceanic crust. As shown in Fig. 12.4, the increase in T_C for the present section is not caused by this process.

Figure 12.5 further illustrates this fact: If the T_C increase was caused by increasing z then one would expect that the data points in the T_C versus z plot are distributed along the upwards curved line for an initial

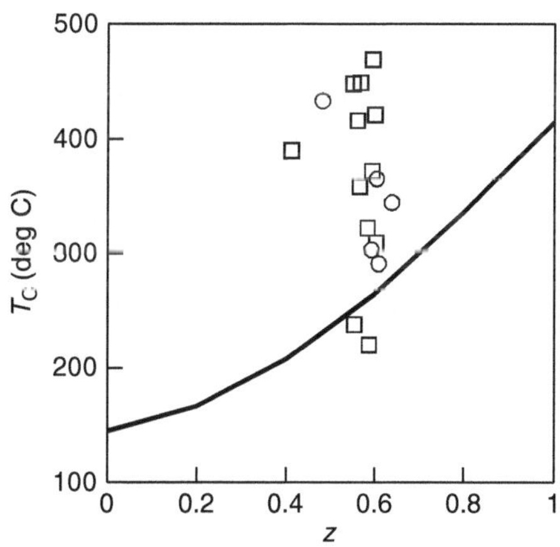

Fig. 12.5 Plot of Curie temperature versus titanomagnetite composition parameter z for the samples from the extrusive part of the section. This plot shows that z stays essentially constant throughout this part of the section despite the change in Curie temperature. The black line shows the trend expected for an initial titanomagnetite composition of $x = 0.6$ if the increase in T_C was caused by low temperature oxidation and an associated increase in the titanomagnetite oxidation parameter z

titanomagnetite composition with $x = 0.6$. Instead, z is essentially constant for all studied samples.

Wang et al. (2006) noted that coercivity generally increases with increasing oxidation for their suite of ocean crust pillow basalts. Coercivity is roughly constant (or decreases very slightly) down the entire extrusive section at Site 1256 (Chapter 11). This supports the EDS results that oxidation is constant throughout the extrusive section.

If low-temperature oxidation cannot account for the increase in T_C, then some other mechanism must be responsible. When studying the downcore variation of rock magnetic parameters for this section (Chapter 11), one finds that with increasing depth within the extrusive part of Hole 1256D, M_{RS}/M_S is decreasing while B_{CR}/B_C is increasing. Such a characteristic evolution of rock magnetic parameters together with an increase in T_C was also found for titanomaghemite subjected to increasing temperatures in laboratory heating experiments (Krása and Matzka 2007). The authors of this study attributed these findings to titanomaghemite inversion.

Titanomaghemite inversion is a mineralogical process whereby the unstable crystal lattice of titanomaghemite inverts into a fine intergrowth of a Ti-poor spinel phase close to magnetite on the one hand, and a mixture of Ti-rich phases such as pseudobrookite, ilmenite or anatase (+hematite, depending on the initial composition and oxidation state) on the other hand (Readman and O'Reilly 1970). The exact composition of the Ti-rich phase mixture is dependent on the initial composition and oxidation state of the titanomaghemite. The process itself was found to be gradual, such that the composition of the spinel phase resulting from inversion gradually approaches the final value close to magnetite with increasing laboratory heating temperature.

Due to the similar downcore variation of rock magnetic parameters in the extrusive part of Hole 1256D as compared with the laboratory heating experiments of Krása and Matzka (2007), we hypothesize that this variation is caused by the same phenomenon of titanomaghemite inversion.

Indeed, previous studies have found a general increase in alteration temperatures within the extrusive part of Hole 1256D (Laverne et al. 2006, Wilson et al. 2006, Koepke et al. 2008), from around 100°C at the top of the section to 250°C at the transition between lavas and sheeted dikes. These temperatures are lower than the ones at which titanomaghemite inversion was observed in laboratory experiments. However, the timescales involved in laboratory experiments are also much shorter than what would be expected for the alteration of magnetic minerals in in situ oceanic crust.

This increase in alteration temperature with increasing depth can explain the increase of Curie temperature observed within the lavas of Hole 1256D. The rising Curie temperature would then be due to the gradually proceeding inversion of the original titanomaghemite similar to what is observed in laboratory experiments, however at much longer time scales.

The gradual increase of Curie temperature during inversion points to spinodal decomposition (Putnis 1992) as the underlying mineralogical process, i.e., within the initially homogeneous crystal, sub-micron domains of Ti-poor and Ti-rich phases form. The composition of these gradually approaches the final compositional values, i.e. a composition close to magnetite for the Ti-poor spinel phase, and pseudobrookite, ilmenite, or anatase as the Ti-rich phase.

The size of these intergrowths is too small to be resolved by SEM techniques as shown schematically in Fig. 12.6 and so the measured composition basically corresponds to the composition of the titanomaghemite before inversion. Nevertheless, measurements of thermomagnetic curves are able to determine the Curie temperature of these fine spinel aggregates. Evidence for the presence of very fine intergrowths caused by inversion comes from the fact that M_{RS}/M_S and B_{CR}/B_C values as plotted in the (Day et al. 1977) plot (see Fig. 11.2) deviate from the ideal SD-MD mixing line (Dunlop 2002), and are pushed towards the SP admixture zone. This is evidence for the presence of very fine, super-paramagnetic (SP) particles, consistent with the fine intergrowths caused by inversion.

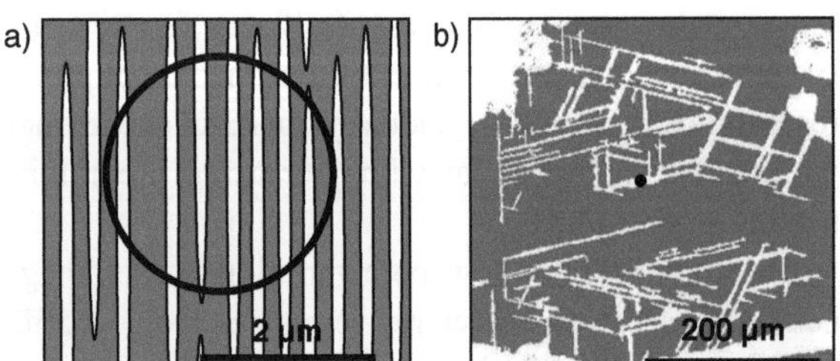

Fig. 12.6 Schematic representation of electron micrographs and EDS analysis volume. (**a**) In sub-micron structures such as the ones caused by titanomaghemite inversion, the EDS analysis volume averages over a volume much larger than the typical length scale of the inversion structure. Therefore, the measured composition basically corresponds to the composition of the titanomaghemite before inversion. (**b**) For larger scale oxy-exsolution structures, such as the ones encountered in the gabbros, EDS analysis is able to resolve the compositional differences

Below the extrusives, the upper part of the sheeted dikes displays Curie temperatures around 570°C, i.e. close to magnetite values, although the titanomagnetite composition and oxidation parameters x and z, respectively, still remain remarkably constant and similar to those observed in the lavas. On the basis of our inversion hypothesis, we interpret this as the presence of fully inverted titanomaghemite, where the resulting ferrimagnetic phase consists of essentially pure magnetite. Alteration temperatures in this part reach values between 250 and 400°C (Wilson et al. 2006). Down to about 1300 m below see floor (mbsf), the intergrowths caused by inversion still cannot be resolved by SEM, therefore the compositional parameters are similar to the extrusives. However, a gradual increase of the dimensions of these intergrowths can be deduced from the fact that the standard deviation of the measurements (as displayed by the error bars in Fig. 12.4) increases significantly.

The zone below 1350 mbsf, down to the dike-gabbro transition, is characterised by the occurrence of the "granoblastic dikes" (Koepke et al. 2008). Here, the sheeted dikes were transformed by thermal metamorphism. Estimates based on various geothermometers imply that temperatures were so high that the dikes were partially remelted. As described above, this part of the core is marked by the occurrence of large, idiomorphic, homogeneous magnetite grains, in association with other secondary minerals. Accordingly, the composition of the magnetic phase here, as witnessed by EDS measurements, is that of almost pure magnetite, with a low degree of oxidation $z < 0.2$.

In the gabbros, magnetic mineralogy is dominated by large titanomagnetite particles, displaying rather coarse oxy-exsolution structures of intergrown lamellae of magnetite and hemo-ilmenite. These structures are so large in scale that the EDS technique is able to distinguish between the two phases. Therefore, the spinel phase is marked by a low compositional parameter $x < 0.2$. However, it shows a certain degree of cation deficiency with z values ranging between 0.2 and 0.4. This might point to a more complex alteration history, with oxy-exsolution followed by further low-temperature oxidation. Such an ongoing alteration is also witnessed by the fairly altered state of the intergrown hemo-ilmenites in the gabbros.

The composition and oxidation state of magnetic minerals in the different parts of the complete section as determined by EDS analysis is shown in the Fe-Ti oxide ternary diagram in Fig. 12.7. The magnetic minerals of the extrusive part and the sheeted dykes all straddle the border between compositional zones 2 and 4 (Readman and O'Reilly 1970), and, as stated before, the inversion process cannot be resolved by EDS analysis. Contrary to that, the Fe-rich lamellae of the high-temperature oxidised magnetominerals of the gabbros are clearly shown to lie close to the compositional field expected for slightly low-temperature oxidised magnetite.

Based on the discussion of magnetic mineralogy, and on observations of rock magnetic parameters (Chapter 11), we assume that the lavas in the upper part of the section acquired a thermoremanent magnetisation during emplacement, which was later affected by low-temperature oxidation and inversion of the primary titanomagnetites. Previous investigations show that such an alteration process alters the record of magnetic palaeointensity, but in general does not alter the directional magnetic record (Krása and Matzka 2007). The same alteration process affected the upper part of the sheeted dikes, however, to a much higher degree, with partial to total replacement of the original titanomagnetite by secondary minerals. Therefore, the origin of the remanence in this part of the section is less clear, and might be significantly influenced by an acquisition of a thermo-chemical remanence (TCRM) in secondary magnetite. This is particularly the case in the lowermost part of the sheeted dikes, the granoblastic dikes. Here, the magnetic grain sizes are larger than elsewhere in the entire crustal section, falling in the pseudo-single domain to multidomain range, and the NRM intensity is reduced by a factor of >2 relative to the rest of the dike section. Together, these observations indicate it less likely that these rocks carry as stable of a primary remanent magnetisation as either the extrusives, upper dikes, or gabbros. The underlying gabbros contain oxy-exsolved titanomagnetite grains, which are able to carry a stable remanence due to the small size and extreme shape anisotropy of the magnetite lamellae. However, the presence of a deeper heat source over time spans extending significantly beyond the time of emplacement as evidenced by the thermal metamorphic overprint in the granoblastic dikes (Koepke et al. 2008), makes it likely that these rocks have acquired their remanence well after the initial emplacement.

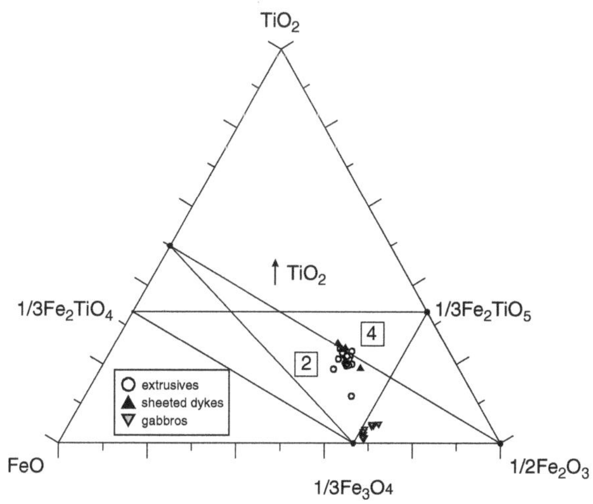

Fig. 12.7 Ternary diagram for Fe-Ti oxides showing the magnetic particle composition of samples of this study. The numbers in boxes designate the titanomaghemite compositional fields (Readman and O'Reilly 1970), which in turn determine the composition of inversion products of titanomaghemite lying within these fields

Overall, the microscopy results and rock magnetic properties indicate that virtually the entire section is capable of carrying a remanent magnetization that could contribute to the marine magnetic anomaly signal. The lowermost dike section probably contributes less per volume than other intervals owing to its alteration history, which has resulted in larger magnetic grain sizes and a relatively less stable magnetization within this relatively thin interval. Knowledge of the timing of acquistion is extremely important to determine the precise contribution of each layer to the marine magnetic anomaly signal. The alteration we document for the extrusive section and upper dikes is similar to what has been documented in laboratory experiments (Krása and Matzka 2007), in which the new magnetic minerals formed by alteration inherit the paleomagnetic direction of the parent minerals. Because parent minerals in these rocks cool and record the ambient geomagnetic field soon after formation at or very near the ridge axis, even after alteration (inversion) they will retain a primary signal, which surely contributes significantly to the marine magnetic anomaly signal. The gabbros (and lowermost dikes) have been altered in ways that indicate they have acquired thermo-chemical remanent magnetization well after emplacement. This does not mean they do not contribute significantly to the anomaly signal but the size of the contribution will depend on how gradually the magnetization was acquired. This is a question that is difficult to answer based on rock magnetic data alone, but can be investigated by combining rock magnetic constraints from studies like this one with marine magnetic anomaly studies that seek to estimate and model the origin of anomalous skewness (e.g., Arkani-Hamed 1988, 1991, Petronotis et al. 1992, Dyment et al. 1997).

Conclusions

Our magnetomineralogical study, together with rock magnetic and petrological data, allows us to further constrain the complex emplacement and alteration conditions throughout a complete section of fast spreading oceanic crust as encountered in ODP/IODP Hole 1256D. This in turn makes it possible to assess the mode of magnetic remanence acquisition, and the reliability of palaeomagnetic data retrieved from this core.

Our study shows that the topmost ~750 m of sheet flows and massive basalts carry a stable TRM whose intensity was later modulated by low-temperature oxidation and inversion of the predominant titanomagnetite. Although the primary magnetic mineralogy of the underlying sheeted dikes was probably very similar to the extrusive part of the section, the magnetic minerals here are affected much more strongly by alteration up to complete replacement by secondary minerals. The alteration degree generally increases with depth. Therefore, while the upper part of the dykes may still carry some primary remanence, a TCRM or secondary TRM completely replaces the primary remanence in the lowermost section of the sheeted dikes, the granoblastic dikes. The gabbros at the bottom of the section carry a stable TCRM, which was acquired significantly after their initial emplacement.

Our study also shows that Curie temperature determinations are able to resolve the changing mineralogical composition of extremely fine intergrowths, which occur on a scale too small to be discerned by electron miscroscopic techniques.

Acknowledgements This research used samples and data provided by the Ocean Drilling Program (ODP) and the Integrated Ocean Drilling Program (IODP). Funding for this research was provided by the National Science Foundation (NSF) through its support of ODP, IODP, and the United States Science Support Program (USSSP) and through NSF grants JOI-T309A4, OCE-0727764, and EAR-IF-0710571 to Herrero-Bervera, and a USSSP Post-Expedition Activity Award and NSF grant OCE-0727576 to Acton. Additional financial support to E. H-B was provided by SOEST-HIGP. D. K. was funded by a Royal Society of Edinburgh BP Trust Research Fellowship. The views expressed are purely those of the authors and may not in any circumstances be regarded as stating an official position of the European Research Council Executive Agency. We are very grateful to the reviewers of this chapter for their very constructive criticisms particularly those of Professor Nikolai Petersen. This is a SOEST contribution number 8147 and HIGP contribution number 1890.

References

Acton GD, Petronotis KE, Cape CD, Ilg SR, Gordon RG, Bryan PC (1996) A test of the geocentric axial dipole hypothesis from an analysis of the skewness of the central marine magnetic anomaly. Earth Planet Sci Lett 144:337–346

Arkani-Hamed J (1988) Remanent magnetization of the oceanic upper mantle. Geophys Res Lett 15:48–51

Arkani-Hamed J (1991) Thermoremanent magnetization of oceanic lithosphere inferred from a thermal evolution model: implications for the source of marine magnetic anomalies. Tectonophysics 192:81–96

Bleil U, Petersen N (1983) Variation in magnetization intensity and low-temperature titanomagnetite oxidation of ocean floor basalts. Nature 301:384–388

Carlut J, Kent DV (2002) Grain-size-dependent paleointensity results from very recent mid-oceanic ridge basalts. J Geophys Res. doi:10.1029/2001JB000439

Day R, Fuller M, Schmidt VA (1977) Hysteresis properties of titanomagnetites: grain-size and compositional dependence. Phys Earth Planetary Inter 13:260–267

Doubrovine PV, Tarduno JA (2004) Self-reversed magnetization carried by titanomaghemite in oceanic basalts. Earth Planet Sci Lett 222:959–969

Doubrovine PV, Tarduno JA (2006) Alteration and self-reversal in oceanic basalts. J Geophys Res. doi:10.1029/2006JB004468

Dunlop DJ (2002) Theory and application of the Day plot (M_{RS}/M_S versus H_{CR}/H_C) 1. theoretical curves and tests using titanomagnetite data. J Geophys Res. doi:10.1029/2001JB000486

Dyment J, Arkani-Hamed J, Ghods A (1997) Contribution of serpentinized ultramafics to marine magnetic anomalies at slow and intermediate spreading centres: insights from the shape of the anomalies. Geophys J R Astron Soc 129:691–701.

Gee JS, Kent DV (2007) Source of oceanic magnetic anomalies and the geomagnetic polarity timescale. In: Kono M (ed) Treatise on geophysics. Elsevier, Amsterdam, pp 455–507

Gee J, Schneider DA, Kent DV (1996) Marine magnetic anomalies as recorders of geomagnetic intensity variations. Earth Planet Sci Lett 144:327–335

Gee JS, Cande SC, Hildebrand JA, Donnelly K, Parker RL (2000) Geomagnetic intensity variations over the past 780 kyr obtained from near-seafloor magnetic anomalies. Nature 408:827–832

Johnson HP, Pariso JE (1993) Variations in oceanic crustal magnetization – systematic changes in the last 160 million years. J Geophys Res 98:435–445

Juarez MT, Tauxe L, Gee JS, Pick T (1998) The intensity of the Earth's magnetic field over the past 160 million years. Nature 394:878–881

Kent DV, Gee J (1994) Grain size-dependent alteration and the magnetization of oceanic basalts. Science 265: 1561–1563

Koepke J, Christie DM, Dziony W, Holtz F, Lattard D, Maclennan J, Park S, Scheibner B, Yamasaki T, Yamazaki S (2008) Petrography of the dike-gabbro transition at IODP Site 1256 (equatorial Pacific): the evolution of the granoblastic dikes. Geochem Geophys Geosyst. doi:10.1029/2008GC001939

Krása D, Matzka J (2007) Inversion of titanomaghemite in oceanic basalt during heating. Phys Earth Planetary Inter 160:169–179

Laverne C, Grauby O, Alt JC, Bohn M (2006) Hydroschorlomite in altered basalts from Hole 1256D, ODP Leg 206: the transition from low-temperature to hydrothermal alteration. Geochem Geophys Geosyst. doi:10.1029/2005GC001180

Matzka J, Krása D (2007) Oceanic basalt continuous thermal demagnetization curves. Geophys J Int 169:941–950

Matzka J, Krása D, Kunzmann T, Schult A, Petersen N (2003) Magnetic state of 10–40 Ma old ocean basalts and its implications for natural remanent magnetization. Earth Planet Sci Lett 206:541–553

Petersen N, Eisenach P, Bleil U (1979) Low temperature alteration of the magnetic minerals in ocean floor basalts. In: Talwani M, Harrison CGA, Hayes D (eds) Deep drilling results in the atlantic ocean: ocean crust. American Geophysical Union, Washington, DC, pp 169–209

Petersen N, Vali H (1987) Observation of shrinkage cracks in ocean floor titanomagnetites. Phys Earth Planetary Inter 46:197–205

Petronotis KE, Gordon RG, Acton GD (1992) Determining paleomagnetic poles and anomalous skewness from marine magnetic anomaly skewness data from a single plate. Geophys J Int 109:209–224

Putnis A (1992) Introduction to mineral sciences. Cambridge University Press, Cambridge

Readman PW, O'Reilly W (1970) The synthesis and inversion of non-stoichiometric titanomagnetites. Phys Earth Planetary Inter 4:121–128

Readman PW, O'Reilly W (1972) Magnetic properties of oxidized (cation-deficient) titanomagnetites (Fe, Ti, $\square)_3O_4$. J Geomagnetism Geoelectricity 24:69–90

Schneider DA (1988) An estimate of the long-term non-dipole field from marine magnetic-anomalies. Geophys Res Lett 15:1105–1108

Smith PPK (1979) Identification of single-domain titanomagnetite particles by means of transmission electron-microscopy. Can J Earth Sci 16:375–379

Smith GM, Banerjee SK (1986) Magnetic-structure of the upper kilometer of the marine crust at deep-sea drilling project Hole-504b, Eastern Pacific-Ocean. J Geophys Res 91:337–354

Teagle DAH, Alt JC, Umino S, Miyashita S, Banerjee NR, Wilson DS, the Expedition 309/312 Scientists (2006) Proceedings of IODP, vol 309/312. Integrated Ocean Drilling Program Management International, Inc., Washington, DC

Tivey MA, Tucholke BE (1998) Magnetization of 0–29 Ma ocean crust on the Mid-Atlantic Ridge, 25 degrees 30' to 27 degrees 10' N. J Geophys Res 103:17807–17826

Vine FJ, Matthews DH (1963) Magnetic anomalies over oceanic ridges. Nature 199:947–949

Wang D, R. Van der Voo, Peacor DR (2006) Low-temperature alteration and magnetic changes of variably altered pillow basalts. Geophys J Int 164:25–35

Wilson DS, Teagle DAH, Acton GD, the Shipboard Scientific Party (2003) An in situ section of upper oceanic crust formed by superfast seafloor spreading at site 1256. Proceedings of ODP, initial Reports, vol 206. Ocean Drilling Program, College Station, TX

Wilson DS, Teagle DAH, Alt JC, Banerjee NR, Umino S, Miyashita S, Acton GD, Anma R, Barr SR, Belghoul A, Carlut J, Christie DM, Coggon RM, Cooper KM, Cordier C, Crispini L, Durand SR, Einaudi F, Galli L, Gao YJ, Geldmacher J, Gilbert LA, Hayman NW, Herrero-Bervera E, Hirano N, Holter S, Ingle S, Jiang SJ, Kalberkamp U, Kerneklian M, Koepke J, Laverne C, Vasquez HLL, Maclennan J, Morgan S, Neo N, Nichols HJ, Park SH, Reichow MK, Sakuyama T, Sano T, Sandwell R, Scheibner B, Smith-Duque CE, Swift SA, Tartarotti P, Tikku AA, Tominaga M, Veloso EA, Yamasaki T, Yamazaki S, Ziegler C (2006) Drilling to gabbro in intact ocean crust. Science 312:1016–1020

Xu WX, Peacor DR, Dollase WA, Van der Voo R, Beaubouef R (1997) Transformation of titanomagnetite to titanomaghemite: a slow, two-step, oxidation-ordering process in MORE. Am Mineral 82:1101–1110

Zhou WM, Peacor DR, Van der Voo R, Mansfield JF (1999a) Determination of lattice parameter, oxidation state, and composition of individual titanomagnetite/titanomaghemite grains by transmission electron microscopy. J Geophys Res 104:17689–17702

Zhou WM, Van der Voo R, Peacor DR (1999b) Preservation of pristine titanomagnetite in older ocean-floor basalts and its significance for paleointensity studies. Geology 27:1043–1046

Zhou WM, Van der Voo R, Peacor DR, Wang DM, Zhang YX (2001) Low-temperature oxidation in MORB of titanomagnetite to titanomaghemite: a gradual process with implications for marine magnetic anomaly amplitudes. J Geophys Res 106:6409–6421

Zhou WM, Van der Voo R, Peacor DR, Zhang YX (2000) Variable Ti-content and grain size of titanomagnetite as a function of cooling rate in very young MORB. Earth Planet Sci Lett 179:9–20

Absolute Paleointensities from an Intact Section of Oceanic Crust Cored at ODP/IODP Site 1256 in the Equatorial Pacific

13

Emilio Herrero-Bervera and Gary Acton

Abstract

We have investigated the magnetic mineralogy and absolute paleointensity of basalt samples from Site 1256 cored during ODP Leg 206 and IODP Expeditions 309 and 312. The site is located on the Cocos Plate 5 km east of the transition zone between marine magnetic anomalies 5Bn.2n and 5Br (~15 Ma). The deepest hole, Hole 1256D, extends 250 m through sediments and 1257 m into the igneous upper oceanic crust generated by superfast seafloor spreading (>200 mm/yr) along the East Pacific Rise. This is the first drill site to penetrate an in situ and intact section of crust. The section consists of about 811 m of basaltic sheet flows and massive lavas, 346 m of sheeted-dike complex, and 100 m of gabbros and granoblastic dikes. Rock magnetic investigations included thermomagnetic analyses, alternating field, thermal demagnetization, saturation IRM, magnetic grain-size and coercivity analyses. Curie points identified titanomagnetites and titanomaghemites as the magnetic carriers and grain-size studies indicate that the carriers are mixtures of single domain (SD) and pseudosingle domain (PSD) grains. Using the Thellier-Coe method, we have attempted paleointensity determinations for 82 specimens sampled from different "stratigraphic" levels of the core. Partial thermal remanent magnetization (pTRM) checks were performed systematically one temperature step down from the last pTRM acquisition in order to document magnetomineralogical changes. The determinations were obtained from the slope of the pTRM gained vs. natural remanent magnetization lost in the Arai diagrams. Only about 6% of the samples (i.e. 5 samples) yielded marginally acceptable results. The paleofield estimated ranges from 16 to 28 μT and has a mean virtual axial dipole moment (VADM) of 5×10^{22} A/m^2, which is concordant with the average intensity for the period between 0 and 160 Myr ($4 \pm 2 \times 10^{22}$ A/m^2) and is about 2/3 of the strength of the present field (~8×10^{22} A/m^2).

E. Herrero-Bervera (✉)
Paleomagnetics and Petrofabrics Laboratory, School of Ocean & Earth Science & Technology (SOEST), Hawaii Institute of Geophysics and Planetology (HIGP), 1680 East West Road, Honolulu, Hawaii 96822, USA
e-mail: herrero@soest.hawaii.edu

13.1 Introduction

In order to understand the formation and evolution of the geodynamo, the geomagnetic field, and Earth's core, paleomagnetists have attempted to reconstruct

E. Petrovský et al. (eds.), *The Earth's Magnetic Interior*, IAGA Special Sopron Book Series 1,
DOI 10.1007/978-94-007-0323-0_13, © Springer Science+Business Media B.V. 2011

the history of the strength of the geomagnetic field from the geologic record (e.g., Juarez et al. 1998, Selkin and Tauxe 2000, Valet 2003, Channell et al. 2009). Sedimentary rocks deposited continuously over time have proven valuable in determining relative paleointensity variations but do not provide absolute values (e.g., Tauxe 1993, Guyodo and Valet 1999, Channell et al. 2009). Generally, successful absolute paleointensity determinations are obtained from samples (rocks and man-made materials) that have thermal remanent magnetizations (TRMs) and that contain magnetic minerals with relatively small grain sizes, preferably in the single domain (SD) size range (e.g., Valet 2003, Dunlop et al. 2005). Furthermore, the magnetic minerals should not alter upon heating because the most commonly accepted absolute paleointensity methods, such as the Thellier-Thellier and derivative methods, require that a sample be heated multiple times. This limits absolute paleointensity measurements primarily to select man-made materials, such as fired ceramics, bricks, and kilns, and to igneous extrusive rocks and glass.

Oceanic crust is by far the largest source of igneous extrusive rocks and is formed relatively continuously at seafloor spreading centers, with a 180 million year record preserved in the ocean basins. It is also known to be a high fidelity recorder of the geomagnetic field (e.g., Vine and Mathews 1963, Gee and Kent 2007). Furthermore, the thick sections of lavas that comprise the upper crust provide a chance to investigate the short-term variations of the geomagnetic field, with the possibility that mean paleomagnetic directions and paleointensities at a site could average geomagnetic secular variation. Thus, basalts and glass of the oceanic crust would seem to have the potential to provide a long history of absolute paleointensity.

Unfortunately, past studies of the upper oceanic crust have shown that the magnetic mineralogy of the extrusives is not ideal for determining paleointensities: multidomain grain size and alteration upon heating are limiting factors, except for a few cases (e.g. Dunlop and Hale 1976, Gromme, et al. 1979) and for the very finest grained units and basaltic volcanic glass (e.g. Pick and Tauxe 1993a, b, Mejia et al. 1996, Carlut and Kent 2000, 2002). Few studies, however, have had the opportunity to study much more than the uppermost portion of the extrusive layer of oceanic crust because of difficulties related to sampling basement in the ocean basins. Possibly the lower extrusives of the upper crust, dikes of the middle crust, or gabbros of the lower crust play a significant role in producing marine magnetic anomalies, and possibly they could yield viable paleointensity estimates.

Site 1256, cored during Ocean Drilling Program (ODP) Leg 206 and Integrated Ocean Drilling Program (IODP) Expeditions 309 and 312, provides an opportunity to test these possibilities. It is the first drill site to penetrate an in situ and intact section of crust, extending through oceanic crustal layer 1 (sediments), layer 2 (extrusives and dikes), and into the top of layer 3 (gabbros) (Wilson et al. 2003, 2006, Teagle et al. 2006). The site is located on the Cocos Plate 5 km east of the transition zone between marine magnetic anomalies 5Bn.2n and 5Br (~15 Ma) (Fig. 13.1). Given the location of the site relative to the transition zone and the identification of some lavas with steep inclinations during Leg 206 (Wilson et al. 2003), another goal of our paleointensity study is to test whether some of the lavas had been extruded while the geomagnetic field was transitioning from a reversed polarity interval at the end of Chron C5Br to a normal polarity interval at the beginning of Chron C5Bn.2n. Such a scenario might be expected for lavas transported from the spreading axis, like the massive lavas that comprise the upper part of the extrusive section. If some of the lavas had recorded a transitional field, we would expect them to have very low paleointensities.

In this paper, we present absolute paleointensity determinations from a suite of ocean crust samples from Site 1256 which include extrusives, dikes, and gabbros. We also provide relevant rock magnetic background information, with more extensive rock magnetic results presented elsewhere (Herrero-Bervera et al. 2011, Chapter 11; Krása et al. 2011, Chapter 12).

13.2 Site Description and Sampling

Site 1256 (6.736°N, 91.934°W) is located in crust that formed at a superfast spreading rate (~220 mm/yr full rate) in the East Pacific Rise about 15 million years ago (Fig. 13.1). The site consists of four drill holes (Wilson et al. 2003, 2006, Teagle et al. 2006, Alt et al. 2007): Holes 1256A and B were dedicated to recovering the ~250 m of sedimentary overburden (Layer 1), with Hole 1256B bottoming out in basement

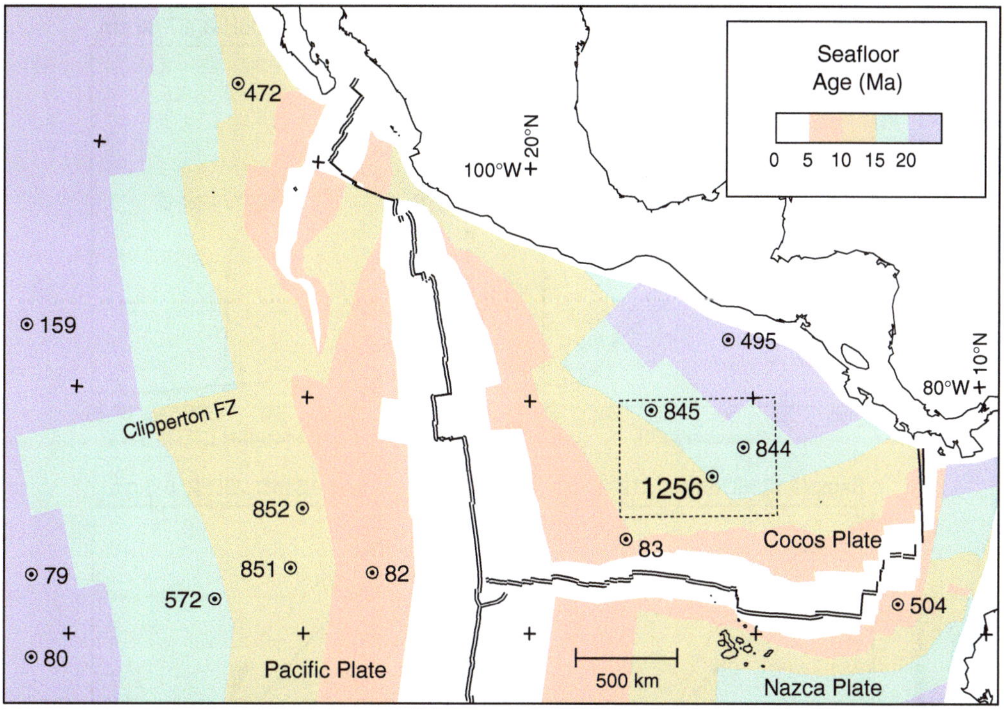

Fig. 13.1 Location of Site 1256 (from Alt et al. 2007)

where a few centimeters of basalt were recovered. Hole 1256C was cored through the lower 32 m of sediment and about 88.5 m into basement. Hole 1256D is the deep basement hole, where coring extended from 276 meters below seafloor (mbsf) down to 1507 mbsf. Together Holes 1256C and D sample about 1257 m of basement.

We collected 82 small specimens (1 cm^3 cubes and smaller-size rock chips) for absolute paleointensity determinations from Holes 1256C and D, with the bulk of these coming from Hole 1256D. Sample intervals were selected to span the different "stratigraphic" levels from the top of basement to the bottom of the drilled hole. We generally avoided intervals with obvious alteration of the fine grain matrix and we specifically targeted some of the finer grained intervals and intervals with glass. Glassy samples are rare and primary occur as sub-millimeter rinds on top of very fine-grained basalt. Most of the samples come from intervals that were described by the shipboard petrologists as cryptocrystalline or microcrystalline basalt, except for the samples from Site 312 characterized by gabbros with different grain sizes (i.e. larger crystals).

13.3 Methods and Results

13.3.1 Magnetic Mineralogy, Grain Size, and Thermomagnetic Characteristics

We have performed systematic thermomagnetic experiments using twin specimens from the samples studied for the determination of the absolute paleointensity (Fig. 13.2). The magnetic moment of a specimen from each sample was measured in strong fields during heating typically up to ~700 °C and then during cooling to room temperature using a Variable Field Translation Balance (VFTB) instrument. Additionally, the susceptibility of representative samples was measured during heating and cooling using a Mini-KappaBridge KLF-3 with a thermostat.

The magnetic mineralogy of most specimens is complex as evidenced by the many samples with two Curie temperatures. The first Curie point occurs between 250 and 350°C and is representative of Ti-rich titanomagnetite and titanomaghemite, while the second one occurs between 500 and 580°C and is representative of almost pure magnetite or titanomagnetite with very low titanium content. These two

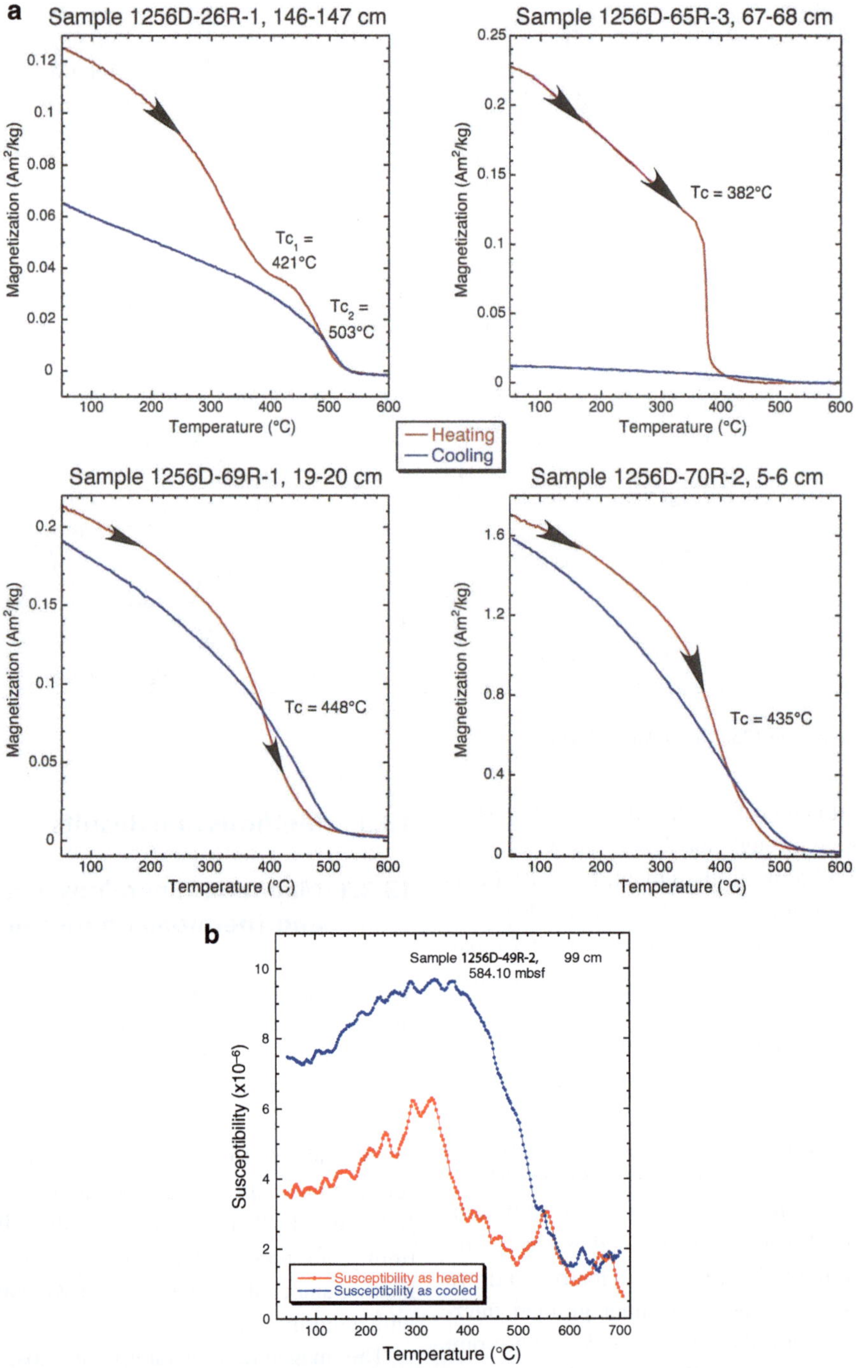

Fig. 13.2 (**a**) Thermomagnetic curves showing that titanomagnetites carry the natural remanence in four of the samples that gave acceptable results. (**b**) Low-field susceptibility versus temperature from one specimen showing the complexity of the magnetic mineralogy with two Curie temperatures of 306 and 410°C. The magnetic carriers for this sample are titanomaghemite and titanomagnetite. The alteration sustained by this sample is more typical of that seen in samples that gave unacceptable paleointensity determinations. The alteration was apparently at temperatures somewhat higher than 530°C in this sample, whereas it occurred at lower temperatures in many other samples

mineralogical phases are common in deep oceanic rocks. The Site 1256 samples also exhibit thermomagnetic behavior commonly associated with submarine basalts, including inversion of titanomaghemite to magnetite either during past heating in situ or during laboratory heating (e.g., Johnson and Merrill 1974, Petersen et al. 1979, Bleil and Petersen 1983, Matzka et al. 2003). The irreversibility of the thermomagnetic curves are thus the result of a composition and texture of the FeTi oxide minerals contained in oceanic rocks, which are sensitive to cooling rate and temperature changes mainly due to the presence of geochemically unstable ferrous ions within these minerals.

13.3.2 Magnetic Granulometry from Hysteresis Experiments

Magnetic hysteresis parameters were performed on small chips of rocks with a Petersen's Variable Field Translation Balance (VFTB) located at the SOEST-HIGP Paleomagnetics and Petrofabrics Laboratory of the University of Hawaii. Saturation remanent magnetization (Mr), saturation magnetization (Ms), and coercive force (Hc) were calculated after removing the paramagnetic contribution. Here we summarize the pertinent behavior, focusing mainly on the five samples with paleointensity determinations. The hysteresis results from an extensive suite of Site 1256 samples, including all 82 paleointensity samples, are presented elsewhere (Chapter 11, this volume).

The ratios of hysteresis parameters for the five paleointensity samples are plotted in a Day diagram (Fig. 13.3) following recent modifications by Dunlop (2002) for grain-size regions that have been defined for pure magnetite. The grain sizes are scattered within the pseudo-single domain (PSD) range for these five samples, as well as for most other samples from Site 1256 (Chapter 11, this volume). The results are not without ambiguity as the bulk PSD grain size estimates can be indicative of grains that either are truly PSD size or are a mixture of SD, PSD, and possibly multidomain (MD) grains. The majority of samples studied also had bulk coercivities of remanence (Hcr) that were low (<15 mT), with a median value of about 10 mT. Similar to the bulk grain size data, the bulk Hcr values do not necessarily indicate that only low coercivity grains are present. Instead, coercivity distributions from single rock chips typically have coercivities that range

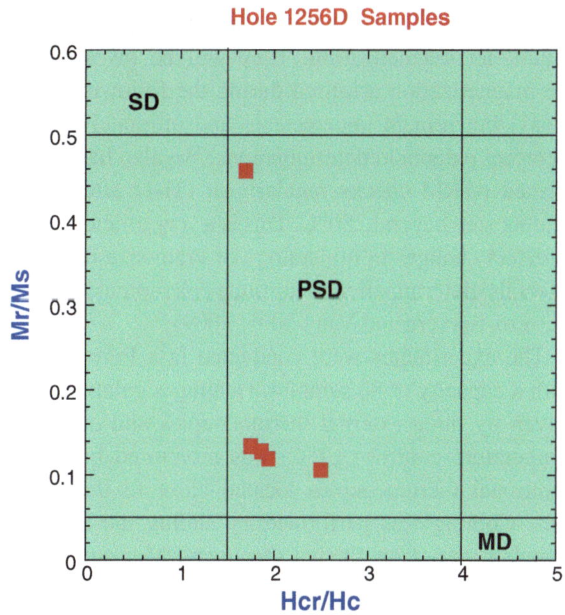

Fig. 13.3 Plot of the hysteresis parameters, Mrs/Ms (ratio of remanent saturation moment Mrs, to saturation moment Ms) against Hcr/Hc (ratio of remanent coercive force, Hcr, to coercive force, Hc). Single domain (SD), pseudo single domain (PSD), multidomain (MD) grain sizes, after Day et al. (1977) and corrected according to Dunlop (2002)

from <1 mT up to about 80 mT. Overall, the hysteresis data are most consistent with an interpretation in which nearly all samples have a range of grain sizes and coercivities, even though the bulk values indicate virtually all samples have PSD grain size and low coercivity.

13.4 Absolute Paleointensity and Remanent Magnetization

Paleointensity experiments were conducted using the Coe version (Coe 1967) of the Thellier experiments but with a different protocol. Instead of measuring the NRM first, we preferred to apply a pTRM before heating the sample in zero field (Aitken et al. 1988, Valet et al. 1998, Herrero-Bervera and Valet 2005, 2009). In this case, magnetomineralogical transformations occur in the presence of a field, resulting in a chemical remanent magnetization (CRM) component that is detected by a deviation of the NRM toward the oven field direction. The opposite situation (demagnetization in zero field first) does not allow one to detect remagnetization components with unblocking temperatures higher than the prior temperature step

Ti (Valet 2003) because the remagnetized grains have a null net magnetization. They will be involved in the magnetization acquired during the following step (Ti+1) but remain undetectable and ultimately yield incorrect paleofield determinations. We also have performed pTRM checks regularly at (Ti-1) after each heating step beyond 280°C. This is a very efficient way to detect changes in mineralogy or grain size that can affect the determination of absolute paleointensity (e.g. Herrero-Bervera and Valet 2005, 2009).

The experiments were conducted in a Pyrox oven with a capacity of 80 samples. Heating regulation was driven by three external thermocouples and accurate temperature control (±1°C) was monitored by three additional thermocouples located close to the samples. Cooling was performed by sliding the heating chamber away from the hot specimens. Measurements were done using a JR-5 Spinner magnetometer in the shielded room of the SOEST-HIGP Paleomagnetics and Petrofabrics laboratory. Each series of experiments were carried out on the 82 samples, which were divided into batches that heated independently. Generally, each batch contained 20–40 samples positioned a few millimeters away from each other within the oven. The magnetization level of the samples was too low to expect significant interactions between adjacent samples.

13.5 Selection of Data and Calculation of Absolute Paleointensity

Various techniques have been proposed to select appropriate data for paleointensity measurements, although much debate surrounds what constitutes an acceptable determination. Ideally, we would like samples to have reversible thermomagnetic curves, which provide evidence that no apparent mineralogical changes have happened during heating, and for samples to have SD magnetic grain sizes. While many Site 1256 samples may have a distribution of magnetic grains that do have these properties, they also generally contain a more significant distribution of grains that do not. Indeed, few igneous rocks have ideal paleointensity properties and so criteria are used to exclude results that are unreliable.

We use three primary criteria here: First, the characteristic (primary) component of the NRM must decay to the origin of a vector demagnetization diagram; second, the slope of the line in Arai plots must be linear over the temperature range where the primary TRM is resolved; and, third, the pTRM checks must not deviate significantly from the initial TRM on sequential temperature steps.

The first criterion is used in various forms in current paleointensity studies (e.g., Perrin and Shcherbakov 1998, Zhu et al. 2001). Any deviation from the demagnetization path from the origin of vector demagnetization diagrams introduces a significant uncertainty in the determination of paleointensity as it reflects failure to isolate the primary TRM. When plotted in sample coordinates, the evolution of the NRM direction also gives crucial information about possible deviations caused by CRMs that have unblocking temperatures higher than the last heating step. Samples that exhibited remagnetization during heating in the presence of the field are characterized by large directional swings of the remaining NRM towards the direction of the applied field. In the present study this was the case for over 90% of the samples. This criterion is probably particularly effective as a consequence of our protocol that requires that laboratory TRM acquisition precede demagnetization in zero field.

The demagnetization diagrams are also important for determining the temperature range over which the primary TRM has been isolated. Samples in which a significant part of the stable TRM is carried by SD grains of magnetite generally are less affected by large viscous components, low-temperature overprints, or drilling overprints. In such cases, a significant percentage of the total NRM intensity remains after removing any viscous components or overprints, allowing the primary TRM to be resolved along linear demagnetization paths that trend to the origin of vector demagnetization diagrams and allowing the viscous or overprinted temperature range to be avoided in Arai plots.

The second criterion retains only Arai plots with a single linear slope within the temperature range that lie above the viscous domain. We have required that the fit of the line must be better than R = 0.98 as depicted in the Arai plots of Fig. 13.4 for the five successful paleointensity determinations. Due to our selection rules, the averaged number of points used for slope determination amounts to 10 and never less than 6 and the "f" (NRM fraction of Coe et al. 1978) parameters are always greater than 45 percent (see Table 13.1).

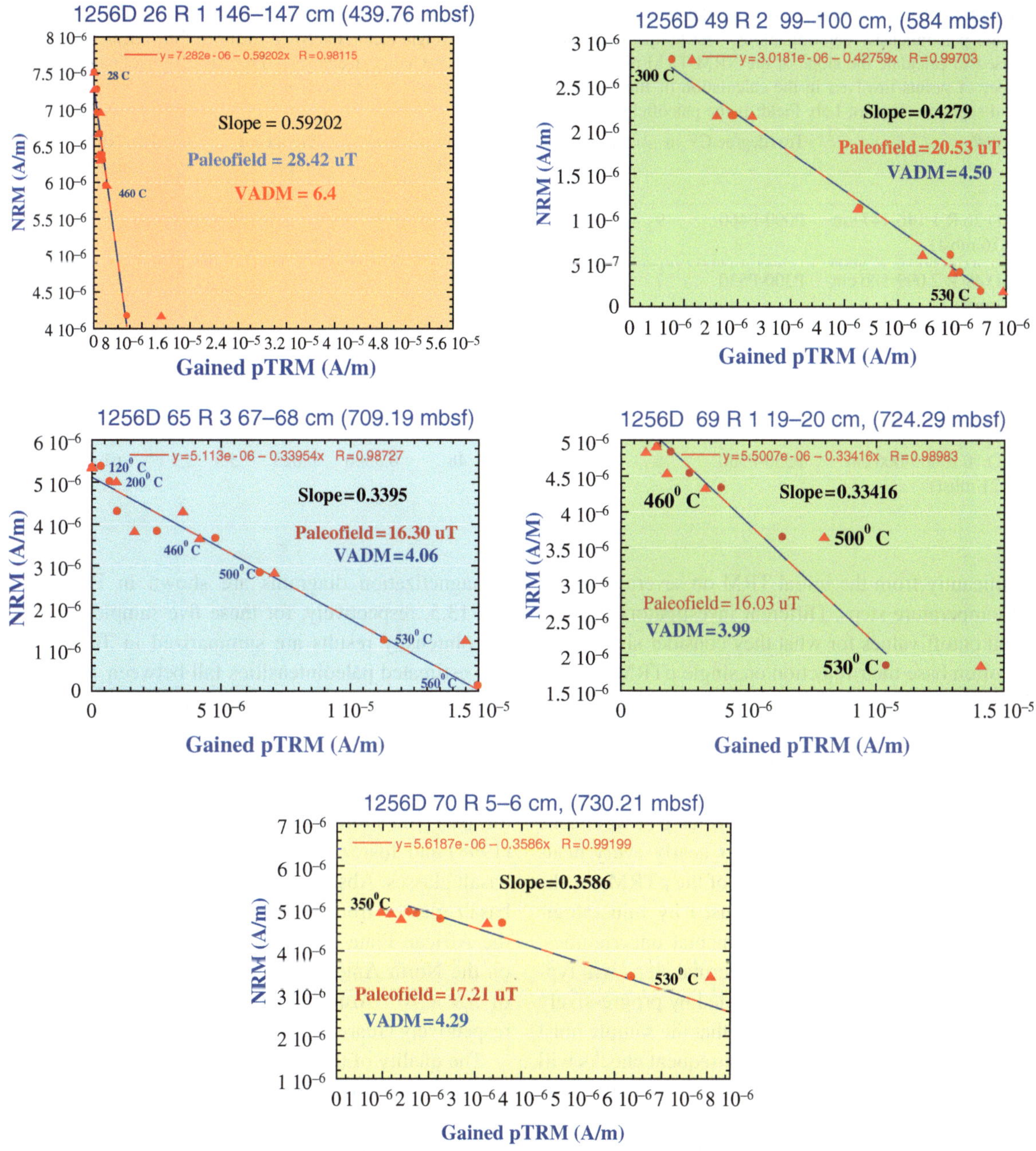

Fig. 13.4 Arai plots of samples from 5 successful determinations of absolute paleointensity. The pTRM checks are shown by triangles

In our case the range of the "f" parameter is from 45.8 up to 88.2. It is important to point out that this parameter is meaningless when there is a strong viscous low temperature component with an opposite direction to the characteristic remanence, which is not the case for the 5 samples in question and when the average quality factor lies much beyond the limits of 1 or 2 that are usually required. The mean value of the gap ratio (see Table 13.1, gap factor of Coe et al. 1978) measuring the relative uncertainty on the definition of the best-fit line is almost twice weaker than the acceptable limit of 0.1 (see Selkin and Tauxe 2000).

The third criterion rejects results if the pTRM checks where consistently negative: that is they differ

Table 13.1 Sample number refers to the IODP sample used for absolute paleointensity determinations, Tre: temperature interval used to calculate the slope of the NRM-TRM diagrams, n: number of points involved in the calculation of the slope and their dispersion (sigma), Lab. Field: is the paleofield estimated for each sample reported in μT; Q, w, f, w are the quality factor, the weight factor, the NRM fraction of the linear TRM-NRM portion that has been considered for calculation of paleointensity and the gap factor, respectively (after Coe et al. 1978)

Sample#	Tre (degree C)	n	Field (uT)	Sigma	Lab Field (uT)	slope	Sigma slope	q	Weight	NRM fraction	g
1256D 26 R 1 146-147 cm (439.76 mbsf)	P000-P460	9	28.42	1.8	48	0.592	0.044	4.2	1.6	45.8	0.67
1256D 49 R 2 099-100 cm (584 mbsf)	P300-P530	7	20.539	0.6	48	0.4275	0.015	18	8.1	86.2	0.73
1256D 65 R 3 067.68 cm (709.19 mbsf)	P120-P530	10	16.296	2.5	48	0.3395	0.062	3.3	1.2	88.2	0.71
1256D 69 R 1 019-020 cm (724.29 mbsf)	P300-P530	7	16.03	2	48	0.33416	0.051	4	1.8	72.5	0.71
1256D 70 R 005-006 cm (730.21 mbsf)	P350-P530	6	17.21	0.9	48	0.3586	0.023	5.3	2.7	61.6	0.55

significantly from the initial TRM on several sequential temperature steps. Different investigators use different cutoff values for what they consider significant, and often base their rejection on single pTRM checks. Here, pTRM checks were considered to be negative when they deviated by more than 5% from the initial TRM, and they did so on sequential temperature steps. We believe that such cutoffs applied to single temperature steps are somewhat arbitrary when pTRM checks are performed systematically after nearly every heating step. Indeed, small deviations of the pTRM checks from the initial TRM can be caused by mineralogical transformations or by experimental uncertainties. In the first case, the subsequent thermal steps will typically be systematically accompanied by progressively larger deviations, which confirms that the sample must be rejected. In the second case, subsequent checks will not be negative and hence a reliable paleointensity determination may be obtained over the temperature interval. Generally, samples that failed the second or third criterion had already notably failed the first criterion, which was the most effective data selection criterion.

13.5.1 Acceptable Absolute Paleointensities

Of the 82 paleointensity experiments, only five samples from the extrusives of the upper crust gave marginally acceptable results. Arai plots and vector demagnetization diagrams are shown in Figs. 13.4 and 13.5, respectively, for these five samples and the paleointensity results are summarized in Table 13.1. The estimated paleointensities fall between 16 and 28 μT. This gives virtual axial dipole moments (VADMs) of 4–7 $\times$ 10^{22} A/m^2, which are about two thirds of the current dipole moment and are comparable to the average VADM of 4 $\pm$ 2 $\times$ 10^{22} A/m^2 for the past 160 million years obtained by Juarez et al. (1998) and Juarez and Tauxe (2000) from submarine basalt glasses. Absolute paleointensities obtained from basalt glasses from DSDP Hole 520 (Chron 5Br) on the African Plate and DSDP Hole 470A (Chron 5B) on the North American plate yielded paleointensities of 5.9 $\times$ 10^{22} A/m^2 and 6.6 $\times$ 10^{22} A/m^2 VADM, respectively (Juarez et al. 1998).

The quality of the new results is questionable. Both the Arai plots and the vector demagnetization diagrams illustrate less than ideal behavior over part of the thermal unblocking spectrum. In particular, the declination in the vector demagnetization diagrams does not decay linearly to the origin of the plot. Instead it has a kink in the trend at medium unblocking temperatures, indicating that the samples contain at least two components of remanent magnetization. The diagrams indicate that the low and high temperature unblocking components have similar directions, whereas the medium unblocking temperature component is similarly shallow but has a declination that makes an obtuse angle to that of the low and high temperature

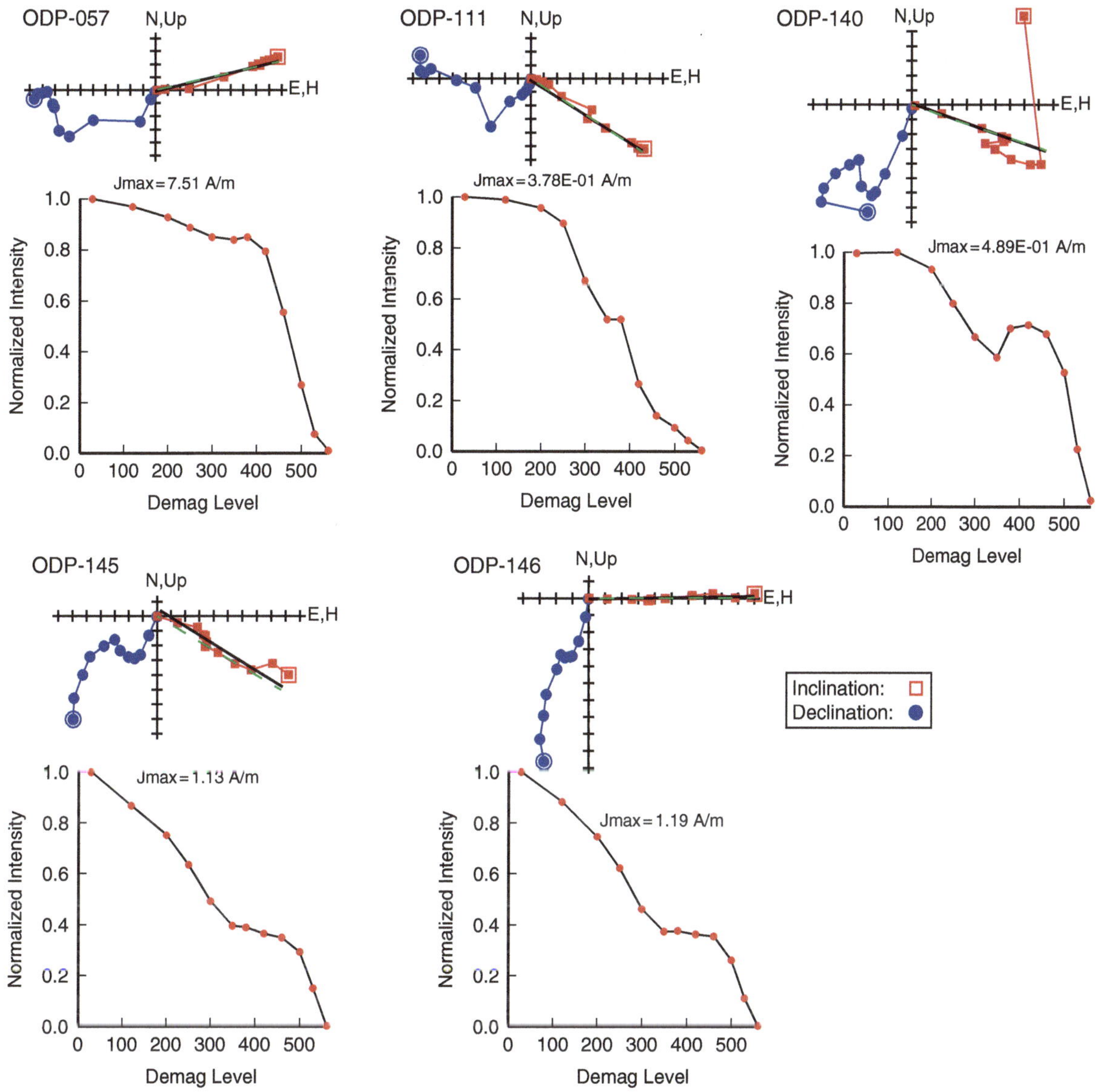

Fig. 13.5 Thermal demagnetization results from five Site 1256 samples. For each sample, the top diagram shows vector end points of paleomagnetic directions on orthogonal demagnetization diagrams or modified Zijderveld plots (squares are inclinations and circles are declinations). The best-fit lines from principal component analysis (PCA) are shown for the FREE PCA option (solid black line) and ANCHORED PCA option (dashed green line) for the vertical component only. For each sample, the lower diagram shows the normalized intensity variation with progressive demagnetization

component. Part of this could be attributed to drilling overprints, but they generally affect magnetic minerals with low-to-medium unblocking temperatures and low coercivities and typically have a very steep inclination. Alternatively, the multiple components could have been acquired in situ as the oceanic crust was altered shortly after formation, within a geomagnetic field that was transitioning from reversed to normal polarity. Oxidized titanomaghemite is also known to sustain self-reversal or partial self-reversal (e.g., Krása and Matzka 2007, Matzka and Krása 2007), which may occur in situ or from mineralogical transformations during laboratory heating. Although the medium unblocking temperature component is not fully resolved, it does trend toward a direction that is antipodal to the low and high unblocking temperature component. Other forms of anomalous behavior can arise from experimental protocol, but we think these are irrelevant to the overall low success rate of paleointensity determinations. In the experimental protocol, all samples were always carefully positioned at the same location within the oven in order to provide consistent temperature and field control from one thermal step to the next. Furthermore, the same protocol and instrumentation has been used repeatedly on other samples with high success rate (e.g. Herrero-Bervera and Valet 2005, 2009; Valet et al. 2010).

The thermomagnetic curves provide evidence that mineralogical transformation occurs to some degree in all the samples. In those samples that did not provide viable absolute paleointensities, the alteration that occurs during heating is generally large, such that the cooling and heating curves are even more offset than shown for Sample 1256D-49R-2, 99 cm, in Fig. 13.2b. This increase in magnetization, susceptibility, and unblocking temperature that occurs after heating is the result of titanomaghemite being converted to low-Ti magnetite, as has been noted in many prior studies of oceanic crust samples (e.g., Petersen et al. 1979, Pariso and Johnson 1991, Xu et al. 1997, Krása and Matzka 2007).

Multidomain effects are also likely a factor. Data in the Arai plots have significantly curved or S-shaped paths (Fig. 13.4), which can be symptomatic of multidomain grains (e.g., Coe et al. 2004). An interpretation of the Arai slopes for data from either above or below temperatures of about 300 to 450°C would yield paleointensity estimates that are about 30–80% higher or lower, respectively, than the slopes of the of the lines connecting the endpoints of the Thellier-Coe experiments.

In summary, most samples failed to produce reliable paleointensity estimates. We suspect that the most likely explanations for the complex demagnetization behavior of the samples is that either they record dual polarities and/or are affected by mineralogical transformation that include partial self-reversal. In addition, some samples display multidomain-like behavior.

13.6 Discussion

Our study finds that the igneous rocks of the oceanic crust are generally ill suited for absolute paleointensity determination because Ti-rich titanomagnetite and titanomaghemite is converted to low-Ti magnetite during laboratory heating. With only 5 of 82 samples giving even marginally acceptable results, the success rate is only 6%, which is exceptionally low even for paleointensity studies. The failure of ocean crust rocks to provide viable absolute paleointensities has been noted before, but our study extends this observation from extrusives of the upper crust down to gabbros of the very top of the lower crust.

As bleak as this result may seem, there is hope. For example, for the upper oceanic crust, one potential workaround is to focus only on glass, where success rates and quality of results can be much higher (e.g., Pick and Tauxe 1993, Selkin and Tauxe 2000, Carlut and Kent 2000). Unfortunately, glass was relative rare at Site 1256, and, that which was recovered, was in demand for geochemical studies as well. We did, however, attempt paleointensity determinations on small pieces of glass and these performed no better than other cryptocrystalline basalts. We are unsure why this is the case for Site 1256. For any ocean drill site, finding enough glass from enough intervals to average geomagnetic secular variation will be difficult. Adding to this difficulty, is the need for a sufficient number of samples from each flow to ensure that discrepancies, such as those observed by Carlut and Kent (2000) and that were thought to be caused by local magnetic anomalies, can be identified.

Deeper oceanic crustal gabbros may offer a more promising target for absolute paleointensity determinations. Thermomagnetic curves for gabbros give Curie temperatures near 580°C, indicating magnetite with little or no titanium is the remanence carrier, and the curves are often reversible, indicating that alteration during laboratory heating is negligible (Pariso and Johnson 1993a, Gee et al. 1997, Gee and Kent 2007). Unlike the extrusive layers, where rapid quenching forms Ti-rich iron oxides, the gabbros cool slowly and are subjected to hydrothermal alteration prior to migrating away from the ridge axis, both of which facilitate the formation of magnetite (e.g., Pariso and Johnson 1993a). Thermal modeling of Site 1256 granoblastic dikes implies the gabbros were at temperatures well above the Curie temperature of magnetite for 10^3–10^4 years (Koepke et al. 2008). How rapidly the gabbros cool below the Curie temperature and block in a TRM is uncertain, but it is probably sufficiently long that individual samples partially average geomagnetic secular variation (Gee and Kent 2007). Although somewhat counterintuitive, hysteresis data from gabbros give PSD magnetic grain sizes similar to the grain sizes of the cryptocrystalline basalts of the upper crust. Pariso and Johnson (1993a) suggested the small grain size was caused by the formation of very fine ilmenite lamellae within larger magnetite grains, effectively subdividing the grains, and by deformation that increases crystal defects. The TRM of gabbros may also be carried dominantly by small magnetite grains that form as inclusions within olivine, pyroxene, and plagioclase or along cracks in olivine grains (Pariso and Johnson 1993a, Gee and Meurer 2002, Gee and Kent 2007). In our case the rejection of the 77 specimes we experimented with was caused by the acquisition of a Chemical Remanent Magnetization (CRM) in the oven during the Thellier-Coe experiments, the presence of multidomain grains (this is the case of the Dikes and Gabbros at Hole 1256), negative pTRM checks and the absence of full linearity of the NRM-TRM diagrams.

Future drilling that is planned for Site 1256 should penetrate deep into the gabbros and provide a much larger sample of gabbros to assess this possibility (ie. Dikes and Gabbros with small magnetite Single Domain (SD) and/or SD and Pseudosingle Domain grains suitable for absolute paleointensity determinations).

Given the low success rate, it is natural to question the meaning of the five acceptable results. Are they real estimates of the field strength at the site when the crust formed or are they merely results from samples with sufficiently atypical mineralogy to marginally pass the various rejection criteria but give unreliable paleofield estimates? Although the five results are somewhat scattered, they give a mean VADM of 5.0×10^{22} A/m^2 that is comparable to the mean for the past 160 Ma. Much of the scatter probably arises from noise caused by the less than ideal behavior of the samples whereas the agreement of the VADM with the long-term field average probably indicates the samples are accurately recording the field. The overall result is therefore probably accurate but has very low precision.

With only five results, we are also unable to examine temporal changes in the field or to average geomagnetic secular variation sufficiently. Because so few units gave reliable result, we are also unable to test adequately the hypothesis that part of the crust at Site 1256 records a transitional field. Three of the samples did give VADMs that are about half of the present field but the paleointensities are too large to be considered transitional. If transitional fields were recorded by some of the 82 samples, it is highly likely that those samples would have been rejected given the difficulties we had in resolving the primary remanence. Basically, even a small amount of alteration would be sufficient to mask a sample that recorded a very weak transitional field.

Conclusions

The primary magnetic mineralogy of the Site 1256 basaltic flows appears to be typical of other mid-ocean ridge basalts (MORB's), which are characterized by fast cooling of primary titanomagnetite (i.e., a solid solution of magnetite and ulvospinel, with an ulvospinel content near 60%) followed by low temperature alteration/oxidation that results in the formation of titanomaghemite (e.g., Petersen et al. 1979, Pariso and Johnson 1991, Krása and Matzka 2007). Likewise, the Site 1256 gabbros display magnetic properties similar to those sampled at outcrops along fracture zones in which fine grained magnetite dominates the magnetic signal (e.g., Pariso and Johnson 1993a, b, Gee and Kent 2007).

Although typical of oceanic crust, the igneous rocks from Site 1256 are generally ill suited for absolute paleointensity determination because Ti-rich titanomagnetite and titanomaghemite is converted to low-Ti magnetite during laboratory heating. Five of 82 samples did, however, give marginally acceptable paleointensity estimates that range from 16 to 28 μT, with a corresponding mean VADM of 5×10^{22} A/m^2. This is concordant with other Chron 5 estimates and with the 0–160 Ma mean VADM from basaltic glass from the ocean basins.

Acknowledgements We are grateful to Mr. James Lau for his laboratory assistance and help with the laboratory measurements. We thank the two anonymous referees for their constructive criticism. We also give special thanks to the participating scientists and crew members of JOIDES Resolution for their help and support during the scientific cruises. This research used samples and data provided by the Ocean Drilling Program (ODP) and the Integrated Ocean Drilling Program (IODP). Funding for this research was provided by the National Science Foundation (NSF) through its support of ODP, IODP, and the United States Science Support Program (USSSP) and through NSF grants JOI-T309A4, OCE-0727764, and EAR-IF-0710571 to Herrero-Bervera and grant OCE-0727576 to Acton. Additional financial support to E. H-B was provided by SOEST-HIGP. This is HIGP and SOEST contributions 1891 and 8148, respectively.

References

Aitken MJ, Allsop AL, Bussel GD, Winter MB (1988) Determination of the intensity of the Earth's magnetic field during archeological times: reliability of the thellier technique. Rev Geophys 26:3–12

Alt JC, Teagle DAH, Umino S, Miyashita S, Banerjee NR, the IODP Expeditions 309 and 312 (2007) Scientists, and the ODP Leg 206 scientific party. Sci Drilling 4:4–10. doi:10,2204/iodp.sd.4.012007

Bleil U, Petersen N (1983) Variations in magnetization intensity and lowtemperature titanomagnetite oxidation of ocean floor basalts. Nature 301:384–388

Carlut J, Kent DV (2000) Paleointensity record in a zero-age submarine basalt glasses: testing a new dating technique for recent MORB's. Earth Planet Sci Lett 183:389–401

Carlut J, Kent DV (2002) Grain-size dependent paleointensity results from very recent mid-ocean ridge basalts. J Geophys Res 107. doi:10.1029/2001JB000439

Channell JET, Xuan C, Hodell DA (2009) Stacking paleointensity and oxygen isotope data for the last 1.5 Myr (PISO-1500). Earth Planet Sci Lett 283:14–23

Coe RS (1967) Paleointensities of the earth's magnetic field determined from tertiary and quaternary rocks. J Geophys Res 72:3247–3262

Coe RS, Gromee S, Mankinen EA (1978) Geomagnetic paleointensities from radiocarbon dated lava flows on Hawaii and the question of the Pacific nondipole low. J Geophys Res 83:1740–1756

Coe RS, Riisager J, Plenier G, Leonhardt abd R, Krása D (2004). Multidomain behavior during Thellier paleointensity experiments: results from the 1915 Mt. Lassen flow. Phys Earth Planet Int 147:141–153

Day R, Fuller M, Schmidt VA (1977) Hysteresis properties of titano-magnetites: grain size and compositional dependence. Phys Earth Planet Ints 13:260–267

Dunlop DJ (2002) Theory and application of the Day plot (Mrs/Ms vs. Hcr/Hc) 1. Theoretical curves and tests using titanomagnetite data. J Geophys Res 107: EM 4-1–EPM 4-22

Dunlop DJ, Hale CJ (1976) A determination of paleomagnetic field intensity using submarine basalts drilled near the Mid-Atlantic Ridge. J Geophys Res 81:4166–4172

Dunlop DJ, Zhang B, Ozdemir O (2005) Linear and nonlinear Thellier paleointensity behavior of natural minerals. J Geophys Res 110. doi:10.1029/2004JB003095

Gee JS, Kent DV (2007) Source of oceanic magnetic anomalies and the geomagnetic polarity timescale. In: Kono M (ed) Treatise on geophysics, vol 5, Geomagnetism. Elsevier, Amsterdam, pp 455–507

Gee JS, Lawrence RM, Hurst SD (1997) Remanence characteristics of gabbros from the mark area: implications for crustal magnetization. In: Karson JA, Cannat M, Miller DJ, Elthon D (eds) Proceedings of ODP, Scientific results, vol 153. Ocean Drilling Program, College Station, TX, pp 429–436

Gee J, Meurer WP (2002) Slow cooling of middle and lower oceanic crust inferred from multicomponent magnetizations of gabbroic rocks from the Mid-Atlantic Ridge south of the Kane fracture zone (MARK) area. J Geophys Res 107:18

Gromme S, Mankinen EA, Marshall M, Coe RS (1979) Geomagnetic Paleointensities by the Thellier's method from submarine pillow basalts: effect of seafloor weathering. J Geophys Res 84:3553–3575

Guyodo Y, Valet J-P (1999) Global changes in geomagnetic intensity during the past 800 thousand years. Nature 399:249–252

Herrero-Bervera E, Valet JP (2005) Absolute paleointensity and reversal records from the Waianae sequence (Oahu, Hawaii, USA). Earth Planet Sci Lett 234:279–296

Herrero-Bervera E, Valet JP (2009) Testing determinations of absolute paleointensity from the 1955 and 1960 Hawaiian flows. Earth Planet Sci Lett 287:420–433

Johnson HP, Merrill RE (1974) Low-temperature oxidation of a single domain magnetite. J Geophys Res 79:5533–5534

Juarez MT, Tauxe L, Gee JS, Pick T (1998) The intensity of the Earth's magnetic field over the past 160 million years. Nature 394:878–881

Juarez MT, Tauxe L (2000) The intensity of the time-averaged geomagnetic field: the last 5 Myr. Earth Planet Sci Lett 175:169–180

Koepke J, et al (2008) Petrography of the dike-gabbro transition at IODP Site 1256 (equatorial Pacific): the evolution of the granoblastic dikes. Geochem Geophys Geosyst 9:Q07O09. doi:10.1029/2008gc001939

Krása D, Matzka J (2007) Inversion of titanomaghemite in oceanic basalt during heating. Phys Earth Planetary Inter 160:169–179

Matzka J, Krása D (2007) Oceanic basalt continuous thermal demagnetization curves. Geophys J Int 169:941–950

Matzka J, Krása D, Kunzmann T, Schult A, Petersen N (2003) Magnetic state of 10–40 Ma old ocean basalts and its implications for natural remanent magnetization. Earth Planet Sci Lett 206:541–553

Mejia V, Opdyke ND, Perfit MR (1996) Paleomagnetic field intensity recorded in submarine basaltic glass from the East Pacific Rise, the last 69 Ka. Geophys Res Lett 23: 475–478

Pariso JE, Johnson HP (1991) Alteration processes at Deep Sea Drilling Project/Ocean Drilling Program Hole 504I at the Costa Rica Rift: implications for magnetization of oceanic crust. J Geophys Res 96:11,703–11,722

Pariso JE, Johnson HP (1993a) Do lower crustal rocks record reversals of the earth's magnetic field? Magnetic petrology of oceanic gabbros from ocean drilling program hole 735B. J Geophys Res 98:16013–16032. doi:10.1029/93jb00933

Pariso JE, Johnson HP (1993b) Do layer 3 rocks make a significant contribution to marine magnetic anomalies? In situ magnetization of gabbros at ocean drilling program hole 735B. J Geophys Res 98:16033–16052. 10.1029/93jb01097

Perrin M, Shcherbakov V (1998) Paleointensity database updated. Eos 79:198

Petersen N, Eisenach P, Bleil U (1979) Low temperature alteration of the magnetic minerals in ocean floor basalts: in Deep Drilling Results in the Atlantic Ocean. In: Talwani M, Harrison CGA, Hayes DE (eds) Ocean *Crust* Maurice Ewing Series, vol 2.. AGU, Washington, DC, pp 169–209

Pick T, Tauxe L (1993a) Holocene paleointensity: Thellier experiments on submarine basaltic glass from the East Pacific Rise. J Geophys Res 98:17949–17964

Pick T, Tauxe L (1993b) Geomagnetic palaeointensities during the Cretaceous Normal superchron measured using submarine basaltic glass. Nature 366:238–242

Selkin PA, Tauxe L (2000) Long-term variations in palaeointensity. Philos Trans R Soc London A 358:1065–1088

Tauxe L (1993) Sedimentary records of relative paleointensity of the geomagnetic field: theory and practice. Rev Geophys 31:319–354

Teagle DAH, Alt JC, Umino S, Miyashita S, Banerjee NR, Wilson DS, the Expedition 309/312 Scientists (2006) Superfast spreading rate crust 2 and 3. Proceeding of IODP, vol 309/312. Integrated Ocean Drilling Program Management International, Inc., Washington, DC. doi:10.2204/iodp. Proc.309312.2006 pp

Valet JP (2003) Time variation in geomagnetic intensities. Rev Geophys 41:1/1004

Valet JP, Herrero-Bervera E (2000) Paleointensity experiments using alternating Field demagnetization. Earth Planet Sci Lett 177:43–58

Valet JP, Herrero-Bervera E, Carlut J, Kondopoulou D (2010) A selective procedure for absolute paleointensity in lava flows. Geophys Res Lett 37. doi:10.1029/2010GL044100, 2010

Valet JP, Tric E, Herrero-Bervera E, Meynadier L, Lockwood JP (1998). Absolute paleointensity from Hawaiian lavas younger than 35 ka. Earth Planet Sci Lett 161:19–32

Vine FJ, Matthews DH (1963) Magnetic anomalies over oceanic ridges. Nature 199:947–949

Wilson DS, Teagle DAH, Acton GD et al (2003) Proceedings of ODP, initial reports, vol 206. doi:10.2973/odp.proc.ir.206.2003

Wilson DS, et al (2006) Drilling to gabbro in intact ocean crust. Science 312:1016–1020

Xu W, Van der Voo R, Peacor DR, Beaubouef RT (1997) Alteration and dissolution of fine-grained magnetite and its effects on magnetization of the ocean floor. Earth Planet Sci Lett 151:279–288

Zhu R, Pan Y, Shaw J, Li D, Li Q (2001) Geomagnetic palaeointensity just prior to the cretaceous normal superchron. Phys Earth Planetary Inter 128:207–222

Paleointensities of the Hawaii 1955 and 1960 Lava Flows: Further Validation of the Multi-specimen Method

14

Harald Böhnel, Emilio Herrero-Bervera, and Mark J. Dekkers

Abstract

The Kilauea 1955 and 1960 lava flows (Big Island of Hawaii, USA), both emplaced in a field of ~36 μT, were studied using the multi-specimen parallel differential partial thermoremanent magnetization (pTRM) paleointensity (PI) method. In nineteen specimens from an upper cooling unit of the 1955 flow, the pTRMs were acquired at a temperature of 450°C and a PI of 34.3 ± 1.5/1.6 μT was obtained, while 13 specimens of the lower cooling unit heated to 230°C resulted in a PI of 38.5 ± 3.3 μT. The 1960 flow was studied at various temperatures of 400, 440, 480, 500 and 550°C. At 400 and 550°C overestimates of the PI were obtained: 47.3 μT and 41.5 μT respectively. The other temperatures yielded PI values ranging from 32.2 to 34.2 μT, just 6–11% lower than the expected field and similar to the 1955 flow results. The 550°C PI result is biased because of mineral alteration as indicated by an 18% susceptibility decrease which reduces the pTRM capacity leading to an overestimate of the PI. The overestimate of the PI at 400°C may be due to the comparatively small pTRM acquired and thus to larger high-temperature pTRM tails. Results obtained with the multi-specimen method for both flows compare favorably with the best other PIs obtained by the Thellier–Coe, Thellier–Thellier and microwave methods. The success rate is high with more than 90% of specimens contributing to a PI. Only the Thellier–Coe method used on single plagioclase crystals and the microwave method have similar success rates.

14.1 Introduction

Traditionally, the study of geomagnetic field variations involves the determination of the paleodirection in a sequence of lavas or sediments. Sediments may potentially provide a continuous record, but climatic interferences and post-depositional processes make the interpretation not always straightforward. Volcanic rocks offer a more accurate record, but each lava flow corresponds to one instant in geologic time, so the record is inherently discontinuous. Paleointensity (PI) determinations of the ancient Earth's magnetic field are required to obtain the full vector record, which is of fundamental importance in developing models of global field variations. Unfortunately the determination of a PI record is not easy; a series of checks

H. Böhnel (✉)
Centro de Geociencias, Universidad Nacional Autónoma de México, Querétaro 76230, México
e-mail: hboehnel@geociencias.unam.mx

E. Petrovský et al. (eds.), *The Earth's Magnetic Interior*, IAGA Special Sopron Book Series 1,
DOI 10.1007/978-94-007-0323-0_14,

is required to ascertain its meaningfulness. Sediments allow retrieving a scaled paleointensity record, referred to as relative PI. In contrast, lavas allow determination of the PI in μT which are referred to as absolute PIs. Here we focus on the latter.

For weak magnetic fields like the Earth's magnetic field the thermoremanent magnetization (TRM) is proportional to the inducing field, in case of negligible anisotropy, allowing a comparison of the (unknown) ancient field with a known laboratory field. Most absolute PI methods used today are derivatives of the Thellier and Thellier (1959) method and require one or multiple heating steps of a rock sample, ideally to up to temperatures close to the Curie temperature of the dominant magnetic mineral. In many cases, however, repeated heating induces chemical alteration of TRM carriers invalidating a comparison between the laboratory partial TRM (pTRM) with the natural TRM acquired during the initial cooling of the rock. Additionally, only single domain (SD) particles yield meaningful results (e.g., Valet 2003). Hence, most protocols in use today involve (time-consuming) preselection experiments, alteration and reproducibility checks during the PI experiment itself, and tests for the presence (and effects) of multi-domain (MD) particles. However, high quality PI values are shown to be reasonably frequently incorrect for historic lava flows, where the Earth's magnetic field intensity is known with certainty, despite this array of selection tools and tests (e.g., Biggin and Thomas 2003, Biggin et al. 2007). This is prompting a methodological research into the veracity of absolute PI determination techniques and the development of new protocols (e.g., Valet 2003). The present contribution involves testing a recently proposed absolute PI protocol (Dekkers and Böhnel 2006) on the Kilauea lava flows emplaced during 1955 and 1960 on the Big Island of Hawaii (USA).

The Dekkers and Böhnel (2006) technique is based on a multi-specimen approach. All specimens are heated only one time and to the same temperature but in a different laboratory field, whereby the natural remanent magnetization (NRM) of each specimen is oriented parallel to the furnace field, which is present during heating and cooling. Consequently, a part of the characteristic remanent magnetization (ChRM) is demagnetized and replaced by a pTRM with an intensity depending linearly on the applied furnace field. When this applied laboratory field is smaller (larger) than the ancient field, the pTRM is negative (positive), and if both fields are equal, no pTRM is induced. As a pTRM is proportional to small fields and approximately independent of the magnetic domain state, this method is less affected by the presence of MD particles than the Thellier-methods (Michalk et al. 2008). By using a laboratory field parallel to the NRM, the MD tail effects are minimized. Finally, the one-time heating and selection of a relatively low temperature reduces thermal alteration and also the total experiment time.

The use of multiple specimens ensures that every single specimen experiences exactly the same treatment so that magnetic history effects so typical for MD particles are avoided. On the other hand it requires that the different specimens have similar cooling rates and very similar magnetic properties, e.g. the same intensity decay at the selected temperature of the PI experiment. It is hard to fulfil this requirement entirely, but it seems reasonable to assume that minor variations amongst specimens are averaged out by using seven or more specimens, in particular if these are closely spaced. Furthermore, differences among specimens would become less important when approaching the paleofield intensity. Then no pTRM is produced, independent of the variations of the unblocking temperatures.

14.2 Previous Paleointensity Results from the 1955 and 1960 Lava Flows from the Big Island of Hawaii

We chose the Hawaii 1955 and 1960 flows erupted from Kilauea volcano because the field intensity for that period is well known: 36.47 μT measured at the Honolulu observatory, located about 370 km NWN, and 36.2 μT calculated for the Kilauea area according to the Definitive Geomagnetic Reference Field (DGRF) 1965 global field model (Tanaka and Kono 1991). Coe and Grommé (1973) estimated that magnetic anomalies caused by underlying lava flows at Kilauea could account for local ± 2 μT deviations from these field values. Recent field intensity measurements reported by Herrero-Bervera and Valet (2009) around their 1960 and 1955 flow sites show similar variations, and they suggested a possible deviation of about ± 3 μT. Many PI studies already have embarked on these flows, offering a kind of benchmark for testing different PI methods. The previous results are listed in Table 14.1 and show quite a large variation of PI

Table 14.1 Existing paleointensity determinations for the Big Island of Hawaii 1955 and 1960 lava flows. The standard deviation (s.d.) refers to the PI (paleointensity) uncertainty for an individual sample or to the number n of the samples, if $n > 1$. When samples were rejected for the mean calculation, their number is given under "Rej."

Reference	PI (μT)	s.d. (μT)	n	Rej.	PI method	Comment
Reference values for field intensity						
Tanaka and Kono (1991)	36.47	–			Honolulu Observatory	measured intensity
Tanaka and Kono (1991)	36.2	–			DGRF 1965	modelled intensity
Hawaii 1955 lava flow						
Chauvin et al. (2005)	42.4	6.9	4	4	Thellier–Thellier	high-T component
	39.3	3.7	7	1	Thellier–Thellier	with weighing
Herrero-Bervera and Valet (2009)	27.4	4.4	18	5	Thellier–Coe	all successful
	34.8	1.3	4	19	Thellier–Coe	only samples with max DRAT<5% and average DRAT<3%
Cottrell and Tarduno (1999)	33.8	3.7	10		Thellier–Coe	plagioclase crystals from one sample
Coe and Grommé (1999)	33.9	1.5	1		Wilson-method	for details also see Cottrell and Tarduno (1999)
	33.1	1.3	1		Van Zijl-method	
	37.4				Thellier–Thellier	
Hawaii 1960 lava flow						
Abokodair (1977)	47.0				Thellier	in air
	42.3				Thellier	in vacuum
Tanaka and Kono (1991)	40.3	8.1	5	1	Thellier–Coe	all samples
	37.0	3.9	4		Thellier–Coe	one excluded
Tsunakawa and Shaw (1994)	34.5	0.9			Shaw	accepted
	47.8	0.8			Shaw	rejected
	51.5	1.5			Shaw	rejected
Tanaka et al. (1995)	36	4			Thellier–Coe	
	43	12			Thellier–Coe	
	44	3			Thellier–Coe	
	40	4			KTT	
	36	7			Kono–Ueno	
	47	7			Kono–Ueno	
Valet and Herrero-Bervera (2000)	41.1	3.7	7	3	Shaw (modified)	no ARM_2 check
	35.2	5.8	3	2	Shaw (modified)	300°C demagnet.
	–	–		1	Thellier–Coe	no success
Hill and Shaw (2000)	33.9	5.6	70	1	Microwave	average slope
	46.8	10.3	22		Microwave	low-T slope
	26.8	5.9	18		Microwave	high-T slope
Yamamoto et al. (2003)	41.9	4.2	4	2	Thellier–Coe	group A
	56.0	7.6	9		Thellier–Coe	group B (TCRM ?)
	40.6	3.2	4		Thellier–Coe	group C
	49.0	9.6	17		Thellier–Coe	all samples
	39.4	7.9	9		LTD-DHT Shaw	all samples
	35.7	3.3	7		LTD-DHT Shaw	2 high PI excluded
Chauvin et al. (2005)	44.7	9.5	7	1	Thellier–Thellier	high-T component
	33.6	4.9	6		Thellier–Thellier	with weighing
Herrero-Bervera and Valet (2009)	36.9	7.4	36	2	Thellier–Coe	all successful
	36.7	1.8	7	31	Thellier–Coe	only samples with max DRAT<5% and average DRAT<3%

values, between about 17 and 58 μT with a clear tendency to overestimate the ancient field.

By far the most detailed study is that by Hill and Shaw (2000), who reported 70 successful PI data sets obtained from 71 samples from two parallel vertical profiles of the 1960 flow, all determined by the microwave technique (e.g., Walton et al. 1996). Individual PIs vary in a wide range between 24 and 52 μT, but the overall average PI values for these two profiles is 33.9 ± 5.6 μT and thus closer to the 1960 field intensity of 36.5 μT than most other studies. We note here that to obtain these averages Hill and Shaw (2000) calculated for 40 samples the PI from the whole interval of applied microwave energies, even when two segments with different slopes were apparent. This procedure is uncommon, and adopting the more accepted use of only one of the measured slopes would result in erroneously high or low PI values: 46.8 ± 10.3 μT for the low temperature slope and 26.8 ± 5.9 μT for the high temperature slope (Table 14.1; Hill and Shaw 2000). A small data set characterized only by two-slope Arai curves would then yield an erroneous PI. This pleads for the recommendation of Biggin et al. (2003) to employ a larger number of samples per lava flow than is usually used. According to their three cases it should be based on at least 6 samples but preferentially the number should be larger than 15.

An important constraint for PI studies is that the NRM is a pure TRM, i.e. not affected by chemical processes occurring during initial cooling, e.g. at temperatures well below 580°C. Yamamoto et al. (2003) studied 19 cores from the 1960 flow using the Thellier–Coe method (Coe 1967) and obtained from 17 successful experiments an average PI of 49.0 ± 9.6 μT. Samples came from 4 clusters within the outcrop (assigned to three groups, A to C) and it was argued that the PI values for their group B samples (Table 14.1) would have recorded a thermochemical remanent magnetization (TCRM) envisaged to have occurred by grain growth and/or exsolution of the titanomagnetite particles during high temperature oxidation but below the Curie temperature. Yamamoto (2006) addressed this question further by analyzing geo-thermometers, indicating that a part of the titanomagnetite and titanohematite minerals from this B group possibly achieved equilibrium at temperatures as low as ~300°C. This may have created – under certain cooling, temperature and compositional circumstances – suitable conditions for Fe-Ti diffusion allowing building up a TCRM and as a consequence a PI overestimation of up to ~70% (Yamamoto 2006). Depending on the remanence contribution of such grains to the bulk NRM, this could explain the much higher PI values obtained for samples of group B compared to that of groups A and C (Table 14.1). At the outcrop sampled by Yamamoto et al. (2003) the 1960 lava flow has a thickness of ~4 m, while at the location sampled by Hill and Shaw (2000) the accessible top part of the flow is thinner than one meter. This considerable difference in thickness probably resulted in different cooling histories and thus may explain that in the microwave PI study less variation was observed.

Yamamoto et al. (2003) also used a PI method, which combines low temperature demagnetization (cycling to 77 Kelvin) with the double heating Shaw method (LTD-DHT method; Tsunakawa and Shaw 1994). This gave PIs much closer to the expected value than their Thellier-style determinations, with a total mean of 39.4 ± 7.9 μT ($n = 9$). A mean of 35.7 ± 3.3 μT results when two high PI values are excluded, but we note that one of these excluded high PI values came from group B suspected to be affected by TCRM acquisition processes and the other was from group C. Also, both high PI samples have a similar technical quality as the remainder, making their exclusion difficult without knowledge of the paleofield. Again this emphasizes the requirement to determine a mean PI value for a rock unit from a reasonably large number of individual determinations, to reduce the detrimental effect of a few divergent results in a small data set. Only 3 out of 10 LTD-DHT Shaw experiments were successful for the group B samples.

Valet and Herrero-Bervera (2000) used the Shaw (1974) method but eliminating the ARM_2 step which is usually carried out to check for thermal alteration effects. For the 1960 flow a mean of 41.1 ± 3.7 μT was based on 7 out of 10 samples. In another experiment a 300°C thermal demagnetization step was added to eliminate potential unstable secondary components, which yielded a better mean of 35.2 ± 5.8 μT but this was based on only 3 samples. Finally, a Thellier–Coe type experiment on one sample failed and did not deliver a PI. The success rate was much lower for the 1955 flow with only two good results out of 9 samples: 39 and 42 μT, but here sister samples from the same drill cores provided acceptable Thellier–Coe PIs of 36.6 and 29.6 μT.

Cottrell and Tarduno (1999) tested the Thellier–Coe PI method using single plagioclase crystals with magnetite inclusions from the Hawaii 1955 lava flow. Such inclusions are single domain like and protected against thermally induced alterations. The paleointensity results obtained from 10 plagioclase crystals extracted from one drill core defined a mean paleofield value of 33.8 ± 3.7 μT, which is only 5% lower than the expected field. A notable aspect is that these field determinations have been obtained only from the 75 to 350°C temperature interval, as at higher temperatures the remaining NRM intensity was too week to be reliably measured.

Chauvin et al. (2005) used the original Thellier–Thellier method on samples from various historic Hawaii lava flows. Seven samples were studied from the 1960 flow, apparently collected from a location close to the one we have sampled and eight samples from the 1955 flow. When combining all accepted Thellier data of the two flows in one diagram each to simulate a multi-specimen approach, paleointensities of 28.2 (1955) and 36.3 μT (1960) were obtained. Averages for the high-temperature fractions of pTRM were 42.4 ± 6.9 μT (1955 flow, $n=4$) and 44.7 ± 9.5 μT (1960 flow, $n=7$). A best estimate was based mainly on the quality factor q (Coe et al. 1978) but especially on the requirement that the temperature interval for calculating the PI had to represent at least 40% of the original NRM intensity. With this criteria, values of 39.3 ± 3.7 μT (flow 1955, $n=7$) and 33.6 ±4.9 μT (flow 1960, $n=6$) were obtained.

Herrero-Bervera and Valet (2009) recently sampled both the 1955 and 1960 lava flows in great detail, with 23 and 39 cores at locations very close to our sampling sites. The Thellier–Coe protocol, but with in-field heating first followed by zero-field heating, was used to determine the PI. Values between 17.5 and 57.9 μT were obtained, with generally lower values for the 1955 flow. Overall averages were 27.4 ± 4.4 μT (1955 flow, $n=18$ out of 23) and 36.9 ± 7.4 μT (1960 flow, $n=36$ out of 38). By selecting only those samples that had pTRM checks with a repeatability of better than 5% (maximum DRAT; Selkin and Tauxe 2000) and average DRAT < 3%, these values enhanced dramatically to 34.8 ± 1.3 μT (1955 flow, $n=4$) and 36.7 ± 1.8 μT (1960 flow, $n=7$). But on the other hand, this reduced significantly the number of accepted PI data to about 18%. It was also observed that samples with high unblocking temperatures produced PI closer to the expected intensity than samples with lower unblocking temperatures.

Other PI studies (Abokodair 1977, Tanaka and Kono 1991, Tsunakawa and Shaw 1994, Tanaka et al. 1995) of the 1960 flow involved much less samples and resulted in PI values between 33.6 and 53.5 μT, thus often overestimating the expected field intensity (see Table 14.1).

Table 14.2 Expected field directions and observed paleomagnetic mean directions for the 1955 and 1960 Hawaii lava flows according to the references shown. Dec (Inc) site-mean declination (inclination); α_{95}, 95% confidence level and n, number of samples used for site-mean calculations, or for the number of sites where indicated

Reference	Flow	Dec (°E)	Inc (°)	α_{95} (°)	n	Comment
DGRF	1955	11.0	37.6	–	–	expected field direction
	1960	11.0	37.7	–	–	expected field direction
Castro and Brown (1987)	1960	11.4	36.0			
Hagstrum and Champion (1994)	1960	14.7	34.5	3.0		
	1960	10.9	35.2	2.2		
	1955–1960	12.5	35.2	1.7	6	average from 6 sites
Hill and Shaw (2000)	1960	10.4	30.1	2.8	17	profile H6001
	1960	14.0	34.4	2.7	14	profile H6002
Chauvin et al. (2005)	1955	10.3	29.7	2.1	15	AF and thermal demagnetization
	1960	13.2	37.0	2.0	12	
	1950–1982	11.8	34.5	3.1	6	average from 6 flows
Herrero Bervera and Valet (2009)	1955	7.8	36.8	2.2	21	AF demagnetization
	1955	13.6	37.7	3.7	20	thermal demagnetization
	1960	12.2	36.7	2.1	31	AF demagnetization
	1960	12.0	36.3	2.3	36	thermal demagnetization

Some of the studies mentioned above also reported directions for these two lava flows. Table 14.2 shows that these in general coincide within ~2–3° with the expected field direction according to the DGRF (declination 11.0°, inclination 37.6° to 37.7°), which was also the typical α_{95} confidence angle of these site mean directions. Hagstrum and Champion (1994) studied the 1955 and 1960 flows in a total of 6 locations and found an average direction (declination 12.5°, inclination 35.2°) close to the expected direction. The deviation at single sampling sites is larger and was interpreted to correspond to local anomalies of the magnetic field during lava emplacement.

14.3 Field and Laboratory Work

Samples from the 1955 flow were collected at the same outcrop described by Herrero-Bervera and Valet (2009), located at 19°23.918'N 154°55.147'W (Fig. 14.1). Here two cooling units were clearly recognizable and 11 (9) un-oriented drill cores were taken from the upper (lower) unit (Fig. 14.2). The 1960 flow was sampled at an outcrop located at 19°30.954'N 154°48.710'W (Fig. 14.1) and chosen because of the close proximity to the sampling localities of Herrero-Bervera and Valet (2009) and Hill and Shaw (2000). Here the accessible massive (top) part of the 1960 lava flow has a thickness of ~90 cm, which is overlain by a scoriacious part of similar thickness (Fig. 14.2). The scoracious part represents lava that cooled rapidly on the top of the flow, while the more massive interior part represents lava flowing in sheets or tubes for some extended time. Unoriented drill cores were taken at three levels: the upper two at about 5–15 cm and the third at about 40–45 cm below the interface between the massive and scoracious lavas (Fig. 14.2). Drill cores were cut into 12–25 mm long specimens for paleomagnetic and paleointensity experiments, and the cutting remnants were preserved for rock magnetic studies. For more details about the two flows we refer to Herrero-Bervera and Valet (2009).

A few samples from each flow were AF and/or thermally demagnetized to establish the presence of potential secondary NRM components and to determine the unblocking temperature spectrum of the magnetic minerals. Examples of orthogonal vector plots for both AF and thermal demagnetization and remanence decay are shown in Fig. 14.3. Both methods are successful in determining a stable remanence direction

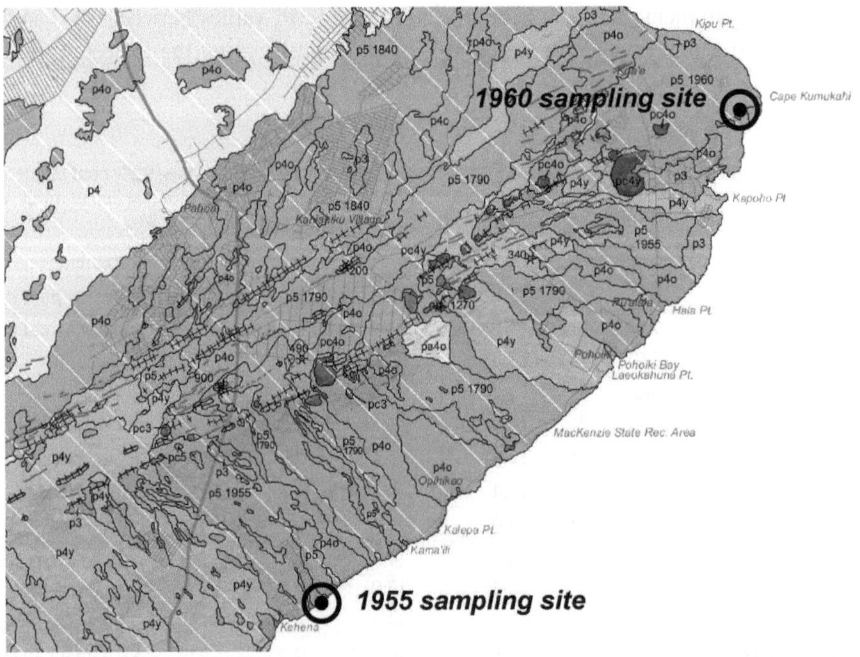

Fig. 14.1 Geological map of the south-eastern part of the Big Island of Hawaii with sampling localities of the Kilauea volcano 1955 flow (19°23.918'N 154°55.147'W) and the 1960 flow (19°30.954'N 154°48.710'W) (modified from Wolfe and Morris 1996). Different lava flows are distinguished by grey shades and codes

Fig. 14.2 Sampled outcrop of the Kilauea 1955 lava flow, with two cooling units; Kilauea 1960 lava flow cut along the Kapoho-Kumakahi Lighthouse Rd. with drill holes at three levels of the massive part overlain by scoriaceous lava

as no secondary magnetization components appeared to be present, confirming the results already obtained by Herrero-Bervera and Valet (2009). The 1955 flow showed different demagnetization behavior for the two cooling units: the lower unit lost more remanence below 300°C than the upper unit, and for both the maximum unblocking temperatures were close to 580°C. The data shown in Fig. 14.3g are from the work of Herrero-Bervera and Valet (2009), as because of limited sample material no thermal demagnetization was done on the NRM of the samples described here. Their sampling sites and ours of the two cooling units (cf. Fig. 14.2) are very closely spaced with drill holes side by side. Demagnetization characteristics of the 1960 flow resembled those of the 1955 upper cooling unit (Fig. 14.3). These data suggest for both lava flows the presence of magnetite with an additional component of titanomagnetite in the lower cooling unit of the 1955 flow.

14.3.1 Rock Magnetic Experiments

Hysteresis properties were determined with a Petersen Instruments variable field translation balance (VFTB) at the SOEST-HIGP Paleomagnetics and Petrofrabrics Laboratory of the University of Hawaii and a Princeton Instrument alternating gradient force magnetometer MicroMag model 2900 in Utrecht. For the MicroMag measurements samples of 4–10 mg were mounted on the sample probe, a maximum field of 2 T was utilized and the field increment was 4 mT (averaging time 0.1 s); samples appeared to be saturated below 1 T (pole shoe saturation correction between 1 and 1.5 T). Backfield curves of the SIRM were acquired to 0.1 T to determine the remanent coercive force B_{cr} (waiting time 1 s, averaging time 0.1 s). The instrumental noise level is $\sim 2 \cdot 10^{-10}$ Am^2, typical sample intensities were at least four orders of magnitude higher. Hysteresis parameters were determined after various thermal treatments and are based on at least three chips per treatment. For the VFTB measurements larger samples of up to 200 mg were used, in a maximum field of 1 T. First, the hysteresis curve and the backfield demagnetization of SIRM were determined, followed by the thermomagnetic heating and cooling curves in a field of ~720 mT. The VFTB has a measurement range of $10^{-8} - 10^{-2}$ Am^2.

To check for the absence of thermochemical alteration low-field susceptibility vs temperature runs were performed with an AGICO (Brno, Czech Republic) KLY3 susceptibility bridge (operating frequency 875 Hz, field strength 300 A/m r.m.s.) equipped with a CS3 heating unit. Typically ~200 mg of crushed rock chips were weighed in the furnace and heated in air which is inside the tube. Typical susceptibilities were at least two orders of magnitude higher than the instrumental noise level of $\sim 4 \cdot 10^{-8}$ SI. At the

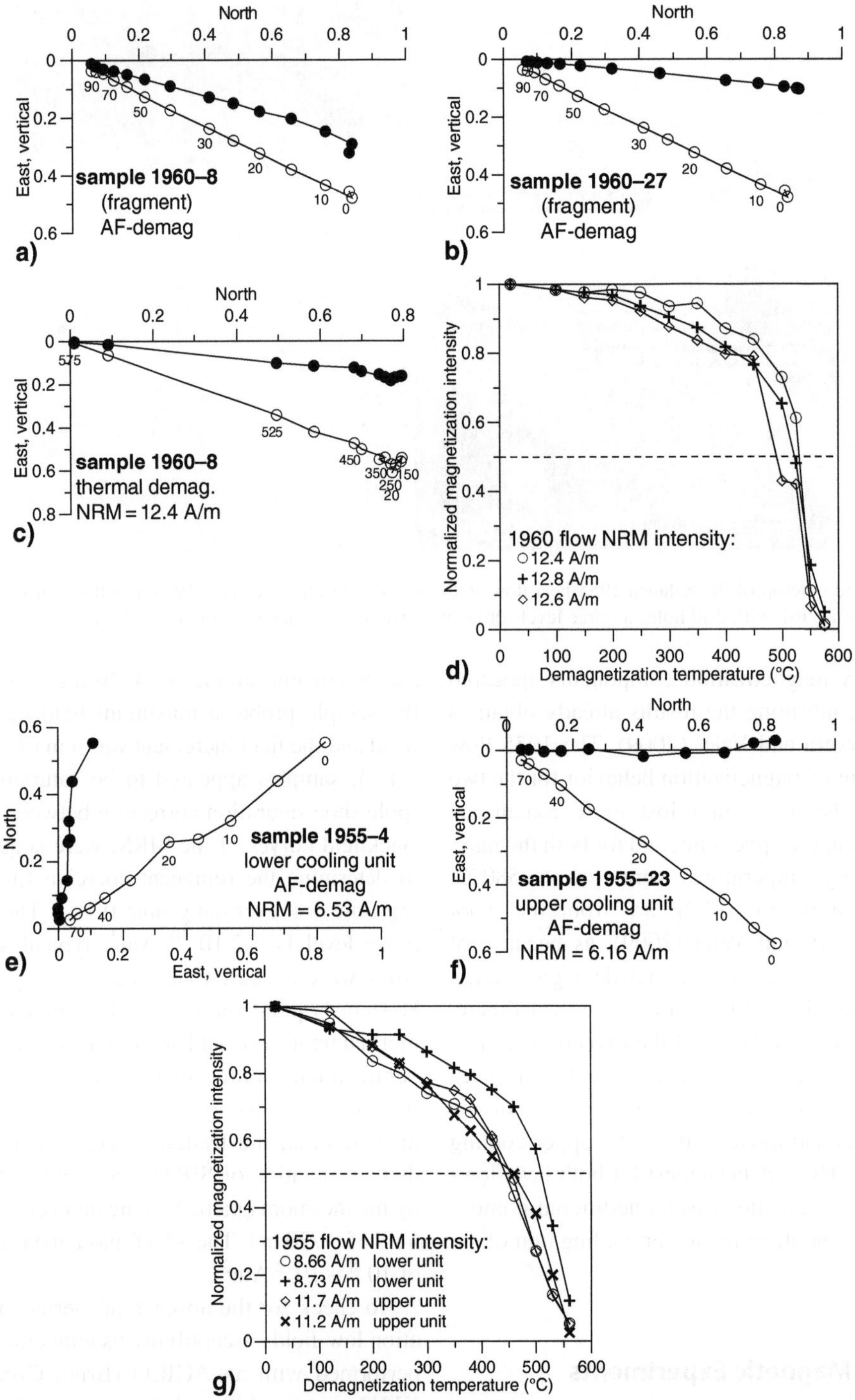

Fig. 14.3 Typical examples of Zijderveld plots (normalized intensity; in specimen coordinates) with labels along curves indicating AF or thermal demagnetization steps in mT or temperature in °C; *black (white)* dots correspond to horizontal (vertical) components. (**a**) alternating field demagnetization of a fragment of sample 1960–8 showing univectorial decay to the origin; (**b**) Zijderveld plot for a fragment of sample 1960–27; (**c**) Zijderveld plot of thermal demagnetization of sample 1960–8; (**e**) Zijderveld plot for sample 1955–3; (**f**) Zijderveld plot for sample 1955–24; (**d**) three examples of normalized NRM intensity curves during stepwise thermal demagnetization for samples from the 1960 flow, and (**g**) four examples for the 1955 flow

desired maximum temperature (several temperatures between 400 and 700°C) a lingering time of 10 min was selected to better simulate the heating trajectory in the PI experiments. For each temperature a fresh sample was weighed.

14.3.2 Rock Magnetic Properties

Hysteresis data are shown in Fig. 14.4 in a Day et al. (1977) plot as magnetization and coercivity ratios. For the 1955 flow, only samples from the lower cooling unit and one sample from the upper unit were available, and for the general hysteresis behavior of the upper cooling unit, the reader is referred to Herrero-Bervera and Valet (2009). All data obtained here indicate pseudo single domain (PSD) behavior, with M_{rs}/M_S values clustering between about 0.3 and 0.45. Coercivity ratios B_{cr}/B_C also are well clustered around 1.3–1.7, with a few data points above 1.75. While the exception of these, all samples could be interpreted as a distinct PSD grain ensemble or by a mixture of single domain (SD) and multidomain (MD) particles (Dunlop 2002).

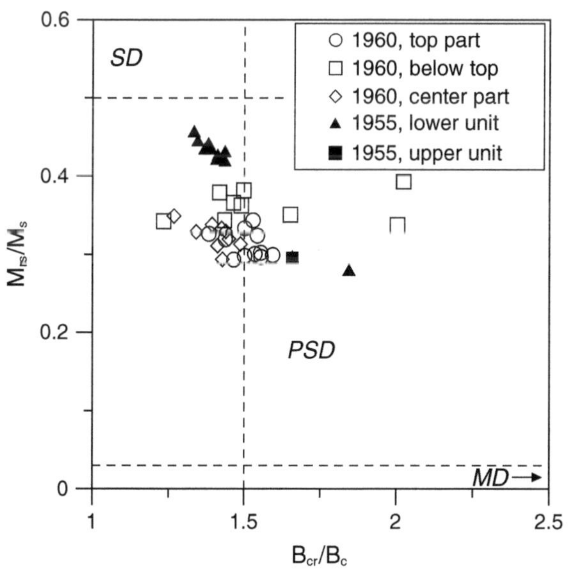

Fig. 14.4 Hysteresis parameters of Hawaii 1955 and 1960 flow samples. SD and PSD denote the single domain and pseudo single domain fields according to Dunlop (2002). The multidomain field (MD) is located outside the graph at $B_{cr}/B_c > 5$, as indicated by the *arrow*

Examples of the high-field thermomagnetic curves are shown in Fig. 14.5. The 1955 flow lower cooling unit exhibits two Curie temperatures around 200°C and 550°C, with near-reversible heating and cooling branches (Fig. 14.5a), suggesting the presence of a mixture of high-Ti and low-Ti titanomagnetite. Samples from the upper cooling unit are characterized by only one Curie temperature close to 500°C; they are almost reversible as well (Fig. 14.5a), suggesting a clear dominance of only low-Ti titanomagnetite. Samples from the 1960 flow show not much variability among the three sampled levels, with very similar thermomagnetic curves (Fig. 14.5b). Curie temperatures are between ~520 and 560°C, and after cooling the magnetization generally increased by about 10 and 20%. The higher Curie temperatures compared to the 1955 flow indicate an even lower Ti content, and the increased magnetization after cooling may be explained by the inversion of titanomagnetite to magnetite.

Samples from the 1960 flow were heated to different maximum temperatures while measuring the low-field magnetic susceptibility. Virtually no alteration occurred up to 510°C as shown by reversible heating and cooling curves, only after cycling to 730°C the susceptibility was somewhat higher (a few per cent) after return to room temperature (Fig. 14.6). This effect was similar to the high-field magnetization curves (Fig. 14.5). Hysteresis parameters showed no change after heating to 500°C. After heating at higher temperatures B_C and B_{cr} tended to increase slightly as did the M_{rs}/M_s ratio, and B_{cr}/B_c went up as well. This could indicate oxidation along preferred planes within particles thereby creating volumes that do not accommodate remanence. This would be in line with the observed decrease in pTRM capacity which leads to erroneously high PI estimates for pTRM acquisition temperatures of 550°C (to be discussed later).

14.3.3 Scanning Electron Microscope Analysis

Polished sections from the 1955 and 1960 flow drill cores were prepared for microscope observations. The reflected light microscope was a LEICA model DMLP with an attached Olympus DP11 digital camera and scanning electron microscope (SEM) observations were done with a JEOL Ltd. Model JSM-6060LV.

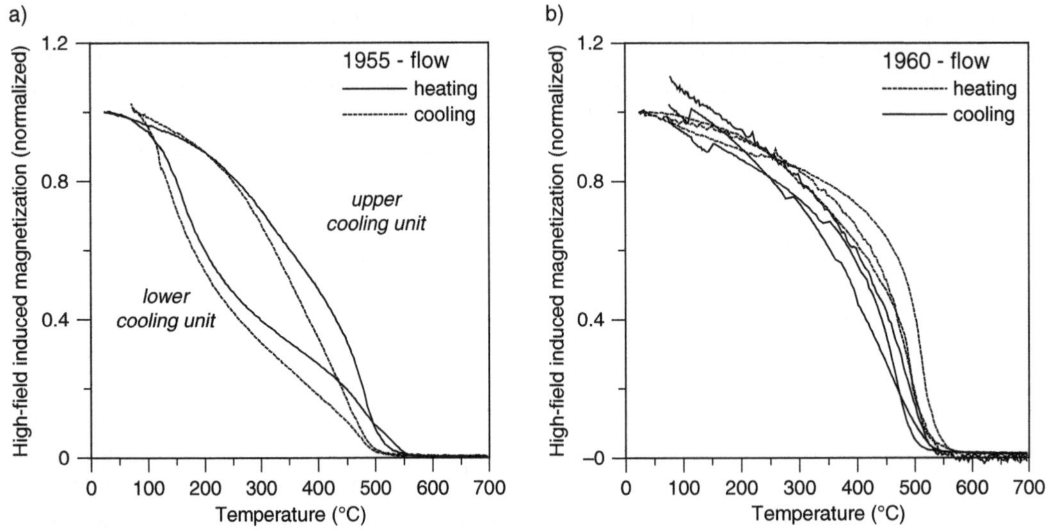

Fig. 14.5 Typical thermomagnetic curves for samples from (**a**) the two cooling units of the 1955 lava flow and (**b**) the three sampling levels of the 1960 lava flow

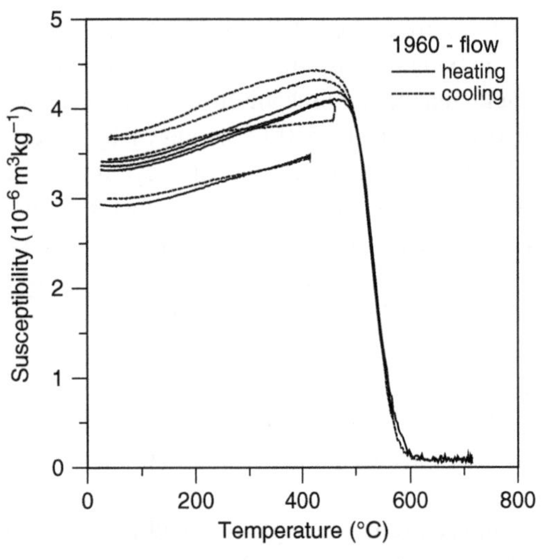

Fig. 14.6 Temperature variation of magnetic susceptibility for sister specimens from one drill core of the 1960 lava flow, up to ~410, 470, 570 and 730°C. Note the small change in susceptibility after heating to 410°C and 470°C indicating only small thermal alteration of magnetic minerals at these temperatures, and larger changes for higher temperatures

For the SEM polished sections were carbon coated. Typical settings were 20 keV, and standard-less semi-quantitative analysis was obtained with a beryllium window EDS detector at a counting time of up to 100 s.

Both the optical micrograph (Fig. 14.7a) as well as the SEM image (Fig. 14.7b) indicates that for the 1960 flow samples the iron oxides are predominantly of needle or skeletal form, which are up to 50 μm long and a few μm wide. At the optical resolution the grains looked homogeneous, but the SEM was capable to resolve sandwich-like intergrowths of high- and low-titanium titanomagnetite (Fig. 14.7c). Parts of the analyzed grains were entirely composed of either low- or high-titanium titanomagnetite (Fig. 14.7d). All studied samples were characterized by this combination of titanomagnetite grains in variable proportion, which coincides with the observed thermomagnetic curves showing sometimes a combination of low and high Curie temperature components. Two rock samples were also studied after heating to 480°C, and there was no evidence for significant compositional changes of titanomagnetite grains.

Samples from the two cooling units of the 1955 lava flow are also characterized by titanomagnetite of variable Ti content (Fig. 14.7e–j). In both cooling units, a mixture of up to 30 μm large and small titanomagnetite (<3 μm) particles was observed. Particles in the upper cooling unit have a larger length/width ratio and there is a larger amount of small grain sizes present (Fig. 14.7e, f), suggesting a faster cooling of this flow unit. The upper cooling unit is dominated by low-Ti titanomagnetite grains (Fig. 14.7h). The lower cooling unit contains more high-Ti titanomagnetite than the upper unit, either as homogeneous grains or as grains with a sandwich-type intergrowth texture of

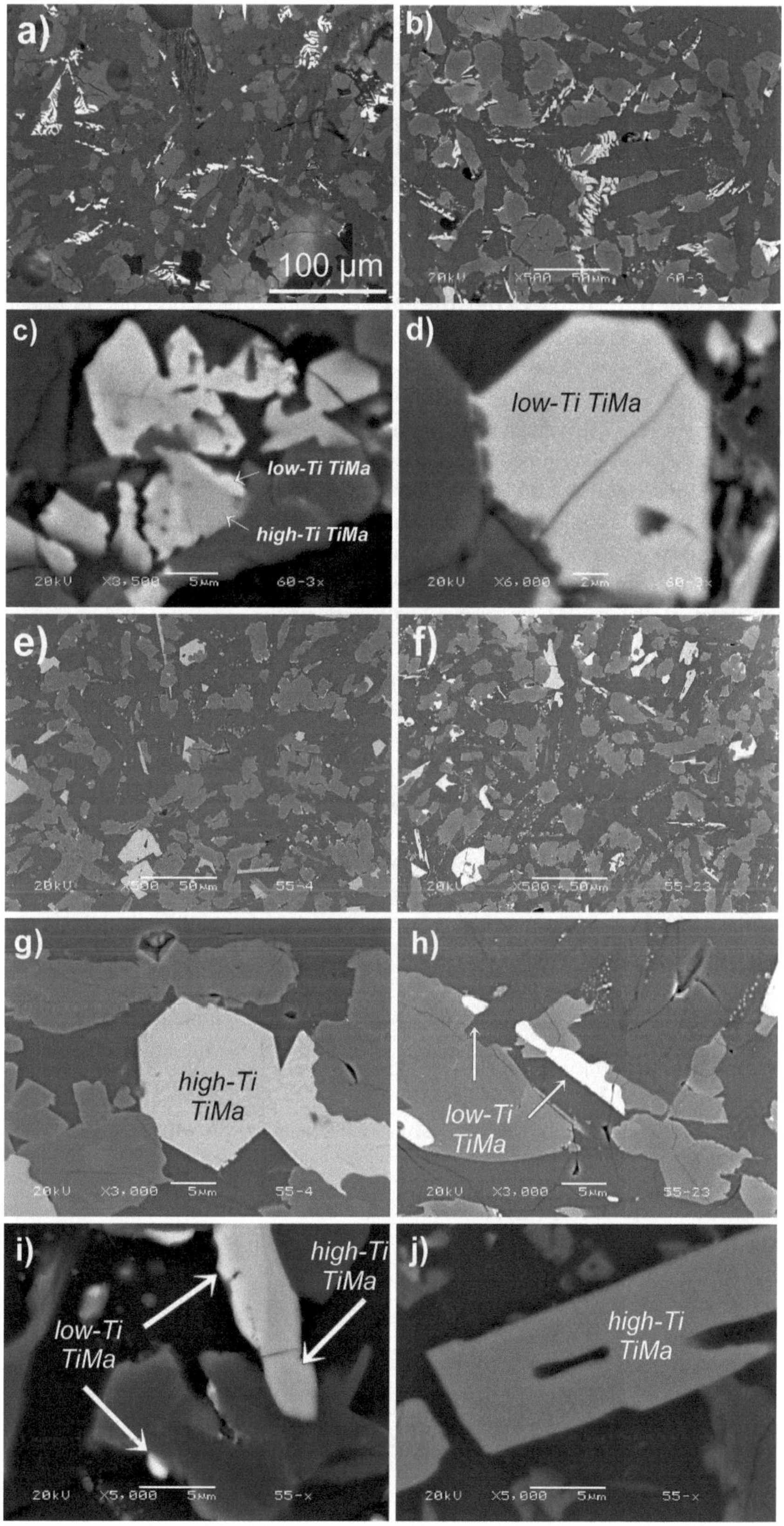

Fig. 14.7 Overview micrographs obtained by (**a**) the reflected light microscope and (**b**) scanning electron microscope (SEM) for a sample from the 1960 lava flow; (**c**) details of inhomogeneous and (**d**) homogeneous titanomagnetite (TiMa) grains. Overview SEM images from the (**e**) 1955 lower and (**f**) upper cooling units; details of (**g**) homogeneous high-Ti and (**h**) low-Ti titanomagnetite grains from the lower and upper cooling units, respectively; (**i**) example of sandwich type texture of titanomagnetite grain with high-Ti and low-Ti parts in a sample from the 1955 lower cooling unit, smaller homogeneous grains appear to be of low-Ti content; (**j**) homogeneous high-Ti titanomagnetite in the same sample from the 1955 lower cooling unit

low- and high-Ti titanomagnetite (Fig. 14.7g, i, j). The SEM analyses and the thermomagnetic experiments therefore support each other, indicating that the lower cooling unit contains titanomagnetite minerals with low and high Curie temperatures, while the upper cooling unit is dominated by minerals with high Curie temperature only.

14.3.4 Paleointensity Experiments

PI experiments were done according to the Dekkers and Böhnel (2006) method. The NRM was measured using an AGICO JR5 magnetometer, and magnetic susceptibility was determined with an AGICO KLY3 or MiniKappabridge instrument before and after heating. One set of samples was alternating field (AF) demagnetized in an AGICO LDA instrument at 5 mT maximum amplitude before measurements to test the effect of the pTRM difference; such a demagnetization would remove minor viscous components. Samples were oriented on a special sample holder so that their NRM direction was parallel to within about 3° precision to the field inside an ASC or Magnetic Measurements Ltd. thermal demagnetizer, and the pTRM acquisition field was maintained constant during heating and cooling. Samples from the lower and upper cooling units of the 1955 flow were exposed to temperatures of 230 and 450°C, respectively, while for the 1960 flow different sample sets were studied at pTRM acquisition temperatures of 400, 440, 480, 500 and 550°C. The remanence intensity after heating was compared to the NRM intensity to calculate the differential pTRMs (ΔpTRM), expressed as a percentage of the NRM intensity of the respective specimen to take into account spatial variations of the magnetic mineral content. Remanence directions after heating were compared to the NRM direction and in case of differences larger than 10° the sample was eliminated from further analysis. These differences may occur due to the secondary magnetization components superimposed on the TRM of a particular specimen or (more probably) due to orienting errors with respect to the field in the oven.

The PI for a given sample set was determined by calculating the linear best-fit to the data points in the ΔpTRM vs. applied field plot as the value where the best-fit line crosses the ΔpTRM = 0 line. Standard deviations (s.d.) were calculated using the 68% confidence limits to this best-fit, which is asymmetric in case that the data points are not distributed symmetrically around this crossing point.

14.3.5 Paleointensity Results from the 1960 Flow

The outcome of the PI experiments is listed in Table 14.3 and the pTRM vs. laboratory field plots are shown in Fig. 14.8. PI values obtained at the two most extreme temperatures, 400 and 550°C, deviate most from the expected field intensity of 36.2 μT. The high PI value obtained at 400°C of 47.3 ± 3.9/3.5 μT is characterized by a small ΔpTRM of <±10% (Fig. 14.8a), which is related to the high unblocking temperature spectrum in these specimens. At 550°C, again an overestimate of 41.5 ± 1.7/1.6 μT was

Table 14.3 Paleointensity results obtained using the multi-specimen parallel differential method. n = the specimen number used for PI calculation, r = rejected specimens. Sample set F is C+D+E. Sample sets labels correspond to panel labels in Fig. 14.8. The error is one standard deviation; n.d., not determined

Sample set	T (°C)	n	r	PI (μT)	± error (μT)
1960-A	400	7	0	47.3	3.9/3.5
1960-B	440	10	0	34.2	1.7/1.9
1960-C	480	7	0	33.8	3.7/3.8
1960-D	480	8	0	33.6	0.7
1960-E	480	7	1	33.8	n.d.
1960-F	480	22	1	33.5	0.7/0.8
1960-G	500	7	0	32.2	1.5
1960-H	550	8	0	41.5	1.7/1.6
1955-I	230	13	1	38.5	3.3
1955-J	450	19	2	34.3	1.5/1.6

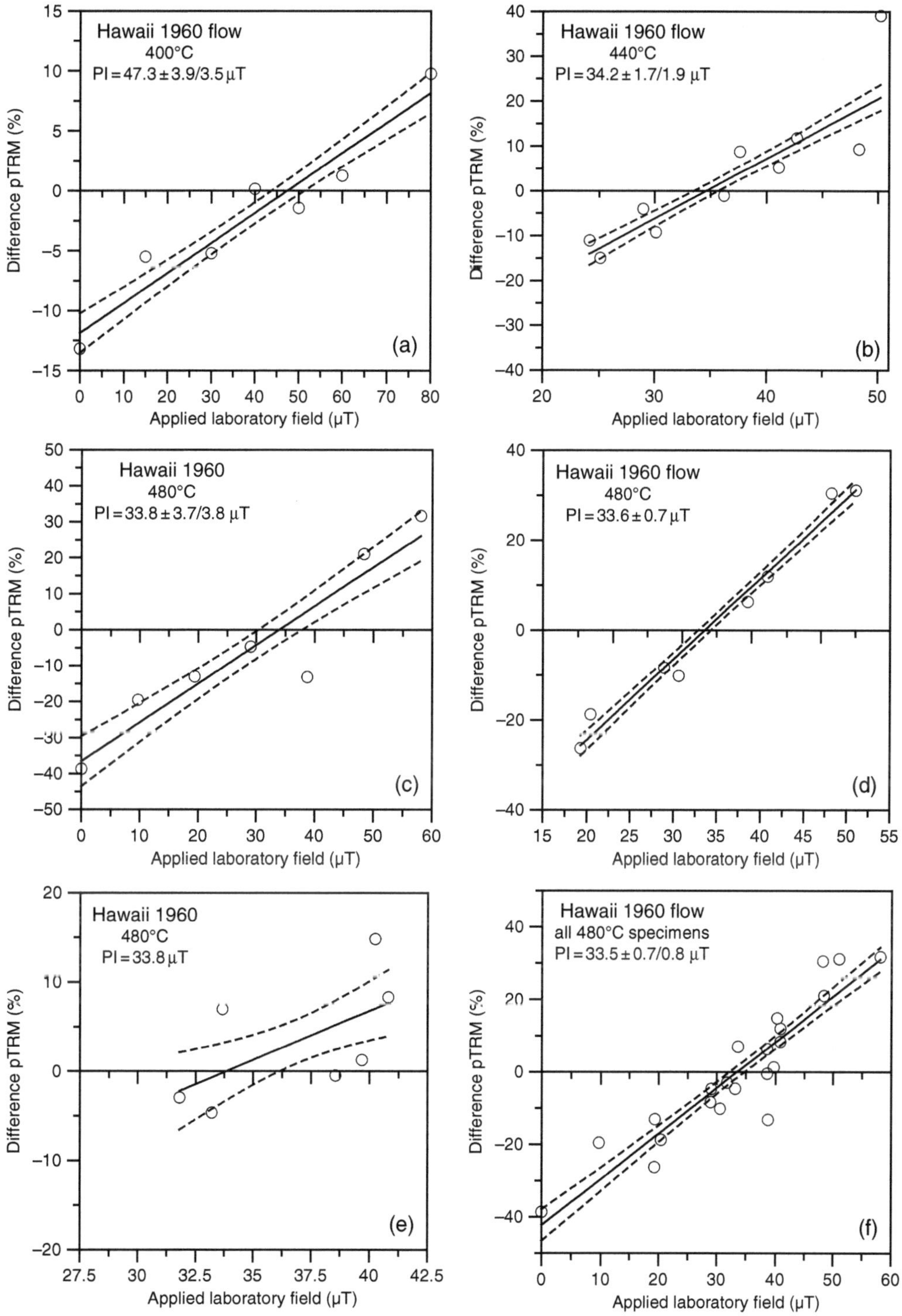

Fig. 14.8 Paleointensity experiments according to the multi-specimen parallel differential pTRM protocol carried out at several pTRM acquisition temperatures for the Kilauea 1960 flow (panels **a–h**) and the Kilauea 1955 flow (panel **i** and **j**). Panel (**a**) pTRM acquisition temperature 400°C, (**b**) 440°C, (**c**), (**d**) and (**e**) 480°C, (**f**) results from (**c**) to (**e**) plotted together, (**g**) 500°C plus AF demagnetization at 5 mT (**h**) 550°C. Panel (**i**) 1955 flow lower cooling unit with pTRM acquisition temperature at 230°C. Panel (**j**) 1955 flow upper cooling unit with pTRM acquisition temperature at 450°C

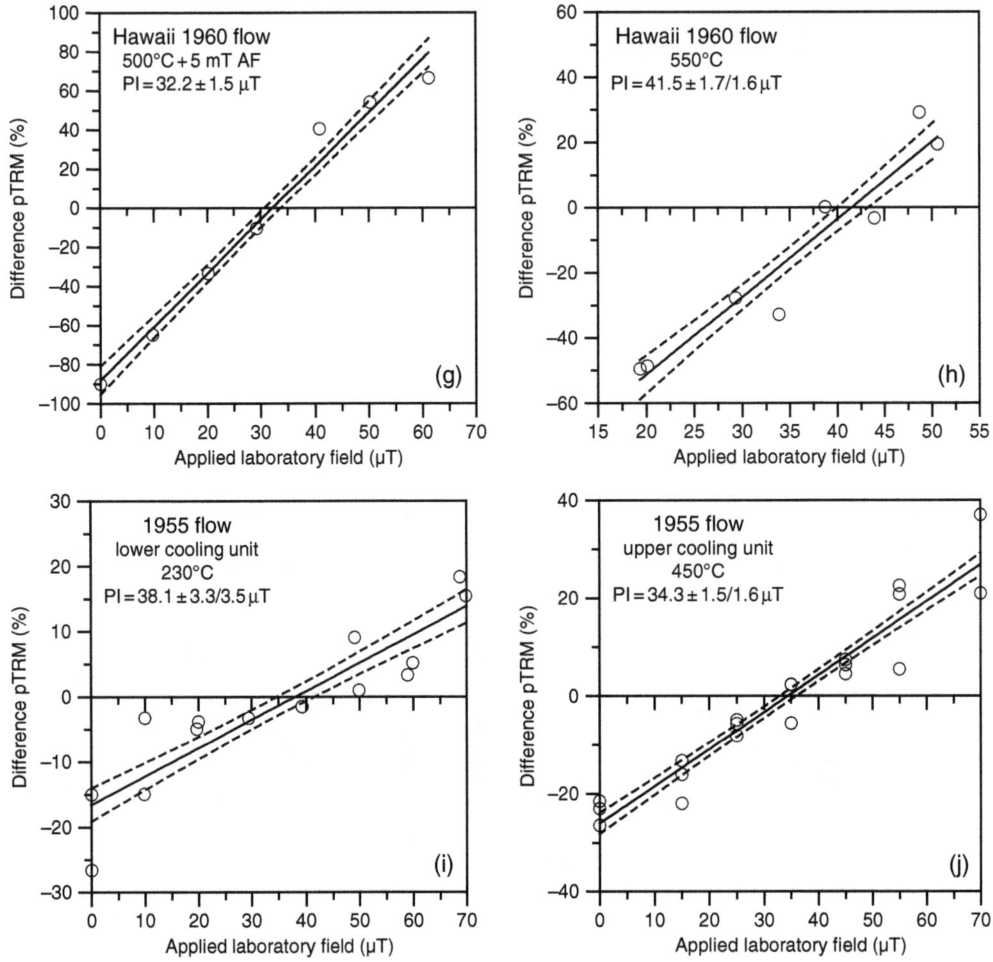

Fig. 14.8 (continued)

obtained (Fig. 14.8h). After the pTRM acquisition at this temperature the initial susceptibility on average dropped by ~18% in average, while at all lower temperatures only minor changes of <2% occurred. We therefore interpret the 550°C PI result to be affected by alteration processes, probably oxidation of titanomagnetite grains that reduced the susceptibility as well as the pTRM capacity. A reduced pTRM capacity would require higher laboratory fields to produce the same pTRM as during the initial natural cooling, which leads to the observed overestimate.

For temperatures between 440 and 500°C paleointensities vary between 32.2 and 34.2 μT, with s.d. values between 0.7 and 3.9 μT (Fig. 14.8b–g). These values are only slightly smaller than the expected field intensity of 36.2 μT. No systematic tendency of PI with temperature is seen, but we point out that all PI values for samples treated at 480°C scatter closely around 33.6 μT.

Experiments at 400°C and one set at 480°C were carried out as described above, with every specimen heated in a particular laboratory field. In other two experiments at 440 and 480°C two specimens were heated in the same laboratory field for each field step (in practice there are small field variations along the oven axis) to check for differences in their pTRM acquisition. Differences between sample pairs appear to be more pronounced at 440°C (Fig. 14.8b) than at 480°C (Fig. 14.8d). We interpret this to be produced by the smaller and relatively more variable pTRM acquisition at 440°C, which produces more scatter of the data points around the best-fit line. This scatter seems to be random, as the PI values obtained from the experiments at these two temperatures are indistinguishable. Therefore, the number of 10 specimens used in the 440°C experiment was adequate to average out the differences produced by the smaller pTRM acquisition.

Another experiment (Table 14.3, set E, Fig. 14.8e, at 480°C) was designed to test the idea that close to the expected field intensity the ΔpTRM should be small. Seven specimens (plus one rejected) were exposed to laboratory fields straddling closely 36 μT. Indeed small ΔpTRM values of +15/–5% were observed. The still observable variability of ΔpTRM is thought to be due to the variability of the unblocking temperature spectra of the specimens used in this experiment. Despite their dispersed distribution, the data points in (Fig. 14.8e) still give a very reasonable PI of 33.8 μT. However, because of the clustering of the data points the quality of the best-fit line is rather poor with large confidence limits. Thus, we consider this not a useful "stand-alone" experiment. Putting together the results of all individual experiments at 480°C, which is a welcome possibility of the multi-specimen method, produces a line based on 22 data points with a PI of 33.5 ± 0.7/0.8 μT (Fig. 14.8f). Again this underestimates the expected field intensity of ~36 μT only slightly.

One set of 7 specimens were AF demagnetized at 5 mT prior to any remanence measurement to simulate the removal of unstable secondary magnetization components. The PI experiment itself was done at 500°C (Table 14.2, set 1960-G, Fig. 14.8g) and a PI of 32.2 ± 1.5 μT was obtained, which is similar to the PI at 480°C. The AF-demagnetization hardly reduced the remanence intensity; therefore marginal difference from the other PI results would be in line with expectations.

14.3.6 Paleointensity Results from the 1955 Flow

According to available thermal demagnetization data from Herrero-Bervera and Valet (2009) as well as the thermomagnetic curves discussed above, the PI experiment was carried out at different temperatures for the two cooling units: 230°C for the lower unit and 450°C for the upper unit. Paleointensity results are listed in Table 14.3. At 230°C a total of 13 specimens were available, and at 450°C 21 specimens. In both groups one respectively two specimens were identified as an outliers, because of strongly deviating ΔpTRM values. Their exclusion is justified by the following experiment: after the pTRM acquisition step all samples were demagnetized at the same treatment temperature, and the rejected specimens showed substantially different remanence decays than the others.

Samples from the upper cooling unit defined the best fit shown in Fig. 14.7j, and the obtained PI of 34.3 ± 1.5/1.6 μT is indistinguishable from that of the 1960 flow and only marginally different from the expected field intensity. Three specimens were used at each field step to check for the reproducibility of the ΔpTRM, and indeed differences were observed: e.g., three samples exposed to a field of 55 μT produced ΔpTRM values between about 5 and 22% (Fig. 14.8j). Under unfavorable circumstances the use of a single specimen per field step could thus produce a bias of the obtained PI. From the lower cooling unit fewer specimens were available, so at every field step only one or two specimens could be used. The reproducibility of ΔpTRM is therefore slightly harder to evaluate. Where two specimens were available per field step, the corresponding data points in Fig. 14.8i are more dispersed around the best fit line, and therefore the standard error of the obtained PI is larger: 38.5 ± 3.3 μT. This PI slightly overestimates the expected field intensity, but due to the larger uncertainty ranges the difference is statistically not significantly different from the expected value of 36.2 μT. The selected treatment temperature of 230°C happened to be low, with relatively small maximum ΔpTRM values around ±18%. In this respect the results resemble those from the 400°C experiment carried out on the 1960 flow, which provided an even larger PI overestimate. In the case of the 1955 flow it was not possible to use a higher pTRM acquisition temperature that would have resulted in larger ΔpTRM values, as after heating to 230°C the susceptibility already increased on average by 10%, indicating that thermal alteration had already started but seemingly was not yet detrimental to the outcome of the PI experiment.

Conclusions

The multi-specimen parallel differential pTRM method (Dekkers and Böhnel 2006) has been used to determine the paleointensities for the 1955 and 1960 Kilauea lava flows, from the Big Island of Hawaii, with an expected PI value of 36.2 μT.

Nineteen specimens from 11 drill cores from the 1955 flows lower cooling unit were treated at 450°C, defining a PI of 34.3 ± 1.5/1.6 μT. Thirteen specimens from the 1955 upper cooling unit were heated to 230°C providing a PI of 38.5 ± 3.3 μT. Compared with the Thellier–Coe PI data reported by Herrero-Bervera and Valet (2009) from these same cooling units, the multi-specimen method yields similar paleointensities in the case of the high unblocking temperature samples, and a result closer to the expected field intensity in the case of the low unblocking temperature samples. These results are based on more than 90% of the specimens used in our experiments, while in the Thellier–Coe study only a small fraction of samples of ~16% were considered to be reliable.

Specimens from 23 drill cores from the centre and upper part of a 90 cm vertical profile through the 1960 flow were used for pTRM acquisition at temperatures between 400 and 550°C. An experiment at 550°C produced a high PI of 41.5 μT concomitant with a significant drop in the magnetic susceptibility of 18% after heating. Apparently thermal alteration processes reduce the pTRM capacity in the specimens yielding a too high PI estimate. Therefore, PI results accompanied by strong susceptibility changes should be rejected. Based on our experience, we currently suggest that susceptibility changes up to 10% are acceptable; however, future studies will have to define a more exact cut-off level. At 400°C pTRM acquisition temperature, the obtained PI of 47.3 μT was too high as well, possibly related to the small maximum ΔpTRM produced (<10%). This result bears resemblance to the overestimate obtained at low ΔpTRM-values (maximum of ~15%) obtained at 230°C for the 1955 flow. The overestimate for 400°C experiment in the 1960 flow is larger: it may thus point to a systematic bias towards high PI values when the ΔpTRM is small. Experiments between 440 and 500°C provided very consistent and well defined PI values between 32.2 and 34.2 μT. Therefore, most PI values obtained from both lava flows are just ~5–11% smaller than the expected field intensity of 36.2 μT as calculated from the 1965 Definitive Geomagnetic Reference Field (DGRF).

Exposing two or even more specimens to each laboratory field provides a useful test to evaluate the between-specimen variation of the pTRM acquisition. Thermal demagnetization at the same temperature as that used for the pTRM acquisition similarly provides useful information about such between-specimen variations. If the demagnetization data of suspect specimens do not compare with the bulk, this could be used as rejection criterion for anomalous data points. The results obtained by the multi-specimen method compare well with the best results provided by other PI methods, e.g. using single plagioclase crystals (Cottrell and Tarduno 1999), and applying very strict selection criteria in Thellier–Coe experiments (e.g. Herrero-Bervera and Valet 2009), or by using the microwave method (Hill and Shaw 2000). The often observed slight underestimate of the field intensity is also seen in those studies and may indeed indicate that at the sampling localities the field intensity may have been ~34 μT and thus lower than the expected intensity of 36.2 μT, due to local terrain effects.

Acknowledgements This work was supported by project IN107809 (PAPIIT, UNAM). We appreciate the help of Marina Vega and Alicia del Real in the SEM analysis of samples. Financial support to E.H-B was provided by SOEST-HIGP and by the National Science Foundation grants EAR-0510061, EAR-0710571, EAR-1015329, and NSF EPSCoR Program. This is a SOEST 8149 and HIGP 1892 contribution.

References

Abokodair AA (1977) The accuracy of the Thelliers technique for the determination of paleointensities of the Earth's magnetic field. PhD thesis, University of California, Santa Cruz

Biggin AJ, Thomas DN (2003) Analysis of long-term variation the geomagnetic poloidal field intensity and evaluation of their relationship with global geodynamics. Geophys J Int 152:392–415

Biggin AJ, Böhnel HN, Zuniga FR (2003) How many paleointensity determinations are required from a single lava flow to constitute a reliable average? Geophys Res Lett 30:1575. doi:10.1029/2003GL017146

Biggin AJ, Perrin M, Shaw J (2007) A comparison of a quasi-perpendicular method of absolute palaeointensity determination with other thermal and microwave techniques. Earth Planet Sci Lett 257:564–581

Castro J, Brown L (1987) Shallow paleomagnetic directions from historic lava flows, Hawaii. Geophys Res Lett 14:1203–1206

Chauvin A, Roperch P, Levi S (2005) Reliability of geomagnetic paleointensity data: the effects of the NRM fraction and concave-up behavior on paleointensity determinations by the Thellier method. Phys Earth Planetary Inter 150: 265–286

Coe RS (1967) Paleointensities of the earth's magnetic field determined from tertiary and quaternary rocks. J Geophys Res 72:3247–3262

Coe RS, Grommé CS (1973) A comparison of three methods of determining geomagnetic paleointensities. J Geomagnetism Geoelectricity 25:415–435

Coe RS, Grommé S, Mankinen EA (1978) Geomagnetic paleointensities from radiocarbon-dated lava flows on Hawaii and the question of the pacific nondipole low. J Geophys Res 83:1740–1756

Cottrell RD, Tarduno JA (1999) Geomagnetic paleointensity derived from single plagioclase crystals. Earth Planet Sci Lett 169:1–5

Day R, Fuller M, Schmidt VA (1977) Hysteresis properties of titanomagnetites: grain-size and compositional dependence. Phys Earth Planetary Inter 13:260–267

Dekkers MJ, Böhnel HN (2006) Reliable absolute paleointensity independent of magnetic domain state. Earth Planet Sci Lett 248:508–517

Dunlop DJ (2002) Theory and application of the Day plot (M_{rs}/M_S versus H_{cr}/H_C) 1. Theoretical curves and tests using titanomagnetite data. J Geophys Res 107:2056. doi:10.1029/2001JB000486

Hagstrum JT, Champion DE (1994) Paleomagnetic correlation of lateQuaternary lava flows in the lower east rift zone of Kilauea volcano, Hawaii. J Geophys Res 99: 21,679–21,690

Herrero-Bervera E, Valet JP (2009) Testing determinations of absolute Paleointensity from the 1955 and 1960 Hawaiian flows. Earth Planet Sci Lett 287:420–433

Hill MJ, Shaw J (2000) Magnetic field intensity study of the 1960 Kilauea lava flow, Hawaii, using the microwave paleointensity technique. Geophys J Int 142:487–504

Michalk DM, Muxworthy AR, Böhnel HN, Maclennan J, Nowaczyk N (2008) Evaluation of the multispecimen parallel differential pTRM method: a test on historical lavas from Iceland and Mexico. Geophys J Int 173:409–420

Selkin PA, Tauxe L (2000) Long-term variation in paleointensity. Philos Trans R Soc London 358:1065–1088

Shaw J (1974) A new method of determining the magnitude of the palaeomagnetic field: application to five historic lavas and five archaeological samples. Geophys J R Astron Soc 39:133–141

Tanaka H, Kono M (1991) Preliminary results and reliability of paleointensity studies on historical and 14C dated Hawaiian lava flows. J Geomagnetism Geoelectricity 43: 375–388

Tanaka H, Athanassopoulos JDE, Dunn JR, Fuller M (1995) Paleointensity determinations with measurements at high temperature. J Geomagnetism Geoelectricity 45:103–113

Thellier E, Thellier O (1959) Sur l'intensité du champ magnétique terrestre dans le passé historique et géologique. Ann Geophys 15:285–376

Tsunakawa H, Shaw J (1994) The Shaw method of paleointensity determinations and its application to recent volcanic rocks. Geophys J Int 118:781–787

Valet JP (2003) Time variations in geomagnetic intensity. Rev Geophys 41:4-1–4-44. doi:10.1029/2001RG000104

Valet JP, Herrero-Bervera E (2000) Paleointensity experiments using alternating field demagnetization. Earth Planet Sci lett 177:43–58

Walton D, Snape S, Rolph TC, Shaw J, Share J (1996) Application of ferromagnetic resonance heating to palaeointensity determinations. Phys Earth Planetary Inter 94: 183–186

Wolfe W, Morris J (1996) Geologic map of the Island of Hawaii, Miscellaneous Investigations Series, Published by the U.S. Geological Survey

Yamamoto Y (2006) Possible TCRM acquisition of the Kilauea 1960 flow, Hawaii: failure of the Thellier paleointensity determination inferred from equilibrium temperature of the Fe-Ti oxide. Earth Planets Space 58:1033–1044

Yamamoto Y, Tsunakawa H, Shibuya H (2003) Palaeointensity study of the Hawaiian 1960 lava flow: implications for possible causes of erroneously high intensities. Geophys J Int 153:263–276

Archaeomagnetic Research in Italy: Recent Achievements and Future Perspectives

15

Evdokia Tema

Abstract

During the last two decades, important advances in archaeomagnetic research in Italy have been made, both in the acquisition of new data and in the improvement of methodologies and data elaboration techniques. Nowadays, 73 directional and 23 intensity results are available, mainly obtained from archaeological sites situated in southern Italy. Most of the data come from the study of ancient kilns and ceramics and their ages range from 1300 BC to 1600 AD. The quantity and quality of the available Italian directional data have permitted the construction of reference secular variation (SV) curves for declination and inclination, using the latest improvements on curve building techniques. These curves describe, with reasonable accuracy, the variations of the Earth's magnetic field in Italy for the 600 BC to 600 AD period, for which many data are available. For older BC periods and for the Medieval times data are still very scarce. Archaeomagnetic dating of in situ archaeological materials is now possible but still caution is needed for the time periods where the reference SV curves are accompanied by large error envelopes. Certainly more new, high quality directional and intensity data of well dated archaeological material are necessary to better describe the variations of the Earth's magnetic field during the past and to make reliable archaeomagnetic dating possible, based on the full description of the Earth's magnetic field vector.

15.1 Introduction

Archaeomagnetic observations in Italy were initiated almost a century and half ago with the early work of Silvestro Gherardi (1866), Professor of Physics at the University of Torino, who noticed important deviations of the magnetic needle when placed close to building walls at Torino. He attributed these deviations to the magnetic polarity of the bricks in the structure itself and stated that the bricks and baked soils are magnetized. However, the Italian Giuseppe Folgerhaiter (1856–1913) is considered as the 'father' of archaeomagnetism, since he was the first to establish the archaeomagnetic research as an experimental method and archaeomagnetism as an independent field of study.

Folgerhaiter (1899) studied the remanent magnetization of ancient bricks and found that they were able

E. Tema (✉)
Dipartimento di Scienze della Terra, Università degli Studi di Torino, Via Valperga Caluso 35, 10125 Torino, Italy
e-mail: evdokia.tema@unito.it

E. Petrovský et al. (eds.), *The Earth's Magnetic Interior*, IAGA Special Sopron Book Series 1,
DOI 10.1007/978-94-007-0323-0_15,

to preserve their original magnetization. He also systematically studied the inclination of Etruscan, Greek and Pompeian ceramic vases concluding that they recorded the inclination of the Earth's magnetic field at the time of their fabrication. According to his studies the inclination of the geomagnetic field in Italy in the 8th century BC was negative. Nowadays, we know that such negative inclination is not correct; nevertheless Folgerhaiter's pioneer studies still remain fundamental. He was the first to propose an inclination variation curve from 800 BC to 100 AD (Fig. 15.1) and he underlined the importance of studying in situ burnt material in order to extend archaeomagnetic studies including the determination of declination.

These promising archaeomagnetic studies in Italy were not followed up for almost a century, until the late 20th century. In the 1980's, the first systematic archaeomagnetic studies were published, mainly regarding directional results from single kilns and in situ burnt bricks (Tanguy et al. 1985, Evans and Mareschal 1986). More attention on archaeomagnetism has been focused in the 1990's and an important number of archaeomagnetic results have been presented (Abdeldayem et al. 1992, Màrton et al. 1992, Nardi et al. 1995, Chiosi et al. 1998, La Torre et al. 1998). However, the greatest progress on archaeomagnetic research in Italy was actually done at the beginning of the 21st century mainly thanks to the EU-funded AARCH project 'Archaeomagnetic Applications for the Rescue of Cultural Heritage'. In the framework of this project, many new archaeomagnetic data for Italy have been produced and advances on data processing and secular variation curve building have been done (Tema et al. 2006).

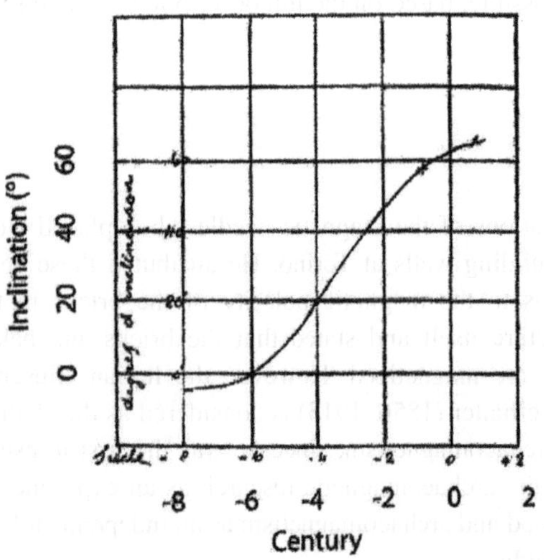

Fig. 15.1 Folgerhaiter's (1899) original figure displaying the inclination variation in Italy for the period 800 BC – 100 AD

Although archaeomagnetic research most commonly implies the use of archaeological materials for the study of the Earth's magnetic field in the past, volcanic rocks from historically dated eruptions have been often used to increase the reference secular variation data. In Italy, lava flows from Etna and Vesuvius have been widely studied and used for archaeomagnetic dating (e.g. Hoye 1981, Rolph and Shaw 1986, Incoronato et al. 2002, Tanguy et al. 2003). Nevertheless, the reliability of the volcanic data is often disputable. Problems related to their age determination, difficulties in the identification in the field of the eruptive products of specific events and/or a possible distortion of the geomagnetic field due to the magnetization of the whole volcanic edifice may be some of the sources that can cause a bias between the palaeomagnetic record and the Earth's magnetic field (Lanza et al. 2005). Thorough palaeomagnetic studies of lava flows from Italian volcanoes show that in many cases the historically attributed age of such rocks is not consistent with their palaeomagnetic direction and thus they have been assigned a new archaeomagnetic age (Tanguy et al. 2003, 2007, Principe et al. 2004).

In this paper a thorough presentation of the archaeomagnetic research in Italy during the last decades and a critical review of the quantity and quality of the available archaeomagnetic data is presented, considering as 'archaeomagnetic' only data coming from archaeological materials. Extensive discussion on archaeomagnetic data from volcanic rocks may be found in Incoronato et al. (2002), Tanguy et al. (2003, 2007) and references therein.

15.2 Archaeomagnetic Directional Results

The first direct measurement of both declination and inclination of the Earth's magnetic field in Italy was done in 1640 AD, at Rome, by Athanasius Kircher (Cafarella et al. 1992). For older periods, in order to establish the variation of the geomagnetic field, archaeomagnetic results are necessary. Even though Italy is a country with rich archaeological history, archaeomagnetic data are still sparse and often difficult to find as they have been published in a variety of journals or only in excavation reports.

Evans and Hoye (2005) have recently published the first systematic compilation of directional archaeomagnetic results, updating the ones previously published by Evans and Mareschal (1989) and including 12 new data. In this paper the authors summarized the results of their archaeomagnetic investigations in Italy over the last 20 yrs and published a total of 29 archaeodirectional results from kilns and other baked structures, all of them coming from southern Italy. Apart from this study, around ten other papers dealing with single sites and published by various authors can be found in the literature.

Tema et al. (2006) published a systematic compilation of all available Italian directional results from the literature, supplemented by 28 new data from the Genève and Torino laboratories. This database includes 67 directions from archaeological sites with ages covering the time period from 5900 BC to 1600 AD and is the only existing compilation of Italian archaeomagnetic data including metadata information such as material studied, laboratory treatment and dating techniques that are nowadays necessary for data quality estimation. Recently, two more studies have been published including directional results from Italian archaeological sites (Hill et al. 2007, Tema and Lanza 2008). Hill et al. (2007) studied a roman amphorae workshop in Albinia (Tuscany) presenting the directional results from five contemporanous Roman kilns and Tema and Lanza (2008) published the archaeomagnetic direction of a 3rd–4th century AD circular lime kiln excavated at Bazzano (northern Italy).

Nowadays, a total of 73 directional results from dated archaeological structures are available for Italy and are summarized here in Table 15.1. The geographical distribution of the data is shown in Fig. 15.2. Most of directional data come from southern Italy whilst central and northern Italy are poorly covered. The age of the studied sites is in almost all cases based on archaeological information and only in few cases dating results from independent experimental dating techniques are available (only 4 thermoluminescence and 3 radiocarbon results, Table 15.1). This is often a problem when dealing with the Italian archaeomagnetic results. Archaeologically proposed ages usually correspond to the period of the use of a archaeological site and can rarely precisely define the date of the abandonment or last use of a structure (e.g. the last firing of a kiln). Thus, in the Italian dataset (Table 15.1), well defined archaeomagnetic directions are often accompanied by poor dating information and therefore their contribution on the reconstruction of the SV of the Earth's magnetic field in the past is limited.

The time distribution of the Italian archaeodirections (Fig. 15.3) shows that the majority of data come from the Roman period with a maximum concentration at the 1st century BC. The time interval 800 BC to 1200 AD is generally continuously covered with at least one archaeodirection per century (the only exceptions are the 3rd and the 7th century AD). For other periods only few data are available while no data exist for 1200 to 1500 AD and 800 to 1100 BC. For prehistoric times, only three results have been published; one coming from a hearth from Grotta dell'Edera di Aurisina (radiocarbon dated: 2500–4300 BC) and two results from ancient hearths studied at the location Laghetti del Crestoso, dated around 5900–5480 BC (Tema et al. 2006).

The materials used for archaeodirection determination are in most cases kilns and small hearths (Fig. 15.4a). In some cases in situ fragments of bricks from burnt walls, mural paintings and funeral pyres have been studied. The laboratory treatment used to derive the archaeomagnetic directions is mainly AF and/or thermal demagnetization. However, in some early archaeomagnetic studies only NRM measurements and Thellier viscosity tests have been performed. The quality of the archaeodirections is generally good with α_{95} being <4° for the 76% of the data (Fig. 15.4b) but it is important to note that information about the magnetic mineralogy and rock magnetic experiments of the studied material is often missing. Such information is crucial for a better understanding of the recorded archaeodirections and is necessary for quality assessment of the obtained results. Data obtained without full demagnetization procedures and accompanied by poor rock-magnetic information, included in Table 15.1, should be cautiously used for further archaeomagnetic applications and adapted selection criteria should be applied according to the target of any future elaboration of the Italian database.

Table 15.1 Italian directional archaeomagnetic data. Columns: No = site reference number; Location of the sampling sites; Lat/Long = site latitude/longitude; Age range; Mean age; Method of dating (A: archaeological estimate, TL: thermoluminescence, ^{14}C: radiocarbon); Material studied; Code name of the structure; (*n*)/*N* = (number of specimens)/number of independently oriented samples; laboratory treatment (AF: alternating field demagnetization, Th: thermal demagnetization, ac: anisotropy correction); *D*, *I* = declination, inclination; α_{95} = 95% semi-angle of confidence; *k* = Fisher's precision parameter; D_r, I_r = declination, inclination relocated to Viterbo (42.45°N, 12.03°E); Reference

No	Site	Lat (°)	Long (°)	Age (years)	Mean age	Method	Material	Code	(n)	N	Treatment	D (°)	I (°)	α_{95} (°)	k	D_r (°)	I_r (°)	Reference
1	Rome, Bibliotheca Apost.	41.90	12.45	1610–1615 AD	1612	A	mural painting	D- Paul	10	10	AF	0.8	67.4	2.4	414	0.9	67.8	Chiari and Lanza (1999)
2	Castelseprio	45.72	8.85	1506–1574 AD	1540	A	hearth	US1020		7	Tv	7.8	59.0	3.1	365	7.9	56.3	Tema et al. (2006)
3	Rome, Foro Traiano 1	41.90	12.45	1480–1520 AD	1500	A	kiln		33	11	AF, Th	359.0	62.4	2.8	258	359	62.9	Tema et al. (2006)
4	Dosso Castello, Cremona	45.13	10.37	1120–1270 AD	1195	TL	hearth	H		7	Tv	22.0	56.0	5.0	145	21.6	54	Tema et al. (2006)
5	Monreale, Palermo	38.07	13.29	1172–1185 AD	1178	A	bricks from building			17	Th	*	51.2	2.3	*	–	–	Tanguy et al. (1985)
6	Dosso Castello, Cremona	45.13	10.37	1080–1250 AD	1165	TL	hearth	C		10	Tv	19.1	55.5	5.7	72	18.8	53.3	Tema et al. (2006)
7	Cefalu, Sicily	38.03	14.07	1132–1148 AD	1140	A	bricks from church			17	Th	*	53.2	1.8	*	–	–	Tanguy et al. (1985)
8	Palazzo Reale, Palermo	38.11	13.33	1130–1150 AD	1140	A	bricks from building			47	NRM	*	50.5	1.3	*	–	–	Tanguy et al. (1985)
9	St.Salvatore d'Alunzio, Sicily	38.73	14.70	1105 AD	1105	A	bricks from church			31	NRM	*	51.8	1.7	*	–	–	Tanguy et al. (1985)
10	Santa Maria di Mili, Sicily	38.13	15.49	1092 AD (or 1542?)	–	A	bricks from church			25	NRM	*	60.7	2.1	*	–	–	Tanguy et al. (1985)
11	St Pietro e Paolo di Agro	37.96	15.32	1000–1060 AD	1030	A	bricks from church			41	AF	*	53.0	1.4	*	–	–	Tanguy et al. (1985)
12	San Vincenzo, Galliano	45.73	9.13	700–1200 AD	950	A	hearth	US24		7	NRM	14.9	62.6	2.6	508	14.3	60.4	Tema et al. (2006)

Table 15.1 (continued)

No	Site	Lat (°)	Long (°)	Age (years)	Mean age	Method	Material	Code	(n)	N	Treatment	D (°)	I (°)	α_{95} (°)	k	D_r (°)	I_r (°)	Reference
13	Dosso Castello, Cremona	45.13	10.37	750–950 AD	850	TL	hearth	E		10	Tv	27.3	62.2	2.2	463	26.4	60.6	Tema et al. (2006)
14	Rome, Foro Traiano 2	41.90	12.45	600–800 AD	700	A	bricks from limekiln		26	8	AF, Th	8.4	61.8	2.3	573	8.5	62.2	Tema et al. (2006)
15	Canosa, Puglia	41.22	16.07	500–600 AD	550	A	bricks from kiln		26	9	AF, Th	359.4	51.3	3.1	279	358.6	52.7	Tema et al. (2006)
16	Ruoti	40.73	15.68	460–540 AD	500	A	oven	F446		16	AF, Th	0.6	58.1	4.2	67	0.5	59.6	Evans and Hoye (2005)
17	Pioppi, Pratola Serra	40.98	14.85	400–600 AD	500	A	bricks from a furnace		18	6	AF	1.6	56.7	3.6	363	1.4	58	Nardi et al. (1995)
18	Santa Maria	39.50	16.93	400–600 AD	500	A	oven			6	AF, Th	359.4	57.6	3.8	318	359.2	60.3	Evans and Hoye (2005)
19	Carlino, Aquileia	45.80	13.22	380–430 AD	405	A	kiln	kiln 1		10	NRM	359.4	58.5	3.0	255	359.3	55.3	Tema et al. (2006)
20	Carlino, Aquileia	45.80	13.22	380–430 AD	405	A	kiln	kiln 2		7	NRM	357.7	55.8	3.1	357	357.6	52.4	Tema et al. (2006)
21	Segesta, Sicily	37.94	12.83	300–500 AD	400	A	kiln	SG2	8	8	Th	2.1	62.3	3.7	222	2.4	65.9	Màrton et al. (1992)
22	Segesta, Sicily	37.94	12.83	300–500 AD	400	A	kiln			5	AF, Th	359.9	63.3	2.8	769	0	66.8	Evans and Hoye (2005)
23	Sibari, Cavallero	39.60	16.50	300–400 AD	350	A	fired structure			10	AF, Th	0.3	55.1	3.2	224	359.9	57.8	Evans and Hoye (2005)
24	Vagnari, Puglia	40.83	16.27	300–400 AD	350	A	tiles from kiln		23	9	AF, Th	2.1	50.8	2.8	343	1.3	52.4	Tema et al. (2006)
25	Ascoli Satriano, Puglia	41.21	15.56	350–400 AD	375	A	bricks from kiln		16	5	AF, Th	2.3	54.4	2.9	701	1.9	55.5	Tema et al. (2006)
26	Bazzano	44.50	11.08	200–400 AD	300	A	kiln		35	12	AF, Th	0.4	60.6	3.1	197	0.5	58.8	Tema and Lanza (2008)
27	Cassano	41.10	14.50	50–300 AD	175	A	kiln		7	7	AF	6.4	51.6	9.2	33	6	52.8	Chiosi et al. (1998)

Table 15.1 (continued)

No	Site	Lat (°)	Long (°)	Age (years)	Mean age	Method	Material	Code	(n)	N	Treatment	D (°)	I (°)	α_{95} (°)	k	D_r (°)	I_r (°)	Reference
28	Lonato, Brescia	45.47	10.48	146 AD–74 BC	36	TL	tile kiln	A		20	NRM	355.8	55.2	1.7	369	356.3	52	Tema et al. (2006)
29	Pompei, Porta Nocera	40.80	14.50	79 AD	79	A	kiln			13	AF, Th	355.0	59.1	1.6	581	354.8	60.7	Evans and Hoye (2005)
30	Pompei, Thermae Stab.	40.80	14.50	62–79 AD	70	A	mural painting		18	18	AF	1.2	58.0	5.5	40	1.1	59.5	Zanella et al. (2000)
31	San Giovanni	40.73	15.68	0–100 AD	50	A	kiln			106	AF, Th	350.5	58.8	0.9	232	350.1	60.7	Evans and Hoye (2005)
32	Pizzica	40.38	16.78	50 AD-50 BC	0	A	kiln			14	AF, Th	356.4	51.4	2.1	367	355.5	53.8	Evans and Hoye (2005)
33	Segesta, Sicily	37.90	12.80	50 AD–50 BC	0	A	kiln	SG1	5	5	AF	355.5	58.2	7.0	120	355.2	62.2	Màrton et al. (1992)
34	Marzabotto, Bologna	44.35	11.20	20 AD–50 BC	−15	A	furnace	A		7	pTD, 0v	356.8	63.4	2.7	498	356.9	61.8	Tema et al. (2006)
35	Capocolonna	39.02	17.07	20 AD–80 BC	−30	A	kiln			10	AF, Th	357.1	59.8	1.5	980	357.1	62.9	Evans and Hoye (2005)
36	Albinia	42.50	11.20	200 BC–100 AD	−50	A	kiln	kiln A		21	Th, ac	358.6	63.4	1.2	723	358.6	63.3	Hill et al. (2007)
37	Albinia	42.50	11.20	200 BC–100 AD	−50	A	kiln	kiln B		24	Th, ac	356.2	66.3	1.1	754	356.1	66.2	Hill et al. (2007)
38	Albinia	42.50	11.20	200 BC–100 AD	−50	A	kiln	kiln C		15	Th, ac	354.6	63.5	1.9	420	354.6	63.4	Hill et al. (2007)
39	Albinia	42.50	11.20	200 BC–100 AD	−50	A	kiln	kiln D		6	Th, ac	358.8	63.2	3.7	386	358.8	63.1	Hill et al. (2007)
40	Albinia	42.50	11.20	200 BC–100 AD	−50	A	kiln	kiln E		16	Th, ac	352.5	62.7	1.8	471	352.5	62.6	Hill et al. (2007)
41	Morgantina	37.38	14.37	0–100 BC	−50	A	kiln			5	AF, Th	5.1	60.0	4.6	188	5.7	64	Evans and Hoye (2005)
42	Apani, Brindisi	40.72	17.90	50–100 BC	−75	A	fired structure			7	AF, Th	356.7	58.0	1.7	1203	356.4	59.8	Evans and Hoye (2005)

Table 15.1 (continued)

No	Site	Lat (°)	Long (°)	Age (years)	Mean age	Method	Material	Code	(n)	N	Treatment	D (°)	I (°)	α_{95} (°)	k	D_r (°)	I_r (°)	Reference
43	Ordona	41.32	15.63	50–150 BC	−100	A	kiln			60	AF, Th	4.8	63.0	1.7	115	5.2	63.7	Evans and Hoye (2005)
44	Metaponto	40.42	16.75	0–200 BC	−100	A	iron smelter			11	AF, Th	353.1	58.0	3.9	120	352.7	60.2	Evans and Hoye (2005)
45	Bassento Destra	40.41	16.17	0–200 BC	−100	A	kiln			10	AF, Th	353.6	61.5	1.9	652	353.6	63.4	Evans and Hoye (2005)
46	St Angelo, Vecchio	40.39	16.72	100–200 BC	−150	A	fired structure			13	AF, Th	358.3	57.9	1.7	617	358.1	59.9	Evans and Hoye (2005)
47	Insula, Metapontum	40.42	16.75	250–300 BC	−275	A	fired structure			8	AF, Th	348.7	61.4	2.5	499	348.5	63.6	Evans and Hoye (2005)
48	Velia	40.16	15.17	250–300 BC	−275	A	kiln			8	AF, Th	355.6	60.2	3.4	271	355.5	62.3	Evans and Hoye (2005)
49	Locri	38.21	16.24	250–300 BC	−275	A	kiln	B		7	AF, Th	358.5	58.6	2.1	807	358.5	62.3	Evans and Hoye (2005)
50	Bisignano	39.51	16.28	250–350 BC	−300	A	kiln			10	AF, Th	9.4	64.3	2.7	333	10.6	66.1	Evans and Hoye (2005)
51	Roccagloriosa	40.10	15.43	300–350 BC	−325	A	funeral pyre			7	AF, Th	–	–	–	4	–	–	Evans and Hoye (2005)
52	Metapontum, Kerameikos	40.42	16.75	325–350 BC	−338	A	fired structure			12	AF, Th	347.0	62.0	2.7	251	346.8	64.2	Evans and Hoye (2005)
53	St Angelo Grieco	40.40	16.79	300–400 BC	−350	A	kiln			10	AF, Th	3.2	63.4	2.9	273	3.9	64.8	Evans and Hoye (2005)
54	Metapontum	40.42	16.75	300–400 BC	−350	A	funeral pyre			9	AF, Th	–	–	–	7	–	–	Evans and Hoye (2005)
55	Motya	37.83	12.50	350–400 BC	−375	A	kiln			7	AF, Th	352.2	61.3	2.2	737	351.6	65	Evans and Hoye (2005)
56	Heraciea	40.21	16.67	350–400 BC	−375	A	kiln			–	AF, Th	–	–	–	–	–	–	Evans and Hoye (2005)
57	Locri	38.21	16.24	400–500 BC	−450	A	kiln	A		9	AF, Th	0.7	61.3	5.2	98	1.2	64.7	Evans and Hoye (2005)
58	Poggioreale	37.48	13.02	430–500 BC	−465	A	burnt walls			6	AF, Th	4.3	60.7	3.9	293	4.8	64.7	Evans and Hoye (2005)
59	Pozzuolo del Fruili, Udine	45.97	13.20	500–550 BC	−525	A	oven	No 8		4	Tv	359.9	74.8	11.7	63	0.5	72.6	Tema et al. (2006)

Table 15.1 (continued)

No	Site	Lat (°)	Long (°)	Age (years)	Mean age	Method	Material	Code	(n)	N	Treatment	D (°)	I (°)	α_{95} (°)	k	D_r (°)	I_r (°)	Reference
60	Metapontum, Temple Area	40.42	16.75	500–550 BC	−525	A	kiln			10	AF, Th	0.6	66.4	1.8	711	1.7	67.8	Evans and Hoye (2005)
61	Chiesa St Francesco, Udine	46.03	13.14	500–600 BC	−550	A	hearth	Tv, AF		7	Tv, AF	6.5	71.1	2.4	642	6.3	68.6	Tema et al. (2006)
62	Treglia, Pontelatore	41.25	14.17	500–600 BC	−550	A	furnace		9	6	AF	353.3	53.7	5.8	79	352.9	55.1	La Torre et al. (1998)
63	Naxos	37.85	15.30	525–575 BC	−550	A	kiln			7	AF, Th	5.6	60.3	1.7	1233	6.3	63.9	Evans and Hoye (2005)
64	Este-Casa di Ricovera	45.42	11.88	600–625 BC	−612	A	hearth	US169		8	Tv, AF	352.7	63.9	9.6	34.3	353.1	61.5	Tema et al. (2006)
65	Salapia	41.41	15.93	700–900 BC	−800	A	kiln			10	AF, Th	7.1	63.5	4.4	121	7.5	64	Evans and Hoye (2005)
66	Arcora, Campomarino, Molise	41.95	15.05	700–900 BC	−800	A	hearth	US2		22	Tv	18.1	62.9	1.0	888	18.3	62.7	Tema et al. (2006)
67	Montagnana, San Zeno	45.23	11.47	700–900 BC	−800	A	hearth	US137		6	AF	36.6	60.1	4.7	200	35.3	58.3	Tema et al. (2006)
68	Scarceta, Manciano	42.57	11.57	1000–1300 BC	−1150	A	hearth	1, D5-3		16	pTD	11.5	62.9	1.5	578	11.5	62.9	Tema et al. (2006)
69	Scarceta, Manciano	42.57	11.57	1000–1300 BC	−1150	A	hearth	2, D5-4		20	Tv	5.4	61.5	1.7	393	5.4	61.4	Tema et al. (2006)
70	Scarceta, Manciano	42.57	11.57	1000–1300 BC	−1150	A	hearth	3, D6-3		10	Tv	9.9	59.7	3.6	177	9.9	59.6	Tema et al. (2006)
71	Grotta dell′ Edera di Aurisina	45.75	13.70	2500–4300 BC	−3400	^{14}C	hearth	2		6	Tv	3.9	60.6	2.9	524	3.6	57.6	Tema et al. (2006)
72	Laghetti del Crestoso	45.85	10.44	5900–5480 BC	−5690	^{14}C	hearth	5		10	0v	4.5	65.7	11.6	18.3	4.2	63.1	Tema et al. (2006)
73	Laghetti del Crestoso	45.85	10.44	5900–5480 BC	−5690	^{14}C	hearth	1		6	Tv, 0v	346.0	64.7	17.7	15.2	346.8	61.8	Tema et al. (2006)

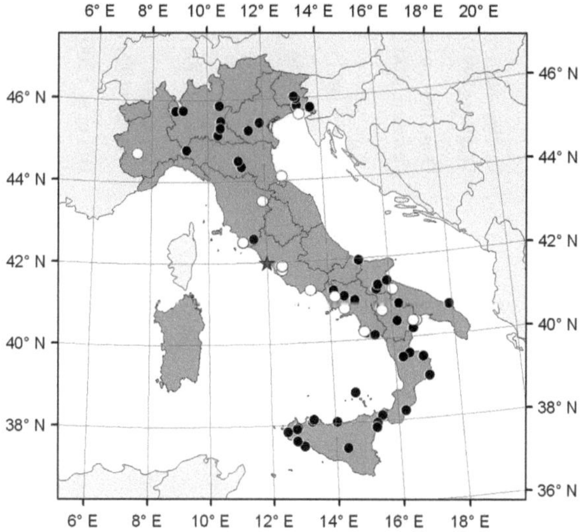

Fig. 15.2 Geographic distribution of the Italian archaeomagnetic directional (*black dots*) and intensity (*white dots*) data. The star shows the Viterbo reference site

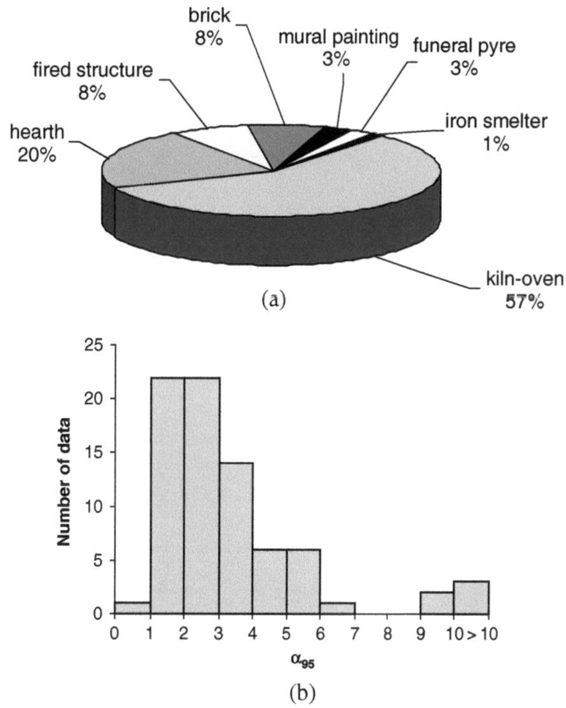

Fig. 15.4 (**a**) Pie diagram of the type of material used for archaeodirection determination; (**b**) histogram of the distribution of the α_{95} semi-angle of confidence

15.3 Archaeomagnetic Intensity Results

The intensity of the Earth's magnetic field in Italy was first measured at the observatory of Pola (Istria Peninsula, now Croatia) in 1880, more that two centuries after the first directional measurement at Rome. Although archaeointensity measurements do not require in situ sampling, reliable archaeointensity results in Italy are still scarce and less numerous than the directional data. In 1986, Evans first presented the archaeointensity results from three Italian archaeological sites and up to now, only seven more studies from archaeological material (Evans 1991, Aitken et al. 1988, Hedley and Wagner 1991, Donadini and Pesonen 2007, Hill et al. 2007 2008, Tema et al. 2010) have been published yielding a total of only 23 determinations. These results are summarised in Table 15.2.

Most of the archaeointensity data come from central and southern Italy (Fig. 15.2). Northern Italy is very poorly covered while, contrary to the directional data, no results exist for Calabria and Sicily, which are two regions with very rich archaeological heritages. Time distribution of the data is shown in Fig. 15.5a. Some data exist for the period from 300 BC to 200 AD but it is impressive that for all other periods data are scarce or

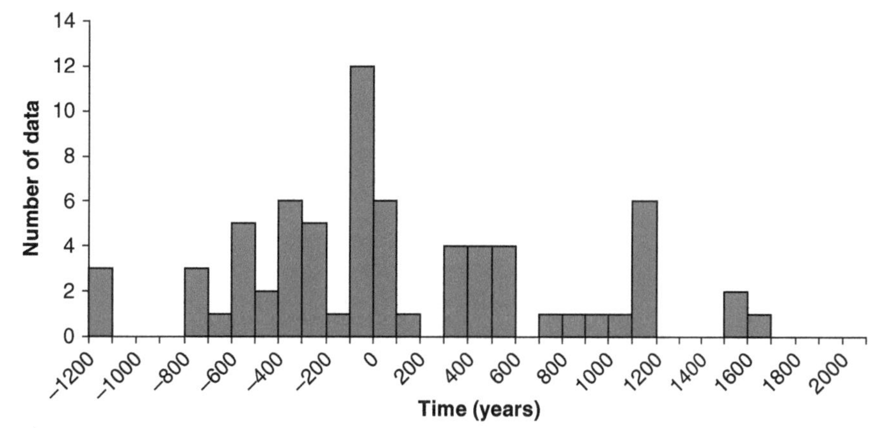

Fig. 15.3 Time distribution of the Italian directional data

Table 15.2 Italian archaeointensity data. Columns: No = site reference number; Location of the sampling sites; Lat/Long = site latitude/longitude; Age range; Mean age; Method of dating (A: archaeological estimate, ^{14}C: radiocarbon); Material studied; Code name of the structure; Cross-reference to Table 1 for directional results from the same structure; $(n)/N$ = (number of specimens)/number of independently oriented samples; Intensity laboratory treatment; corrections applied (ac: anisotropy correction, cr: cooling rate correction, Kc: Kono (1978) correction); $F \pm \sigma$, palaeointensity estimate and its standard deviation; VADM, virtual axial dipole moment; Reference

No	Site	Lat (°)	Long (°)	Age (years)	Mean Age	Dating method	Material	Code	Directional results (Table 15.1)	(n)	N	Intensity method	Corrections	F (μT)	± σ	VADM (10^{22} A m^2)	Reference
1	Bernalda	40.4	16.7	1960 AD	1960	A	kiln				4	Shaw	Kc	45.7	2.2	7.9	Evans (1986)
2	Bernalda	40.4	16.7	1960 AD	1960	A	kiln	1686 b 1				classical Thellier		47.6	1.1	8.2	Aitken et al. (1988)
3	Saluzzo	44.7	7.5	1500–1700 AD	1600	A	displaced brick			15	5	Thellier modified by Coe	ac, cr	46.0	3.4	7.5	Tema et al. (2010)
4	Rome, Foro Traiano 1	41.9	12.5	1480–1520 AD	1500	A	kiln		*No (3)*	10	8	Thellier modified by Coe	ac, cr	53.6	1.2	9.07	Tema et al. (2010)
5	Rome, Foro Traiano 2	41.9	12.5	600–800 AD	700	A	kiln		*No (14)*	4	4	Thellier modified by Coe	ac, cr	73.4	3.5	12.4	Tema et al. (2010)
6	Canosa	41.2	16.1	500–600 AD	550	A	kiln		*No (15)*	11	8	Thellier modified by Coe	ac, cr	61.1	3.2	10.4	Tema et al. (2010)
7	Carlino	45.6	13.3	380–430 AD	405	A	kiln				1	classical Thellier	ac, cr	62.4	1.3	10.1	Hedley and Wagner (1991)
8	Rome	41.8	12.5	105–115 AD	110	A	displaced brick			3	1	Thellier modified by Coe		66.4	4.9	11.2	Donadini and Pesonen (2007)
9	Pompeii	40.8	14.5	79 AD	79	A	kiln		*No (29)*		6	Shaw	Kc	61.2	1.0	10.5	Evans (1991)
10	Southern Italy	40.7	15.7	0–100 AD	50	A	kiln	site K			5	Shaw	Kc	71.3	8.2	12.2	Evans (1986)
11	Southern Italy	40.4	16.8	0–100 AD	50	A	kiln	site PK			4	Shaw	Kc	75.3	2.5	12.9	Evans (1986)
12	Arezzo	43.5	11.9	15–20 BC	−18	A	ceramic	DP1/18			2	classical Thellier	ac, cr	64.1	1.3	10.6	Hedley and Wagner (1991)
13	Calès	41.1	14.2	50 BC	−50	A	ceramic				2	classical Thellier	ac, cr	77.3	1.6	13.2	Hedley and Wagner (1991)
14	Albinia	42.5	11.2	200 BC–100 AD	−50	A	kiln	kiln A	*No (36)*		21	classical Thellier	ac, cr	69.8	3.1	11.7	Hill et al. (2007)

Table 15.2 (continued)

No	Site	Lat (°)	Long (°)	Age (years)	Mean Age	Dating method	Material	Code	Directional results (Table 15.1)	(n)	N	Intensity method	Corrections	F (μT)	± σ	VADM (10^{22} A m^2)	Reference
15	Albinia	42.5	11.2	200 BC–100 AD	−50	A	kiln	kiln B	*No (37)*		24	classical Thellier	ac, cr	66.5	3.7	11.2	Hill et al. (2007)
16	Albinia	42.5	11.2	200 BC–100 AD	−50	A	kiln	kiln C	*No (38)*		15	classical Thellier	ac, cr	62.2	4.2	10.4	Hill et al. (2007)
17	Albinia	42.5	11.2	200 BC–100 AD	−50	A	kiln	kiln D	*No (39)*		5	classical Thellier	ac, cr	62.8	3.1	10.5	Hill et al. (2007)
18	Albinia	42.5	11.2	200 BC–100 AD	−50	A	kiln	kiln E	*No (40)*		15	classical Thellier	ac, cr	64.5	3.3	10.8	Hill et al. (2007)
19	Albinia	42.5	11.2	200 BC–100 AD	−50	A	amphorae				37	classical Thellier	ac, cr	63.8	3.2	10.7	Hill et al. (2007)
20	Mt San Biagio	41.3	13.4	120–180 BC	−150	A	ceramic	XI 78			2	classical Thellier	ac, cr	52.7	1.1	9.0	Hedley and Wagner (1991)
21	Rimini	44.1	12.6	220–260 BC	−240	A	ceramic	CAM336			2	classical Thellier	ac, cr	70.2	1.5	11.6	Hedley and Wagner (1991)
22	Southern Italy	40.2	15.1	200–300 BC	−250	A	kiln	site ZK			3	Shaw	Kc	79.7	2.4	13.7	Evans (1986)
23	Metaponto	40.4	16.8	800–600 BC	−700	A, C^{14}	displaced brick				38	classical Thellier	ac, cr	85.0	5.0	14.6	Hill et al. (2008)

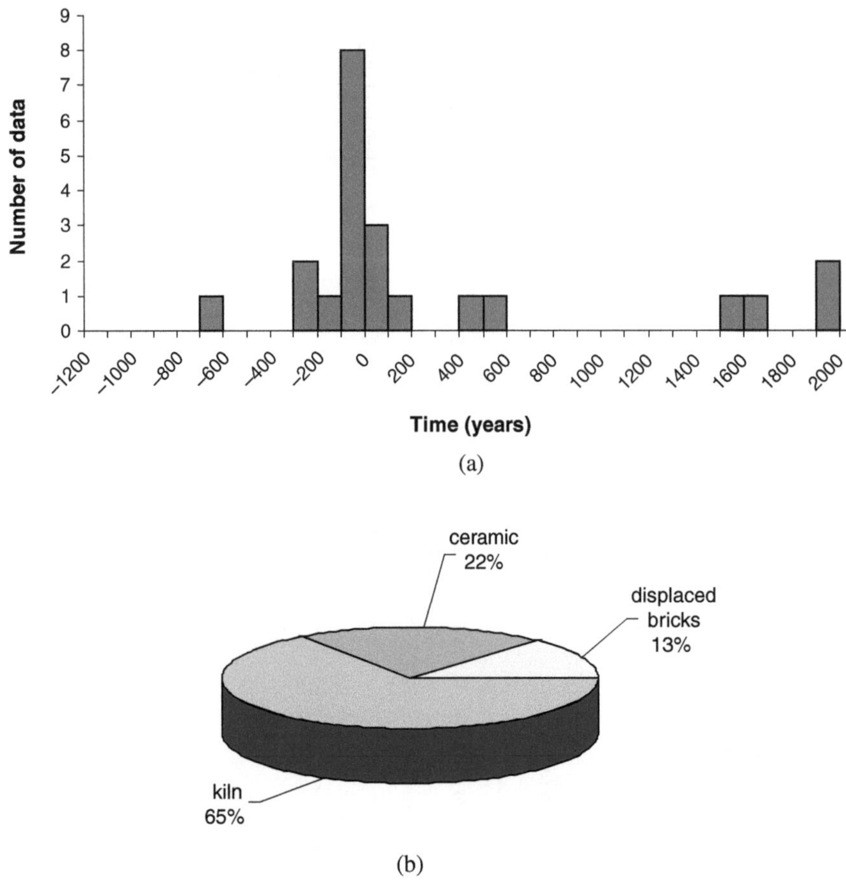

Fig. 15.5 (**a**) Time distribution of the Italian intensity data; (**b**) pie diagram of the type of material used for archaeointensity determination

completely absent! The 65% of the results come from the archaeointensity determination of kilns, 22% from ceramics and in three cases (13%) displaced bricks have been studied (Fig. 15.5b). In the case of kilns, only in very few studies have directional measurements been done in the same material and they are noted in Table 15.2 (Hoye 1982, Tema et al. 2006, Hill et al. 2007). Various laboratory techniques have been used for the archaeointensity determination. In most cases classical Thellier and Thellier (1959) or Thellier modified by Coe (Coe 1967, Coe et al. 1978) methods were used. The rest of the results were obtained using the Shaw (1974) method. Evans (1986), (1991) has investigated the mineralogical and magnetic changes during intensity laboratory experiments and has applied Kono's correction (Kono 1978) to his results. This variety of methods used often makes difficult the evaluation of the reliability of the results, particularly when analytical rock-magnetism information is missing in the original papers. Cooling rate corrections and magnetic anisotropy investigations are only applied in the most recently published works (Hill et al. 2007, 2008, Tema et al. 2010).

15.4 Recent Achievements in Archaeomagnetic Studies

In the early days of palaeomagnetism, the main concern was to develop instruments, techniques and procedures in order to improve the precision of the palaeomagnetic analysis. Nowadays, the archaeomagnetic experimental techniques, that are basically the same as those used in palaeomagnetic studies on rocks, are well established. Therefore, the challenge in archaeomagnetic research is now mainly focused on identifying new materials that could be suitable for archaeomagnetic studies, better understanding the magnetization processes and the causes that can bias the recorded magnetization from that of the external field, and reconstructing the secular variation of the Earth's magnetic field in the past. In Italy, during the last few years, important achievements have been accomplished towards these directions.

15.4.1 Pictorial Remanent Magnetization (PiRM)

Apart from the archaeological materials traditionally used in archaeomagnetic studies, the mural paintings may be an alternative source of information about the geomagnetic field secular variation. Chiari and Lanza (1997) studied the remanent magnetization of Italian mural paintings and demonstrated that the red colour of many of them contains haematite grains as pigment. When this pigment is applied to the wall, the grains are free to move and they align their magnetic moment with the Earth's magnetic field (Fig. 15.6). Once the painting is dried, the magnetic grains maintain their orientation acquiring a 'pictorial' remanent magnetization (PiRM). The mean directions from mural paintings of known date sampled in the neighbourhood of Torino (Chiari and Lanza 1997), painted between 1740 and 1954, were well defined and consistent with the direction of the Earth's magnetic field at the time of their painting, as deduced from historical direct measurements. Chiari and Lanza (1999), following their previous paper, investigated the PiRM of five mural paintings from the *Bibliotheca Apostolica* (Vatican, Roma) that were painted in different ages. The remanence directions from two of those murals were close to the expected values for the time they were painted. However, the directions from the other three murals, which were close to each other, were significantly different from the direction of the Earth's magnetic field expected for their historical age but similar to the Earth's field direction measured in Rome in the first half of the 19th century. According to Chiari and Lanza (1999), a possible explanation to these anomalous directions could be a restoration of the murals at the beginning of 19th century, as also supported by historical sources. Anyhow, a misalignment of the direction recorded through the PiRM with respect to the Earth's magnetic field is not excluded by the authors. This conclusion is very interesting as it brings to light another important contribution of the archaeomagnetic research to the cultural heritage studies (the authenticity and restoration of art pieces).

Zanella et al. (2000) studied the magnetic remanence from mural paintings in Pompeii and compared their directions to the archaeomagnetic direction of a nearby kiln (Evans and Mareschal 1989) and the directions of the pyroclastic rocks from the Vesuvius 79 AD eruption. This comparison showed that the PiRM directions of the red coloured Pompeii murals are statistically indistinguishable from that obtained from other, traditionally studied, materials (kiln, pyroclastic rocks) and that the Pompeii murals have retained their remanent magnetization for nearly two thousand years. Thus, red-coloured paintings can be included among the materials suitable for archaeomagnetic research, and can significantly contribute to our knowledge of the secular variation of the Earth's magnetic field in the past, particularly in Italy, where numerous well-dated paintings are available.

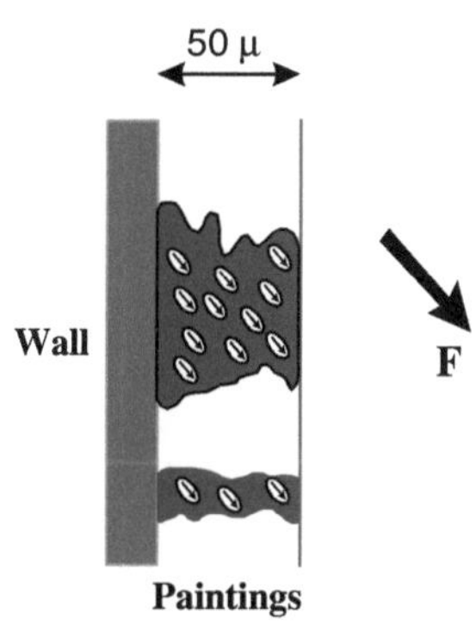

Fig. 15.6 Representation of the magnetization of a mural painting. Symbols: white spots = haematite grains; h = colour film thickness (some tens of μm); F = Earth's magnetic field, (redrawn from Chiari and Lanza, 1997)

15.4.2 Magnetic Anisotropy

Archaeomagnetic studies are based on the principle that archaeological artefacts, under certain circumstances, acquire a remanent magnetization parallel to the local geomagnetic field at the time of their production. In the early archaeomagnetic studies this assumption was considered a priori true. Nowadays, it is known that various mechanisms can cause a bias of the recorded magnetic direction, one of which is the magnetic anisotropy. Fabrication procedures used for the production of several archaeological artefacts such as tiles, bricks and ceramics often cause the preferential alignment of the magnetic grains which results in a deviation of the remanence direction with respect to the external field (Aitken et al. 1981, Veitch et al. 1984, Sternberg 1989, Chauvin et al. 2000, Hedley 2001, Hus et al. 2002 2004).

In Italy, even if many of the archaeomagnetic results come from kilns, bricks and ceramics (Figs. 15.4a and 15.5b) that are often strongly anisotropic, magnetic anisotropy is still rarely investigated. Among the 73 directional results included in the Italian database (Table 15.1), in only 5 cases the magnetic anisotropy correction was applied, although in some of the original papers a possible important magnetic anisotropy effect was mentioned. The complicated and time consuming laboratory procedure for the full TRM tensor determination is probably the main reason for the lack of such investigations. Recently, Tema (2009) systematically investigated the magnetic anisotropy of bricks from five archaeological kilns excavated in southern and central Italy. Measurements of the anisotropy of the magnetic susceptibility (AMS), isothermal remanent magnetization (AIRM) and anhysteretic remanent magnetization (AARM) showed a well developed planar magnetic fabric that matches the flat shape of the bricks. In the case where the bricks lay horizontally within a kiln, the planar fabric resulted in an inclination shallowing of the archaeomagnetic direction with respect to that of the Earth's magnetic field at the time of their last cooling. Tema (2009) showed that the directions of the principal axes of the anisotropy ellipsoid are almost the same, irrespectively of the type of anisotropy measured. She proposed a simplified method to estimate the TRM anisotropy degree, involving only two sample heatings, once the orientation of the magnetic remanence ellipsoid is defined by the easier and faster AARM measurements. In her study, the estimation of the magnetic anisotropy effect in the inclination record on the grounds of ATRM measurements yielded a shallowing that varied from 4° to 10° for individual samples. Such inclination difference may significantly bias the archaeomagnetic dating and for further archaeomagnetic inferences it is important to consider that most of the directional results included Table 15.1 are not corrected for the magnetic anisotropy effect. Only in recent Italian directional studies (Table 15.1: Hill et al. 2007, Tema and Lanza 2008) and intensity studies (Hill et al. 2007, 2008, Tema et al. 2010) had the magnetic anisotropy been systematically investigated and the results corrected when it was necessary.

15.4.3 Secular Variation Curve Building

Once sufficient archaeomagnetic data of well-dated archaeological artefacts are available for a certain region, it is possible to determine a reference secular variation (SV) curve that describes the detailed local variations of the Earth's magnetic field in the past. The establishment of reliable reference curves is, however, one of the most complex parts of archaeomagnetic research, mainly due to the uneven distribution of raw data in both space and time. Early reference curves were drawn by interpolating data points by hand (e.g. Clark et al. 1988). Nowadays, recent advances on curve building involve more sophisticated techniques based on the statistical elaboration of the reference data. The most frequently used statistical approaches are: the moving window technique (Sternberg and McGuire 1990, Batt 1997, Kovacheva et al. 1998), the bivariate Le Goff statistics (Daly and Le Goff 1996, Le Goff et al. 2002) and the latest proposed hierarchical Bayesian modelling (Lanos 2004, Lanos et al. 2005).

Tema et al. (2006) proposed a preliminary Italian directional SV curve based on a compilation of 64 Italian archaeomagnetic results, supplemented with 10 directional results from volcanic rocks from Vesuvius and Ischia, of unquestionable age. The authors used the hierarchical Bayesian modelling based on a roughness penalty (Lanos 2004) to build the Italian SV curves for declination and inclination. Bayesian modelling allows the fitting of a spherical spline function based on the roughness penalty to the data in three dimensions (declination, inclination and time), automatically adjusting the window width to the density of the points along the time axes and allowing the points to move within their dating error ranges. The obtained curve is accompanied by an error envelope calculated at the 95% confidence level.

For the calculation of the Italian SV reference curve, Tema et al. (2006) relocated all Italian data to the Viterbo repeat station (42.45°N, 12.03°E) using the virtual geomagnetic pole (VGP) method (Noel and Batt 1990). Lanza and Zanella (2003) proposed Viterbo (about 70 km from Rome) as the optimum reference point for relocating directional Italian secular variation data. They used direct measurements from 113 repeat stations of the Italian Geomagnetic Network (Coticchia et al. 2001) and calculated the relocation error via the VGP method at each of them. They concluded that the Viterbo repeat station is the one characterized by the smallest relocation error in

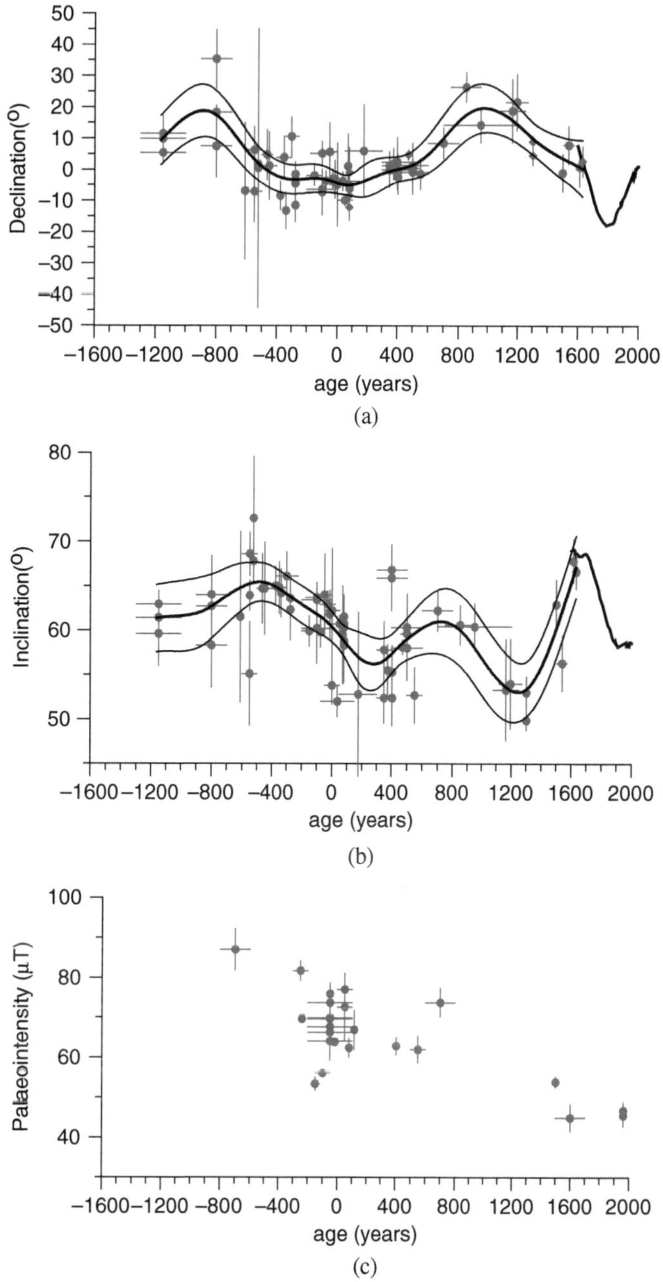

Fig. 15.7 Smoothed SV curves for (**a**) declination and (**b**) inclination for Italy surrounded by the 95% error envelope as obtained from the Bayesian modelling (from Tema et al. 2006). Together are plotted the raw data and the historical measurements for the last four centuries (Jackson et al. 2000). (**c**) Italian archaeointensity data plotted versus age. All directions are reduced to the reference site Viterbo (42.45°N, 12.03°E)

both declination and inclination (± 0.3°), wherever the original site is situated in Italy. To keep the uniformity between the Italian directional and intensity data, Tema et al. (2010) have also chosen the Viterbo latitude to reduce the Italian intensity results. They showed that reducing at Viterbo intensity data from the two most distant localities from Viterbo in the Italian intensity dataset (Saluzzo and Etna) introduces an error less than 1.2 μT. This error is of the same order or even lower than the usual error associated with the palaeointensity determinations (usually expressed as standard deviation) and therefore the Viterbo station can be established as the common reference point for all SV studies in Italy.

The Italian SV curves proposed by Tema et al. (2006) describe the variation of the direction of the Earth's magnetic field in Italy for the time period from 1300 BC to 1700 AD (Fig. 15.7a, b). For more recent periods, the D and I at Viterbo can be calculated using the historical geomagnetic field model of Jackson et al.

(2000). For the time periods for which many data exist (e.g. 79 AD), the curves are well defined with narrow error envelopes while for those time periods with few or no data the curves are only roughly estimated (Fig. 15.7a, b). As Bayesian calculations take into consideration both the measurement and time uncertainties, it is probable that for those periods poorly covered by the data the curves are affected by an over-smoothing effect (Tema et al. 2006). Intensity Italian results are too few to build an intensity SV reference curve (Fig. 15.7c) even when the intensity data from volcanic rocks are considered (Tema et al. 2010).

15.5 Archaeomagnetic Dating in Italy

Once a well-established SV reference curve is available for a region, archaeomagnetic dating can be obtained by comparing the magnetic direction (and/or intensity) of the feature to be dated with the reference curve. In the last 5 years, great progress in the statistical treatment for the comparison of the archaeomagnetic directions with the reference curves (Le Goff et al. 2002, Lanos 2004) has provided powerful dating tools. Archaeomagnetic dating in Italy, however, is only possible based on the directional data and can give reliable dates for the time periods for which the Italian directional SV curves are well established and accompanied by narrow error envelopes (Fig. 15.7). As for now, and as previously discussed, no intensity reference curve is available for Italy. Intensity results can only be used for direct comparison with the other data from the Italian archaeointensity dataset (Tema et al. 2010) or compared with the reference curves of nearby countries.

The Italian reference curves have been calculated using the Bayesian method, and therefore archaeomagnetic dating can also be done using the last advances of the Bayesian statistic approach (Lanos 2004, Lanos et al. 2005). This dating technique allows the estimation of the calendar date interval of an archaeological feature by calculating the probability densities separately for each geomagnetic field element (declination, inclination and intensity when available) after comparison with the reference SV curves. The final dating interval is obtained by combining the separate probability densities in order to find the most probable solution (Lanos 2004).

Tema and Lanza (2008) used the Italian SV to date a lime kiln excavated at Bazzano (northern Italy). This circular kiln was unearthed during a rescue excavation. Sparse archaeological information based on some ceramic fragments found in the nearby area suggests that the kiln was in use during 3rd–4th century AD but no further scientific or archaeological dating evidence about the time of its abandonment was available. The kiln was dated using the RENDATE software that is based on Bayesian statistics (Lanos 2004, RENDATE available on line at the web site: http://www.meteo.be/CPG/aarch.net/download en.html). The calculated age at the 95% level of confidence is 99 BC–650 AD. This age interval is very wide, even though the kiln's experimental mean archaeomagnetic direction is well defined, characterised by a small α_{95} semi-angle of confidence ($\alpha_{95} = 3.1°$). According to the authors this is due to the fact that the declination value of Bazzano intersects the Italian D curve at a part of the curve described by minor changes of the D values versus time. Also, taking into account the measurement's error (dD) and the curve's uncertainty band, the dating interval obtained at 95% of probability is inevitably wide. Tema and Lanza (2008) have also calculated the possible date of the kiln's last firing with 68% of probability and they found a much shorter time interval, 281–603 AD.

Hill et al. (2007) used the Italian SV curves to compare the directional results from 5 contemporaneous kilns excavated at Albinia (central Italy). The archaeomagnetic dating interval calculated at 95% confidence level using the mean value from the five kilns is: 499 BC to 45 BC (Fig. 15.8a). Even if this is wide, it is still in good agreement with the archaeologically given age for the site, which is 200 BC to 100 AD. It clearly suggests that the most probable age of the kilns is towards the older part of the archaeological estimation. Because of the lack of an intensity SV curve for Italy, comparison of the intensity from these kilns and a collection of amphorae samples from the same site, was only possible using individual Italian data and the Greek reference curve (De Marco et al. 2008). According to Hill et al. (2007), the intensity data also suggest that the date of the Albinia archaeological site is closer to 200 BC.

Recently, great progress in archaeomagnetic dating has been done based on geomagnetic field models that predict the variations of the Earth's magnetic field in a certain region. Lodge and Holme (2008) reviewed the

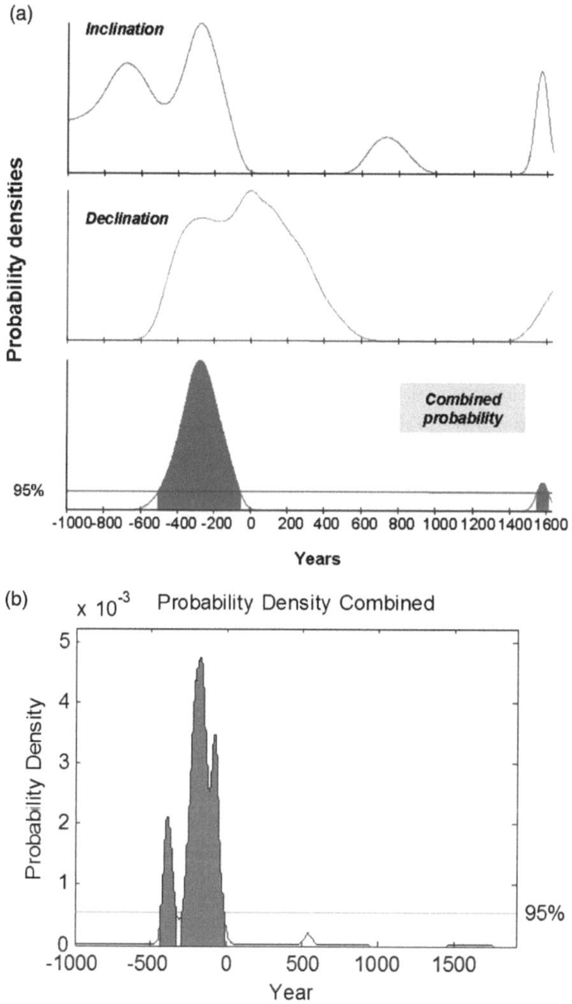

Fig. 15.8 (**a**) Archaeomagnetic dating for Albinia kilns and amphorae collection (data from Hill et al. 2007) calculated at 95% of probability level, after comparison with: (**a**) the Italian SV curves (Tema et al. 2006). The calibrated date is obtained by combining the probability densities of inclination and declination; (**b**) the reference curves generated by the regional SCHA.DIF.3K model (Pavón-Carrasco et al. 2009). The calibrated dating interval is obtained by combining the probability densities of inclination, declination and intensity. The probability densities are illustrated as the *grey areas*

published secular variation curves for five European localities (France, Germany, Hungary, Iberia, England) and introduced a new approach to produce secular variation curves for archaeomagnetic dating. They proposed a global model (GMADEK2K.1), calculated using physical constraints and the same modelling procedure as in the CALS family global models (Korte and Constable 2005), and they suggested that the GMADEK2K.1 should be used as a reference curve for the European archaeomagnetic dating for the last two millennia.

Pavón-Carrasco et al. (2009) have recently proposed a regional archaeomagnetic model that produces the geomagnetic field variations in Europe for the last 3000 years, modelling together the three geomagnetic elements: declination, inclination and intensity. This new model, SCHA.DIF.3K, has been obtained by least sums of absolute deviation inversion of archaeomagnetic data using spherical cap harmonics for the spatial representation of the field and sliding windows in time. In the model's input database (essentially based on the database of Korte et al. 2005) only archaeological material has been used and no lake sediment and lava flow data were considered (Pavón-Carrasco et al. 2009). The SCHA.DIF.3K model directly predicts the geomagnetic field at the site of interest, avoiding, in this way, any eventual relocation error.

To check the use of the regional SCHA.DIF.3K model for archaeomagnetic dating in Italy, we have used it to date again the Albinia kilns and amphorae collection (Hill et al. 2007) that were previously dated using the classical Italian SV curve. Archaeomagnetic data (declination, inclination and intensity) from these structures are of high quality (corrected for anisotropy and cooling rate effect) and have not been included either in the construction of the Italian reference curve or in the generation of the regional model. They therefore constitute an independent set of data. The archaeomagnetic dating has been carried out according to the Bayesian mathematical method of Lanos (2004), but in this case the reference curves that have been used are those generated by the regional SCHA.DIF.3K model, calculated directly at the geographic coordinates of Albinia. From the mean archaeomagnetic direction calculated from the five kilns ($D = 356.1°, I = 63.8°, \alpha_{95} = 1.8°$) and the mean intensity ($F = 64.9 \pm 2.8\ \mu T$, calculated from the kilns and the amphorae data), a probability density function has been calculated separately for the declination, inclination and intensity. The final date obtained combining these probability functions at 95% of probability (Fig. 15.8b), suggests that the production of the amphorae and the abandonment of the kilns had occured around 295 BC to 15 BC. This time interval is in very good agreement with the archaeological age of the site and is narrower than the one obtained by the Italian SV curve. The SCHA.DIF.3K model as an archaeomagnetic dating tool represents an improvement because it is build with an in situ archaeomagnetic database, eliminating, in this way, relocation errors. Moreover it offers, for the moment, the only possibility for archaeomagnetic dating in Italy based on a full description of the geomagnetic field vector. However, the SCHA.DIF.3K model shows some short term geomagnetic field variations that are not always supported by the Italian data (Tema et al. 2010) and are probably an artificial effect of the modelling procedures. The Albinian dating example clearly shows that even when a precise archaeomagnetic record is obtained (small measurement errors, α_{95} and σ) accurate archaeomagnetic dating is still not always possible.

15.6 Conclusions and Future Perspectives

Archaeomagnetic dating is a relative dating technique that relies on two physical phenomena: the secular variation of the Earth's magnetic field and the ability of certain archaeological features to become permanently magnetized according to the magnetic field pertaining at the time of their last firing. Archaeomagnetic dating is, thus, possible only if a robust reference SV curve is available for a certain region and precise measurement of the archaeomagnetic direction and intensity of the archaeological material can be done. During last decades, in Italy important advances in both these directions have been achieved. Improvements on instrumentation, methodologies and laboratory techniques have been done and in the last years new, high quality Italian directional and intensity data, corrected for magnetic anisotropy and cooling rate effects, have been published. The Italian directional dataset has been elaborated using the Bayesian statistical approach and preliminary SV curves for the declination and inclination that describe the variation of the Earth's magnetic field in Italy during the last three millennia have been drawn.

Nevertheless, a great amount of work still remains to be done in order to improve the quantity and quality of the reference data that would permit the establishment of reliable SV curves, necessary for accurate archaeomagnetic dating based on a full description of the geomagnetic field vector. An archaeomagnetic database is a «live organism» and a continuous update is necessary. Re-evaluation of the reliability of the existing archaeomagnetic data is required, particularly for the cases where data come from a study with a limited number of samples, have poorly constrained archaeological age and where necessary measurement corrections are missing. Much future work should be undoubtedly focused on the acquisition of new reliable data of well dated archaeomagnetic artefacts. The Italian archeaointensity database is very poor and a large amount of work should be done in this direction in the future. Directional data for the 600–1600 AD period are still missing, and data for periods older than 700 BC are needed to extend our knowledge of the past geomagnetic field variations. In the existing directional SV curve (Fig. 15.9), even though it is not so detailed, some general features of the variation of the Earth's magnetic field can be still noticed. Based on the rule that the magnetic field recorded at an individual site should trace a segment of an ellipse in a clockwise direction if the underlying azimuthal motion of the non-dipole field is westward, and conversely, eastward motion should result in an anti-clockwise ellipse segment (Runcorn 1959, Dumberry and Finlay 2007), in the Italian SV curve we can observe eastward motion between 700 AD and 1400 AD and westward motion between 800 BC and 200 AD. Some abrupt directional changes can be also observed. A sharp change in direction is clearly noticed at the 9th century BC while a less abrupt change may be seen around 250 AD (low inclination peak). These directional changes could be correlated with the archaeomagnetic jerks observed in France (Gallet et al. 2003, 2009) at 800 BC and 200 AD. However, the transition between westward and eastward motion in the Italian SV curve (Fig. 15.9) seems not to occur so sharply as in the French SV curve (Gallet et al. 2003) and no archaeomagnetic jerk is observed around 1350 AD. Nevertheless, the poor data, the time uncertainties and the oversmoothing of the Italian SV curve make difficult any clear interpretation of these directional changes that are at the moment constrained by only 3–4 data points (Fig. 15.7a, b).

The big challenge of the future archaeomagnetic studies, however, still is to better understand the mineralogical background of the burnt clay materials which is essential for a reliable determination of the TRM of archaeological materials. These kinds of studies would

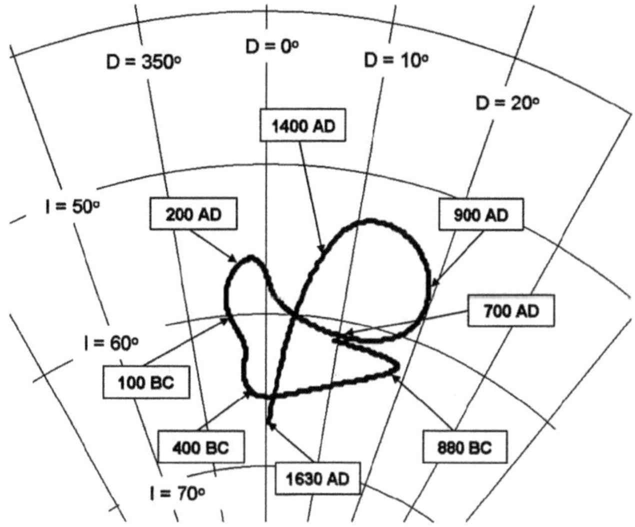

Fig. 15.9 Archaeomagnetic directional secular variation curve for Italy plotted in an inclination-declination diagram (Bauer plot)

also contribute to the application of archaeomagnetism to a larger variety of archaeological problems, such as material provenance, reconstruction of firing temperatures and restoration of archaeological artefacts, and can help in the recognition of archaeomagnetism as a dating method and an important tool in archaeological studies. The establishment of a good collaboration with the archaeological community is essential, particularly nowadays that the fast economic development often threatens the archaeological heritage, and especially in Italy where the cultural heritage is so rich and the archaeological sites are so numerous.

Acknowledgments Elena Zanella is highly acknowledged for her useful comments on an early version of the manuscript. I would also like to sincerely thank Roberto Lanza for his precious advices and for continuously encouraging me to work on archaeomagnetism. Two anonymous reviewers are acknowledged for their useful remarks and Emilio Herrero-Bervera is greatly thanked for inviting me to participate in this IAGA Special Series Book.

References

Abdeldayem A, Tarling DH, Marton P, Nardi G, Pierattini D (1992) Archaeomagnetic study of some kilns and burnt walls in Selinunte archaeological township, Sicily. Sci Technol Cult Heritage I:129–141

Aitken MJ, Alcock PA, Bussel GD, Shaw CJ (1981) Archaeomagnetic determination of the past geomagnetic intensity using ancient ceramics: allowance for anisotropy. Archaeometry 23(1):53–64

Aitken MJ, Allsop AL, Bussell GD, Winter MB (1988) Determination of the intensity of the Earth's magnetic field during archaeological times: reliability of the Thellier technique. Rev Geophys 26:3–12

Batt CM (1997) The British archaeomagnetic calibration curve: an objective treatment. Archaeometry 39:153–168

Cafarella L, De Santis A, Meloni A (1992) The historical Italian geomagnetic data catalogue. Publication ING, Rome, p 160

Chauvin A, Garcia Y, Lanos, Ph., Laubenheimer F (2000) Palaeointensity of the geomagnetic field recovered on archaeomagnetic sites from France. Phys Earth Planetary Inter 120:111–136

Chiari G, Lanza R (1997) Pictorial remanent magnetization as an indicator of secular variation of the Earth's magnetic field. Phys Earth Planetary Inter 101:79–83

Chiari G, Lanza R (1999) Remanent magnetization of mural paintings from the Bibliotheca Apostolica (Vatican, Rome). J Appl Geophys 41:137–143

Chiosi E, La Torre M, Nardi G, Pierattini D (1998) Archaeomagnetic data from a kiln at Cassano (South Italy). Sci Technol Cult Heritage 7(2):13–17

Clark AJ, Tarling DH, Noël M (1988) Developments in archaeomagnetic dating in Britain. J Archaeol Sci 15:645–667

Coe RS (1967) Paleo-intensities of the Earth's magnetic field determined from tertiary and quaternary rocks. J Geophys Res 72(12):3247–3262

Coe RS, Grommé S, Mankinen EA (1978) Geomagnetic palaeointensities from radiocarbon-dated lava flows on Hawaii and the question of the Pacific nondipole low. J Geophys Res 83(B4):1740–1756

Coticchia A, De Santis A, Di Ponzio A, Dominici G, Meloni A, Pierozzi M, Sperti, M (2001) Italian magnetic network and geomagnetic field maps of Italy at year 2000. Boll di geodesia e scienze affini 4(Anno LX):261–291

Daly L, Le Goff M (1996) An updated and homogeneous world secular variation data base. 1. Smoothing the archaeomagnetic results. Phys Earth Planetary Inter 93:159–190

De Marco E, Spatharas V, Gómez-Paccard M, Chauvin A, Kondopoulou D (2008) New archaeointensity results from archaeological sites and variation of the geomagnetic field intensity for the last 7 millennia in Greece. Phys Chem Earth 33:578–595

Donadini F, Pesonen L (2007) Archeointensity determinations from Finland, Estonia and Italy. Geophysica 43(1–2): 3–18

Dumberry M, Finlay CC (2007) Eastward and westward drift of the Earth's magnetic field for the last three millennia. Earth Planet Sci Lett 254:146–157

Evans ME (1986) Paleointensity estimates from Italian kilns. J Geomagnetism Geoelectricity 38:1259–1267

Evans ME (1991) An archaeointensity investigation of a kiln at Pompeii. J Geomagnetism Geoelectricity 43:357–361

Evans ME, Hoye GS (2005) Archaeomagnetic results from southern Italy and their bearing on geomagnetic secular variation. Phys Earth Planetary Inter 151:155–162

Evans E, Mareschal M (1986) An archaeomagnetic example of polyphase magnetization. J Geomagnetism Geoelectricity 38:923–929

Evans E, Mareschal M (1989) Secular variation and magnetic dating of fired structures in southern Italy. In: Maniatis Y (ed) Archaeometry, Proceedings of the 25th international symposium. Elsevier, Amsterdam, pp 59–68

Folgerhaiter G (1899) Sur les variations séculaires de l'inclination magnétique dans l'antiquité. Arch Sci Phys Nat (Genève) 8:5–16

Gallet Y, Genevey A, Courtillot V (2003) On the possible occurrence of archaeomagnetic jerks in the geomagnetic field over the past three millennia. Earth Planet Sci Lett 214:237–242

Gallet Y, Hulot G, Chulliat A, Genevey A (2009) Geomagnetic field hemispheric asymmetry and archaeomagnetic jerks. Earth Planet Sci Lett 284:179–186

Gherardi S (1866) Sunto di altre sperienze ed osservazioni sul magnetismo dei mattoni, terre cotte, certi minerali e terreni ferriferi e di una intravenuta cagione fin qui non avvertita di variamenti nell'azione del magnetismo del globo da un punto all'altro anche prossimi della sua superficie. Il Nuovo Cimento 23–24(1):5–17

Hedley I (2001) New directions in archaeomagnetism. J Radioanal Nucl Chem 247(3):663–672

Hedley I, Wagner GC (1991) A magnetic investigation of Roman and pre-Roman pottery. In: Pernicka E, Wagner GC (eds) Archaeometry '90. Birkhaeuser, Basel, pp 275–284

Hill M, Lanos, Ph., Chauvin A, Vitali D, Laubenheimer F (2007) An archaeomagnetic investigation of a Roman amphorae workshop in Albinia (Italy). Geophys J Int 169:471–482

Hill M, Lanos Ph, Denti M, Dufresne, Ph. (2008) Archaeomagnetic investigation of bricks from the VIIIth–VIIth century BC Greek–indigenous site of Incoronata (Metaponto, Italy). Phys Chem Earth 33:523–533

Hoye GS (1981) Archaeomagnetic secular variation record of Mount Vesuvius. Nature 291:216–218

Hoye GS (1982) A magnetic investigation of kiln wall distortion. Archaeometry 24:80–84

Hus J, Ech-Chakrouni S, Jordanova D (2002) Origin of magnetic fabric in bricks: its implications in archaeomagnetism. Phys Chem Earth 27:1319–1331

Hus J, Geeraerts R, Plumier J (2004) On the suitability of refractory bricks from a mediaeval brass melting and working site near Dinant (Belgium) as geomagnetic field recorders. Phys Earth Planetary Inter 147:103–116

Incoronato A, Angelino A, Romano R, Ferrante A, Sauna R, Vanacore G, Vecchione C (2002) Retrieving geomagnetic secular variations from lava flows: evidence from Mount Arso, Etna and Vesuvius (southern Italy). Geophys J Int 149:724–730

Jackson A, Jonkers ART, Walker MR (2000) Four centuries of geomagnetic secular variation from historical records. Philos Trans R Soc London Ser A 358:957–990

Kono M (1978) Reliability of palaeointensity methods using alternating field demagnetization and anhysteretic remanence. Geophys J R Astron Soc 54:241–261

Korte M, Constable CG (2005) Continuous geomagnetic field models for the past 7 millennia: 2. CALS7K. Geochem Geophys Geosyst 6:Q02H16

Korte M, Genevey A, Constable C, Frank U, Schnepp E (2005) Continuous geomagnetic field models for the past 7 millennia: 1. A new global data compilation. Geochem Geophys Geosyst 6:Q02H15. doi:10.1029/2004GC000800

Kovacheva M, Jordanova N, Karloukovski V (1998) Geomagnetic field variations as determined from Bulgarian archaeomagnetic data. Part II: the last 8000 years. Surv Geophys 19:431–460

La Torre M, Livadie Arbore C, Nardi G, Pierattini D (1998) Archaeomagnetic study of the Late Archaic furnace of Treglia (Campania, Southern Italy). Sci Technol Cult Heritage 7(2):7–12

Lanos Ph (2004) Bayesian inference of calibration curves: application to archaeomagnetism. In Buck CE, Millard AR (eds) tools for constructing chronologies, crossing disciplinary boundaries. Series: Lecture Notes in Statistics, Vol 177. Springer-Verlag, London, pp 43–82

Lanos Ph, Le Goff M, Kovacheva M, Schnepp E (2005) Hierarchical modelling of archaeomagnetic data and curve estimation by moving average technique. Geophys J Int 160:440–476

Lanza R, Meloni A, Tema E (2005) Historical measurements of the Earth's magnetic field compared with remanence directions from lava flows in Italy over the last four centuries. Phys Earth Planetary Inter 148:97–107

Lanza R, Zanella E (2003) Palaeomagnetic secular variation at Vulcano (Aeolian Islands) during the last 135kyr. Earth Planet Sci Lett 213:321–336

Le Goff M, Gallet Y, Genevey A, Warmé N (2002) On archeomagnetic secular variation curves and archeomagnetic dating. Phys Earth Planetary Inter 134:203–211

Lodge A, Holme R (2008) Towards a new approach to archaeomagnetic dating in Europe using geomagnetic field modelling. Archaeometry 50(3). doi:10.1111/j.1475-4754.2008.00400.x

Màrton P, Abdeldayem D, Tarling DH, Nardi G, Pierattini D (1992) Archaeomagnetic study of two kilns at Segesta, Sicily. Sci Technol Cult Heritage I:123–127

Nardi G, Pierattini D, Talamo P (1995) Archaeomagnetic data from Campania (Southern Italy): the "Medieval" furnace of Pratola Serra, Avellino. Sci Technol Cult Heritage 4(I):71–77

Noel M, Batt CM (1990) A method for correcting geographically separated remanence directions for the purpose of archaeomagnetic dating. Geophys J Int 102:753–756

Pavón-Carrasco FJ, Osete ML, Torta JM, Gaya-Pique L.R. (2009) A regional archaeomagnetic model for Europe for the last 3000 years, SCHA.DIF.3K: applications to archaeomagnetic dating. Geochem Geophys Geosyst 10:Q03013. doi:10.1029/2008GC002244

Principe C, Tanguy JC, Arrighi S, Paiotti A, Le Goff M, Zoppi U (2004) Chronology of Vesuvius' activity from AD 79 to 1631 based on archaeomagnetism of lavas and historical sources. Bull Volcanol 66:703–724

Rolph TC, Shaw J (1986) Variations of the geomagnetic field in Sicily. J Geomagnetism Geoelectricity 38:1269–1277

Runcorn SK 1959. On the theory of the geomagnetic secular variation. Ann Geophys 15:87–92

Shaw J (1974) A new method of determining the magnitude of the palaeomagnetic field: application to five historic lavas and five archaeological samples. Geophys J R Astron Soc 39:133–141

Sternberg RS (1989) Archaeomagnetic palaeointensity in the American Southwest during the past 2000 years. Phys Earth Planetary Inter 56:1–17

Sternberg RS, McGuire RH (1990) Techniques for constructing secular variation curves and for interpreting archaeomagnetic dates. In: Eighmy JL, Sternberg RS (eds) Archaeomagnetic dating. University of Arizona Press, Tucson, pp 109–134

Tanguy JC, Bucur I, Thompson JFC (1985) Geomagnetic secular variation in Sicily and revised ages of historic lavas from Mount Etna. Nature 318(6045):453–455

Tanguy JC, Condomines M, Le Goff M, Chillemi V, La Delfa S, Patanè G (2007) Mount Etna eruptions of the last 2750 years: revised chronology and location through archaeomagnetic and ^{226}Ra-^{230}Th dating. Bull Volcanol 70:55–83

Tanguy JC, Le Goff M, Principe C, Arrighi S, Chillemi V, Paiotti A, La Delfa S, Patanè G (2003) Archaeomagnetic dating of Mediterranean volcanics of the last 2100 years: validity and limits. Earth Planet Sci Lett 211:111–124

Tema E (2009) Estimate of the magnetic anisotropy effect on the archaeomagnetic inclination of ancient bricks. Phys Earth Planetary Inter 176:213–223

Tema E, Lanza R (2008) Archaeomagnetic study of a lime kiln at Bazzano (northern Italy). Phys Chem Earth 33:534–543

Tema E, Hedley I, Lanos P (2006) Archaeomagnetism in Italy: a compilation of data including new results and a preliminary Italian secular variation curve. Geophys J Int 167:1160–1171

Tema E, Goguitchaichvili A, Camps P (2010) Archaeointensity determinations from Italy: new data and the Earth's magnetic field strength variation over the past three millennia. Geophys J Int 180:596–608

Thellier E, Thellier O (1959) Sur l'intensité du champ magnétique terrestre dans le passé historique et géologique. Ann Geophys 15:285–376

Veitch RJ, Hedley IG, Wagner JJ (1984) An investigation of the intensity of the geomagnetic field during Roman times using magnetically anisotropic bricks and tiles. Arch Sci (Geneva) 37(3):359–373

Zanella E, Gurioli L, Chiari G, Ciarallo A, Cioni R, De Carolis E, Lanza R (2000) Archaeomagnetic results from mural paintings and pyroclastic rocks in Pompeii and Herculaneum. Phys Earth Planetary Inter 118:227–240

The Termination of the Olduvai Subchron at Lingtai, Chinese Loess Plateau: Geomagnetic Field Behavior or Complex Remanence Acquisition?

16

Simo Spassov, Jozef Hus, Friedrich Heller,
Michael E. Evans, Leping Yue, and Tilo von Dobeneck

Abstract

We present a detailed investigation of the geomagnetic polarity transition that terminated the Olduvai subchron as recorded by loess/paleosol sediments at Lingtai in the central Chinese Loess Plateau (CLP). The polarity transition occurs within loess layer L25, where mineral magnetic parameters show considerable variations and sedimentation rate changes occur. The magnetic record obtained after thermal cleaning exhibits more than twenty apparent polarity flips, most of which occur within a stratigraphic distance corresponding to no more than ~15,000 years. We argue that these results do not represent the actual behavior of the geomagnetic field. Instead, we propose that the combined effect of detrital and pedogenic remanences—which almost always co-exist on the central CLP—are responsible. These cause significant, lithologically-controlled, delays in the acquisition of the total remanence. In effect, the sediments act as a filter that generates noisy magnetic output from possibly simple input. We conclude that loess/paleosol sediments from the central CLP are poor candidates for tracking short-term geomagnetic field behavior such as polarity transitions, geomagnetic excursions and paleosecular variation.

16.1 Introduction

The possibility of deciphering geomagnetic field behavior during polarity reversals is offered by the remanent magnetization encoded in volcanic and sedimentary sequences formed at appropriate times, although the remanence acquisition process is very different in these two rock types. Volcanics become magnetized during cooling and thus possess a thermoremanent magnetization (TRM) that is usually acquired within a very short time compared to the timescale of polarity reversals. However, this advantage is offset by the problem that volcanic rocks are created sporadically and do not provide continuous high-resolution time coverage. Sediments usually carry a post-detrital remanent magnetization (pDRM) acquired after mechanical locking of the remanence carriers in the sediment matrix. The time interval between gravitational settling and mechanical locking depends critically on the environment, which involves many chemical and physical factors. It generally takes

S. Spassov (✉)
Section du Magnétisme Environnemental, Centre de Physique du Globe de l'Institut Royal Météorologique de Belgique, B-5670 Dourbes (Viroinval), Belgium
e-mail: simo.spassov@meteo.be

E. Petrovský et al. (eds.), *The Earth's Magnetic Interior*, IAGA Special Sopron Book Series 1,
DOI 10.1007/978-94-007-0323-0_16, © Springer Science+Business Media B.V. 2011

much longer than TRM acquisition in volcanics. On the other hand, sediments may provide essentially continuous records of the geomagnetic field.

It is well established that Chinese loess/paleosol sediments provide excellent records of past climatic variations that can be tied to absolute time via major geomagnetic polarity boundaries (Heller and Liu 1982, 1984) which serve as control points for more detailed chronostratigaphies (Heller et al. 1987, Lu et al. 1999, Zhou and Shackleton 1999, Heslop et al. 2000, Sun et al. 2006). In recent years several high-resolution paleomagnetic records from the CLP have been published that apparently demonstrate the ability of loess/paleosol sediments to record shorter-term geomagnetic variations, such as excursions (Zhu et al. 1999, Guo et al. 2002, Pan et al. 2002, Yang et al. 2004, 2005, 2007), polarity transitions (Rolph et al. 1989, Zhu et al. 1993, 1994, McIntosh et al. 1996) and secular variation (Heslop et al. 1999, Zhu et al. 2000). In view of the complex acquisition processes of natural remanent magnetization in loess/paleosol sediments (Hus and Han 1992, Zhou and Shackleton 1999, Evans and Heller 2001, Spassov et al. 2003a), it is essential that the resolving power of these sediments be closely examined in order to check the validity of these claims. Well-sampled sections containing many apparent polarity flips offer an ideal opportunity to carry out this exercise. Here, we analyze such a record covering the termination of the Olduvai subchron at Lingtai in the central CLP.

16.2 Field and Laboratory Techniques

The section studied is located about 15 km south of the town of Lingtai (Gansu province) at 34.98°N, 107.56°E and ~1340 m above sea level. It contains 33 pairs of loess/paleosol units that extend down to 175 m depth and has an average sedimentation rate of 6.7 cm/kyr (Spassov et al. 2003b). The entire section was sampled continuously by carving out vertical block samples about 30 cm in height that were subsequently cut into ~2×2×2 cm specimens. The depth interval of interest here is centered on loess layer L25, the upper part of which is a yellowish brown silty clay overlain by carbonate nodules, and the lower part of which shows clear signs of weathering (Ding et al. 1999). Taking account of these lithological differences, Sun et al. (2006) derive sedimentation rates of 11.5 cm/kyr and 4.0 cm/kyr for the upper and lower parts of L25, respectively. On this basis, the upper part of L25 (131.62–132.76 m) represents 9.9 kyr, whereas the lower part (132.76 and 134.46 m) represents 42.5 kyr.

A subset of 144 specimens was selected for paleomagnetic measurements. Almost all of them exhibit an initial normal polarity of the natural remanent magnetization (NRM) due to a secondary overprint. Therefore, all 144 specimens were stepwise thermally demagnetized up to 520°C in an ASC thermal demagnetizer installed in a magnetically shielded room. The remanence remaining after each heating step was measured with a 2G Enterprises SQUID cryogenic magnetometer housed in the same magnetically shielded room. The overprint component was removed at temperatures between 250 and 300°C, and the characteristic remanence direction (ChRM) was then determined using principal component analysis (Kirschvink, 1980). The quality of the ChRM direction is assessed by the maximum angular deviation (MAD) of the principal eigenvector which is defined as:

$$\mathrm{MAD} = \cot\sqrt{\frac{\tau_2^2 + \tau_3^2}{\tau_1}} \qquad (16.1)$$

with τ_i being the eigenvalues of the principal eigenvector, i.e. the best fit line (see Kirschvink 1980).

Measurements of the magnetic low-field susceptibility were performed on a KLY-2 susceptibility bridge, and the frequency dependence of low-field magnetic susceptibility was measured with a Bartington susceptometer equipped with a dual frequency sensor operating at 0.465 and 4.65 kHz.

16.3 Results and Discussion

ChRM declination and inclination yield a very complicated pattern during transition from the normal Olduvai to reversed Matuyama polarity zone (Fig. 16.1). A short reversed polarity interval, comprising 8 specimens, is observed just below 134 m with slightly elevated MAD values (mean = 11°). Between 134.04 and 132.98 m, there are 42 specimens with normal polarity and generally lower MAD values (mean = 8°). The zone between 132.98 and 131.80 m exhibits fluctuating remanence directions with variable MAD values, mainly between 10 and 25°

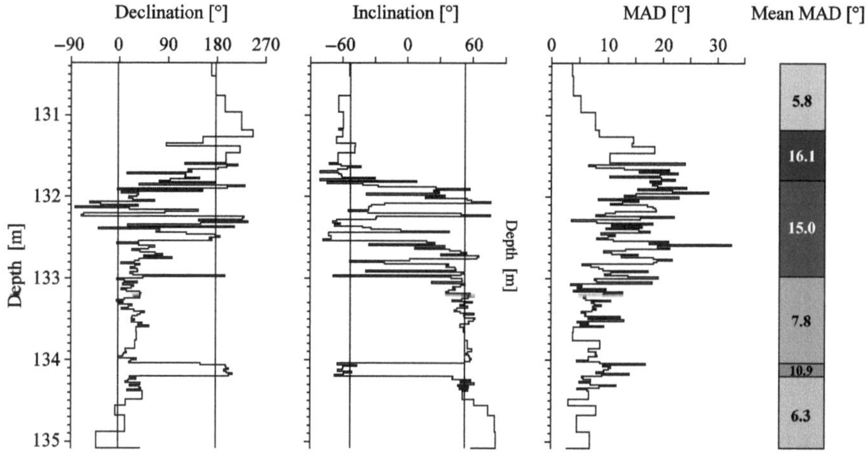

Fig. 16.1 Declination, inclination and maximum angular deviation (MAD) of the characteristic remanent magnetization of thermally demagnetized samples from the terminal Olduvai transition at Lingtai. Depths are given in meters from the *top* of the section. The lines represent the present-day field at Lingtai according to the International Geomagnetic Reference Field (IGRF). The column to the right gives the mean MAD's of the characteristic remanence (ChRM) directions for depth intervals with reversed, normal and mixed polarities. The stronger the gray shade the higher the MAD

(mean = 15°, $N = 59$). Above 131.80 m, only reversed polarity is observed but the MAD values remain high (mean = 16°, $N = 15$). Generally, they do not fall below 10° until 131.18 m, while above 131.18 m, the MAD's do not exceed 9° (mean = 5.8°, $N = 5$).

In the approximately 15,000 years represented between 132.98 and 131.80 m, inclination changes sign 19 times. This implies polarity episodes of about 800 years, separated by transitions lasting no more than a century or two. These are not characteristics normally regarded as typical for real geomagnetic field behavior. In their summary, Merrill and McFadden (1999), as well as Clement (2004) and Valet (2003) conclude that, although the time taken for a polarity reversal to take place is not well known, it probably lies between 1000 and 8000 years.

In what follows, we investigate an alternative mechanism that appeals to the mineralogical and sedimentological factors that control the acquisition of remanence on the central CLP. This involves a critical assessment of the relevant mineral magnetism parameters, to which we now turn.

Low susceptibility values of about 30×10^{-8} m^3/kg are observed in the calcareous horizon near the top of paleosol S25 and in the lower part of paleosol S24 (Fig. 16.2). Within loess layer L25 the susceptibility is enhanced in the lower part, reaching values of almost 80×10^{-8} m^3/kg. The F-factor, which is the percental difference between measured low and high frequency susceptibility, relative to the low frequency measurement, shows a similar pattern, varying between 3 and 11%.

The observed magnetic enhancement in the lower part of L25 (F-factor >7%, susceptibility $>50 \times 10^{-8}$ m^3/kg) indicates that weak pedogenesis took place, in agreement with the field evidence. Comparing Figs. 16.1 and 16.2, we note that this pedogenesis is associated with lower MAD values, whereas in the less-weathered upper part of L25 (F-factor <7%, susceptibility $<50 \times 10^{-8}$ m^3/kg) fluctuating—but generally higher—MAD values occur. Post-depositional processes created stratigraphic variations in lithology and mineral magnetism that impose themselves on the observed paleomagnetic pattern. The result is that the majority of the observed polarity changes become associated with higher MAD values, i.e. with samples whose remanence is less well-defined during demagnetization. A similar situation is found at Baoji (Yang et al. 2007), where several suggested geomagnetic excursions—as well as the Olduvai termination itself—are associated with higher MAD values (see their Fig. 9). Although MAD values have been used for many years as a general guide to the quality of remanence vectors obtained from progressive demagnetization, formal statistical confidence thresholds have never been established. McElhinny and McFadden (2000) state that MAD values ≥15°

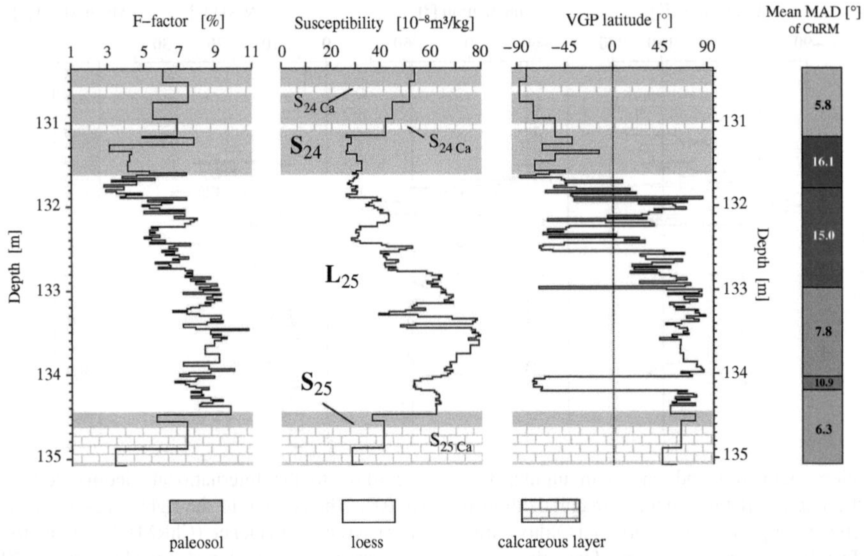

Fig. 16.2 Stratigraphic variation of frequency dependence of low-field susceptibility (F-factor), low-field susceptibility and VGP latitude for the Olduvai termination in loess layer L25 at Lingtai. Label S indicates a paleosol horizon and index Ca a calcareous horizon in the paleosol. The column to the right gives the mean MAD's of the characteristic remanence (ChRM) directions (see Fig. 16.1) for depth intervals with reversed, normal and mixed polarities. The stronger the gray shade, the higher the MAD

"are often considered ill defined and questionable", but MAD $\leq 10°$ "would be considered to be reasonably good".

Starting from the full Lingtai data set considered here, which exhibits 21 polarity changes in the meter or so immediately above 133 m (Figs. 16.1 and 16.3a), the elimination of samples with MAD > 15° reduces the number of changes to nine (Fig. 16.3b). Removing samples with MAD > 10° further reduces this to only five (Fig. 16.3c). Clearly, the apparently rapid fluctuations of the geomagnetic field are based largely on samples whose remanence record is compromised.

We suggest that the Lingtai samples with higher MAD values possess two remanence components of opposite polarity acquired (see upper row in Fig. 16.4) in the manner proposed by Spassov et al. (2003a), a point to which we return below. For the moment, we take the conservative view that Fig. 16.3c is more likely to represent the actual geomagnetic field behavior. The simplest interpretation is that the uppermost normal-to-reversed transition (near 131.80 m) represents the real end of the Olduvai subchron, with two earlier reversed intervals centered on 132.4 and 134.1 m, respectively.

16.3.1 Comparison with Other Records

In the following we compare our reduced VGP pattern (Fig. 16.3c) with other records of the Olduvai/Matuyama boundary. The loess/paleosol deposits at Baoji (Shaanxi Province) have been studied by different authors. The section is located 5 km north of the city of Baoji (34.4°N, 107.1°E) at an elevation of approximately 970 m, about 120 km south of Lingtai. A total of 37 paleosols has been identified by Rutter et al. (1991) in the 159 m thick section. Due to the absence of orographic barriers, the climatic and sedimentary environment is very similar to that of Lingtai. Magnetostratigraphic investigations were performed by Rutter et al. (1991) and recently by Yang et al. (2005). The thickness of loess L25 at Baoji is about 2.70 m (Rutter and Ding 1993) corresponding to an average sedimentation rate of 8.1 cm/kyr for L25,

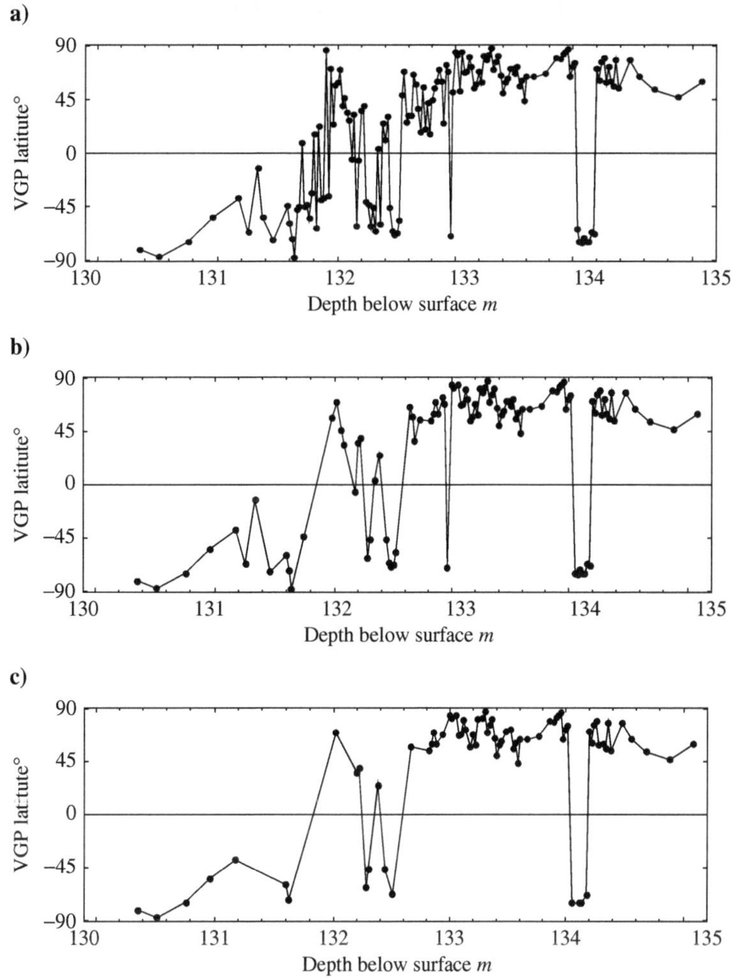

Fig. 16.3 VGP profiles of the UOB at Lingtai. (**a**) original data including all 144 samples. Note here that the intervals [131.60; 133.60] and [133.92; 134.38] are continuously represented. (**b**, **c**) Samples rejected with MAD > 15° and MAD > 10°, respectively

which is similar to that of Lingtai. The micromorphological description by Rutter and Ding (1993) revealed indications of relatively strongly weathering of L25.

The paleomagnetic investigations of Rutter et al. (1991) gave no hint of transitional geomagnetic field directions. The obtained VGP latitudes of the upper Olduvai boundary (UOB) transition are almost all larger than 50°N or S as expected for a geomagnetic dipole and only a single polarity change is observed. As their sampling interval is about 20 cm, direct comparison of polarity patterns is impossible.

A much more detailed paleomagnetic study of the UOB at Baoji was undertaken by Yang et al. (2005). It contains a zone 1.4 m thick containing 15 ChRM polarity changes, which is different to the number found at Lingtai (cf. Table 16.1), despite having the same sampling density of 2 cm. Also the polarity pattern is different: at Baoji the apparent "reversal" frequency increases from top to bottom (Fig. 8c in Yang et al. 2005), while at Lingtai the opposite is observed (cf. Fig. 16.3). Because of the short distance between the two sites, the same transitional field morphology and timing can be expected and the same pattern should be observed in both sections.

Reversal duration and transitional field morphology depend on geographic position (e.g. Clement

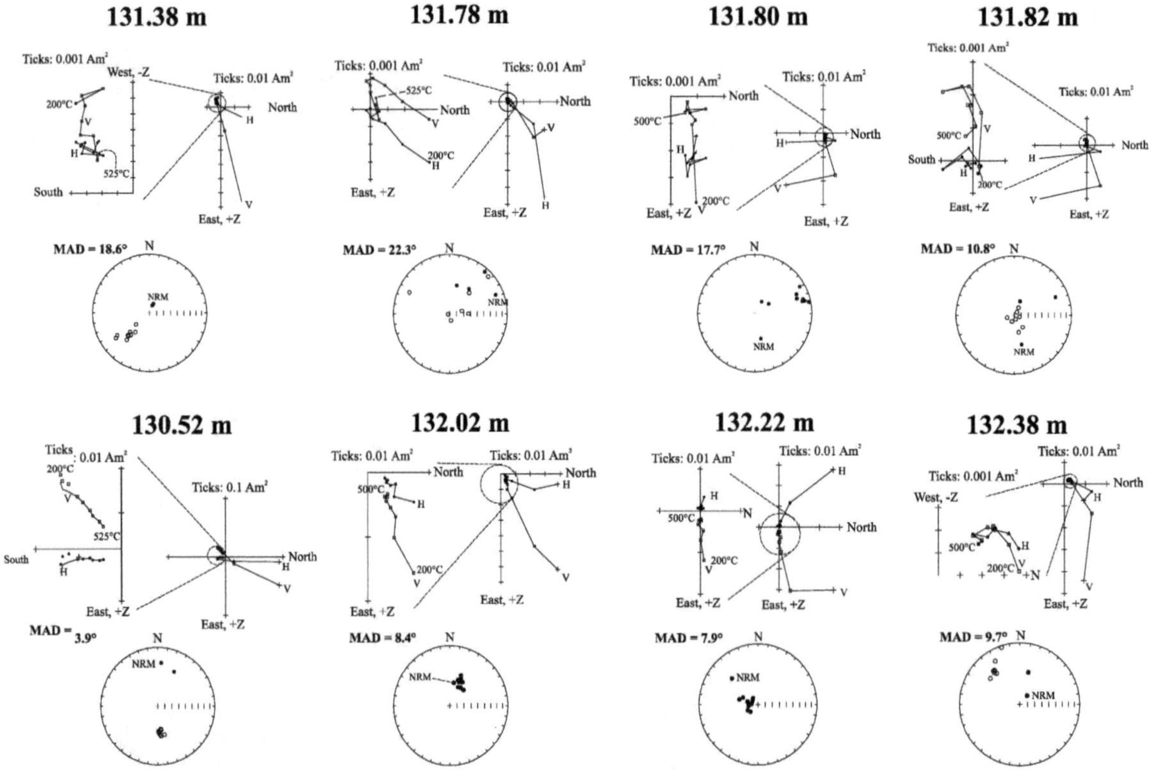

Fig. 16.4 Zijderveld diagrams (Zijderveld 1967) and corresponding equal area plots of samples with maximum angular deviations (MAD) >15° (*upper row*) and <10° *lower row*. H and V denote the *horizontal* and the *vertical* component of the remanence vector during progressive thermal demagnetization, respectively

2004). Hence, records from different regions of the globe will show different transitional field behavior. Nevertheless, different UOB records are compared in the following to seek further observational evidence for a possibly complicated UOB behavior in other types of sediments.

Table 16.1 lists 15 marine and 2 lacustrine UOB records from different parts of the world. The comparison shows that far fewer polarity changes were observed, which may be an artifact due to insufficient sampling intervals for certain records. However, it is much more important that records from the same geographic locality show considerable differences in the number of polarity changes (e.g. Holt and Kirschvink 1995, Channell and Lehmann 1999) indicating that transitional field behavior is not faithfully recorded.

Summarizing, there is not enough evidence to interpret the multiple ChRM polarity changes at the UOB in Lingtai as geomagnetic field behavior, neither from theoretical aspects nor from comparisons with other UOB records.

16.3.2 Lock-in Aspects

Here, we investigate to what extent the polarity profile of Fig. 16.3c arises from delayed remanence lock-in controlled by lithological variations. Instead of performing the lock-in model calculations in the depth domain, as in Spassov et al. (2003a), we adapted their model for the time domain. This was necessary, because the sedimentation rate is not constant within L25, as is clear from the susceptibility variations shown in Fig. 16.5. The astronomically derived sedimentation rates of Sun et al. (2006) were used for transforming depth into time. Consequently, the lock-in function λ of the postdetrital (pDRM) and pedogenic (CRM) remanence is expressed in the time domain (τ):

Table 16.1 Observation of the upper Olduvai boundary in different sedimentary environments

Origin	Location	Latitude	Longitude	SR [cm/kyr]	Number of VGP or ChRM inclination polarity changes	Reference
marine	ODP site 981	55.48°N	14.65°W	10.0	1	Channell et al. (2003)
marine	ODP site 982	57.52°N	15.87°W	2.5	1	Channell and Guyodo (2004)
marine	ODP site 981A	~56°N	~14°W	5.6	1	Channell and Lehmann (1999)
marine	ODP site 981B	~56°N	~14°W	6.3	1	Channell and Lehmann (1999)
marine	ODP site 983A	~56°N	~14°W	15.8	3	Channell and Lehmann (1999)
marine	ODP site 983B	~56°N	~14°W	15.8	5	Channell and Lehmann (1999)
marine	ODP site 983C	~56°N	~14°W	16.8	5	Channell and Lehmann (1999)
marine	ODP site 984B	~56°N	~14°W	10.8	1	Channell and Lehmann (1999)
marine	ODP site 984C	~56°N	~14°W	10.6	1	Channell and Lehmann (1999)
marine	ODP site 984D	~56°N	~14°W	11.0	3	Channell and Lehmann (1999)
marine	DSDP	49.86°N	24.23°W	8.3	3	Clement and Kent (1987)
marine	ODP site 1101	64.37°S	70.27°W	6.9	1	Guyodo et al. (2001)
lacustrine	Death March Canyon	35°N	116°W	33.1	1	Holt and Kirschvink (1995)
lacustrine	Confusion Canyon	35°N	116°W	28.1	7	Holt and Kirschvink (1995)
marine	K7501	37.37°N	179.60°W	1.6	5	Herrero-Bervera and Khan (1992)
marine	MD972143	15.87°N	124.65°W	~1.5	1	Horng et al. (2003)
marine	ODP site 983	60.40°N	23.64°W	16.4	3	Mazaud and Channell (1999)
loess	Lingtai, central CLP	34.98°N	107.56°E	8.5	23	this study
loess	Baoji, central CLP	34.41°N	107.12°E	9.1*	15**	*calculated from Rutter et al. (1991)
						**Yang et al. (2005)

$$\lambda_{p\text{DRM}+\text{CRM}}(\tau, e(t), SR(t)) = \begin{cases} 0 & \text{for} \quad \tau < \dfrac{d_0}{SR(t)} \\[2ex] e(t)\dfrac{\tau - \dfrac{d_0}{SR(t)}}{\dfrac{d_1}{SR(t)} - \dfrac{d_0}{SR(t)}} & \text{for} \quad \dfrac{d_0}{SR(t)} \leq \tau < \dfrac{d_1}{SR(t)} \\[2ex] e(t) & \text{for} \quad \dfrac{d_1}{SR(t)} \leq \tau < \dfrac{d_2}{SR(t)} \\[2ex] e(t) + (1 - e(t))\dfrac{\tau - \dfrac{d_2}{SR(t)}}{\dfrac{d_3}{SR(t)} - \dfrac{d_2}{SR(t)}} & \text{for} \quad \dfrac{d_2}{SR(t)} \leq \tau < \dfrac{d_3}{SR(t)} \\[2ex] 1 & \text{for} \quad \tau \geq \dfrac{d_3}{SR(t)} \end{cases} \tag{16.2}$$

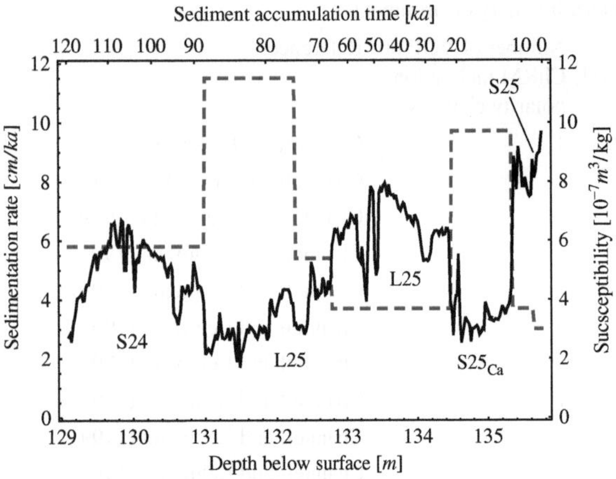

Fig. 16.5 Magnetic susceptibility variation (*black solid*) and sedimentation rates (*gray shaded*) around the upper Olduvai boundary at Lingtai. The latter were taken from Sun et al. (2006) and were used to transform the susceptibility record from the depth into the time domain. The model time scale is given on the top of the graph and increased from *right* to *left*. L and S denote loess and paleosol layers, respectively and the index Ca qualifies a calcareous layer

The parameter function $e(t)$ is the magnetic low-field susceptibility as a function of time scaled to vary between 0 and 1, and $SR(t)$ is the sedimentation rate as a function of time, which was taken from Sun et al. (2006), see Fig. 16.5. Variable τ is the elapsed time interval during which a layer at the sediment surface is progressively buried. It is the time analogue of the burial depth ζ in Bleil and v. Dobeneck (1999) or Spassov et al. (2003a). Variables d_0 to d_3 are lock-in depths, indicating the different stages of the detrital pedogenic lock-in process. No magnetization is acquired until d_0, pDRM is gradually locked until d_1. Pedogenic CRM acquisition starts at d_1 and is completed at d_3. For reasons of simplicity linear pDRM and CRM acquisition is assumed. The reader is referred to Fig. 3 in Spassov et al. (2003a) for further explanations. However, as our model is established in the time domain, the lock-depths are divided by the sedimentation rate.

The modeled remanence at the upper Olduvai boundary was then calculated using the following formula:

$$M(t) = -1 + \int_{t=0}^{t} \lambda_{\mathrm{PDRM+CRM}}(\tau, e(t), SR(t)) \frac{\partial H(t-\tau)}{\partial \tau} d\tau \qquad (16.3)$$

with $H(t)$ being the time varying geomagnetic field. We used two different field behaviors which are contained in two forms of the following equation:

$$H(t) = \tanh\left(\frac{c_2+c_1}{c_2-c_1}\left(t-\frac{c_2+c_1}{2}\right)\right) \times \tanh\left(\frac{c_4+c_3}{c_4-c_3}\left(t-\frac{c_4+c_3}{2}\right)\right) \times \tanh\left(\frac{c_6+c_5}{c_6-c_5}\left(t-\frac{c_6+c_5}{2}\right)\right) \qquad (16.4)$$

Simple behavior was simulated by omitting the first and second factors, while for complicated behavior, which assumes a precursor before the reversal, all three factors were used. Parameters c_1 to c_6 control temporal position and duration of the field polarity switches. We have chosen values of 1, 100, 1, 120, 1, 55 kyr for c_1 to c_6, respectively in order to simulate the Olduvai/Matuyama reversal occurring in the lower part of S24 at is presumed from its stratigraphic position (see Heslop et al. 2000). In case of complicated field behavior, the values of c_1 to c_4 permit the simulation of 10 ka long reversal precursor occurring about 20 ka before the actual reversal; a situation such as observed for the Brunhes-Matuyama boundary, (see Coe et al. 2004, Valet et al. 2005). The intensity decrease of geomagnetic field has been chosen to change smoothly during the reversal as we place emphasis on lithologenic and sedimentary variations influencing the magnetic remanence record. However, geomagnetic field variations during polarity transition may not be that smooth, which would result in a much more complicated remanence record.

Figure 16.6 summarizes our model calculations for simple and complicated field behavior. Assuming a simple reversal occurring in the lower part of S24, the rather large lock-in depths estimated by Spassov et al. (2003a) for the Matuyama-Brunhes boundary at Lingtai do not explain the observed VGP polarity flips at the upper Olduvai boundary (compare Figs. 16.3c and 16.6a). Indeed there are three lithology-induced polarity flips but they occur in the lower part of L25. In contrast, with much shallower lock-in depths, the detrital-pedogenic lock-in is completed at 140 cm, the multiple polarity zone moves towards the upper part of L25 but the reversed polarity interval observed in the beginning of L25 is missing (Fig. 16.6b). Because further variation of the lock-in depths did not improve

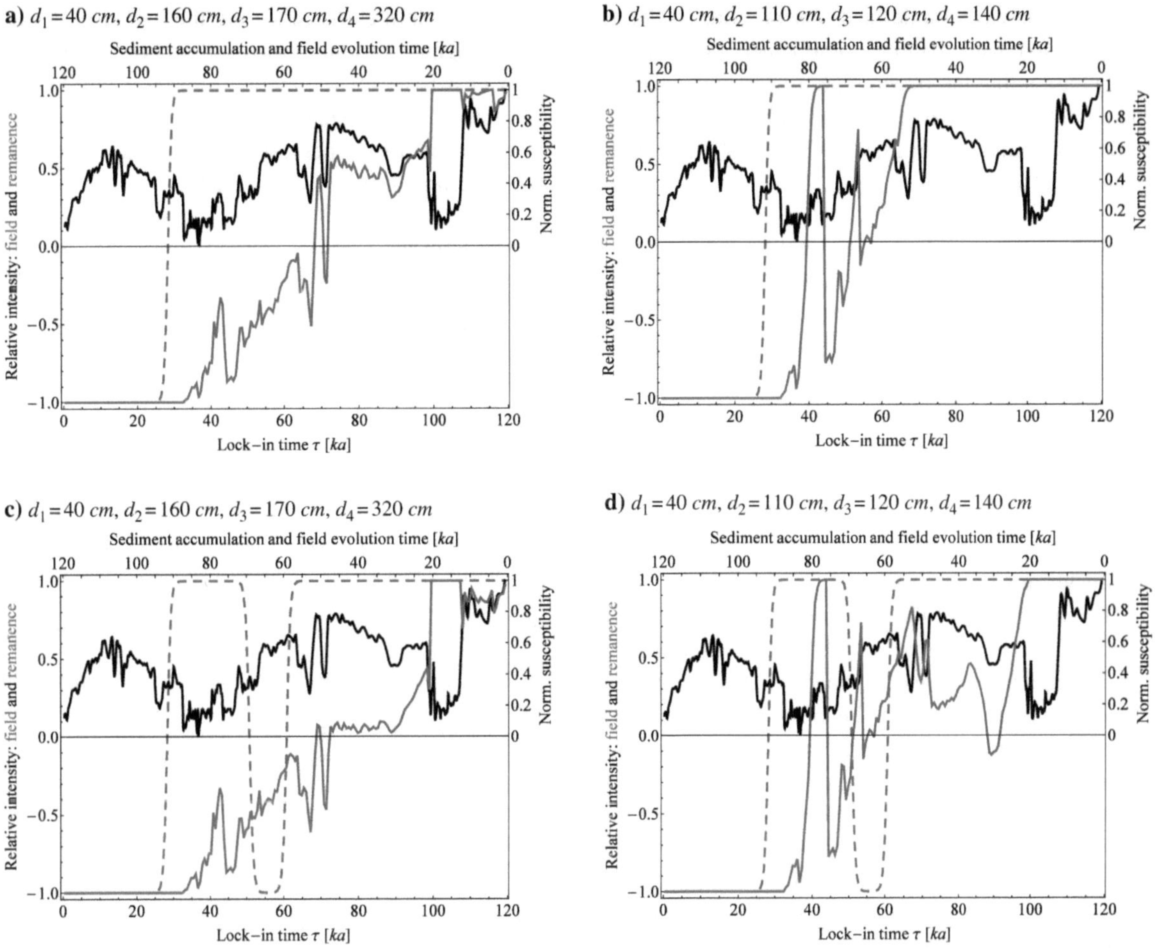

Fig. 16.6 Modeled remanence record (*grey solid*) for two different lock-in depth scenarios and two different assumed transitional field behaviors (*gray dashed*). Only scenario (d) is in fair agreement with the VGP pattern (Fig. 16.3c). Parameters d_1 to d_3 are lock-in depths, indicating different stages of the detrital-pedogenic lock-in process (see text and Fig. 3 in Spassov et al. 2003a). In order to calculate a time dependent lock-in function they are divided by the sedimentation rate function $SR(t)$, see equation (2). Note that the modeled intensity changes are solely due to magnetic property and sedimentation rate changes, as the field is assumed to decrease smoothly

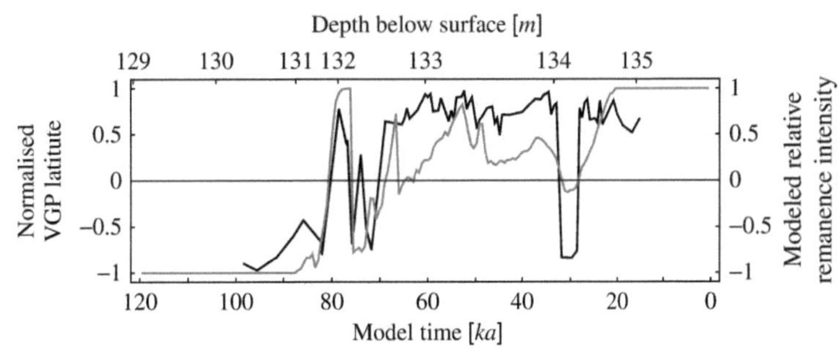

Fig. 16.7 Comparison between observed VGP latitudes (MAD < 10°) normalized to 1 (*black*) and modeled relative remanence intensity (*gray*) of the scenario in Fig. 6d. The observed equatorial VGP crossings agree fairly well with the modeled polarity changes

agreement between model and observation, modeling was repeated with a more complicated field behavior, i.e. assuming the occurrence of a precursor about 20 ka before the actual reversal. Again, the large lock-in depths found by Spassov et al. (2003a) cannot explain the observations (Figs. 16.3c and 16.6c), but model calculations with shallower lock-in depths are in fair agreement with the observations (Figs. 16.6d and 16.7). We interpret the multiple polarity zone centered at 132.4 m (see Fig. 16.3c) as a lithology-induced artifact, while the other reversed interval at 134.1 m (Fig. 16.3c) may be interpreted as a precursor. This is supported by Holt and Kirschvink (1995) and Mazaud and Channell (1999) who observed an excursion at about 7500 and 13500 years, respectively, before the Olduvai termination.

Conclusions

The termination of the Olduvai subchron at Lingtai falls in a loess layer which is characterized by magnetic property and sedimentation rate changes. Consequently, many of the observed VGP directions have high MAD values. Although omitting data points with large errors, a multiple polarity zone was still identified. Its polarity pattern, however, differs from nearby as well as from other global records. With the aid of a detrital-pedogenic lock-in model we were able to explain a part of the observed VGP signal as an artifact of delayed remanence acquisition and another part as an apparently true field behavior. The following conclusions are drawn:

1. Loess layer L25 at Lingtai shows variable sedimentation rates and is strongly affected by weathering/pedogenesis, resulting in lithogenic variations.
2. The occurrence of intermediate remanence directions and high MAD's and hence the reliability of the characteristic remanence are related to and influenced by lithogenic variations.
3. The observed upper Olduvai boundary polarity pattern in the upper part of L25 at Lingtai does not represent true transitional geomagnetic field behavior but reflects lithogenic variations.
4. The detrital/pedogenic lock-in model is generally able to explain downward shifts of geomagnetic polarity boundaries and the occurrence of mixed polarity zones. However, appropriate input parameters, reflecting the whole range of lithogenic variations have still to be sought.
5. Loess/paleosol sediments from the central CLP cannot be considered as trustworthy geomagnetic field recorders, when timescales of a few thousand years are concerned.

Acknowledgements Financial support from the Belgian Science Policy and the Natural Sciences and Engineering Research Council of Canada is gratefully acknowledged, as well as the interest and the effort of two anonymous reviewers for their constructive comments.

References

Bleil U, v Dobeneck T (1999) Geomagnetic events and relative paleointensity records – clues to high-resolution paleomagnetic chronostratigraphies of Late Quaternary marine sediments? In: Fischer G, Wefer G (eds) Use of proxies in paleoceanography: examples from the South Atlantic. Springer, Heidelberg, pp 635–654

Channell JET, Lehmann B (1999) Magnetic stratigraphy of North Atlantic sites 980–984. In: Raymo ME, Jansen E, Blum P, Herbert TD (eds) Proceedings of the ocean drilling program, scientific results, vol 162. College Station, TX (Ocean Drilling Program), pp 113–130. doi:10.2973/odp.proc.sr.162.002.1999

Channell JET, Labs J, Raymo ME (2003) The Reunion Subchronozone at ODP Site 981 (Feni Drift, North Atlantic). Earth Planetary Sci Lett 215:1–12

Channell JET, Guyodo Y (2004) The Matuyama Chronozone at ODP Site 982 (Rockall Bank): evidence for decimeter-scale magnetization lock-in depths. In: Chapman Conference on Timescales of the Internal Geomagnetic Field, Geophysical monograph, vol 145. American Geophysical Union, Washington, DC, pp 205–219

Coe RS, Singer BS, Pringle MS, Zhao XX (2004) Matuyama–Brunhes reversal and Kamikatsura event on Maui: paleomagnetic directions, $^{40}Ar/^{39}Ar$ ages and implications. Earth Planetary Sci Lett 222:667–684

Clement BM (2004) The dependence of geomagnetic polarity reversal durations on site latitude. Nature 428:637–639

Clement BM, Kent DV (1987) Geomagnetic polarity transition records from five hydraulic piston core sites in the North Atlantic. Init Rep Deep Sea Drilling Project 94:831–852

Ding ZL, Xiong SF, Sun JM, Yang SL, Gu ZY, Liu TS (1999) Pedostratigraphy and paleomagnetism of a ~7.0 Ma eolian loess-red clay sequence at Lingtai, Loess Plateau, north-central China and the implications for paleomonsoon evolution. Paleogeogr Paleoclimatol Paleoecol 125:49–66

Evans ME, Heller F (2001) Magnetism of loess/paleosol sequences: recent developments. Earth Sci Rev 54:129–144

Guo B, Zhu RX, Florindo F, Ding ZL, Sun JM (2002) A short, reverse polarity interval within the Jaramillo subchron: evidence from the Jingbian section, northern Chinese Loess Plateau. J Geophys Res 107. doi 10.1029/2001JB000706

Guyodo Y, Acton GD, Brachfeld S, Channell JET (2001) A sedimentary paleomagnetic record of the Matuyama chron from the Western Antarctic margin (ODP Site 1101). Earth Planetary Sci Lett 191:61–74

Heller F, Liu TS (1982) Magnetostratigraphical dating of loess deposits in China. Nature 300:431–433

Heller F, Liu TS (1984) Magnetism of Chinese loess deposits. Geophys J R Astron Soc 77:125–141

Heller F, Meili B, Wang JD, Li HM, Liu TS (1987) Magnetization and sedimentation history of loess in the central loess plateau of China. In: Liu TS (ed) Aspects of Loess research. China Ocean Press, Beijing, pp 147–163

Herrero-Bervera E, Khan MA (1992) Olduvai termination: detailed palaeomagnetic analysis of a north central Pacific core. Geophys J Int 108:535–545

Heslop D, Langereis CG, Dekkers MJ (2000) A new astronomical timescale for the loess deposits of northern China. Earth Planetary Sci Lett 184:25–139

Heslop D, Shaw J, Bloemendal J, Chen F, Wang J, Parker E (1999) Sub-millennial scale variations in east Asian monsoon systems recorded by dust deposits from the north-western Chinese Loess Plateau. Phys Chem Earth A 24:785–792

Holt JW, Kirschvink JL (1995) The upper Olduvai geomagnetic field reversal from Death Valley, California: a fold test of transitional directions. Earth Planetary Sci Lett 133:475–491

Horng C-S, Roberts AP, Liang W-T (2003) A 2.14-Myr astronomically tuned record of relative geomagnetic paleointensity from the western Philippine Sea. J Geophys Res 108. doi: 10.1029/2001JB001698

Hus J, Han J (1992) The contribution of loess magnetism in China to the retrieval of past global changes – some problems. Phys Earth Planetary Inter 70:154–168

Kirschvink JL (1980) The least-squares line and plane and the analysis of paleomagnetic data. Geophys J R Astron Soc 62:699–718

Lu HY, Liu XD, Zhang FQ, An ZS, Dodson J (1999) Astronomical calibration of loess-paleosol deposits at Luochuan, central Chinese Loess Plateau. Paleogeogr Paleoclimatol Paleoecol 154:237–246

Mazaud A, Channell JET (1999) The top Olduvai polarity transition at ODP Site 983 (Iceland Basin). Earth Planetary Sci Lett 166:1–13

McElhinny MW, McFadden PL (2000) Paleomagnetism: continents and oceans. Academic, San Diego, CA, p 386

McIntosh G, Rolph TC, Shaw J, Dagley P (1996) A detailed record of normal-reversed polarity transition obtained from a thick loess sequence at Jiuzhoutai, near Lanzhou, China. Geophys J Int 127:651–664

Merrill RT, McFadden PL (1999) Geomagnetic polarity transition. Rev Geophys 37:201–226

Pan YX, Zhu RX, Liu QS, Guo B, Yue LP, Wu HN (2002) Geomagnetic episodes of the last 1.2 Myr recorded in Chinese loess. Geophys Res Lett 29. doi10.1029/2001GL014024

Rolph TC, Shaw J, Derbyshire E, Wang JT (1989) A detailed geomagnetic record from Chinese loess. Phys Earth Planetary Inter 56:151–164

Rutter N, Ding ZL (1993) Paleoclimates and monsoon variations interpreted from micromorphologic features of the Baoji paleosols, China. Quaternary Sci Rev 12:853–862

Rutter N, Ding ZL, Evans ME, Liu TS (1991) Baoji-type pedostratigraphic section, Loess Plateau, North-Central China. Quaternary Sci Rev 10:1–22

Spassov S, Heller F, Evans ME, Yue LP, Ding ZL, v. Dobeneck T (2003a) A lock-in model for the complex Matuyama-Brunhes boundary record of the loess/paleosol sequence at Lingtai (Central Chinese Loess Plateau). Geophys J Int 155:1–17

Spassov S, Heller F, Kretzschmar R Evans ME, Yue LP, Nourgaliev DK (2003b) Detrital and pedogenic magnetic mineral phases in the loess/paleosol sequence at Lingtai (Central Chinese Loess Plateau). Phys Earth Planetary Inter 140:255–275

Sun YB, Clemens SC, An ZS, Yu ZW (2006) Astronomical timescale and paleoclimatic implication of stacked 3.6-Myr monsoon records from the Chinese Loess Plateau. Quaternary Sci Revi 25:33–48

Valet J-P (2003) Time variations in geomagnetic intensity. Rev Geophys 41. doi:10.1029/2001RG000104

Valet J-P, Meynadier L, Guyodo Y (2005) Geomagnetic dipole strength and reversal rate over the past two million years. Nature 435:802–805

Yang TS, Hyodo M, Yang ZY, Fu JL (2004) Evidence for the Kamikatsura and Santa Rosa excursions recorded in eolian deposits from the southern Chinese Loess Plateau. J Geophys Res 109. doi: 10.1029/2004JB002966

Yang TS, Hyodo M, Yang ZY, Sun JM (2005) A first paleomagnetic and rock magnetic investigation of calcareous nodules from the Chinese Loess Plateau. Earth Planets Space 57:29–34

Yang TS, Hyodo M, Yang ZY, Ding L, Fu JL, Mishima T (2007) Early and middle Matuyama geomagnetic excursions recorded in the Chinese loess-paleosol sediments. Earth Planets Space 59:825–840

Zhou LP, Shackleton NJ (1999) Misleading positions of geomagnetic reversal boundaries in Eurasian loess and implications for correlation between continental and marine sediment sequences. Earth Planetary Sci Lett 168: 117–130

Zhu RX, Laj C, Mazaud A (1994) The Matuyama-Brunhes and Upper Jaramillo transitions recorded in a loess section at Weinan, north-central China. Earth Planetary Sci Lett 125:143–158

Zhu RX, Pan YX, Liu QS (1999) Geomagnetic excursions recorded in Chinese loess in the last 70,000 years. Geophys Res Lett 26:505–508

Zhu RX, Ding ZL, Wu HN, Huang BC, Jiang L (1993) Details of magnetic polarity transitions recorded in Chinese loess. J Geomagnetism Geoelectricity 45: 289–299

Zhu RX, Bin G, Pan YX, Liu QS, Zeman A, Suchy V (2000) Reliability of geomagnetic secular variations recorded in a loess section at Lingtai, north-central China. Sci China D 43:1–9

Zijderveld JDA (1967) AC demagnetization of rocks: analysis of results. In: Collinson DW, Creer KM, Runcorn SK (eds) Methods in palaeomagnetism. Elsevier, New York, NY, pp 254–286

Magnetic Fabric of the Brazilian Dike Swarms: A Review

17

M. Irene B. Raposo

Abstract

In Brazil, there are many Precambrian and Phanerozoic mafic dike swarms of variable length, chemistry and structural trend. Phanerozoic dike swarms are more abundant and widely distributed than those of Precambrian ages. The longest dikes and the densest swarms are Mesozoic, e.g. the Ponta Grossa swarm in Paraná State. It is well accepted that magnetic anisotropies yield the most efficient methods to determine petrofabric orientation in rocks, particularly where standard petrofabric techniques are inadequate or inefficient, and even in rocks that are visually isotropic. Magnetic fabrics, also called magnetic anisotropies, can be determined using either low-field anisotropy of magnetic susceptibility (AMS), which is the most popular method, or anisotropy of magnetic remanence (AMR). These are powerful tools for structural geology and have been applied in many geological situations. Magnetic fabrics were determined in many dike swarms of Precambriam and Mesozoic ages together with extensive rock magnetic studies. The main magnetic fabric for these swarms is related to magma flow, and the relative position between magma sources and emplacement fractures could be inferred. In some swarms AMS and AMR tensors are coaxial, whereas in others the AMS fabric is primary in origin but the AMR fabric is tectonic and acquired after dike emplacement.

17.1 Introduction

One of the most important features of dikes is that they represent former conduits for the passage of magma from deeper levels of the Earth to the surface. Mafic dike swarms are important tectonic features of the continental crust since they give information about the tectonic processes that deformed the lithosphere. The investigation of dike emplacement processes helps us in understanding how continental swarms are developed. Traditional methods to study the emplacement of dikes are mainly based on petrographic analyses (e.g. Greenough et al. 1988), oriented vesicles (e.g. Coward 1980), and fingers, grooves and lineations (e.g. Baer and Reches 1987). However, these methods are not useful when the petrographic fabric is poorly defined or absent. Even when the petrofabric is strongly developed, observations made on oriented thin sections under microscope are tedious, and methods based on field observations of flow structures must be used

M.I.B. Raposo (✉)
Institute of Geosciences, São Paulo University, 05508-080, São Paulo, SP, Brazil
e-mail: irene@usp.br

E. Petrovský et al. (eds.), *The Earth's Magnetic Interior*, IAGA Special Sopron Book Series 1,
DOI 10.1007/978-94-007-0323-0_17, © Springer Science+Business Media B.V. 2011

with care since the distinction between lineations and grooves is not always evident. These indicators are generally absent. One of the first applications of magnetic techniques (anisotropy of magnetic susceptibility, AMS) was an attempt to determine flow directions in mafic dikes (e.g. Khan 1962). The use of AMS as a proxy indicator of flow direction became widespread after the important work performed on Hawaiian dikes by Knight and Walker (1988), on which they described the empirical relationship between outcrop flow indicators and AMS. It is nowadays well accepted that magnetic anisotropies yield the most efficient methods to determine the petrofabric orientation in rocks even when these are visually isotropic.

In Brazil, the most extensive and widespread dike swarms are those of the Mesozoic age (Fig. 17.1, Sial et al. 1987) that were emplaced during the opening of the South Atlantic Ocean. Some of them are coast-parallel (Fig. 17.1) and might represent part of the main ocean-forming rift (Almeida 1986). Others are inland and represent more limited extension along zones of crustal arching (e.g. Ponta Grossa Arch, Fig. 17.1). Proterozoic dike swarms occur mainly within the Amazon and São Francisco Cratons (Fig. 17.1) where dikes show variable structural trends. Magnetic fabric has been determined for both Mesozoic and Proterozoic Brazilian swarms, and the dike emplacement mechanisms were investigated. In this paper we show how magnetic methods may help to investigate magma flow, to provide information on the mode of emplacement of the dikes, and to determine the relative position of magma sources and fractures. Secondary magnetic fabrics which overprint the primary ones will be also discussed. Rock-magnetic studies for each swarm will not be shown in this paper then the reader is invited to go to the original papers.

17.2 Magnetic Fabric

Magnetic fabrics, also called magnetic anisotropies, can be determined using either low-field anisotropy

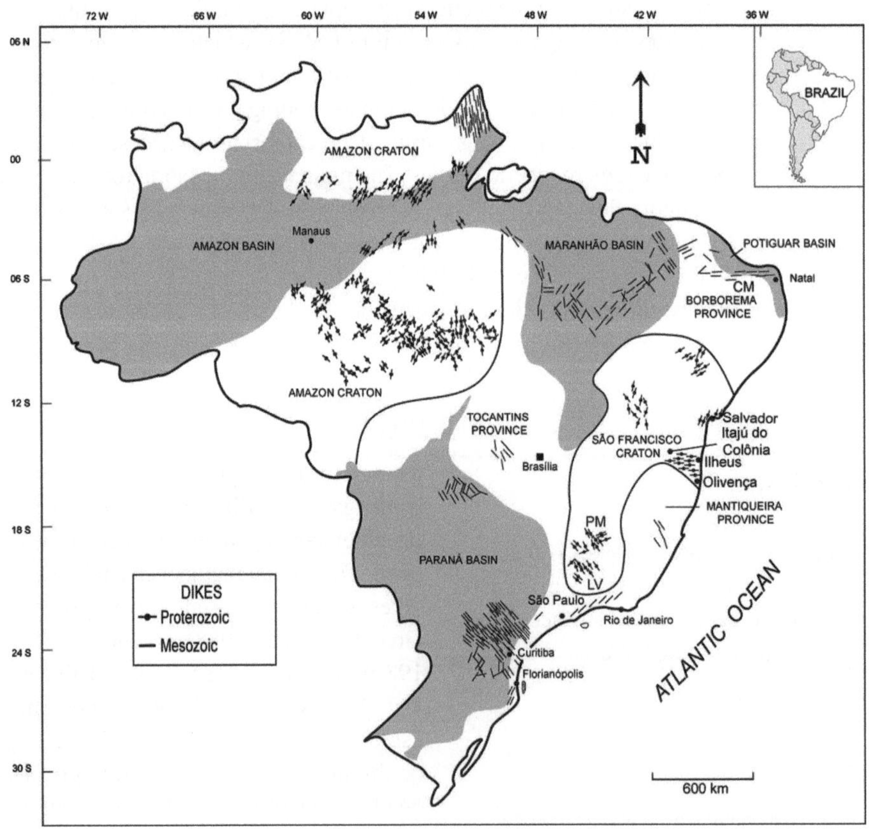

Fig. 17.1 Simplified map of the main Brazilian mafic dike swarms (Modified from Sial et al. 1987)

of magnetic susceptibility (AMS), anisotropy of magnetic remanence (AMR), or high field anisotropy. AMS is the most popular because it can be quickly obtained, and therefore it is widely applied in many geological situations than the other options. Magnetic fabric is a symmetrical second-rank tensor whose square matrix is given either by the magnetic susceptibility (Km) for AMS fabric, or by the remanent magnetic susceptibility for AMR anisotropy (Jackson 1991). This tensor is expressed by its principal eigenvectors (their orientations) and eigenvalues (their magnitudes) $K_{max} > K_{int} > K_{min}$ (for AMS), or $AMR_{max} > AMR_{int} > AMR_{min}$ (for AMR) representing the maximum, intermediate, and minimum axes of susceptibility, respectively. The maximum axis (K_{max}) represents the magnetic lineation and the minimum axis (K_{min}) the pole of the magnetic foliation, which is the plane formed by maximum and intermediate axes ($K_{max} - K_{int}$). Magnetic fabrics are usually determined using oriented cylindrical 2.5 × 2.2 cm specimens.

A normal fabric is expected when the magnetic fabric and the petrofabric have the same shape and orientation. In this sense, the minimum axis is perpendicular to bedding, flow or flattening plane whereas the maximum axis is parallel to current direction, flow direction or mineral stretching (Rochette 1988). In such a case, the magnetic axes correspond one by one to the petrofabrics axes.

17.2.1 Anisotropy of Low-Field Magnetic Susceptibility (AMS)

AMS describes the variation of magnetic susceptibility with direction within a sample, and represents the contribution of all the rock-forming minerals whether these are dia-, para- or ferromagnetic. In rocks in which AMS is carried by either paramagnetic Fe-bearing silicate matrix minerals or (titano) hematite or pyrrhotite, the AMS is due to the preferred crystallographic orientation of these minerals (magnetocrystalline anisotropy, Tarling and Hrouda 1993). On the contrary, in rocks in which AMS is carried by ferrimagnetic minerals, such as titanomagnetite or magnetite, the origin of the AMS is related to the grain shape (shape-anisotropy), and to the magnetic grain interaction (Cañon-Tapia 1996), or their distribution (distribution anisotropy; Hargraves et al. 1991) within the rock.

In dikes three types of AMS fabric have been defined according to the eingenvector orientations with respect to the dike plane. These fabric types were defined for basaltic dikes from Oman ophiolite by Rochette et al. (1991), and are called as "normal", "intermediate" and "reverse". The latter was termed "inverse" by Rochette et al. (1992), therefore we will use the term *inverse*. These fabric types have been found in many swarms worldwide (Knight and Walker 1988, Raposo and Ernesto 1995, Tauxe et al. 1998, Herrero-Bervera et al. 2002, Borradaile and Gauthier 2003 among many others). Figure 17.2 shows an example of these fabrics for dikes from Brazilian swarms.

In the *normal* fabric the AMS foliation ($K_{max} - K_{int}$ plane) is parallel to the dike plane and the magnetic foliation pole (K_{min}) is nearly perpendicular to it. This fabric has been usually interpreted as a flow fabric, and the K_{max} inclination has been used to infer the relative position between magma source and fractures. This assumption is based on previous research performed in dikes elsewhere which show a good concordance between K_{max} and flow indicators obtained from field evidence (Knight and Walker 1988; Ernst and Baragar 1992; Tauxe et al. 1998; Varga et al. 1998; Callot et al. 2001; Raposo et al. 2004 among many others). Therefore *normal* AMS fabric for Brazilian dikes is similarly interpreted as magma flow.

The *intermediate* fabric type is defined by K_{max} and K_{min} axes clustering close to the dike plane while K_{int} axes are nearly perpendicular to this plane. This fabric can be either tectonic or primary in origin. If primary its interpretation has been considered as due to the vertical compaction of a static magma column with the minimum stress along the dike strike (Park et al. 1988). Such interpretation is supported by data from Emerman and Marret (1990) and Kratinová et al. (2006). Therefore primary *intermediate* AMS fabric for Brazilian dikes has the same interpretation.

The *inverse* fabric type displays K_{int} and K_{min} axes forming a plane parallel to the dike, with K_{max} perpendicular to dike wall. This fabric is usually attributed to a single-domain (SD) effect if small, prolate (titano)magnetite grains carry the AMS (Stephenson et al. 1986). Such grains have maximum susceptibility axis (K_{max}) perpendicular to the long axis of the grains whereas the minimum susceptibility axis (K_{min}) is parallel to it. This occurs because elongated SD grains have a zero susceptibility along their long axis and maximum susceptibility along to their

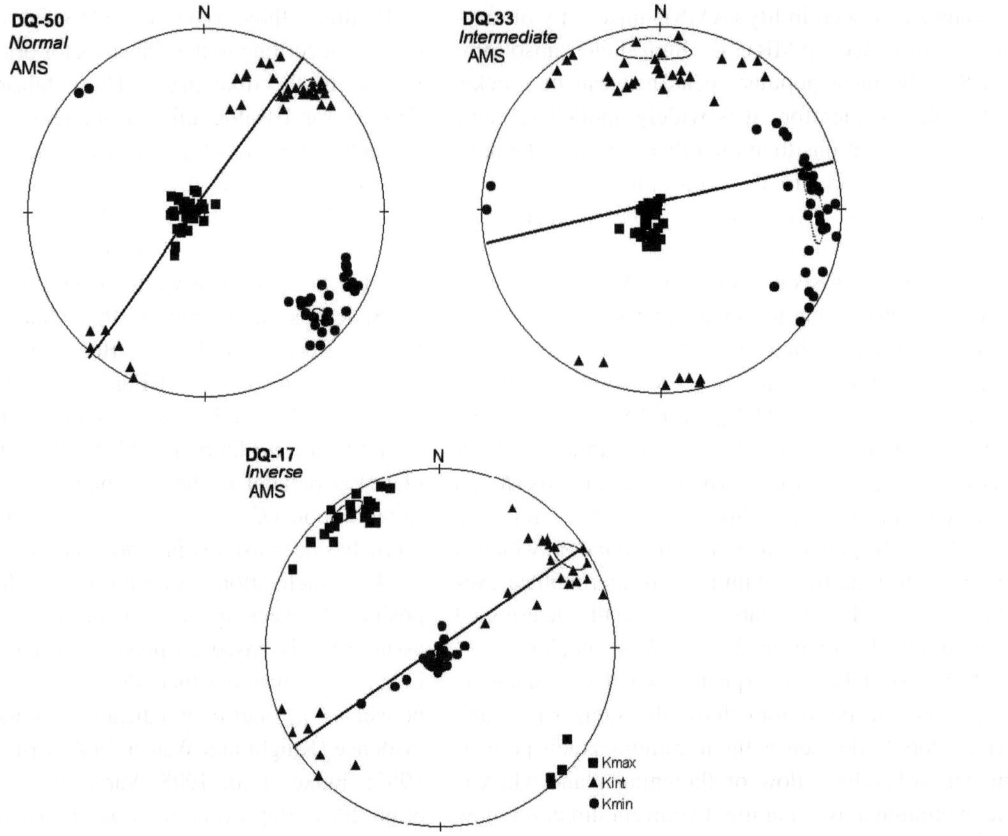

Fig. 17.2 Lower hemisphere, equal area projection showing some examples of AMS fabrics usually found in mafic dike swarms. Dashed line ellipses are 95% confidence ellipses, calculated using the bootstrap method of Constable and Tauxe (1990). The full line represents the dike plane with vertical dip. These sites are from São Sebastião e IlhaBela (coastline of of São Paulo State) swarms

short axis (Stephenson et al. 1986). This is referred to as an "inverse" magnetic anisotropy. In contrast, multi-domain (MD) prolate grains of such minerals have K_{max} axis parallel to their long axis and K_{min} is normal to it. To know if SD (titano)magnetite grains are present a detailed rock magnetism study must be performed in all rocks in which one wants to determine the magnetic fabric (for example; granites and sediments). The inverse fabrics in Brazilian dikes are not related to the SD effect as will be shown in the following sections.

17.2.2 Anisotropy of Magnetic Remanence Magnetization (AMR)

AMR isolates the magnetic fabric arising from the contributions of remanence-bearing grains that usually occur mostly as iron oxides (Jackson 1991). Although these oxides are low-abundance accessory minerals, their "subfabric" may represent a different portion of the crystallization or strain history from that of the main rock-forming minerals that contribute only to AMS (Borradaile 2001). Determination of AMR requires the application and measurement of an artificial permanent magnetization along different directions through the sample, and is exclusively due to the alignment of the remanence carriers. Samples should be demagnetized by an alternating field after each measurement and before remagnetizing in a new orientation, except for anisotropies of thermo-remanent and viscous magnetizations. Application of the AMR technique requires specific instruments, and a predetermined scheme of the sequential positions on which the magnetization is induced. The time involved in the experiment is usually much longer than the one required for AMS determination, and a computer program must be used to process the data. The AMR can be total or partial depending on the remanent coercivity spectra of the samples. In the past the AMR was generally used as an auxiliary method for detailed magnetic studies as, for example, to verify whether there is a single domain (SD) effect in an anomalous AMS fabric because the SD effect will never occur in AMR since the particle magnetization (i.e. remanence) always orients itself parallel to the long axis of such a particle (Jackson 1991). Studies of pseudo-single domain (PSD) (titano)magnetite grains in Brazilian dikes have shown that their fabric is similar to that found in MD (titano)magnetite-bearing dikes (Raposo and D'Agrella-Filho 2000; Raposo et al. 2004, 2007; Raposo and Berquó 2008). However, nowadays the interest in determining AMR tensor is increasing because, in some cases, both AMS and AMR tensors might have been acquired at different times or by different processes as it will be shown later.

Among the artificial remanences that can be used to calculate the AMR tensors, the anhysteretic remanence is the most used, and it is preferable since it is acquired under a weak field. These guarantees that the magnetization is linearly related to the induced field and the ARM tensors are of second-rank. However, the isothermal remanent magnetization (IRM) may also be successfully used if a careful choice of applied fields is made, such as in weak fields corresponding to Rayleigh's law (Daly and Zinsser 1973). For a strong field, the remanence is a nonlinear function of the applied field, and the remanence and the applied field are therefore not related by a second-rank tensor (Jackson 1991) making correlation between these tensors and the orientation of the petrofabric rather difficult in rocks.

In Brazil, the measurements of AMR started with the installation of instruments during mid 1997 at the Magnetic Anisotropies and Rock Magnetism Laboratory of the Geosciences Institute – São Paulo University. The first mafic dikes studied corresponded to Ilhéus-Olivença swarms (Raposo and D'Agrella-Filho 1998; 2000).

For determining the IRM anisotropy (AIRM) we use a pulse magnetometer (Magnetic Measurements). The procedure involves applying a magnetic field for a short period of time, the measurement of the IRM, and subsequent demagnetization to destroy the imposed magnetization before the next IRM acquisition, with repetition along different positions for each specimen. Before IRM determinations, samples should be demagnetized by AF tumbling at 100 mT to establish the base level.

For the anisotropy of anhysteretic remanence magnetization (AARM) determinations we use a Molspin alternating field demagnetizer as the source of the alternating magnetic field (AF field). Superimposition of a steady field (DC field) is attained by a small coil (home-made) inside and coaxial to the demagnetizer, which is controlled by a Molspin apparatus. With this apparatus we are also able to determine both total and partial (pAARM) AARM. The computer program (also home-made) that we use to calculate the AARM and AIRM (mainly for AARM) tensors, allows magnetization to be imposed in any position; all that is necessary to know are the coordinates of the position in which the magnetization is induced. In our Laboratory we have position schemes which permit induction of magnetization in more than 22 positions.

It is worth noting that in AARM studies it is usually expected that samples are affected by gyroremanent magnetization (GRM), which is a magnetization acquired perpendicular to the AF field at each remanence acquisition step if small grains of titanomagnetite, magnetite or maghemite are present (Stephenson 1993). To verify the GRM influence the GRM tensor should be determined experimentally for the same specimens using Stephenson's method (Stephenson 1993), and the AARM and pAARM tensors should be corrected.

17.3 Mesozoic Swarms in Brazil

Even though Mesozoic dikes swarms are widespread in Brazil (Fig. 17.1) here we report only those whose magnetic fabric was determined. These swarms are located in Southern and Northeastern Brazil. The Mesozoic magmatism in Southern Brazil (Fig. 17.3) is represented mainly by the basaltic flows of the Serra Geral Formation (Paraná Basin), the dikes swarms from the Ponta Grossa Arch, the Florianópolis region, dikes along the coast between São Paulo and Rio de Janeiro, and by several alkaline complexes that lie along tectonic features associated with the evolution of the Paraná Basin. The emplacement of the dike swarms

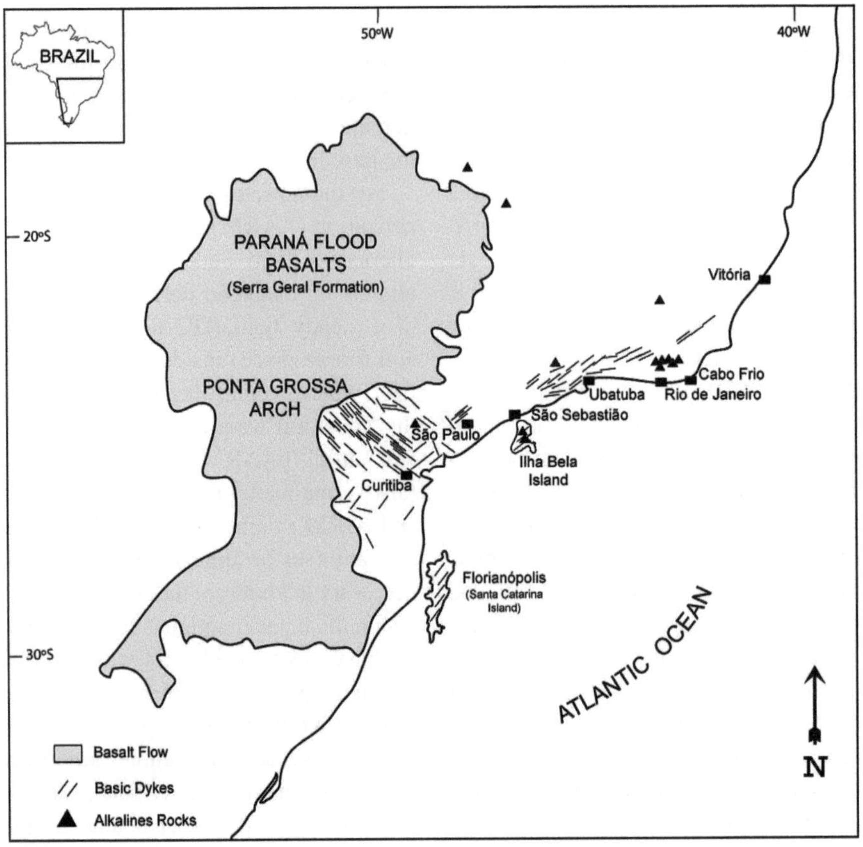

Fig. 17.3 Simplified map (out of scale) showing Paraná Basin and the Mesozoic dike swarms in the Southeastern of Brazil. Modified from Almeida, (1986)

and the alkaline complexes was related to the processes responsible for the opening of Atlantic Ocean (Almeida 1986).

17.3.1 Ponta Grossa Dike Swarm – (PG dikes)

The Ponta Grossa Arch is a large (~134,000 km^2) tectonic feature on the eastern border of the Paleozoic-Mesozoic Paraná Basin (Figs. 17.1 and 17.3). This arch was most active during the Jurassic and Lower Cretaceous, when its structures were enhanced and the arch took its present configuration (Almeida 1986). Associated with this structure is one of the largest Phanerozoic mafic dike swarms in Brazil: the Ponta Grossa (PG) dike swarm. The first AMS studies in Brazil were performed on PG dikes, and their preliminary results were presented at the Brazilian Geological Congress in 1990 (Raposo and Ernesto 1990). Dikes intrude both Paleozoic sedimentary rocks of the Paraná Basin (in the west) and Precambrian rocks of the crystalline basement (in the east). Dike thicknesses vary from tens to hundreds of meters, and they are often tens of kilometers in length. They trend mainly NW-ward with subvertical dips; however, NE-trending dikes are also found mainly in the eastern part of the arch. The PG dikes as well as the Mesozoic volcanic rocks of Paraná Basin – or Paraná Province (Fig. 17.3) – are related to the breakup of western Gondwanaland, having been emplaced during the initial stages of the South Atlantic rifting (Almeida 1986). Dikes are mainly tholeiitic basalts and more rarely rhyolite (Piccirillo et al. 1990). They are younger than the flood volcanics of the Paraná Basin whose $^{40}Ar/^{39}Ar$ ages fall in the range 131.4 ± 0.4 to 129.2 ± 0.4 Ma, with a peak at 130.5 Ma (Renne et al. 1996).

AMS fabric in the PG dikes (total of 95 dikes) was determined using the Minisep instrument (Molspin Ltda). Results from this study reveal that 51% have *normal* AMS fabric; 38% have *intermediate* AMS fabric; 4% have *inverse* AMS fabric; and in 7% the data are too scattered for any axes to be defined (Raposo and Ernesto 1995). *Normal* AMS fabric was interpreted as due to magma flow (as addressed in the previous section) and, based on the K_{max} inclination, it was inferred that the majority of dikes (58%) were fed by horizontal or sub-horizontal ($K_{max} < 30°$) magma flow whereas 42% were fed by inclined to vertical flow. K_{max} inclinations are steepest in the Curitiba (Figs. 17.1 and 17.3) region, from which geochemical data show that dikes in this area are the most primitive (Piccirillo et al. 1990). This suggests that this area could have been closer to a magma source. However, it is not plausible to predict only one magma source acting inside the Ponta Grossa arch because contemporaneous dikes were fed by compositionally distinct magmas (Raposo 1992). Therefore other magmas sources were inferred (for more details see Raposo and Ernesto 1995). *Intermediate* AMS fabric was interpreted as the result of crystallization from magma under stress. *Inverse* AMS fabric is not related to a SD effect since the rock magnetism (Raposo 1992) showed that the carrier of AMS is MD titanomagnetite. Therefore the origin of this fabric is probably linked to some local tectonic event.

17.3.2 Florianópolis Dike Swarm – (FL Dikes)

The FL dikes are well exposed in and around Florianópolis city (Fig. 17.3). They cut the Proterozoic crystalline basement rocks, and their thicknesses vary from 0.1 to 70 m although values between 0.5 and 10 m are most frequent. In contrast to PG dikes, the majority of sampled dikes (54 dikes) trend N5 – 80E which is coincident with the main structures of the crystalline basement, with the direction of swarms along the Brazilian coast (Fig. 17.3), and with the eastern members of the PG dikes (Fig. 17.3). A few dikes (eight dikes) trend N10–80 W. Locally the NW-trending system cuts NE-trending dikes, indicating that the former is younger than the main system. All dikes are vertical or subvertical in dip. Geochemical data indicate that the FL dikes are similar in composition to the Paraná flood basalts (Serra Geral Formation) and to the PG dikes (Piccirillo et al. 1990). $^{40}Ar/^{39}Ar$ ages of the FL dikes concentrate in two intervals ~119–122 Ma and ~126–128 Ma with an average of 124 ± 4 Ma (Raposo et al. 1998). These ages are consistent with the intrusion of the dikes during the development of the East Brazilian continental margin, a passive margin formed as a consequence of crustal breakup on the late Aptian (Chang et al. 1992).

AMS fabric in the FL dikes (total of 62 dikes) was also determined using the Minisep instrument. Around 94% of dikes show *normal* fabric (56 dikes) whereas 4 dikes display *intermediate* fabric. No inverse fabric was found (Raposo 1997). *Normal* fabric was interpreted as magma flow and the majority of the dikes were fed by inclined to vertical flows (K_{max} inclination is greater than 30°). Only 12 dikes in which K_{max} inclination does not exceed 30° were fed by horizontal flow. This pattern indicates that Santa Catarina Island (Florianóplois) overlies the magma source. Since PG and FL dikes have similar geochemical data (Piccirillo et al. 1990), and as the PG dikes in the Curitiba area also have vertical flows, the same geochemical source can be inferred for both swarms. However, this inferred magma chamber must have been active at different times or another similar chamber was involved. During this time the South American plate had moved, as indicated by the ages of the dikes, and by the paleomagnetic poles from both swarms (Raposo et al. 1998).

17.3.3 Dikes from Coastline of São Paulo and Rio de Janeiro States

Dike swarms are widespread along the Serra do Mar between São Paulo and Rio de Janeiro in Southern Brazil (Fig. 17.3). They crosscut polymetamorphosed Archean and Proterozoic rocks of the Costeiro Complex (Almeida 1986). The dikes are mainly basalts and lamprophyres, and they crop out side by side in the beaches. It is believed that the basaltic activity occurred during the Early Cretaceous and was then partly coeval with PG and FL dikes (e.g. Almeida 1986); the lamprophyric dikes are, however, younger than the basalts (Almeida 1986). Width of dikes ranges from a few centimeters up to 2 m for the lamprophyres, and up to >10 m for the basalts. Their trend is predominately N40–50E (Fig. 17.3) with vertical dips. In São Paulo state, swarms are concentrated mainly in São Sebastião, Ubatuba, and IlhaBela (Fig. 17.3). These swarms are presently being studied by magnetic methods. Preliminary results for the Ubatuba swarms (Raposo et al. 2009) show that *normal* AMS fabric is dominant on both basalt and lamprophyres dikes. The analysis of the K_{max} inclination permitted to infer that all lamprophyre dikes were fed by horizontal flow ($K_{max} < 30°$), suggesting that they were locate far from the magma source. On the other hand, basic dikes were fed by both horizontal and inclined ($K_{max} > 30°$) flows, suggesting that some of them were far and the others were close to the magma source. This indicates movement of the South American plate at the time of intrusions, or that more than one magma source was involved. An interesting point to be observed in these dikes is that the rock-magnetic properties show that both swarms have very similar magnetic properties, and the magnetic mineral present in the dikes is pseudo-single-domain (2–5 μm magnetite grains) independent of dike compositions. However, six (out of nine) lamprophyric dikes show an unusual magnetic behavior mainly found in low *KxT* curves. The magnetic susceptibility of these dikes is dependent on the field intensity, whereas in the other three lamprophyres and in the basic dikes *K* is field-independent. The *K* variation with field intensity suggests that titanomagnetite could be present in the six lamprophyre dikes. Magnetic measurements at liquid helium temperature (transition around 38 K) and Mössbauer spectroscopy data for these dikes suggest the presence of the iron carbonate siderite. There is no difference in the magnetic fabrics between anomalous ($n = 6$) and non-anomalous ($n = 3$) lamprophyre dikes (Raposo et al. 2009).

17.3.4 Ceará Mirim Dike Swarm – (CM Dikes)

The CM is a 350-km-long, Early Cretaceous (145–125 Ma) tholeiitic dike swarm (Fig. 17.1) emplaced during the early stages of rifting and opening of the equatorial Atlantic Ocean (Archanjo et al. 2002, and reference therein). CM is the only swarm from northeastern Brazil from which the magnetic fabric was determined in addition to image analyses of rock thin sections (Archanjo et al. 2002). The CM dikes intruded the Precambrian basement of the Borborema Province (including mylonites formed along NNE-trending shear zones). The dikes trend E-W near the southern border of the Potiguar basin, but in the western part of the province these dikes become nearly parallel to a NE-SW trending Eo-creataceous rift system in the Potiguar basin (Fig. 17.1). Individual dikes crop out as vertical bodies <1 km long, and from 1.2 m up to 150 m wide. Dikes 20–100 m wide are relatively common. A few sets form an *en echelon* pattern suggesting that the emplacement occurred at a shallow crustal level (Archanjo et al. 2002).

AMS fabric was determined in 50 dikes from CM swarm (Archanjo et al. 2002). Rock-magnetic properties and petrofabric studies reveal that the shape-preferred orientation of the opaque grains (PSD-MD magnetite with low Ti content) controls the magnetic fabric. The AMS fabrics of 58% of the dikes are abnormal (*intermediate* and *inverse*); the remaining 42% of the dikes show a *normal* fabric regionally characterized by steep magnetic foliations and subhorizontal lineations (Archanjo et al. 2002). However, in the central eastern part of the swarm lineations plunge subvertica/vertical, suggesting a magmatic feeder zone. The shape alignment of plagioclase prisms supports the inferred regional flow pattern, whereas petrofabric studies indicate that the abnormal magnetic fabric is unrelated to flow (Archanjo et al. 2002). Abnormal AMS fabric is attributed to crystallization of magnetic oxides in residual magma volumes accommodated into spaces opened normal to the stretching directions. A high-temperature oxidation stage during cooling of the dikes recrystallized nearly pure magnetite which could locally grow following the regional stress field (Archanjo et al. 2002).

17.4 Precambrian Swarms in Brazil

There are many Proterozoic dike swarms in Brazil (Fig. 17.1), yet we focus on those cropping out in eastern Brazil (in the São Francisco Craton; SFC, LV, PM, Ilhéus, and Salvador, Fig. 17.1), for whose magnetic fabrics were determined. The SFC is situated in the central-eastern part of South America, and it is perhaps the most important of the four Brazilian cratons. These cratons (see Teixeira et al. 2000, for review and reference therein) are underlain by Archean to Paleoproterozoic continental crust (granite-greenstone belts; high-grade terranes, and Early and Middle Proterozoic supracrustal sequences). The SFC is the westerly extension of the Congo Craton (W-Africa), according to the pre-Atlantic continental fits, and both behaved as a single plate in Western Gondwana (Teixeira et al. 2000). It covers almost all Bahia State and a large part of Minas Gerais State. The SFC borders are surrounded by fold belts formed during the Brasiliano/Pan-African orogeny (~680–550 Ma; Teixeira et al. 2000). The studied areas for magnetic fabric are part of the north-eastern (Ilhéus and Salvador dikes), and southern (Para de Minas (PM) and Lavras (LV)) parts of the SFC which were formed by the accretion of Archean and Paleoproterozoic terranes mainly during the Transamazonian collisions (2.24–1.94 Ga; Teixeira et al. 2000). Several important mafic dike swarms with different ages and compositions occur in the SFC. These swarms are related to Early Proterozoic tensional episodes that affected the entire SFC (Teixeira et al. 2000).

17.4.1 Ilhéus-Olivença Dikes

There are three dikes swarms in the Ilhéus region (Fig. 17.1). In Brazil these swarms were the first from which we determined the AMS with a Kappabridge instrument (Agico, Czech Republic), and applied the AMR techniques (Raposo and D'Agrella-Filho 2000).

Magnetic fabrics were measured on 81 dikes from Ilhéus, Olivença and Itaju do Colônia swarms located in SE Bahia state. The dikes intrude high-grade metamorphic terrains as old as 3.2 Ga (Teixeira et al. 2000). The dikes are tholeiitic, having vertical to subvertical dip-angles trending mainly E-W-ward. Their thicknesses vary from few centimeters to about 20 m (average of 3 meters). $^{40}Ar/^{39}Ar$ data (Renne et al. 1990) indicate ages of 1.012 (Ilhéus dikes) and 1.078 Ga (Olivença dikes). Rock magnetic properties shows that Ti-poor titanomagnetite to pure magnetite with PSD/MD grain sizes carry the magnetic fabrics (Raposo and D'Agrella-Filho 2000). *Normal* AMS fabric is the most common fabric type within the studied swarms (64 dikes). *Intermediate* fabric (7 dikes) is interpreted in the same matter as for Mesozoic Brazilian swarms. *Inverse* AMS fabric was found in 9 dikes. AARM measurements were performed on three dikes; two of them with *inverse* and one with *normal* AMS fabrics. AARM was found to be coaxial for *normal* AMS fabric and it resulted in *intermediate* and *normal* for *inverse* AMS fabrics. A combination of AMS and AARM fabrics suggests that the flow-induced magmatic fabric for these dikes was overprinted after dike emplacement by some local event, probably related to processes associated to the Brasiliano Orogeny.

Normal AMS fabric is interpreted as due to magma flow (as stresses in the previous section). The analysis of K_{max} inclination allows to infer that the dikes from Ilhéus were fed dominantly by horizontal fluxes ($K_{max} < 30°$) whereas those from Olivença and Itaju do Colônia were mainly by inclined to vertical fluxes ($K_{max} > 60°$, (Raposo and D'Agrella-Filho 2000). Dikes from Itaju do Colônia region have the steepest K_{max} suggesting that this region could be closer to a magma source. On the other hand, a magma source near the Olivença area could be also inferred since the majority of the dikes were fed by inclined flow. The existence of more than one source is supported by geochemical (Bellieni et al. 1991), geochronological (Renne et al. 1990) and paleomagnetic (D'Agrella-Filho et al. 1990) data which show that contemporaneous dikes were fed by compositionally distinct magmas. AMS data also suggest that the majority of the dikes were fed by upward flow coming from west to east since K_{max} is steepest in the west of the studied area (near Itaju do Colônia, Fig. 17.1), and only few of them were fed from upward flow from east to west (for more detail see Raposo and D'Agrella-Filho 2000).

17.4.2 Pará de Minas Dikes – (PM Dikes)

This swarm is located to the southern part of the SFC (Figs. 17.1 and 17.4). These dikes intrude granitoids and poly-phase, high-grade metamorphic terrains and greenstone belts, as well banded iron formations (BIFs) and other mafic dike swarms (Teixeira et al. 2000). The dikes strike dominantly N50–60 W with vertical to subvertical dip-angles. Their thicknesses are variable (10–100 m) although values around 50 m are more frequent. They are mainly tholeiitic basalts, andesi- basalts and lati-basalts in composition (Chaves 2001). Some dikes are porphyritic with large plagioclase phenocrysts (>5 cm) oriented along the same trend of the dike. Rb-Sr and U-Pb data indicate intrusion ages in the range of 1.740–1.714 Ga (Chaves 2001). Structural observations at a few outcrops, such as bayonet asymmetric (bayonet) branching or the presence of small apophyses near the dike contact with the country rocks, suggest that the dikes were fed from NW to SE.

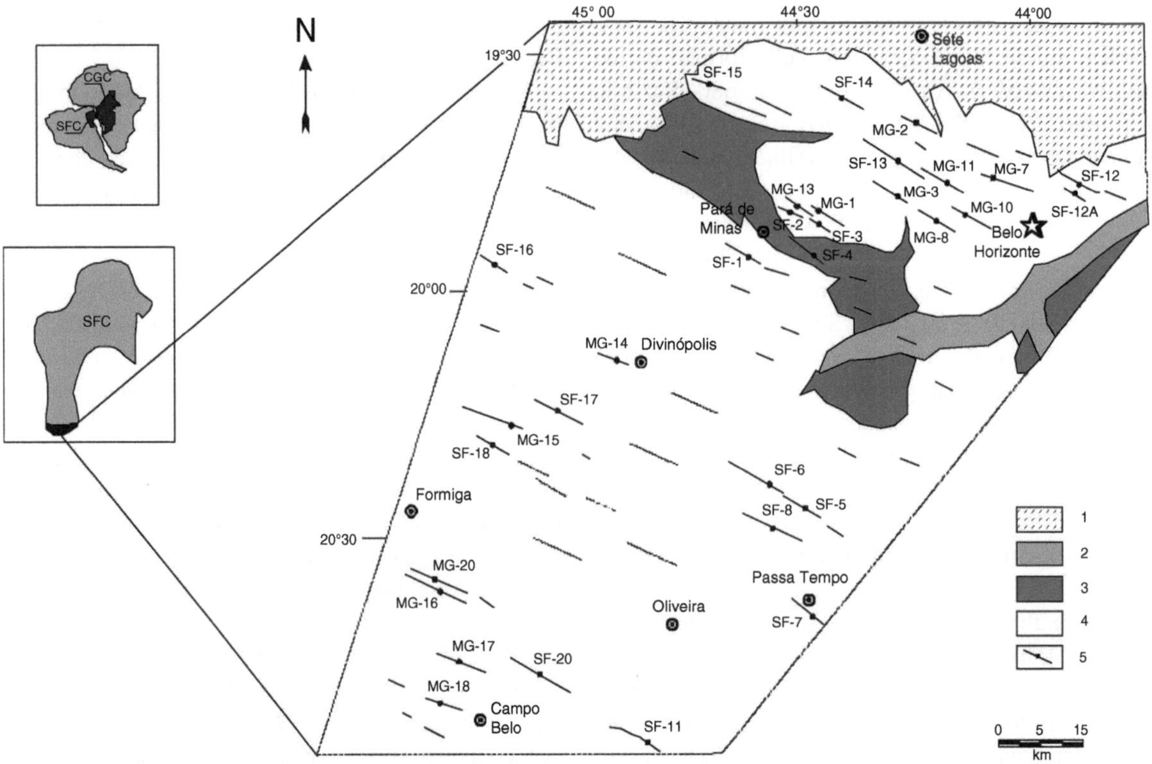

Fig. 17.4 Simplified geological map of southern portion of São Francisco Cratons showing the Pará de Minas Dikes. 1- Bambui Group; 2- BIFs (Minas Supergroup); 3- Greenstone belt (Rio das Velhas Supergroup); 4-granitoids and high-grade metamorphic terrains (Rio das Velhas Supergroup); 5- sampled dikes. CGC and SFC are Congo and São Francisco Cratons, respectively. After Raposo et al. (2004)

Magnetic fabrics were determined by AMS and AMR on 32 dikes from this swarm (Raposo et al. 2004). The latter was performed imposing both anhysteretic (total (AARM) and partial (pAARM) and isothermal remanence magnetizations (AIRM). At most sites AMS is dominantly carried by ferromagnetic minerals, yet at some sites the paramagnetic contribution exceeds 70% of bulk susceptibility. Rock magnetism and thin section analysis allows dikes to be classified as non-hydrothermalized and hydrothermalized (Raposo et al. 2004). pAARM was measured for hydrothermalized dikes since the remanent coercivity spectra from a pilot specimen of each site reveal Ti-poor titanomagnetite with three different coercivities and consequently different grain sizes suggesting that magnetic phases were formed during the hydrothermal alteration. According to the spectra, pAARM was determined from 0 to 30 mT ($pAARM_{0-30}$), from 30 to 60 mT ($pAARM_{30-60}$) and from 60 to 90 mT ($pAARM_{60-90}$) windows. Magnetic measurements show that the mean magnetic susceptibility is usually lower than 5×10^{-3} (SI), which is not common for mafic dikes. Ti-poor titanomagnetite to pure magnetite with pseudo-single-domain grain sizes carry the majority of magnetic fabrics for the non-hydrothermalized dikes, whereas coarse to fine-grained Ti-poor titanomagnetite carries the majority of magnetic fabrics in hydrothermalized dikes (see Raposo et al. 2004 for details).

Three primary AMS fabrics are recognized which are coaxial with ARM fabric, except for two dikes, from both non-hydrothermalized and hydrothermalized dikes. *Normal* AMS fabric surprisingly is not dominant (31%). This fabric is interpreted as magma flow in which the analysis of K_{max} inclination permitted the inference that the dikes were fed by horizontal or subhorizontal fluxes ($K_{max} < 30$ °). The similarity between *normal* AMS, $pAARM_{0-30}$, $pAARM_{30-60}$ and $pAARM_{60-90}$ fabrics from hydrothermalized dikes indicates that magnetic grains formed due to late-stage crystallization or to remobilization of iron oxides during hydrothermal alteration after dike emplacement acquired a mimetic fabric coaxial to the primary fabric given by coarse-grained, early crystallized Ti-poor titanomagnetite (given by AMS). *Intermediate* AMS fabric is the most important (41%) in the investigated swarm. It is interpreted as due to vertical compaction of a static magma column with the minimum stress along the dike strike (Raposo et al. 2004). AMR determinations for these sites also are *intermediate* except for two dikes. In one of them, AIRM fabric resulted in *normal* AMS fabric whereas for the other the AARM fabric resulted in an *inverse* AMS fabric. A combination of AMS and AMR fabrics suggest that magmatic fabric for both dikes were overprinted by some late local event, probably related to Brasiliano orogenic processes after dike emplacement. *Inverse* AMS fabric is a minority (4 dikes). AMR determinations also are *inverse* suggesting a primary origin for *inverse* AMS fabric.

17.4.3 Lavras Dikes (LV Dikes)

Magnetic fabric measurements and rock magnetic studies were carried (Raposo et al. 2007) on three mafic dike swarms (total of 38 dikes). They are located in Minas Gerais State, SE Brazil, and they represent the southernmost swarms in the São Francisco Craton (Fig. 17.5). Dikes cut Archaean granite-gneiss-migmatite and Paleoprototerozoic terrains. Petrographic, geochemical and isotopic studies classify the swarms as basic-noritic, basic and metamorphic suites (Pinese et al. 1995, 1997). The basic-noritic swarm is the oldest with a model Sm-Nd age ~2.65 Ga, whereas the basic and metamorphic suites have a Rb-Sr age of ~1.87 Ga. Dikes from the metamorphic suite still preserve igneous characteristic, which implies that they were metamorphosed in low grade and practically undeformed. Chemical similarities (major, minor, trace elements-REE) and Rb-Sr isotope data suggest that this suite may have had a similar mantle source and magmatic evolution to those of the basic suite (Pinese et al. 1995, 1997). On the other hand, the geochemical and isotopic data (Sr and Nd) suggest that the basic-noritic and basic/metamorphic suites were derived from two different sources and had different magmatic evolutions (Pinese et al. 1995, 1997).

Magnetic fabrics were determined by applying both AMS and AARM techniques. Three primary AMS fabrics are recognized which are all coaxial with to the AARM fabric. *Normal* AMS fabric is dominant in the basic suite (16 out of 20 analyzed dikes) and occurs in 4 and 3 dikes from the basic-noritic and metamorphic suites, respectively. Rock magnetic studies reveal that coarse and fine magnetite grains or Ti-poor titanomagnetite are the main magnetic mineral in the dikes (Raposo et al. 2007). However, for some specimens there is also a significant contribution (>50%) from paramagnetic minerals to the bulk magnetic susceptibility, suggesting that these minerals may have a strong contribution to the AMS fabric. Such behavior is not common on dike swarms but it has also been found in a swarm north of the investigated area at Pará de Minas dikes (PM, Fig. 17.4). The AMS is related to all rock-forming minerals whereas the AARM is related to coarse and fine-grained magnetite. The abnormal magnetic fabrics (*intermediate* and *inverse* types) found in our study cannot be explained by grain size effect (SD or PSD particles), since in all cases the AMS and AARM tensors are coaxial, suggesting a primary origin for the AMS fabrics.

The basic suite is the most important swarm in the studied region and the main magnetic fabric found for these dikes is related to magma flow, in which 66% of the dikes were fed by inclined flow, whereas the remaining ones were fed by lateral flow. Based on the K_{max} inclination and the geographic position of the dikes it was possible to infer that the dikes were fed by mantle sources NW of Lavras and/or SW of Bom Sucesso cities (Fig. 17.5). If these dikes had similar ages to those from PM, this magma source could have fed also PM dikes. However, this hypothesis must be check with geochemical and geochronological data.

Even though the basic-noritic and metamorphic suites are represented by a relatively small number of dikes, the AMS fabric related to magma flow is as important as the intermediate AMS fabric. The flow fabric (K_{max} inclination) also shows that they were fed by inclined flows which allow to infer that mantle sources (or magma chambers) were located N of Lavras city for basic-noritic suite, and SW of Bom Sucesso city for metamorphic suite (Fig. 17.5).

The magnetic fabrics recognized for the three studied swarms are primary in origin, and are independent of the magnetic mineralogy, the geochemical composition, the strikes of the dikes, and the ages of

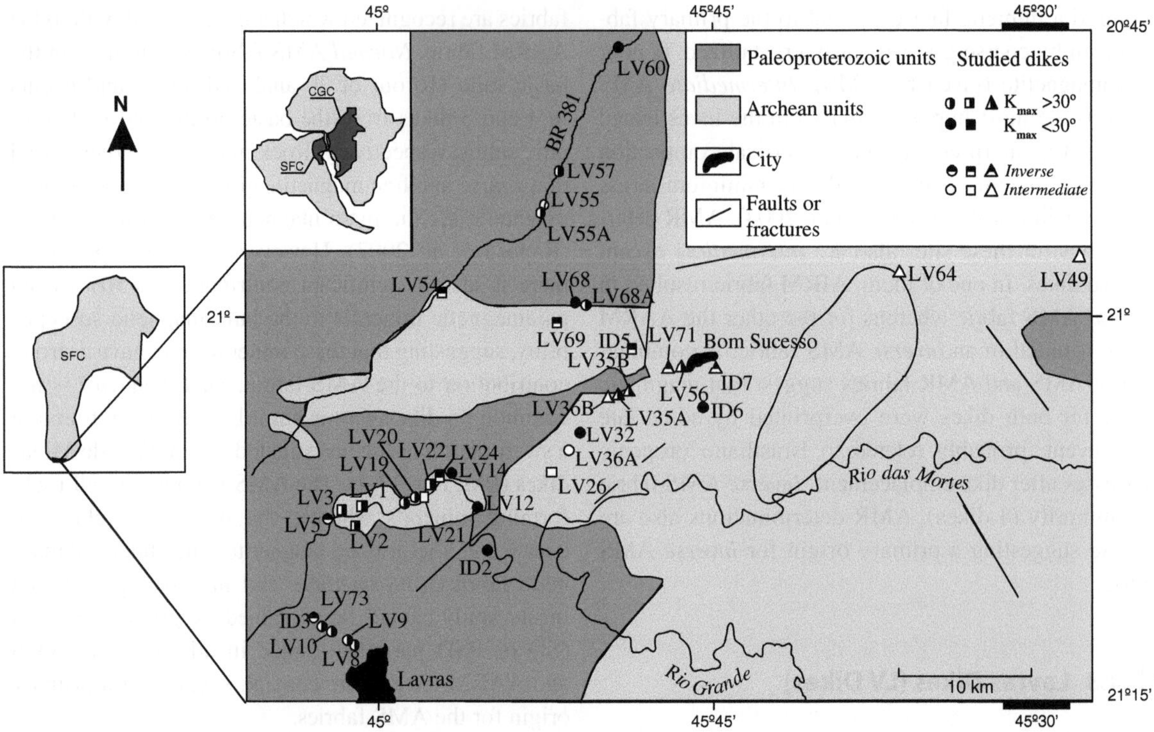

Fig. 17.5 Simplified geological map of the Lavras region. Archean units are: granulites, granite-gneisse-migmatites, anphibolites, ultramafic rocks and TTG gneisses. Paleoproterozoic units are: granites, orthogneisses, gneisse-monzonitic-migmatites and metassediments from Minas Supergroup. CGC and SFC are Congo and São Francisco Cratons, respectively. Strike of dikes were omitted to clarify the figure. Spheres, squares and triangles are the basic, basic-noritic and metamorphic suites, respectively. After Raposo et al. (2007)

the swarms since the same magnetic minerals and magnetic fabric types were found in the dikes from all suites. Geochemical and isotopic data indicate that two mantle sources were involved. On the other hand, the flow fabric suggests that at least three source regions (or magma chambers) were involved in the emplacement of the dike swarms

17.4.4 Salvador Dikes

Magnetic fabric and rock magnetic studies were carried (Raposo and Berquó 2008) on 25 unmetamorphosed mafic dikes of the Meso-Late Proterozoic (~1.02 Ga) dike swarm from Salvador city (Bahia State, NE Brazil). This area lies in the north-eastern part of the São Francisco Craton (Figs. 17.1 and 17.6). Dikes crop out along the beaches and in quarries around Salvador city, and cut across both amphibolite dikes and granulites. Their widths range from a few centimeters up to 30 m with an average of ~4 m, and show two main trends of N140–190 and N100–120 with vertical dip-angles.

Magnetic fabrics were determined using AMS and AARM methods. The magnetic mineralogy was investigated by many experiments (Raposo and Berquó 2008) including remanent magnetization measurements at variable low temperatures (10–300 K), Mössbauer spectroscopy, high temperature magnetization curves (25–700°C) and Scanning Electron Microscopy (SEM). Results from all experiments show that the magnetite grains found in these dikes are large and we discard the presence of single domain grains. The composition is close to stoichiometric with low Ti substitution, and its Verwey transition occurs around 120 K.

Two AMS fabrics were found for this swarm; *normal* and *intermediate*. Rock magnetism showed that there is no difference in the magnetic mineral phases for dikes showing both types of AMS fabric, and that PSD-MD magnetite grains are the carriers of both magnetic susceptibility and remanences (see Raposo

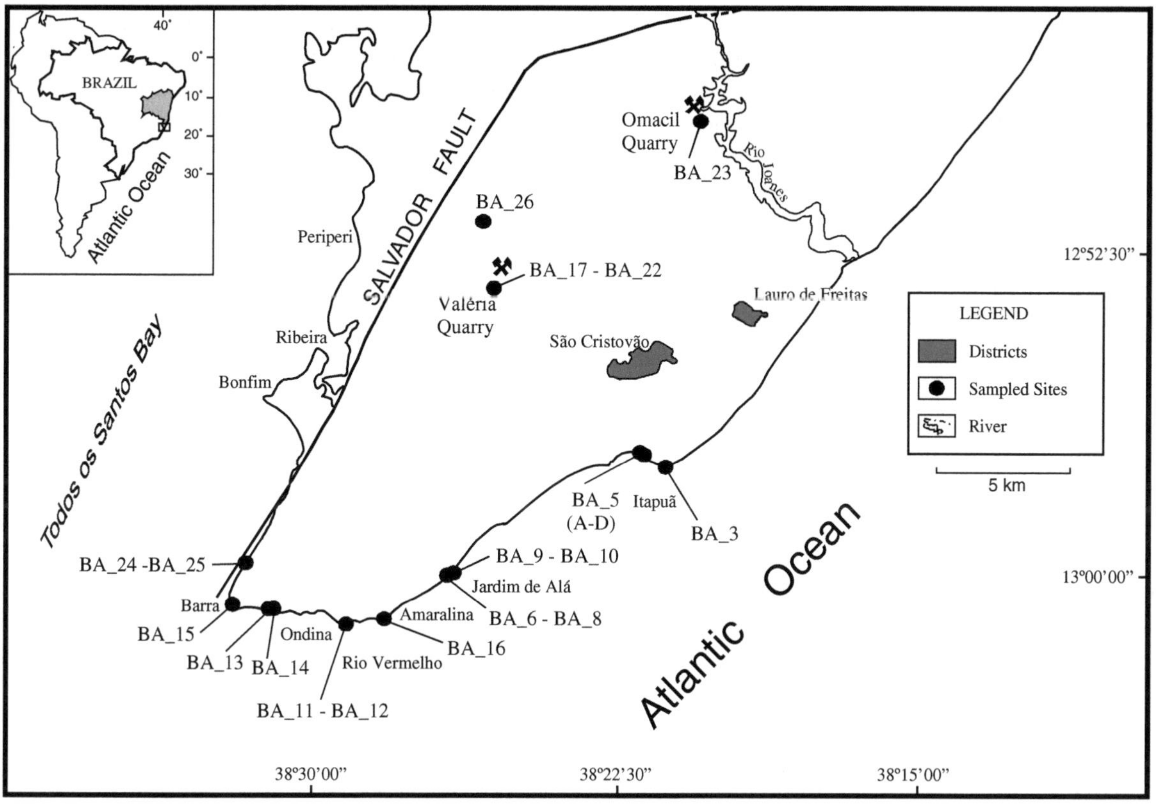

Fig. 17.6 Simplified map with the localization of the sampled dikes from Salvador swarm

and Berquó 2008 for details). The *intermediate* AMS fabric was found in only four dikes. Rock magnetic studies for these dikes show that this fabric cannot be attributed to a grain size effect. Therefore, we deduced that the four dikes acted as stress conduits, on where continuous compression forced material along the dike direction. The *normal* AMS fabric is the most important in the investigated swarm. This fabric was interpreted as due to magma flow in accordance with field evidence found by Corrêa-Gomes (1992). The analysis of K_{max} inclination permitted to infer that ~80% of dikes were fed dominantly by gently plunging fluxes ($K_{max} < 30°$) whereas only four dikes were fed by inclined ($30° < K_{max} < 60°$) up to vertical fluxes ($K_{max} > 60°$). Such K_{max} plunge values allow inferring two geographic positions for the mantle source. One of them, which horizonntally fed the majority of dikes, could be located in southern Bahia, and might be the same as the one that fed the E-W dikes from Ilhéus-Olivença according to geochemical (Bellieni et al. 1991) and geochronologic data (Renne et al. 1990). The other source, which fed dikes with the steepest K_{max} values would have underlain the Valéria quarry area (Fig. 17.6), where these dikes out crops. This source also could have horizontally fed the other dikes.

Magnetic fabric of the dikes was also studied by AARM technique. Both AMS and AARM fabrics are not coaxial for all investigated dikes. For the majority of dikes with *normal* AMS fabric the AARM tensor is *inverse* with respect to the AMS fabric, in a few dikes it is *intermediate* AMS fabric, and in just one dike it is *normal* AMS fabric. For three dikes with *intermediate* AMS fabric the AARM fabric is also *intermediate*, and for one dike it is *inverse* AMS fabric. However, in all dikes the AARM magnetic lineation ($AARM_{max}$) is oriented to N30–60E. Magnetite grains are the carriers of AARM fabrics and are responsible for the magnetic susceptibility (Raposo and Berquó 2008). However, the non-parallelism between AARM and AMS fabrics suggest that magnetite is not the carrier of the AMS probably because it is weakly oriented, and its contribution to low field susceptibility is not sufficiently anisotropic to deflect the AMS fabric (i.e. to become AMS parallel to AARM). This is also verified in high field anisotropy and by normal and cathode-luminesce petrography (Raposo, McReath and Hirt, in preparation). The N30–60E orientation given by AARM lineation ($AARM_{max}$) is similar to the orientation of many faults found in the coastal area of Bahia State from which the Salvador normal fault is the most important. These faults were formed during Cretaceous rifting in the Recôncavo-Tucano-Jatobá system formed during the opening of the South Atlantic (Milani and Davision 1988). The similarity between the orientation of AARMmax and the Salvador fault suggests that the dikes were affected by the fault system, and also suggests that magnetite grains were rotate clockwise from dike plane. This is supported by the presence of many small NE-striking faults found in the dikes. Therefore we concluded that the AMS fabric of the Salvador swarm is primary in origin whereas the AARM fabric is tectonic and younger than AMS and it was acquired during the break-up of the Gondwanaland (see Raposo and Berquó 2008 for details).

Another important result from this swarm is related to the widest dike which has distinct fabrics at the margins and in its interior. Specimens from both margins have *normal* AMS fabric whereas specimens from the center and those about 12 m away from one of the contacts show *intermediate* AMS fabrics (see Raposo and Berquó 2008). However, the three parts of the dike have similar K_{max} orientations. This pattern suggests that a stress field played an important role in the emplacement of this dike. In addition, it could be explained by pure shear which is dominant in the center of the dike and it might be responsible for the inversion of the K_{int} and K_{min} AMS axes (Féménias, et al. 2004). AARM fabrics for this dike are also not coaxial with AMS tensors. For the specimens with *normal* AMS fabric (both margins) the AARM fabric is "intermediate" AMS fabric whereas for the specimens with *intermediate* AMS fabric the AARM fabrics became *normal* AMS fabric. In these three cases, the maximum remanence axes ($AARM_{max}$) are very similar and, once again, have the N30–60E orientation with steep plunges.

17.5 Conclusions

In this paper we have shown the application of magnetic fabric studies on Brazilian mafic swarms with different compositions from Mesozoic as well as Proterozoic. It has been shown that magnetic methods gave good results which permitted to infer the emplacement and source position of dikes. The results also allowed detecting tectonic fabric given either by the abnormal either AMS or AARM.

Acknowledgements The author thanks the Brazilian agency FAPESP (mainly the Grants No. 95/8399-0 which she could make the Laboratory) for its financial support. I also thank all my students for their help in both field and laboratory works. The comments of the manuscript from both Eugenio Veloso and an anonymous referee are also thanked.

References

Almeida FFM (1986) Distribuição regional e relações tectônicas do magmatismo pós-Paleozoico no Brasil. Rev Bras Geosci 16:325–349

Archanjo CJ, Araúújo MGS, Launeau P (2002) Fabric of the Rio Ceará-Mirim mafic dike swarm (northeastern Brazil) determined by anisotropy of magnetic susceptibility and image analysis. J Geophys Res 107(B3):2046. doi: 10.1029/2001JB000268

Baer G, Reches Z (1987) Flow pattern of magma in dikes, Makhtesh Romon, Israel. Geology 15:569–572

Bellieni G, Petrini R, Piccirillo EM, Cavazzini G, Civetta L, Comin-Chiaramonti P, Melfi AJ, Bertolo S, De Min A (1991) Proterozoic mafic dyke swarms of the São Francisco Craton (SE-Bahia State, Brazil): petrology and Sr-Nd isotopes. Eur J Mineral 3:429–449

Borradaile GJ (2001) Magnetic fabrics and petrofabrics: their orientation distribution and anisotropies. J Struct Geol 23:1581–1596

Borradaile GJ, Gauthier D (2003) Interpreting anomalous magnetic fabrics in ophiolite dikes. J Struct Geol 25:171–182

Callot J-P, Geoffroy L, Aubourg C, Pozzi JP, Mege D (2001) Magma flow directions of shallow dykes from the East Greenland volcanic margin inferred from magnetic studies. Tectonophysics 335:313–329

Cañon-Tapia E (1996) Single-grain versus distribution: a simple three-dimensional model. Phys Earth Planetary Inter 94: 149–158

Chang HK, Kowsmann RO, Figueiredo AMF, Bender AA (1992) Tectonic and stratigraphy of the East Brazilian rift system: an overview. Tectonophysics 213:97–138

Chaves AO (2001) Enxames de diques máficos do Setor Sul do Craton do São Francisco-MG. PhD Thesis, Instituto de Geociências da Universidade de São Paulo, 152 pp

Constable C, Tauxe L (1990) The bootstrap for magnetic susceptibility tensor. J Geophys Res 95:8383–8395

Corrêa-Gomes LC (1992) Diques máficos: Uma reflexão teórica sobre o tema e o seu uso no entendimento prático da geodinâmica fissural. MSc Dissertation, Instituto de Geociências da Universidade Federal da Bahia, 196p

Coward MP (1980) The analysis of flow profilies in a basltic dyke using strained vesicles. J Geol 137:605–615

D'Agrella-Filho MS, Pacca IG, Renne PR, Onstott TC (1990) Paleomagnetism of Middle Proterozoic (1.01 to 1.08 Ga) mafic dykes in southeastern Bahia State-São Francisco Craton. Earth Planetary Sci Lett 101:332–348

Daly L, Zinsser H (1973) Étude comparative des anisotropies de susceptibilité et d'aimantation rémanente isotherme: Conséquences pour l'analyse structurale et le paléomagnétisme. Ann Géophys 29:189–200

Emerman SH, Marret R (1990) Why dikes? Geology 18:231–233

Ernst R.E, Baragar WRA (1992) Evidence from magnetic fabric to the flow pattern of magma in the Mackenzie giant radiating dyke swarm. Nature 356:511–513

Féménias O, Diot H, Berza T, Gauffriau A, Demaiffe D (2004) Asymmetrical to symmetrical magnetic fabric of dikes: paleo-flow orientations and paleo-stress recorded on feeder-bodies from the Motru dike swarm (Romania). J Struct Geol 26:1401–1418

Greenough JD, Ruffman A, Owen JV (1988) Magma injection directions inferred from fabric study of the Popes Harbour dike, eastern shore, Nova Scotia, Canada. Geology 16:547–550

Hargraves RB, Johnson D, Chan CY (1991) Distribution anisotropy: the cause of AMS in igneous rocks? Geophys Res Lett 18:2193–2196

Herrero-Bervera E, Cañon-Tapia E, Walker GPL, Guerrero-Garcia JC (2002) The Nuuanu and Wailau Giant Landslides: insights from paleomagnetic and anisotropy of magnetic susceptibility (AMS) studies. Phys Earth Planetary Inter 129:83–98

Jackson M (1991) Anisotropy of magnetic remanence: a brief review of mineralogical sources, physical origins and geological applications, and comparison with susceptibility anisotropy. Pure Appl Geophys 136:1–28

Khan AM (1962) The anisotropy of magnetic susceptibility of some igneous and metamorphic rocks. J Geophys Res 67:2873–2885

Knight MD, Walker GPL (1988) Magma flow direction in dikes of the Koolau Complex, Oahu, determined from magnetic fabric studies. J Geophys Res 93:4301–4319

Kratinová Z, Závada P, Hrouda F, Schulmann K (2006) Non-scaled analogue modeling of AMS development during viscous flow: a simulation on diaper-like structures. Tectonophysics 418:51–61

Milani JE, Davison I (1988) Basement control and transfer tectonics in the Recôncavo-Tucano-Jatobán rift, Northeast Brazil. Tectonophysics 154:41–70

Park K, Tanczyk EI, Desbarats A (1988) Magnetic fabric and its significance in the 1400 Ma Mealy diabase dykes of Labrador, Canada. J Geophys Res 93:13689–13704

Piccirillo EM, Bellieni G, Cavazzini G, Comin-Chiaramonti P, Petrini R, Melfi AJ, Pinese JPP, Zantedeschi, De Min A (1990) Lower Cretaceous tholeiitic dyke swarm from the Ponta Grossa Arch (Southeast Brazil): petrology, Sr-Nd isotopes and genetic relationship with the Paraná flood volcanics. Chem Geol 89:19–48

Pinese JPP (1997) Geoquímica, geologia isotópica e aspectos petrológicos dos diques máficos Pré Cambrianos da região de Lavras (MG), porção sul do Cráton do São Francisco. PhD thesis, Instituto de Geociências da Universidade de São Paulo, 178 pp

Pinese JPP, Teixeira W, Piccirillo EM, Quemeneur JJG, Bellieni G (1995) The Precambrian Lavras mafic dykes, southern São Francisco Craton, Brazil: preliminary geochemical and geochronological results. In: Baer and Heimann (eds) Physics and chemistry of Dykes, Balkema. Rotterdam, Netherlands, pp 205–218

Raposo MIB (1992) Paleomagnetismo do enxame de diques do Arco de Ponta Grossa. PhD thesis, University of São Paulo, 105 pp

Raposo MIB (1997) Magnetic fabric and its significance in the Florianópolis dyke swarm, southern Brazil. Geophys J Int 31:159–170

Raposo MIB, Berquó TS (2008) Tectonic fabric revealed by AARM of the Proterozoic Mafic Dike Swarm in the Salvador City (Bahia State): São Francisco Craton, NE Brazil. Phys Earth Planetary Inter 167:179–194

Raposo MIB, D'Agrella-Filho MS (2000) Magnetic fabrics of dike swarm from SE Bahia state (Brazil): their significance and implications for Mesoproterozoic basic magmatism in the São Francisco Craton. Precam Res 9:309–325

Raposo MIB, D'Agrella-Filho MS (1998) Magnetic fabric of Mesoproterozoic dike swarms from Bahia State (Brazil)-São Francisco Craton: preliminary results. International conference on Precambrian and Craton tectonics (14th International Conference on Basement Tectonics). Ouro Preto (MG), Brazil, Extended Abstract

Raposo MIB, Ernesto M (1990) Anisotropia de suscetibilidade magnética de diques máficos do Arco de Ponta Grossa e algumas implicações quanto ao seu modo de colocação. Boletim de resumos do XXXVI Congresso Brasileiro de Geologia, Natal, Brasil

Raposo MIB, Ernesto M (1995) Anisotropy of magnetic susceptibility in the Ponta Grossa dyke swarm (Brazil) and its relationship with magma flow direction. Phys Earth Planet Inter 87:183–196

Raposo MIB, Chaves AO, Lojkasek-Lima P, DÁgrella-Filho MS, Teixeira W (2004) Magnetic fabrics and rock magnetism of Proterozoic dike swarm from the southern São Francisco Craton, Minas Gerais State, Brazil. Tectonophysics 378:43–63

Raposo MIB, D'Agrella-Filho MS, Pinese JPP (2007) Magnetic Fabrics and Rock Magnetism of Archaean and Proterozoic dike swarms in the southern São Francisco Craton, Brazil. Tectonophysics 443:53–71

Raposo MIB, Ernesto M, Renne PR (1998) Paleomagnetism and ^{40}Ar/^{30}Ar dating of the early Cretaceous Florianópolis dike swarm (Santa Catarina Island), Southern Brazil. Phys Earth Planet Inter 108:275–290

Raposo MIB, Mello IJS, Berquó TS (2009) Magnetic fabrics and rock-magnetism studies of Early-Late Cretaceous mafic dike swarms from Ubatuba (São Paulo State, Brazil): Preliminary results. IAGA, Sopron, Hungary

Renne PR, Deckart K, Ernesto M, Féraud G, Piccirillo EM (1996) Age of the Ponta Grossa dike swarm (Brazil), and implications to Paraná flood volcanism. Earth Plan Sci Lett 144:199–211

Renne PR, Onstott TC, D'Agrella-Filho MS, Pacca IG, Teixeira W (1990) ^{40}Ar/^{39}Ar dating of 1.0–1.1 Ga magnetizations from São Francisco and Kalahari cratons: tectonic implications for Pan-African and Brasiliano mobile belts. Earth Planetary Sci Lett 101:349–366

Rochette P (1988) Inverse magnetic fabric in carbonate-bearing rocks. Earth Planetary Sci Lett 90:229–237
Rochette P, Jackson M, Aubourg C (1992) Rock magnetism and the interpretation of anisotropy magnetic susceptibility. Rev Geophys 30:209–226
Rochette P, Jenatton L, Dupy C, Boudier F, Reuber I (1991) Diabase dikes emplacement in the Oman Ophiolite: a magnetic fabric study with reference to geochemistry. In: Peten TJ, Nicolas A, Coleman R (eds) Ophiolite genesis and evolution of the oceanic lithosphere. Kluwer Academic, Dordrecht
Sial AN, Oliveira EP, Choudhuri A (1987) Mafic dyke swarms of Brazil. Geol Assoc Can Spec Paper 34:467–481
Stephenson A (1993) Three-axis static alternating field demagnetization of rocks and the identification of natural remanent magnetization, gyroremanent magnetization and anisotropy. Geophys J Int 98:373–381
Stephenson A, Sadikum S, Potter DK (1986) A theoretical and experimental comparison of the anisotropies of magnetic susceptibility and remanence in rocks and minerals. Geophys J R Astron Soc 84:185–200
Tarling DH, Hrouda F (eds) (1993) The magnetic anisotropy of rocks. Chapman and Hall, London
Tauxe L, Gee JS, Staudigel H (1998) Flow directions in the dikes from anisotropy of magnetic susceptibility data: the bootstrap way. J Geophys Res 103:17775–17790
Teixeira W, Sabaté P, Barbosa J, Noce CM, Carneiro MA (2000) Archean and Paleoproyerozoic tectonic evolution of the São Francisco Craton, Brazil. In: Cordani UG, Milani EJ, Thomaz-Filho A, Campos DA (eds) Tectonic evolution of South America. Rio de Janeiro, Brazil, pp 287–310
Varga RJ, Gee JS, Staudigel H, Tauxe L (1998) Dike surface lineations as magma flow indicators within the sheeted dike complex of the Troodos ophiolite, Cyprus. J Geophys Res 103:5241–5256

AMS in Granites and Lava Flows: Two End Members of a Continuum?

18

Edgardo Cañón-Tapia

Abstract

Significant differences between granites and lava flows can require different basic assumptions when interpreting AMS results. Among the differences between both types of rocks, perhaps the earliest in being recognized was the wider range of mineral compositions found in granites. Such difference can result in complex mineral assemblages that, in turn, can complicate the interpretation of AMS results in granites relative to the AMS measured in lava flows. Closely linked to this mineralogic effect is the distinction between "primary" flow fabrics and "secondary" effects. Such distinction is a matter of concern in most granites whereas the AMS of lava flows is usually considered "primary" without further examination. As the increasing evidence obtained from lava flows shows, however, the AMS of lavas is not as simple as the general model of hydrodynamic alignment of particles would suggest and much can be learned from lava flows that can be applied directly to the interpretation of AMS in granites. In this work, the better understanding of the fabric of lava flows that has been obtained in recent years is used as the basis for a reassessment of the basic assumptions needed for the correct interpretation of AMS in both lava flows and granitic rocks.

18.1 Introduction

The anisotropy of magnetic susceptibility (AMS) has proved repeatedly to be an important tool that can be used to study the emplacement process of rocks with much detail. The technique has experienced an exponential growth starting in the 1990's, and such trend seems to continue until present. An important trigger of such rapid growth is associated to the development of instruments capable to measure AMS with high accuracy and relatively little effort. In addition, the number of interested workers has also increased, as has also increased the breadth of cases that are studied with this technique.

At the time of the first monograph devoted to AMS (Tarling and Hrouda 1993), the applications of this method to the study of sedimentary and igneous rocks seemed to have received less attention from the community than within the realm of metamorphic rocks. Also, at the time of publication of such monograph it seemed convenient to make a distinction between

E. Cañón-Tapia (✉)
Departamento de Geología, CICESE, Ensenada, BC 92143, Mexico; CICESE, Geology Department, PO Box 434843, San Diego, CA 92143, USA
e-mail: ecanon@cicese.mx

E. Petrovský et al. (eds.), *The Earth's Magnetic Interior*, IAGA Special Sopron Book Series 1,
DOI 10.1007/978-94-007-0323-0_18, © Springer Science+Business Media B.V. 2011

the processes and approaches necessary to study the AMS of sedimentary, igneous and metamorphic rocks. Both situations have changed dramatically over the years.

On the one hand, the AMS of igneous rocks experienced a bloom of attention from the scientific community. This is revealed by the extremely large number of studies of individual granites that use AMS measurements almost in a routinely form at present. As an example of the magnitude of such growth an internet search will produce more than 50 hits of published articles containing the words "granites" and "magnetic fabric" for a given year after 2000, whereas less than 20 hits are found if the year is 1996 or before.

On the other hand, as progress has been made concerning the origin of AMS of rocks, it has become clear that an approach to AMS based on a general classification of rock types presents some inconvenient features. Examples of these problems can be found when comparing the petrofabric of different types of rocks, but also when comparing different subtypes of the same basic group of rocks. For instance, while undoubtedly many aspects of sedimentary rocks are not equal to those of metamorphic rocks, it is also true that the processes of emplacement and other characteristics of many sedimentary rocks and many pyroclastic deposits are essentially the same. Thus, a distinction between sedimentary and igneous rocks is not always justified. In contrast, using the same approach in the interpretation of all igneous rocks might also be unjustified in some cases. This becomes clearer if one considers the numerous coincidences in the form in which petrofabric is acquired in some volcanic and igneous rocks and contrasts them with the acquisition of petrofabric in pyroclastic rocks, even when all three types are of igneous origin.

An alternative approach that is based more on processes of petrofabric acquisition rather than in specific rock types might be more convenient to fully exploit the potential that AMS has as a petrofabric method. For example, as pointed out by Cañón-Tapia (2005) the acquisition of a mineral fabric in lava flows and dykes can be visualized as the result of the movement of rigid particles immersed in a viscous fluid, whereas the fabric acquisition in volcaniclastic and some sedimentary rocks (e.g., turbidites) seems to be better described by a particulate flow model where turbulence might be important as well as other depositional and post emplacement events. Consequently, it would seem that although the traditional distinction between rock types still remains valid as a classification scheme, within the context of AMS studies is more advantageous to adopt an approach that focuses in processes of fabric acquisition.

For these reasons, and further considering (a) the large number of published works that have used AMS for the study of granites since the publication of the seminal Tarling and Hrouda (1993) monograph, (b) the fact that many of those works simply present AMS results as a routine source of information and therefore do not present new insights concerning the origin of AMS, and (c) the advantages offered by adopting a process-oriented approach to explain the acquisition of a mineral fabric in general, it would seem that a traditional review paper summarizing works dealing with AMS in igneous rocks published in the past fifteen years would become very impractical. Therefore, this work departs from the traditional review-paper and puts emphasis on the examination of processes of fabric acquisition rather than attempting to provide specific examples of AMS studies. More specifically, attention is focused here on the common processes shared by two types of igneous rocks (lava flows and small tabular intrusions on the one hand and large intrusive rocks, mainly granites, in the other), attempting to summarize some of the most recent developments made in the understanding of the acquisition of a magnetic fabric relevant for these cases. By following this approach, the many common aspects of AMS shared by the rock types examined in this work are highlighted, therefore allowing us to better appreciate the various forms in which the still unresolved issues in both types of rocks might be approached. Behind this approach is the belief that fostering such an "interdisciplinary" perspective should contribute to a more rapid development of the field.

18.2 Granites and Lava Flows: How Are They Related to Each Other?

Although in a general classification scheme of rocks granites, small tabular intrusions and lava flows are all considered to be igneous rocks, in practice the many differences distinguishing each of these rock subtypes prevails. For this reason, specialized studies of each of these subtypes commonly make little reference to the literature reporting discoveries or advances made in the context of the other two rock subtypes.

The differences between the three rock subtypes considered here start to become evident when the definition of each group is considered. For an intrusive rock to be classified as a granite geologists would need to estimate its content of quartz, alkali feldspars, plagioclase and feldespathoids plotting their results in a QAFP diagram (e.g., Best 1982). Thus, from the mineralogical point of view the term "granite" has well defined boundaries that distinguish it from other types of intrusive rocks. In contrast, the term "lava flow" is a rather descriptive term that applies to an igneous extrusive rock and that does not make any reference to the mineral or chemical compositions of the erupted magma, although in practice some bias exists towards the usage of this term to describe the effusive products of rather mafic (basaltic to andesitic) composition. Similarly, "tabular intrusion" or "dyke" is a rather descriptive term that applies to intrusive rocks that although does not make reference to the mineral or chemical composition of the magma that gives place to them, has a bias towards the more mafic compositions.

Other differences are related to the cooling and stress environments associated to the emplacement of each of these rock subtypes. Due to the larger time of cooling associated with granites, these rocks may acquire their petrofabric as a result of external (tectonic) stresses, while lava flows and dykes usually will crystallize much more rapidly and on the absence of a tectonic stress, this being valid for most dykes despite their intrusive character due to their relatively small dimensions at least in comparison with the dimensions of granites. Consequently, it would seem that "granites" and "lava flows"/"dykes" designate entirely different rocks that have very little in common.

Actually, this appreciation has prevailed in the literature relevant for the study of AMS, and studies made on one of these rock types rarely take advantage of studies made on the other two types of rocks. Furthermore, such independence of the fields of study sometimes might give the impression that any attempt to compare the AMS properties of these three types of rocks might be unjustified. Undoubtedly, if attention is limited to their definitions or to the different ranges of viscosity and/or stress associated with them it would be impossible to find similarities between these types of rocks. If attention is focused on the fundamental physical processes controlling fabric acquisition, however, a different picture emerges.

Indeed, despite the differences in the numerical values of the associated viscosities and /or stresses, lava flows, dykes and granites are the result of similar physical processes that involve the viscous flow of a liquid phase that has a certain concentration of solid particles in suspension. Whether such flow takes place entirely beneath the surface of the Earth, or over that surface, does not introduce a significant difference in the basic aspects of the physical processes involved. Actually, the differences are more related to the boundary and initial conditions imposed to those processes than to the processes themselves (Fig. 18.1). Undoubtedly such differences in the boundary conditions are important in controlling the end result of a particular situation, but it is stressed that such

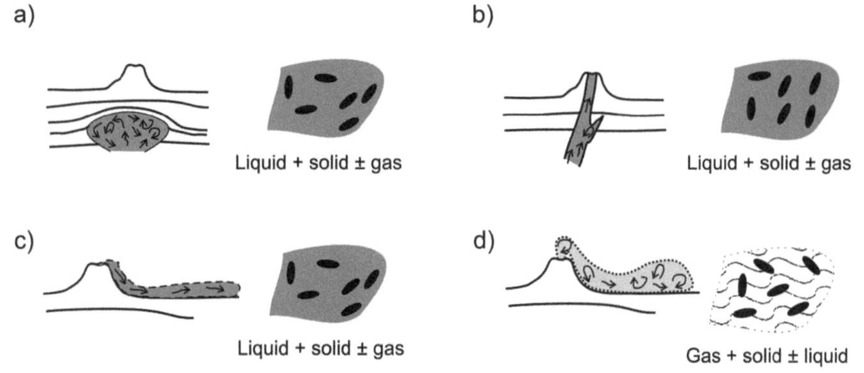

Fig. 18.1 Diagram showing four types of igneous rocks that can be defined attending to their rheological properties and characteristics of emplacement. The *left* of each diagram shows the environment of emplacement and the typical flow regime (arrows within shadowed zone), whereas the *right side* illustrates the typical array of particles at a cm scale. (**a**) Large size plutons, (**b**) tabular intrusions, (**c**) lava flows, and (**d**) pyroclastics

differences are not related to the processes themselves. As those processes are related to the acquisition of a mineral fabric (magnetic or not), it is therefore justified attempting to establish a connection between both rock types. Thus, the underlying trend followed in this work is based in such a similarity of physical processes, although differences in the boundary conditions are pointed out as necessary. Also, in the remainder of this work lava flows and dykes are considered to belong to the same rock category, and any reference to "lava flows" also implies a reference to "dykes" (and vice versa) even if the latter are not explicitly mentioned. The explicit mention of lava flows and dykes will be made only when a clear distinction between both types of rocks is strictly necessary.

18.3 The Spectrum of AMS in Lava Flows and Granites

18.3.1 Bulk Susceptibility

The bulk susceptibility of granites varies over more than three orders of magnitude (Fig. 18.2), its lower end being as low as 1×10^{-5} and the upper end as high as 4×10^{-2} SI (e.g., Chlupácová et al. 1975, Hrouda and Lanza 1989, Tarling and Hrouda 1993, Zák et al. 2005, Bouchez 1997). The low end of this range corresponds to the paramagnetic, or magnetite-free granites; the upper part corresponding to granites that contain magnetite. For this reason, Bouchez (1997) considered that the distribution of bulk susceptibilities in granites is bimodal, corresponding to the broad subdivision between ilmenite and magnetite granites proposed by Ishihara (1977). A similar conclusion had been reached much earlier in the former U.S.S.R. by Dortman (1984), who based their conclusions in observations made on more than 10,000 specimens. Consequently, the bimodal distribution of susceptibilities in granites is a well established observation.

When examined in more detail, bulk susceptibility might be used to produce a much finer classification of granitic rocks. For instance, Gleizes et al. (1993) suggested a correspondence between different pluton types and bulk susceptibility that could be used as a proxy for rock classification. Accordingly, Leucogranites have bulk susceptibilities $< 10 \times 10^{-5}$, monzogranites have a bulk susceptibility between 10×10^{-5} and 19×10^{-5}, granodiorites have a bulk susceptibility larger than that of monzogranites but lower than 31×10^{-5} and quartz diorites have the largest values of bulk susceptibility exceeding 30×10^{-5}. Similar results were reported by Aydin et al. (2007), leading these authors to suggest that the variation of bulk susceptibility could actually be useful as a proxy for geochemical differentiation. These results would suggest that variation of bulk susceptibility with rock types might be correlated with the total content of iron of the rock, as measured from geochemical analyses. The existence of S-type and I-type granodiorites with virtually the same chemical and mineral compositions, yet with bulk susceptibilities that differ by as much as

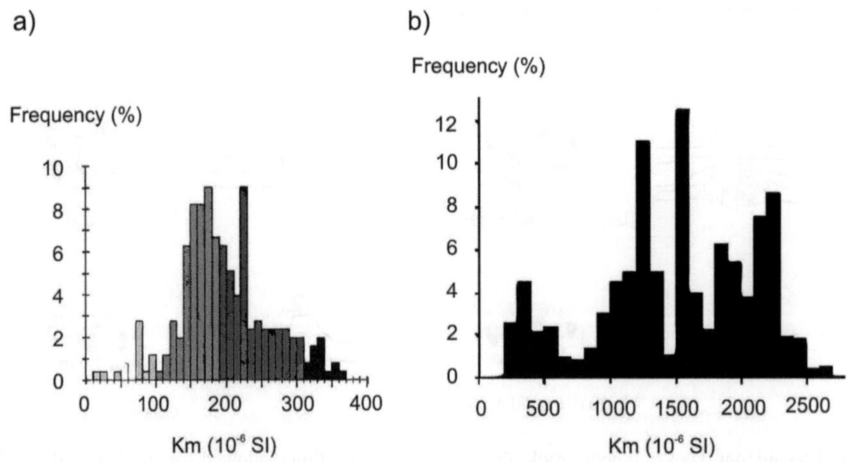

Fig. 18.2 Histograms showing the variation of bulk susceptibility within granites. Note the much larger range in values on the histogram in (**b**). The different grey shades on (**a**) correspond to different lithological components of the same intrusion. (Modified from Gleizes et al. 1993, Aydin et al. 2007)

two orders of magnitude, however, indicates that the possible correlations between magnetic susceptibility and rock type needs to be further examined in a more general scenario.

In the case of lava flows, the conditions of oxygen fugacity during their crystallization commonly favor the formation of titanomagnetites. Consequently, the ferromagnetic minerals commonly will dominate the bulk susceptibility of this type of rocks, resulting in bulk susceptibilities usually in the order of 10^{-2} SI (Fig. 18.3). Nevertheless, when examined in more detail the bulk susceptibility of lava flows is not unimodal, and similarly to the case of granites it seems to display some trends associated with rock type. At the lower end of the range, phonolites and rhyolites have bulk susceptibilities $\sim 10^{-4}$–10^{-3} SI (Cañón-Tapia and Castro 2004, Hrouda et al. 2005), although some mafic dykes with bulk susceptibilities $<10^{-2}$ have also been reported (e.g., Raposo et al. 2004). Nevertheless, most basaltic lavas usually will have bulk susceptibilities exceeding 6×10^{-2} SI. At a finer scale, there are some differences concerning the range and distributions of

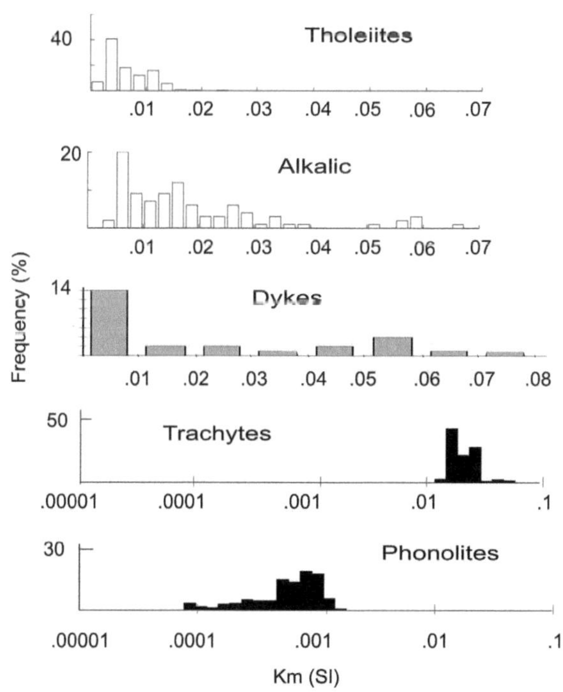

Fig. 18.3 Histograms showing the variation of bulk susceptibility within lava flows and tabular intrusions. Note the different range of Km on the phonolites, and the various distributions displayed as a function of rock type. (Modified from Cañón-Tapia 2004a, Raposo et al. 2004, Hrouda et al. 2005)

bulk susceptibilities reported for different rock types that to some extent are reminiscent of the associations proposed to exist in the case of granites. For example, Cañón-Tapia (2004b) found that basaltic tholeiites tend to have lower bulk susceptibilities than basalt-alkalic lavas (average values $<1 \times 10^{-2}$ and $\sim 2 \times 10^{-2}$, respectively). Similar results were reported by Hrouda et al. (2005) who showed that within the same volcanic region the bulk susceptibility of phonolites is nearly two orders of magnitude lower than that of trachytes (average values $\sim 5 \times 10^{-4}$ and 3×10^{-2}, respectively). Consequently, despite the predominance of magnetites, the bulk susceptibility of lava flows can vary by as much as an order of magnitude.

An additional degree of complexity found when attempting to establish a correlation between bulk susceptibilities and different types of lava flows is brought forward by the occurrence of middle to low temperature changes that can alter the composition of the ferromagnetic minerals (e.g., Ade-Hall et al. 1968a, b, c, 1971). These alterations might decrease magnetic susceptibility and could overshadow differences originally related to chemical composition. Consequently, similarly to the case of granites, the proposed association of bulk susceptibility and composition of lava flows also needs to be further explored before it can be universally applied.

In any case, considering that the bulk susceptibilities of granites and lava flows vary by more than an order of magnitude, and that there are some trends that can be associated to subtle variations in composition within the realm of each rock type, it would seem justified to conclude that the spectra of bulk magnetic susceptibilities of granites and lava flows have essentially the same general characteristics.

18.3.2 Degrees of Anisotropy and Shape of Mineral Fabrics

Quantifying the degree of anisotropy and the shape of the magnetic fabric is not a simple task. Over the years several parameters attempting to yield a quantitative estimate of these aspects of the susceptibility tensor have been proposed. Also, several criteria for the selection of those parameters have been proposed based on aspects concerning the method of measurement, similarities with other forms of quantifying mineral fabrics and/or geometrical considerations (e.g., Hrouda

1982, Ellwood et al. 1988, Cañón-Tapia 1994, 2007). Nevertheless, until present there is no consensus concerning which parameters should be used to analyze the degree of anisotropy and shape of the magnetic fabrics in each case.

Unfortunately, the relationship between the different parameters used to report results of magnetic fabric measurements made both in granites and lava flows is not linear. Consequently, it is difficult to make a precise comparison of results, and any attempt to assess a range of values of these aspects of the magnetic fabric is therefore restricted to a semi-quantitative analysis. Nevertheless, some generalizations concerning these aspects of the magnetic fabric can be made; generalizations that are valid both for granites and lava flows alike. First, many studies reveal that the degree of anisotropy of a single rock unit is not uniform (e.g., Archanjo et al. 1995, Cañón-Tapia et al. 1995, 1996, 1997, Cañón-Tapia and Coe 2002, Ferré et al. 2004, Hrouda et al. 1999, Román-Berdiel et al. 1998, Táborská and Breiter 1998). The most homogeneous rocks will have a variation of ~2 % (this number is more or less the same regardless of the exact definition of the parameter used to determine the degree of anisotropy), but most commonly variations of up to 4% are found (Fig. 18.4). Second, within a single rock unit it is also common to find magnetic fabrics that are lineated (prolate) and some that are foliated (oblate). The amount of development of each of these two types of fabrics is variable, and there is no clear correlation with rock type, composition or size of the rock unit. Third, although there are cases in which the degree of anisotropy does not exceed the intrinsic anisotropy of the mineral grains present in a rock, there are many examples in which this mineralogical limit is not observed (e.g., Bouchez 1997, Cañón-Tapia and Coe 2002, Cañón-Tapia and Castro 2004, Rochette et al. 1994). Fourth, within a given rock unit the observed variations in these two types of parameters sometimes display some correlation with structural features of the rock. These structural features include rock boundaries, enclave or vesicle deformation, vesicle or mineral abundances, etc. For all of these reasons, it seems reasonable to conclude that some of the variations documented to take place in the degree of anisotropy and/or in the shape of the susceptibility tensor are related to specific aspects of the emplacement history of the studied rock unit. This conclusion is valid both for granites and lava flows, and therefore, it would seem that in both of these rock types the range of values of the degree of anisotropy and the shape of the magnetic fabric has essentially the same origins.

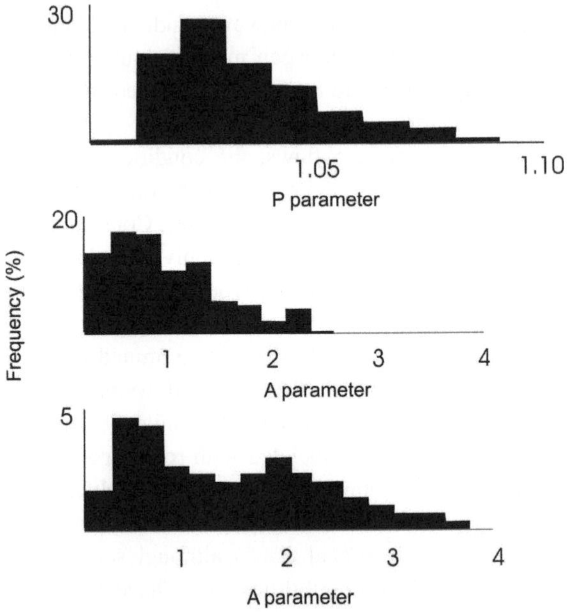

Fig. 18.4 Histograms showing the variation of the degree of anisotropy of various igneous rocks. The upper histogram corresponds to granites and reported degree of anisotropy using a different parameter than that used in the lower two diagrams and that correspond to lava flows of different compositions. Despite the differences in numerical values of the two parameters, note the similitudes in the shape of the distribution, especially between the two uppermost examples. (Modified from Bouchez 1997, Cañón-Tapia 2004b)

18.4 The Origin of the Magnetic Anisotropy

The range of bulk susceptibilities typical of granites and lava flows has influenced the form in which magnetic anisotropy is commonly interpreted. In the case of the paramagnetic granites it is clear that the magnetic anisotropy will be controlled by the fabric of the paramagnetic minerals, most commonly being that of biotite (Bouchez 1997). In contrast, the magnetic anisotropy of the ferromagnetic granites, and that of all lava flows, is commonly considered to be controlled by the fabric of magnetite due to their relatively high bulk susceptibility. As shown by Borradaile and Gauthier (2003) and Raposo and Berquó (2008), however, the bulk susceptibility may be controlled by a mineral phase whereas the flow-related anisotropy may be controlled by a different mineral phase with lower bulk susceptibility. This might occur because the contribution of a mineral population to the anisotropy of the rock is a function of several factors that include the intrinsic susceptibility of the mineral, its intrinsic anisotropy and the degree of preferred orientation of the individual mineral grains. Consequently, the orientation (and intensity) of the magnetic anisotropy of ferromagnetic granites and lava flows is not necessarily associated to the fabric of the magnetite responsible for the large bulk susceptibility. Thus, it is convenient to examine in some detail the origin of the AMS of rocks, even in the cases where the mineral source controlling the bulk susceptibility is well established.

Concerning the intensity of the magnetic anisotropy, Bouchez (1997) noted that the paramagnetic granites tend to be less anisotropic than the ferromagnetic ones. He attributed this difference to the action of magnetic interactions between the ferromagnetic grains because the intrinsic anisotropy of the ferromagnetic minerals tends to be extremely small. Consequently, according to him, the only form in which a petrofabric dominated by ferromagnetic minerals could achieve large degrees of anisotropy would be through magnetic interactions of neighboring grains.

Although such interpretation is very reasonable, it is based on three underlying assumptions that may not be valid at all times. One of these assumptions is that the mineral phases involved have a well constrained degree of anisotropy. The second assumption is that such intrinsic anisotropy constitutes an upper limit for the anisotropy of the rock, and the third assumption considers that large degrees of anisotropy can be achieved only in the cases of a perfect alignment of the mineral grains. The first of these assumptions is justified because due to their cubic structure magnetite grains tend to be equant, therefore having an intrinsic low degree of anisotropy. Nevertheless, some grains that could be considered equant when optically inspected (aspect ratio of ~1.2) could yield a degree of anisotropy of nearly 2%, which is higher than the degree of anisotropy found in many rocks, and in particular of S-Type granites. Furthermore, there are some cases where magnetite grains can be found to be present in the form of elongated microlites or well oriented magnetite aggregates. For instance, Cañón-Tapia and Castro (2004) made optical and AMS measurements on rock samples in which elongated microlites of magnetite promoted a large degree of anisotropy unrelated to the action of magnetic interactions. Also, Grégoire et al. (1995) showed that the shape anisotropy arising from the preferred orientation of mineral aggregates dominated the anisotropy arising from magnetic orientations, yet the degree of anisotropy of the rock was not directly associated to the shape of individual minerals.

As for the second and third assumptions, limiting the anisotropy of the rock to the intrinsic anisotropy of the minerals, Cañón-Tapia and Castro (2004) showed that very large degrees of anisotropy exceeding those of a single grain could be achieved depending on the distribution of the particles. Furthermore, these authors showed that it is not necessary to have a perfect alignment of the grains to produce a large degree of anisotropy (Fig. 18.5). Consequently, it is not entirely justified to conclude that large degrees of anisotropy are related to a composite fabric that involves magnetic interactions, even if the fabric is dominated by magnetite grains. For this reason, the large degrees of anisotropy found in some ferromagnetic granites (e.g., Archanjo 1993) should not be associated with

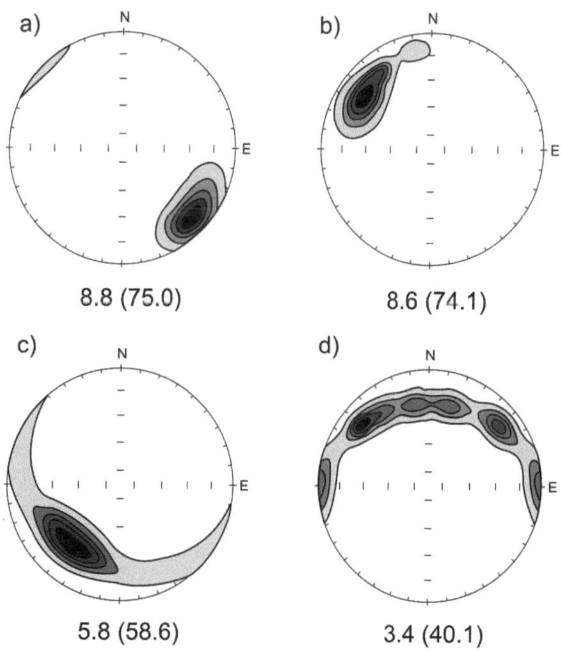

Fig. 18.5 Diagrams showing a synthetic distribution of particles and the degree of anisotropy obtained from the same distribution as a function of particle aspect ratio. The numbers on parenthesis display extreme cases of particles with an aspect ratio of 0.2, whereas the other numbers correspond to aspect ratios of 0.9. (Modified from Cañón-Tapia and Castro 2004)

magnetic interactions unless other sources of information supporting such association are available.

The above statement does not deny the possibility that magnetic interactions can indeed substantially increase the degree of anisotropy of a group of ferromagnetic grains. Such an effect has been well documented by theory and experiments (Stephenson 1994, Grégoire et al. 1995, Cañón-Tapia 1996, 2001) and therefore it should be considered as a reasonable explanation for some of the observations made both in lava flows and granites. Other possible departures from a simple relation between mineral and magnetic fabrics are associated to the single domain (SD) effect, in which the largest and shortest dimensions of a grain correspond to the minima and maxima susceptibility axes, respectively (e.g., Rochette et al. 1991, 1999).The extent to which magnetic interactions and the SD effect are widespread, however, is still debatable.

In summary, it is established that despite the apparent differences between the magnetic fabrics of granites and lava flows, the variability of the magnetic fabric in both types of rocks is equally large. To some extent such variability is associated to the wide diversity of compositions that are encompassed by the terms "granite" and "lava flows". Nevertheless, such variability also reflects the fact that the underlying processes controlling the acquisition of the magnetic fabric in both instances might be complex. In the following sections attention is focused on the mechanical processes responsible for the acquisition of a mineral fabric in both types of rocks.

18.5 Fabric Acquisition in Igneous Rocks: How It Is the Expected AMS?

Despite the apparent differences in the definitions of lava flows and granites, the rocks that are described by each of these terms share many features, especially when examined from the point of view of the mechanisms controlling their emplacement. Indeed, both lava flows and granites are formed as the result of movement and/or deformation of a viscous liquid that contains solids and probably gaseous phases in suspension. Consequently, the mechanical aspects of their emplacement can be described by resorting to fluid dynamic approaches, and many aspects of the emplacement of both lava flows and granites can be described by using terms that are common to the realm of plastic flow. Furthermore, as pointed out by Cañón-Tapia (2005), the similarities between lava flows and granites are much more marked in this sense than between two different types of extrusive igneous rocks such as lava flows and the deposits of pyroclastic activity. The reason for such similarities are based on the facts that for lava flows and granites the continuum phase is a viscous liquid, whereas in the case of pyroclastic rocks the continuum phase has an almost negligible viscosity because it is a gas. In addition, the liquid phase in lava flows and granites usually will move following a laminar regime, whereas in the case of pyroclastic currents very often flow takes place in a turbulent form. Thus, from a mechanical point of view it is entirely justified to compare some physical characteristics of both types of rocks, leaving aside (temporarily) their inherent differences in chemical or mineralogic composition.

18.5.1 Cyclicity of the Fabric

Underlying most models of petrofabric acquisition is the mechanical model of movement of an ellipsoid immersed in a moving viscous fluid proposed by Jeffery (1922). Jeffery's equations of motion of rigid grains subject to forces exerted by a moving fluid along its surface can be used to estimate the angular velocity of the grain (ω) as a function (1) of the ellipsoid axes (A, B, C), (2) of the components of the strain rate tensor (d, f, g) and (3) of the components of the vorticity tensor (ξ, β, ζ):

$$\begin{aligned}(B^2 + C^2)\omega_x &= B^2(\xi + f) + C^2(\xi - f) \\ (C^2 + A^2)\omega_y &= C^2(\beta + g) + A^2(\beta - g) \\ (A^2 + B^2)\omega_z &= A^{2\sim}(\zeta + d) + B^{2\sim}(\zeta - d)\end{aligned} \quad (18.1)$$

These equations must be solved numerically (Freeman 1985, Hinch and Leal 1979, Iezzi and Ventura 2002, Jezek 1994, Jezek et al. 1994, 1996) unless some simplifying assumption is introduced, in which case it is possible to find analytical solutions. Some of these simplifications, however, yield results that eliminate the cyclic behavior inherent in the complete equations of movement, or that are valid to describe only very special cases of deformation (e.g., Gay 1966, 1968, Owens 1974).

Of particular interest for the study of AMS in igneous rocks is the work of Dragoni et al. (1997). These authors solved Jeffery's equations for the special case of an axi-symmetric particle (i.e., A = C in Eq. (18.1)). They expressed the solutions in terms of Euler angles θ and ϕ:

$$\tan\phi = r\tan\left[\frac{r\dot{\gamma}t}{r^2+1} + \arctan\left(\frac{1}{r}\tan\phi_0\right)\right]$$
$$\tan^2\theta = \frac{r^2\cos^2\phi_0 + \sin^2\phi_0}{r^2\cos^2\phi + \sin^2\phi}\tan^2\theta_0 \qquad (18.2)$$

where r = A/B is the elongation ratio of the grain, t is time, and θ_0, ϕ_0 denote the initial orientation of the particle. Thus, it is possible to use Eq. (18.2) to find the orientation of any axi-simmetric particle as a function of the strain experienced by the moving fluid and of the initial orientation of the particle.

Notably, neither the original formulation of Jeffery (1922) nor the special case examined by Dragoni et al. (1997) makes an explicit mention of the role played by the viscosity of the fluid phase. Such apparent omission is explained by considering that the viscosity is essentially a parameter that relates the speed of the deformation with the stress acting on the fluid. Thus, an increase in viscosity only implies that in order to reach a certain amount of deformation it is necessary to wait longer than it would be necessary if the viscosity was lower. In the terms of Eq. (18.2), this implies that a given state of orientation will be reached very soon in some cases, whereas in other cases it may take much longer, but nevertheless such state of deformation will be reached as long as the stresses continue to act in the fluid. In practice, this implies that despite the large differences in viscosity likely to characterize lava flows on the one hand and granites on the other, the same fluid equations are likely to be valid to describe processes of emplacement applicable to both types of rocks.

Based on the theoretical model expressed by Jeffery's equations, there are several forms to calculate the fabric of a rock that experienced plastic deformation (flow) during its creation (e.g., Cañón-Tapia and Chávez-Álvarez 2004a, Fernandez 1987, Freeman 1985, Gay 1966, Ildefonse et al. 1997, Hrouda et al. 1994, Jezek et al. 1994, Manga 1998). One approach is to constrain the evolution of a multiparticle system in which each particle follows the equations of movement described above, and in particular as expressed by Eq. (18.2). To some extent, this was the approach followed by Dragoni et al. (1997), although the initial distribution of particles that they considered was not random. Due to the symmetry in the initial distribution of the system of particles considered by those authors, the resultant fabric (i.e., the average orientation of all the particles in the system) remained always within a given plane. Consequently, their results were not very realistic as there is no a priori reason that could serve to justify an initial distribution of particles with a marked symmetry about the plane of flow.

To address this issue, Cañón-Tapia and Chávez-Álvarez (2004b) studied cases in which the particles had an initial orientation that was randomly distributed (Fig. 18.6). As these authors showed, due to the cyclic movement of each particle during deformation, the fabric of a rock also has a cyclic behavior (Fig. 18.7). Nevertheless, the more elongated (or flattened) particles can produce the illusion of a stable fabric orientation because the system of particles spends much time in a given orientation. In addition, the amount of deformation required to complete one cycle in the movement of the particles is also extremely

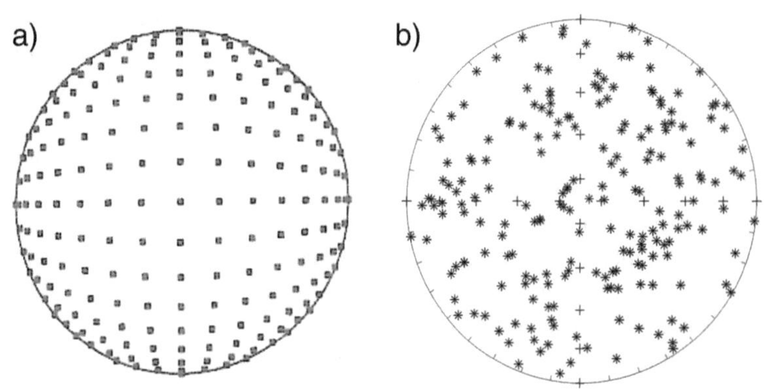

Fig. 18.6 Two examples of initial distributions used to calculate the evolution of a mineral fabric as a function of deformation during flow. Note that the symmetry of the uniform distribution on (**a**) is not present in the random distribution illustrated on (**b**). (Modified from Dragoni et al. 1997, Cañón-Tapia and Chávez-Álvarez 2004b)

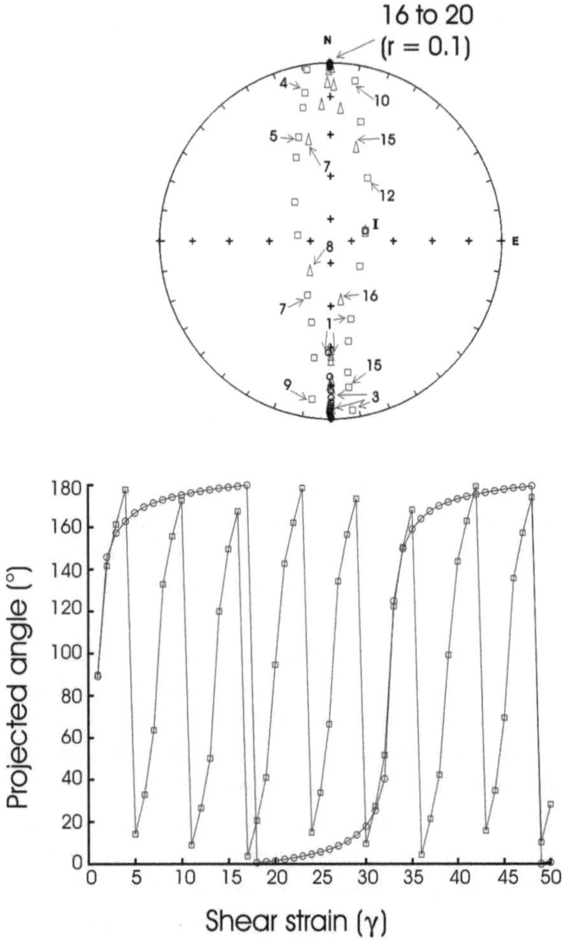

Fig. 18.7 Diagrams showing the evolution of a mineral fabric as a function of deformation during flow. The different symbols on the equal area projection represent the average mineral orientation obtained from particles with different aspect ratios and the numbers indicate selected values of the deformation. Note that for very elongated particles (aspect ratio = 0.1) a quasi-stable fabric is defined for a relatively large range of deformation. The lower diagram displays the results for two different aspect ratios that nonetheless display a cyclic behaviour of the mineral fabric. (Modified from Cañón-Tapia and Chávez-Álvarez, 2004b)

large, therefore contributing to create the illusion of a stable fabric in many cases of practical interest. Furthermore, Cañón-Tapia and Chávez-Álvarez (2004b) also found examples of mineral fabrics that are perpendicular to the direction of shear, therefore suggesting that in some cases the relationship between mineral fabric and shear orientation is not as simple as it had been assumed until then.

In summary, one of the characteristics of incorporating the full model of particle motion developed by Jeffery (1922) in models of mineral fabric acquisition is that the mineral fabric will have a cyclic behavior. Although this result implies a complex relationship between fabric and flow regime, it is possible to take advantage of the predicted variation of mineral fabric as a function of deformation to extract more information concerning the details of emplacement of a particular rock type if the appropriate sampling scheme is followed, as was later demonstrated by Cañón-Tapia and Herrero-Bervera (2009).

18.5.2 Causes for Non-cyclic Behavior

At this point in the discussion it is important to note that the cyclic behavior described above is valid only if each particle in the system is mechanically isolated from its neighbors during fluid motion. If there are mechanical interactions between particles, for example due to a large particle concentration, then a stable fabric orientation can be achieved even for particles with intermediate values of aspect ratios. Some of the relevant thresholds of particle concentration that can stop the cyclic movement of the mineral fabric have been determined experimentally (Arbaret et al. 1996, 1997, Fernandez et al. 1983, Ildefonse et al. 1992). These works have shown that below a threshold concentration between 15 and 20% the particles behave cyclically, and it is only for particle concentrations above 20% that collisions between particles force the mineral fabric to achieve a nearly flow-parallel orientation almost independently of the specific particle shape (Arbaret et al. 1996, Ildefonse et al. 1992). Consequently, by paying attention to the orientation of the fabric along the suspected direction of flow, it would seem possible (at least in principle) to assess the concentration of particles present in a particular magma at the time of its emplacement. If the fabric is shown to be cyclic the concentration of minerals most likely was small, whereas a stable fabric orientation would be indicative of a much larger particle concentration at the time of the rock emplacement/ deformation.

18.5.3 Implications of the Theoretical Models of Fabric Acquisition

In practice, the cyclicity vs. non-cyclicity of the mineral fabric predicted by theory suggests that the AMS of lava flows and many small dykes or other similar tabular intrusions should differ from that of most granites, even if fabric acquisition is controlled by the same physical mechanisms in both cases. Indeed, flow of lava (magma) in extrusive rocks or small tabular intrusions can take place in conditions of relatively low mineral concentration (e.g., Cashman et al. 1999, Polacci et al. 1999). In contrast, deformation affecting many granites will take place after the emplacement of the magma at the crustal level of the intrusion, and in conditions favoring a much larger concentration of minerals. Consequently, according to theory, the mineral fabric of granites will tend to be more stable for longer distances, whereas in lava flows and tabular intrusions the fabric might change rapidly in short distances.

Actually, the theoretically expected mineral fabric in either lava flows and granites is in excellent agreement with many observations made on both types of rocks. This agreement is still valid even if the mineral phase present during deformation was not the ferromagnetic mineral controlling the AMS. This is possible because the preferred orientation of magnetite grains, created after flow (deformation) had ceased, might be controlled by the highly anisometric intergranular spaces between the silicate minerals present during flow. Indeed, such process has been document for both lava flows and granites (Hrouda et al. 1971, Hargraves et al. 1991).

18.6 The Influence of Late Deformation

One of the characteristics of earlier measurements of the AMS of lava flows was its relatively large variability in intensity of bulk susceptibility and on other parameters and orientations of principal susceptibility. Such variability led earlier workers to suggest that AMS was of little use to infer flow directions in lava flows (see Cañón-Tapia 2004a, for a complete list of references). As explained in the previous section, the reasons for such variability may not be related to noise, but rather reflect the cyclic behavior of a group of particles embedded in a deforming (flowing) matrix, although in fact there are other sources of complexity in the process of fabric acquisition that also can contribute to blur the AMS signal.

For the case of lava flows, the AMS has usually been interpreted to be primary (and hence simple) in every instance. Nevertheless, as shown by Cañón-Tapia and Coe (2002) the AMS of some lava flows might be very complex (Fig. 18.8). In particular the process known as endogenous growth, relatively common in lava flows (e.g., Hon et al. 1994), can modify the pattern of AMS selectively. Although in the strictest sense this process can not be called post emplacement (because it actually is an integral part of the emplacement of the finally formed lava flow), it produces important deviations from the fabric acquired during emplacement of lava flows or dykes formed in a single stage.

In the case of granites, post emplacement processes are also of concern. The most commonly invoked process responsible for the deviation of an original magmatic fabric is related to later tectonic events. In these cases, the original fabric experiences a rotation towards the direction of principal post-emplacement strain. Such fabric overprinting may be partial or completely obliterate the original magmatic fabric (Benn 1994, Bouchez 1997, Hrouda et al. 2002). Usually, a distinction between original magmatic fabric and post-emplacement acquired fabric can be made in these cases by paying attention to the pattern of fabrics relative to the walls of the pluton, or to the surrounding country rock. If the orientation of the principal susceptibilities parallels the walls of the pluton, and/or has a different orientation than the AMS of the country rock, then a magmatic deformation can be inferred. If the pattern of principal susceptibilities, however, cuts along the walls of the intrusion and is parallel to the orientation outside the pluton, then the case for tectonic overprint can be justified.

In some other cases, however, the complexity of the fabric within a single pluton can be associated to successive kinematic events without leaving a trace of solid-state deformation. In these cases, use of other magnetic techniques, such as the anisotropy of anhysteretic remanent magnetization (AARM), can provide additional clues concerning the origin of the mineral fabrics (e.g., Trinidade et al. 1999, Fig. 18.9). In addition, some variations of susceptibility values can be helpful to identify a cryptic zoning within a particular pluton. In these cases, the mineral fabric might have been acquired as the result of the progressive injection of different pulses of magma (e.g., Méndez-García 2005).

In summary, it is clear that we can find a range of post emplacement alterations of the AMS in both granites and lava flows alike. In some cases, it might be

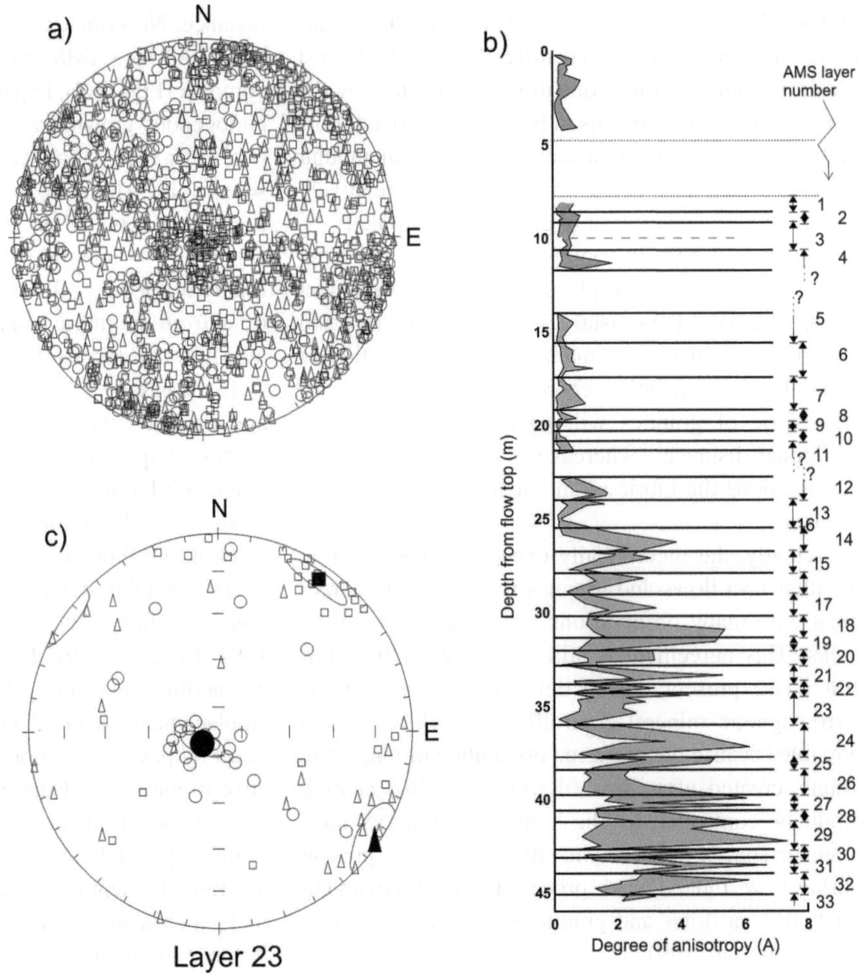

Fig. 18.8 Diagrams showing the variation of the AMS results obtained from a single lava flow. When all of the samples are plotted in a single diagram, the distribution of the principal susceptibilities is extremely noisy as shown in (**a**), but if attention is given to the various layers defined by fluctuations of the degree of anisotropy shown in (**b**), the principal susceptibilities define relatively well grouped susceptibility axes, as illustrated in (**c**). (Modified from Cañón-Tapia and Coe 2002)

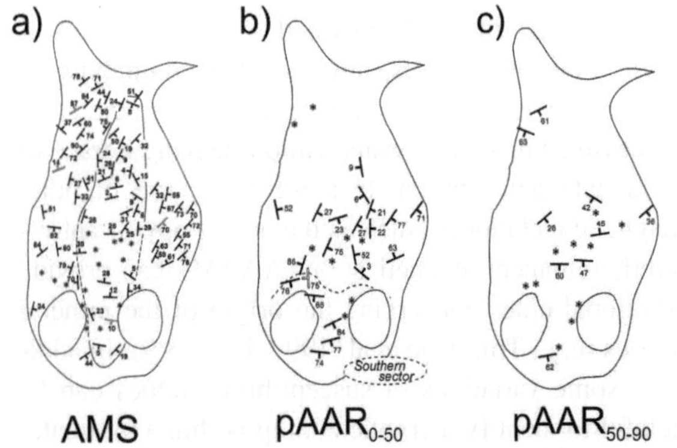

Fig. 18.9 Differences between the orientation of the magnetic fabric deduced from (**a**) the anisotropy of magnetic susceptibility and (**b**), (**c**) the anisotropy of anhysteretic remanence. (**b**) and (**c**) differ from each other in the field interval used to induce the remanent magnetization. In all cases, the asterisks mark fabrics with orientations poorly determined by the corresponding method. (Modified from Trindade et al. 1999)

possible to make an interpretation of the measured fabric based on additional sources of observation, including field and labotratory techniques. Nevertheless, in general, various possible scenarios will need to be carefully evaluated in a case by case basis.

18.7 How Does the AMS of a Lava Deformed on Laboratory Conditions Relate to the Magmatic Fabric of Granites?

Motivated by the wide diversity of magnetic fabric signatures that can be found to exist in lava flows, Cañón-Tapia and Pinkerton (2000) completed a series of experiments aiming to establish the form in which flow-related deformation relates to the measured magnetic anisotropy of a rock. These authors showed that several aspects of the AMS of lava flows are a function of both the thermal and shearing history of the volume of rock sampled. In particular, these experiments revealed that the bulk susceptibility was very large if the lava was allowed to reach a temperature close to 1,000°C before being quenched, whereas the lava of the same composition quenched from temperatures around 1,300°C yielded low values of bulk susceptibility. All of these samples were holohyaline and therefore the variations of bulk susceptibility are due to the microlites of ferromagnetic grains rather than to minerals of large shape. Unfortunately these microlites are too small to make a statistically significant observation concerning their orientation to compare with the AMS. Nevertheless, due to the lack of other mineral phases in the samples, the correspondence between the mineral and the magnetic fabric is straightforward.

As for other characteristics of the magnetic fabric, it was found that the direction of deformation could be preserved only if the deformation took place immediately before quenching the samples, although the more anisotropic samples were obtained when the deformation occurred at the lower experimental temperatures. Such results indicate only the last phases of deformation are detectable by using the AMS method, and that high strain rates do not result in high degrees of anisotropy if either the deformation ends while lava is still fluid (i.e., it has low viscosity) or if the orientation of maximum shear stress varies with time.

When applied to natural rocks, taking into consideration that an increment in crystal content of lava (1) increases the viscosity of the fluid and (2) promotes the acquisition of a more stable fabric due to enhanced particle-particle mechanical interaction during deformation, the experimental results of Cañón-Tapia and Pinkerton (2000) suggest that flow related fabrics in magmas with a high crystal content are more likely to be preserved in nature than those of lavas with a lower crystal content. An exception to this rule would be provided if the lavas with low crystal content are chilled relatively rapidly. Another implication of these experimental results is that the magnetic fabric of even the most viscous magmas (i.e. fluid systems with a high content of minerals in solution) can be relaxed if the magma remains at an elevated temperature for a long time after the deformation had ceased. In consequence, even when a pluton might favor the acquisition of a stable fabric due to its larger mineral concentration relative to that found in most lava flows, such mineral fabric might be relaxed if the pluton comes to rest while magma is still at a relatively high temperature, or has a crystal content below the threshold marking the change from a Newtonian to a Non-Newtonian rheology. Such threshold is still a matter of debate, but some authors estimate it to be as high as 80% (Fig. 18.10). Consequently, at least in theory it would seem possible that the mineral fabric of some granites can be relaxed in the same form implied by the experimental results of Cañón-Tapia and Pinkerton (2000).

For these reasons, it is concluded that even if the grain-scale deformation models discussed above provide a very accurate description of the process of acquisition of a "primary flow fabric", the fabric that is finally measured in the laboratory may be unrelated to those processes. The key factor that needs to be taken into consideration involves the relative timing of the end of the flow and the achievement of a critical viscosity that prevents the relaxation of the acquired fabric. Consequently, these experiments indicate that the concept of a "primary flow fabric" in the case of lava flows is relatively unconstrained, suggesting that a careful evaluation of results is needed in a case by case basis.

Interestingly, as noted by Paterson et al. (1998) the concept of a "primary flow fabric", as applied to fabric patterns in plutons, also contains some ambiguities. The source of such ambiguities resides in the blending of three different conditions that might alter the final fabric measured in the pluton. Thus, the interpretation of magmatic fabrics in plutons commonly assumes that (1) the hydrodynamic alignment of mineral grains suspended in the melt dominates the fabric, (2) that

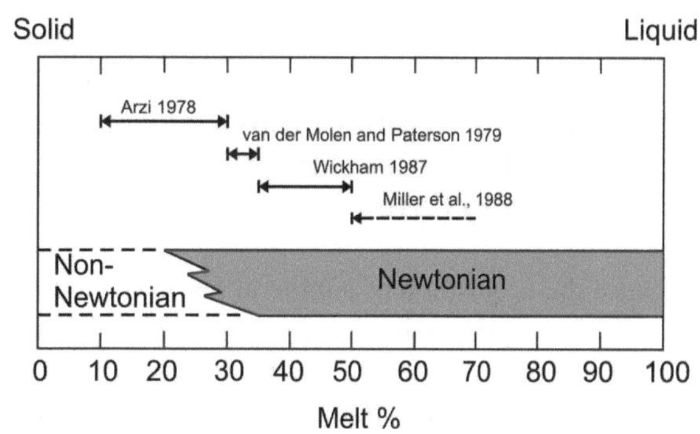

Fig. 18.10 Diagram showing the position of the critical melt percentage required to induce a significant change in the rheological behavior of a silicate fluid according to various authors. Also, the transition between a Newtonian and Non-Newtonian regime from Petford (2003) is indicated

planar and linear mineral mineral fabrics define planes and lines of flow, respectively, and (3) that map-scale fabric patterns form exclusively by processes inside the pluton when the fabric parallels the boundaries that separate the intrusion and the country rock. All of these three assumptions are shared by the common interpretation of magnetic fabrics in lava flows, but as the experimental results of Cañón-Tapia and Pinkerton (2000) showed, those conditions may not be entirely satisfied in every circumstance. Furthermore, the analysis made by Paterson et al. (1998) led these authors to emphasize the need to assess the timing relationships prevalent during emplacement of the pluton and the lock-in of the mineral fabric before making a geologic interpretation of the observed fabric patterns because those patterns are very likely to preserve only the last increment of strain before magma crystallization. Again, this conclusion is identical to the conclusion reached by the examination of the results of Cañón-Tapia and Pinkerton (2000) obtained while examining the behavior of the magnetic fabric in laboratory deformed lavas, even if the crystal content of both types of samples is entirely different. For these reasons, it is considered here that both studies reflect processes that are fundamental in controlling the emplacement of both types of rock, and that therefore, the AMS of lava flows and granites is not as different as it might seem at first sight. Actually, to some extent it would be justified to say that the results obtained by Cañón-Tapia and Pinkerton (2000) provide an experimental justification to the conclusions reached by Paterson et al. (1998), even when both of these studies were motivated by entirely different reasons, and were completed in very different settings.

18.8 Should Magmatic and Magnetic Fabrics be Considered Equivalent in Every Case?

Because the acquisition of a mineral fabric (including a magnetic one) depends very strongly on the deformation history of a rock, and to some extent is insensitive to the chemical composition of the melt (which only influences a numerical value associated to the same physical property), it turns out that the study of the AMS of lava flows and granites has many things in common. Nevertheless, from the specific point of view of magnetic measurements there are some significant differences between most lava flows and granites that need to be considered to make a proper interpretation of results. The two most significant sources of differences in the magnetic fabric of granites and lava flows are:

18.8.1 Magnetite Content

In lava flows the magnetic fabric is commonly related to the presence of a Ti-magnetite either directly crystallized from the magma or formed as the result of alteration of a ferrosilicate, or as inclusions in that silicate. In contrast, in some granites Ti-magnetites might be completely absent. As discussed above, this difference controls the range of bulk susceptibilities that are expected to be found in each type of rock. Most importantly, the larger content of ferromagnetic minerals in lava flows makes more probable the occurrence of a magnetic fabric containing a component associated to the magnetic interactions between the ferromagnetic grains, rather than being associated to the preferred orientation of the minerals contained in the rock. Thus, in some cases it might be necessary to complete an optical examination of selected specimens to assess the possibility of a mineral fabric that is not coaxial with the measured magnetic fabric.

18.8.2 Fabric Homogeneity

The crystal content in lava flows at the time of plastic flow might be relatively small whereas in granites such process is very likely to take place when crystal content is relatively high. Because particle concentration in most granites is thought to be larger than 20%, it is expected that the fabric of these rocks remains stable for long distances. In contrast, particle concentration can be well below the critical threshold values in many dykes and lava flows (e.g., Cashman et al. 1999, Polacci et al. 1999). Consequently, in these types of rocks the petrofabric might be cyclic and therefore have a more complex character than for most granites. Consequently, in lava flows fabric variations might take place in the scale of centimeters whereas in granites the mineral fabric might be homogeneous at scales of even kilometers.

Nevertheless, as shown by Olivier et al. (1997) the magnetic fabric of many granites may display variations at small scales comparable to those observed in many lava flows. Such variations can be considered to be a function of heterogeneous grain size and local inhomogeneities of the magma that can take place within a pluton. These inhomogeneities can even influence the detection of a significant anisotropy at a specimen scale, and therefore it is important to pay attention to the significance tests that indicate how confidently two of the magnetic axis were distinguished from each other during the measuring of the anisotropy of a single specimen (Pueyo et al. 2004). Such criteria are equally applicable to the studies of AMS made in lava flows.

Conclusions

It has been shown in this work that despite the differences in the definition of the terms "granites" and "lava flows" these two types of rocks display many common features when attention is focused on their AMS signature. Nevertheless, the similarities in the process of acquisition of the AMS of these two rock types should not be interpreted as indicative of a similar AMS signature under every circumstance. Actually, due to variations in some parameters controlling the acquisition of any mineral fabric, and in particular of mineral concentration during the late stages of plastic flow, there is a range of possible AMS signatures that can be found both in lava flows and granites alike. Consequently, rather than visualizing the AMS of lava flows and granites as two completely unrelated types of rocks it is much better to visualize their AMS signatures as controlled by a common process, that due to the special conditions prevalent in a case by case basis ultimately determine the measured fabric (whether optically determined or with the use of magnetic methods). This situation indicates that the AMS of lava flows, dykes and granites can be considered to be part of a continuum that also reflects the conditions of aquisition of any other type of mineral fabric (Fig. 18.11). Awareness of such circumstances would result in a much better interpretation of the measured AMS signature, and a sounder reconstruction of the emplacement history of that particular rock.

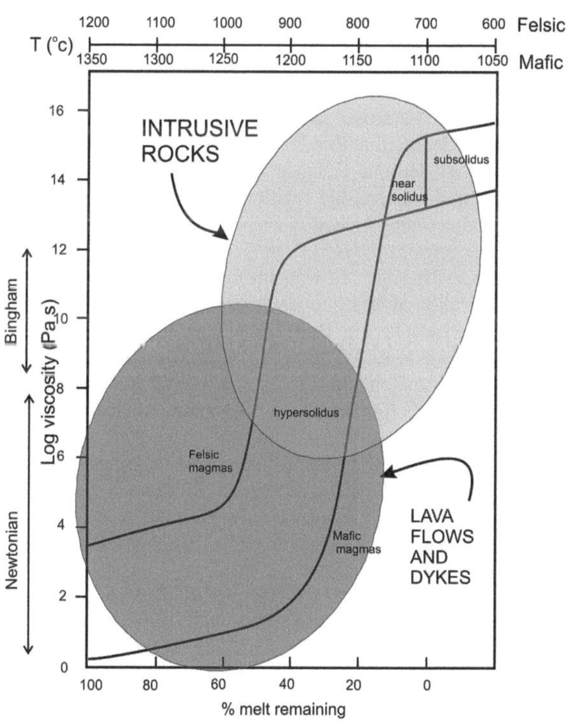

Fig. 18.11 Diagram showing the relationship between temperature, % of melt remaining and rheological properties of both felsic and mafic magmas. The curves delimiting the three types of deformation (subsolidus, near solidus and hypersolidus) are taken from Paterson et al. (1998). The shaded areas include the different regimes discussed in this work, and are only illustrative

Acknowledgements The comments made by two anonymous reviewers helped to clarify many of the ideas presented in the text are greatly appreciated. I also express my thanks to E. Herrero-Bervera for the encouragement that led me to give a more formal expression to some of the ideas presented here and that had been informally discussed several times.

References

Ade-Hall J, Palmer HC, Hubbard TP (1971) The magnetic and opaque petrological response of basalts to regional hydrothermal alteration. Geophys J Roy Astronomical Soc 24:137–174

Ade-Hall J, Khan MA, Dagley P, Wilson RL (1968a) A detailed opaque petrological and magnetic investigation of a single Tertiary lava flow from Skye, Scotland – I Iron-titanium oxide petrology. Geophys J R Astron Soc 16:375–388

Ade-Hall J, Khan MA, Dagley P, Wilson RL (1968b) A detailed opaque petrological and magnetic investigation of a single Tertiary lava flow from Skye, Scotland – II Spatial variations of magnetic properties and selected relationships between magnetic and opaque petrological properties. Geophys J R Astron Soc 16:389–399

Ade-Hall J, Khan MA, Dagley P, Wilson RL (1968c) A detailed opaque petrological and magnetic investigation of a single Tertiary lava flow from Skye, Scotland – III Investigations into the possibility of obtaning the intensity of the ambient

magnetic field (Fanc) at the time of the cooling of the flow. Geophys J Roy Astronomical Soc 16:401–415
Arbaret L, Diot H, Bouchez JL (1996), Shape fabrics of particles in low concentration suspensions: 2D analogue experiments and applications to tiling in magma. J Struct Geol 18:941–950
Arbaret L, Diot H, Bouchez JL, Lespinasse P, de Saint-Blanquat, M. (1997) Analogue 3D simple-shear experiments of magmatic biotite subfabrics. In: Bouchez JL, Hutton DHW, Stephens WE (eds) Granite: from segregation of melt to emplacement fabrics. Kluwer, Dordrecht, pp 129–143
Archanjo CJ (1993) Fabriques de plutons granitiques et déformation crustale du Nord-Est du Brésil: une étude par l'anisotropie de la susceptibilité magnétique de granites ferromagnétiques. Thesis University, Paul-Sabatier, Toulouse 167 pp
Archanjo CJ, Launeau P, Bouchez JL (1995) Magnetic fabric vs. magnetite and biotite shape fabrics of the magnetite-bearing granite pluton of Gameleiras (northeast Brazil) Phys Earth Planetary Inter 89:63–75
Aydin A, Ferré EC, Aslan Z (2007) The magnetic susceptibility of granitic rocks as a proxy for geochemical differentiation: example from the Saruhan granitoids, NE Turkey. Tectonophysics 441:85–95
Benn, K. 1994. Overprinting of magnetic fabrics in granites by small strains: numerical modeling. Tectonophysics 233:153–162
Best MG (1982) Igneous and metamorphic petrology. W.H. Freeman and Company, New York, NY, 630 pp
Borradaile GJ, Gauthier D (2003) Interpreting anomalous magnetic fabrics in ophiolite dikes. J Struct Geol 25:171–182
Bouchez JL (1997) Granite is never isotropic: an introduction to AMS studies of granitic rocks. In: Bouchez JL, Hutton DHW, Stephens WE (eds) Granite: from segregation of melt to emplacement fabrics. Kluwer, Dordrecht, pp 95–112
Cañón-Tapia E (1994) AMS parameters: guidelines for their rational selection. Pure Appl Geophys 142:365–382
Cañón-Tapia E (1996) Single-grain versus distribution anisotropy: a simple three-dimensional model. Phys Earth Planetary Inter 94:149–158
Cañón-Tapia E (2001) Factors affecting the relative importance of shape and distribution anisotropy in rocks: theory and experiments. Tectonophysics 340:117–131
Cañón-Tapia E (2004a) Anisotropy of magnetic susceptibility of lava flows and dykes: an historical account. In: Martín-Hernández F, Lüneburg C, Aubourg MC, Jackson M (eds) Magnetic fabric. Methods and applications. Geological Society, London, pp 205–225
Cañón-Tapia E (2004b). Flow direction and magnetic mineralogy of lava flows from the central parts of the Peninsula of Baja California, Mexico. Bull Volcanol 66:431–442
Cañón-Tapia E (2005) Uses of anisotropy of magnetic susceptibility in the study of emplacement processes of lava flows. In: Manga M, Ventura G (eds), Kinematics and dynamics of lava flows. Geological Society of America, Boulder, Colorado, USA, pp 29–46
Cañón-Tapia E (2007) Susceptibility, parameters, anisotropy. In: Gubbins D, Herrero-Bervera E (eds), Encyclopedia of Geomagnetism and Paleomagnetism. Springer, Dordrecht, pp 937–939
Cañón-Tapia E, Castro J (2004) AMS measurements on obsidian from the Inyo Domes, CA: a comparison of magnetic and mineral preferred orientation fabrics. J Volcanol Geothermal Res 134:169–182
Cañón-Tapia E, Chávez-Álvarez MJ (2004a) Theoretical aspects of particle movement in flowing magma: implication for the anisotropy of magnetic susceptibility of dykes. In: Martín-Hernández F, Lüneburg C, Aubourg MC, Jackson M (eds), Magnetic fabric. Methods and applications. Geological Society, London, pp 227–249
Cañón-Tapia E, Chávez-Álvarez MJ (2004b) Rotation of uniaxial ellipsoidal particles during simple shear revisited: the influence of elongation ratio, initial distribution of a multiparticle system and amount of shear in the acquisition of a stable orientation. J Struct Geol 26:2 073–2 087
Cañón-Tapia E, Coe R (2002) Rock magnetic evidence of inflation of a flood basalt lava flow. Bull Volcanol 64:289–302
Cañón-Tapia E, Herrero-Bervera E (2009), Sampling strategies and the anisotropy of magnetic susceptibility of dykes. Tectonophysics 466:3–17
Cañón-Tapia E, Pinkerton H (2000) The anisotropy of magnetic susceptibility of lava flows: an experimental approach. J Volcanol Geothermal Res 98:219–233
Cañón-Tapia E, Walker GPL, Herrero-Bervera E (1995) Magnetic fabric and flow direction in basaltic pahoehoe lava of Xitle volcano, Mexico. J Volcanol Geothermal Res 65:249–263
Cañón-Tapia E, Walker GPL, Herrero-Bervera E (1996) The internal structure of lava flows – insights from AMS measurements I: near vent aa. J Volcanol Geothermal Res 70:21–36
Cañón-Tapia E, Walker GPL, Herrero-Bervera E (1997) The internal structure of lava flows – insights from AMS measurements II: Hawaiian pahoehoe, toothpaste lava and 'a'a. J Volcanol Geothermal Res 76:19–46
Chlupácová M, Hrouda F, Janák F, Rejl L (1975) The fabric, genesis and relative age relations of the granitic rocks of the Cistá-Jesenice Massif (Czechoslovakia) as indicated by magnetic anisotropy. Gerl Beitr Geophys 84:487–500
Cashman KV, Thornber, C, Kauahikaua JP (1999) Cooling and crystallization of lava in open channels, and the transition of Pahoehoe Lava to 'a'a. Bull Volcanol 61:306, 323
Dragoni M, Lanza, R, Tallarico A (1997) Magnetic anisotropy produced by magma flow: theoretical model and experimental data from Ferrar dolerite sills (Antarctica) Geophys J Int 128:230–240
Dortman NB (1984). Physical properties of rocks and mineral deposits. Nedra, Moscow, 455 pp. (in Russian)
Ellwood BB, Hrouda F, Wagner JJ (1988) Symposia on magnetic fabrics: introductory comments. Phys Earth Planetary Inter 51:249–252
Ferré EC, Martín-Hernández F, Teyssier, C, Jackson M (2004) Paramagnetic and ferromagnetic anisotropy of magnetic susceptibility in migmatites: measurements in high and low fields and kinematic implications. Geophys J Int 157:1119–1129
Fernandez A (1987) Preferred orientation developed by rigid markers in two-dimensional simple shear strain: a theoretical and experimental study. Tectonophysics 136:151–158
Fernandez A, Feybesse JL, Mezure JF (1983) Theoretical and experimental study of fabrics developed by different shaped markers in two-dimensional simple shear. Bull de la Soc geol de France 25:319–326

Freeman B (1985) The motion of rigid ellipsoidal particles in slow flows. Tectonophysics 113:163–183

Gay NC (1966) Orientation of mineral lineation along the flow direction in rocks: a discussion. Tectonophysics 3:559–564

Gay NC (1968) The motion of rigid particles embedded in a viscous fluid during pure shear deformation of the fluid. Tectonophysics 5:81–88

Gleizes G, Nédélec A, Bouchez JL, Autran, A, Rochette P (1993) Magnetic susceptibility of the Mount-Louis Andorra ilmenite-type granite (Pyrenees): a new tool for the petrographic characterization and regional mapping of zoned granite plutons. J Geophys Res 98:4317–4331

Grégoire V, de Saint-Blanquat M, Nédélec A, Bouchez JL (1995) Shape anisotropy versus magnetic interactions of magnetite grains: experiments and application to AMS in granitic rocks. Geophys Res Lett 20:2765–2768

Hargraves RB, Johnson D, Chan CY (1991) Distribution anisotropy: The cause of AMS in igneous rocks? Geophys Res Lett 18:2193–2196

Hinch EJ, Leal LG (1979) Rotation of small non-axisymmetric particles in a simple shear flow. J Fluid Mech 92:591–608

Hon K, Kauahikaua J, Denlinger, R, Mackay K (1994) Emplacement and inflation of pahoehoe sheet flows: observations and measurements of active lava flows on Kilauea Volcano, Hawaii. Bull Geol Soc Am 106:351–370

Hrouda F (1982) Magnetic anisotropy of rocks and its application in geology and geophysics. Geophysical Survey 5: 37–82

Hrouda, F, Lanza R (1989) Magnetic fabric in the Biella and Traversella stocks (Periadriatic Line): Implications for the mode of emplacement. Phys Earth Planet Interiors 56:337–348

Hrouda F, Chlupacova M, Rejl L (1971) The mimetic fabric of magnetite in some foliated granodiorites, as indicated by magnetic anisotropy. Earth Planet Sci Lett 11:381–384

Hrouda F, Chlupacova M, Schulmann K, Smid J, Zavada P (2005) On the effect of lava viscosity on the magnetic fabric intensity in alkaline volcanic rocks. Studia Geophys Geod 49:191–212

Hrouda F, Melka R, Schulmann K (1994) Periodical changes in fabric intensity during simple shear deformation and its implications for magnetic susceptibility anisotropy of sedimentary and volcanic rocks. Acta Univ Carol 38:37–56

Hrouda F, Putis M, Madarás J (2002) The Alpine overprints of the magnetic fabrics in the basement and cover rocks of the Veporic Unit (western Carpathians, Slovakia) Tectonophysics 359:271–288

Hrouda F, Táborská S, Schulmann K, Jezek J. Dolejs D (1999) Magnetic fabric and rheology of co-mingled magmas in the Nasavrky plutonic complex (E Bohemia): implications for intrusive strain regime and emplacement mechanism. Tectonophysics 307:93–111

Iezzi, G, Ventura G (2002) Crystal fabric evolution in lava flows: results from numerical simulations. Earth Planetary Sci Lett 200:33–46

Ildefonse B, Arbaret, L, Diot H (1997) Rigid particles in simple shear flow: is their preferred orientation periodic or steady-state? In: Bouchez JL, Hutton DHW, Stephens WE (eds) Granite: from segregation of melt to emplacement fabrics. Petrology and structural geology. Kluwer, Dordrecht, pp 177–185

Ildefonse B, Launeau P, Bouchez JL, Fernandez A (1992) Effect of mechanical ointeractions on the development of shape preferred orientations: a two-dimensional experimental approach. J Struct Geol 14:73–83

Ishihara S (1977) The magnetite-series and ilmenite-series granitic rocks. Mining Geol 27:293–305

Jeffery GB (1922) The motion of ellipsoidal particles immersed in a viscous fluid. Proc R Soc London 102:161–179

Jezek J (1994) Software for modelling the motion of rigid triaxial ellipsoidal particles in viscous flow. Comput Geosci 20:409–424

Jezek J, Melka R, Schulmann K, Venera Z (1994) The behaviour of rigid triaxial ellipsoidal particles in viscous flows – modeling of fabric evolution in a multiparticle system. Tectonophysics 229:165–180

Jezek J, Schulmann K, Segeth K (1996) Fabric evolution of rigid inclusions during mixed coaxial and simple shear flows. Tectonophysics 257:203–221

Manga M (1998) Orientation distribution of microlites in obsidian. J Volcanol Geothermal Res 86:107–115

Méndez-García CH (2005) Mecanismo de emplazamiento del plutón El Testerazo, Baja California, deducido a través de mediciones de ASM. MSc thesis, CICESE, Ensenada, 94 pp

Olivier P, de Saint-Blanquat M, Gleizes G, Leblanc D (1997) Homogeneity of granite fabrics at the metre and dekametre scales. In: Bouchez JL, Hutton DHW, Stephens WE (eds) Granite: from segregation of melt to emplacement fabrics. Kluwer, Dordrecht, pp 113–127

Owens WH (1974) Mathematical model studies on factors affecting the magnetic anisotropy of deformed rocks. Tectonophysics 24:115–131

Paterson SR, Fowler TK, Schmidt KL, Yoshinobu AS, Yuan ES, Miller RB (1998) Interpreting magmatic fabric patterns in plutons. Lithos 44:53–82

Petford N (2003) Rheology of granitic magmas during ascent and emplacement. Ann Rev Earth Planet Sci 31:399–427

Polacci M, Cashman KV, Kauahikaua JP (1999) Textural characterization of the pahoehoe – 'a'a transition in Hawaiian basalt. Bull Volcanol 60:595–609

Pueyo EL, Román-Berdiel MT, Bouchez JL, Casas AM, Larrasoaña JC (2004) Statistical significance of magnetic fabric data in studies of paramagnetic granites. In: Martín-Hernández F, Lüneburg C, Aubourg MC, Jackson M (eds), Magnetic fabric: methods and applications. The Geological Society, London, pp 395–420

Raposo MIB, Berquó TS (2008) Tectonic fabric revealed by AARM of the proterozoic mafic dyke swarm in the Salvador City (Bahia State); São Francisco Craton, NE Brazil. Phys Earth Planetary Inter Syst 167:179–194

Raposo MIB, Chaves AO, Lojkasek-Lima P, DÁgrella-Filho MS, Teixeira W (2004) Magnetic fabrics and rock magnetism of Proterozoic dike swarm from the southern São Francisco Craton, Minas Gerais state, Brazil. Tectonophysics 378:43–63

Rochette P, Aubourg C, Perrin M (1999) Is this fabric normal? A review and case studies in volcanic formations. Tectonophysics 307:219–234

Rochette P, Jackson M, Aubourg C (1991) Rock magnetism and the interpretation of anisotropy of magnetic susceptibility. Rev Geophys 30:209–226

Rochette P, Scaillet B, Guillot S, Pêcher A, Le Fort P (1994) Magnetic mineralogy of the High Himalayan leucogranites: structural implications. Earth Planetary Sci Lett 126; 217–234

Román-Berdiel T, Aranguren A, Cuevas J, Tubía JM (1998) Compressional granite-emplacement model: structural magnetic study of the Trives Massif (NW Spain). Lithos 44; 37–52

Stephenson A (1994) Distribution anisotropy: two simple models for magnetic lineation and foliation. Phys Earth Planetary Inter 82:49–53

Táborská S, Breiter K (1998) Magnetic anisotropy of an extremely fractionated granite: the Podlesí Stock, Krusné hory Mts., Czech Republic. Acta Univ Carolinae Geol 42:147–149

Tarling D, Hrouda F (1993) The magnetic anisotropy of rocks. Chapman and Hall, London, 217 pp

Trindade RIF, Raposo MIB, Rnesto M, Siqueira R (1999) Magnetic susceptibility and partial anhysteretic remanence anisotropies in the magnetite-bearing granite pluton of Tourão, NE Brazil. Tectonophysics 314:443–468

Zák J, Schulmann K, Hrouda F (2005) Multiple magmatic fabrics in the Sázava pluton (Bohemian Massif, Czech Republic): a result of superposition of wrench-dominated regional transpression on final emplacement. J Struct Geol 27:805–822

Anisotropy of Magnetic Susceptibility in Variable Low-Fields: A Review

19

František Hrouda

Abstract

Theory of the Anisotropy of Magnetic Susceptibility (AMS) assumes field-independent rock susceptibility in the low fields used by common AMS meters. This is valid for rocks whose AMS is carried by diamagnetic and paramagnetic minerals and also by pure magnetite, while rocks with pyrrhotite, hematite or titanomagnetite may show significant variation of susceptibility in common measuring fields. Consequently, the use of the contemporary AMS theory is in principle incorrect in these cases. Fortunately, it has been shown by practical measurements and mathematical modelling of the measuring process that the variations of the principal directions and of the AMS ellipsoid shape with field are very weak, which is important in most geological applications. The degree of AMS, however, may show conspicuous variation with field and, if one wants to make precise quantitative fabric interpretation, it is desirable to work with the AMS of the field-independent component. Three methods exist for simultaneous determination of the field-independent and field-dependent AMS components, all based on standard AMS measurement in variable fields within the Rayleigh Law range. The field-dependence of the AMS can be used in solving some geological problems. For example, in volcanic and dyke rocks with inverse magnetic fabric, one can decide whether this inversion has geological (special flow regime of lava) or physical (SD vs. MD grains) causes. In rocks consisting of two magnetic fractions, one with field-independent susceptibility (magnetite, paramagnetic minerals) and the other possessing the field-dependent susceptibility (titanomagnetite, hematite, pyrrhotite), one can separate the AMS of the latter fraction and in favourable cases also of the former fraction.

19.1 Introduction

The theory of low-field Anisotropy of Magnetic Susceptibility (AMS) assumes a linear relationship between magnetization and magnetizing field, resulting in field-independent susceptibility. This is valid for diamagnetic and paramagnetic minerals by definition and also for pure magnetite, while in titanomagnetite,

F. Hrouda (✉)
AGICO Inc., Ječná 29a, CZ-621 00 Brno, Czech Republic;
Institute of Petrology and Structural Geology, Charles University, CZ-128 43 Praha, Czech Republic
e-mail: fhrouda@agico.cz

E. Petrovský et al. (eds.), *The Earth's Magnetic Interior*, IAGA Special Sopron Book Series 1,
DOI 10.1007/978-94-007-0323-0_19, © Springer Science+Business Media B.V. 2011

pyrrhotite and hematite the susceptibility may be clearly field-dependent even in low fields used in common AMS meters (Figs. 19.1 and 19.2) (Worm et al. 1993, Markert and Lehmann 1996, Jackson et al. 1998, de Wall 2000, Hrouda 2002, de Wall and Nano 2004, Hrouda et al. 2006).

Markert and Lehmann (1996) treated the field variation of the AMS theoretically for the Rayleigh Law region of the magnetization curve. They generalized the originally scalar Rayleigh Law to three dimensions and developed a technique for the simultaneous measurement of both the tensors using vibrating sample magnetometer.

De Wall (2000) investigated the field-dependent AMS of titanomagnetite-bearing dyke rocks, Hrouda (2002) studied pyrrhotite-bearing rocks and single crystals of hematite, Pokorný et al. (2004) measured titanomagnetite-bearing volcanic rocks and pyrrhotite-bearing schists and Hrouda et al. (2009a, b) investigated ultramafic rocks and host granulite both with pyrrhotite. In all these rocks, virtually no field variations were found in the orientations of the principal susceptibilities and in the shape of the susceptibility ellipsoid, while in the degree of AMS clear variations with field were revealed.

Hrouda (2009) developed methods for resolving the field-dependent AMS into the field-independent (e.g. due to paramagnetics or magnetite) and field-dependent (e.g. due to pyrrhotite or titanomagnetite) components on the basis of standard AMS measurement in variable low fields and wrote a program (ANIFIELD) to perform this resolution practically.

The present paper summarises the results obtained untill now and tries to illustrate advantages and

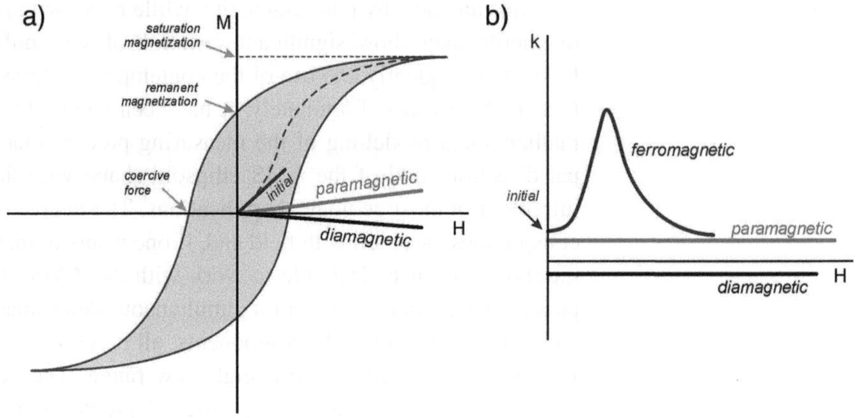

Fig. 19.1 Idealized relationship between magnetization and intensity of magnetizing field for diamagnetic, paramagnetic and ferromagnetic substances (**a**) and field variation of susceptibility for the same substances (**b**). Adapted from Brož (1966)

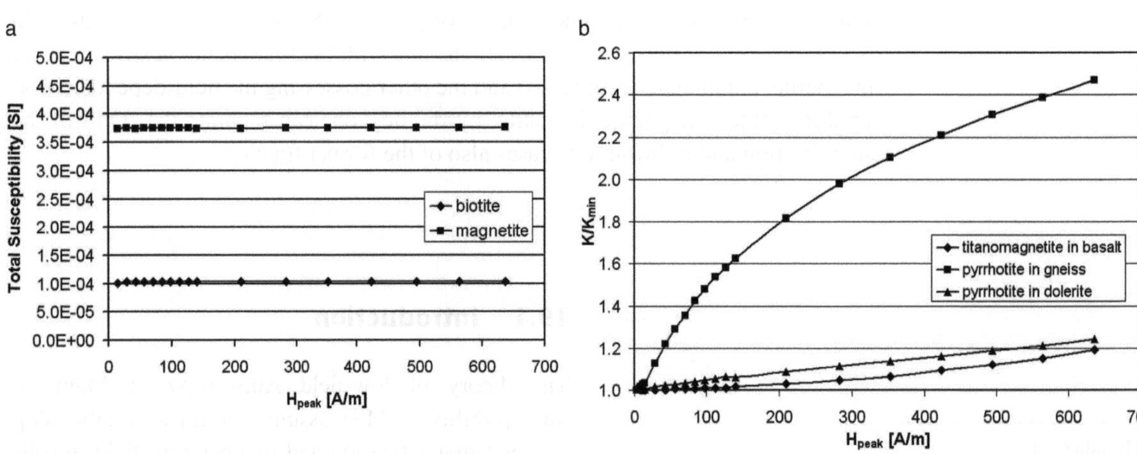

Fig. 19.2 Examples of field variation of susceptibility in various minerals. (**a**) biotite and magnetite, no field variation of total susceptibility (unnormalized by specimen volume), (**b**) pyrrhotite and titanomagnetite in various rocks, susceptibility normalized by the minimum value strongly varies with field. Adapted from Hrouda (2002) and Hrouda et al. (2006)

disadvantages of the field variation of the AMS of rocks in the structural geology research.

19.2 Theoretical Considerations

Theory of the low-field AMS is based on the assumption of the linear relationship between magnetization and magnetizing field

$$\boldsymbol{M} = \mathbf{K}\boldsymbol{H} \tag{19.1}$$

where $\boldsymbol{M}$ is the magnetization vector, $\boldsymbol{H}$ is the field intensity vector, and $\mathbf{K}$ is the symmetric second-rank tensor of magnetic susceptibility. (The SI of units is used throughout the paper.)

It is usual to represent the susceptibility tensor by convenient parameters derived from principal susceptibilities (e.g. Nagata 1961, Jelínek 1981), for instance

$$\begin{gathered} K_m = (K_1 + K_2 + K_3)/3 \\ P = K_1/K_3 \\ T = (2\,\eta_2 - \eta_1 - \eta_3)/(\eta_1 - \eta_3) = 2\,ln\,F/ln\,P - 1 \end{gathered} \tag{19.2}$$

where $K_1 \geq K_2 \geq K_3$ are the principal susceptibilities, $\eta_1 = ln\,K_1$, $\eta_2 = ln\,K_2$, $\eta_3 = ln\,K_3$, and $F = K_2/K_3$. The parameter K_m is called the mean susceptibility and characterizes the qualitative and quantitative content of magnetic minerals in a rock. The parameter P, called the degree of AMS, indicates the intensity of the preferred orientation of magnetic minerals in a rock. The parameter T, called the shape parameter, characterizes the symmetry or shape of the AMS ellipsoid. If $0 < T \leq +1$ the AMS ellipsoid is oblate, if $-1 \leq T < 0$ the AMS ellipsoid is prolate.

Magnetization of multi-domain materials in low fields (less than coercivity) can be described by the empirical Rayleigh Law (e.g. Néel 1942)

$$M = \kappa H + \alpha H^2 \tag{19.3}$$

where M is magnetization, H is field intensity, κ is initial susceptibility, and α is Rayleigh coefficient (see Fig. 19.3).

For anisotropic materials, Markert and Lehmann (1996) suggested the Rayleigh Law in the following form

$$\boldsymbol{M} = \boldsymbol{\kappa}\boldsymbol{H} + \boldsymbol{\alpha}H\boldsymbol{H}, \tag{19.4}$$

where $\boldsymbol{\kappa}$ is the second rank initial susceptibility tensor, $\boldsymbol{\alpha}$ is the second rank Rayleigh tensor, and H is modulus of the vector $\boldsymbol{H}$.

From the purely theoretical point of view, the Markert and Lehmann (1996) approach working with tensors of the second rank is not fully correct, because the relationship between magnetization and squared field should be described by the tensor of the higher rank than rank two. However, Hrouda (2009) checked linearity between individual susceptibility tensor components and field on mathematical models obtaining excellent fits deserving the above approach valid at least from the practical point of view. In addition, the Markert and Lehmann (1996) approach is analogous to the approach used in the anisotropy of magnetic remanence, which works with non-linear relationship between the remanent magnetization and magnetizing field, and the anisotropy is in numerous cases successfully represented by the second rank tensor, too (Cox and Doell 1967, Daly and Zinsser 1973, Stephenson et al. 1986, Jackson 1991, Jelínek 1993, Hrouda 2002a).

19.3 Modelling AMS Variation with Field Within the Rayleigh Law Region

Hrouda (2007, 2009) investigated the AMS variation with field within the Rayleigh Law region on mathematical models that simulated the measuring process. In the modelling, three groups of rocks were considered, viz. titanomagnetite-bearing rocks, pyrrhotite-bearing rocks, and haematite-bearing ones.

First, the initial susceptibility tensor was constructed for each rock group specified by the mean susceptibility, degree of AMS, shape parameter, and orientations of principal susceptibilities. Then, 15 directional initial susceptibilities corresponding to the Jelínek (1977) rotatable design were calculated as well as 15 directional Rayleigh Law susceptibilities. These data served as input data for AMS calculation using the Jelínek (1977) method based on linear theory. In addition to calculation of standard AMS parameters, this method also evaluates the error in fitting the tensor to the measured data as well as the errors in determining the principal susceptibilities and principal directions. The results can be summarized as follows.

In all models, the mean susceptibility linearly increases with field (Fig. 19.4a). The curves for

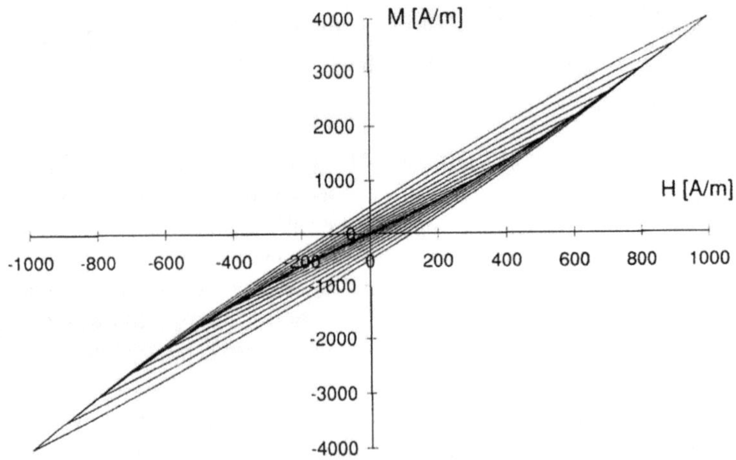

Fig. 19.3 Low-field Rayleigh hysteresis loop for a model material with the initial susceptibility of 3 [SI] and a quadratic coefficient of 2 × 10^{-3} m/A. Adapted from Jackson et al. (1998)

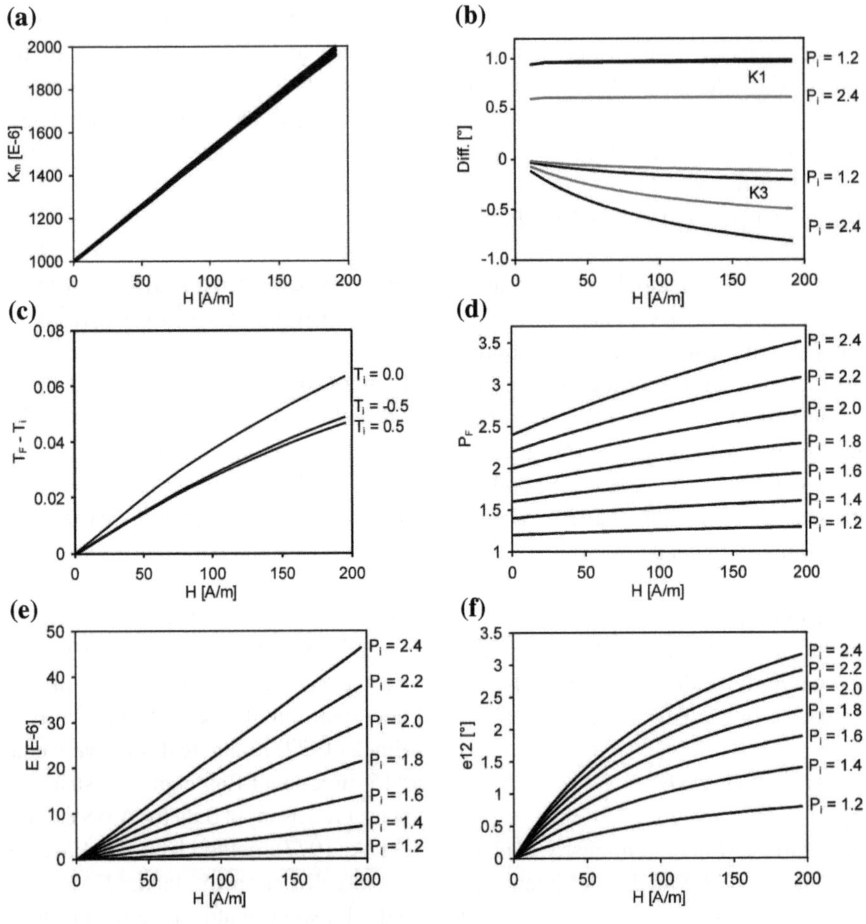

Fig. 19.4 Variations of AMS parameters with field for the model of pyrrhotite-bearing rocks as obtained using linear theory of AMS calculation (P_i indicates the degree of AMS of the initial susceptibility tensor), (**a**) mean susceptibility (K_m), (**b**) differences between model value and initial susceptibility tensor value for azimuth (*black line*) and plunge (*grey line*) for maximum (K_1) and minimum (K_3) susceptibility directions, (**c**) difference in shape parameters (T_f -T_i, between model and initial susceptibility tensor values), (**d**) degree of AMS (P_F), (**e**) fitting error (*E*, mean value of the absolute values of the difference between "measured" and fit susceptibilities), (**f**) confidence angle (*e12*, between K_1 and K_2 directions). Adapted from Hrouda (2007)

the individual modelling runs are very near one to another, but they do not coincide precisely as indicated by progressive widening of the line, which evidently indicates the effect of linear fit to non-linear data within the Rayleigh Law region. The orientations of the principal directions are in all fields very near to those of the initial susceptibility tensor, the angles between the respective directions being mostly less than 1° (Fig. 19.4b). The shape parameter changes with field only very slightly (Fig. 19.4c), while the degree of AMS may increase with field considerably (Fig. 19.4d). The fitting error and the confidence angles are very low for the low fields, increasing with field strongly (Fig. 19.4e, f). The values of the confidence angles increase with field evidently due to the linear fit to non-linear data. Consequently, high values obtained in practical measurements need not necessarily indicate poor measurement accuracy.

19.4 Methods for AMS Resolving into Field-Independent and Field-Dependent Components

19.4.1 Directional Susceptibilities Method

Using the model by Henry (1983) and Henry and Daly (1983), the rock susceptibility within the Rayleigh Law region can be described with sufficient accuracy as

$$k_r = \Psi + A_{md}H \tag{19.5}$$

where k_r is the rock susceptibility, $\Psi = K_d + K_p + K_{ma} + K_{sd} + K_{md}$, and $A_{md}H$ is the Rayleigh member. K_d, K_p, K_{ma}, K_{sd}, K_{md} are the susceptibility contributions of diamagnetic fraction, paramagnetic fraction, pure magnetite, SD grains of all ferromagnetics, initial susceptibility of MD ferromagnetic grains showing field-dependent susceptibility, respectively. This is the equation of the straight line whose intercept represents the field-independent susceptibility and the slope characterizes the field-dependent susceptibility.

The principle of the AMS determination lies in measuring the susceptibility of a specimen in at least six independent directions, the measured susceptibilities being called the directional susceptibilities (Janák 1965), and subsequent fitting the susceptibility tensor to these data using the least squares method (e.g. Jelínek 1977). For separating the field-independent and field-dependent AMS components, the AMS is measured in several fields within the Rayleigh Law region, straight lines are fit to the susceptibility vs. field intensity data for each measuring direction, determining the field-independent and field-dependent directional susceptibilities. From these data, the tensors of the field-independent susceptibility and of the initial susceptibility of the MD ferromagnetic fraction can be calculated, both tensors obeying conditions of the Eq. (19.1). (For details see Hrouda 2009.)

19.4.2 Tensor Elements Method

The model of rock susceptibility described above can also be written in tensor terms (Hrouda 2009), the tensor components being

$$k_{r(ij)} = \Psi_{ij} + A_{md(ij)}H. \tag{19.6}$$

Realizing that Ψ_{ij} are field-independent, it is obvious that this is the equation of the straight line whose intercept represents the field-independent susceptibility. From the components of the $A_{md(ij)}$ tensor, the AMS parameters of the initial susceptibility of the MD ferromagnetic fraction are

$$\begin{aligned} P_I &= \sqrt{(A_{md(1)}/A_{md(3)})} \\ F_I &= \sqrt{(A_{md(2)}/A_{md(3)})} \\ T_I &= 2\ ln\ F_I/ln\, P_I - 1. \end{aligned} \tag{19.7}$$

The mean initial susceptibility cannot be calculated. Determination of the field-independent susceptibility tensor and the initial susceptibility tensor of MD fraction from standard AMS measurements in variable fields is easy provided that the $\mathbf{k}_r$ tensor can be obtained by measurement. However, the methods for the AMS measurement by the most frequently used instruments, i.e. the inductance bridges, are not able to directly measure $\mathbf{k}_r$. They use the linear theory, calculating $\mathbf{K}$ in Eq. (19.1). Fortunately, mathematical modelling by Hrouda (2007, 2009) has shown that the differences between $\mathbf{k}_r$ and $\mathbf{K}$ are so small that these two tensors can be regarded as equal from the practical point of view.

19.4.3 Tensor Subtraction Method

If the AMS is measured in two different fields ($^AH < {}^BH$) within the Rayleigh Law range and the respective susceptibility tensors are subtracted ($^B\mathbf{k}_r - {}^A\mathbf{k}_r$), the Rayleigh contribution tensor can be determined as follows

$$\mathbf{A}_{md} = ({}^B\mathbf{k}_r - {}^A\mathbf{k}_r)/({}^BH - {}^AH). \quad (19.8)$$

The field-independent susceptibility tensor can then be calculated

$$\boldsymbol{\Psi} = {}^A\mathbf{k}_r - \mathbf{A}_{md}\,{}^AH. \quad (19.9)$$

19.5 Potential Geological Implications

The method of AMS resolution into field-independent and field-dependent components is relatively new, developed and applied to solving geological problems by the present author. Even though there are not too many applications, the existing experience suggests that the method can have interesting implications outlined below.

19.5.1 Determination of Orientation Tensor

Preferred orientation of minerals can be in addition to graphical methods characterized purely mathematically by the orientation tensor, defined as (Scheidegger 1965)

$$\mathbf{E} = (1/N)\begin{vmatrix} \Sigma\, l_i^2 & \Sigma\, l_i m_i & \Sigma\, l_i n_i \\ \Sigma\, m_i l_i & \Sigma\, m_i^2 & \Sigma\, m_i n_i \\ \Sigma\, n_i l_{il} & \Sigma\, m_i n_i & \Sigma\, n_i^2 \end{vmatrix} \quad (19.10)$$

where l_i, m_i, n_i are the direction cosines of the i-th linear element represented by unit vector and N is the number of the linear elements considered. The principal values of this tensor ($E_1 \geq E_2 \geq E_3$) have the following property $E_1 + E_2 + E_3 = 1$. Consequently, $E_1 > E_2 = E_3$ represents a cluster type of distribution, whereas $E_1 = E_2 > E_3$ corresponds to a girdle type pattern. The advantage of the orientation tensor is that it can be used for characterization of the preferred orientation of minerals determined by both non-magnetic and magnetic methods.

A straightforward relationship exists between the orientation tensor and the susceptibility tensor of the rock, the AMS of which is carried by single magnetic mineral (Ježek and Hrouda 2000)

$$\mathbf{K} = k\mathbf{I} + \Delta\mathbf{E} \quad (19.11)$$

where $\mathbf{K}$ is the rock susceptibility tensor, $\mathbf{I}$ is the identity matrix. For magnetically rotational prolate grains, $k = k_2 = k_3$, $\Delta = k_1 - k$ (k_1, k_2, k_3 are the grain principal susceptibilities) and $\mathbf{E}$ is the orientation tensor of grain maximum susceptibility axes. For magnetically rotational oblate grains, $k = k_1 = k_2$, $\Delta = k - k_3$ and $\mathbf{E}$ is the orientation tensor of grain minimum susceptibility axes. For the grains displaying so called perfectly triaxial AMS ($k_1 - k_2 = k_2 - k_3$), $k = k_2$, $\Delta = k_1 - k_2 = k_2 - k_3$ and $\mathbf{E} = \mathbf{E}_x - \mathbf{E}_z$ is the Lisle orientation tensor ($\mathbf{E}_x$, $\mathbf{E}_z$ are the Scheidegger orientation tensors of the maximum and minimum axes, respectively).

As the susceptibility tensor $\mathbf{K}$ in Eq. (19.1) is field-independent, the relationship (19.11) can be used for correct calculation of the orientation tensor of a paramagnetic mineral or pure magnetite. In case of mineral with field-dependent AMS (pyrrhotite or hematite), the rock susceptibility would be

$$\mathbf{k}_r = \boldsymbol{\Psi}_{md} + \mathbf{A}_{md}H. \quad (19.12)$$

Using the susceptibility tensor measured in a standard way would result in overestimation of the orientation tensor. Correct way would be using the initial susceptibility tensor ($\boldsymbol{\Psi}_{md}$), which is field-independent.

Figure 19.5a shows the field variation of the degree of AMS for several specimens of ultramafite embedded in granulite from the Bory locality (Moldanubian Zone, Bohemian Massif). In all specimens, the degree of AMS increases with field. Evaluation of several tens of specimens, whose field-dependent AMS is carried by titanomagnetite and/or pyrrhotite, has shown that the degree of AMS always increases with field if the mean susceptibility increases, even though the former increase is not as rapid as the latter increase.

Figure 19.5b shows model effect of the field variation of the degree of AMS on the determination of the orientation tensor. The model considers the degree of the anisotropy of initial susceptibility ranging from 1.1 to 2.0 and MD pyrrhotite as the only AMS carrier having the grain $P_c = 100$. The abscissa plots the E_1/E_3 ratio of pyrrhotite c-axes calculated from

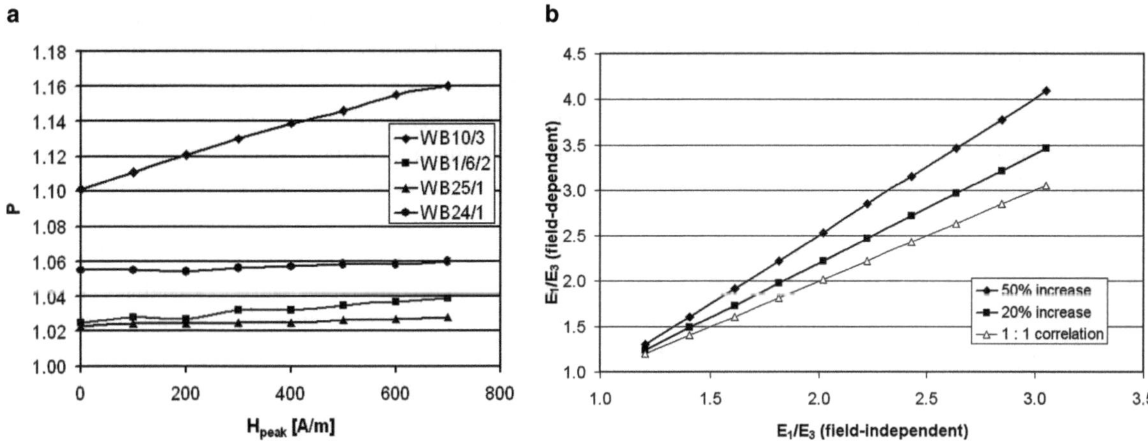

Fig. 19.5 The effect of field variation of AMS on the determination of the orientation tensor. (**a**) variation of the degree of AMS with field on example of ultramafic rocks from the Bory locality (Moldanubian Zone, Bohemian Massif), (**b**) mathematical model of the effect of the field variation of AMS on the determination of the orientation tensor (for details see the text)

the field-independent initial susceptibility, while the ordinate plots the same ratio calculated from the field-dependent whole-rock AMS. One curve shows the 1:1 correlation, while the other two represent the field-dependent AMS exhibiting the increase in *P* by 20 and 50%, respectively. It is obvious that the effect of the field variation is remarkable and using the initial susceptibility tensor, which is field-independent, is therefore strongly recommended.

19.5.2 Strain from AMS

Strain analysis is one of the most laborious techniques of structural analysis being confined to rocks containing convenient strain indicators (oolites, concretions, reduction spots, lapilli, fossils). For this reason, many attempts have been made to use the AMS as a strain indicator. One of the methods is that developed by Ježek and Hrouda (2007) consisting of modelling the development of the AMS during rock deformation and subsequent matching with measured data. In the modelling, standard (field-independent) AMS is used. However, in rocks whose AMS is carried by a mineral with field-dependent susceptibility, the strain estimation made on the basis of measured AMS is incorrect, because the measured degree of AMS is higher than that considered in modelling. Figure 19.6 shows the model effect of the field variation of the degree of AMS on determination of strain from the AMS. The model considers pure shear strain, the degree of the

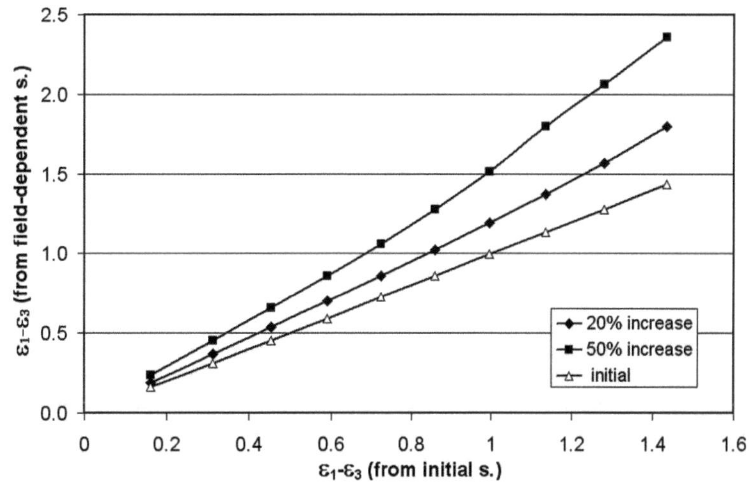

Fig. 19.6 The effect of field variation of AMS on the determination of strain using the Ježek and Hrouda (2007) method (for details see the text)

anisotropy of initial susceptibility ranging from 1.1 to 2.0 and MD pyrrhotite as the only AMS carrier having the grain $P_c = 100$. The abscissa plots the ε_1–ε_3 difference (ε_1 and ε_3 are maximum and minimum natural strains) calculated from the initial susceptibility, while the ordinate plots the same difference calculated from the field-dependent whole-rock AMS. One curve shows the 1:1 correlation, while the other two represent the field-dependent AMS exhibiting the increase in P by 20 and 50%, respectively. In this case, using field-independent initial susceptibility tensor would be more correct.

19.5.3 Inverse Magnetic Fabric in Volcanic and Dyke Rocks

In lava flows, sills, and dykes, whose AMS is carried by MD titanomagnetite, in which the magnetic fabric resembles the grain shape fabric, the magnetic foliation is often found to be near the flow plane and the magnetic lineation is mostly parallel to the lava flow direction (Ernst and Baragar 1992, Canon-Tapia et al. 1994, Raposo and Ernesto 1995, Hrouda et al. 2002). Less frequently, the magnetic foliation can also be roughly perpendicular to the flow plane and the magnetic lineation perpendicular to the lava flow direction. This magnetic fabric is called the inverse fabric being due to special lava flow regime. The inverse fabric can also exist in rocks with normal lava flow regime if their AMS is carried by SD grains, because in these grains the maximum and minimum susceptibilities are inverse with respect to grain shape (Fig. 19.7a) (Potter and Stephenson 1988). This dualism can have unpleasant consequences for the interpretation of a particular magnetic fabric (see Fig. 19.7b).

Provided that the AMS is dominantly carried by single magnetic mineral consisting of the mixture of MD and SD particles, the rock susceptibility is

$$\mathbf{k}_r = (\boldsymbol{\Psi}_{sd} + \boldsymbol{\Psi}_{md}) + \mathbf{A}_{md} H \quad (19.13)$$

and the field-independent AMS component is controlled by both SD and MD particles, while the

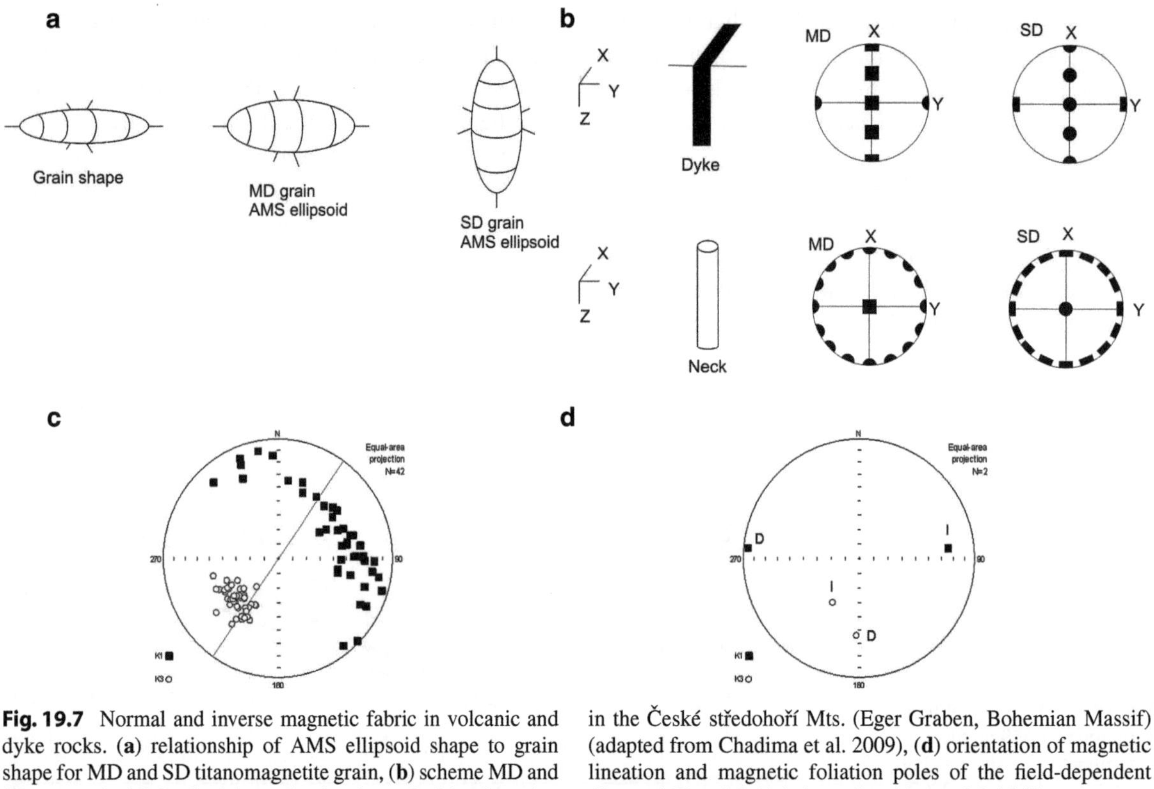

Fig. 19.7 Normal and inverse magnetic fabric in volcanic and dyke rocks. (**a**) relationship of AMS ellipsoid shape to grain shape for MD and SD titanomagnetite grain, (**b**) scheme MD and SD magnetic fabrics in dyke and volcanic neck, (**c**) orientation of magnetic lineation and magnetic foliation poles of the whole-rock AMS with respect to dyke plane in a trachybasalt dyke in the České středohoří Mts. (Eger Graben, Bohemian Massif) (adapted from Chadima et al. 2009), (**d**) orientation of magnetic lineation and magnetic foliation poles of the field-dependent (denoted *D*) and field-independent (denoted *I*) AMS components of one specimen in a trachybasalt dyke in the České středohoří Mts. (Eger Graben, Bohemian Massif)

field-dependent component is controlled solely by MD grains. In the coordinate system of principal directions of MD grain fabric, the field-independent component is

$$\Psi = \begin{vmatrix} \Psi_{md(1)} + \Psi_{sd(3)} & 0 & 0 \\ 0 & \Psi_{md(2)} + \Psi_{sd(2)} & 0 \\ 0 & 0 & \Psi_{md(3)} + \Psi_{sd(1)} \end{vmatrix} \quad (19.14)$$

Then, if $(\Psi_{md(1)} + \Psi_{sd(3)}) > (\Psi_{md(3)} + \Psi_{sd(1)})$, which can be also written as $(\Psi_{md(1)} - \Psi_{md(3)}) > (\Psi_{sd(1)} - \Psi_{sd(3)})$, the field-independent and field-dependent AMS components are coaxial. In this case, the AMS is predominantly carried by MD particles and the eventual inverse magnetic fabric has geological causes. On the other hand, if the field-independent and field-dependent components are anti-coaxial, the AMS is predominantly carried by SD particles and the inverse magnetic fabric has very likely physical causes.

An example can be presented from a trachybasalt dyke in the České středohoří Mts. (Eger Graben, NW Bohemian Massif), where clearly inverse magnetic fabric exists, with magnetic foliation perpendicular to the dyke plane (Fig. 19.7c, locality CS27 in Chadima et al. 2009). The field-dependent AMS component, shown on example of one specimen, is roughly coaxial with the whole-rock AMS and with the field-independent component (Fig. 19.7c, d) indicating that this inverse magnetic fabric has geological causes (probably static compaction within the dyke as suggested by Raposo and Ernesto 1995). This

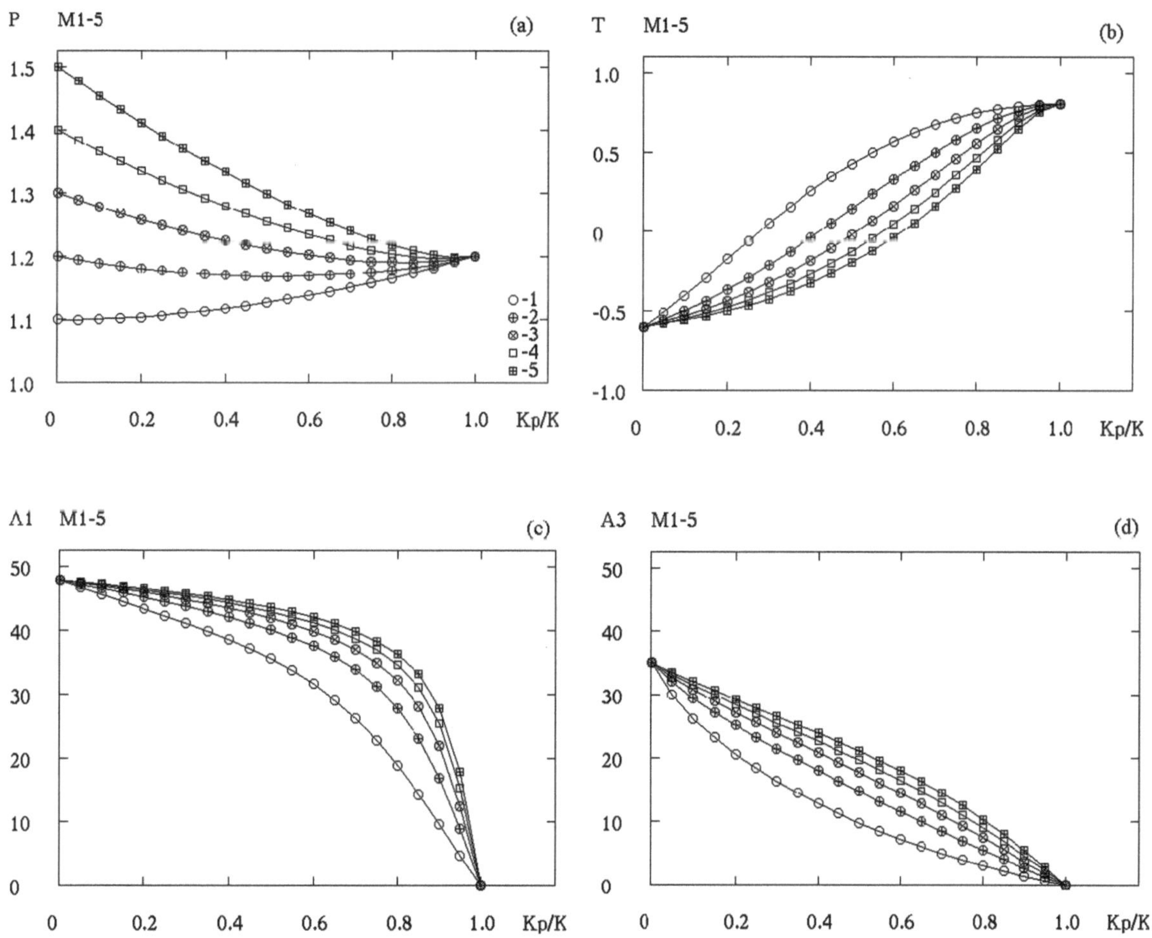

Fig. 19.8 Model of the effect of ferromagnetic and paramagnetic fractions on the whole-rock AMS. K_p/K is the ratio of paramagnetic to the whole-rock bulk susceptibility. (**a**) degree of AMS, (**b**) shape parameter, (**c**) angle between whole-rock and paramagnetic magnetic lineations, (**d**) angle between whole-rock and paramagnetic magnetic foliation poles. After Hrouda (2010)

conclusion is confirmed also by the results of the anisotropy of magnetic remanence (for details see Chadima et al. 2009).

19.5.4 Separation of Paramagnetic and Field-Dependent Ferromagnetic AMS Components

Let us consider a rock whose AMS is controlled by paramagnetic fraction and field-dependent MD ferromagnetic fraction, for example, originally weakly magnetic schist mineralized by pyrrhotite. The susceptibility of such a rock is

$$\mathbf{k}_{\mathrm{r}} = (\boldsymbol{\Psi}_p + \boldsymbol{\Psi}_f) + \mathbf{A}_{\mathrm{f}} H, \qquad (19.15)$$

where $\boldsymbol{\Psi}_p$ is the contribution tensor of susceptibility of the paramagnetic fraction, $\boldsymbol{\Psi}_f$ is contribution tensor of initial susceptibility of ferromagnetic fraction and $\mathbf{A}_{\mathrm{f}}$ is contribution Rayleigh tensor of the ferromagnetic fraction. It is obvious that while the field-dependent component is controlled solely by ferromagnetic fraction, the field-independent component is controlled by both the paramagnetic fraction and the initial susceptibility of the ferromagnetic fraction. Modelling by Hrouda (2009, 2010) showed that $\boldsymbol{\Psi}_f$ and $\mathbf{A}_{\mathrm{f}}$ tensors are virtually coaxial, while the $\boldsymbol{\Psi}_p$ tensor can show in general different orientation. If the $\boldsymbol{\Psi}_p$ tensor dominates over the $\boldsymbol{\Psi}_f$ tensor, the AMS resolution into field-dependent and field-independent components can serve as separate determination of the paramagnetic mineral fabric and the ferromagnetic mineral fabric. On the other hand, if the $\boldsymbol{\Psi}_f$ tensor dominates over the $\boldsymbol{\Psi}_p$ tensor, one can only determine the ferromagnetic mineral fabric and resign on determining the paramagnetic mineral fabric (see Fig. 19.8).

19.6 Problems of the Linear Theory

There are a few examples when the linear theory does not seem to satisfactorily describe the AMS of rocks with field-dependent susceptibility. For example, hematite single crystals were measured by the KLY-3S Kappabridge modified in such a way that 64 directional susceptibilities may have been measured in each of three perpendicular planes and recorded

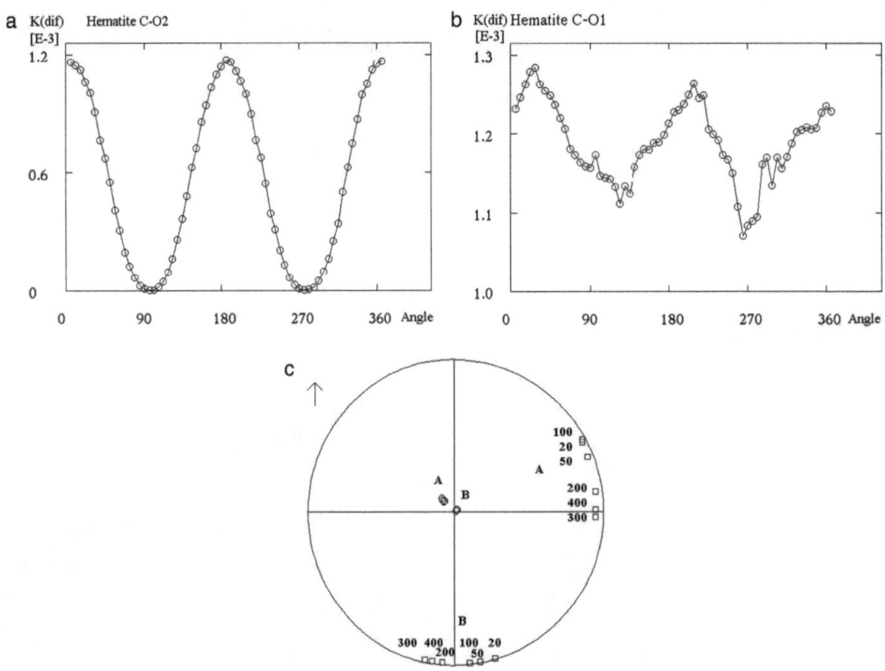

Fig. 19.9 Variation of AMS of single crystals of hematite with field. (**a**) susceptibility of a single crystal measured in a plane perpendicular to the basal plane (**b**) susceptibility of a single crystal measured within basal plane (**c**) orientations of principal directions in two hematite single crystals (numbers at magnetic lineations indicate intensity of measuring field). Legend: square – magnetic lineation, circle – magnetic foliation pole, numbers at magnetic lineations denote the intensity of measuring field in A/m rms values. Adapted from Hrouda (2002)

on disk (Hrouda 2002). Figure 19.9a shows these susceptibilities measured in the plane perpendicular to the basal plane. The susceptibility curve follows very well the sinusoidal curve, which means that the AMS in this plane is well represented by an ellipse. Figure 19.9b shows the susceptibilities measured parallel to the basal plane. It is clear that the susceptibility curve no longer follows the sinusoidal curve, which means that the AMS in the basal plane is represented neither by an ellipse nor by a circle, which would be expected, following the Neumann's principle (cf. Nye 1957, Hrouda 1973), if the relationship between the magnetization and field were linear. High-field magnetic anisotropy of hematite crystals, as summarized by Stacey and Benerjee (1974), indicates that the magnetization within the basal plane is not isotropic, showing more complex pattern (three-fold or six-fold) controlled by the crystal lattice. Similar reasons may apply to the separation of lower field maximum susceptibilities from the stronger field ones (Fig. 19.9c).

A possible way in solving the problem may be respecting the non-linearity and measuring the susceptibility in so many directions that a contour diagram of the directional susceptibility can be presented instead of the susceptibility tensor. Unfortunately, this is laborious and time consuming. In addition, the beauty and elegance would be lost of the AMS presentation in simple terms of principal susceptibilities and parameters derived from them and orientations of magnetic foliation and magnetic lineation. On the other hand, this approach would be efficient if more detailed results were obtained than those arising from the simple susceptibility tensor. Of course, the best solution would be development of the AMS theory respecting the non-linear relationship between the magnetization and field. This theory is still waiting for its inventor.

19.7 Conclusions

The theory of the AMS of rocks assumes that the rock susceptibility is field independent in low fields used by common AMS meters. This assumption is valid in rocks the AMS of which is carried by paramagnetic minerals and/or by pure magnetite. In rocks, whose AMS is carried by titanomagnetite, hematite or pyrrhotite, the AMS may be field dependent giving rise to problems in measurement and mainly in interpretation. Fortunately, as shown by practical measurements and mathematical modelling of the measuring process, the variations of the principal directions and of the AMS ellipsoid shape with field are very weak, negligible with respect to common measuring errors, which is important in most geological applications. On the other hand, the degree of AMS may show conspicuous variation with field and, if one wants to make precise quantitative fabric interpretation, as for example the determination of the orientation tensor from AMS or the interpretation the AMS in quantitative terms of strain, it is desirable to work with the AMS of the field-independent component.

The field-dependence of the AMS can be used in solving some geological problems. For example, in volcanic and dyke rocks with inverse magnetic fabric, one can decide whether this inversion has geological (special flow regime of lava) or physical (SD vs. MD grains) causes. In rocks consisting of two magnetic fractions, one with field-independent susceptibility (magnetite, paramagnetic minerals) and the other possessing the field-dependent susceptibility (titanomagnetite, hematite, pyrrhotite), one can separate the AMS of the latter fraction and in favourable cases also of the former fraction.

Acknowledgements Dr. Martin Chadima is thanked for providing the field-dependent AMS of the dike rock from the locality of CS27. The research was partly supported financially by the Ministry of Education and Youth of the Czech Republic (Scientific Program MSM0021620855).

References

Brož J (1966) Modern problems of ferromagnetism (in Czech). NČSAV, Praha 189 pp

Chadima M, Cajz V, Týcová P (2009) On the interpretation of normal and inverse magnetic fabric in dikes: examples from the Eger Graben, NW Bohemian Massif. Tectonophysics 466:47–63

Canon-Tapia E, Herrero-Bervera E, Walker GPL (1994) Flow directions and paleomagnetic study of rocks from the Azufre volcano, Argentina. J Geomagnetism Geoelectricity 46:143–159

Cox A, Doell RR (1967) Measurements of high coercivity magnetic anisotropy. In: Collinson DV, Creer KM, Runcorn SK (eds) Methods in palaeomagnetism. Elsevier, Amsterdam, pp 477–482

Daly L, Zinsser H (1973) Etude comparative des anisotropies de susceptibilité et d'aimantation rémanente isotherme. Conséquences pour l'analyse structurale et le paléomagnétisme. Ann Géophys 29:189–200

De Wall H (2000) The field dependence of AC susceptibility in titanomagnetites: implications for the anisotropy of magnetic susceptibility. Geophys Res Lett 27:2409–2411

De Wall H, Nano L (2004) The use of field dependence of magnetic susceptibility for monitoring variations in titanomagnetite composition – a case study on basanites from the Vogelsberg 1996 Drillhole, Germany. Studia Geophys Geod 48:767–776

Ernst RE, Baragar WRA (1992) Evidence from magnetic fabric for the flow pattern of magma in the Mackenzie giant radiating dyke swarm. Nature 356:511–513

Henry B (1983) Interprétation quantitative de l'anisotropie de susceptibilité magnétique. Tectonophysics 91:165–177

Henry B, Daly L (1983) From qualitative to quantitative magnetic anisotropy analysis: the prospect of finite strain calibration. Tectonophysics 98:327–336

Hrouda F (1973) A determination of the symmetry of the ferromagnetic mineral fabric in rocks on the basis of the magnetic susceptibility anisotropy measurements. Gerl Beitr Geophys 82:390–396

Hrouda F (2002) Low-field variation of magnetic susceptibility and its effect on the anisotropy of magnetic susceptibility of rocks. Geophys J Int 150:715–723

Hrouda F (2002a) The use of the anisotropy of magnetic remanence in the resolution of the anisotropy of magnetic susceptibility into its ferromagnetic and paramagnetic components. Tectonophysics 347:269–281

Hrouda F (2007) Anisotropy of magnetic susceptibility of rocks in the Rayleigh Law region: modelling errors arising from linear fit to non-linear data. Studia Geophys Geod 51:423–438

Hrouda F (2009) Determination of field-independent and field-dependent components of anisotropy of susceptibility through standard AMS measurements in variable low fields I: theory. Tectonophysics 466:114–122

Hrouda F (2010) Modelling relationship between bulk susceptibility and AMS in rocks consisting of two magnetic fractions represented by ferromagnetic and paramagnetic minerals – implications for understanding magnetic fabrics in deformed rocks. J Geol Soc India 75:254–266

Hrouda F, Chlupáčová M, Mrázová Š (2006) Low-field variation of magnetic susceptibility as a tool for magnetic mineralogy of rocks. Phys Earth Planetary Inter 154:323–336

Hrouda F, Chlupáčová M, Novák JK (2002) Variations in magnetic anisotropy and opaque mineralogy along a kilometer deep profile within a vertical dyke of the syenogranite porphyry at Cínovec (Czech Republic). J Volcanol Geotherm Res 113:37–47

Hrouda F, Faryad SW, Chlupáčová M, Jeřábek P, Kratinová Z (2009a) Determination of field-independent and field-dependent components of anisotropy of susceptibility through standard AMS measurements in variable low fields II: An example from the ultramafic body and host granulitic rocks at Bory in the Moldanubian Zone of Western Moravia, Czech republic. Tectonophysics 466:123–134

Hrouda F, Faryad SW, Jeřábek P, Chlupáčová M, Vitouš P (2009b) Primary magnetic fabric in an ultramafic body (Moldanubian Zone, European Variscides) survives exhumation-related granulite-amphibolite facies metamorphism. Lithos (2008), 111:95–111

Jackson M (1991) Anisotropy of magnetic remanence: a brief review of mineralogical sources, physical origins, and geological applications, and comparison with susceptibility anisotropy. PAGEOPH 136:1–28

Jackson M, Moskowitz B, Rosenbaum J, Kissel C (1998) Field-dependence of AC susceptibility in titanomagnetites. Earth Planet Sci Lett 157:129–139

Janák F (1965) Determination of anisotropy of magnetic susceptibility of rocks. Studia Geophys Geod 9:290–301

Jelínek V (1977) The statistical theory of measuring anisotropy of magnetic susceptibility of rocks and its application. Geofyzika n.p. Brno

Jelínek V (1981) Characterization of magnetic fabric of rocks. Tectonophysics 79:T63–T67

Jelínek V (1993) Theory and measurement of the anisotropy of isothermal remanent magnetization of rocks. Travaux Geophys 37:124–134

Ježek J, Hrouda F (2000) The Relationship bBetween the Lisle orientation tensor and the susceptibility tensor. Phys Chem Earth A 25:469–474

Ježek J, Hrouda F (2007) SUSIE: A program for inverse strain estimation from magnetic susceptibility. Comput Geosci 33:749–759

Markert H, Lehmann A (1996) Three-dimensional Rayleigh hysteresis of oriented core samples from the German Continental Deep Drilling Program: susceptibility tensor, Rayleigh tensor, three-dimensional Rayleigh law. Geophys J Int 127:201–214

Nagata T (1961) Rock magnetism. Maruzen, Tokyo

Néel L (1942) Theory of Rayleigh's law of magnetization. Cahier Phys 12:1–20

Nye JF (1957) Physical properties of crystals. Clarendon Press, Oxford

Pokorný J, Suza P and Hrouda F (2004) Anisotropy of magnetic susceptibility of rocks measured in variable weak magnetic fields using the KLY-4S Kappabridge. In: Martín-Hernández F, Lüneburg CM, Aubourg C, Jackson M (eds) Magnetic fabric: methods and applications. , Special Publications, vol 238. Geological Society, London, pp 69–76

Potter DK, Stephenson A (1988) Single-domain particles in rocks and magnetic fabric analysis. Geophys Res Lett 15:1097–1100

Raposo MIB, Ernesto M (1995) Anisotropy of magnetic susceptibility in the Ponta Grossa dyke swarm (Brazil) and its relationship with magma flow direction. Phys Earth Planetary Inter 87:183–196

Scheidegger AE (1965) On the statistics of the orientation of bedding planes, grain axes, and similar sedimentological data. US Geol Surv Prof Paper 525-C:164–167

Stacey FD and Benerjee SK (1974) The physical principles of rock magnetism. Development in solid earth geophysics. Elsevier, Amstredam, 195 pp

Stephenson A, Sadikun S, Potter DK (1986) A theoretical and experimental comparison of the anisotropies of magnetic susceptibility and remanence in rocks and minerals. Geophys J R Astron Soc 84:185–200

Worm H-U, Clark D, Dekkers MJ (1993) Magnetic susceptibility of pyrrhotite: grain size, field and frequency dependence. Geophys J Int 114:127–137

A Multi-Function Kappabridge for High Precision Measurement of the AMS and the Variations of Magnetic Susceptibility with Field, Temperature and Frequency

20

Jiří Pokorný, Petr Pokorný, Petr Suza, and František Hrouda

Abstract

A new MFK1-FA Kappabridge is introduced that precisely measures the magnetic susceptibility of rocks and the anisotropy of susceptibility. The instrument has the following features: separation of the in-phase (real) and out-of-phase (imaginary) components, auto-ranging and auto-zeroing, automated measurement of the field variation of the bulk susceptibility, AMS measurements using the spinning specimen method, built-in circuitry for controlling the non-magnetic furnace (CS-4) and cryostat (CS-L), full instrument control by an external computer, sophisticated hardware and software diagnostics. Examples are shown to illustrate the capability of the instrument for rock magnetic and palaeomagnetic research.

20.1 Introduction

The magnetic susceptibility of rocks is one of the most frequently investigated parameters in rock magnetism and palaeomagnetism. It can be measured both in the field on rock outcrops using various hand-held susceptibility meters, and especially very precisely, in the laboratory on rock specimens. Modern laboratory instruments measure not only the susceptibility at room temperature, but also the variation of susceptibility with temperature, the variation of susceptibility with measurements in low fields, the susceptibility at variable operating frequencies, and the anisotropy of susceptibility. All of these measurements have wide applications in geology, geophysics, physics, palaeoclimatology and the environmental sciences.

The purpose of the present paper is to introduce the recently developed multi-function MFK1-FA Kappabridge, which is able to measure the rock susceptibility in all the above mentioned regimes. The paper is devoted as our tribute to the late Dr. Vít Jelínek, CSc., the principal designer of the Kappabridges of the KLY series, who passed away on October 14, 2007. After his graduation in Physics at the Masaryk University in Brno, Czech Republic, in 1959, and working a short engagement in telecommunications, Dr. Jelínek co-developed the KT-1 hand-held susceptibility meter (Bartošek and Jelínek 1961), and then he developed the JR-1 Spinner Magnetometer for measuring the remanent magnetization of rocks (Jelínek 1966). Even though constructed with electronic tubes, the last instrument had an outstanding sensitivity of 5 μA/m, which was unparalleled to other rock magnetic instruments working with the classical (non-cryogenic) principles of the time. The next

F. Hrouda (✉)
AGICO Inc., Ječná 29a, CZ-621 00 Brno, Czech Republic;
Institute of Petrology and Structural Geology, Charles University, CZ-128 43 Praha, Czech Republic
e-mail: fhrouda@agico.cz

E. Petrovský et al. (eds.), *The Earth's Magnetic Interior*, IAGA Special Sopron Book Series 1,
DOI 10.1007/978-94-007-0323-0_20, © Springer Science+Business Media B.V. 2011

important development by Dr. Jelínek was the KLY-1 Kappabridge for measuring rock susceptibility and its anisotropy (Jelínek 1973). This instrument, based on germanium transistors, also had an outstanding sensitivity of 4×10^{-8} [SI], which was relatively soon innovated to the KLY-2 Kappabridge (sensitivity 4×10^{-8} [SI]; Jelínek 1980) that became one of the most popular instruments for measuring the anisotropy of magnetic susceptibility (AMS) of rocks, with more than 150 pieces manufactured and sold the world over. The AMS was measured using the rotatable design of 15 directional susceptibilities, also originally developed by Dr. Jelínek (Jelínek 1977). In addition, Dr. Jelínek worked out the original method for the statistical evaluation of the second rank tensors such as AMS (Jelínek 1978), and introduced rational parameters to characterize the AMS of rocks quantitatively (Jelínek 1981). In order to positively respond to the rapidly growing interests of structural geologists in AMS, the KLY-3S Kappabridge was developed (Jelínek and Pokorný 1997), which measured the slowly spinning specimen in a considerably shortened measuring time. For purposes of identification of magnetic minerals in rocks, the non-magnetic furnace and the cryostat apparatuses were developed (Parma and Zapletal 1991) that enabled to be measured, the susceptibility variation with temperature, in cooperation with the Kappabridges of the KLY series. Methods for resolving the temperature-dependent rock susceptibility into ferromagnetic and paramagnetic components were also elaborated (Hrouda 1994, Hrouda et al. 1997). The KLY-3 Kappabridge was then innovated to the KLY-4 Kappabridge which measures the susceptibility and the AMS in variable low-fields (Pokorný et al. 2004, Hrouda 2009).

In order to satisfy the demands of environmental magnetists for measurement of frequency-dependent susceptibility, the MFK1-FA Kappabridge was developed that enabled the susceptibility at three operating frequencies to be measured. In this instrument, several circuit solutions developed for the Kappabridges of the KLY series were exploited, being extended by newly developed composite pick-up coils and circuits controlling the bridge performance at different operating frequencies. This approach enables frequency-dependent susceptibility to be measured with all of the successful features of the KLY series of Kappabridges retained.

20.2 Instrumentation

The Multi-Function Kappabridge (MFK1-FA) is a precision fully-automatic inductivity bridge equipped with automatic compensation of the bridge's unbalance drift as well as automatic switching to the appropriate measuring range, both working over the entire measuring range. Automatic zeroing compensates both real and imaginary components, the zeroing circuits being digitally controlled by firmware. The instrument is fully controlled by an external computer. The measuring coils for the basic frequency of 976 Hz are designed as 6th-order compensated solenoids with remarkably high field homogeneity. The output signal from the pick-up coils is amplified, filtered and digitized, with the raw data being transferred directly to the computer, which controls all the instrument functions. Special diagnostics is embedded, which monitors important processes during measurement. A simplified block diagram of the instrument is shown in Fig. 20.1.

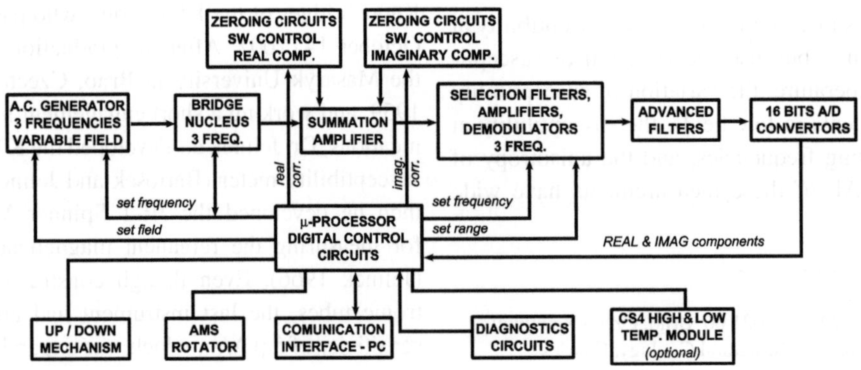

Fig. 20.1 Simplified block diagram of the MFK1-FA Kappabridge

The measuring fields in the MFK1-FA Kappabridge are presented in terms of peak values, following the convention used in most susceptibility meters, in contrast to the Kappabridges of the KLY series in which the measuring field was presented in terms of the *rms* values (sometimes called the effective values), the relationship between them being $K_{peak} = \sqrt{2}\, K_{rms}$.

The main features of the MFK1-FA Kappabridge are:

- Susceptibility measurement at 3 operating frequencies
- Susceptibility measurement in variable measuring fields
- High Sensitivity

Operating Frequency [Hz]	Measuring Field [A/m]	Measuring Range [SI]	Sensitivity [$\times 10^{-8}$]
976	2–700	0.9	4
3904	2–350	0.3	6
15616	2–200	0.7	12

- Separation of the in-phase (real) and out-of-phase (imaginary) components
- Automatic compensation of the real and imaginary components
- Auto-ranging and auto-zeroing over the entire measurement range
- Automated measurement of the field variation of bulk susceptibility
- AMS measurement using the spinning specimen method (specimen slowly rotates at 0.4 r.p.s., sequentially about three perpendicular axes)
- Built-in circuitry for controlling the non-magnetic furnace (up to 700°C) and the cryostat (from −196°C)
- Full instrument control by the external computer
- Sophisticated hardware and software diagnostics

Calibration is performed for both bulk susceptibility and spinning mode AMS at the reference field of 200 A/m, which is available at all operating frequencies.

20.3 Accuracy of Susceptibility Measurement in Variable Field and at Three Frequencies

The accuracy of measurement was investigated on 8 artificially made specimens containing multidomain magnetite mixed with powder of plaster of Paris. Bulk susceptibility ranged from 1×10^{-5} to 5×10^{-2}, the susceptibility difference between individual specimens being half an order. The whole collection was measured 10 times in three fields (50, 100, 200 A/m peak) and at three frequencies (976, 3904 and 15616 Hz). Before each measurement of the collection, the instrument was calibrated. The specimens were measured in one and the same orientation in order to avoid possible effects of magnetic anisotropy. Then, the arithmetical mean, root-mean-square error and root-mean-square error divided by the arithmetical mean and expressed in % (relative error), were calculated for each field and frequency.

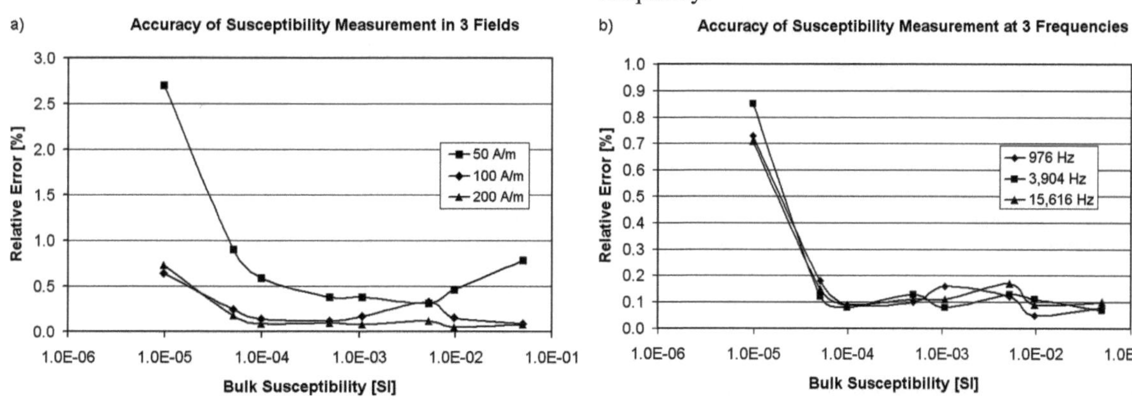

Fig. 20.2 Accuracy of susceptibility measurements in variable fields and at variable operating frequencies. (**a**) relative error of measurements in three fields (50,100, 200 A/m), all at the operating frequency of 976 Hz. (**b**) relative error of measurement at three operating frequencies (976, 3904, 15616 Hz), all in the measuring field of 200 A/m

Figure 20.2a shows the variation of the relative error with field for the measurements made at the frequency of 976 Hz. The relative error is similar in all three fields. In the magnetically weakest specimen with bulk susceptibility 1×10^{-5}, it is 1–3%. In all other specimens it is less than 0.2%. Figure 20.2b shows the variation of the relative error with the operating frequency, the magnetizing field being 200 A/m. The relative error is very similar at all three frequencies. In the magnetically weakest specimen, it is slightly less than 1%. In all other specimens, it is less than 0.2%.

20.4 Accuracy in the Determination of Frequency-Dependent Susceptibility

The frequency-dependent susceptibility is characterized by the following commonly accepted parameter (Dearing et al. 1996):

$$\chi_{FD} = 100(\chi_{LF} - \chi_{HF})/\chi_{LF},$$

where χ_{LF}, χ_{HF} are the mass susceptibilities at the lower and higher frequencies, respectively.

The precision in the determination of the χ_{FD} parameter was investigated through repeated measurement of a collection of cave sediments, 6 times on different days (the measuring field was 200 A/m). The

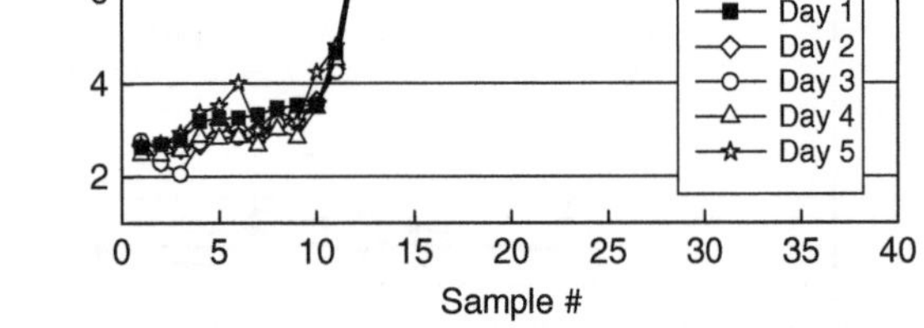

Fig. 20.3 Results of repeated measurements of the frequency-dependent mass susceptibility of cave sediments (provided by Drs. M. Chadima and J. Kadlec). The collection was measured 6 times on different days; the instrument was newly calibrated on each measuring day and for each frequency (for details see the main text)

results are in Fig. 20.3. Then, the data of the first day were ordered according to the increasing χ_{FD} parameter. The data of the other days were ordered in the order of the first day. If the line connecting the individual points parallels the first day line, the measuring precision is better than the differences between neighbouring points. On the other hand, if the lines cross, the precision is lower. It is obvious that the variation in the χ_{FD} parameter in the order of 1% is well reproducible, and the measurements can be interpreted in terms of the magnetic granulometry, even in those weakly magnetic sediments.

20.5 Variation of Bulk Susceptibility with Measuring Field

Diamagnetic and paramagnetic minerals show no variation of susceptibility with field by definition, and pure magnetite, even though ferrimagnetic, also shows virtually no field variation of susceptibility, while titanomagnetite, hematite and pyrrhotite may show very strong field variation (e.g. Worm et al. 1993, Jackson et al. 1998, Hrouda 2002, Hrouda et al. 2006).

The MFK1 Kappabridge is equipped for fully automated measurement of the susceptibility variation with field. Figure. 20.4a–c show examples of field variation of susceptibility of magnetite, pyrrhotite, hematite, and titanomagnetites. In magnetite, the susceptibility curve is more or less parallel to the abscissa, indicating no susceptibility variation with field within the field range used. In pyrrhotite, the susceptibility increases with field strongly and relatively rapidly. The susceptibility difference between the weakest and strongest field may be several hundred per cent. The susceptibility increase curve is rapid and linear in the field range of 80–120 A/m. Afterwards, the susceptibility change becomes slower, the curve becoming slightly curved and convex in shape. In hematite, the susceptibility variation with field is varied. In some cases it can be very strong, as strong as in pyrrhotite. However, the interval with linear dependence is no doubt wider than in pyrrhotite. In titanomagnetites, the susceptibility also increases with field, but the increase is not as strong and rapid as in pyrrhotite and hematite. In addition, the field interval with linear susceptibility increase may be much wider, its upper limit being typically more than 400 A/m and can be well over the whole instrument range. The

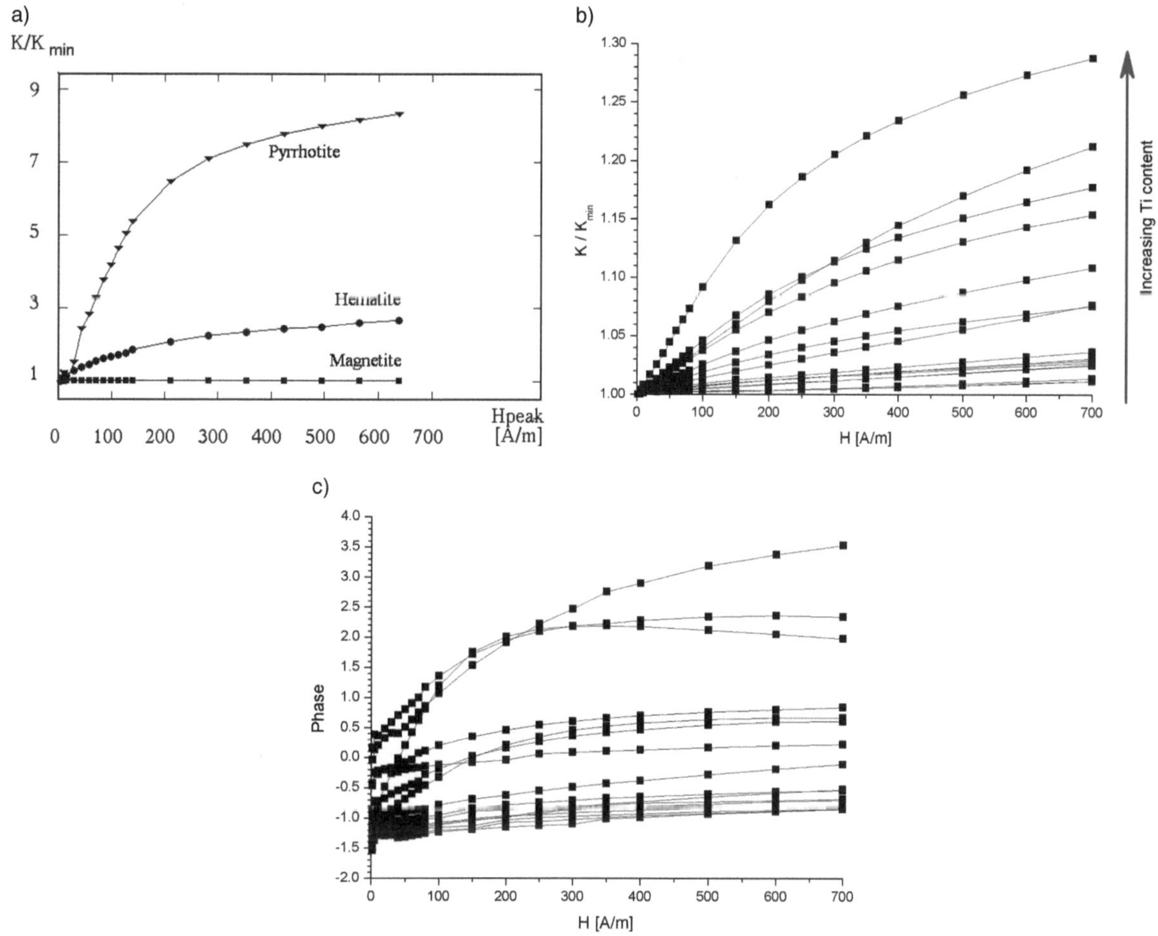

Fig. 20.4 Variation of bulk susceptibility of minerals and rocks with measuring field. (**a**) typical curves for magnetite, hematite, pyrrhotite, (**b**) curves for Fe-Ti oxides with variable contents of Ti, dyke rocks of the České středohoří Mts. (for more information see Chadima et al. 2009), (**c**) curves of the phase angle for Fe-Ti oxides with variable contents of Ti for the same specimens as in (**b**)

intensity of the field variation depends on the Ti content in the mineral. In titanomagnetite with very low Ti content approaching magnetite, the field variations of both the in-phase and out-of-phase susceptibilities (presented in terms of phase angle) are very weak, while in the Fe – Ti oxides with high Ti content, the variations are strong (see Fig. 20.4b, c).

As the field variation of susceptibility is different in different minerals and the automated measurement is very rapid (one specimen measurement takes about 10 minutes), it is recommended to do these measurements in all specimens prior to selecting the pilot specimens for the time-consuming investigation of the temperature variation of susceptibility.

20.6 Variation of Bulk Susceptibility with Temperature

The investigation of the variation of bulk magnetic susceptibility with temperature is a powerful technique for the identification of magnetic minerals in rocks and environmental materials, because the susceptibility vs. temperature curves show features typical of individual magnetic minerals or groups of minerals. For example, the curves of paramagnetic minerals are always represented by a hyperbola, while in ferromagnetic minerals they are more complex, characterized by an acute decrease of the susceptibility at the Curie temperature that can be very different (by 100°C or more) in individual minerals. In addition, the investigation of the temperature variation of magnetic susceptibility is also useful in investigating the compositional and/or phase changes taking place during the heating and cooling processes. Knowledge of these changes is important in the studies of the paleointensity of the Earth's magnetic field (e.g. Coe 1967) and in the thermal enhancing of the rock magnetic fabric (e.g. Henry et al. 2003).

The MFK1-FA Kappabridge measures the temperature variation of the bulk susceptibility of powdered specimens in cooperation with the recently developed non-magnetic CS-4 Furnace Apparatus in the temperature interval between room temperature and 700°C (and exceptionally to 800°C), and with CS-L Cryostat Apparatus in the temperature interval between the temperature of liquid nitrogen (–196°C) and room temperature. Figure 20.5 presents examples of such measurements. The susceptibilities are presented in terms of total susceptibility, which directly informs us of the measured signal. The relationship between the total and bulk susceptibility is as follows:

$$k_{\text{bulk}} = (V_o/V)\, k_{\text{total}},$$

where V_o is the nominal volume of the instrument ($V_o = 10\ \text{cm}^3$ in MFK1-FA), and V is the actual volume of the specimen measured (typically 0.2–0.3 cm^3).

Figure 20.5a shows the heating curves of a weakly magnetic granite specimen. The most conspicuous feature is the paramagnetic hyperbola in the low-temperature part of the curve continuing to about 200°C of the high-temperature part. The other important feature is the susceptibility increase above 400°C followed by an acute susceptibility decrease between

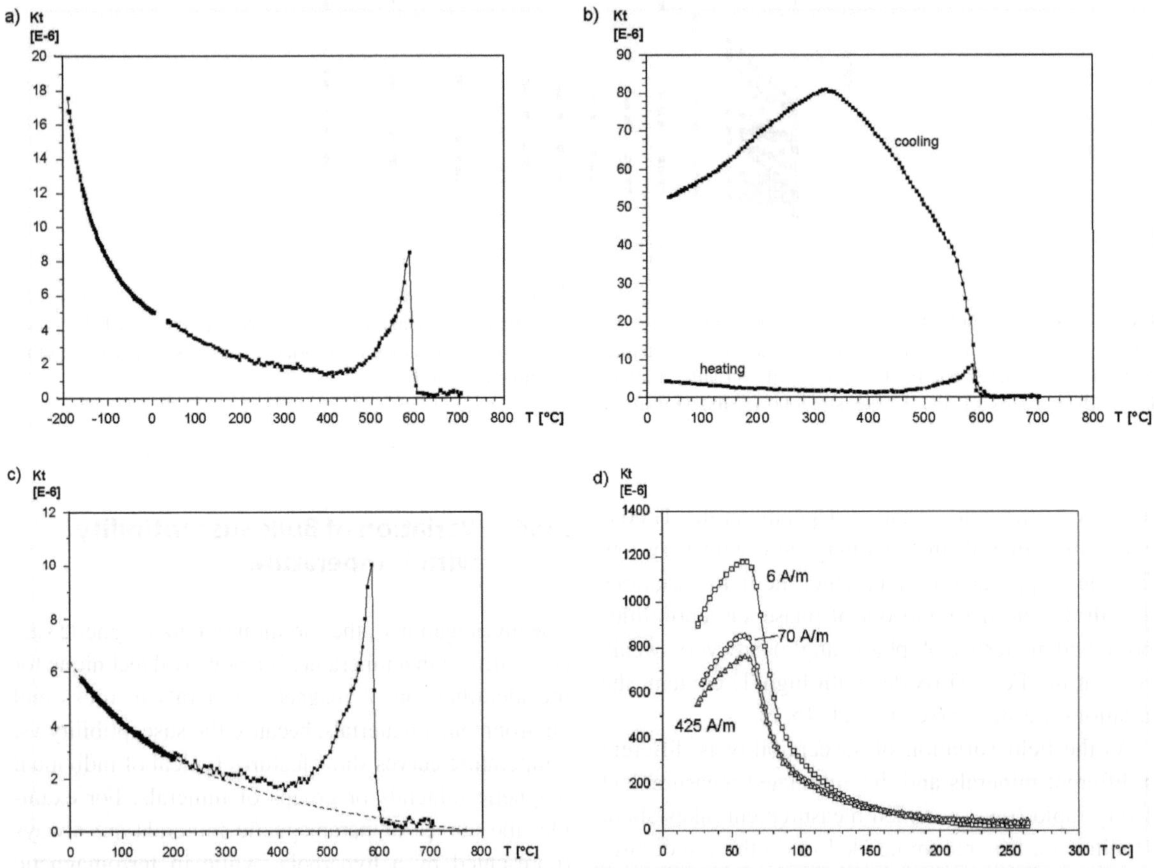

Fig. 20.5 Various susceptibility vs. temperature curves for weakly magnetic granite and basalt specimens (provided by Drs. J. Lehmann and C. Vahle). (**a**) low- to high-temperature heating curves for a granite specimen, (**b**) heating and cooling curves in a high temperature measurement for a granite specimen, (**c**) heating curve with hyperbola fit in the interval between 30 and 200°C for a granite specimen, (**d**) heating curves for a basalt specimen measured in different fields

590 and 600°C due to the Curie temperature of magnetite. Figure 20.5b shows both the heating and cooling curves in the high-temperature measurement. The cooling curve shows much higher susceptibilities than the heating curve, indicating the creation of new magnetite during specimen heating. Figure 20.5c shows the cooling curve with the hyperbola fit to the curve in the interval between 30 and 200°C using the method by Hrouda (1994) and extended to high temperatures. This hyperbola is very near the susceptibilities measured above 600°C, indicating almost no magnetite present in the specimen before its heating.

Figure 20.5d shows the heating curves of a basalt specimen measured in 3 different measuring fields. The peak just before the susceptibility decrease due to the Curie temperature is the most pronounced in the weakest field (6 A/m) considered, decreasing with increasing field.

20.7 AMS Field Variation

The theory of the low-field AMS is based on the assumption of a linear relationship between the magnetization and magnetizing field, resulting in field-independent susceptibility. This is valid for diamagnetic and paramagnetic minerals by definition, and also for pure magnetite. On the other hand, titanomagnetite, hematite and pyrrhotite may show a very strong field variation, making the above assumption invalid (for more details see Hrouda 2002, 2009). Methods were developed for determining the field-independent and field-dependent AMS components through standard measurement of the AMS in variable fields within the Rayleigh Law range (Hrouda 2009).

Figure 20.6 shows the variations of the bulk susceptibility $Km = (k_1 + k_2 + k_3)/3$, where $k_1 \geq k_2 \geq k_3$ are the principal susceptibilities), the degree of AMS ($P = k_1/k_3$), the shape parameter ($T = ln(k_2/k_3)/ln(k_1/k_3) - 1$) and the principal directions with field in a specimen of ultramafite. The bulk susceptibility and the degree of AMS clearly increase with field (Fig. 20.6a, b), while the shape parameter and principal directions vary only very mildly (Fig. 20.6c, d). The degree of AMS of the field-independent component ($P = 1.11$) is much lower than that of the field-dependent component ($P = 1.65$). The principal directions of both the components are coaxial with the whole-rock AMS. The former component

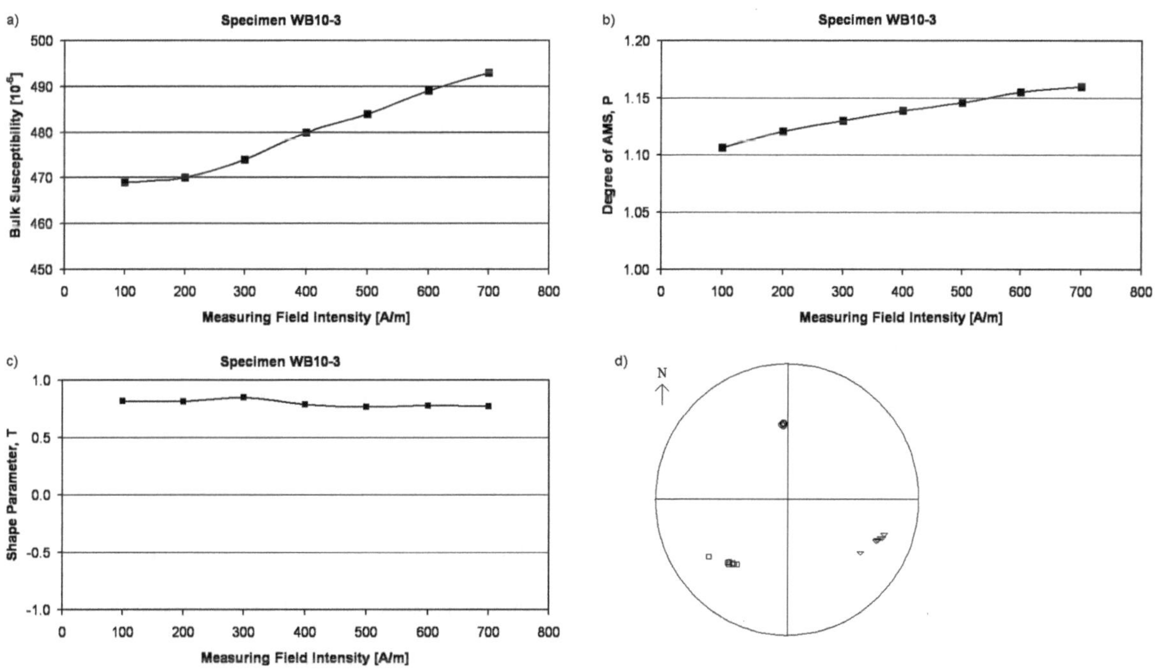

Fig. 20.6 Variations of the bulk susceptibility and the AMS with field in a specimen of ultramafite. (**a**) bulk susceptibility ($Km = (k_1 + k_2 + k_3)/3$), (**b**) degree of AMS ($P = k_1/k_3$), (**c**) shape parameter ($T = ln(k_2/k_3)/ln(k_1/k_3) - 1$), (**d**) the principal directions

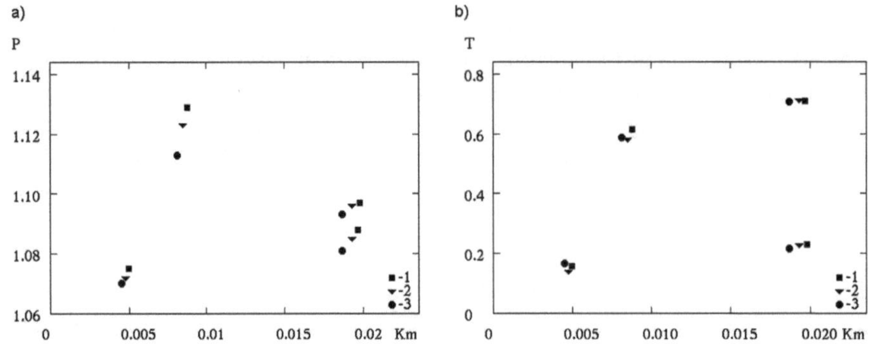

Fig. 20.7 Frequency-dependent susceptibility and the AMS in Celtic and medieval ceramics (provided by Dr. Z. Petáková). (**a**) *P* vs. *Km* plot, (**b**) *T* vs. *Km* plot. *Legend*: 1 – frequency of 976 Hz, 2 – frequency of 3904 Hz, 3 – frequency of 15616 Hz

is due to less anisotropic paramagnetic minerals, while the latter component is due to pyrrhotite possessing strong magnetocrystalline anisotropy.

20.8 Frequency-Dependent AMS

In some geological processes, such as for example very low-grade metamorphism, new very fine-grained magnetic minerals may originate. Their fabric can be investigated by the anisotropy of magnetic remanence (e.g. Trindade et al. 1999). As the frequency-dependent magnetic susceptibility is traditionally interpreted as resulting from interplay between superparamagnetic (SP) and stable single domain (SSD) or even multidomain (MD) magnetic particles, the frequency-dependent AMS can in principle help us in investigating the fabric of very fine-grained SP particles. Figure 20.7a shows the *P* vs. *Km* plot for selected specimens of the ceramics measured by the MFK1-FA Kappabridge in three frequencies. In all specimens, both the bulk susceptibility and the degree of AMS are the highest at the lowest frequency and in turn they are the lowest at the highest frequency. On the other hand, the values of the shape parameter *T* do not correlate with the bulk susceptibility (Fig. 20.7b). The orientations of the directions of the principal susceptibilities are highly coaxial at all three frequencies.

Acknowledgements Drs. Martin Chadima, Jaroslav Kadlec, Jeremie Lehmann, Zdeňka Petáková and Carsten Vahle are thanked for providing us with specimens illustrating the abilities of the MFK1-FA Kappabridge.

References

Bartošek J, Jelínek V (1961) Portable susceptibility meter of rocks in situ (κ-meter) (in Czech). Geologický průzkum 12:374–376

Chadima M, Cajz V, Týcová P (2009) On the interpretation of normal and inverse magnetic fabric in dikes: examples from the Eger Graben, NW Bohemian Massif. Tectonophysics 466:47–63

Coe RS (1967) The determination of paleointensities of the Earth's magnetic field with emphasis on mechanisms which could cause non-ideal behavior in Thellier's method. J Geomagnetism Geoelectricity 19:157–179

Dearing JA, Dann RJL, Hay K, Lees JA, Loveland PJ, Maher BA, O'Grady K (1996) Frequency-dependent susceptibility measurements of environmental materials. Geophys J Int 124:228–240

Henry B, Jordanova D, Jordanova N, Souque C, Robion P (2003) Anisotropy of magnetic susceptibility of heated rocks. Tectonophysics 366:241–258

Hrouda F (1994) A technique for the measurement of thermal-changes of magnetic- susceptibility of weakly magnetic rocks by the CS-2 apparatus and KLY-2 Kappabridge. Geophys J Int 118:604–612

Hrouda F (2002) Low-field variation of magnetic susceptibility and its effect on the anisotropy of magnetic susceptibility of rocks. Geophys J Int 150:715–723

Hrouda F (2009) Determination of field-independent and field-dependent components of anisotropy of susceptibility through standard AMS measurements in variable low fields I: Theory. Tectonophysics 466:114–122

Hrouda F, Chlupáčová M, Mrázová Š (2006) Low-field variation of magnetic susceptibility as a tool for magnetic mineralogy of rocks. Phys Earth Sci Planetary Inter 154:323–336

Hrouda F, Jelínek V, Zapletal K (1997) Refined technique for susceptibility resolution into ferromagnetic and paramagnetic components based on susceptibility temperature-variation measurement. Geophys J Int 129:715–719

Jackson M, Moskowitz B, Rosenbaum J, Kissel C (1998) Field-dependence of AC susceptibility in titanomagnetites. Earth Planet Sci Lett 157:129–139

Jelínek V (1966) High sensitivity spinner magnetometer. Studia Geophys Geod 10:58–77

Jelínek V (1973) Precision A.C. bridge set for measuring magnetic susceptibility of rocks and its anisotropy. Studia Geophys Geod 17:36–48

Jelínek V (1977) The statistical theory of measuring anisotropy of magnetic susceptibility of rocks and its application. Geofyzika, National Enterprise, Brno, 77 pp

Jelínek V (1978) Statistical processing of magnetic susceptibility measured on groups of specimens. Studia Geophys Geod 22:50–62

Jelínek V (1980) Kappabridge KLY-2. A precision laboratory bridge for measuring magnetic susceptibility of rocks (including anisotropy). Leaflet Geofyzika, Brno

Jelínek V (1981) Characterization of magnetic fabric of rocks. Tectonophysics 79:T63–T67

Jelínek V, Pokorný J (1997) Some new concepts in technology of transformer bridges for measuring susceptibility anisotropy of rocks. Phys Chem Earth 22:179–181

Parma J, Zapletal K (1991) CS-1 apparatus for measuring the temperature dependence of low-field susceptibility of minerals and rocks (in co-operation with the KLY-2 Kappabridge). Leaflet Geofyzika, Brno

Pokorný J, Suza P, Hrouda F (2004) Anisotropy of magnetic susceptibility of rocks measured in variable weak magnetic fields using the KLY-4S Kappabridge. In: Martín-Hernández F, Lüneburg CM, Aubourg C, Jackson M (eds) Magnetic fabric: methods and applications. Special publications, vol 238. Geological Society, London, pp 69–76

Trindade RIF, Raposo MIB, Ernesto M, Siqueira R (1999) Magnetic susceptibility and partial anhysteretic remanence anisotropies in the magnetite-bearing granite pluton of Tourao, NE Brazil. Tectophysics 314:443–468

Worm H-U, Clark D, Dekkers MJ (1993) Magnetic susceptibility of pyrrhotite: grain size, field and frequency dependence. Geophys J Int 114:127–137

Rema6W – MS Windows Software for Controlling JR-6 Series Spinner Magnetometers

21

Martin Chadima, Jiří Pokorný, and Miroslav Dušek

Abstract

Rema6W is a Microsoft Windows software package that provides full control of the AGICO JR-6 series dual speed spinner magnetometers (models JR-6 and JR-6A) together with instant data visualization. The main features include measurements in two speeds of rotation and three different acquisition times using automatic, semi-automatic, or manual specimen holders. In addition to a fully graphical user interface, the main functions of the program can be controlled using simple keyboard shortcuts. Acquired data are stored in an ASCII text data file allowing for easy viewing, editing, and further processing using Remasoft, PMGSC Paleomagnetism Data Analysis, or SuperIAPD programs. Acquired data are automatically sorted according to the specimen names and/or magnetic states, thus enabling an instant control on paleomagnetic demagnetization process. Plots for each specimen or magnetic state can be directly printed, copied into the clipboard, or exported as a Windows metafile graphics.

21.1 Introduction

A spinner magnetometer is a laboratory instrument for measuring the remanent magnetization of rocks, minerals, environmental samples, and synthetic materials. Due to the relatively simple construction, reliability, and uncomplicated maintenance, various types of spinner magnetometers have been routinely applied to solving problems in paleomagnetism, archeomagnetism, magnetic mineralogy, magnetic fabric, and magnetometry. The measuring principle is based on the law of electromagnetic induction. A specimen of defined size and shape rotating at a constant speed in a vicinity of a nearby detector induces an AC voltage signal. The amplitude and phase of the induced signal depend on the magnitude and direction of the vector of magnetic remanence of the specimen. For further information on spinner magnetometers refer to Pokorný (2007).

AGICO manufactures several models of spinner magnetometers that are commonly used in paleomagnetism labs, with the latest series (JR-6) including an automated model (JR-6A) and manual model (JR-6). Prior to this study, the instrument was controlled by an external PC computer using a Microsoft (MS) DOS based program called *Rema6*. Recent advances in operational systems and computer hardware together with the need for an instant visualization of acquired data, signaled the need for improvement in the software interface. The purpose of

M. Chadima (✉)
AGICO Inc., Ječná 29a, Brno, Czech Republic; Institute of Geology AS CR, v.v.i., Prague, Czech Republic
e-mail: chadima@agico.cz

E. Petrovský et al. (eds.), *The Earth's Magnetic Interior*, IAGA Special Sopron Book Series 1,
DOI 10.1007/978-94-007-0323-0_21, © Springer Science+Business Media B.V. 2011

the present paper is to introduce the main features of *Rema6W* software as a MS Windows based computer control of the JR-6 series spinner magnetometers.

21.2 JR-6 Series Spinner Magnetometers

The JR-6 series dual speed spinner magnetometer is based on the "classical" design where a specimen rotates about a single axis inside a pair of the Helmholtz coils (Agico 2004). In such an arrangement only the components of magnetic remanence vector in the plane perpendicular to the axis of rotation are measured; the component of the vector parallel to the rotation axis induces no signal. To measure the full vector of magnetic remanence, the specimen must be at least once manually re-positioned, thus two measuring positions are needed at the minimum. The spinner magnetometer JR-6A, equipped with an automatic holder manipulator, enables the automatic positioning of three successive measuring positions by turning the specimen around the imaginary body diagonal of a cube without any manual interference from the user. The standard measurement design consists of successive measurements in three, four, or six positions to reduce the measurement errors due to the inaccurate shape of specimen and by instrument noise, and, in case of manual holder, to eliminate any residual non-compensated components of remanence vector of the holder (Fig. 21.1).

The JR-6 series spinner magnetometer features two speeds of rotation: the high speed (87.7 revolutions per second), and the low speed (16.7 rps). The low speed of rotation increases the possibility of measuring fragile specimens, soft specimens, and specimens with considerable deviations in size and shape. The acquisition time depends on the strength of the specimen and, if desired, may be shortened or prolonged by a factor of two. The repeat mode enables the repetition of measurement in the current specimen position without stopping the spinning. The vendor-estimated sensitivity is 2.4×10^{-6} Am^{-1} when using high rotation speed and the 10 s integration time. The measuring range is from about 2.4×10^{-6} Am^{-1} up to 12,500 Am^{-1}.

21.3 Rema6W Program

Rema6W (ver. 6.1.0.) was developed in MS Visual Basic 6 with the frequent use of the Win32 API (Application Programming Interface) graphical functions stored in the dynamic link libraries of MS Windows. The program is intended to be user-friendly by using an uncomplicated and straightforward graphical user interface. The interface is based on the main window containing three tabs associated with different view panes and several auxiliary windows (Fig. 21.2).

21.3.1 Setting, Calibration, and Correction for Holder

The *Settings* window can be launched from the main window (*Tools* → *Settings*). In the *Settings* window one can set holder type (automatic, semi-automatic, or manual); number of positions for manual holder (two, four, or six); specimen type (cylinder or cube); acquisition time (short, normal, or long); speed of rotation (high or low); switch on/off the repeat mode. In addition, the user may set a type of spherical projection,

Automatic or semi-automatic holder

Position	M(x)	M(y)	M(z)
1	+X		+Z
2		+Y	+Z
3	+X	+Y	

Manual holder

Position	M(x)	M(y)	M(z)
1	-X		+Z
2		-Y	+Z
3	+X		-Z
4		+Y	-Z
5	+X	-Y	
6	-X	+Y	

Fig. 21.1 Position design for automatic, semi-automatic; and manual specimen holder. Positive and negative values refer to the components measured relative to the orientation of the holder (not applicable for the automatic or semi-automatic holder)

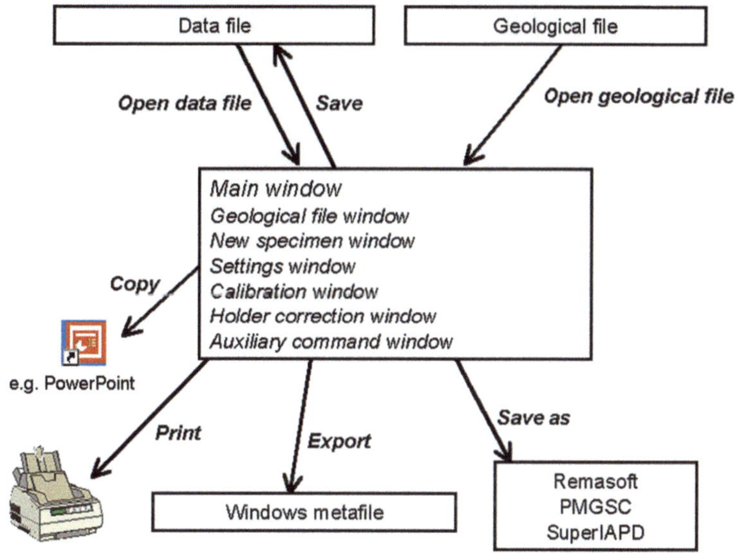

Fig. 21.2 Simplified flow chart of *Rema6W* showing the main windows and features of the program

displayed axes of the Zijderveld diagram, actual specimen volume, and the orientation parameters describing the sampling scheme.

Instrument calibration is controlled from the *Instrument calibration* window (*Tools* → *Calibration*). Calibration routine is performed by measuring the calibration standard of nominal magnetization and volume provided with the instrument. Calibration coefficients are expressed in the terms of gain and phase. Separate calibrations must be done for the high speed and the low speed.

The correction for the empty holder can be done in the *Holder correction* window (*Tools* → *Holder correction*). Two components of magnetization perpendicular to the axis of rotation are measured. In the case of automatic holder, due to its assymetric construction, two components of magnetization must be measured in all three successive positions. After changing the speed of rotation, the holder correction values are cleared and new holder correction measurement should be performed.

21.3.2 Data Acquisition

The *Data acquisition* view pane of the main window enables the user to make measurements according to the pre-set position design, holder type, measuring speed, and acquisition time (Fig. 21.3). Typical measurement cycle can be summarized as follows: Inputting new specimen name and magnetic state (plus, optionally, the orientation data) → Executing measurements (each position can be repeatedly remeasured, if necessary) → Saving or rejecting the results. The measurement cycle can be interrupted and data rejected anytime before completing. Current measurement can be immediately stopped in the case of any mechanical problem, e.g., strong vibration, specimen getting loose from the holder. After fixing up the problem the measurement cycle may be resumed. The above-described cycle can be operated either by clicking on the respective buttons, or by simple keyboard shortcuts; the shortcuts considerably facilitate the operation during a routine measurement of the extensive sets of specimens.

Orientation data for a new specimen can be input either manually from the memo book, or using information stored in a pre-created file, referred to as the geological file. Specimen orientation data include orientation angles plus, optionally, orientation of the mesoscopic foliation and lineation (in paleomagnetism usually bedding and fold axis, respectively). After saving, the specimen information is stored in the data file so it is not necessary to input the specimen information again when the same specimen in repeatedly measured during the course of demagnetization/magnetization process.

After completing the current measurement, two measured components of magnetization are recalculated using the calibration coefficients gain and

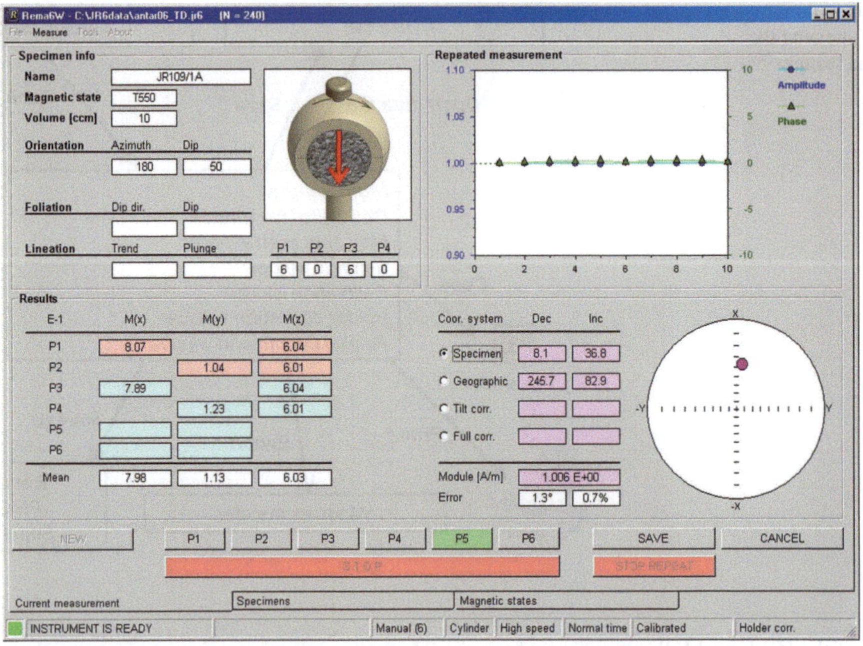

Fig. 21.3 A typical screen shot of the *Data acquisition* pane of the main window of the program

phase, and corrected for the actual volume of the specimen and remanent magnetization of the empty holder.

It is worth noting that after completing the measurement using the four- or six-position manual holder design, the holder residual values are fully eliminated even if the correction for the holder has not been performed. Such a correction is possible due to the design of specimen positioning with respect to the holder orientation; each component of magnetization is measured twice both in normal and reverse orientation with respect to the holder orientation (see Fig. 21.1). Calculated residual values of holder are automatically subtracted from the displayed components. This does not hold true for the two-position manual holder design and three-position automatic or semi-automatic holder design where one must fully rely on the holder values measured during holder correction procedure.

After completing measurements in all the pre-set positions, the mean values of three orthogonal components of magnetization are calculated together with precision of the measurement. The precision is expressed as an angle error of the direction and also as a relative error of magnitude of the magnetic remanence vector. The magnitude of the remanence vector is calculated together with declination and inclination in the specimen, geographic, tilt-corrected and full-corrected coordinate systems according to the specimen orientation data and the sampling scheme as described by four orientation parameters.

If desired, up to ten repeated measurements can be performed in the current position without stopping the specimen spinning. This feature is particularly useful in monitoring the effects of the viscous magnetization on the measurement. The results are displayed as the amplitude normalized by the amplitude of the first measurement and the phase difference from the first measurement. Repeated measurement can be stopped anytime before completing the default number of repetitions, the average of the last three measurements is calculated when possible.

21.3.3 Displaying of Measured Data

Specimens view pane of the main window displays the list of all the specimens stored in the currently opened data file (Fig. 21.4). After clicking on the desired specimen name, a table is presented showing the magnetic remanence in all respective magnetic states (e.g., natural remanent magnetization, individual steps of alternating field or thermal demagnetization) expressed as the magnitude together with declination and inclination in the currently used coordinate system. The

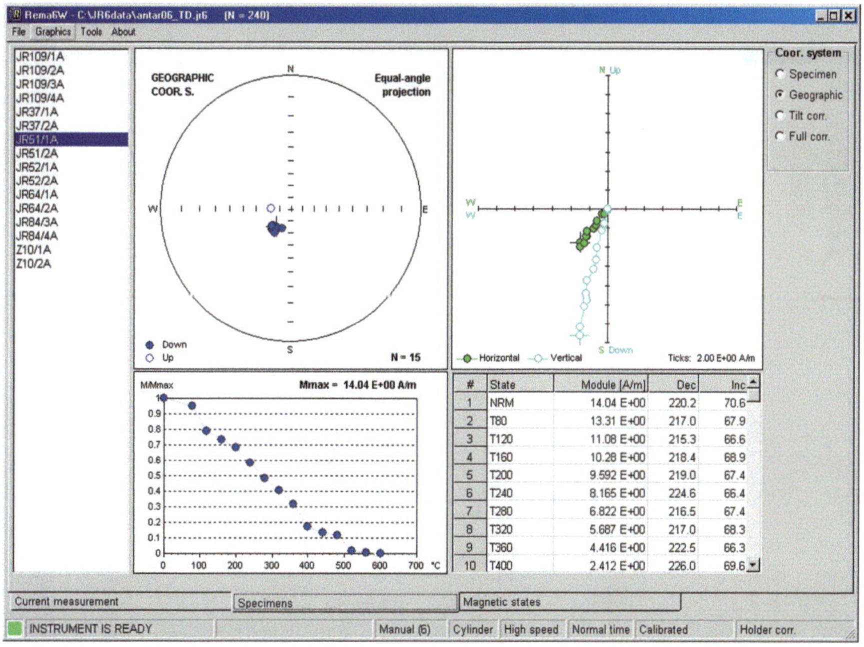

Fig. 21.4 A typical screen shot of the *Specimen* pane of the main window of the program

end-points of specimen remanence vectors are graphically displayed in the spherical projection, and as two perpendicular orthogonal projections (the Zijderveld diagram) together with the normalized magnitude as a function of demagnetization or magnetization values. Such diagrams are standard visualizations of the alternating field or thermal demagnetization treatment employed in paleomagnetic research. Data for each specimen are automatically refreshed whenever a new measurement is appended into the currently opened data file. Specimen data can be saved as the specimen files ready to be processed by Remasoft30 (Chadima and Hrouda, 2006), PMGSC Paleomagnetism Data Analysis (Enkin et al., 2003), or SuperIAPD (Torsvik et al. 1999). The graphics can be directly printed, copied into to clipboard, or exported into the Windows metafile format.

Magnetic states view pane of the main window displays the list of all the magnetic states stored in the currently opened data file (Fig. 21.5). After clicking on the desired magnetic state, a table is presented showing the magnetic remanence of all respective specimens expressed as the magnitude together with declination and inclination in the currently used coordinate system. The end-points of specimen remanence vectors are graphically displayed in the spherical projection. Such diagram facilitates visualization of the progress of demagnetization or magnetization of the large set of specimens. The graphics can be directly printed, copied into to clipboard, or exported into the Windows metafile format.

21.3.4 Auxiliary Commands

The auxiliary commands can be executed from the *Auxiliary commands* window (*Tools → Auxiliary commands*). These commands serve mainly for the basic test and maintenance of the automatic holder manipulator.

21.3.5 Data file

The acquired data are stored in the simple ASCII data file (extension *.jr6, see Agico, 2004 for the exact description of the format) in columns of the fixed width; each line represents one measurement. Each record contains specimen name, magnetic state, and three orthogonal components of the magnetic remanence vector in the specimen coordinate system together with their order of magnitude, specimen orientation angles, orientation of foliation and lineation, orientation parameters, and the precision.

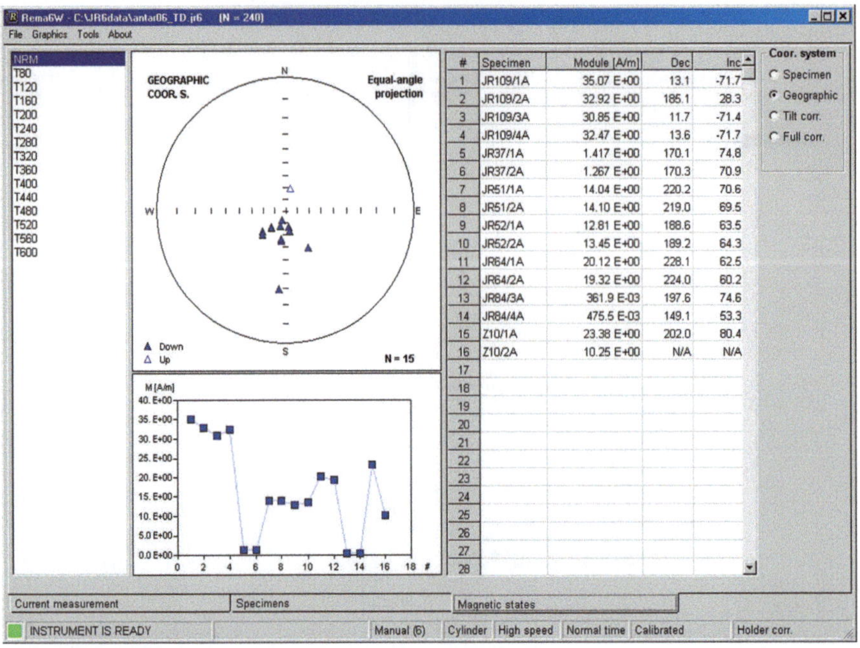

Fig. 21.5 A typical screen shot of the *Magnetic states* pane of the main window of the program

Transformation into the geographic, tilt-corrected, or full-corrected coordinate systems according to the specimen orientation data is done repeatedly before displaying each specimen or magnetic state. This enables further corrections of the specimen orientation data, if necessary. In addition to the data file, the measurement report is stored as a simple ASCII text file.

21.4 Conclusions

It is worth noting that the program works as a data viewer even if no instrument is connected or switched on. Acquired data stored in the *.jr6 files can be, e.g., viewed "off-line" after all the measurements has been completed, electronically sent for discussion to a colleague who does not own the JR-6 magnetometer.

The user interface consisting of data acquisition followed by sorting and graphical display is aimed for the future inclusion of basic data processing into the program. This would eliminate a need for the use of a successive data processing program and the software would provide a bridge between data acquisition and data processing programs. Another future outlook is to include the measuring scheme of magnetizing positions for obtaining the tensor of anisotropy of magnetic remanence (AARM, AIRM).

The program is available free of charge at: www.agico.com.

Any feedback, error reports and suggestions are highly appreciated.

Acknowledgements Illustrative figures depicting individual measuring positions were drawn by Jiří Vysloužil. Roman Veselovskiy is acknowledged for his effort in testing the program. Constructive review of Gary Acton significantly improved the clarity of the paper. The software was partly developed for the purpose of solving the Project #A300130612 of the Grant Agency of the Academy of Sciences of the Czech Republic, and the Academic Research Plan #AV0Z30130516.

References

Agico, Inc. (2004) Spinner magnetometer JR6/JR6A user's manual. Brno, Czech Republic. http://www.agico.com

Chadima M, Hrouda F (2006) Remasoft 3.0 – a user-friendly paleomagnetic data browser and analyzer. Travaux Géophys XXVII:20–21. http://www.agico.com/software/remasoft/

Enkin R, Wuolle K, McCann C, Carretero M, Voroney M, Baylis T, Morton K, Jaycock D, Baker J, Beran L (2003) PMGSC – Paleomagnetism Data Analysis, ver. 4.2. http://gsc.nrcan.gc.ca/sw/paleo_e.php

Pokorný J (2007) Spinner magnetometer. In: Gubbins D, Herrero-Bervera E (eds) Encyclopedia of geomagnetism and paleomagnetism. Springer, The Netherlands, 1054pp

Torsvik TH, Brinden JC, Smethurst MA (1999) SuperIAPD1999 – software package. Geological Survey of Norway, Trondheim, Norway. http://www.geodynamics.no/software.htm

Experimental Study of the Magnetic Signature of Basal-Plane Anisotropy in Hematite

22

Karl Fabian, Peter Robinson, Suzanne A. McEnroe, Florian Heidelbach, and Ann M. Hirt

Abstract

The crystal symmetry of hematite in the basal plane predicts three easy magnetization axes for the sublattice spin orientation above the Morin transition. Spin canting then leads to three preferred magnetization axes perpendicular to these easy axes. By combining detailed crystallographic orientation by EBSD measurements with dense magnetic hysteresis and remanence curves as a function of rotation angle, the relation between crystallography and magnetic properties has been experimentally verified for the basal plane of a natural hematite crystal. The measurements lead to a better understanding of the interplay between spin canting, remanence and magnetic susceptibility at different field strengths. The measurement results coincide qualitatively with theoretical predictions, and provide experimental evidence for quantitative evaluation by more complex micromagnetic modeling.

22.1 Introduction

Hematite, α-Fe_2O_3, is one of the most common natural magnetic minerals. It occurs in metamorphic and igneous rocks, commonly with exsolution intergrowths of ilmenite, as high-grade ore bodies in metamorphic rocks due to contact metasomatism, in hydrothermal veins, and, due to its high oxidation state, it is a widespread product of diagenesis of weathered sediments. Fine-grained hematite is found in almost all types of continental and marine sediments. Its magnetic properties result from a subtle balance of antiferromagnetism (AF) and spin-orbit-coupling leading in ideal cases to a spin-canted antiferromagnetic (CAF) moment at room temperature. In natural hematite, impurities, defects, and residual moments of incomplete AF compensation in fine particles may add to, or even dominate, the total magnetic moment (De Grave and Vandenberghe, 1990). The unraveling of the complex magnetic properties of hematite is still a topic of ongoing research (Morrish, 1994). Here, we verify experimentally the ideal magnetocrystalline structure of pure monocrystalline hematite by means of direct magnetic measurements on a sample on which the crystal structure orientation has been verified by electron backscatter diffraction (EBSD). We thereby confirm the magnetic consequences of the CAF structure and provide a geometrical discussion of the response of the CAF spin-configuration to different applied fields.

K. Fabian (✉)
Norwegian Geological Survey, 7491 Trondheim, Norway
e-mail: karl.fabian@ngu.no

E. Petrovský et al. (eds.), *The Earth's Magnetic Interior*, IAGA Special Sopron Book Series 1,
DOI 10.1007/978-94-007-0323-0_22, © Springer Science+Business Media B.V. 2011

22.1.1 Hematite Structure

The crystal structure of hematite, also common to Al_2O_3, Cr_2O_3 and V_2O_3, is in the rhombohedral system, though most conveniently indexed using hexagonal axes a_1, a_2, and a_3 in the (0001) basal plane and c normal to it. The basic unit consists of layers of Fe^{3+} ions coordinated by six oxygens in distorted octahedrons. The octahedrons share edges to form sheets parallel to the basal plane (Fig. 22.1), where 1/3 of the octahedrons remain vacant. The layers are stacked on top of each other along the c-axis (pink layer above gray layer in Fig. 22.1) in such a way that each octahedron shares one face (shaded green where visible) with an octahedron either above or below, with one vacancy either above or below. The Fe^{3+} ions across these shared faces are repelled from each other. This has the effect of shortening the edges of the shared faces, and the repulsion creates an up and down puckering of cations in the (0001) layers (indicated by dark = down, and light = up shading in each layer), which relates to the distances between cations in adjacent layers, and ultimately in turn to the intensities of different magnetic interactions. The ordered sequence of octahedral layers has a repeat every six layers. The space group symbol is $R\bar{3}c$, where $\bar{3}$ denotes a threefold axis of rotary inversion and c a glide plane parallel to the c-axis. When magnetized at room temperature the magnetic moments of individual Fe^{3+} ions are parallel to each other within an octahedral layer, and $\approx 180°$ different in adjacent layers. Thus, in this magnetic form (see below) the arrangement of magnetic moments (sublattice magnetization directions) violate rhombohedral symmetry, but agree with the c glide. Early studies by Besser et al. (1967), using natural crystal forms and ferro-magnetic resonance spectra, implied that the sublattice magnetizations in hematite are in fact preferentially oriented parallel to a-crystallographic axes which happen also to be normal to edges of Fe^{3+} octahedrons. Using EBSD and intensive magnetic measurements together here, we confirm these implications and tie magnetic properties directly to crystal structure.

22.1.2 Spin Canting in Hematite

According to Dzialoshinskii (1957), the atomic rhombohedral space group D_{3d}^6 of hematite is consistent with three possible magnetic point groups corresponding to three possible states:

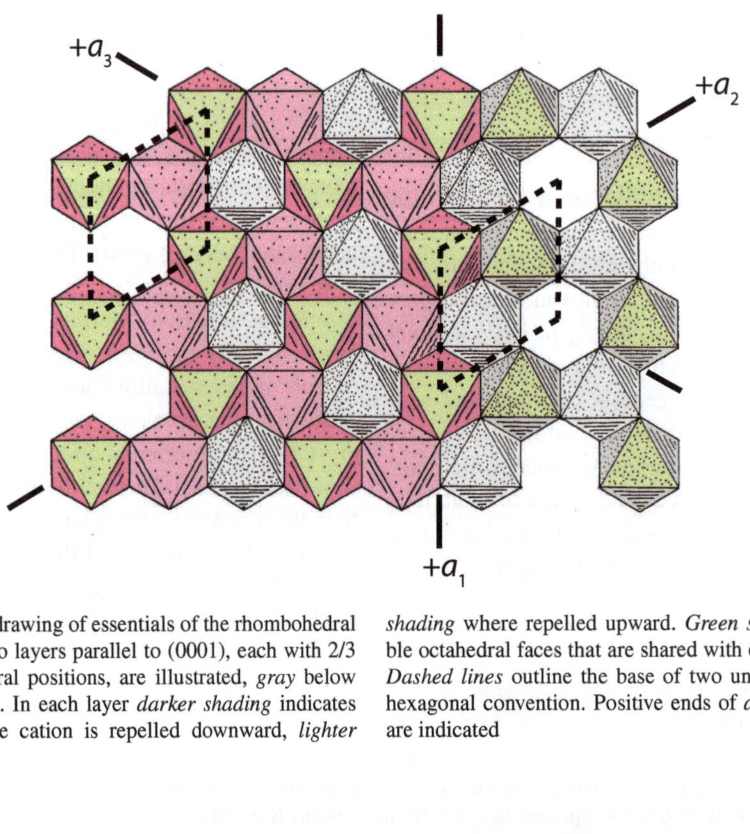

Fig. 22.1 Schematic drawing of essentials of the rhombohedral hematite structure. Two layers parallel to (0001), each with 2/3 occupancy of octahedral positions, are illustrated, *gray* below and *pink*, fixed on *top*. In each layer *darker shading* indicates octahedrons where the cation is repelled downward, *lighter shading* where repelled upward. *Green shading* indicates visible octahedral faces that are shared with overlying octahedrons. *Dashed lines* outline the base of two unit cells indexed in the hexagonal convention. Positive ends of a-crystallographic axes are indicated

1. State I: $(2\,C_3, 3\,U_2, l, 2\,S_6, 3\,\sigma_d)$ Highest symmetry, corresponding to the largest symmetry group, where all spins are exactly aligned with the c-axis.
2. State II: (U_2, l, σ_d) Lower symmetry, where a residual moment is confined to lie in the basal plane, perpendicular to an a-axis. Such a moment is possible by spin canting within the basal plane.
3. State III: $(U_2 R, l, \sigma_d R)$ Low symmetry, where a residual moment lies in an a-c-plane. This could be achieved by canting out of the c-axis.

The low symmetry of state II permits a spontaneous magnetic moment along an axis in the basal plane coincident with the c-glide plane. Dzialoshinskii (1957) proposed that this moment in hematite occurs as an intrinsic moment generated by spin canting described by an exchange energy

$$J\,\mathbf{m_1}\cdot\mathbf{m_2} + \mathbf{D}\cdot(\mathbf{m_1}\times\mathbf{m_2}), \qquad (22.1)$$

where $\mathbf{m_{1,2}}$ are the two sublattice-magnetization vectors, J is the classical isotropic exchange constant, and $\mathbf{D}$ the Dzialoshinskii vector. Later, Moriya (1960) analyzed the superexchange coupling in hematite and demonstrated that the physical origin of spin canting is spin-orbit coupling. The order of magnitude of $|\mathbf{D}|$ therefore is $|\mathbf{D}| \approx \Delta g/g\,J$, where g is the gyromagnetic ratio and $\Delta g = g - 2$. The canting angle γ is then

$$\gamma \approx \frac{1}{2}\,|\mathbf{D}/J| = \frac{1}{2}\,|\Delta g/g| \approx 10^{-3}, \qquad (22.2)$$

in agreement with the observed value of $\gamma \approx 1.14 \cdot 10^{-3} \approx 0.065^{\circ}$.

22.1.3 Morin Transition and Anisotropy

By heating through the Morin-transition, $T_M \approx 260$ K, the magnetic symmetry of hematite changes from state I below T_M to state II above T_M, such that at room temperature hematite is a weak ferromagnet due to spin canting. Above T_M, a strong magnetic anisotropy effectively confines the sublattice magnetization of hematite to the basal plane (Morrish, 1994). Along the c-axis susceptibility is low, coercivity is extremely high, and magnetization can be moved out of the basal plane only under the influence of very strong fields. Within the basal plane, the threefold atomic symmetry creates three equivalent a-axes (Fig. 22.2) corresponding to three equivalent magnetic states II. The magnetic anisotropy energy in the basal plane accordingly is supposed to show sixfold symmetry. In first order the assumed magnetocrystalline anisotropy is described by

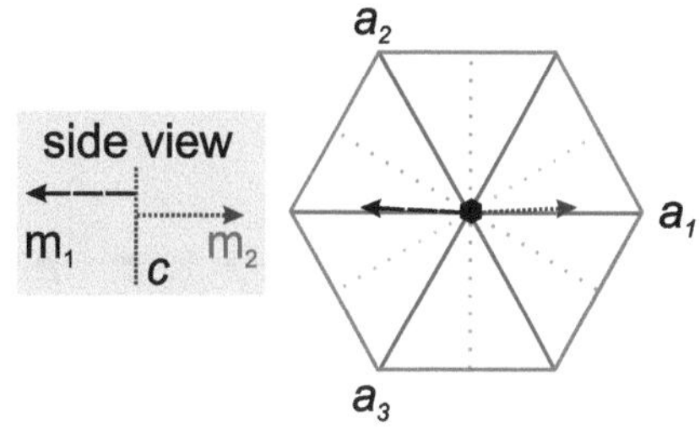

Fig. 22.2 The *left* sketch shows the antiferromagnetic spin configuration as viewed from the c-axis in hematite and shows that within the basal plane the canted magnetization vectors m_1 (*dashed*) and m_2 (*dotted*) are oriented along one of the three easy axes a_1, a_2, a_3

$$e_B = -K_B\,(\cos 6\theta_1 + \cos 6\theta_2)\,, \qquad (22.3)$$

where K_B is a basal plane anisotropy constant and $\theta_{1,2}$ are the azimuths with respect to an easy axis of the sublattice-magnetization vectors $\mathbf{m_{1,2}}$, which are supposed to lie in the basal plane.

22.1.4 Basal-Plane Geometry in Hematite

The topic of this study is the direct detection of magnetic variations within the basal plane. Using experimental magnetic measurements it should be possible to confirm the relationship, and the theoretically assumed relation, between the orientations of magnetic sublattices and the structure of Fe^{3+} octahedrons in the basal plane related to the positions of a-crystallographic axes. This subject has become a recent focus of interest in the study of single hemo-ilmenite crystals (Robinson et al., 2006a, b). Here, the natural remanent magnetization (NRM), strictly confined to the basal plane, appears to derive from a vector sum of the lamellar magnetization parallel to the direction of the sublattice magnetizations, and the spin-canted magnetization of the hematite, which is normal to the mean of the sublattice magnetizations. Satisfactory magnetic measurements of the anisotropy of hematite in the basal plane apparently require very pure and undistorted crystals to minimize

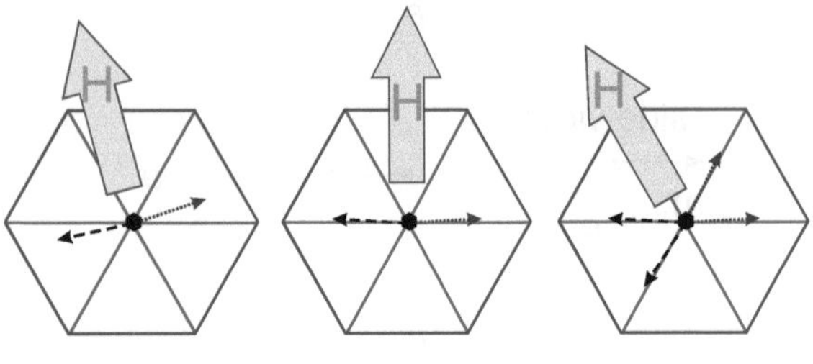

Fig. 22.3 Some possible responses of hematite's CAF spin configuration in the basal plane to different applied external field vectors *H*. *Left*: rotation of the canted moment against basal-plane anisotropy. *Middle*: Increasing the canting angle with field strength. *Right*: Increasing the canting-angle while keeping the individual spins close to the easy *a*-axes

magnetostrictive energies (Banerjee, 1963). The early successful study by Flanders and Schuele (1964) found widely varying estimates for the anisotropy constant K_B from torque measurements on a range of hematite samples. They also performed magnetic measurements of remanence as a function of angle, moment, coercive force, susceptibility and rotational hysteresis. Besser et al. (1967) used ferromagnetic resonance spectra to show that the sublattice magnetizations in the basal plane are parallel to one of the three crystallographic *a*-axes, such that the spin-canted magnetic moment is perpendicular to one of the *a*-axes. Interestingly, the study of Porath and Raleigh (1967) associates triaxial anisotropy in hematite with multiple rhombohedral twins due to high strain in natural crystals. The present study tries to relate the magnetic response of a pure hematite crystal to a precise crystallographic characterization. Depending on field direction and field strength, the possible spin-deflection mechanisms may vary as sketched in Fig. 22.3. Detailed measurements will help to resolve which processes occur.

22.2 Experimental Results

22.2.1 Sample Orientation

The present study was conducted on a natural hematite crystal approximately 2 × 3 × 4 mm in size. On one shiny face of this crystal, parallel to the basal plane, the crystallographic orientation was determined by electron backscatter diffraction (EBSD) in the scanning electron microscope (SEM) (e.g. Dingley, 1984). This technique allows for determination of the complete orientation of all parts of a crystal with a spatial resolution of less than 1 μm^2. In a fully automated version it is possible to map out the orientations in a crystalline surface (Adams et al., 1993) in a regular grid with variable step size and dimensions. In our case the crystal was mounted with the assumed basal plane parallel to the sample holder surface, and was scanned with a step size of 20 μm and a grid of 134 × 108 steps. The crystal surface was measured as-is without any further preparation. A Zeiss Gemini 1530 FEG-SEM with a Schottky emitter, and equipped with a EBSD detector by HKL technology (now Oxford Instruments) was used, employing an accelerating voltage of 20 keV and a beam current of 9 nA.

A pixel map of the 11318 data points measured on the shiny basal-plane surface of the hematite crystal is shown in Fig. 22.4. The vertical edge on the right was used as a reference for both the magnetic and EBSD measurements. Two other straight edges at 120-degree angles are also crystallographically significant. The predominant medium orange color of the image indicates surface areas with a common crystallographic orientation. Small areas of brighter green, blue or red, indicate areas where crystallographic orientation deviates very slightly from the mean (see Fig. 22.4b). However, they are all within less than 1° of the orientations colored in orange. Three lower-hemisphere equal-area diagrams in Fig. 22.4 show orientations of selected electron diffraction data. The zone axis {001} shows that crystallographic *c*-axes, with rare exceptions, are exactly normal to the shiny (001) basal plane. The zone axis {100} shows reflections from planes with indices (100). Such planes are exactly parallel to *a*-axes, so the spots indicate the positions of directions more or less exactly parallel to the flat direction of the crystal, and also exactly perpendicular either to the right-hand reference edge, or to one of the other straight edges shown in Fig. 22.4 keyed directly to the magnetic data. The zone axis {110} shows reflections from planes with indices (110). Such planes are exactly normal to *a*-axes, so the spots indicate the positions of *a*-axes more or less exactly parallel to the flat direction of the crystal, and also exactly parallel either to the right-hand reference edge, or to one of the other straight edges shown in Fig. 22.4 that are keyed directly to the magnetic data. In summary, then, we can conclude that each of the straight edges of the hematite crystal platelet is exactly parallel to one of the three crystallographic *a*-axes.

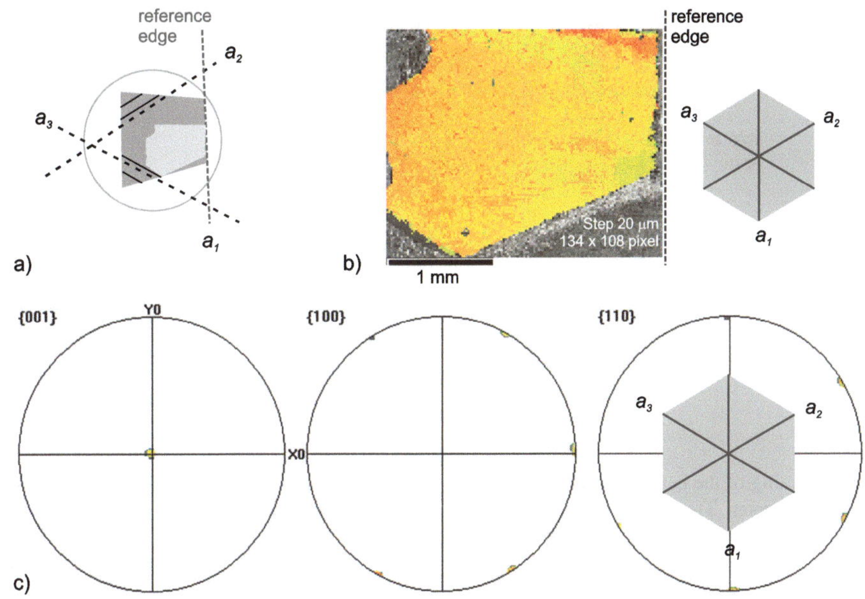

Fig. 22.4 EBSD measurement of local crystallographic directions. The investigated area with respect to the magnetic measurement orientation is sketched in (**a**). The reference edge refers to the alignment of the crystal in the vibrating-sample magnetometer. The electron backscatter image in (**b**) shows the color coded orientation of *c*-crystallographic axes of all 11318 pixels in the scanned area. The homogenous color indicates a uniform orientation without visible twinning. In (**c**) the same information is displayed by pole figures in equal-area projection of the lower hemisphere. The fact that all data points fall on equivalent locations indicates perfect alignment of *c*-axes and *a*-axes within the studied sample area. The scatter of orientations within the pole figures ($\approx 1.5°$) is smaller than the symbol size used

22.2.2 VSM Measurements in the Basal Plane

The angular-dependent magnetic measurements were performed using a vibrating sample magnetometer (PMC Micromag 3900) equipped with a computer controlled continuous sample rotation about its *Z*-axis, allowing a maximum applied field of 1.4 T. As sketched in Fig. 22.5, the hematite crystal was mounted with the basal plane aligned with the bottom flat surface of a cylindrical sample holder. Because the field is applied along the instrument's *X*-axis, rotation around the *Z*-axis leaves the field inside the basal-plane. By manual rotation, the reference edge of the sample was rotated approximately in line with the instrument's *Y*-axis. Optical inspection showed that this succeeded within 6–9° (Fig. 22.5). The sample was positioned so that when viewed from top as sketched in Fig. 22.5, it lay as close as possible symmetrically around the rotation axis, in order to have its 'magnetic center' on the *Z*-axis. Centering is important, because the four VSM pick-up coils would record a spurious $2\,\theta$ signal from an off-centered sample. This is best understood by imagining a paramagnetic sphere, with volume V and magnetic susceptibility χ mounted by a distance Δr off-center in a homogenous field H. By rotating the sample, the X coordinate of the sample varies between $-\Delta r$ and Δr, while its magnetization $\chi\, V H$ stays constant. Because the induced voltage from a constant vibrating magnetization depends non-linearly on the distance from the coils, the X variation leads to a $2\,\theta$ periodicity in the apparent magnetization. By optimally centering the sample on the rod, this spurious $2\,\theta$ signal can be minimized.

The instrument was then programmed to perform a series of measurements for each rotation angle θ from $\theta = 0°$ to $\theta = 358°$ in steps of 2°. These measurements were

1. High-field hysteresis branch from zero field to +1 T in steps of 10 mT,
2. Low-field branch from zero field to +10 mT in steps of 0.5 mT,
3. Back-field curve from zero field to −100 mT in steps of 10 mT.

From the first measurement it is possible to infer a slope-corrected estimate for $M_s(\theta)$, the value of

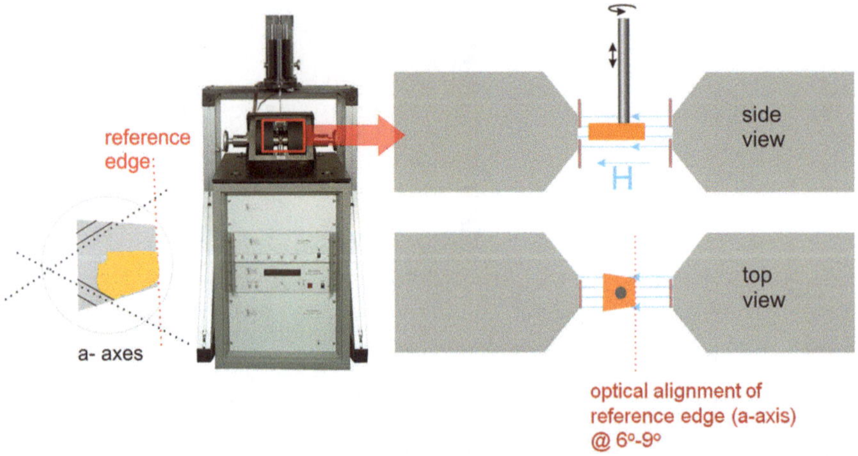

Fig. 22.5 Oriented mounting of the hematite-crystal inside the vibrating-sample magnetometer. The a-axis corresponding to the reference edge is visually aligned to lie approximately perpendicular to the applied field. The instrument position of 0° corresponds to an angle of 6–9° between reference edge and true perpendicular

$M_{rs}(\theta)$ and the high-field, as well as the low-field slope of the hysteresis branch. The second measurement mainly gives a more precise determination of the low-field slope, which yields a reliable approximation to the saturation-remanence-aligned susceptibility $\chi_r(\theta)$. The back-field curves determine the values of $H_{cr}(\theta)$.

22.2.3 Remanence Coercivity in the Basal Plane

The back-field curve is obtained by applying first a large positive field of 0.5 T, then a smaller negative field $-H$ which is switched off, and then the remanent magnetization $M_{bf}(H)$ is measured in zero field. The field value H_{cr}, for which $M_{bf}(H_{cr}) = 0$, is the coercivity of remanence. Figure 22.6a shows two extreme back-field curves in the basal plane, one measured at 8°, approximately perpendicular to an a-axis, and one at 22°, at an intermediate peak of H_{cr}. The values of $H_{cr}(\theta)$ as a function of rotation angle θ are of special interest, because they do not depend on the position of the 'magnetic center' of the sample. At $H_{cr}(\theta)$ the induced voltage from the sample magnetization is zero, and the zero position is not influenced by a 2 θ periodicity in signal amplification. Therefore, we expected $H_{cr}(\theta)$ to show most clearly the 6 θ periodicity of basal-plane anisotropy. The right-hand side of Fig. 22.6 indeed clearly shows this 6 θ periodicity, and also that $H_{cr}(\theta)$ is minimal whenever an a-axis is perpendicular to the field along X, and maximal if an a-axis is aligned with X. However the data show inexplicable peaks at symmetric rotations between these extreme positions, which may be related to twinning, although EBSD measurements showed no evidence for twins on the surface measured.

22.2.4 Saturation Magnetization and Saturation Remanence in the Basal Plane

Saturation magnetization is determined by fitting a straight line to the last 18 points of the high-field curve above 0.85 T, as indicated in Fig. 22.7a. This yields $M_s(\theta)$ as well as the high-field susceptibility $\chi_{hf}(\theta)$. For $M_{rs}(\theta)$ we use the zero-field value of the high-field curve. While the result in Fig. 22.7c shows that $M_s(\theta)$ is almost constant, $M_{rs}(\theta)$ varies with 6 θ periodicity, in phase with $H_{cr}(\theta)$. The ratio of minimal to maximal M_{rs} is 0.89, which agrees well with the theoretical value of $\cos 30° \approx 0.87$. The latter assumes that after applying a field, the sublattice moments in the remanence state are aligned with that a-axis, which allows the smallest possible angle of the canted moment with respect to the direction vector of the previously applied field (Fig. 22.7b). This angle is always smaller or equal to 30°, and therefore the minimal remanence measured has to be greater or equal to $\cos 30°$ of the maximal remanence.

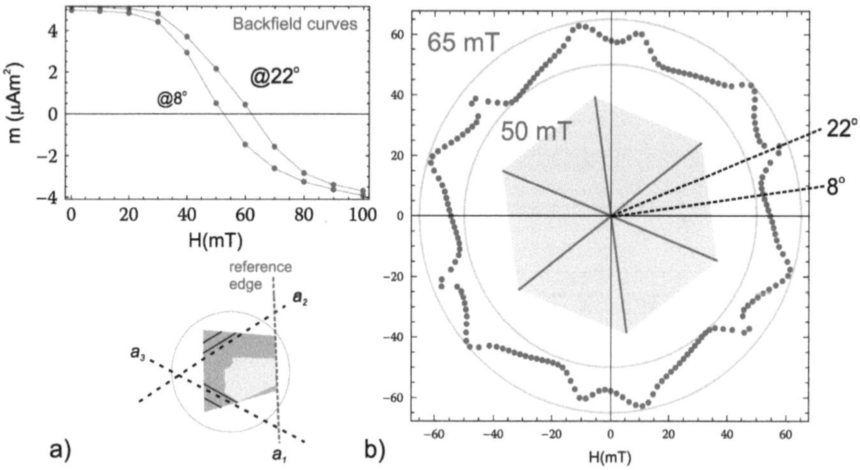

Fig. 22.6 Remanence coercivity H_{cr} as a function of rotation angle θ. (**a**) shows sample orientation and two typical backfield curves at $\theta = 8°$ and $\theta = 22°$. At the position of $\theta = 8°$ the applied field is approximately perpendicular to the reference edge, which corresponds to a crystallographic a-axis. (**b**) is a polar plot of all H_{cr} measurements versus θ in relation to the crystallographic a-axes. The absolute values of H_{cr} vary between 50 mT and 65 mT. The origin of the symmetric intermediate peaks, e.g. at 22°, is unclear

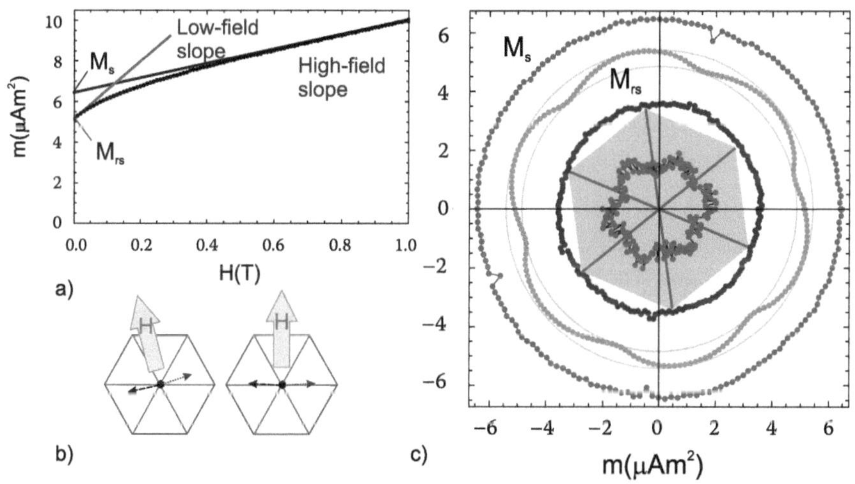

Fig. 22.7 From measurements of hysteresis branches between zero field and 1 T it is possible to infer M_s, M_{rs} and high-field as well as low-field slopes as shown in (**a**). The sketch in (**b**) shows the geometrical situation of the spin-canted moments within a field applied in different directions within the basal plane. The angular dependent measurements are plotted in (**c**). The two *innermost curves* represent high-field (*dark gray*) and low-field (*light gray*) slopes in relative units, whereby the values for the low-field slope must be multiplied by 10 to yield the same units as for the high-field slope. The variation of M_s (*outer dark gray*) and M_{rs} (*outer light gray*) is plotted in units of magnetic moment measured. The *two grey circles* indicate minimal and maximal values of M_{rs} of 4.85 and 5.42 μAm^2, respectively

A crude estimate of magnetic susceptibility is given by the slope of the high-field curve at zero as drawn in Fig. 22.7a. This results in a rather noisy data set in Fig. 22.7c (light green), which varies in anti-phase with $M_{rs}(\theta)$, because a smaller value of M_{rs} coincides with a steeper descend near zero.

22.2.5 Weak-Field Susceptibility in the Basal Plane

The susceptibility signal from the initial slope of the high-field curve at zero is resolved in more detail by the subsequent low-field curve in Fig. 22.8a. This curve

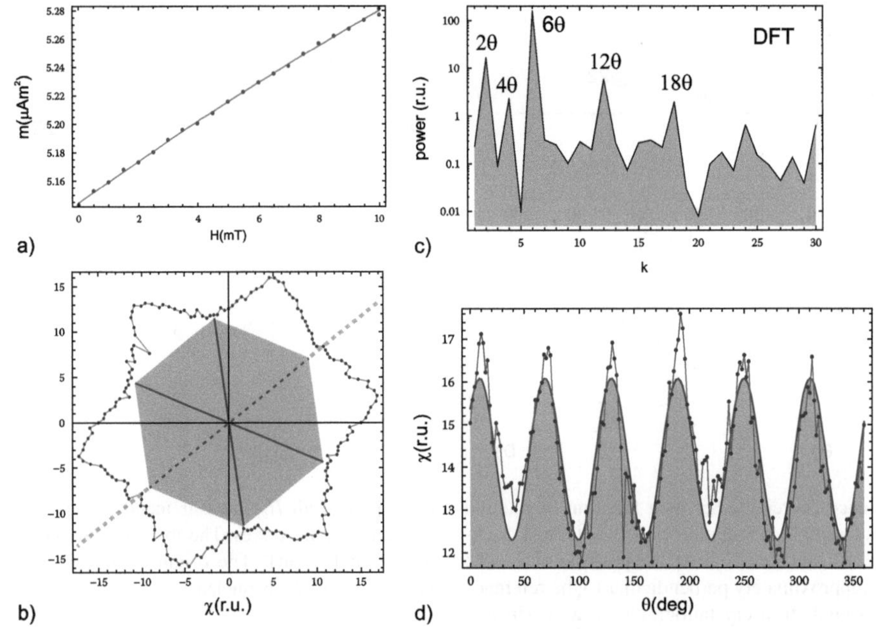

Fig. 22.8 (**a**) Weak-field susceptibility χ is estimated from the initial slope of hysteresis branch 0–10 mT measured after the high-field curve in the same direction. The slope is determined from a second-order polynomial fit to the full data set. (**b**) Polar diagram of weak-field susceptibility. High values of χ correspond to directions perpendicular to the three a-axes in the basal plane. (**c**) Power spectrum of the discrete Fourier transform of $\chi(\theta)$. While most energy is localized in the 6 θ-peak, there are also clear contributions from higher harmonics and from a 2 θ variation. (**d**) Plot of susceptibility versus angle. Sinusoidal 6 θ variation shows the inverse transform of the 6 θ-component in (**c**)

starts from the state of saturation remanence, where the spins should be aligned with the a-axis closest to the perpendicular of the field direction. The initial slope of this low-field curve as a function of θ is plotted in a polar diagram in Fig. 22.8b. It clearly shows the same six-fold symmetry as the innermost curve in Fig. 22.7c, but with a better signal-to-noise ratio. Comparison to the crystallographic orientation verifies that maximum susceptibility is observed perpendicular to the a-axes. Besides the predominant six-fold symmetry, the diagram exhibits a slight but visible elongation along the a-axis marked by a dotted line in Fig. 22.8b. A discrete Fourier transform of susceptibility versus rotation angle in Fig. 22.8c gives a quantitative confirmation that the predominant 6 θ variation has some non-sinusoidal distortion, visible by the relative high energy of the multiples 12 θ, and 18 θ. The direct comparison of the original data and the filtered 6 θ component in Fig. 22.8d shows that, besides the 2 θ variation noted above, the deviation from the 6 θ signal mainly comes from the more pointed positive peaks.

22.3 Discussion

22.3.1 Mechanisms of Magnetization Change in the Basal Plane

There are two possible magnetic mechanisms of magnetization change in the basal plane. First, an increase in H can increase the canting angle γ leading to increased net magnetization without rotating $\mathbf{m}_{1,2}$ far away from their a-axes. Second, when the field is not perpendicular to the a-axis close to $\mathbf{m}_{1,2}$, the individual sublattice moments can rotate out of this easy axis, keeping the canting angle γ constant. By just studying the angular variation of $M_{rs}(\theta)$ these mechanisms cannot be distinguished, because $M_{rs}(\theta)$ is measured in zero field, where the sublattice moments are aligned with an a-axis. The observation that neither M_s nor the high-field slope (dark gray) in Fig. 22.7c depend on θ, indicates, that above 0.8 T the second mechanism is essentially saturated, which means that $\mathbf{m}_1 - \mathbf{m}_2$ lies perpendicular to the field direction, and further magnetization change occurs exclusively by field-induced spin canting. It is therefore of interest, which mechanism dominates the magnetization change at low field values, a question best treated by investigating the low-field susceptibility data in Fig. 22.8.

If only the first mechanism occurs, it is easy to estimate geometrically the angular variation of weak-field susceptibility in the basal plane. Maximum susceptibility χ_{max} then occurs along a perpendicular p to an a-axis. When the field axis is rotated by an angle θ away from p, the effective field along p is reduced by a factor $\cos\theta$, but also the measured magnetization component is reduced by the same factor. Because the largest angle away from an a-axis is 30°, the maximal decrease in susceptibility with respect to a field perpendicular to the a-axis would be

$$\cos^2 30° = \frac{3}{4}, \tag{22.4}$$

corresponding to a relative reduction from $\chi_{max} = 16$ to $\chi_{min} = 12$ units in Fig. 22.8d. This reduction coincides astonishingly well with the experimental result, although the observed minimum values are slightly higher ≈ 12.4 and the signal shape is more pointed than a pure $\cos^2\theta$ dependence. These minor deviations indicate that besides the spin-canting mechanism, there is a second contribution to the low-field susceptibility, which we assign to the second mechanism of spin rotation away from the a-axis. Because this mechanism is of minor importance at low fields < 2 mT, and saturated in high fields > 0.8 T, it must contribute significantly to the magnetization change at intermediate fields.

22.3.2 Which Coercivity Mechanism?

The observed symmetric variation of $H_{cr}(\theta)$ between ≈ 50 mT and ≈ 65 mT shows that orientation with respect to the a-axis also controls the coercivity mechanism which creates this change in remanence. As in the case of low-field susceptibility, there are two main coercivity mechanisms which might contribute. The first is *spin rotation*, where the sublattice magnetizations $\mathbf{m}_{1,2}$ with increasing backfield $-H$ rotate in the basal plane, while keeping the canting angle γ approximately constant. By this mechanism, differential rotation of neighboring regions could create domain walls, which can be pinned at defects thereby leading to the observed coercivity. A second possible mechanism is reduction of the canting angle γ by the backfield $-H$ to a point where after reducing the field to zero the canting angle changes from positive to negative values. This hypothetical process shown in Fig. 22.9 is here denoted as *spin flip*, referring to a small angle opposite movement of the sublattice magnetizations $\mathbf{m}_{1,2}$. The term *spin flop* is commonly used for a field induced transition between state I and state II (Morrish, 1994). In contrast, a spin flip occurs entirely within state II, but would be marked by a change of sign of the Dzialoshinskii vector **D**.

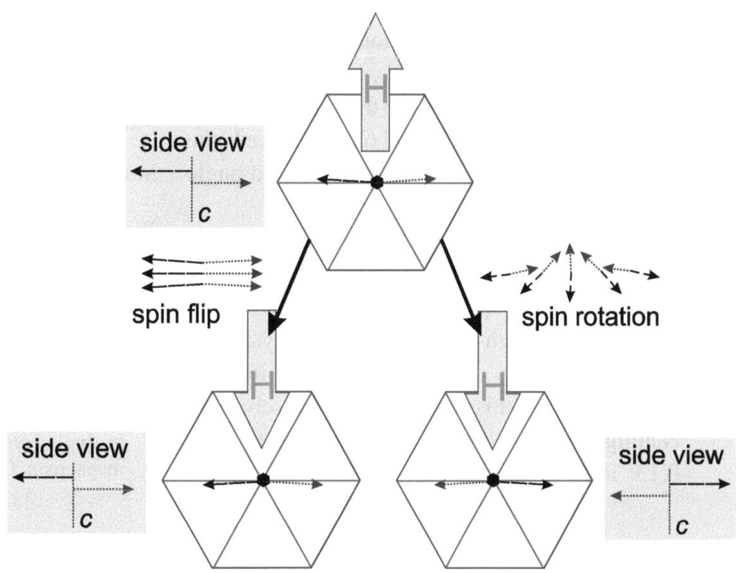

Fig. 22.9 Two possible magnetization-reversal mechanisms to explain the variation of H_{cr} with rotation angle. Spin flip requires small angular movement of the individual spins into a new inverse spin-canting state (*left*). Spin rotation (*right*) requires large angle movements of the individual spins, but allows retention of the spin canting

22.3.3 Why Not Spin Flip

The main observation in favor of the spin-flip hypothesis is that the minima of $H_{cr}(\theta)$ lie perpendicular to the a-axes. If spin-rotation would be the most important mechanism, one would expect this to be more effective when the field acts perpendicular to the remanent moment, i.e. along the a-axes. On the other hand, in case of spin-flip the angular variation of H_{cr} should be proportional to $\cos\theta$ leading to a ratio of minimal to maximal H_{cr} of $\cos 30° = 0.87$. However the observed value is ≈ 0.77 which is closer to $\cos^2 30° = 0.75$, and much smaller than expected for the spin-flip scenario. Another problem with spin flip is that the high-field slope of the hysteresis curves show that to create an induced magnetization of 2 M_s a field of $H = 1.6$ T is required. If the same canting energy is necessary for negative canting, it would require at least $H = -1.3$ T to annihilate M_{rs} by the spin-canting mechanism. If the above estimates are valid, spin rotation would seem to be the more likely coercivity mechanism. Yet, the energy barrier between positive and negative canting could be considerably smaller, as indicated by low-field susceptibility, and further investigation will be needed to resolve this question.

Conclusions

1. By combining EBSD and magnetic hysteresis measurements on a hematite crystal, it is confirmed that the basal plane contains three magnetically equivalent a-axes, which are easy axes for the sublattice magnetization.
2. Remanence coercive force H_{cr} is maximal along and minimal perpendicular to the a-axes. Its relative variation between 50 and 65 mT exceeds the variation expected from a pure $\cos\theta$ field-angle dependence.
3. M_s is independent of field direction indicating sufficiently easy rotation of spins out of the easy a-axes below the applied maximum field of 1 T.
4. The observed high-field slope of the hysteresis curves most likely originates from field-induced spin canting. In this interpretation the zero-field canting-angle $\gamma \approx 0.065°$ is doubled to $\gamma \approx 0.13°$ in an applied field of 1.6 T.
5. M_{rs} as a function of rotation angle is maximal along and minimal perpendicular to the a-axes. Its relative variation agrees with the theoretically expected $\cos\theta$ field-angle dependence.
6. Weak-field susceptibility after saturation is minimal along and maximal perpendicular to the a-axes. Its variation also approximately agrees with a $\cos^2\theta$ field-angle dependence.
7. If high-field extrapolation is valid, spin flip is an unlikely coercivity mechanism below 1.5 T

Acknowledgements We gratefully acknowledge funding of KF, PR, and SM through the Norwegian Research Council (MATERA, Nanomat). Background measurements for sample selection have been performed at the Institute for Rock Magnetism, University of Minnesota, and were funded by NSF.

References

Adams BL, Wright SI, Kunze K (1993) Orientation imaging: the emergence of a new microscopy. Metall Transact 24A:819–833

Banerjee S (1963) An attempt to observe the basal plane anisotropy of hematite. Philos Mag 8:2119–2120

Besser P, Morrish A, Searle C (1967) Magnetocrystalline anisotropy of pure and doped hematite. Phys Rev 153:632–640

De Grave E, Vandenberghe R (1990) Mössbauer effect study of the spin structure in natural hematites. Phys Chem Minerals 17:344–352

Dingley D (1984) Diffraction from sub-micron areas using electron backscattering in a scanning electron microscope. Scan Electron Microsc 11:569–575

Dzialoshinskii IE (1957) Thermodynamic theory of "weak" ferromagnetism in antiferromagnetic substances. Soviet Phys JETP 5:1259–1272

Flanders PJ, Schuele WJ (1964) Anisotropy in the basal plane of hematite single crystals. Philos Mag 9:485–490

Moriya T (1960) Anisotropic superexchange interaction and weak ferromagnetism. Phys Rev 120:91–98

Morrish AH (1994) Canted antiferromagnetism: hematite. World Scientific Publishing, Singapore

Porath H, Raleigh C (1967) An origin of the triaxial basal–plane anisotropy in hematite crystals. J Appl Phys 38:2401–2402

Robinson P, Heidelbach F, Hirt AM, McEnroe SA, Brown L (2006a) Crystallographic-magnetic correlations in single crystal haemo-ilmenite: New evidence for lamellar magnetism. Geophys J Int 165:17–31

Robinson P, Heidelbach F, Hirt AM, McEnroe SA, Brown LL (2006b) Correction – crystallographic-magnetic correlations in single crystal haemo-ilmenite: new evidence for lamellar magnetism. Geophys J Int 165:431

Anorthosites as Sources of Magnetic Anomalies

23

Laurie L. Brown, Suzanne A. McEnroe, William H. Peck, and Lars Petter Nilsson

Abstract

Magnetic anomalies provide information about location, size and composition of earth structures, ore bodies and tectonic features even in bodies containing only a few percent magnetic minerals. Here we investigate the magnetic properties and oxide mineralogy of anorthosites, rocks rich in plagioclase (>90%), and compare their magnetic signatures to aeromagnetic anomaly maps of the regions. Two of the anorthosite complexes have large negative anomalies associated with them; both have low susceptibility and high remanence related to hemo-ilmenite mineralogy and remanent directions antiparallel to the present field. One complex has appreciable natural remanent magnetization quasi-parallel to the present field, and strong susceptibility, creating an enhanced positive anomaly. The fourth anorthosite has little or no magnetic anomaly over much of its area, in accordance with the weak remanence, low susceptibility and variable magnetic mineralogy observed. The anorthosite samples producing significant anomalies, and maintaining strong and stable natural remanent magnetization over geologic time all contain oxides of the hematite-ilmenite series. This study adds support to 'lamellar magnetization' whereby exsolved phases in the ilmenite-hematite system produce strong and stable magnetization with only minor amounts of oxide material.

23.1 Introduction

Information on the magnetization of the Earth's crust comes from the study of anomalies measured at ground, aeromagnetic and satellite levels, as well as the study of rocks exposed at the surface. Magnetic anomalies have long been used as an ideal way to "discover" high-grade ore deposits, to outline shallow and deep geologic structures, and to identify buried magnetic objects from 55-gallon drums to batholiths. In many studies the assumption prevails that observed magnetization is the result of an induced field, parallel to the present field, that is controlled in magnitude by the susceptibility of the source material. The possibility of remanent magnetization contributing to terrestrial anomalies is commonly ignored, although remanence is well accepted as the dominant component in marine surveys. The Koenigsberger ratio, comparing remanent magnetization (NRM) to induced magnetization, provides a key to contribution of the

L.L. Brown (✉)
Department of Geosciences, University of Massachusetts, Amherst, MA 01003, USA
e-mail: lbrown@geo.umass.edu

E. Petrovský et al. (eds.), *The Earth's Magnetic Interior*, IAGA Special Sopron Book Series 1,
DOI 10.1007/978-94-007-0323-0_23, © Springer Science+Business Media B.V. 2011

amount of NRM and induced magnetization which is needed to successfully interpret magnetic anomalies from crustal rocks.

Anorthosites are plutonic igneous rocks composed almost entirely of plagioclase feldspar. Mafic minerals, usually clinopyroxene, orthopyroxene, olivine and iron oxides make up no more than 10% of the rock, with magnetite and ilmenite as the predominant oxides present. Anorthosites are commonly found in association with a series of other igneous rocks, namely mangerite, charnockite, and granite, and in these cases are referred to as AMCG complexes. Many of the known large anorthosite bodies are Proterozoic in age, and are also associated with layered intrusions, and are found in both orogenic and non-orogenic environments. The origin of anorthosites has been debated for many years, and although now a magma source is well accepted, the location of that source in the mantle or crust is still in question (Ashwal 1993, Longhi 2005). Besides their ancient existence on Earth and enigmatic problems as to their origin, anorthosites host economic deposits, including ilmenite, magnetite, chromite, and platinum group elements, making them objects of continued interest to the geological and geophysical community. Anorthosite is a common rock on the Moon (Papike et al. 1998, Hawke et al. 2003, Takeda et al. 2006) and has been suggested as a possible rock type on other planets such as Mars and Mercury (Ashwal 1993, Blewett et al. 2002).

Due to the typical composition of over 90% plagioclase feldspar, anorthosites in general have largely been ignored by the magnetic community, and considered to be "non-magnetic" or treated as 'paramagnetic crust'. Previous work on anorthosites from southern Norway (Brown and McEnroe 2008, McEnroe et al. 2001) has shown anorthosite can have strong remanent-dominated magnetic anomalies associated with them. To expand on this work we have sampled other Proterozoic anorthosites of varying ages and geologic environments. The accessory oxide mineralogy reflects the general oxidation state of the magma, with the more oxidized anorthosites containing rhombohedral members of the hematite and ilmenite solid solution, and the more reduced anorthosites have magnetite as an accessory phase. Both hemo-ilmenite phases (ilmenite host with hematite exsolution) and ilmeno-hematite phases (hematite host with ilmenite exsolution) are associated with the more oxidized anorthosites.

In this paper we investigate the magnetic properties of anorthosites ranging in ages and geologic setting that document a wide variety of magnetic properties. Two of the anorthosites, the Morin Complex in southern Canada and the Marcy Anorthosite in the Adirondack Mountains, NY state, are late Proterozoic bodies metamorphosed during the Grenville orogeny (1.1 Ga). A third anorthosite is from Flakstadøy, Lofoten islands, Norway, an Archean body metamorphosed at 1.8 Ga. The Rogaland Anorthosite Province of southern Norway, emplaced after peak Sveconorwegian metamorphism, provides three large anorthosite bodies (Egersund-Ogna, Håland-Helleren, and Åna-Sira) and the associated smaller Garsaknatt body. Here we examine the magnetic properties of these anorthosites, discuss their oxide mineralogy and evaluate their contributions to magnetic anomalies on Earth, and as a possible sources for anomalies on other planets.

23.2 Methods

Most of the samples used in this study were collected as parts of larger paleomagnetic studies of the anorthosite complexes, allowing for a number of separate samples collected as various sites distributed over the bodies. Cores were drilled and oriented at the outcrop, or oriented blocks were collected and later drilled in the lab. Samples from the Morin anorthosite were collected initially for geochemical studies and thus were not oriented.

Measurements of susceptibility and natural remanent magnetization (NRM) were measured on all samples using a Sapphire susceptibility bridge and either on a 2G cryogenic magnetometer at the University of Massachusetts, or a JR-6A spinner magnetometer at the Norwegian Geological Survey. Optical observations on polished thin sections were made using a reflecting light microscope, or a Scanning Electron Microscope (SEM).

Values of susceptibility and NRM for each anorthosite body are compared by plotting induced magnetization (Ji), calculated from the product of susceptibility and the strength of the present magnetic field at the location of the anorthosite, versus the natural remanent magnetization (Jr). The ratio of remanent magnetization (NRM) to induced magnetization, known as the Koenigsberger ratio (Q), is used as a measure of the importance of remanent magnetization in a paleofield direction to the induced magnetization in the direction of the present field. In general, Q values greater than 0.5 indicate the contribution of a remanent component, while values greater than 2 indicate a significant NRM component, and values > 5 show the predominance of the NRM direction.

23.3 Morin Anorthosite

23.3.1 Geology

The ca. 1.15 Ga Morin Anorthosite-Mangerite-Charnockite-Granite (AMCG) suite (Martignole and Schrijver 1970, Emslie 1975, Doig 1991) intrudes orthogneiss and paragneiss of the Morin terrane in the Grenville Province of Quebec (Fig. 23.1). The Morin Anorthosite massif consists of a western lobe of anorthosite and leucogabbro, and is surrounded on the west side by contemporaneous jotunite, mangerite, and monzonite plutons. The eastern lobe of the massif is a separate intrusion (Peck and Valley 2000), and it and part of the western lobe underwent deformation and recrystallization in the Morin Shear Zone (Zhao et al. 1997). The Morin terrane is polymetamorphic; the AMCG suite intruded high-grade gneisses and subsequently was metamorphosed and variably deformed under granulite-facies conditions (Peck et al. 2005). In the western lobe of the anorthosite igneous textures are preserved, and cm-scale plagioclase megacrysts are surrounded by plagioclase, clinopyroxene, and orthopyroxene with minor alkali feldspar, quartz, hornblende, and opaques. Besides the grey color of preserved igneous plagioclase common to other massif anorthosites, igneous plagioclase from the Morin anorthosite also can have a distinctive maroon color attributed to fine-grained hematite inclusions (Emslie 1975). Anorthosite of the eastern lobe in the Morin Shear Zone has white plagioclase and minor clinopyroxene, orthopyroxene, and opaques. This anorthosite is recrystallized to a 1–2 mm polygonal mosaic with

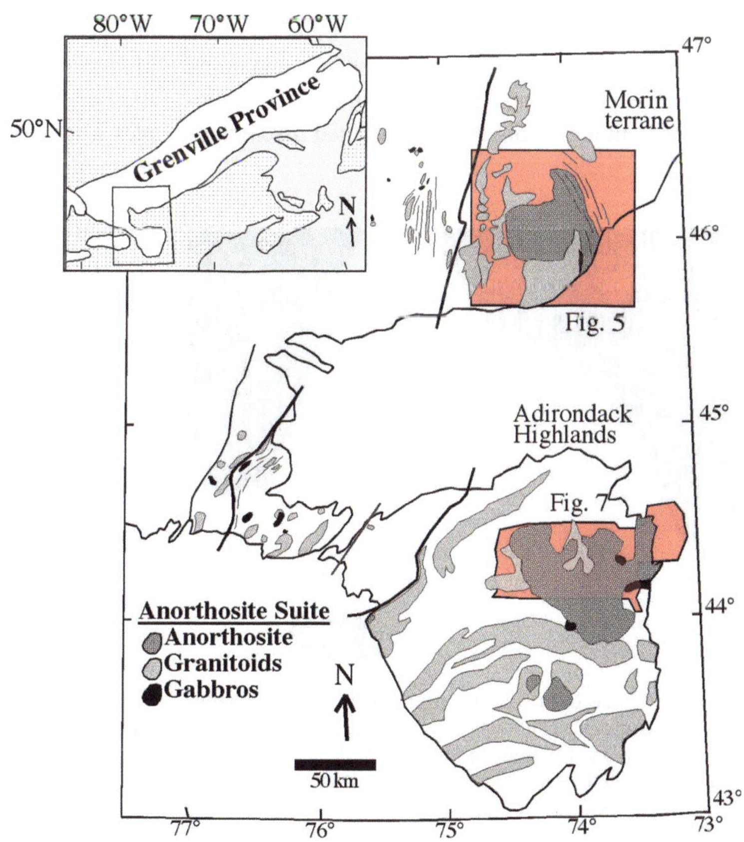

Fig. 23.1 Location map for the Morin Anorthosite, Quebec and the Marcy Anorthosite, New York, both part of the Grenville province of eastern North America

relict pyroxene augen. Jotunite (pyroxene monzodiorite) has sharp contacts with anorthosite, and is made up of plagioclase, clinopyroxene, orthopyroxene, and opaques with minor quartz, alkali feldspar, hornblende, biotite, apatite, and garnet. Mangerite (pyroxene quartz monzonite) has gradational contacts with jotunite, and is composed of plagioclase, alkali feldspar, and quartz with minor clinopyroxene, orthopyroxene, hornblende, biotite, apatite, and garnet.

23.3.2 Magnetic Mineralogy

Opaque minerals in anorthosite (0–5%), jotunite (1–20%), and mangerite (1–5%) are dominated by iron-titanium oxides (Emslie 1975). Anorthosite and leucogabbro of the west lobe contain both exsolved hemo-ilmenite (Fig. 23.2) and ilmeno-hematite; both commonly have secondary lamellae within the primary exsolved phases (Pogue 1999, Petrogenesis of the Morin anorthosite, Grenville Province, Quebec, unpublished MA thesis, Washington University St. Louis). Some samples also contain minor discrete magnetite grains. In the eastern lobe oxides are less abundant than in the western lobe, and metamorphic assemblages of hemo-ilmenite + titanite containing ilmenite lamellae and composite grains of hemo-ilmenite + titanite. This is a more altered assemblage than is found in the western zone. In contrast, jotunite and mangerite contain ilmenite and magnetite as separate oxide phases.

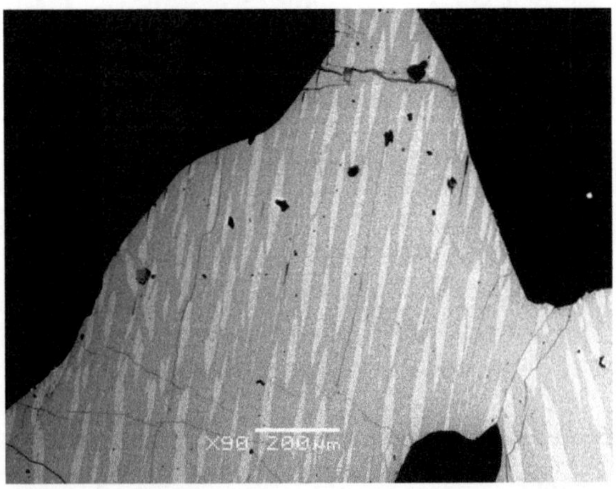

Fig. 23.2 Oxide grain from the Morin anorthosite containing exsolved hemo-ilmenite, typical of samples from the western lobe (this sample from the Ivry Fe-Ti deposit). The host (*grey*) is ilmenite with ubiquitous hematite lamellae (*white*). The hematite exsolution produced lamellae from a few microns to 25 μm thick

23.3.3 Magnetic Properties

Magnetic properties were measured on 38 samples from the Morin Complex, including 25 samples from the anorthosite (20 from the West lobe, 5 from the East lobe), 7 samples from the jotunites and 5 samples from the mangerite. The anorthosite samples have magnetic susceptibilities that range from 2×10^{-4} to 3×10^{-1}, although only two samples have high susceptibilities of 1×10^{-1} or larger. Without these oxide-rich samples the average susceptibilities for the anorthosites is 6.5×10^{-3}. Samples from the mylonitized east lobe all have very low susceptibilities, averaging 6×10^{-4}. Both the jotunite and mangerite samples have higher susceptibilities in general than the anorthosites, ranging from 2×10^{-3} to 1.0×10^{-1} and averaging 6.0×10^{-2}. NRM values for the anorthosites also show considerable divergence, with the lowest values being from the east lobe and the highest values (10 and 12 A/m) from the two samples with very high susceptibility. Average NRM value, excluding the two high samples, is 1.0 A/m while the 5 samples from the more altered east lobe average 0.1 A/m. Jotunite and mangerite samples have stronger magnetization, with average values of 2.7 and 2.1 A/m respectively.

A plot of Ji, calculated using the local field of 43.7 A/m, versus Jr for the Morin samples is shown in Fig. 23.3a. A majority of the anorthosite samples fall above the Q = 1 line, indicating that the remanent magnetization is stronger than the induced magnetization. Our samples from Morin were not oriented, therefore information about the direction of the NRM direction is lacking and can not yet combined with the intensity data. A detailed paleomagnetic study of the Morin anorthosite was made by Irving et al. (1974) where 24 sites (117 samples) from the West Lobe of the Morin were studied for directional data. All samples were demagnetized using alternating field methods and only final directions are reported. However, the discussion of demagnetization behavior indicates that many samples had uni-component directions, and that the initial NRM directions were similar to the final demagnetization directions. A plot of these final directions, with

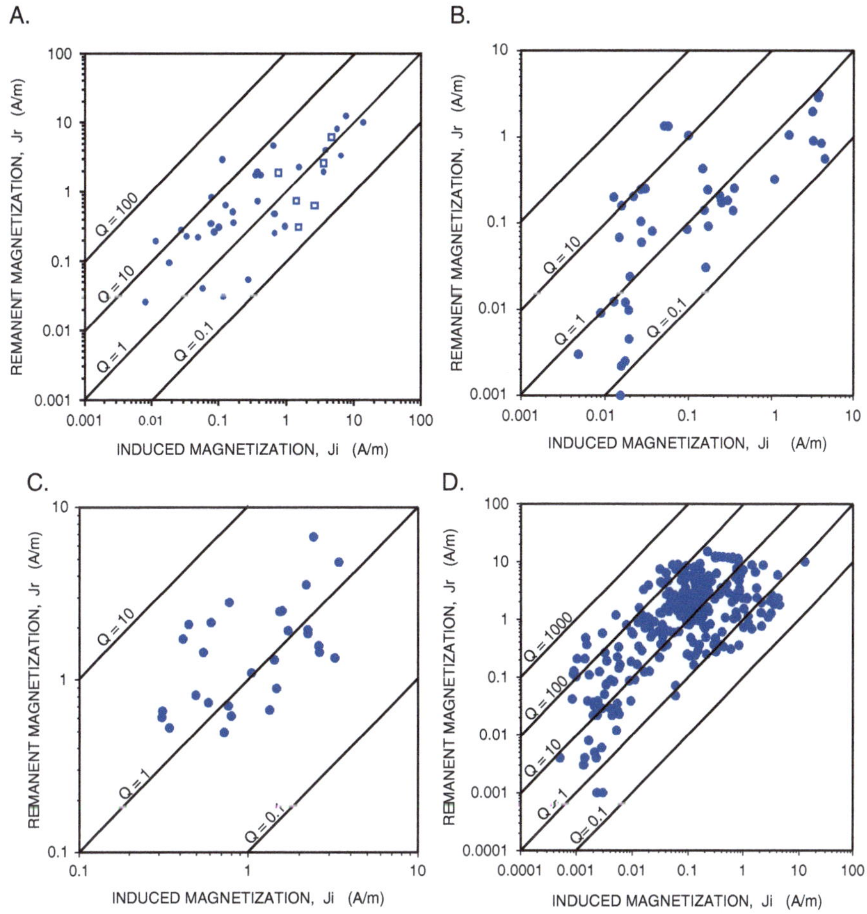

Fig. 23.3 Plots of induced magnetization (Ji) versus remanent magnetization (Jr) for the studied anorthosites, with diagonal lines indicating respective Q values. **a**. Samples from the Morin Complex, closed circles, anorthosites; open circles, jotunites; open squares, mangerites. **b**. Data from samples of the Marcy anorthosite, Adirondack Mountains. **c**. Samples from the Flakstadøy Complex, Lofoten, Norway. **d**. Samples from the anorthosites of the Rogaland Complex, Norway; see Figure 12 for plots of separate anorthosite bodies from Rogaland

steep, negative inclination, and westerly declinations is shown in Fig. 23.4a.

23.3.4 Magnetic Map

Both regional (e.g. Ludden and Hynes 2000) and local (Fig. 23.5) magnetic anomaly maps over the Morin complex have distinct features. In both cases strong negative anomalies are associated with the area directly around the Morin complex. Figure 23.5, which is centered on the Morin Complex, also highlights the numerous positive anomalies in this region. The north and western part of the map is over outcrops of granulite terrain associated with Grenville basement in this part of southern Quebec. To the south and east the anomalies are more subdued and less well defined as Paleozoic sediments in the St. Lawrence Valley cover the Precambrian basement. The two pronounced negative regions, one of nearly 2000 nT below background and one area of lesser but still negative values correspond directly to the west and east lobes of the Morin anorthosite. The associated mangerites and jotunites, encircling the anorthosite to the west and forming both the septum between the two lobes, and the narrow excursions into the west lobe, are marked by strong positive anomalies as expected from rocks dominated by coarse-grained magnetite. Although some of the other lows on this map are related to specific positive anomalies (see lower southwest corner of the map), the

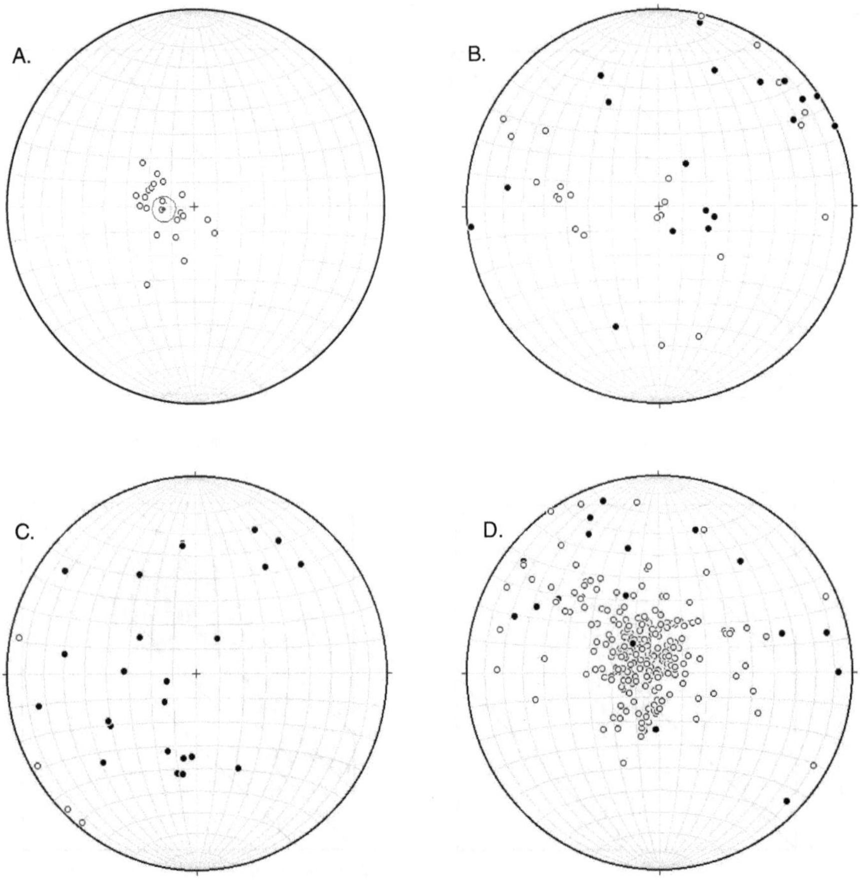

Fig. 23.4 Plots of remanent magnetic directions from oriented samples of anorthosite. Open circles are projections on the upper hemisphere; closed circles are projections on the lower hemisphere. **a**. Morin anorthosite showing sites after demagnetization, replotted from Irving et al. (1974). **b**. Marcy anorthosite, individual samples, NRM. **c**. Lofoten anorthosite, individual samples, NRM, **d**. All anorthosite samples from the Rogaland Complex, NRM; see Fig. 23.12 for magnetic directions from separate bodies from Rogaland

lows over the Morin anorthosite delineate the mapped limits of the anorthositic rocks.

23.4 Marcy Anorthosite, Adirondack Mountains, NY

23.4.1 Geology

The Adirondack Highlands in New York, lying some 200 km south of the Morin terrane (Fig. 23.1), is also part of the Grenville Province and is geologically similar to the Morin terrane. The Marcy anorthosite massif underlies the High Peaks region of the Highlands, which with other satellite anorthosite bodies is part of a ca. 1.15 Ga AMCG suite (McLelland et al. 2004). Migmatitic paragneiss and orthogneiss country rocks are intruded by AMCG-suite plutons, and were variably re-crystallized by granulite facies metamorphism during the 1.10–1.03 Ga Ottawan orogeny (Bohlen et al. 1985, Heumann et al. 2006)

The Marcy massif is a composite intrusion and contains coarse anorthosite and leucogabbro, with more mafic compositions occurring near the borders of the massif (15 to >20% ferromagnesian minerals; Davis 1971). In general, the interior of the massif (the 'Marcy facies') preserves igneous textures overprinted by metamorphic mineral growth, and is composed of coarse blue-gray plagioclase in a matrix of lighter plagioclase, clinopyroxene, and orthopyroxene with minor opaques, hornblende, and garnet. The margins of the massif ('Whiteface facies') commonly include

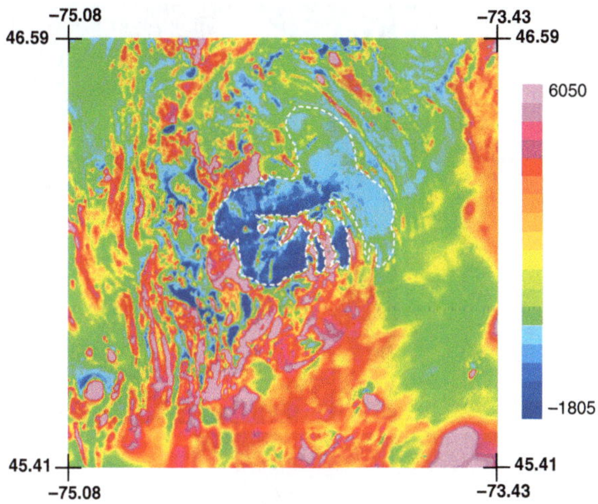

Fig. 23.5 Magnetic anomaly map of the Morin Complex region, southern Quebec, Canada. Data supplied by Natural Resources Canada. White dashed line is outline of Morin anorthosite from Emslie (1975)

hybrid lithologies formed by assimilation of country rocks, and are often recrystallized, foliated, and more mafic than the Marcy facies. Whiteface facies rocks are made up of white plagioclase, clinopyroxene, orthopyroxene, and hornblende with minor opaques, garnet, quartz, potassium feldspar, and biotite (Buddington 1939). Some skarns developed during anorthosite contact metamorphism have low oxygen isotope ratios caused by interaction with heated meteoric water, indicating that these plutons intruded at relatively shallow depths (<10 km, Valley and O'Neil 1982).

23.4.2 Magnetic Mineralogy

In reflected light, the anorthosite and leucogabbro samples are shown to contain between 0.5 and 1% oxides, with magnetite commonly subordinate to ilmenite. Discrete magnetite grains are typically large, from 0.5 to 1.0 mm in size, and minor pseudo-single-domain magnetite grains are present as inclusions in plagioclase. Commonly the large magnetite grains have abundant spinel needles, and minor oxidation-exsolution lamellae of ilmenite. Spinel rods, or needles commonly form at the edges of the ilmenite lamellae. Numerous anorthosite samples also contain hemo-ilmenite (ilmenite host with hematite exsolution). The hematite lamellae are typically less than a micron thick. In many samples the hemo-ilmenite shows an alteration, with ilmenite altered to a rutile + hematite. Other anorthosite samples have magnetite, as described above, with coexisting ilmenite without hematite exsolution lamellae. Many samples contain pyrite as an accessory phase.

Numerous samples contain plagioclase grains, which have well-preserved inclusions of hemo-ilmenite, and/or magnetite (Fig. 23.6a, c). The hemo-ilmenite inclusions show at least two generations of hematite exsolution (Fig. 23.6b). The magnetite grains typically show oxidation- exsolution of ilmenite. The oxidation-exsolution process results in smaller grain sizes for the magnetite, here typically in the pseudo-single-domain size region. Shown in Fig. 23.5c is a magnetite grain with a large ilmenite 'sandwich lamellae'. Though many Marcy anorthosite samples contain oxide inclusions the abundance of these inclusions is very low.

23.4.3 Magnetic Properties

Seven sites of the Marcy anorthosite (42 samples) provide a wide range of magnetic properties. Susceptibility measurements range over three orders of magnitude from 1×10^{-4} to 1×10^{-1}; although individual sites show much less scatter. Two sites have uniformly strong susceptibilities with averages greater than 1.0×10^{-2}, two sites have intermediate susceptibilities averaging $2\text{–}4 \times 10^{-3}$, and three sites have very low susceptibilities with averages less than 7×10^{-4}. As to be expected, NRM values for the samples also show considerable variability, but again consistent within separate sites. Initial intensity ranges from a high of 3.1 A/m to a low of 0.001 A/m with two sites having NRM averages greater than 1 A/m, and the remaining 5 sites having averages less than 0.3 A/m. Using a local present field value of 43.2 A/m, the induced magnetization incurred by these rocks is calculated, and plotted against the NRM values (Fig. 23.3b). Although over 38% of the samples have Q values greater than 1, only a few samples ($N = 4$) have both Ji and Jr values greater than 1 A/m. The rocks appear to have some samples with dominant remanence and some areas with dominant induced field, but both are of relatively small amounts likely due to alteration of the oxides in most of our samples.

Directional data for the samples from the Marcy anorthosite are plotted in Fig. 23.4b. Large scatter in

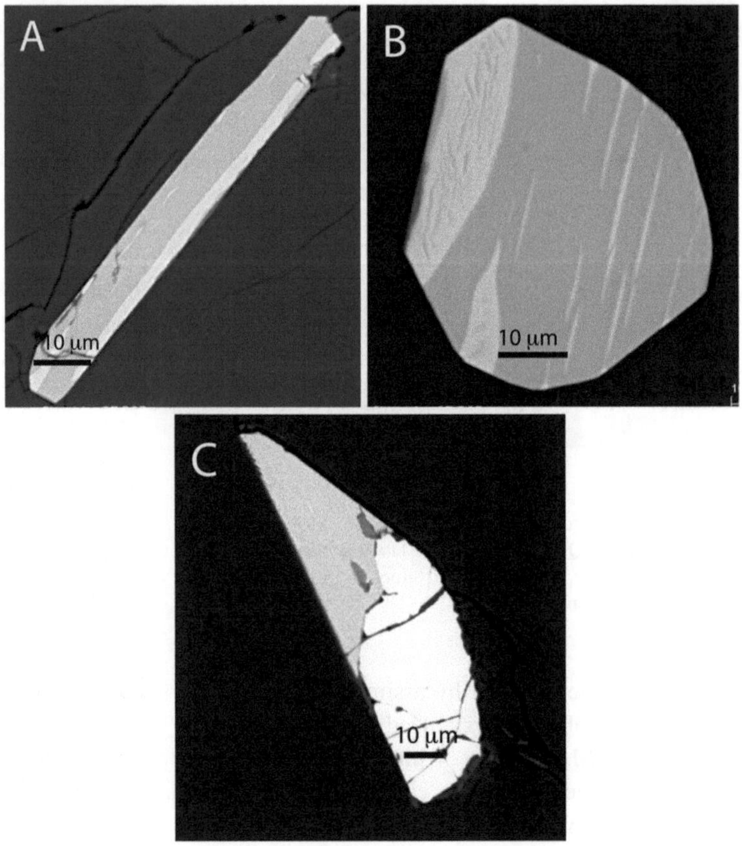

Fig. 23.6 Scanning electron back-scatter image of oxide inclusions in plagioclase grains from the Marcy anorthosite. (**a**) Ilmenite inclusion (I) with two generations of hematite (H) lamellae. The larger hematite lamella contains small ilmenite lamellae. (**b**). Hemo-ilmenite grain with ilmenite host (I) and multiple generations of exsolution lamellae of hematite (H). (**c**) Magnetite with oxy-exsolution of ilmenite

both inclinations and declinations are observed, with a full range of steep to shallow inclination, both negative and positive. Two sites show internal consistency in NRM directions, the rest of sites have considerable scatter in NRM directions.

23.4.4 Magnetic Map

Magnetic anomaly data over part of the Marcy anorthosite is presented in Fig. 23.7. Although the range of anomalies seen on the map go from a high of over 3200 nT to a low of –742 nT, the majority of map is represented by anomalies of only –200 to + 200 nT. The Marcy anorthosite, which extends across the mapped area from west to east, is represented by negative anomalies, up to –500 nT in the west, to the magnetically quiet area of –100 to +100 nT anomalies in the east. Samples used in this study come only from the central and eastern part of the massif. Several distinct positive anomalies exist within the Marcy terrain. These are related to iron-rich deposits in the massif, as the Tahawus Fe-Ti deposit, located in the south-central part of the map.

23.5 Flakstadøy Anorthosite, Lofoten, Norway

23.5.1 Geology

The Lofoten islands are an ancient piece of lower crustal rocks now up thrust through the Cretaceous offshore continental margin in the North Sea off the northwestern coast of Norway (Fig. 23.8). Multiphase tectonic and rifting events in the late Paleozoic through the early Cenozoic has resulted in faulting and uplift of basement blocks now exposed as the spectacular bedrock of the islands (Bergh et al. 2007). Archean rocks are preserved in the Vesterålan islands to the north, while the Lofoten are primarily Paleoproterozoic gneisses, metamorphosed around 1830 Ma, since intruded by AMCG suites, predominately charnockites and mangerites in the late

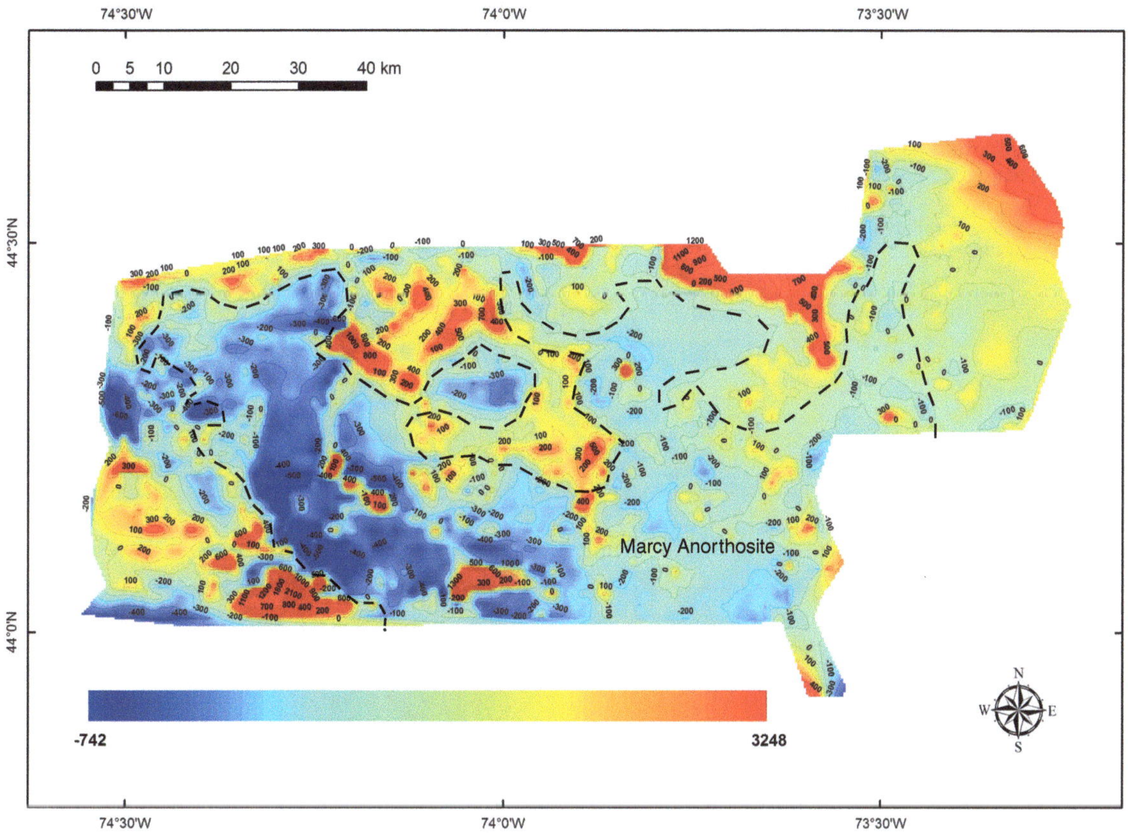

Fig. 23.7 Magnetic anomaly map of the northeastern part of the Adirondack Mountains, New York State, USA. Data replotted from the United States Geological Survey (2001). Contour lines labeled in nT. Black dashed line is outline of Marcy anorthosite body from McLelland et al. (2004)

Paleoproterozoic (Griffin et al. 1978). It appears the many plutonic bodies were emplaced over a short time (1800–1790 Ma) during the Svecofennian orogeny at a time of plate reorganization (Corfu 2004). Anorthosites and associated gabbroic rocks are limited, with the largest body being the Flakstadøy anorthosite on the east side of the island in the vicinity of Nusfjord (Romey 1971).

Samples for this study come from the Flakstadøy Basic Complex (FBC), which contains anorthosite sensu-stricto, leucogabbro, and leuco-troctolite (Romey 1971, Markl et al. 1998). These related rocks appear to have undergone polybaric crystallization at pressures of 0.95–0.4 GPa and temperatures between 1180 and 1120°C, putting emplacement depths of around 12 km in the mid crust (Markl et al. 1998). The anorthosite is composed of >90% plagioclase of An57 to An47 with additional constituents of magnetite, ilmenite, ilmeno-hematite, clinopyroxene and occasional olivine and orthopyroxene.

23.5.2 Magnetic Mineralogy

The magnetic mineralogy from the FBC samples is diverse. However, magnetite (Fig. 23.9a) is common in all samples. The rhombohedral ilmeno-hematite is found only at two localities. The ilmeno-hematite grains are large (Fig. 23.9b, c), many > 100 microns in size and contain at least two generations of ilmenite exsolution lamellae parallel to (0001). The ilmenite lamellae contain second-generation hematite lamellae also parallel to (0001). Near end member magnetite grains with no visible lamellae are subordinate to the ilmeno-hematite grains with ilmeno-hematite (Fig. 23.9c) making up less that 0.5% of the rock.

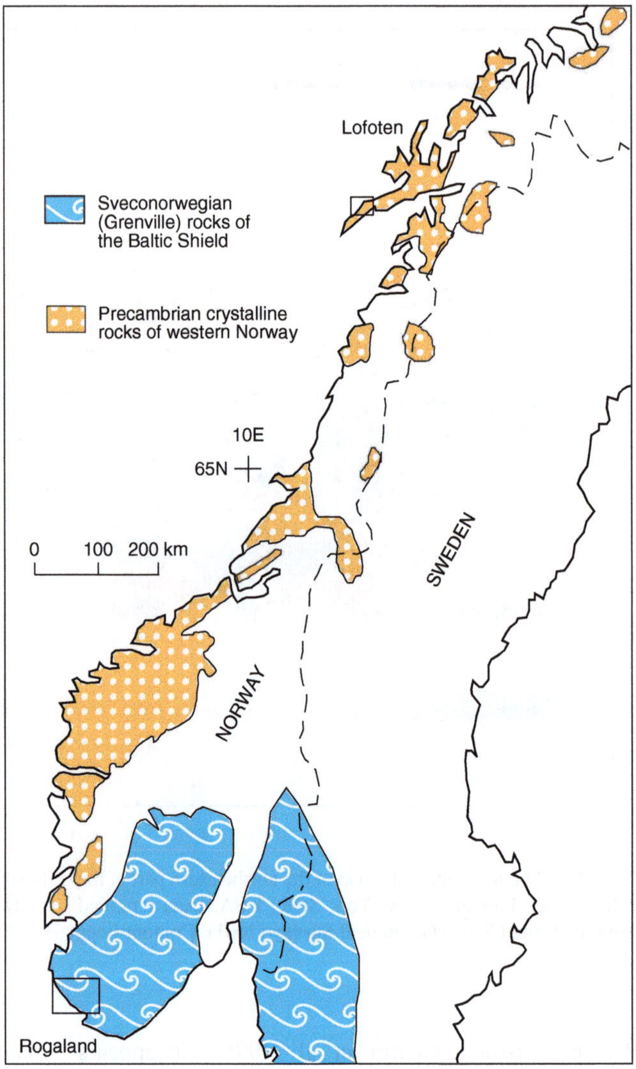

Fig. 23.8 Simplified map of western Scandinavia showing the location of anorthosites studied in the Lofoten, part of the Precambrian crystalline rocks of western Norway, and the Rogaland Complex, intruded into the Sveconorwegian metamorphic terrane of southern Norway

Most of the observed magnetite grains were >100 μm, well into the multi-domain size range for magnetite. These samples have very high coercivity with 90% of the magnetization remaining above an AF demagnetization of 100 mT with the bulk of the magnetization carried by the ilmeno-hematite.

Anorthosite samples, which contain only magnetite, and no rhombohedral oxide, contain magnetite that is relatively rich in aluminum. These magnetites have oxidation exsolution lamellae of ilmenite oriented along {111} and the hercynite ($FeAl_2O_4$) lamellae that are parallel {100}. The first generation of hercynite lamellae are large lamellae and in many areas crosscut the ilmenite lamellae. The ilmenite lamellae commonly have small exsolutions of hercynite. The trellis of ilmenite and the hercynite lamellae, which would reduce the effective magnetic grain size, extensively breaks up these large magnetite grains. There is coexisting ilmenite present as discrete grains and these grains at the optical scale appear to have no exsolution.

23.5.3 Magnetic Properties

Five sites (28 samples) were collected from Nusfjord westward to Storvatnet sampling the anorthosite and leucogabbro (leuco-troctolite). Samples have high susceptibilities, with a range from 7.4 × 10^{-3} to 6.3 × 10^{-2} with all sites having mean susceptibilities >1.2×10^{-2}. NRM values for the leucogabbros are also strong, averaging 1.2 A/m for the entire population, with a range of 0.5 A/m to 6.8 A/m. Systematic

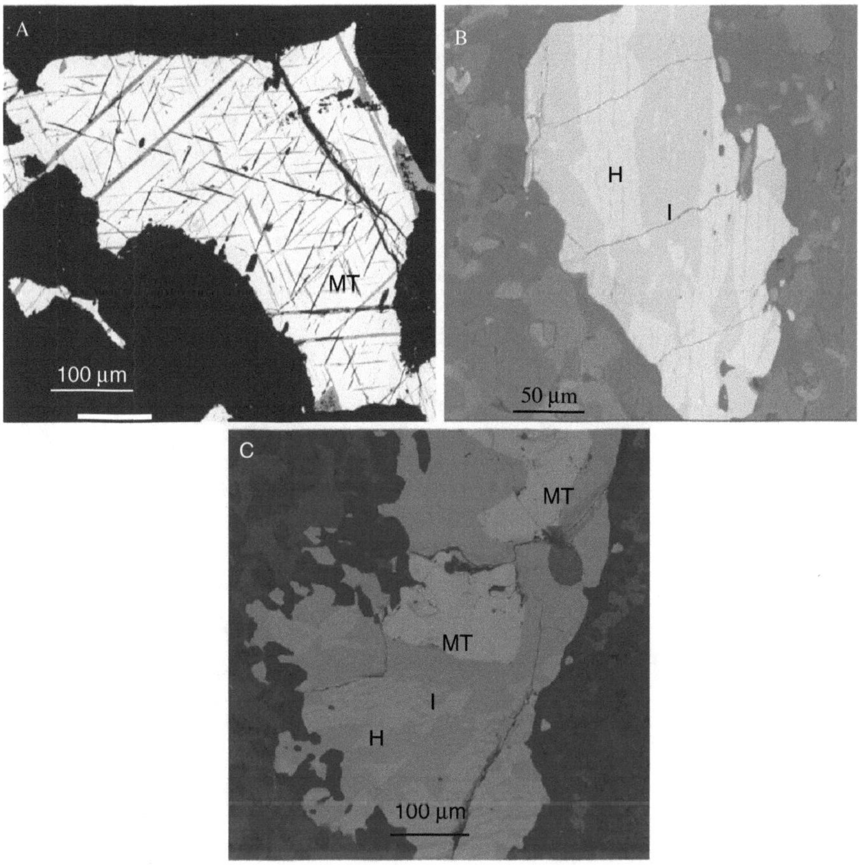

Fig. 23.9 Scanning electron back-scatter image of discrete oxide grains from the Lofoten anorthosites. (**a**) Large magnetite grain (*white*) with oxidation exsolution of ilmenite (*grey*) and hercynite lamellae (*black*). (**b**) Ilmeno-hematite grain with multiple generations of exsolution lamellae of ilmenite (*gray*) in a titanohematite host (*white*). First generation lamellae of ilmenite, 50 μm, appear to have only one set of hematite lamellae. The titanohematite host (*white*) appears to have two to three generations of ilmenite lamellae. (**c**) Large grains of hemo-ilmenite, ilmeno-hematite and magnetite (MT). The ilmenite host material (I) contains far fewer lamellae, than the titanohematite hosts (H), which show multiple generations of ilmenite lamellae. Magnetite grains are all mutli-domain in size

variations are observed from site to site, with the strongest sites along the shore of Nusfjord and the weaker sites to the west towards Storvatnet, with the eastern sites averaging 2–3 A/m and the two western sites averaging 1.4 and 0.7 A/m respectively.

Calculations of induced magnetization, made using an average magnetic field value of 41.93 A/m for Flakstadøy, are compared to the remanent magnetizations in Fig. 23.3c. Over 60% of the samples lie above the $Q=1$ line, indicating the importance of remanence. Due to the high susceptibility, induced magnetization values are generally high, and nearly half of the samples have the unique characteristic of having both induced and remanent magnetization > 1 A/m.

Directions of natural remanent magnetization for the FBC cores are scattered, but nearly all the directions are positive with intermediate to steep inclinations (Fig. 23.4c). Over 45% of the samples yield inclinations greater than 45°, and only four samples produced negative inclinations, all of which are extremely shallow.

23.5.4 Magnetic Map

Both regional and local maps of the present magnetic field in the Lofoten indicate a region of magnetic high over the island of Flakstadøy (Olesen et al. 2002). A section of the Svolvaer quadrangle

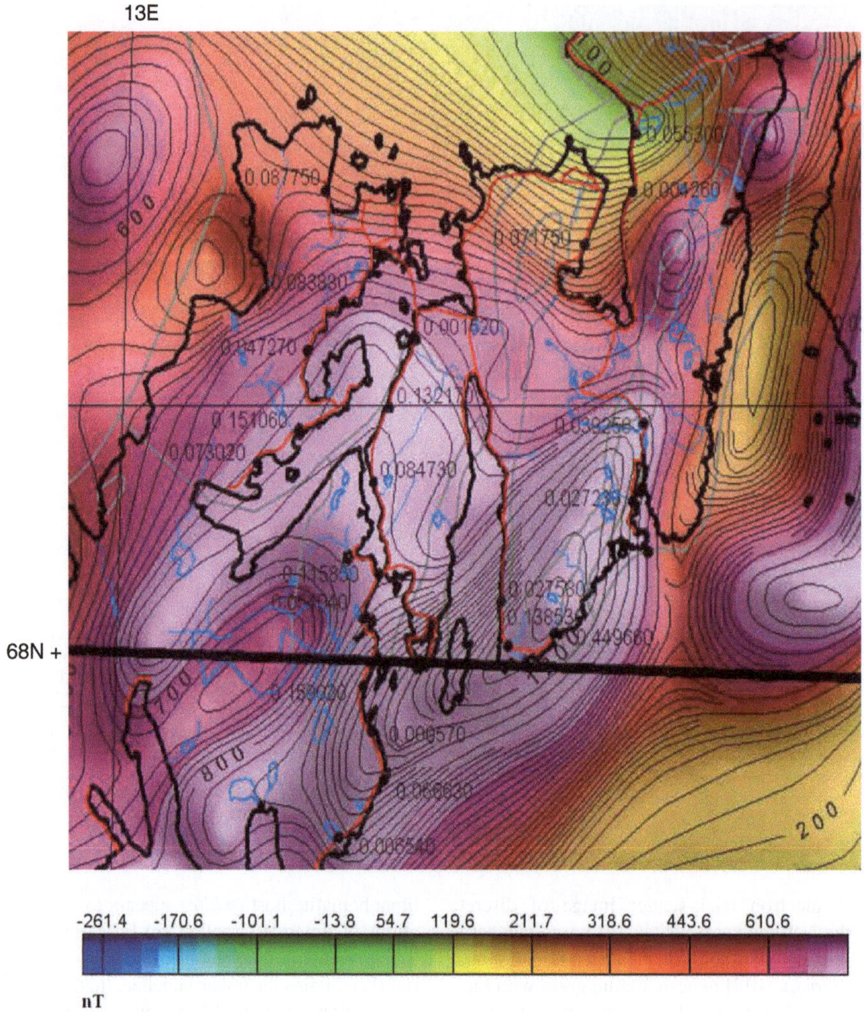

Fig. 23.10 Magnetic anomaly map of part of the Lofoten Islands, northern Norway, from Gellein (2007). Flakstadøy Basic Complex (FBC) is outlined in the dashed white line on Flakstadøy Island. Contour intervals are 20 and 100 nT; labels on contours are in nT. Peak anomaly in southern part of FBC is greater than 1400 nT

anomaly map (Gellein 2007) including Flakstadøy portrays the high anomalies present in this area (Fig. 23.10), particularly on the east side of the island where the FBC outcrops. The area of studied samples (Nusfjord to Storvatnet) lies on the north side of a large anomaly high with a maximum of 1400 nT above background. The anomaly is stronger in the south, east and north part of the FBC, which is primarily leucogabbro and leuco-troctolite, than in the eastern part, which is predominantly anorthosite, although the entire island shows a positive anomaly.

23.6 Rogaland Anorthosite Complex

23.6.1 Geology

The early Neoproterozoic Rogaland Anorthosite Province (RAP) is a composite massif-type anorthosite province consisting of a number of anorthositic plutons, mafic, dominantly jotunitic or noritic intrusions as well as a collection of granitoid bodies (Michot and Michot 1969). The RAP is hosted by Mid-Proterozoic ortho- and paragneisses along the coastline of south-westernmost Norway (Fig. 23.8). The offshore extension of the RAP under the North Sea is approximately the same size as the 1700 km^2 on-land portion. This offshore portion of the province is traceable through the younger sediments from airborne and surface geophysical surveys (McEnroe et al. 2001, Olesen et al. 2004).

23.6.1.1 The Egersund-Ogna Anorthosite

The c. 18 × 28 km Egersund–Ogna Anorthosite is the largest, and oldest of the individual anorthosite plutons in the RAP. The c. 930 Ma body (Schärer et al. 1996) is a dome-shaped, concentric lithologic structure that is concordant with the foliation in the Sveconorwegian gneiss that envelops the anorthosite. The central part of the Egersund-Ogna Anorthosite body consists of medium-grained homogeneous anorthosite that, in its center contains phenocrysts and megacrysts of orthopyroxene and plagioclase. Here, labradorising plagioclase also occurs most extensively in the province. Some of the zones are strongly foliated whereas others are not deformed. Petrographically and chemically, the anorthosite is monotonous and made up of granulated, equal-sized (1–3 cm), homogeneous plagioclase ($An_{40–50}$) with locally some megacrysts of orthopyroxene and plagioclase (Duchesne and Maquil 1987).

23.6.1.2 The Håland- Helleren Anorthosite

The Håland Anorthosite body is located to the southeast of the Egersund-Ogna massif and separated from the latter by a strongly foliated contact zone that includes a septum of similarly highly strained Sveconorwegian gneisses. The Håland body is composed of anorthosites and leuconorite showing complex spatial relationships. Varying degrees of superimposed deformation is making is making the relationship even more complex. The northern part of the Håland Anorthosite is dominated by anorthosites with pseudo-enclaves of leuconorite. The southern part consists of interlayered anorthosite and leuconorite with enclaves of phases showing modal-like layering. The smaller, and undeformed, Helleren anorthosite is intruding and cutting the Håland body in its western part. The Helleren anorthosite consists of massive, rather coarse-grained anorthosite and leuconorite. As is custom with workers in the region, these two associated bodies are referred to as the Håland-Helleren anorthosite (Marker et al. 2003a).

23.6.1.3 The Åna-Sira Anorthosite

The Åna-Sira Anorthosite is a fairly homogeneous anorthosite massif, which consists essentially of anorthosite and minor oxides, which is irregularly interspersed with anorthositic leuconorite. Plagioclase chemistry reveals a NW-SE trending pattern with the most primitive compositions in the central part of the anorthosite body ($An_{40–57}$) compared to the rim in the SW and NE respectively ($An_{39–45}$) (Zeino-Mahmalat and Krause 1976). The Åna-Sira Anorthosite hosts the two most important ilmenite ore bodies in the RAP, the Storgangen and Tellnes deposits as well as a number of smaller deposits (Schiellerup et al 2003).

23.6.1.4 The Garsaknatt Anorthositic-Leuconoritic Intrusion

The Garsaknatt is a small, dominantly anorthositic to leuconoritic body, intruded into the gneissic envelope slightly to the northeast of the main complex. The Garsaknatt body also hosts segments of modally layered norites as well as containing regions of iridescent plagioclase-bearing anorthosite.

23.6.2 Magnetic Mineralogy

All the anorthosite samples investigated contained oxide minerals, although usually less than a few percent. Discrete hemo-ilmenite grains, ranging in size from tens of microns to 0.5 millimeters, are common in the Åna-Sira anorthosite. However oxides rarely make up more than 1–2% of the sample. The hemo-ilmenite grains contain very fine exsolution of hematite lamellae parallel to (0001), with the first generation of hematite lamellae are rarely wider than 1–2 μm (Fig. 23.11a). Subsequent generations of lamellae are much smaller and the exsolution is expected to continue down to approximately unit cell size of hematite (1.2 nm). An extensive chemical and magnetic study on the exsolution lamellae and microstructures in hemo-ilmenite was made by electron microprobe and transmission electron microscope (TEM) on a sample from the Frøytlog hemo-ilmenite deposit, which is a small historic deposit in the Åna-Sira anorthosite (McEnroe et al 2002). TEM observations showed that the hematite lamellae continued down to a few nanometers in thickness.

Given that the emplacement and cooling conditions for the Åna-Sira anorthosite and the Frøytlog

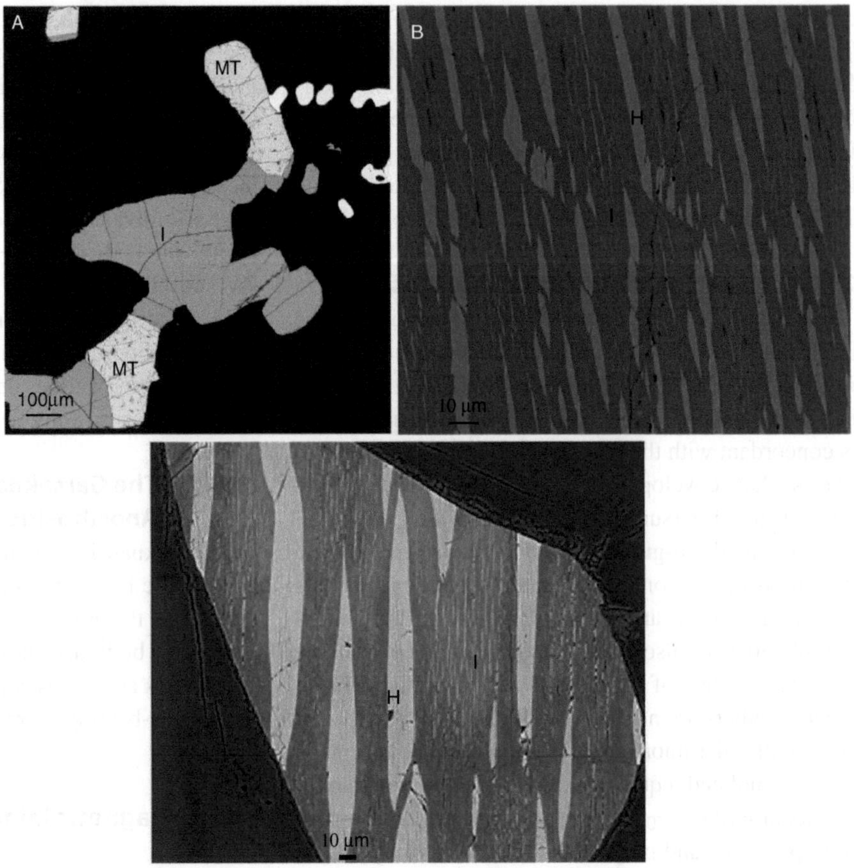

Fig. 23.11 Scanning electron back-scatter image of discrete oxide grains from the Rogaland anorthosites. (**a**). Hemo-ilmenite from the Åna-Sira anorthosite contains abundant exsolution of hematite (H) lamellae. First generation hematite lamellae are < 3 μm thick, subsequent generations of hematite lamellae are much smaller. (**b**). Large grains of magnetite (MT) adjacent to large composite hemo-ilmenite grain from the Håland-Helleren anorthosite. The hemo-ilmenite contains fine-scale hematite lamellae. (**c**). Close up of a hemo-ilmenite grain from the Garsaknatt leuconorite showing fine-scale exsolution lamellae of hematite (H) in ilmenite (I) host

hemo-ilmenite are similar, there should be strong similarities in the exsolution history. Scanning electron microscope observations show numerous generations of lamellae in the hemo-ilmenite in the Åna-Sira anorthosite (Fig. 23.11a). Minor inclusions of sulfides and rutile are also found within the plagioclase grains. Locally large magnetite grains with few microstructures and minor of oxidation exsolution lamellae of ilmenite are also found, but are not abundant.

The Egersund-Ogna body is the most oxidized of the anorthosites discussed here contains the fewest oxides, much less than 0.5%. Both hemo-ilmenite as described above, and ilmeno-hematite, with multiple generations of both ilmenite and hematite lamellae are found. Rare orthopyroxene grains with magnetite and aluminous spinel needles are occur in the Egersund-Ogna anorthosite. The oxides are rare and some hemo-ilmenite grains are altered.

Samples from the Håland-Helleren anorthosite typically contain more magnetite than the other anorthosite bodies in Rogaland. The magnetite grains are large, usually >50 mm and contain only minor oxidation-exsolution lamellae of ilmenite and spinel blades or needles (Fig. 23.11b). Hemo-ilmenite grains are more abundant than magnetite. Hemo-ilmenite has the same exsolution features as described above for the Åna-Sira anorthosite. The distinction here is that the Håland-Helleren samples on average contain more magnetite than the other anorthosites. Given the large grains size of the magnetites and limited exsolution lamellae and microtextures many magnetites should behave as multi-domain grains and dominantly contribute to the inducing signal.

The Garsaknatt leuconorite contains large grains of both magnetite and hemo-ilmenite. The hemo-ilmenite contain multiple generations of hematite lamellae (Fig. 23.11c). Almost all samples observed in the optical microscope contained hemo-ilmenite and at least half of the samples contained magnetite. The magnetite has fine oxidation-exsolution of ilmenite, both as trellis lamellae, and sandwich type. Aluminous spinel forms needles and plates. When the hemo-ilmenite and magnetite are in contact there is typically a spinel simplicity at the boundary of the two phases. Though the hemo-ilmenite grains are large these contains the same abundant hematite exsolution feature as described above. However the Garsaknatt leuconorite has much less oxide than the Håland-Helleren or the Åna-Sira bodies.

A late alteration with the breakdown of hemo-ilmenite to ilmenite to hematite + rutile can be found in some samples from all the anorthosite bodies.

23.6.3 Magnetic Properties

Brown and McEnroe (2008) investigated the magnetic properties of the three largest anorthosites in the Rogaland Complex, Egersund-Ogna (E-O), Håland-Helleren (H-H), and Åna-Sira (Å-S). In addition, a detailed paleomagnetic study was carried out on the Egersund-Ogna anorthosite (Brown and McEnroe 2004). Although the anorthosites are spatially connected, and similar in age and general composition, the magnetic properties of the three bodies vary widely. It is interesting to note that even with a great variation in susceptibility and remanence properties, all of the anorthosites yield Q values close to, or greater than 1 (Fig. 23.3d). Included in the study here are samples from the Garsaknatt anorthosite-leuconorite, a smaller body close to but not contiguous on the surface with the Åna-Sira body.

The Egersund-Ogna body has the lowest susceptibilities, ranging from 1×10^{-5} to 5×10^{-3}, with an average value of 5×10^{-4}, indicating a material with very little or magnetite. The induced magnetization generated by this body is very low, with no cores having Ji >0.2 A/m. Despite this, samples from the Egersund-Ogna have an average remanence of 0.6 A/m, with a range from 2 mA/m to 4 A/m. As seen in Fig. 23.12a all samples from the E-O have Q values greater than 1, indicating that even though the remanence is relatively low, the NRM dominates completely over the limited induced magnetization.

Samples from the Åna-Sira anorthosite have a range of susceptibilities from 8×10^{-4} to 8×10^{-2} and a mean susceptibility of 6×10^{-3}. This does not represent a bimodal distribution, but rather 80% of the Åna-Sira samples have low susceptibilities in the 10^{-3} range. The resulting Ji values for this body average 0.2 A/m, and only 5% of samples are greater than 1.0 A/m (Fig. 23.12c). The lowest Jr measured equals 0.3 A/m while the highest remanence is 9.4 A/m. Over 90% of the Åna-Sira samples have Jr >1.0 A/m with an average of 3.3 A/m. The Q values for 95% of Åna-Sira rocks are greater than 1.0, and over 70% of the samples have Q >10 indicating the dominance of the NRM over induction in these rocks, resulting in the large distinct negative anomalies.

Both Håland-Helleren and Garsaknatt samples have higher average susceptibility values than other Rogaland anorthosites, reflecting the presence of magnetite in these bodies. Although both bodies have average susceptibilities of 2×10^{-2}, the Garsaknatt body has a much large range of values from 4×10^{-5} to 3×10^{-1}. Induced magnetization from both bodies averages less than 1 A/m (Fig. 23.12b, d), whereas the remanence for these bodies averages over 1 A/m. The Håland-Helleren, in particular, has the highest NRM values in the Rogaland complex with a range from 0.8 A/m to 15 A/m. As to be expected with strong remanence the Q values for Håland-Helleren samples are uniformly large with 95% >1 and nearly 60% greater than 10.

The directions of NRM in all the Rogaland samples are dominated by steep negative inclinations (Fig. 23.4d). When directions for the individual bodies are plotted it is evident that both the Egersund-Ogna body (Fig. 23.13a) and the Åna-Sira body (Fig. 23.13c) are almost entirely represented by this very steep negative inclination with a northwest declination. As the present field in Rogaland has an inclination of 72° and a declination of 0°, the remanence directions here are nearly antiparallel to the induced directions. For Håland-Helleren and Garsaknatt bodies the NRM directions are a slightly scattered, though overall steep negative inclinations and northwestern declinations prevail (Fig. 23.13b, d). The negative anomalies in these regions are a direct result of the negative inclinations of the natural remanent magnetization values.

23.6.4 Magnetic Map

The Rogaland Igneous Province has distinct aeromagnetic signatures with large negative and sporadic positive anomalies over the anorthosite bodies (McEnroe et al. 1996, 2001), as well as banded positive and negative anomalies over the Bjerkreim-Sokndal layered (BKS) intrusion (Fig. 23.14). Distinct negative

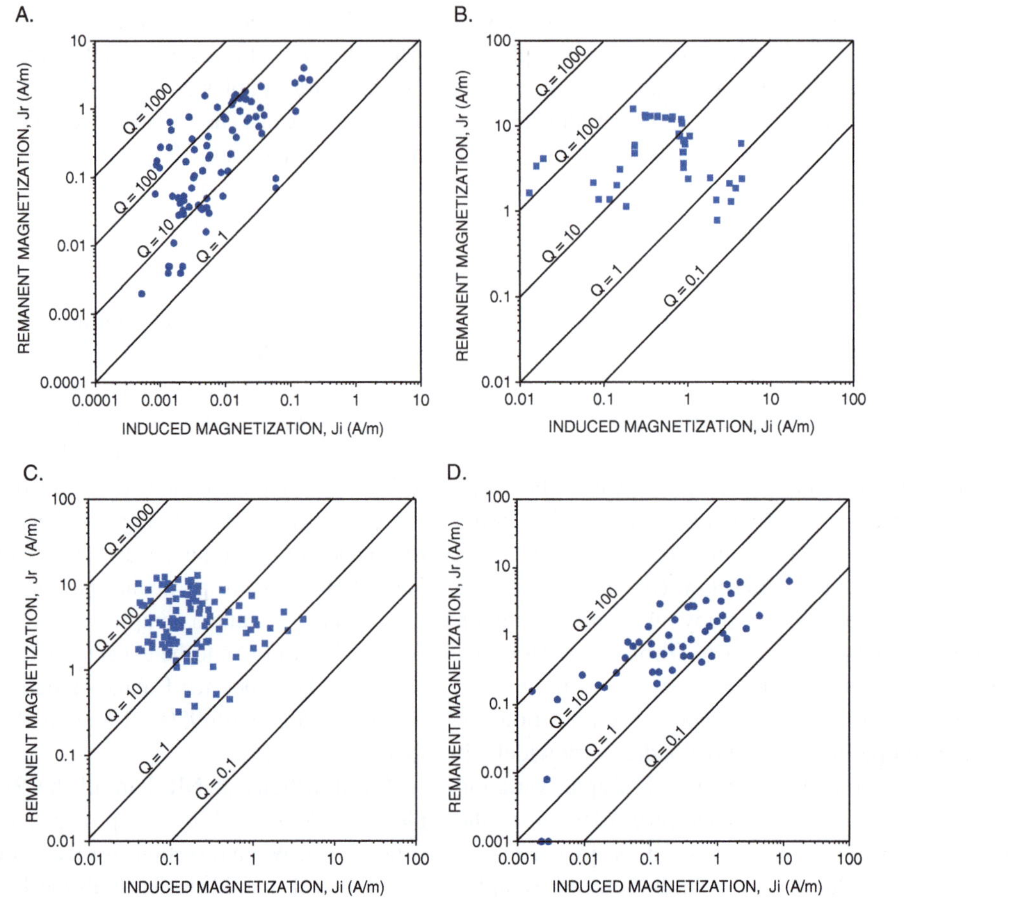

Fig. 23.12 Plots of induced magnetization (Ji) versus remanent magnetization (Jr) for separate anorthosite bodies from the Rogaland Complex, with diagonal lines indicating respective Q values. **a**. Egersund-Ogna anorthosite **b**. Håland-Helleren anorthosite **c**. Åna-Sira anorthosite **d**. Garsaknatt anorthosite

anomalies within the Bjerkreim-Sokndal layered intrusion with ranges greater than 10,000 nT are discussed by McEnroe et al. (2004a, b), and an overview of the magnetic properties of the BKS are given by McEnroe et al. (2009). In the Rogaland anorthosites the Åna-Sira anorthosite has the largest negative anomaly of nearly 3000 nT below background, while the Håland-Helleren and Egersund-Ogna show both negative and positive anomalies of a more subdued nature. Most positive anomalies in the region of Egersund appear are related to either shallow mafic bodies, or areas of alteration. The Garsaknatt body, northwest of the Åna-Sira and the eastern limb of the Bjerkreim-Sokndal shows a very distinct oval shape negative anomaly, which delineates the area of the intrusion. The surrounding metamorphic gneisses are dominated by positive anomalies.

23.7 Discussion

23.7.1 Magnetic Mineralogy and Magnetic Properties of Anorthosites

Observations of oxide mineralogy and measurement of magnetic properties on the four anorthosite complexes described here provide important information on the magnetic behavior of these interesting rocks. It is obvious that some anorthosites have strong and stable magnetizations that are capable of influencing, or even dominating the observed magnetic anomalies. Magnetic properties, represented by susceptibility and NRM intensity, show a range of values spanning five orders of magnitude over the seven different anorthosites bodies studied, while NRM directions

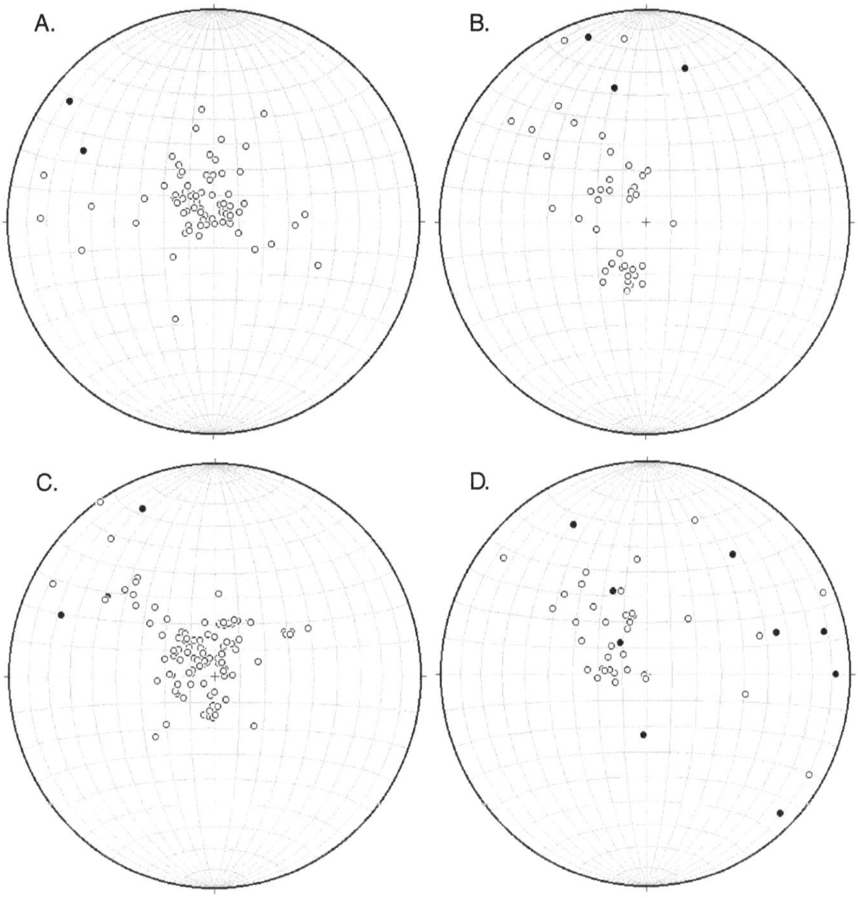

Fig. 23.13 Plots of natural remanent magnetism (NRM) directions from oriented samples of anorthosite from separate bodies from the Rogaland Complex. Open circles are projections on the upper hemisphere; closed circles are projections on the lower hemisphere. **a**. Egersund-Ogna anorthosite **b**. Håland-Helleren anorthosite **c**. Åna-Sira anorthosite **d**. Garsaknatt anorthosite.

range from steeply negative, nearly antiparallel to the inducing field, to steeply positive. The wide range of values in both intensity and susceptibility is emphasized in Fig. 23.15, where all the samples from this study are plotted together. The characteristics of Ji (directly dependent on susceptibility) and Jr depend on the mineral phase, composition, concentration, and grain size of the oxides present in the material. In these samples we see three different possible oxide populations: magnetite, hemo-ilmenite, and mixtures of both these magnetic minerals. The presence or absence of magnetite is discernable in the susceptibility measurements, with values in the 1×10^{-2} range indicative of magnetite and those $<5 \times 10^{-3}$ range indicative of hemo-ilmenite. The susceptibility interpretations are born out by the optical observations, where high susceptibility samples, like those in the Flakstadøy complex, all contain magnetite. On the other hand, hemo-ilmenite grains are the only oxides present in samples with very low susceptibility, like the Egersund-Ogna body, and many samples in the Åna-Sira anorthosite.

Likewise, the Jr values are directly related to the magnetic minerals present in the rocks, but here the relationship is more complicated. The anorthosites dominated by magnetite as the only oxide present generally have moderate NRM values, such as the Flakstadøy complex with an average Jr of 2 A/m or the Garsaknatt body with an average of 1.5 A/m. But the strongest remanences come from the anorthosites with mixed mineralogy, or only hemo-ilmenite populations, such as the Åna-Sira body in the Rogaland complex with an average remanence value greater than 3 A/m, and values up to 9 A/m, or the Håland-Helleren body with an average remanence of 6 A/m. The magnetic mineralogy in these bodies is dominantly

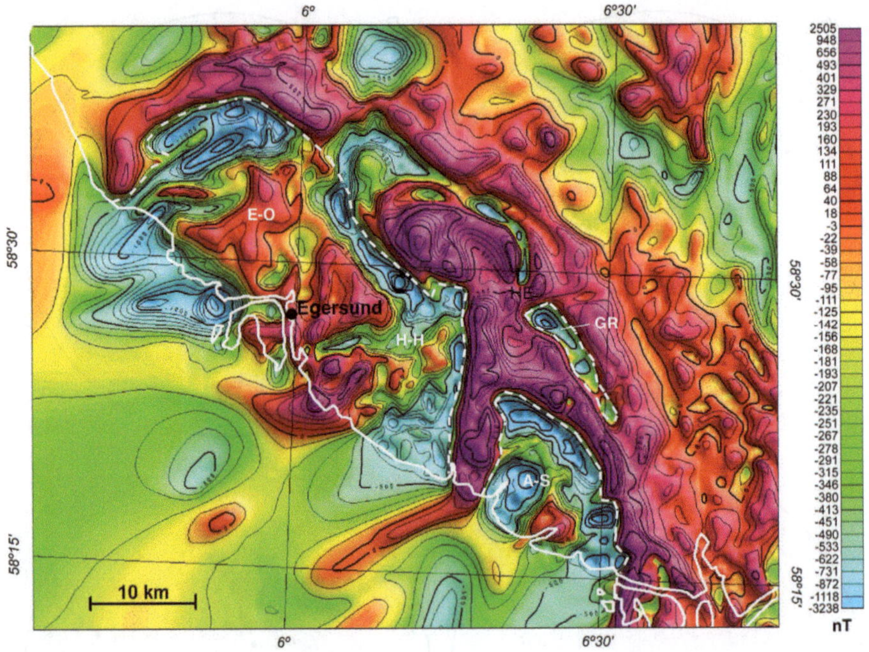

Fig. 23.14 Magnetic anomaly map of the Rogaland Anorthosite Province, southern Norway (from McEnroe et al. 2004a). Solid white line is the coastline, dashed white line is the extent of mapped anorthosite, taken from Marker et al. (2003b). E-O, Egersund-Ogna anorthosite; H-H, Håland-Helleren anorthosite; A-S, Åna-Sira anorthosite, GR, Garsaknatt anorthosite

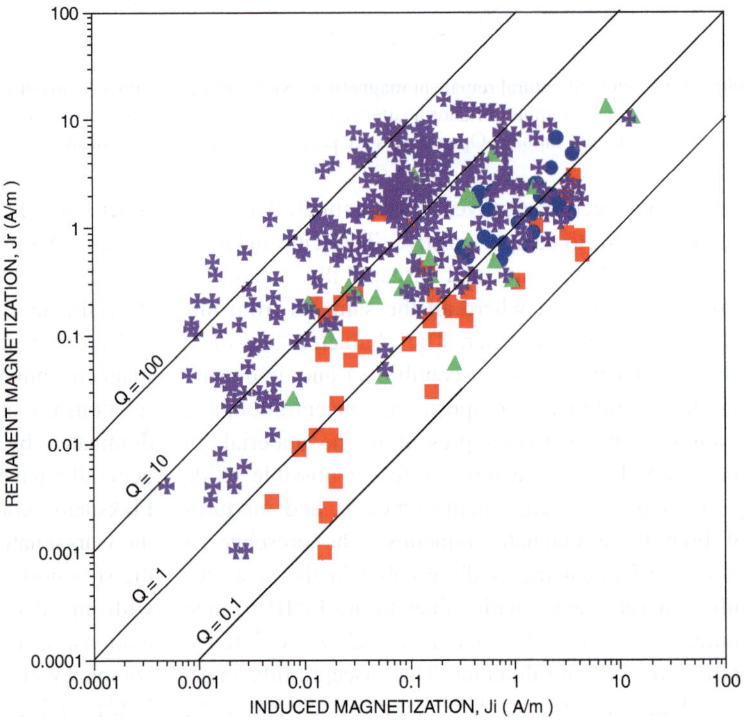

Fig. 23.15 Plot of induced magnetization (Ji) versus remanent magnetization (Jr) for all the samples discussed in this paper by bodies. Morin Anorthosite, triangles; Marcy Anorthosite, squares; Flakstadøy Complex, circles; Rogaland Complex, crosses

hemo-ilmenite, producing a strong and very stable lamellar magnetization (Robinson et al. 2002, 2004). The lamellae are found on a range of sizes, down to only a few nanometers in thickness. It is the abundance of these lamellae that will control the amount of magnetization in the hemo-ilmenite samples (Robinson et al. 2002, 2004, McCammon et al. 2009). Though these structures are very small they are stable due to the exchange coupling between the lamellae, the contact layer and the host (Fabian et al. 2008, McEnroe et al. 2008).

The predominance of hemo-ilmenite results in low average susceptibility values with Ji values much lower than expected. If considerable magnetite is also present the remanence is a combination of both magnetic phases present, but the induced component will increase, because it will be dominated by the much large susceptibility of magnetite.

23.7.2 Connection Between Magnetic Anomalies and Anorthosites

Magnetic anomalies derived from aeromagnetic data are a product of the induced field at the location combined with the remanent field of the rocks in the region. It has generally been considered that remanent properties are vastly over shadowed by the induced components, and commonly only susceptibility of the source material has been considered important. In this paper we present a number of anorthosite complexes that have distinct aeromagnetic anomalies associated with them. Because all four anorthosite complexes studied are located at mid to high latitudes, they all have present field directions with steep positive inclinations ($\sim$70°) and northerly declinations ($\sim$0°). If induced components are dominant, all four regions should show positive anomalies with the size of the anomalies related to the strength of the local susceptibility. Only one region studied here, the Flakstadøy Basic Complex in the Lofoten Islands, is identified with a strong positive anomaly (Fig. 23.10). The other three anorthosite complexes are associated with negative anomalies of varying size, strongly implying that something other than susceptibility influences magnetic field distortions in the area.

Two of the anorthosite complexes, the Morin complex and the Rogaland complex, show strong negative anomalies directly linked to the surface outcrops of anorthosite. In the case of the Morin complex, 80% of the anorthosite samples have Q values > 1 and Jr values that are generally 2–3 times larger than the Ji values (Fig. 23.3a). Since the NRM directions from the anorthosites are steeply negative (Fig. 23.4a), they tend not only to overwhelm the induced component, but also to provide additional magnetization nearly anti-parallel to the present field. This is particularly evident in the western lobe of the Morin anorthosite where strong remanent magnetizations are found. The eastern lobe samples from the zone of mylonitization, have lower susceptibility resulting in a smaller induced field, but also possess very little remanence so the entire anomaly is more subdued.

The Rogaland region shows a strong correlation between the magnetic anomaly pattern and the mapped locations of anorthosites (Fig. 23.14), again with strong negative anomalies present. In this case 92% of the anorthosite samples have Q values > 1, with over 60% of the total having Q values >10, indicating that the remanent components far outweigh the induced components. This is evident in Fig. 23.3d where 65% of the samples have NRM intensities > 1 A/m but only 8% of the samples have Ji components >1 A/m. It appears from this consideration that the anorthosites in the Rogaland area should show strong negative anomalies. This is definitely true for the Åna-Sira and Garsaknatt bodies (Fig. 23.14), but only partially true for the Egersund-Ogna and Håland-Helleren where some positive anomalies are observed. These anomalies are associated with prominent jotunite dikes intruded into the anorthosite as well as regions of known Fe-Ti deposits, especially south of Egersund (Duchesne and Schiellerup 2001, Schiellerup et al. 2003)

In the Lofoten Islands the Flakstadøy Complex lies in a region of positive anomalies, with the southern part of the complex associated with the largest magnetic anomaly in the region (Fig. 23.10). The large positive anomaly of nearly 1400 nT (Gellein 2007) is not surprising when one considers the magnitude and direction of both the Jr and Ji components. Over half of the Lofoten sample indicate Jr > 1, with many of these also having Ji >1 (Fig. 23.3c). As the dominate NRM direction for the Lofoten are vectors with positive inclinations, many of which are steep, the Ji component is enforced by the Jr magnetization, and the larger than expected positive anomaly results.

The only set of samples where the association between a distinct negative, or positive anomaly, and the outcropping anorthosite is not well correlated is from the Adirondacks (Fig. 23.8). The Marcy Anorthosite appears to have a strong negative anomaly in the western part, but much of the body has only variations of –200 to 100 nT. This pattern agrees with our magnetic data where over 80% of both Jr and Ji components are <1 A/m (Fig. 23.3b). In addition, the NRM directions form the Adirondack samples are scattered, being evenly divided between negative and positive inclinations (Fig. 23.4b), so no consistent combination of induced and remanent components exists. More than half of our sample collection from the Marcy anorthosite showed alteration to the oxides with only the oxide inclusions in the plagioclase that were predominately unaltered. However these inclusions make up a very small fraction of the oxide in the rock resulting in a limited contribution to the NRM. The result is a magnetic anomaly of only small and scattered variation.

Conclusions

This study on the magnetic mineralogy and magnetic properties of four different anorthosite complexes yields considerable information on the magnetic behavior of these plagioclase rich rocks. Despite their predominant mono-mineralic nature, anorthosites can be strongly magnetic due to the presence of stable lamellar magnetization residing in hemo-ilmenites. Two complexes, the Morin in southern Quebec and the Rogaland in southern Norway reflect the presence of hemo-ilmenite oxides with low susceptibility but typically large NRM values and directions nearly antiparallel to the present field. These rocks produce large negative anomalies over large regions of the anorthosites. In the Flakstadøy complex in the Lofoten Islands both hemo-ilmenite and magnetite are present, and the rocks have both strong remanence and large induced components. Because the remanent direction is roughly parallel to the present field the anorthosites produce an enhanced positive anomaly. In the Marcy anorthosite in the Adirondack Mountains of New York State the oxide mineralogy is sparse and variable. Both susceptibility and NRM values are low and little discernable anomaly results. The anorthosites here are nearly 1 byr, or older, and many samples have preserved their magnetization over that time on a constantly changing Earth. They have shown they can be important sources for magnetic anomalies on our planet and certainly should be considered when investigating magnetic anomalies on other planetary bodies.

Acknowledgements Parts of this research have been funded by NSF (USA), NFR (Norway) and NGU. Thanks to Peter Robinson for field assistance, Chris Koteas and Weining Zhu for GIS assistance, John Valley for support of past petrologic studies of the Morin Complex, and to Institute for Rock Magnetism, University of Minnesota, supported by a NSF instrument and facility grant, for use of their instruments.

References

Ashwal LD (1993) Anorthosites. Springer, Berlin

Bergh SG, Eig K, Kløvjan OS et al (2007) The Lofoten-Vesterålan continental margin: a multiphase Mesozoic-Paleogene rifted shelf as shown by offshore-onshore brittle fracture analysis. Norw J Geol 87:29–58

Blewett DT, Hawke BR, Lucey PG (2002) Lunar pure anorthosite as a spectral analog for Mercury. Meteor Planet Sci 37:1245–1254

Bohlen SR, Valley JW, Essene EJ (1985) Metamorphism in the Adirondacks I: Petrology, pressure, and temperature. J Petrol 26:971–992

Brown LL, McEnroe SA (2004) Paleomagnetism of the Egersund-Ogna anorthosite, Rogaland, Norway, and the position of Fennoscandia in the Late Proterozoic. Geophys J Int 158:479–488

Brown LL McEnroe SA (2008) Magnetic properties of anorthosites: a forgotten source for planetary magnetic anomalies? Geophys Res Lett. doi:10.1029/2007GL032522

Buddington AF (1939) Adirondack igneous rocks and their metamorphism. Geol Soc Am Mem 7:1–354

Corfu F (2004) U-Pb age, setting, and tectonic significance of the Anorthosite-Mangerite-Charnockite-Granite suite, Lofoten-Vesterålen, Norway. J Petrol 45:1799–1819

Davis BTC (1971) Bedrock geology of the St. Regis Quadrangle, New York, NY State Mus Map Chart 16

Doig R (1991) U-Pb zircon dates of Morin anorthosite suite rocks, Grenville Province Quebec. J Geol 99:729–738

Duchesne JC, Maquil R (1987) The Egersund-Ogna massif. In: Maijer C, Padget P (eds) The geology of southernmost Norway – an excursion guide. Special publication, vol 1. Norwegian Geological Survey, Trondheim, Norway, pp 50–56

Duchesne JC, Schiellerup H (2001) The iron-titanium deposits. In: Duchesne JC (ed) The Rogaland Intrusive Massif. Special publication, vol 29. Norwegian Geological Survey, Trondheim, Norway, pp 56–75

Emslie RF (1975) Major rock units of the Morin Complex, southwestern Quebec. Geol Surv Can Paper 74-48:1–37

Fabian K, McEnroe SA, Robinson P et al (2008) Exchange bias identifies lamellar magnetism as the origin of the natural remanent magnetization in ilmeno-hematite from Modum, Norway. Earth Planetary Sci Lett. doi: 10.1016/j.epsl.2008.01.034

Gellein J (2007) Aeromagnetic anomaly map, Svolvær. Scale 1:250,000. Nor Geol Surv, Trondheim, Norway

Griffin WL, Taylor PN, Hakkinen JW et al (1978) Archean and Proterozoic crustal evolution in Lofoten-Vesterålen, N Norway. J Geol Soc London 135:629–647

Hawke BR, Peterson CA, Blewett DT et al (2003) Distribution and modes of occurrence of lunar anorthosite. J Geophys Res. doi: 10.1029/2002JE001890

Heumann MJ, Bickford M, Hill BM et al (2006) Timing of anatexis in metapelites from the Adirondack lowlands and southern highlands: a manifestation of the Shawinigan orogeny and subsequent anorthosite-mangerite-charnockite-granite magmatism. Geol Soc Am Bull 118:1283–1298

Irving E, Park JK, Emslie RF (1974) Paleomagnetism of the Morin Complex. J Geophys Res 79:5482–5490

Longhi J (2005) A mantle or mafic crustal source for Proterozoic anorthosites? Lithos 83:183–198

Ludden J, Hynes A (2000) The Lithoprobe Abitibi-Grenville transect: two billion years of crust formation and recycling in the Precambrian Shield of Canada. Can J Earth Sci 37:459–476

Marker M, Schiellerup H, Meyer G et al (2003a) Introduction to the geological map of the Rogaland Anorthosite Province 1:75 000. In: Duchesne JC, Korneliussen A (eds) Ilmenite deposits and their geological environment. Special publication, vol 9. Norwegian Geological Survey, Trondheim, Norway, pp 109–116

Marker M, Schiellerup H, Meyer G et al (2003b) Geologic map of the Rogaland Anorthosite Province, 1:75,000. Special publication, vol 9. Norwegian Geological Survey, Trondheim, Norway

Markl G, Frost BR, Bucher K (1998) The origin of anorthosites and related rocks from the Lofoten Islands, northern Norway: I. Field relations and estimation of intrinsic variables. J Petrol 39:1425–1452

Martignole J, Schrijver, K (1970) Tectonic setting and evolution of the Morin anorthosite, Grenville Province, Quebec. Geol Soc Finl Bull 42:165–209

McCammon C, McEnroe SA, Robinson P et al (2009) Mössbauer spectroscopy used to quantify natural lamellar remanent magnetization in single-grains of ilmeno-hematite. Earth Planetary Sci Lett 288:268–278

McEnroe SA, Brown LL, Robinson P (2004a) Earth analog for Martian magnetic anomalies: Remanence properties of hemo-ilmenite norites in the Bjerkreim-Sokndal Intrusion, Rogaland, Norway. J Appl Geophys 56:195–212

McEnroe SA, Brown LL, Robinson P (2008) Remanent and induced magnetic anomalies over a layered intrusion: Effects from crystal fractionation and recharge events. Tectonophysics. doi:10.1016/j.tecto.2008.11.021

McEnroe SA, Fabian K, Robinson P et al (2009) Crustal magnetism, lamellar magnetism and rocks that remember. Elements. doi:10.2113/gselements.5.4.241

McEnroe SA, Harrison RJ, Robinson P et al (2002) Nanoscale hematite-ilmenite lamellae in massive ilmenite rock: an example of 'lamellar magnetism' with implications for planetary magnetic anomalies. Geophys J Int 151:890–912.

McEnroe SA, Robinson P, Panish P (1996) Rock magnetic properties, oxide mineralogy, and mineral chemistry in relation to aeromagnetic interpretation and search for ilmenite reserves. Norw Geol Surv Rep 96–060

McEnroe SA, Robinson P, Panish P (2001) Aeromagnetic anomalies, magnetic petrology and rock magnetism of hemo-ilmenite- and magnetite-rich cumulates from the Sokndal Region, South Rogaland, Norway. Am Mineral s86:1447–1468

McEnroe SA, Skilbrei JR, Robinson P et al (2004b) Magnetic anomalies, layered intrusions and Mars. Geophys Res Lett. doi: 10.1029/2004GL020640

McLelland JM, Bickford ME, Hill BM et al (2004) Direct dating of Adirondack Massif anorthosite by U-Pb SHRIMP analysis of igneous zircon; implications for AMCG complexes. Geol Soc Am Bull 116:1299–1317

Michot J, Michot P (1969) The problem of anorthosites: the South-Rogaland Igneous Complex, Southwestern Norway. In: Isachsen YW (ed) The origin of anorthosite and related rocks. NY State Mus Sci Ser Mem 18:399–410

Olesen O, Lundin E, Nordgulen Ø et al (2002) Bridging the gap between the onshore and offshore geology in Nordland, northern Norway. Norw J Geol 82:243–262

Olesen O, Smethurst MA, Torsvik T, Bidstrup T (2004) Sveconorwegian igneous complexes beneath the Norwegian-Danish Basin. Tectonophysics 387:105–130.

Papike JJ, Ryder G, Shearer CK (1998) Lunar samples. In: Papike JJ (ed) Planetary materials. Rev Mineral 36:5-1–5-234

Peck WH, DeAngelis MT, Meredith MT et al (2005) Polymetamorphism of marbles in the Morin terrane (Grenville Province, Quebec). Can J Earth Sci 42:1949–1965

Peck WH, Valley JW (2000) Large crustal input to high $\delta^{18}O$ anorthosite massifs of the southern Grenville Province: new evidence from the Morin Complex, Quebec. Contrib Mineral Petrol 139:402–417

Robinson P, Harrison JR, McEnroe SA et al (2002) Lamellar magnetism in the hematite-ilmenite series as an explanation for strong remanent magnetization. Nature 418: 517–520

Robinson P, Harrison JR, McEnroe SA et al (2004) Nature and origin of lamellar magnetism in the hematite-ilmenite series. Am Mineral 89:725–747

Romey WD (1971) Basic igneous complex, mangerite, and high-grade gneisses of Flakstadøy, Lofoten, northern Norway: I. Field relations and speculations on origin. Norw Geol Tidsskrift 51:33–61

Schärer U, Wilmart E, Duchesne J (1996), The short duration and anorogenic character of anorthosite magmatism: U-Pb dating of the Rogaland complex, Norway. Earth Planetary Sci Lett 139:335–350

Schiellerup H, Korneliussen A, Heldal T et al (2003) Mineral resources in the Rogaland Anorthosite Province, South Norway: origins, history and recent developments. In: Duchesne JC, Korneliussen A (eds) Ilmenite deposits and their geological environment. Norw Geol Surv Spec Publ 9:116–134

Takeda H, Yamaguchi A, Bogard DD et al (2006) Magnesian anorthosites and a deep crustal rock from the farside crust of the Moon. Earth Planetary Sci Lett 247:171–184

United States Geological Survey (2001) Adirondack Mountains North, New York, Digital flight-line aeromagnetic data set. US Geol Surv Open File Rep 02–0361

Valley JW, O'Neil JR (1982) Oxygen isotope evidence for shallow intrusion of Adirondack anorthosite. Nature 300:497–500

Zeino-Mahmalat R, Krause H (1976) Plagioklase im anorthosit-komplex von Åna-Sira, SW-Norwegen. Petrologische und chemische untersuchungen. Norw Geol Tidsskrift 56:51–94.

Zhao X, Ji S, Martignole J (1997) Quartz microstructures and c-axis preferred orientations in high-grade gneisses and mylonites around the Morin anorthosite (Grenville Province). Can J Earth Sci 34:819–832

Magnetic Record in Cave Sediments: A Review

24

Pavel Bosák and Petr Pruner

Abstract

Dating cave sediments by the application of the palaeomagnetic method – magnetostratigraphy – is a difficult and sometimes risky task, as the method is comparative in its principles and does not provide numerical ages. For dating clastic cave sediments and speleothems it is limited by the complex conditions occurring underground so that it is often necessary to combine it with other methods that offer supplementary absolute-, calibrate-, relative- or correlate-ages. Interpretation of magnetostratigraphic results faces other serious problems that may endanger palaeomagnetic studies in given caves if they are not detected. The sedimentary fills of a number of profiles are separated into individual sequences and cycles, divided by breaks in deposition (unconformities). The dynamic character of cave fill deposition is reflected in the start or termination of individual magnetozones at unconformities in a number of profiles. The general character of cave depositional environments with their numbers of post-depositional changes, hiatuses, reworking and re-deposition does not allow precise calculation of the temporal duration of individual interpreted magnetozones. All these factors contribute to the fact that exact calibration of the geometric characteristics of the magnetostratigraphic logs with the GPTS cannot be attained at all or only with problems, if it is not adjusted using results of other dating methods.

24.1 Introduction

The infill of underground caves and surface or near-surface karst forms has been a focus of its students since the development of speleology and karstology as independent scientific disciplines (for review see e.g., Shaw 1992). It was caused by the rich fossil contents of many cave/karst sites, yielding mammal, bird, amphibian, insect, plant and other remains, and archaeological finds. Therefore, karst voids started to be described as conservers of the geological past (cf. Bosák et al. 1989).

The palaeomagnetic method can yield correlated-ages over the entire Phanerozoic time-scale. Nevertheless, if not supported by other dating methods such as detailed palaeontological determinations of fossils, matching with the standard magnetostratigraphic scale (determined by numerical dating of volcanic products) is a truly hard problem. Something easier is the application of palaeomagnetic methods in

P. Pruner (✉)
Institute of Geology AS CR, v.v.i., Rozvojova 269, 165 00 Praha 6, Czech Republic
e-mail: pruner@gli.cas.cz

E. Petrovský et al. (eds.), *The Earth's Magnetic Interior*, IAGA Special Sopron Book Series 1,
DOI 10.1007/978-94-007-0323-0_24, © Springer Science+Business Media B.V. 2011

areas with young block rotations, as in young orogenic belts and their forelands.

The stratigraphic analysis of cave/karst fills is based on a number of methods, some yielding numerical output, some other calibrated-age, relative-age, and correlated-age (Colman and Pierce 2000). The proper and exact dating of karst processes, including filling of cave/karst voids, is most often the only means of reconstructing the evolution of individual karst features, extensive karst regions, speleogenetical or fossilization processes. The application of a number of dating methods in past decades enabled also the more exact dating of processes in the karst (Ford and Williams 1989, 2007; Bosák 2002). Nevertheless many of such methods cannot be applied to karst materials, and/or their span is too short to be able to cover long time ranges of karst/cave evolution (e.g., the Th/U method).

24.2 Characteristics of Cave/Karst Fill

Karst sediments are a special kind of geologic materials. The development of karst and/or part of the karst system can be "frozen" and rejuvenated for a multiplicity of times (Bosák 1989, 2002, 2003, 2008), and the dynamic nature of karst can lead to re-deposition and reworking of classical stratigraphic order. Those processes can make the karst record unreadable and problematic for interpretation (see Osborne 1984). Temporary (e.g., filling by cave sediments) and/or final interruption of karstification (fossilization *s. s.*) is due to the loss of the hydrological function of the karst (Bosák 1989, p. 583). The introduction of new energy (hydraulic head) to the system may cause reactivation of karstification reflected in the polycyclic and polygenetic nature of karst formation.

The karst environment favours both the preservation of palaeontological remains and their destruction. On one hand, karst is well known for its wealth of palaeontological sites (e.g., Horáček and Kordos 1989), but most cave fills are completely sterile on the other hand. The role of preservation is very important because karstlands function as traps or preservers of the geologic and environmental past, especially of terrestrial (continental) history where correlative sediments are mostly missing, but they carry also marine records (Horáček and Bosák 1989).

The methodology applied to obtain correlated-age (magnetostratigraphic) results depends on the nature of the geologic material filling the karst. The fills of exokarst landforms (especially some epikarst forms) offer more possibilities for the preservation of fossil fauna and flora than do cave interiors. Palaeontological finds are rare in interior cave sediments, where phreatic conditions prevail (in main channels the remains are destroyed by gravel and in side passages with slow currents only fine sediments can be deposited). Troglobitic fauna and flora are usually much too small in number and volume to be significant (Ford and Williams 1989, 2007). Therefore, fossil remains within a cave, that come from the surface (carried in by sinking rivers) or from trogloxenes (e.g., cave-using bats, some birds and mammals), are more important. Airborne grains (pollen, volcanic ash) can only be important when favourable air-circulation patterns are developed within a cave. Nevertheless, cave sediments, especially far from the ponor or other entrance, tend to be highly depleted in fossil fauna (Bosák et al.1998) and/or the preservation of the fossils is too poor for precise determinations (Bosák et al. 2000a).

It is generally known that fossils will be preserved mostly in the upper parts of a sedimentary fill and/or in entrance cave facies, and the time ranges of most absolute dating methods applicable to karst deposits is relatively short (cf. Ford and Williams 1989, 2007).

Some fossils in such cave facies can have been redeposited by the erosion of older sediments and/or derived from collapsed near-surface (epikarst) fills of greater age (cf. e.g., Horáček and Kordos 1989) or represent only fragmental and poorly preserved material (Horáček et al. 2007). Paleontological remains which were deposited in entrance parts or in shallow caves are destroyed by the denudation in relatively short time together with the caves they were deposited in. Because of that reason the faunal remains of the cave entrance facies can not be very old.

The cave environment can be divided from the sedimentological point of view into an entrance facies and an interior facies (Kukla and Ložek 1958). The *entrance facies* is developed in the front of cave opening, in cave entrance and in entrance part of the cave, more o less in the photic zone. The principal processes, which take part in the formation of the entrance facies is slope retreat and unroofing causing the cave shortening. It includes fine-grained sediments transported from the vicinity of the cave by wind and water and coarser clasts transported into the cave by slope processes. The entrance facies represents the most valuable section of the cave from a stratigraphic point of view. The cave entrance contains pollen as well as datable archaeological and palaeontological remains that are protected from surface erosion, weathering and biochemical alteration (Ford and Williams 1989, 2007). Fauna remains (both bones and coproliths) occur as thanatocenoses (rests of animal, which died on place) and taphocenoses (rest of animals brought by carnivors). The disadvantage of the entrance facies is (1) the small volume and (2) continuous slope retreat shifting depocenter inside the former cave. Both factors do not enable the deposition and preservation of thicker sedimentary complexes. The results in relatively short stratigraphic record captured in the entrance facies in caves and rock shelters; in the Central Europe mostly only the Last Glacial up to Holocene (Kukla and Ložek 1958, Ložek 1973).

The *interior facies* develops in those parts of the cave that are more remote from the surface. Sedimentary sequences here can be extensive, consisting of fluvial gravels and sands overlain by flood or injecta deposits of laminar silts and clays often intercalated by speleothems. They can also contain dejecta, colluvial material and outer clastic sediments (including marine ones) often redeposited and/or injected for longer distances within the cave (cf. Ford and Williams 1989, 2007). They form in vadose conditions. Due to the dynamic environment of cave interiors and periodicity of events, sedimentary sequences often represent a series of depositional and erosional events (sedimentary cycles). They are separated by unconformities (breaks in deposition), in which substantial time-spans can be hidden (Bosák et al. 2000b, Pruner and Bosák 2001, Bosák 2002, 2003, 2008, Bosák et al. 2003, Zupan Hajna et al. 2008b). The erosional phases can be much longer that depositional events, which can represent single-flood episodes (Zupan Hajna et al. 2008a, b). Relics of phreatic silts and clays are relatively rare and they typically contain no fossils.

The stratigraphic order in sedimentary sequences is usually governed by the law of superposition, according to which the overlying bed is younger than the underlying one under normal tectonic settings. The law is valid for the majority of sedimentary sequences. However, river terraces and karst environment may present exceptions. The succession of processes connected with entrenchment of river systems cause higher levels of sediments to be older than lower ones. Karst, owing to its dynamic nature, polycyclic and polygenetic character carries some other thresholds – the karst records can be damaged by the simple process of erosion and re-deposition. The reactivation of karst processes often mixes karst fill of different ages (collapses, vertical re-depositions in both directions, etc., e.g., Horáček and Bosák 1989). Contamination of younger deposits by re-deposited fossil-bearing sediments has been known elsewhere in caves (Bosák et al. 2003). Well-known are also sandwich structures, described by Osborne (1984): younger beds are inserted into voids in older ones. Those processes degrade the record in karst archives (Horáček and Bosák 1989).

The final accumulation phase has been dated in caves in most cases, i.e. when the cave is in a quasi-stationary state because the input of energy (water) has been interrupted, detaching the cave from the local hydrological regime for different reasons and for highly differing time-spans; the cave becomes fossilised, at least temporarily. The temporary fossilisation of the cave (i.e. fill by cave sediments) and rejuvenation (excavation of sediments) mostly reflect changes in the resurgence area, especially vertical change (in both directions) of base level at the karst springs. The rejuvenation of the karst process can excavate the previous cave fill/fills completely, which is the most common case resulting from the polycyclic nature and dynamics of cave environments (e.g., Panoš 1963, 1964, Kadlec et al. 2001, Zupan Hajna et al. 2008b). Under favourable settings, fills belonging to more infill phases (cycles) separated by distinct hiatuses (unconformities) can occur in one sedimentary profile. Such amalgamation is typical especially in ponor (sinkhole) parts of the cave (e.g., Kadlec et al. 2001).

24.3 Palaeomagnetism of Karst Sediments

Palaeomagnetic studies of cave deposits can serve as helpful tool to interpret the age of karst sediments as well as to understand the evolution of karst. Cave deposits are generally a significant source of information on palaeomagnetic polarities and on rock-magnetic data as well (cf. White 1988, Ford and Williams 1989, 2007). The aim of such studies was to determine the principal magnetic polarity directions both in clastic and chemogenic deposits, to compare them with the GPTS (Cande and Kent 1995), and to prepare data for the stratigraphic correlation of studied sections.

The geomagnetic field has reversed many times in the past as evidenced from independent sources. One problem that may appear is that over long time spans observed reversals of rock magnetization can result from both self-reversal and/or geomagnetic field reversals. Self-reversal requires the coexistence and interaction of two ferromagnetic constituents; it can be a complicating factor in some studies. However, numerous lines of evidence indicate that most reversals of magnetization in rocks are due to reversals of the geomagnetic field. If the field-reversal theory is correct, there must be a precise stratigraphic correlation of N and R magnetized strata throughout the world. The GPTS has been built up from a number of independent numerical dating studies and is consistent with astrochronology (Cande and Kent 1995). Use of records of ancient variations or reversals as a dating tool, not only in cave sediments, relies on matching the curves of declination, inclination and intensity in a given deposits with established GPTS.

24.3.1 Methods

24.3.1.1 Field Procedures

In early work, the sampling of profiles was usually carried out in two steps: (1) collection of so-called 'pilot' samples spaced 10 cm and more apart, followed by (2) detailed sampling along and across the boundaries of magnetozones detected with different polarization. The second step was taken only after laboratory processing and evaluation of palaeomagnetic properties of the samples taken in the first step. Repeated sampling in some profiles from Slovenia has shown that only dense sampling, i.e. a high-resolution approach, can ensure reliable results (Bosák et al. 2003, Zupan Hajna et al. 2008b). Sample spacing of 2–4 cm in clastic sediments and continuous log/core in speleothems allow the detection of short-lived palaeomagnetic excursions (Bosák et al. 2005b, Bella et al. 2007b, Zupan Hajna et al. 2008a, b), which were not expected to be recorded in cave sediments in general, and excludes the necessity of returning to caves or other sites several times, see Fig. 24.1.

24.3.1.2 Laboratory Procedures and Processing

Measured data should be subjected to multi-component analysis of the remanence (Kirschvink 1980). The individual components must be precisely established to determine direction of the characteristic remanent magnetization (ChRM). Mean ChRM directions must be analyzed using the statistics for spheres (Fisher 1953) but small number of samples could not be used for a reliable interpretation. Only the complete step/field apparatus offered by both demagnetization methods (thermal – TD and/or by alternating field – AF) have to be applied. Examples of TD and AF demagnetization results are illustrated in Figs. 24.2 and 24.3 for samples with R polarity directions The application of complete analysis only to pilot samples and shortened, selected field/step approach to other samples did not offer a sufficient data set for reliable interpretation (Bosák et al. 2003).

24.3.2 Applications and Thresholds

Palaeomagnetic studies conducted in caves have been aimed at determining the age of sediments based on magnetic polarity (magnetostratigraphy) and/or palaeo-secular variations, and on palaeoenvironmental applications of mass-specific MS. In caves, palaeomagnetic studies have been applied principally to fine-grained deposits (fine-grained sands, silts, clays) and some calcite and aragonite speleothems. Systematic acquisition of palaeomagnetic data in Divaška jama (Slovenia) allowed the construction of the magnetostratigraphic profile (Fig. 24.4).

The results of palaeomagnetic analyses are usually not applied only to the dating of infilling processes themselves, but are also means of reconstructing landscape evolution and climate changes, especially when combined with other dating methods (cf. Bosák 2002). Some parameters allow the determination of flow direction of underground streams. In young orogenic belts and their forelands the palaeomagnetic method can be applied for the reconstruction of young block movements. Some of the palaeomagnetic parameters allow the determination of angles of rotation: knowing such parameters from dated correlated sediments covering karst or in adjacent basins, approximate dating is then possible.

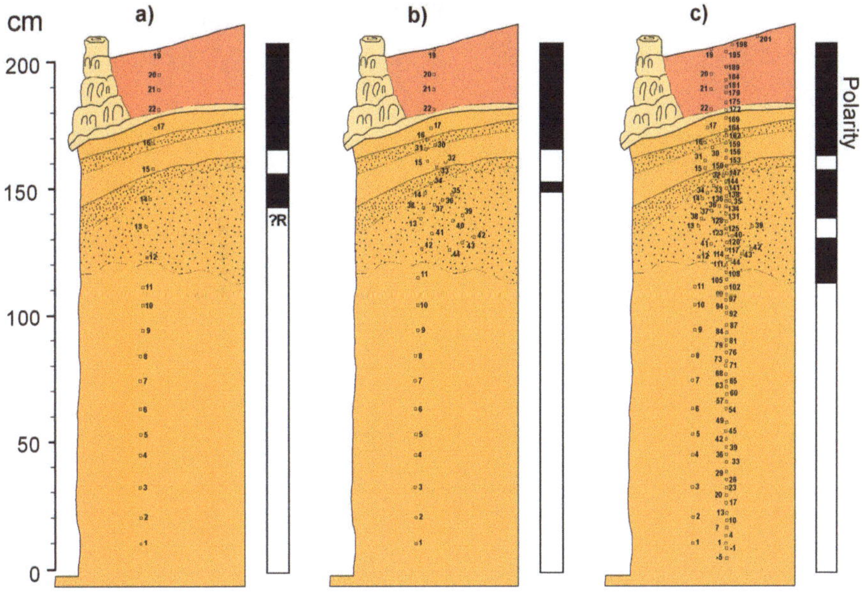

Fig. 24.1 Lithological log of profile in the Divaška jama with position of samples from (**a**) 1997, (**b**) 1998 and (**c**) 2004 (modified from Zupan Hajna et al. 2008b)

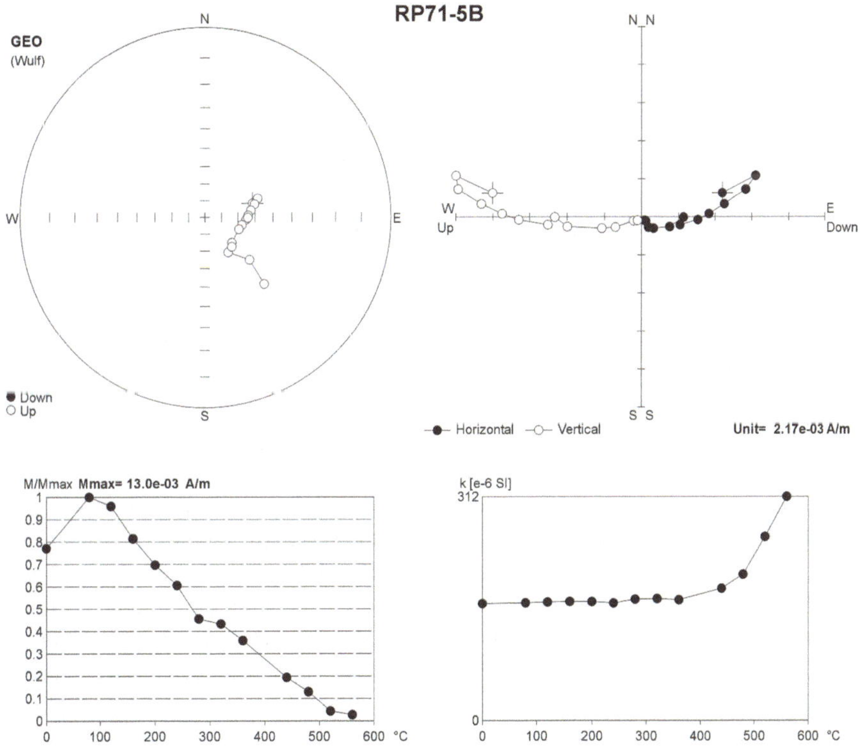

Fig. 24.2 Results of TD demagnetization of flowstone sample with reverse polarity from Račiška pečina cave. A stereographic projection of the natural remanent magnetization of a sample in the natural state (cross section) and after progressive TD and/or AF demagnetization. Zijderveld diagram – *solid circles* represent projection on the *horizontal plane* (XY), *open circles* represent projections on the north-south *vertical plane* (XZ). A graph of normalized values of the remanent magnetic moments versus thermal demagnetizing fields; M – modulus of the remanent magnetic moment of a sample subjected to TD and/or AF demagnetization. A graph of the normalized values of volume magnetic susceptibility versus thermal demagnetizing fields; k – value of volume magnetic susceptibility of a sample subjected to TD demagnetization

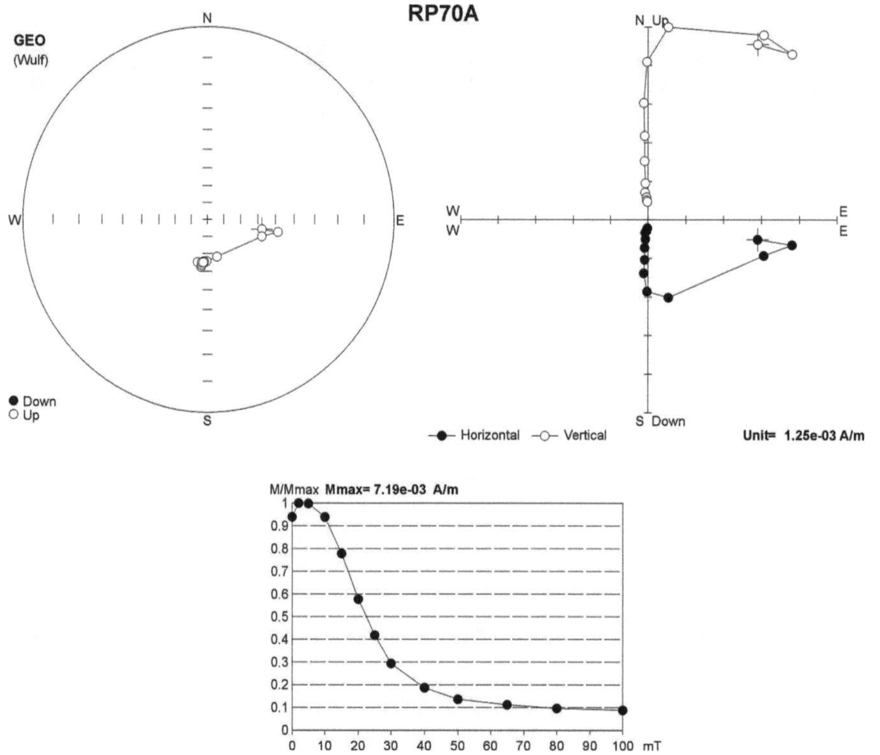

Fig. 24.3 Results of AF demagnetization of flowstone sample with reverse polarity from Račiška pečina cave. For caption see Fig. 24.2

Very shortly after the first applications of paleomagnetism in caves, general comments, definitions of limits and thresholds appeared (e.g., Homonko 1978, Stober 1978, Papamarinopoulos 1978, Noel 1982, 1985, Papamarinopoulos and Creer 1983, Noel and Thistlewood 1989, Ellwood 1999). Verosub (1977) and Noel (1986b) determined that clastic sediments can suffer much post-depositional alteration of the declination and inclination signals, especially if they have drained. Noel and Bull (1982) suggested that bioturbation of sediments may also be a major problem. Williams et al. (1986) emphasised that carriers of remanent magnetisation are usually detrital grains of magnetite.

Bosák et al. (2000b, 2003) and Zupan Hajna et al. (2008b) summarized the principal thresholds in the application of the magnetostratigraphy in cave deposits: e.g., dynamics of cave environment with number of syn- and post-sedimentary changes (sloping, tilting, sediment undercutting, collapses, percolation of fluids especially along wall/sediment interface resulting in so-called Liesegang features, frost and bioturbation alterations, reworking and redeposition), cyclic character of cave fill deposition, unconformities hiding substantial geological time, the general lack of fossils in interior cave facies, redeposition of fossils from older sediments and dejecta, limited number of numerical methods. The described character of deposition resulted that the velocity of sedimentation cannot be calculated in such profiles. The time duration of individual magnetozones cannot be calculated properly and the geometric character of obtained magnetostratigraphic picture cannot be compared with the GPTS.

Repeated sampling in some profiles from Slovenia has shown that only dense sampling, i.e. a high-resolution approach, can ensure reliable results (Bosák et al. 2003, Zupan Hajna et al. 2008b). Sample spacing of 2–4 cm in clastic sediments and continuous log/core in speleothems excludes the necessity of returning to caves or other sites several times. Only the complete step/field apparatus offered by both demagnetization methods have to be applied. The application of complete analysis only to pilot samples (e.g., Šebela and Sasowsky 1999, 2000, Sasowsky

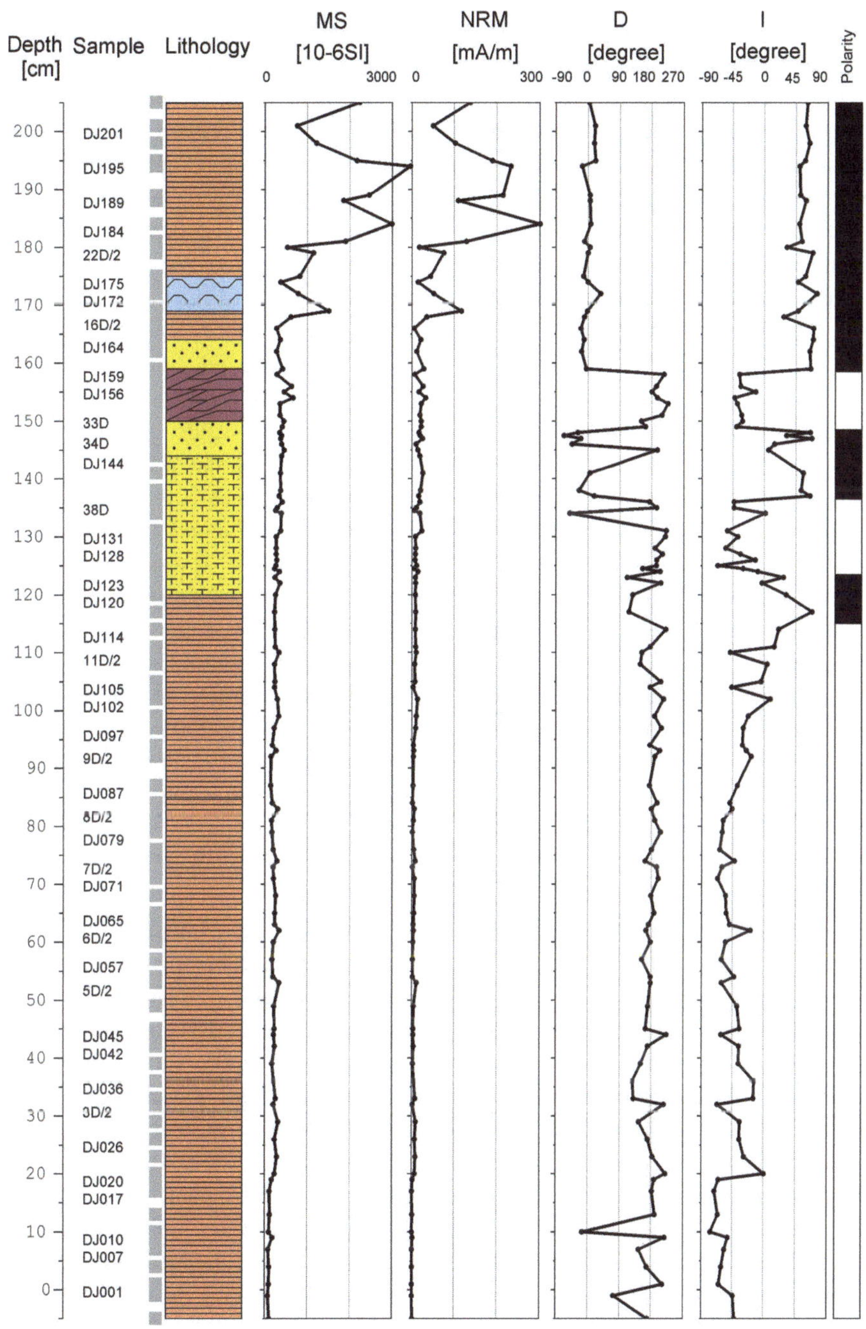

Fig. 24.4 Basic magnetic and palaeomagnetic properties of Divaška jama profile. Legend: Lithology: straight *lines* in *brown* – siltyclay, *dots* in *yellow* – sand, *waves* in *blue* – flowstone, T with *dots* in *yellow* – calcareous silt, collapsed *boxes* in *dark brown* – collapse structure; MS – magnetic susceptibility; NRM – natural remanent magnetization; D – declination; I – inclination

et al. 2003) and shortened, selected field/step approach to other samples (Panuschka et al. 1997) did not offer a sufficient data set for reliable interpretation (Bosák et al. 2003). Measured data should be subjected to multi-component analysis of the remanence (Kirschvink 1980). The individual components must be precisely established to determine the chemical remanent magnetism (CRM) directions. Mean

CRM directions must be analyzed using the statistics for spheres (Fisher 1953) but small number of samples could not be used for a reliable interpretation.

Homogeneity of palaeomagnetic data and lithological character especially of laminated sequences of cave sediments can indicate continuous deposition favourable for record of short-lived palaeomagnetic excursions. Nevertheless, reports on them are scarce in literature (Løvlie and Sandnes 1987, Valen et al. 1997, Bella et al. 2007b, Zupan Hajna et al. 2008a, b).

24.3.3 Dating of Cave/Karst Sediments

The first attempts to apply palaeomagnetic analysis and magnetostratigraphy to cave deposits were carried out by J.S. Kopper and K.M. Creer. Kopper (1975) noted R chrons in two Majorcan caves (Spain). Pons et al. (1979) dated fossil-bearing deposits in Cova de Canet (Esporles, Spain).

Victor A. Schmidt (1982) conducted the pioneering successful large-scale magnetostratigraphic dating in relic and active passages of the Mammoth Cave – Flint Ridge system (Kentucky, USA). He dated the oldest fine-grained sediments back to about 2.1 Ma. This first large-scale attempt to use palaeomagnetism and magnetostratigraphy for dating of cave fill resulted in the spread of the method over the world. Noel and Bull (1982) studied sediments from Clearwater Cave (Mulu, Sarawak, Malaysia). Bosák et al. (1982) sampled an excellent profile of palaeontologically dated Pleistocene sediments in Żabia Cave (Cracow–Wieluń Upland, Central Poland). Measurements were completed but problems with the interpretation of very weak signals resulted in no real results being obtained at that time. Reversal in Early Quaternary deposits was detected by Noel et al. (1984) in Mason Hill (Derbyshire, Great Britain). Williams et al. (1986) interpreted palaeomagnetism of cave sediments from a karst tower at Guilin (China). Schmidt et al. (1984) dated cave sediments in the New South Wales (Australia). Hill (1987) detected the R polarity of silts from Carlsbad Caverns (New Mexico, USA). Kadlec et al. (1992, 1995) dated sedimentary fill in the Aragonitová Cave (Czech Karst, Czech Republic). Müller (1995) dated sediments in Ofenloch Cave (Switzerland). Šroubek and Diehl (1995) indicated the presence of sediments older than 0.78 Ma in caves of the Moravian Karst (Czech Republic). Hercman et al. (1997) used magnetostratigraphy as the control of Th/U dating of speleothems in the Demänovská Ice Cave (Slovakia). Hobléa (1999) dated sediments in Granier System (Chartreuse, France). The study of cave sediments from Črnelsko brezno (Kanin plateau, Julian Alps, Slovenia) by Audra (2000a) indicated age >1 Ma for cave deposits in a deep shaft and much longer speleogenesis of those vadose shafts than expected. Pruner et al. (2000) and Bella et al. (2007b) dated fill of principal caves and cave systems in Slovakia, indicating mostly N magnetization connected with the Brunhes Chron and young fill of pre-existing cave voids. Šebela et al. (2001) dated cave sediments in the Baiyun Cave (Yunnan Karst, China). Kadlec et al. (2000, 2002a, b, 2003, 2004c), and Kadlec and Táborský (2002) dated cave fill and karst sediments in several places of the Moravian Karst, and in the Czech Karst (Czech Republic). Pruner et al. (2002) proved the expected age of principal galleries in Stratenská Cave (Slovakia) to Pliocene. By magnetostratigraphy and Th/U dating of aragonite, Bosák et al. (2002a, 2005a) proved Quaternary age of principal speleogenetical processes in the Ochtinská Aragonite Cave (Slovakia). Audra et al. (in prep.) confirmed the Th/U data from the Grotte de la Clamouse (Hérault, France) by magnetostratigraphy. Bosák et al. (2004a) combined palaeomagnetic and Th/U dating in the Baradla Cave (Hungary), confirming the young age of cave fill (due to multiple cave exhumation). Kadlec et al. (2004a, b) applied magnetostratigraphy and fission track analysis for dating of multi-level cave systems in the Nízké Tatry Mts. (Slovakia) from at least Miocene. Kadlec et al. (2005) contributed to dating of rich fossiliferous cave sediments in Za Hájovnou Cave (Javoříčko Karst, Moravia Czech Republic) to Brunhes and Maruyama Chrons. The MS and magnetization intensity values indicate change in source material derived from catchment area of the sinking creek. Bosák et al. (2005b) dated sediments in the Pocala Cave and unroofed cave in Borgo Grotta Gigante (Trieste region, Italy) obtained by full-core drilling by magnetostratigraphy and Th/U method to pre-Quaternary and Quaternary; unroofing of the Pocala Cave is younger than 197 ka. Bella et al. (2005, 2007a) reconstructed the evolution of hypogenic Belianská Cave (Tatra Mts., Slovakia) down to Upper Miocene. Žák et al. (2007) used palaeomagnetic parameters for dating of old cavity fills in the Czech Karst (Czech Republic) to Early–Middle Triassic. Chess et al. (2010)

used detailed study of cave sediments for dating of event of massive sedimentation around 900 ka.

24.3.4 Dating of Archaeological Sites in Caves

Palaeomagnetic data are also used in archaeological sites. Kopper and Creer (1976) described the application of palaeomagnetic dating for stratigraphic interpretation of archaeological sites. Creer and Kopper (1974) dated cave paintings in Tito Bustillo Cave (Asturias, Spain). Papamarinopoulos (1977), Poulianos (1980) and Papamarinopoulos et al. (1987) dated site with so-called Petralona Man, which is expected to be up to 800 ka old (Petralona Cave, Chalkidiki, Greece). Noel (1990) applied the method in caves of Guanxi (China). The famous South African hominid sites of Makapansgat and Swartkrans were studied by Brock et al. (1977), McFadden et al. (1979), Herries (2003) and Herries et al. (2006a, b).

24.3.5 Reconstructions of Landscape Evolution

Palaeomagnetic analysis and magnetostratigraphy of cave clastic sediments are often applied to the reconstruction of landscape and climate evolution of karst regions. The best are multi-level caves and cave systems. Audra and Rochette (1993) reconstructed glaciations from sediment dating in Valliers Cave (Vercors, France). Sasowsky et al. (1995) determined river incision by magnetostratigraphy of cave sediments in the Appalachian plateaus (USA).

Løvlie et al. (1995) and Valen et al. (1997) studied stratigraphy of Pleistocene partly subglacial cave sediments in several caves from northern Norway. Quaternary evolution of the British South Pennines was reconstructed by Th/U and palaeomagnetic data by Rowe et al. (1988). Audra (1996, 2000b) reconstructed the karst evolution of Dévoluy Massif (Hautes–Alpes, France). Springer et al. (1997) calculated incision rate for Cheat River (West Virginia, USA) from sediments filling maze caves along the river. Audra et al. (1999) applied magnetostratigraphy in Muruk Cave (Nakanaï Mts., New Britain, and East Papua–New Guinea). In the combination with the Th/U dating they proved the very young speleogenesis and uplift. Kadlec et al. (2001, 2002b) reconstructed evolution of semi-blind valleys and their catchment areas and cave systems in the northern part of the Moravian Karst (Czech Republic) with the application of four dating methods. Audra et al. (2001) dated the start of the entrenchment of a deep underground canyon in Aven de la Combe Rajeau (Ardèche, France) to about 2.0 Ma and correlated it with terrace systems of the Ardèche River. Auler et al. (2002) dated, in combination with fission track analysis, multi-level Gruta do Padre (eastern Brazil) to about 34–25 Ma and deduced that fluvial incision rates probably correcpond to isostatic uplift of the cratonic area. Audra et al. (2002) applied palaeomagnetic analysis of cave sediments for analysis of evolution of Tennengebirge karst (Austria). Musgrave and Webb (2004) analyzed Pleistocene landscape and climate evolution of the basis of palaeomagnetic analysis of sediments in the Buchan Cave (southeastern Australia). Sasowsky et al. (2004) interpreted the time-span of the development of Kooken Cave (Pennsylvania, USA). Kadlec et al. (2004a) applied magnetostratigraphy, Th/U, fission track and cosmogenic nuclide methods in cave deposits of multi-level cave systems in the Demänová and Jánská Valleys for the reconstruction of karst landscape evolution and valley incision of the Nízke Tatry Mts. (Slovakia). Stock et al. (2005) applied combination palaeomagnetic and numerical dating methods in landscape evolution studies and Audra et al. (2006, 2007) in the reconstruction of cave genesis in Alps from Miocene. Jurková et al. (2008) interpreted the relict of flowstone found in situ on the top of limestone ridge in the Ještěd Ridge (northern Bohemia, Czech Republic) as conseurence of rapid unroofing caused by tectonic uplift and deep valley incision in last about 0.7 (or 3.6) Ma.

24.3.6 Palaeoclimatic Applications

The most common measured magnetic property, mass-specific MS (c), is a function of concentration, grain size and mineralogy of magnetic minerals found in the sediments. It has proved to be a sensitive detector of the long-term, large-scale terrestrial climatic change in cave deposits (Šroubek et al. 2001). The susceptibility has been applied to a study of palaeo-secular variation (Ellwood 1971, Creer and Kopper 1974, 1976, Papamarinopoulos et al. 1991). Kopper and Creer (1973) fitted the declination and inclination curves to a ^{14}C controlled record in Majorcan caves. Anisotropy of magnetic susceptibility was applied to characterise the sedimentary fabrics and palaeo-flow directions (Noel and St. Pierre 1984, Noel 1983, 1986a; Turner and Lyons 1986, Kostrzewski et al. 1991). Ellwood et al. (1996), Šroubek and Diehl (1995), and Šroubek et al. (1996, 2001) described environmental magnetic properties of cave sediments. Ellwood (1999) compared kappa highs with palaeoclimatic changes in caves of Albania and correlated the data with some other caves of the Mediterranean area. Matching susceptibility curves obtained from cave sediments with known the secular variation curve for a region enables establishment of a chronology for the sediments (e.g., Ellwood 1971, Noel 1983, 1986a, b). Thistlewood and Noel (1991) applied the plots of declination, inclination, intensity of magnetization and magnetic susceptibility to the correlation of several sedimentary sections in Peak Cavern (Great Britain). Kadlec (2003) determined the flow regime from the anisotropy of magnetic susceptibility in the Ochozská Cave (Moravian Karst, Czech Republic).

24.3.7 Speleothems

The palaeomagnetism of calcite and aragonite speleothems has been used to test for N or R polarities where the sample is known to be older than 350 ka or older than about 1.25 Ma (Th/U and U/Pb methods), and to obtain dated, high-resolution curves of recent secular variations (Ford and Williams 1989, 2007).

Latham et al. (1979) made the first analyses, using speleothems from Canada and Great Britain. In spite of very low magnetic intensities, speleothems can carry the NRM either as a chemical precipitate (CRM) or as floodwater or filtrate detrital grains (DRM; Latham 1981, 1989). The carrier of magnetism is magnetite. The R magnetic signature in speleothem was detected by Latham et al. (1982) in British Columbia and by Šroubek and Diehl (1995) in the Moravian Karst. Latham et al. (1986, 1987, 1989) studied secular variation on stalagmites from Mexico and Canada obtaining good secular variation curves and evidences for a drift. Palaeomagnetic data calibrated the reconstruction of palaeoclimatic and depositional conditions between 350 and 700 ka in Rana (northern Norway; Lauritzen et al. 1990). Palaeosecular variations were studied also by Openshaw et al. (1993) on Xingweng stalagmite (China) and Lean et al. (1995) on stalagmites from the Vancouver Island (Canada). Martin (1991) detected palaeoinclination changes in the last 4–20 ka for stalagmite and flowstone in Gardner Cave (Washington, USA). Brooks et al. and Ford (in Hill 1987) applied palaeomagnetic analysis to speleothems from the Carlsbad Caverns. Palaeomagnetic and rock magnetic studies on speleothems were carried out also by Kyle (1990) in Gardner Cave (USA), Morinaga et al. (1989) in Japan, Liu et al. (1988) from Ping Le (China) and Perkins and Maher (1993) in Great Britain. Lauritzen et al. (1994) calibrated isoleucine epimerization from flowstone by palaeomagnetic data in Hamarnesgrotta (northern Norway). Bosák et al. (2002b) dated a 2.5 m thick profile in flowstones in Snežna jama (Slovenia) back to more than 5 Ma and related it to relief evolution of Kamnik–Savinja Alps. Fold tests on dome-like stalagmites from Račiška pečina Cave and Pečina Na Borštu Cave (SW Slovenia) of differing sizes and ages (Pleistocene, Pliocene) indicate the domelike structures are primary (Pruner et al. 2010).

24.3.8 Complex Palaeomagnetic Studies in Slovenia

Intensive palaeomagnetic research in Slovenia in past 14 years has contributed substantially to the understanding of cave sediments in different tectonic and geomorphic settings in the territory (for detailed summary see Zupan Hajna et al. 2008b, 2010, Pruner et al. 2009). The most important result is the discovery that cave fills have substantially older ages than generally expected earlier (max. about 350 ka; see summary in Gospodarić 1988). Palaeomagnetic data in combination with other dating methods, especially biostratigraphy, have shifted the possible beginning not only of the speleogenesis but also of the cave filling processes in Slovenia far below the Tertiary/Quaternary boundary (here assigned its traditional age at 1.8 Ma). The data support and better define the estimated ages of the surface and cave sediments that were based on geomorphic evidences, especially from unroofed caves.

The evolution of the caves (from the start up to their total destruction by denudation) took part within one karstification period (sensu Bosák et al. 1989), which began with the regression of Eocene sea and exposing of limestones at the surface within complicated overthrusted structure, which formed principally during Oligocene to early Miocene. The interpretation of palaeomagnetic data, with some support from palaeontological finds, indicates that karst developed in pulses tightly linked with transgression/regression history, tectonic evolution and changes of the geodynamic regime a related overprint in relief evolution. Individual pulses were not sharply limited, however, and therefore cannot be tied to precisely defined karst phases. Moreover, the complicated geological structure and tectonic/geomorphic evolution makes the picture less clear due to the differing tectonic evolution of individual morpho-structural units, which often have also quite different histories of evolution of the relief and karst.

The reconstruction of evolution of karst and caves is complicated by the lack of surface karst sediments. Correlative deposits in surrounding depositional basins are either too old (Pannonian Basin), or poorly dated (e.g., fills of Plio/Quaternary intermontane basins). Sediments representing the last 6 Ma are preserved only in caves or unroofed caves. They contain records of past climatic, palaeogeographic and tectonic changes, although not continuous. Cave sediments therefore are good tools for reconstructing landscape evolution and tectonic regimes in their respective areas.

Research in the Dinaric, Alpine and Isolated karsts opened new horizons for the interpretation of karst and cave evolution, both of individual geomorphologic units and of extensive areas. The data inform us that a number of common features and evolutionary trends exist in all the studied areas. On the other hand, there are distinct differences of evolution of smaller geomorphic units within the more extensive ones, which result mostly from differential tectonic movements connected with the post-6 Ma counter-clockwise rotation of the Adria block (Vrabec and Fodor 2006). The analysed cave fills indicate some distinct phases of massive deposition in caves, as follows:

Sediments older than 1.2 Ma (numerical age)/ 1.77 Ma (palaeomagnetic age); up to or greater than 5.0 Ma. The interpretation of the upper age limit, if based on the palaeomagnetic data, represents the rough estimate of the alternation of R and N polarized magnetozones typical for the period of Matuyama and older chrons. Ages in this category are adjusted to the age interpreted at the Črnotiče I site and Divača and Kozina profiles or Divaška jama (Bosák et al., 1999, 2000a, 2004b, 2007a, b, Zupan Hajna et al. 2007a). Snežna jama belongs to this period also (Bosák et al. 2002b). Filling can be dated to uppermost Miocene and Pliocene (e.g., Bosák et al. 2000b). The cave with the presumably oldest sediment in our study is Grofova jama. The montmorillonite fill, if derived from intensive weathering of volcanoclastic products, should originate from products of Oligocene to early Miocene volcanic activity in Italy or north-eastern Slovenia. The age of the fill, accroding to the apatite fission track dating is 21.7 ± 6.9 M can represent one oof the oldest phases of cave evolution in the Kras region.

Sediments dated from about 0.78 Ma up to more than 4.0 Ma (palaeomagnetic age). This group contains a succession of detected ages. The base of most profiles can be interpreted as probably not much older than 3.58 Ma, i.e. the datum adjusted by palaeontological finds in the Črnotiče II and Račiška pečina sites (Horáček et al. 2007). It seems that some phases could be distinguished: (a) more than 0.78 up to about 4.2 Ma (palaeomagnetic ages; e.g., Bosák et al. 2005b, c; Pruner et al. 2010), and (b) less than 0.78 to about 2 Ma (palaeomagnetic ages; e.g., Zupan Hajna et al. 2008a), i.e. something between Brunhes/Matuyama boundary (and somewhat younger) and base of Jaramillo and/or Olduvai subchrons (and somewhat older).

Sediments younger than 0.78 Ma. Caves containing sedimentary fill younger than the Brunhes/Matuyama boundary have one common and typical feature – a part of the cave is still hydrologically active, with one or more streams flowing in the lower levels (e.g., Postojnska jama, Planinska jama; Zupan Hajna et al. 2008a; Križna jama). This category with includes also young depositional phase(s) in caves with older fills (e.g., Jama pod Kalom/Grotta Pocala; Bosák et al. 2005b; Divaška jama; Bosák et al. 1998, Račiška pečina; Pruner et al. 2010). We therefore interpreted most of the sediments as being younger than 0.78 Ma, belonging to different depositional events within the Brunhes chron. Nevertheless, the N polarization in some profiles can be linked with N polarized subchrons older than 0.78 Ma.

Conclusions

Dating cave sediments by the application of the palaeomagnetic method – magnetostratigraphy – is a difficult and sometimes risky task, as the method is comparative in its principles and does not provide numerical ages. For dating clastic cave sediments and speleothems it is limited by the complex conditions occurring underground so that it is often necessary to combine it with other methods that offer supplementary absolute-, calibrate-, relative- or correlate-ages. The proper and exact dating of karst processes, including filling of cave/karst voids, is most often possible only by means of reconstruction of evolution of individual karst features, extensive karst regions, speleogenetical or fossilization processes. The application of a number of dating methods in past decades enabled also the more exact dating of processes in the karst. Nevertheless many of such methods cannot be applied to karst materials, and/or their span is too short to be able to cover long time ranges of karst/cave evolution (e.g., the Th/U method). The palaeomagnetic method can yield correlated-ages over the entire Phanerozoic time-scale. Nevertheless, if not supported by other dating methods such as detailed palaeontological determinations of fossils, matching with the standard magnetostratigraphic scale (determined by numerical dating of volcanic products) is a truly hard problem. Something easier is the application of palaeomagnetic methods in areas with young block rotations, as in young orogenic belts and their forelands. Some of the palaeomagnetic parameters allow the determination of angles of rotation: knowing such parameters from dated correlated sediments covering karst or in adjacent basins, approximate dating is then possible.

The application of magnetostratigraphy to both clastic and chemogenic cave sediments can be seemed to be an ideal tool for dating. Nevertheless, our practical experience both with the cave interior deposits and with the palaeomagnetic methodology (sampling, construction and interpretation of magnetostratigraphic profiles) has shown that there are substantial real problems. Some of them have been already considered in some of the previous studies cited, but only in a more limited form and case by case depending on local situations and interpretation problems. There are also other problems with the stratigraphy of cave fills, erosion features, available palaeontology, etc.

Correlation of the magnetostratigraphic results, and the interpretations tentatively placed upon them has shown that in the majority of cases, application of an additional dating method is needed to either reinforce the palaeomagnetic data or to help to match them with the GPTS. The most helpful data are from absolute dating and palaeontological (biostratigraphic) correlative-age estimates. Nevertheless, these also have limits to their utility. Sometimes the nature of cave sediments, does not favour the preservation of fossil remains that could contribute to the dating of cave sediments or to the correlation of magnetozones with the GPTS.

Interpretation of magnetostratigraphic results faces other serious problems that may endanger palaeomagnetic studies in given caves if they are not detected. The sedimentary fills of a number of our profiles were separated into individual sequences and cycles, mostly of fluvial or ponding nature, because they were divided by breaks in deposition (unconformities). Unconformities and/or intercalated precipitates indicate the highly complicated deposition dynamics that can be found in many sedimentary sequences; entire caves and cave systems can be completely filled and emptied several times. Erosion removes parts of the originally deposited logs. Therefore, unconformities within sedimentary profiles can hide substantial amounts of geological time. The long-lasting breaks can also explain some differences in declination values between upper and lower parts of a section that indicate rotations of the particular tectonic block containing the cave. Some of the unconformities were expressed by post-depositional changes, erosion and/or precipitation features. The breaks were often associated with mass movement and sediment re-deposition within caves. Sedimentary sections can be undercut by erosion, resulting in slumping and/or inclination of the fill. Other slope movements could occur within caves due to cold during Pleistocene glacial periods, in respective geographic positions. Compaction, especially of fine-grained deposits can lead to sandwiching. Unconformity surfaces are often emphasized by precipitation of iron-rich substances (limonite crusts, sometimes enriched in Mn-bearing compounds, Liesegang-like features) which can change the original palaeomagnetic record.

The dynamic character of cave fill deposition is reflected in the start or termination of individual magnetozones at unconformities in a number of profiles, which is comparable with situation reported on a number of Quaternary carbonate platforms (McNeil et al. 1988, 1998). The general character of cave depositional environments with their numbers of post-depositional changes, hiatuses, reworking and re-deposition does not allow precise calculation of the temporal duration of individual interpreted magnetozones. All these factors contribute to the fact that exact calibration of the geometric characteristics of the magnetostratigraphic logs with the GPTS cannot be attained at all or only with problems, if it is not adjusted using results of other dating and geomorphic methods.

In spite of a number of theoretical and practical problems mentioned above, the palaeomagnetic and magnetostratigraphy analysis of cave fills and karst sediments represents usefull tool to know the time-evolution sequences in karstogenesis (and speleogenesis, too). Palaeomagnetic analysis yields number of additional data, which can be applied for the interpretation of post-depositional evolution of the respective site, sediment and territory. Palaeomagnetic parameters can indicate rotation of individual tectonic block and broader territory, post-depositional changes (like tilting or changes in geochemical environment, precipitation of minerals, etc.). In a combination with other avaibalbe applicable dating methods, palaeomagnetism and magnetostratigraphy of cave/karst sediments can bring surprising dating results, as documented on number of places, e.g., recently in Slovenia or Slovakia.

Acknowledgments The text was prepared in the frame of Institute Research Plan No. CEZ AV0Z30130516, grant project of the Grant Agency of the Academy of Science of the Czech Republic No. IAA300130701, and program KONTAKT No. MEB 090908.

References

Audra P (1996) L'rapport de l'étude des remplissages à la connaissance de la karstogenèse: le cas du chourum du Goutourier (Massif du Dévoluy, Hautes-Alpes). Revue d'analyse spatiale quantitative et appliquée, Mélanges Maurice Julian "Géomorphologie, risques naturels et aménagement", 38–39, 109–120, Nice

Audra P (2000a) Le karst haut alpin du Kanin (Alpes juliennes, Slovénie-Italie). Etat des connaissances et données récentes sur le fonctionement actuel et lévolution plio-quaternaire des structures karstiques. Karstologia 35:27–38

Audra P (2000b) Pliocene and Quaternary Karst Development in the French Prealps – Speleogenesis and Significance of Cave Fill. In: Klimchouk AB, Ford DC, Palmer AN, Dreybrodt W (eds) Speleogenesis. Evolution of Karst Aquifers. National Speleological Society, Huntsville

Audra P, Bini A, Camus H, Delange P (in prep) Les sédiments de la grotte de Clamouse (Hérault). Mémoires de Spéléo-club du Paris

Audra P, Bini A, Gabrovšek F, Häuselmann P, Hobléa F, Jeannin PJ, Kunaver J, Monbaron M, Šušteršič F, Tognini P, Trimmel H, Wildberger A (2006) Cave genesis in the Alps between the Miocene and today: a review. Z Geomorphol NF 50: 153–176

Audra P, Bini A, Gabrovšek F, Häuselmann P, Hobléa F, Jeannin PJ, Kunaver J, Monbaron M, Šušteršič F, Tognini P, Trimmel H, Wildberger A (2007) Cave and karst evolution in the Alps and their relation to paleoclimate and paleotopography. Acta Carsologica 36:53–67

Audra P, Camus H, Rochette P (2001) Le karst de plateaux jurassique de la moyenne vallée de lÁrdèche: datation par paléomagnetisme des phases dévolution plio-quaternaires (aven de la Combe Rajeau). Bull Soc Géol Fr 172:121–129

Audra P, Quinif Y, Rochette P (2002) The genesis of the Tennengebirge karst and caves (Salzburg, Austria). J Karst Cave Stud 64:153–164

Audra P, Lauritzen SE, Rochette P (1999) Datations de sédiments (U/Th et paléomagnétisme) dún hyperkarst de Papouasie-Nouvelle-Guinée (Montagnes Nakanaï, Nouvelle-Bretagne). Karst 99. Colloque européen. Des paysages du karst au géosysteme karstique: dynamiques, structures et enregistrement karstiques, Etudes de géographie physique, (suppl. XXVIII):43–54, CAGEP, Aix-en-Provence

Audra P, Rochette P (1993) Premières traces de glaciations du Pléistocène inférieur dans le massif des Alpes. Datations par paléomagnetisme de remplissages à la grotte Valliers (Vercors, Isère, France). Compte-rendus à lÁcadémie des Sci 2(11):1403–1409

Auler RS, Smart PL, Tarling DH and Farrant AR (2002) Fluvial incision rates derived from magnetostratigraphy of cave sediments in the cratonic area of eastern Brazil. Z Geomorphol 46(3):391–403

Bella P, Bosák P, Głazek J, Hercman H, Kicińska D, Nowicki T, Pavlarčik S, Pruner P, (2005) The antiquity of the famous Belianská Cave (Slovakia). 14th International Congress of Speleology, Athens-Kalamos. Final Programme & Abstract Book, Athens, pp 144–145

Bella P, Bosák P, Głazek J, Hercman H, Kadlec J, Kicińska D, Komar M, Kučera M, Pruner P (2007a) Datovanie výplní Belianskej jaskyne: geochronologické záznamy jej genézy. Aragonit 12:127–128. Liptovský Mikuláš

Bella P, Bosák P, Pruner P, Hochmuth Z, Hercman H (2007b) Magnetostratigrafia jaskynných sedimentov a speleogenéza Moldavskej jaskyne a spodných častí Jasovskej jaskyne. – Slovenský kras, XLV:15–42. Liptovský Mikuláš

Bosák P (1989) Problems of the origin and fossilization of karst forms. In: Bosák P, Ford DC, Głazek J, Horáček I (eds) Paleokarst. A Systematic and Regional Review, 577–598, Elsevier–Academia, Amsterdam–Praha

Bosák P (2002) Karst processes from the beginning to the end: how can they be dated? In: Gabrovšek F (ed) Evolution of karst: from prekarst to cessation. Carsologica Založba ZRC, Postojna–Ljubljana, pp 191–223

Bosák P (2003) Karst processes from the beginning to the end: how can they be dated? Speleogenesis Evol Karst Aquifers 1(3):24

Bosák P (2008) Karst processes and time. Geologos 14:15–24. Poznań

Bosák P, Bella P, Cílek V, Ford DC, Hercman H, Kadlec J, Osborne A, Pruner P (2002a) Ochtiná Aragonite Cave (Slovakia): Morphology, Mineralogy and Genesis. Geol Carpathica 53(6):399–410

Bosák P, Ford DC, Głazek J (1989) Terminology. In: Bosák P, Ford DC, Głazek J, Horáček I (eds) Paleokarst. A systematic and regional review. Elsevier–Academia, Amsterdam–Praha, pp 25–32

Bosák P, Bella P, Cílek V, Ford DC, Hercman H, Kadlec J, Osborne A, Pruner P (2005a) Ochtiná Aragonite Cave (Slovakia): morphology, mineralogy and genesis. Speleogenesis Evol Karst Aquifers, 3, 2, 16 pp

Bosák P, Głazek J, Horáček I, Szynkiewicz A (1982) New locality of Early Pleistocene vertebrates – Żabia Cave at Podlesice, Central Poland. Acta Geol Pol 32(3–4):217–226, Warszawa

Bosák P, Hercman H, Kadlec J, Móga J, Pruner P (2004a) Palaeomagnetic and U-series dating of cave sediments in Baradla Cave, Hungary. Acta Carsologica 33/2(13):219–238

Bosák P, Hercman H, Mihevc A, Pruner P (2002b) High resolution magnetostratigraphy of speleothems from Snežna Jama, Kamniške–Savinja Alps, Slovenia. Acta Carsologica 31/3(1):15–32

Bosák P, Knez M, Otrubová D, Pruner P, Slabe T, Venhodová D (2000a) Palaeomagnetic research of fossil cave in the highway construction at Kozina, SW Slovenia. Acta Carsologica 29/2(1):15–33

Bosák P, Knez M, Pruner P, Sasowsky I, Slabe T, Šebela S (2007a) Palaeomagnetic research into unroofed caves opened during the highway construction at Kozina, SW Slovenia. – Annales (Anali za istrske in mediteranske študije), Series Historia Naturalis, 17(2):249–260. Koper

Bosák P, Knez M, Pruner P, Sasowsky I, Slabe T . (2007b) Paleomagnetne raziskave jame brez stropa pri Kozini. In: Knez M, Slabe T (eds) Kraški pojavi, razkriti med gradnjo slovenskih avtocest. Založba ZRC. Ljubljana, pp 185–194

Bosák P, Mihevc A, Pruner P (2004b) Geomorphological evolution of the Podgorski Karst, SW Slowenia: contribution of magnetostratigraphic research of the Črnotiče II site with Marifugia sp. Acta Carsologica 33/1(12):175–204

Bosák P, Mihevc A, Pruner P, Melka K, Venhodová D, Langrová A (1999) Cave fill in the Črnotiče Quarry, SW Slovenia: palaeomagnetic, mineralogical and geochemical study. Acta Carsologica 28(2):15–39

Bosák P, Pruner P, Hercman H, Calligaris R, Tremul A (2005b) Paleomagnetic analysis of sediments in Pocala Cave and Borgo Grotta Gigante (Trieste region, Italy). Ipogea 4(2004):37–51, Trieste

Bosák P, Pruner P, Kadlec J (2003) Magnetostratigraphy of cave sediments: application and limits. Studia Geophys Geodaetica 47(2):301–330

Bosák P, Pruner P, Mihevc A, Zupan Hajna N (2000b) Magnetostratigraphy and unconformities in cave sediments: case study from the Classical Karst, SW Slovenia. Geologos 5:13–30, Poznań

Bosák P, Pruner P, Mihevc A, Zupan Hajna N, Horáček J, Kadlec J, Man O, Schnabl P (2005c) Palaeomagnetic and palaeontological research in Račiška pečina Cave, SW Slovenia. 14th International Congress of Speleology, Athens-Kalamos. Final Programme and Abstract Book, 204–205, Athens

Bosák P, Pruner P, Zupan Hajna N (1998) Paleomagnetic research of cave sediments in SW Slovenia. Acta Carsologica 27/2(3):151–179

Brock A, McFadden PL, Partridge TC (1977) Preliminary palaeomagnetic results from Makapansgat and Swartkrans. Nature 266:249–250

Cande SC, Kent DV (1995) Revised calibration of the geomagnetic polarity timescale for the Late Cretaceous and Cenozoic. J Geophys Res 100(B4):6093–6095

Chess DL, Chess CA, Sasowsky ID, Schmidt VA, White WB (2010) Clastic sediments in the Butler Cave – Sinking Creek System, Virginia, USA. Acta Carsologica 39(1):11–26

Colman SM, Pierce KL (2000) Classification of Quaternary geochronologic methods. In: Noller JS, Sowers JM, Lettis WR (eds) Quaternary geochronology. Methods and Applications, American Geophysical Union, Washington, DC, pp 2–5

Creer KM, Kopper JS (1974) Paleomagnetic dating of cave paintings in Tito Bustillo Cave, Asturias, Spain. Science 168:348–350

Creer KM, Kopper JS (1976) Secular oscillations of the geomagnetic field recorded by sediment deposited in caves in the Mediterranean region. Geophys J R Astronomical Soc 45:35–58

Ellwood BB (1971) An archeomagnetic measurement of the age and sedimentation rate of Climax cave sediments, southwest Georgia. Am J Sci 271:304–310

Ellwood BB (1999) Identifying sites and site correlation using electrical and magnetic methods. Program, 64th annual meeting of the society for american archaeology, Chicago, p 101

Ellwood BB, Petruso KM, Harrold FB, Korkuti M (1996) Paleoclimate characterization and intra-site correlation using magnetic susceptibility measurements: and example from Konispol Cave, Albania. J Field Archaeol 23:263–271

Fisher R (1953) Dispersion on a sphere. Proc Roy Soc A 217:295–305

Ford DC, Williams PW (1989) Karst geomorphology and hydrology. Unwin Hyman, London, 601 pp

Ford DC, Williams PW (2007) Karst hydrology and geomorphology. Wiley, Chichester, 562 pp

Gospodarić R (1988) Paleoclimatic record of cave sediments from Postojna karst. Annales de la Société Géologique de Belgique 111:91–95

Hercman H, Bella P, Gradziński M, Glazek J, Lauritzen SE, Lovlie R (1997) Uranium-series dating of speleothems from Demänova Ice Cave: a step to age estimation of the Demänova Cave System (The Nízke Tatry Mts., Slovakia). Ann Soc Geologorum Poloniae 67:439–450

Herries AIR (2003) Magnetostratigraphy of the South African hominid palaeocaves. Am J Phys Anthropol 113(Suppl. 36)

Herries AIR, Adams JW, Kuykendall KL, Shaw J (2006a) Speleology and magnetobiostratigraphic chronology of the GD 2 locality of the Gondolin hominin-bearing paleocave

deposits, North West Province, South Africa. J Hum Evol 51(6):617–631
Herries AIR, Reed KE, Kuykendall KL, Latham AG (2006b) Speleology and magnetobiostratigraphic chronology of Buffalo Cave fossil site, Makapansgat, South Africa. Quaternary Res 66(2):233–245
Hill CA (1987) Geology of Carlsbad Cavern and other caves in the Guadelupe Mountains, New Mexico and Texas. N M Bureau Mines Miner Resour Bull 117:1–152, Socorro
Hobléa F (1999) Contribution à la connaissance et à la gestion environnementale de géosystèmes karstiques montagnards: Etudes savoyarde. PhD Thesis, Université de Lyon, 995 pp
Homonko P (1978) A palaeomagnetic study of cave and lake deposits in Britain. MSc. Thesis, University of Newcastle upon Tyne
Horáček I, Bosák P (1989) Special characteristics of paleokarst studies. In: Bosák P, Ford DC, Głazek J, Horáček I (eds) Paleokarst. A systematic and regional review, Elsevier–Academia, Amsterdam–Praha, pp 565–568
Horáček I, Kordos L (1989) Biostratigraphic investigations in paleokarst. In: Bosák P, Ford DC, Głazek J, Horáček I (eds) Paleokarst. A systematic and regional review, Elsevier–Academia, Amsterdam–Praha, pp 599–612
Horáček I, Mihevc A, Zupan Hajna N, Pruner P, Bosák P (2007) Fossil vertebrates and paleomagnetism update one of the earlier stages of cave evolution in the Classical Karst, Slovenia: Pliocene of Črnotiče II site and Račiška pečina. Acta Carsologica 37/3:451–466
Jurková N, Bosák P, Komar M, Pruner P (2008) Relict flowstone at Machnín (the Ještěd Ridge, North Bohemia, Czech Republic) and its importance for relief evolution. Geomorphol Slovaca Bohemica 2(2007):19–24. Bratislava
Kadlec J (2003) Reconstruction of flow directions by measurements of anisotropy of magnetic susceptibility in fluvial sediments of the Ochozská Cave, Moravan Karst. Geologické výzkumy na Moravě a ve Slezsku v r. X:5–7, Brno
Kadlec J, Chadima M, Pruner P, Schnabl P (2005) Paleomagnetic dating of sediments in the Za Hájovnou Cave in Javoříčko. Přírodovědné studie Muzea Prostějovska, 8:75–82
Kadlec J, Hercman H, Beneš V, Šroubek P, Diehl JF, Granger D (2001) Cenozoic history of the Moravian Karst (northern segment) Cave sediments and karst morphology. Acta Musei Moraviae Sci Geol LXXXVII:111–160, Brno
Kadlec J, Hercman H, Danišik M, Pruner P, Chadima M, Schnabl P, Šlechta S, Grygar T, Granger D (2004a) Dating of cave sediments and reconstruction of karst morphology of the Low Tatra Mts. 3. Národní speleologický kongres, Rozšířené abstrakty, Česká speleologická společnost, Praha, pp 30–32
Kadlec J, Hlaváč J, Horáček I (2002a) Sedimenty jeskyně Arnika v Českém krasu. Český kras XXVIII:13–15, Beroun
Kadlec J, Hercman H, Nowicki T, Głazek J, Vít J, Šroubek P, Diehl JF, Granger D (2000) Dating of the Holštejnská Cave deposits and their role in the reconstruction of semiblind Holštejn Valley Cenozoic history (Czech Republic). Geologos 5:57–64, Poznań
Kadlec J, Jäger, O., Kočí A, Minaříková D (1992) The age of sedimentary fill in the Aragonitová Cave. Český kras XVII:16–23, Beroun
Kadlec J, Jäger O, Kočí A, Minaříková D (1995) The age of sedimentary fill in the Aragonitová Cave. Studia Carsologica 6:20–31, Brno
Kadlec J, Pruner P, Hercman H, Chadima M, Schnabl P, Šlechta S (2004b) Magnetostratigraphy of sediments preserved in caves of the Low Tatra Mts. Výskum, využívanie a ochrana jaskýň (4):15–19, Správa Slovenských jaskýň, Liptovský Mikuláš
Kadlec J, Pruner P, Venhodová D, Hercman H, Nowicki T (2002b) Age and genesis of sediments in Šošůvská Cave (Moravian Karst, Czech Republic). Acta Musei Moraviae Sci Geol LXXXVIII:229–243, Brno
Kadlec J, Pruner P, Venhodová D, Hercman H, Nowicki T (2004c) Age and genesis of sediments in the Ochozská Cave, Moravian Karst. 3. Národní speleologický kongres, Rozšířené abstrakty. Česká speleologická společnost, Praha, pp 33–36
Kadlec J, Schnabl P, Pruner P, Lisá L, Žák K, Hlaváč J (2003) Paleomagnetic dating of cave sediments in the Czech Karst. Český Kras XXIX:21–25, Beroun
Kadlec J, Táborský Z (2002) Tertiary cave sediments in the Malá dohoda Quarry near Holštejn in the Moravan Karst. Geologické výzkumy na Moravě a ve Slezsku v r. 2001(IX):30–33, Brno
Kirschvink JL. (1980) The least-squares line and plane and the analysis of palaeomagnetic data. Geophys J Roy Astronomical Soc 62:699–718
Kopper JS (1975) Preliminary note on the paleomagnetic reversal record obtained from two Mallorcan caves. Endins 2:7–8, Palma de Mallorca
Kopper JS, Creer KM (1973) Cova dets Alexandres, Majorca: paleomagnetic dating and archaeological interpretation of its sediments. Caves Karst Sci 15(2):13–20
Kopper JS, Creer KM (1976) Palaeomagnetic dating stratigraphic interpretation in archaeology. MASCA Newsletter, University of Pennsylvania, pp 12, 1–3
Kostrzewski A, Noel M, Thistlewood L, Zwoliński Z (1991) Cave deposits of the Chocholowska Valley. Geografia (UAM) 50:289–309, Poznań
Kukla J, Ložek V. (1958) K problematice výzkumu jeskynních výplní (To the problem of research of cave fills). Československý Kras 11:19–59, Praha
Kyle M (1990) Paleomagnetism of speleothems in Gardner Cave, Washington. National Speleological Soc Bull 52(2):87–94
Latham AG (1981) The Palaeomagnetism, Rock Magnetism and U-Th Dating of Speleothem Deposits. PhD Thesis, McMaster University, Hamilton
Latham AG (1989) Magnetization of speleothems: detrital or chemical? Proceedings of the 10th international congress of speleology, Budapest, pp 1, 82–84
Latham AG, Ford DC, Schwarcz HP, Birchall T (1989) Secular variations from Mexican stalagmites: their potential and problems. Phys Earth Planetary Inter 56: 34–48
Latham A, Schwarcz HP, Ford DC (1979) Palaeomagnetism of stalagmite deposits. Nature 280(5721):383–385
Latham A, Schwarcz HP, Ford DC (1986) The paleomagnetism and U-Th dating of Mexican stalagmite, DAS2. Earth Planetary Sci Lett 79:195–204

Latham A, Schwarcz HP, Ford DC (1987) Secular variations of the Earths ́ magnetic field from 18.5 to 15.0 ka B.P., as recorded in a Vancouver Island stalagmite. Can J Earth Sci 24:1235–1241

Latham A, Schwarcz HP, Ford DC, Pearce GW (1982) The paleomagnetism and U-Th dating of three Canadian speleothems: evidence for the westward drift, 5.4–2.1 ka B.P. Can J Earth Sci 19:1985–1995

Lauritzen S-E, Haugen JE, Løvlie R, Gilje-Nielsen H (1994) Geochronological Potential of Isoleucine Epimerization in Calcite Speleothems. Quaternary Res 41(1):52–58

Lauritzen S-E, Løvlie R, Moe D, Østbye E (1990) Paleoclimate deduced from a multidiscipluinary study of a half-million-year-old stalagmite from Rana, northern Norway. Quaternary Res 34(3):306–316

Lean C, Latham AG, Shaw J (1995) Palaeosecular variation from a Vancouver Island stalagmite and comparison with contemporary North American records. J Geomagnet Geoelectricity 47(1):71–88

Liu YY, Morinaga H, Horie I, Murayama H, Yakashawa K (1988) Preliminary report on palaeomagnetism of a stalagmite in Ping Le, South China. J Geomagnet Geeolectricity 15:21–22

Løvlie R, Ellingsen KL, Lauritzen SE (1995) Paleomagnetic cave stratigraphy of sediments from Hellemofjord, northern Norway. Geophys J Int 120(2):499–515

Løvlie R, Sandnes A (1987) Palaeomagnetic excursions recorded in mid-Weichselian cave sediments from Skjonghelleren, Valderøy, W. Norway. Phys Earth Planetary Inter 45(4): 337–348

Ložek V. (1973) Příroda ve čtvrtohorách (Nature during Quaternary). Academia, Praha, 372 pp

Martin K (1991) Paleomagnetism of speleothems in Gardner Cave, Washington. National Speleological Society, Bulletin 52(2), (1990):87–94

McFadden PL, Brock A, Partridge TC (1979) Palaeomagnetism and the age of the Makapansgat hominid site. Earth Planetary Sci Lett 12:332–338

McNeil DF, Ginsburg RN, Chang S, Kirschvink JL (1988) Magnetostratigraphic dating of shallow-water carbonates from San Salvador, Bahamas. Geology 16:8–12

McNeil DF, Grammer GM, Williams SC (1998) A 5 My chronology of carbonate platform margin aggradation, southwestern Little Bahama Bank, Bahamas. J Sedimentary Res 68(4):603–614

Morinaga M, Inokuchi H, Yaskawa K (1989) Palaeomagnetism of stalagmites (speleothems) in SW Japan. Geophys J 96:519–528

Müller BU (1995) De Höhlensedimente des Ofenlochs. Stalactite 45(1):25–35

Musgrave RJ, Webb JA (2004) Palaeomagnetic analysis of sediments on the Buchan Caves, Southeastern Australia, provides a pre-Late Pleistocene data for landscape and climate evolution. In: Sasowsky ID, Mylroie J (eds.) Studies of Cave Sediments. Physical and Chemical Records of Paleoclimate, Kluwer Academic/Plenum Publ., New York, NY, pp 47–69

Noel M (1982) Caves, mud and magnetism. Caves Caving 15:28–30

Noel M (1983) The magnetic remanence and anisotropy of susceptibility of cave sediments from Agen Allwedd, South Wales. Geophys J Roy Astronomical Soc 72: 557–570

Noel M (1985) Caves, mud and magnetism: an update. Caves Caving 27:14–15

Noel M (1986a) The paleomagnetism and magnetic fabric of cave sediments from Pwll y Gwynt, South Wales. Phys Earth Planetary Inter 44:62–71, Amsterdam

Noel M (1986b) The paleomagnetism and magnetic fabric of sediments from Peak Cavern, Derbyshire. Geophys J Roy Astronomical Soc 84:445–454

Noel M (1990) Palaeomagnetic and Archaeomagnetic Studies in the Caves of Guanxi. Cave Sci 17(2):73–76

Noel M, Bull PA (1982) The palaeomagnetism of sediments from Clearwater Cave, Mulu, Sarawak. Cave Sci 9(2): 134–141

Noel M, Shaw RP, Ford TD (1984) A palaeomagnetic reversal in Early Quaternary sediments in Mason Hill, Matlock, Derbyshire. Mercian Geolol 9:235–242

Noel M, St. Pierre S (1984) The paleomagnetism and magnetic fabric of cave sediments from Gronligrotta and Jordbrugrotta, Norway. Geophys J Roy Astronomical Soc 78:231–239

Noel M, Thistlewood L (1989) Developments in cave sediment palaeomagnetism. In: Lowes FJ et al (eds.) Geomagnetism and palaeomagnetism. Kluwer, Dordrecht, pp 91–106

Openshaw S, Latham A, Shaw J, Xuewen Z (1993) Preliminary results on recent palaeomagnetic secular variation recorded in speleothems fom Xingweng, Sichuan, China. Cave Science, 22, 3:93–99

Osborne RAL (1984) Lateral facies changes, unconformities and stratigraphic reversals: their significance for cave sediments stratigraphy. Cave Sci 11(3):175–184

Panoš V (1963) On the origin and age of denudation surfaces in the Moravian Karst. Československý kras, 14(1962–1963):29–41, Praha

Panoš V (1964) Der Urkarst in Ostflügel der Böhmishen Masse. Zeitschrift für Geomorphologie NF 8(2):105–162

Panuschka BC, Mylroie JE, Carew JL (1997) Stratigraphic tests of the utilization of paleomagnetic secular variation for correlation of paleosols, San Salvador Island, Bahamas. Proceedings of the 8th symposium on geology of the bahamas and other carbonate regions, bahamian field station. San Salvador, Bahamas, pp 148–157

Papamarinopoulos S (1977) The first known European? Bull Univ Edinb 13(10):1–3

Papamarinopoulos S (1978) Limnomagnetic studies on Greek sediments. PhD Thesis, University of Edinburgh, Edinburgh

Papamarinopoulos S, Creer KM (1983) The palaeomagnetism of cave sediments. In: Creer KM, Tucholka P, Barton CE (eds) Geomagnetism of baked clays and recent sediments. Elsevier, Amsterdam, 243–248

Papamarinopoulos S, Readman PW, Maniatis Y, Simopoulos A (1987) Palaeomagnetic and mineral magnetic studies of sediments from Petralona Cave, Greece. Archaeometry 29: 50–59

Papamarinopoulos S, Readman PW, Maniatis Y, Simopoulos A (1991) Paleomagnetic and mineral magnetic studies of sediment from Balls ́ Cavern, Scholarie, USA. Earth Planetary Sci Lett 102:198–212

Perkins AM, Maher BA (1993) Rock magnetic studies of british speleothems. J Geomagnet Geoelectr 45:143–153

Pons J, Moyá S, Kopper JS (1979) La fauna e mamiferos de la Cova de Canet (Esporles) y su chronologia. Endins 5–6:55–58, Palma de Mallorca

Poulianos AN (1980) A new fossilised inion-parietal bone in Petralona Cave. Anthropos 7:34–39, Athens

Pruner P, Bosák P (2001) Palaeomagnetic and magnetostratigraphic research of cave sediments: theoretical approach, and examples from Slovenia and Slovakia. Proceedings, 13th International Speleological Congress, 4th Speleological Congress of Latin America and the Caribbean, 26th Brazilian Congress of Speleology, Brasilia, 1:94–97, 15–22 July 2001. Brasilia DF

Pruner P, Bosák P, Kadlec J, Venhodová D, Bella P (2000) Paleomagnetic research of sedimentary fill of selected caves in Slovakia. Výskum, využívanie a ochrana jaskýň. 2. vedecká konferencia s medzinárodnou účastóu, 16.–19. novembra 1999, Demänovská Dolina. Zborník referátov. Správa Slovenských jaskýň, Liptovský Mikuláš, pp 13–25

Pruner P, Bosák P, Kadlec J, Man O, Tulis J, Novotný L (2002) Magnetostratigraphy of the sedimentary fill of the IVth cave level in Stratenská Cave. Výskum, využívanie a ochrana jaskýň. 3. vedecká konferencia s medzinárodnou účastóu 2001, Stará Lesná. Zborník referátov. Správa Slovenských jaskýň, Liptovský Mikuláš, pp 3–15

Pruner P, Bosák P, Zupan Hajna N, Mihevc A (2009) Cave sediments in Slovenia: results of 10 years of palaeomagnetic research. Slovenský Kras 47(2):173–186. Liptovský Mikuláš

Pruner P, Zupan Hajna N, Mihevc A, Bosák P, Venhodová D, Schnabl P (2010) Paleomagnetic and rockmagnetic studies of cave deposits from Račiška pečina and Pečina v Borštu caves (Classical Karst, Slovenia). Stud Geophys Geod 54: 27–48

Rowe P, Austin T, Atkinson T (1988) The Quaternary evolution of the British South Pennines from uranium series and palaeomagnetic data. Annales de la Société Géologique de Belgique 111:97–106

Sasowsky ID, Clotts RA, Crowell B, Walko SM, LaRock EJ, Harbert W (2004) Paleomagnetic analysis of a long-term sediment trap, Kooken Cave, Huntingdon County, Pennsylvania, USA. In: Sasowsky ID, Mylroie J (eds) Studies of cave sediments. Physical and chemical records of paleoclimate. Kluwer Academic/Plenum Publ., New York, NY, pp 71–81

Sasowsky ID, Šebela S, Harbert W (2003) Concurrent tectonism and aquifer evolution >100,000 years recorded in cave sediments, Dinaric karst, Slovenia. Environ Geol 44(1): 8–13

Sasowsky ID, White WB, Schmidt VA (1995) Determination of stream incision rate in the Appalachian plateaus by using cave-sediment magnetostratigraphy. Geology 23:415–418

Schmidt VA (1982) Magnetostratigraphy of sediments in Mammoth Cave, Kentucky. Science 217:827–829

Schmidt VA, Jennings J, Bao H (1984) Dating of cave sediments from Wee Jasper, New South Wales, by magnetostratigraphy. Aust J Earth Sci 31:361–370

Šebela S, Sasowsky I (1999) Age and magnetism of cave sediments from Postojnska jama cave system and Planinska jama Cave, Slovenia. Acta Carsologica 28/2(18):293–305

Šebela S, Sasowsky I (2000) Paleomagnetic dating of sediments in caves opened during highway construction near Kozina, Slovenia. Acta Carsologica 29/2, 23:303–312

Šebela S, Slabe T, Kogovšek J, Hong L, Pruner P (2001) Baiyun Cave in Naigu Shilin, Yunnan Karst, China. Acta Geol Sinica (Engl. Edition) 75(3):279–287

Shaw T (1992) The history of cave science, the exploration and study of limestone caves, to 1900. Sydney Speleological Soc, 338 pp, Broadway

Stober JC (1978) Palaeomagnetic secular variation studies on Holocene lake sediments. PhD. Thesis. University of Edinburgh

Stock GM, Granger DE, Sasowski ID, Anderson RS, Finkel RC (2005) Comparison of U-Th, paleomagnetic, and cosmogenic burial methods for dating caves: implications for landscape evolution studies. Earth Planetary Sci Lett 236(1–2):388–403

Springer GS, Kite JS, Schmidt VA (1997) Cave sedimentation, genesis, and erosional history in the Cheat river canyon, West Virginia. Geol Soc Am Bull 109(5):524–532

Šroubek P, Diehl JF (1995) The paleoenvironmental implications of the study of rock magnetism in cave sediments of the Moravian Karst. Knihovna České speleologické společnosti 25:29–30, Praha

Šroubek P, Diehl JF, Kadlec J, Valoch K (1996) Preliminary study on the mineral magnetic properties of sediments from the Kůlna Cave. Studia Geophys Geodaetica 3:301–312

Šroubek P, Diehl JF, Kadlec J, Valoch K (2001) A Late Pleistocene paleoclimate record based on mineral magnetic properties of the entrance facies sediments of Kulna Cave, Czech Republic. Geophys J Int 147:247–262

Thistlewood L, Noel M (1991) A paleomagnetic study of sediments from Maypole Inlet, Peak Cavern. Cave Sci 17, 1:55–58

Turner GM, Lyons RG (1986) A paleomagnetic secular variation record from c. 120,000 yr-old New Zealand cave sediments. Geophys J Roy Astronomical Soc 87:1181–1192

Valen V, Lauritzen SE, Løvlie R (1997) Sedimentation in a hogh-latitude karst cave: Sirijordgrotta, Nordlan, Norway. Norsk Geologisk Tidsskrift 77(4):233–250

Verosub KL (1977) Depositional and post-depositional process in the magnetization of sediments. Rev Geophys Space Phys 15:129–143

Vrabec M, Fodor L (2006) Late Cenozoic tectonics of Slovenia: structural styles at the Northeastern corner of the Adriatic microplate. In: Pinter N, Grenerczy G, Weber J, Stein S, Medak D (eds) The Adria microplate: GPS geodesy, tectonics and hazards. NATO Science Series, IV, Earth Environ Sci 61, Springer, Dordrecht, pp 151–168

White WB (1988) Geomorphology and Hydrology of Karst Terrains. Oxford University Press, Oxford, pp 315–317

Williams PW, Lyons RG, Wang X, Fang L, Bao H (1986) Interpretation of the paleomagnetism of cave sediments from a karst tower at Guilin. Carsologica Sinica 5(2): 113–126

Žák K, Pruner P, Bosák P, Svobodová M, Šlechta S (2007) New type of paleokarst sediments in the Bohemian Karst (Czech Republic), and their regional tectonic and geomorphological relationships. Bull Geosci 82(3):275–290. Praha

Zupan Hajna N, Bosák P, Pruner (2007a) Raziskave jamskih sedimentov iz zapolnjene jame pri Divači. In: Knez M, Slabe T (eds) Kraški pojavi, razkriti med gradnjo slovenskih avtocest. Založba ZRC. Ljubljana

Zupan Hajna N, Mihevc A, Pruner P, Bosák P (2008a) Cave Sediments from Postojnska–Planinska Cave System (Slovenia) Evidence of Multi-Phase Evolution in Epiphreatic Zone. Acta Carsologica 37(1):63–86. Postojna

Zupan Hajna N, Mihevc A, Pruner P, Bosák P (2008b) Palaeomagnetism and Magnetostratigraphy of Karst Sediments in Slovenia. Carsologica 8:1–266. Založba ZRC SAZU, Postojna–Ljubljana

Zupan Hajna N, Mihevc A, Pruner P, Bosák P (2010) Palaeomagnetic research on karst sediments in Slovenia. Int J Speleol 39(2):47–60. Bologna

A Quantitative Model of Magnetic Enhancement in Loessic Soils

25

María Julia Orgeira, Ramon Egli,
and Rosa Hilda Compagnucci

Abstract

We present a quantitative model for the climatic dependence of magnetic enhancement in loessic soils. The model is based on the widely accepted hypothesis that ultrafine magnetite precipitates during alternating wetting and drying cycles in the soil micropores. The rate at which this occurs depends on the frequency of drying/wetting cycles, and on the average moisture of the soil. Both parameters are estimated using a statistical model for the soil water balance that depends on frequency and intensity of rainfall events and on water loss by evapotranspiration. Monthly climatic tables are used to calculate the average soil moisture and the rate of pedogenic magnetite production, which is proportional to a new parameter called magnetite enhancement proxy (MEP). Our model is tested by comparing MEP calculated for known present-day climates with the magnetic enhancement of modern soils. The magnetic enhancement factor, defined as the ratio between a given magnetic enhancement parameter and MEP, is expected to be a site-independent constant. We show that magnetic enhancement differences between soils from the Chinese Loess Plateau and from Midwestern U.S. are explained by our model, which yields similar magnetic enhancement factors for the two regions. Our model is also successful in predicting the mean annual rainfall threshold above which magnetic enhancement declines in a given type of climate.

25.1 Introduction

Magnetic susceptibility records of loess/paleosol sequences have been used to reconstruct climatic changes during the Neogene (e.g. Liu et al. 1992, Heller et al. 1993, Banerjee et al. 1993, Maher et al. 1994, Liu et al. 1995, Maher 1998, Fang et al. 1999, Maher and Thompson 1999, Maher and Hu 2006). Climatic reconstructions are based on the observation that the magnetic susceptibility of the uppermost soil horizons (A and top of B) is higher than that of the underlying layers, as first discovered by Le Borgne (1955), provided that the soil parent material is not excessively magnetic. Magnetic enhancement is caused by superparamagnetic (SP) and single domain (SD) magnetite (Fe_3O_4) or maghemite (γ-Fe_2O_3) particles (Maher 1986, Evans and Heller 1994), commonly referred to as *pedogenic magnetite*,

R. Egli (✉)
Department of Earth and Environmental Sciences,
Ludwig-Maximilians University, 80333 Munich, Germany
e-mail: egli@geophysik.uni-muenchen.de

E. Petrovský et al. (eds.), *The Earth's Magnetic Interior*, IAGA Special Sopron Book Series 1,
DOI 10.1007/978-94-007-0323-0_25, © Springer Science+Business Media B.V. 2011

which are preserved in the loess/paleosol record (see Heller and Evans 1995, Maher and Thompson 1999, Evans and Heller 2003 for reviews). Even if pedogenic magnetite represents only a negligible fraction of pedogenic Fe minerals by mass (Cornell and Schwertmann 2003), its magnetization is much stronger and controls the bulk magnetic properties of the enhanced horizons.

The mechanism of pedogenic Fe minerals formation and the role of climate in it are not yet fully understood. Several hypotheses have been formulated to explain magnetic enhancement. Natural fires cause the partial reduction of weakly magnetic iron oxyhydroxides to magnetite or maghemite in the presence of organic matter (Le Borgne 1960, Kletetschka and Banerjee 1995); however, their effective role as systematic enhancement mechanism is not considered fundamental in most cases (Maher 1986). A widely accepted model for the precipitation of ultrafine magnetite in soils requires the oxidation of $Fe^{2+}(aq)$ at near-neutral pH, which has been shown experimentally to produce a magnetic material that is very similar in chemical composition, morphology and grain size (Taylor et al. 1987, Maher 1988). In a first step, Fe^{2+} ions are released by weathering of Fe-bearing silicates during repeated wetting and drying cycles. Fe^{2+} ions oxidize rapidly to Fe^{3+} in presence of oxygen, and Fe^{3+} hydrolysis induces the precipitation of poorly crystalline oxyhydroxides such as ferrihydrite ($Fe_5HO_8{\cdot}4H_2O$) (Cornell and Schwertmann 2003). Ferrihydrite is easily reduced during episodic anaerobicity caused by organic matter respiration (Fischer 1988), leading to the precipitation of magnetite and other Fe(II) minerals (Tamaura et al. 1983, Tronc et al. 1992). This so-called "fermentation mechanism" has been proposed by Le Borgne (1955), Mullins (1977), and Dearing et al. (1996), and was later refined by Maher (1998). A similar redox mechanism involving direct precipitation of magnetite by metal dissimilatory reducing bacteria under anaerobic conditions (Lovley et al. 1987) has been considered as a possible alternative (Maher 1998, Dearing et al. 2001, Maher et al. 2003, Banerjee 2006). A completely different enhancement path recently proposed by Torrent et al. (2006, 2007) postulates the formation of an intermediate ferrimagnetic phase when ferrihydrite is converted to hematite (α-Fe_2O_3), in which case the reducing environment necessary for magnetite precipitation is not essential.

Heller and Liu (1984) found a significant correlation between magnetic susceptibility of a loess/paleosol profile from the Chinese Loess Plateau (CLP) and oceanic $\delta^{18}O$, providing the first continuous record of Pleistocene glacial/interglacial stages in a continental section. They proposed topsoil decalcification and down-profile carbonate reprecipitation as a means of iron oxide concentration. An alterative explanation of susceptibility variations by Kukla et al. (1988) postulated a constant atmospheric input of iron oxides which has been diluted by weakly magnetic dust at times of rapid loess accumulation. Maher and Thompson (1992) confirmed the correlation with oceanic climate records, which they attributed to the degree of soil formation as seen by the accumulation of secondary ferrimagnetic minerals (magnetic *enhancement*). The magnetic record is thus controlled by temperature and humidity of the regional climate, rather than variations in the deposition rate.

Magnetic susceptibility variations in loess/paleosol sections can be interpreted in terms of detrital and pedogenic mineral fluxes, expressed as accumulated mass per unit surface and time. If each group of minerals is characterized by specific, invariant magnetic properties, the mass fluxes can be converted to magnetic susceptibility fluxes, and vice-versa (Beer et al. 1993). Heller et al. (1993) used ^{10}Be – which is supplied to the sediment through dust accumulation and rain – to calculate susceptibility fluxes, discovering large differences in pedogenic magnetite production rates between loess and paleosol layers. Furthermore, they found a clear match between present-day mean annual rainfall (MAR, see Table 25.1 for a list of symbols and abbreviations) and paleosol susceptibility at different sites on the CLP. Paleo-rainfall reconstructions based on susceptibility *fluxes* and susceptibility *enhancements* differ mainly by the underlying assumptions about the timing of pedogenic processes.

The systematic study of large numbers of modern soils in warm temperate climates (Maher and Thompson 1995, Liu et al. 1995, Han et al. 1996, Porter et al. 2001, Maher et al. 2003, Geiss and Zanner 2007) confirmed the existence of a significant correlation between MAR and susceptibility enhancement. This evidence led to the proposal of a so-called *soil climofunction* (Jenny 1941) linking climate, expressed by MAR, with production of secondary ferrimagnetic minerals, quantified by pedogenic susceptibility (fluxes) (Heller et al. 1993, Maher et al. 1994). A soil

Table 25.1 Definition of important parameters used in the text, sorted by subject: (1) water balance, (2) iron content, and (3) magnetic measurements

Symbol	Name	Definition	Relations
R	Rainfall	Volume of water per unit surface and time from rainfall	$\langle R\rangle = \lambda\zeta$
MAR	Mean annual rainfall	Total R over 1 year	
λ	Rain event rate	Average number of rain events per unit time	
ζ	Rain event mean depth	Mean volume of water per unit surface during a rain event	
I	Infiltration rate	Volume of water per unit surface and time that penetrates the soil	$I \le R$
E	Evapotranspiration	Volume of water per unit surface and time that leaves the soil by evapotranspiration	$E = \mathrm{ET}$
$\hat{E}$	Potential evapotranspiration	Maximum possible value of E reached when $s \ge s_E$	$\hat{E} = \mathrm{PET}$
L	Leakage	Volume of water per unit surface and time that leaves the soil	
ϕ	Porosity	Total pore volume per unit soil volume	
H	Active soil thickness	Thickness of the top layer modeled as a homogeneous soil	
h	Reduced soil thickness	Active soil thickness without pores	$h = \phi H$
s	Moisture	Fraction of pore volume filled with water	
s_E	–	Moisture threshold above which $E = \hat{E}$	
s_K	–	Moisture threshold above which $K = \hat{K}$	
K	Deep infiltration	Volume of water per unit surface and time that leaves the active soil layer by drainage	
$\hat{K}$	Saturated hydraulic conductivity	Maximum possible value of K	
W	Moisture ratio	Ratio between water input R and maximum water loss by evapotranspiration $\hat{E}$	
Fe_t	Total Fe	Total mass concentration of Fe	
Fe_d	CBD-extractable Fe	Total mass concentration of Fe extractable by the citrate-bicarbonate-dithionite method	
χ	Susceptibility	Mass normalized low-field susceptibility in m^3/kg	
χ_{fd}	Frequency dependency of χ	Difference between χ measured at frequencies that differ by one order of magnitude (0.47 and 4.7 kHz)	
$\Delta\chi$	Susceptibility enhancement	Difference between the maximum enhanced horizon (χ_B) and the C horizon (χ_C)	
χ_p	–	Susceptibility of pedogenic minerals, normalized by their mass	
χ_{ARM}	Susceptibility of ARM	Mass-normalized ARM, divided by the DC field H_{DC} used to acquire the ARM, in m^3/kg	

climofunction is the key for reconstructing paleorainfall from magnetic measurements of loess/paleosol sections, under the assumption that the function is invariant over time (e.g. Heller et al. 1993, Maher et al. 1994, Florindo et al. 1999a, b, Evans et al. 2002).

The soil climofunction depends on additional parameters related to the climate itself (e.g. temperature, seasonality), and to soil forming factors such as chemistry, Fe supply, drainage, and time (Jenny 1941, Maher 1998, Hanesch and Scholger 2005). The evolution of soil-forming parameters with time is described by a so-called *chronofunction* (Jenny 1941, Bockheim 1980). The final onset of a stationary regime is known as the *mature stage* in soil evolution models (Jenny 1941). The time required for reaching this stage depends on the weathering intensity. Contrasting views exist on the rapidity of soil development, with estimates spanning from hundreds of years for particle size, pH, and organic matter content of Iowa loess deposits (Hallberg et al. 1978), to >0.5 Myr for the magnetic enhancement of Californian soils formed on stratified aeolian silt and sand terraces (Singer et al. 1992). Contradictory chronofunction reconstructions – expressed as peak magnetic susceptibility vs. paleosol development duration – have been obtained for the CLP. Vidic et al. (2004) report a positive correlation based on three different age models (r^2 = 0.44–0.81), while no correlation is found when soil development duration estimates are calculated from ages measured with optically stimulated luminescence (Maher and Hu 2006). The caveat in reconstructing reliable chronofunctions is the age model used for calculating soil development durations: models based on correlation with reference curves (magnetic polarity or $\delta^{18}O$) (Grimley et al. 2003, Vidic et al. 2004) tend to overlook sedimentation rate discontinuities, overestimating the development duration of the most enhanced soils.

Fe supply is generally not a limiting factor for magnetic enhancement, since the typical Fe content of pedogenic ferrimagnetic minerals (<0.1 wt%) is much lower than the total iron concentration (Fe_t) of most parent materials (2–5 wt%) (Maher 1998). Secondary ferrimagnetic minerals also make an insignificant contribution to the citrate-bicarbonate-dithionite (CBD) extractable iron (Fe_d), which is a proxy for the Fe contained in fine-grained secondary minerals. Fine et al. (1995) found a linear correlation between magnetic susceptibility and Fe_d of loess/paleosol samples from the CLP, with magnetic enhancement starting at Fe_d = 7 g/kg and a slope of 2.6×10^{-4} m^3/kg(Fe_d), which can be interpreted as the susceptibility of Fe_d-bearing minerals normalized by their Fe content. For comparison, much higher values, in the order of 4.3×10^{-3} m^3/kg(Fe_d), would result if Fe_d is assumed to originate *only* from pedogenic magnetite (see the Appendix for a proof). This means that magnetic enhancement is caused by <6% Fe in secondary iron minerals.

The concentration of pedogenic minerals is controlled by several factors, such as production rate, dust inputs during soil formation, mineral dissolution, and vertical mass transport between horizons (Brimhall and Dietrich 1987, Begét et al. 1990, Anderson and Hallet 1996, Maher 1998, Porter et al. 2001). The interplay of all factors becomes important when pedogenic mineral production is slow compared to the other processes. In general, magnetic enhancement depends on soil type, vegetation, and climate (Maher 1986, Dearing et al. 1996, Hanesch and Scholger 2005). The formation of pedogenic magnetite is favored in well drained, not very acidic soils (pH $\approx$ 5.5–8.0), and is correlated with organic matter content and cation exchange capability (Maher 1998). Weathering of primary Fe minerals and precipitation/dissolution of secondary iron oxides is sensitive to pH and Eh. For example, Fe^{2+} is most rapidly removed from Fe-bearing silicates under reducing, acidic conditions (White et al. 1994). The alternation of strongly reducing and oxidizing conditions induced by water table oscillations leads to magnetite dissolution, and – ultimately – to a depletion of secondary ferrimagnetic oxides (Maher 1998, Dearing et al. 2001, Hanesch and Scholger 2005, Fischer et al. 2008). Magnetite dissolution is also promoted by a number of substances that might be present in the upper soil horizons, such as organic ligands (Appelo and Postma 2005), and pore water silica (Florindo et al. 2003).

Site-specific climofunction variations are minimal on the CLP, where loess has exceptionally uniform chemical and physical properties that promote rapid accumulation of pedogenic minerals. Maher et al. (1994) report following empirical relation:

$$\mathrm{MAR} = 222 + 199\,\log_{10}(\chi_\mathrm{B} - \chi_\mathrm{C}), \qquad (25.1)$$

between MAR, expressed in mm/yr, and the susceptibility enhancement $\Delta\chi = \chi_B - \chi_C$ where χ_B and χ_C are the susceptibilities of B and C horizons, respectively, in units of 10^{-8} m^3/kg. The climofunction in Eq. (25.1) has been successfully extended to loessic soils from the Russian steppe (Maher et al. 2003) and other areas of the northern hemisphere temperate zone (Maher and Thompson 1995). These results support the use of magnetic parameters as rainfall proxies for paleosols where chemical conditions and drainage favored the formation and preservation of ferrimagnetic iron oxides.

On the other hand, the implementation of a climofunction that is *universally valid* for a certain category of soils and soil forming conditions is faced with some unresolved problems. Limiting the discussion to loess as parent material, these are: (1) a large scatter of $\Delta\chi$ values (typically a factor of 2) for modern soils from sites with similar MAR; (2) systematic, yet unexplained geographic differences within the CLP (Guo et al. 2001, Bloemendal and Liu 2005) and between CLP and North America (Geiss and Zanner 2007, Geiss et al. 2008); (3) no correlation between $\Delta\chi$ and MAR above a cursive MAR threshold; and (4) apparent lack of magnetic enhancement in soils where the accumulation of pedogenic iron oxides is not suppressed by known causes such as gleyzation (Orgeira et al. 2008). Furthermore, Jahn et al. (2001) report inconsistencies between chemical and magnetic indicators of pedogenesis, which may be explained by selective sensitivity to different pedogenic processes.

Magnetic enhancement depends ultimately on the long-term balance between dissolution of primary and pedogenic Fe minerals on the one hand, and precipitation of magnetic iron oxides on the other. This balance is controlled by several environmental factors – the most important being soil moisture and the chain of chemical reactions related to it (Maher 1998). The importance of soil water balance is demonstrated in laboratory simulations of flooding and drying cycles, which produce a complex sequence of magnetic mineral accumulation and dissolution (Crockford and Willett 1995). MAR is only one of the parameters controlling soil moisture: it represents the water supply rate, which is counterbalanced by *drainage* – in which case soil porosity and thickness play an important role – and by *evapotranspiration* (ET). ET depends on the type of vegetation cover and on climatic parameters such as insolation, wind speed, and temperature. Aspects of Fe oxides accumulation and dissolution driven by the soil water balance, such as seasonal oscillations (Tite and Linington 1975) and excessive drainage (Schwertmann et al. 1982), have been recognized since a long time. One of the most systematic investigations of water balance effects on pedogenesis has been carried out on a sequence of soils located on the flanks of a volcano (Chadwick et al. 2003). While the parent volcanic rocks are identical at all sampling sites, rainfall and *potential evapotranspiration* (PET) vary strongly with altitude. Several geochemical parameters – such as pH, weathering rate, and effective cation exchange capacity – were shown to depend critically on the difference between MAR and PET, which is a measure of the net water balance. Dramatic changes occur at the point where MAR starts to exceed PET: therefore, MAR–PET = 0 was considered a sort of *pedogenic threshold* (Chadwick and Chorover 2001).

The idea of a pedogenic threshold driven by soil water balance was used by Orgeira and Compagnucci (2006) to explain magnetic enhancement differences in loessic soils from Russia, China and Argentina. They observed that soils from sites characterized by negative values of MAR–PET, such as the CLP, are magnetically enhanced, contrary to locations where this difference is ~0 or positive. This trend is also valid for soils with no evidence of waterlogging or gleyzation, as in many Argentinean loess/paleosol profiles.

It is evident from the previous discussion that – even in case of a climofunction that does not depend on soil-specific parameters – MAR is not sufficient to fully describe the climatic control over pedogenic Fe oxides production and preservation. Furthermore, the empirical logarithmic expression for the susceptibility enhancement of loessic soils (Eq. (25.1)) is not of a form that can be obtained from the solution of mass balance and chemical reaction kinetic equations. In this article, we provide a quantitative frame for the magnetic enhancement model proposed by Maher and co-workers (Maher et al. 1994, Maher 1998), which is combined with a stochastic treatment of soil moisture derived from Rodriguez-Iturbe et al. (1999). The result is a physically-derived expression that fits published magnetic enhancement data on modern loessic soils worldwide. Our calculations are able to explain systematic differences between modern soils from the CLP and from North America in terms of climatic factors that are not completely expressed by MAR. Furthermore, pedogenic thresholds limiting the magnetic enhancement of soils formed in cold climates (Alaska and Siberia), or in climates with strong monsoons (Southern China) are correctly predicted. These results demonstrate that soil water balance is one of the most important parameters affecting the formation and dissolution of magnetic minerals – acting not only as a threshold, as originally postulated, but effectively as a regulating factor in a wide range of climates.

25.2 Soil Magnetic Properties in Different Regions of the World

In this section we provide a review of published data on soil magnetic enhancement according to geographical location and climate. Since the focus of this article is on the climatic control of pedogenesis, the variability of other soil forming factors – such as parent material composition and soil age – is minimized by limiting our dataset to modern loessic soils or Pleistocene/Holocene paleosols. The relatively homogeneous composition and texture of loess, its weatherability and pH buffering capability, as well as good drainage properties, provide adequate conditions for pedogenic magnetite accumulation and preservation (Evans and Heller 2003, Maher 1998). Soil development on loess is believed to occur rapidly (Hallberg et al. 1978, Maher and Hu 2006), in which case soil age effects do not need to be considered. For comparison reasons, we also discuss few examples of non-loessic soils. These include heamtitic "terre rosse" from the Mediterranean region (Torrent and Cabedo 1986), and a set of intensively studied Hawaiian volcanic soils (Chadwick et al. 2003).

Soil provenance is divided into five major regions, according to modern climate and magnetic enhancement characteristics. Susceptibility enhancement and annually averaged climatic data for representative soils from these regions are summarized in Table 25.2 and discussed in the following.

Table 25.2 Summary of magnetic and climatic properties of representative loessic soils. The moisture ratio *W* is calculated as the ratio between MAR and yearly PET

Site	Lat.	Long.	$\Delta\chi$ [10^{-8} m^3/kg]	χ_{fd} [10^{-8} m^3/kg]	MAR [mm/yr]	*W*
Asia						
Luochuan (Bloemendal and Liu 2005)	35.4	108.3	+136	–	522	0.939
Duanjiapo (Bloemendal and Liu 2005)	34.2	109.2	+190	–	578	0.831
SE China (Han et al. 1996)	25.0	112.0	~ 25	–	1550	1.957
Karamadian (Forster and Heller 1994, Forster et al. 1994)	38.4	69.4	+35	3.9	326	0.397
North America						
Arbor Cemetery (Geiss and Zanner 2007)	40.55	−100.4	+19	–	494	0.856
Mt. Calvary Cemetery (Geiss and Zanner 2007)	40.87	−95.4	+24	–	890	1.457
South America						
Zárate (Orgeira et al. 2008)	−33.68	−59.68	+40[a]	4.7[a]	1088	1.453
Verónica (Orgeira et al. 2008)	−35.35	−57.28	−65...−30	0.6...4	961	1.297
Alaska and Siberia						
Halfway House (Liu et al. 1999, Lagroix and Banerjee 2002)	64.6	−148.9	−70[b]	1.2[b]	264	0.746
Lozhok (Kravchinsky et al. 2008)	54.45	83.32	+55[b]	5.4[b]	429	0.869
Novokuznetsk (Kravchinsky et al. 2008)	53.72	87.17	~ 0[b]	1.8[b]	561	1.114
Europe and Russia						
Sedlek near Prague (Forster et al. 1996)	50.28	14.39	+29[c]	3.5[c]	455	1.450
Sedlek near Mikulov (Forster et al. 1996)	48.92	16.71	+47[c]	9.0[c]	485	1.710
Coconi (Danube Plain) (Panaiotu et al. 2001)	44.16	26.83	+66	8.1	570	0.811
Stavropol (Maher and Thompson 1995)	44.57	41.00	+32	–	410	0.751
Volgograd (Maher and Thompson 1995)	48.72	44.50	+27	–	450	0.495
Red soils (Mediterranean)						
La Ramba (Torrent et al. 2010)	37.59	−4.65	~ 30	~3.0	600	0.568
Montilla (Torrent et al. 2010)	37.59	−4.65	~ 70	~10	600	0.568

[a]Magnetically enhanced soil profile only.
[b]Most recent paleosol.
[c]Average enhancement of paleosols with respect to loesses.

25.2.1 Central Asia

The Chinese Loess Plateau (CLP) provides the longest and most complete sequence of terrestrial wind-blown sediment (Heller and Liu 1984, 1986), which occur in form of extended and thick Pleistocene to Holocene loess/paleosols sequences overlaying tertiary red clays deposits (Ding et al. 2001). It represents an important terrestrial paleoclimatic archive that has been the subject of extensive research over the last ~20 years (Evans and Heller 2003, Maher 2009). In wintertime, dust from arid regions, including sources beyond proximal deserts on the Siberian-Mongolian High, is transported to the plateau by northwesterly winter monsoon winds, while southeasterly winds dominate in summer (An 2000, Maher et al. 2009). Heavy rainfall (>1000 mm/yr) in the southern part of the CLP is caused by humid air brought by the Indian and the SW East Asian Monsoons (e.g. Han et al. 1996). As a result, the CLP is characterized by a strong climatic gradient between arid regions with large dust accumulation and low weathering rates in the N, and humid regions with opposite dust accumulation and weathering characteristics in the S (Derbyshire et al. 1995).

The high chemical and isotopic homogeneity of Chinese loess suggests that dust is supplied, mixed, and recycled from several source regions (Jahn et al. 2001, Maher et al. 2009). Further transformation of the deposited dust depends on the intensity of the two East Asian Monsoon systems. Increasing summer monsoon intensities raise temperature and rainfall, favoring vegetation, weathering and pedogenesis, as particularly evident in paleosols formed during interglacial stages. Paleosols contain a higher concentration of ultrafine (SP and SD) magnetite and/or maghemite grains (e.g. Maher 1986, Liu et al. 1992, Banerjee et al. 1993, Eyre and Shaw 1994, Evans and Heller 1994, Liu et al. 1995, Maher 1998, Jackson et al. 2006, Liu et al. 2007), which are the main responsible for magnetic enhancement, as recorded by magnetic susceptibility (χ, $\Delta\chi$), its frequency dependence (χ_{fd}), and anhysteretic remanent magnetization (χ_{ARM}). The enhanced SD magnetite fraction has been initially attributed to fossil magnetosomes produced by magnetotactic bacteria (Evans and Heller 1994), a hypothesis supported by living magnetic bacteria findings in the Ah horizon of a low-moor soil (Fassbinder et al. 1990), and fossil magnetosome chains in paleosol samples from the CLP (Maher and Thompson 1995, Maher 1998). Magnetotactic bacteria, however, do not contribute significantly to the soil magnetic signature under conditions typical for the CLP (Egli 2004, Dearing et al. 2001).

Han et al. (1996) conducted a systematic magnetic survey of the top horizon (5–10 cm from the surface) of 160 modern soils over the CLP, finding a systematic magnetic enhancement with an increasing trend towards the southern limit of the Plateau. These results have been confirmed by N-S transects across the CLP (Florindo et al. 1999a, Deng et al. 2000, Evans et al. 2002). A positive and systematic correlation between magnetic enhancement parameters and MAR, as well as the mean annual temperature (MAT), exists for all sites where MAR does not exceed ~1100 mm/yr. Above this limit, which occurs S of the Yangtze river (~32°N), soils are magnetically enhanced, but susceptibility is not correlated with rainfall or temperature. Similar results for modern CLP soils have been obtained by Porter et al. (2001), who discuss the link between χ and several annually averaged climatic parameters, including PET. They found a good correlation with a soil water balance proxy given by MAR − PET; however, MAR × MAT was suggested to be a better proxy of "potential pedogenic activity".

Homogeneous dust composition over the CLP and the weak magnetic susceptibility of loess in comparison to well developed soils provides an ideal situation for modeling the influence of climate on the magnetic record of loess/paleosol sequences (e.g. Heller et al. 1993, Maher et al. 1994, Balsam et al. 2004). On this basis, similar trends could be observed for the spatial distribution of χ during interglacial periods of the last 600 kyr and modern rainfall maps (Hao and Guo 2005). Discrepancies in the relationship between χ and MAR involve sections that are close to the S boundary of the CLP (Evans and Rokosh 2000, Guo et al. 2001, Bloemendal and Liu 2005). For example, S5-S8 paleosols from Duanjiapo (34.2°N, 109.2°E) are magnetically less enhanced than corresponding units from central parts of the CLP, while the opposite trend is observed for present-day MAR. In contrast to magnetic data, geochemical parameters clearly indicate a period of significantly higher weathering at Duanjiapo, which resulted in almost complete decalcification (Bloemendal and Liu 2005). Paleorainfall underestimation by magnetic proxies at the S boundary of the CLP has been attributed to a climatic threshold above which pedogenic magnetite is not preserved (Evans and Rokosh 2000, Guo et al. 2001).

Discrimination between low- and high-coercivity minerals indicates that parts of the Duanjiapo section affected by discrepancies between chemical and magnetic proxies of pedogenesis contain proportionally less ferrimagnetic and more antiferromagnetic iron oxides (Bloemendal and Liu 2005). This is also a typical feature of Tertiary red clays, whose low susceptibility contrasts with chemical indicators of pedogenesis, as for example Fe_d/Fe_t (Ding et al. 2001). Although substantial gleyzation in some portions of the red clay sequence is suggested by the occurrence of dark Fe-Mn films, the pedogenic magnetite/maghemite grain size distribution – expressed by the ratio between magnetic parameters sensitive to the SD and SP fractions, respectively – is not significantly different from that of ordinary paleosols (Nie et al. 2010). This observation is hardly compatible with magnetite dissolution, which is expected to selectively eliminate the smaller grain sizes (Smirnov and Tarduno 2000). Therefore, reduced magnetic enhancement on the S boundary of the CLP is probably caused by a diminished production of ferrimagnetic minerals, rather than by dissolution phenomena.

Outside the CLP, small loess deposits in the Kashmir valley (India) contain paleosol horizons that are magnetically enhanced with respect to loess layers. Mineral magnetic data correlate well with the global marine $\delta^{18}O$ record (Gupta et al. 1991). The region is characterized by MAR $\approx$ 600 mm/yr and MAT $\approx$ 13°C.

25.2.2 North America

Magnetic data for North America (Rousseau and Kukla 1994, Grimley et al. 2003, Geiss et al. 2004, Geiss and Zanner 2006, 2007) refer to modern prairie soils formed mainly on the Peoria loess formation, which has been deposited during the Last Glacial period. The investigated sites are located in the centre of the Great Plains and in the Central Lowlands, following a strong east-west MAR gradient (400–1000 mm/yr). Lower Peoria loess sections show varying degrees of reworking after aeolian deposition. The great local relief in some areas fostered gravity-driven processes that complicate the stratigraphic interpretation of some sections (Bettis et al. 2003a). Early phases of deposition where marked by permafrost (Bettis et al. 2003b), which might have altered the loess deposits. Dust sources of the Peoria formation have been traditionally located in the river valleys that transported melt water from the Laurentide Ice sheet. Other contributions include non-glacial sources for some of these deposits (Bettis et al. 2003a, b). Peoria loess is less homogeneous than CLP loess, both geochemically and in terms of grain size, probably because of dust admixtures from different sources (Muhs and Bettis 2000), and later Holocene inputs (Muhs and Zárate 2001, Muhs et al. 2001, Muhs et al. 2004, Jacobs and Mason 2005). Dust heterogeneities are also reflected by the magnetic properties of modern soil parent material. For example, magnetic susceptibility of underlying loess (χ_C) ranges from 100 to 1000 mm^3/kg, compared to 100–350 mm^3/kg for Chinese loess (Geiss and Zanner 2007).

All modern soils are magnetically enhanced in the top horizons (Ap and A). Holocene loess deposition might contribute to the topsoil magnetization; however, magnetic enhancement is due to ultrafine ferrimagnetic minerals, similarly to Chinese soils, rather than to primary minerals brought by dust (Grimley et al. 2003, Geiss and Zanner 2006, Geiss et al. 2008). The maximum susceptibility enhancement is only 20–50% of χ_C, making pedogenic susceptibility estimates based on $\Delta\chi = \chi_B - \chi_C$ very sensitive to the alteration of primary minerals (weathering) and syn-pedogenic dust inputs. Grimley et al. (2003) observed that magnetic enhancement is best recorded by magnetic parameters that are less sensitive to multidomain (MD) primary minerals, such as χ_{fd} and χ_{ARM}.

A positive correlation exists between magnetic enhancement in the top horizons and MAR (Geiss and Zanner 2007). However, $\Delta\chi$ appears to be influenced by variations in parent material magnetic properties, and the correlation with modern rainfall data is not as good as on the CLP. Furthermore, $\Delta\chi$ is consistently ~3.7 times lower than expected from CLP modern soils with similar MAR. *Relative enhancement*, defined as the ratio between topsoil and parent loess magnetic parameters, appears to provide a better degree of correlation with MAR than *absolute enhancement*, which is based on the difference between soil and parent material (Geiss and Zanner 2007). A possible explanation for the better performance of relative magnetic enhancement parameters is that the formation of pedogenic iron oxides is controlled by climate *and* by Fe supply from primary minerals. Relative enhancement parameters, however, define climofunctions that are even more strongly dependent on the geographic provenance of soils. Because primary ferrimagnetic minerals contain a negligible fraction of total wheatherable Fe, relative magnetic enhancement parameters should be calculated with respect to Fe_t. The total Fe contents of Peoria loess (2–4.5 wt% Fe_2O_3 (Muhs and Bettis 2000)) and of Chinese loess (4.5–5.5 wt% Fe_2O_3 (Jahn et al. 2001)) are too similar to explain the difference in magnetic enhancement of the two regions in terms of Fe availability. This is also the case for other geochemical parameters, which do not provide an obvious explanation for the observed magnetic enhancement differences between CLP and U.S.

25.2.3 South America

Loess deposits and reworked loess are widely distributed in the Pampean Plain (Argentina), including deposits at Pcias de San Luis, Córdoba, Santa Fe, La Pampa, and Buenos Aires (e.g. Teruggi 1957, Muhs and Zárate 2001). These deposits form sequences of Pleistocene loess and paleosols which extend to middle or late Holocene in some cases (Orgeira 1990). The loess outcrops considered in this article belongs to the "Pampeano" unit (former "Buenos Aires Formation" (Ameghino 1909)). They are characterized by a high textural and mineralogical homogeneity with increasing sand content toward SW (Zárate and Blasi 1993). Aeolian deposition during cold dry periods is combined with reworking and redeposition by fluvial processes. Dominant SW winds transported clastic particles mobilized by glacial and fluvial action from the Andes region (Teruggi 1957, Smith et al. 2003). Pampean loess contains high concentrations of volcano-pyroclastic particles, which can reach 60% in ashy horizons. The main constituents are volcanic glass shards and plagioclase, with minor quartz contributions (e.g. Teruggi 1957, Muhs and Zárate 2001). The glass shards appear relatively unweathered; however, trace elements such as As in shallow groundwater indicate ongoing dissolution (Nicolli et al. 2004). The magnetic susceptibility of Argentinean loess is comprised between 600 and 1300 mm^3/kg, and reflects a relatively large concentration of MD low-Ti titanomagnetites (Orgeira et al. 2008), while the total Fe content of 3.6–5.6 wt% Fe_2O_3 is similar to that of other loesses (Gallet et al. 1998).

Here we discuss results relative to modern soils classified as Argiudolls from three sites in the Buenos Aires Province: (1) Verónica and (2) Zárate, both located on the Buenos Aires Formation (Orgeira et al. 2008), and (3) a transect located on sandy loess that is probably more recent than the Buenos Aires Formation (Bartel et al. 2006). Modern soils from Verónica are characterized by a marked depletion of magnetic minerals, while moderate magnetic enhancement characterizes the B and A horizons of modern soils from the other sites. This difference has been interpreted in terms of preservation of detrital and pedogenic magnetite in Zárate, as opposed to extensive dissolution in the other sites (Orgeira et al. 2008). On the other hand, Pleistocene and late Holocene paleosols from the Pampean Plain are characterized by an apparent lack of magnetic enhancement (Orgeira and Compagnucci 2006). Magnetite dissolution – even without waterlogging – could be promoted by volcanic glass weathering, which releases silica in the soil pore water. High silica concentrations have been found to dissolve magnetite in marine siliceous sediments (Florindo et al. 2003). On the other hand, lack of magnetic enhancement could be the result of a complex interplay between weathering of lithogenic magnetic minerals and the formation of secondary iron oxides through pedogenesis, with opposed magnetic signatures that compensate each other.

25.2.4 Northern Loess Boundaries: Alaska and Siberia

Loess from central Alaska has a different mineralogical composition with respect to Peoria loess. Quartz is a dominant mineral, accompanied by unusually high concentrations of Fe minerals and Al_2O_3. The total absence of carbonates and low clay content reflect its origin from granites, metabasalts, and schists (Muhs et al. 2003). Vegetation might have played an important role in modulating loess production, which is maximal during glacial periods, and loess accumulation, which is enhanced during interglacials, because of boreal forests acting like a dust trap (Muhs et al. 2003). Extremely thick loess sediments of aeolian origin, often reworked by secondary processes, cover a large area in southern Siberia. Reworking occurred during the development of a fluvial system, which was triggered by an uplift of the region during Late Pliocene/Early Pleistocene (Zhu et al. 2003). Alaskan loess susceptibilities of 1100 ± 300 mm^3/kg (Lagroix and Banerjee 2002) are comparable with maximum values for North America, and are about one order of magnitude larger than those of Chinese loess. Even larger susceptibility values are characteristic for loesses from southern Siberia (Chlachula et al. 1998).

Loess/paleosols sequences in Alaska and Siberia have opposite magnetic susceptibility signatures with respect to China and North America, with paleosols being less magnetic than loesses (e.g. Begét et al. 1990, Chlachula et al. 1998). This trend has been explained by the so-called *wind-vigor* model with the greater efficiency of atmospheric entrainment and transport of dense (~5000 kg/m^3) magnetic Fe oxide grains during glacial times, when wind action is maximal, with respect to the less dense (e.g. quartz, ~2600 kg/m^3) non-magnetic minerals (Evans 1999, 2001). According to this model, glacial sediments contain higher concentrations of magnetic minerals and are more magnetic. This mechanism is likely active in all loess regions, including the CLP. The reason for the opposite trends of Alaska and Siberia with respect to the CLP is attributed to the different intensities of pedogenic mineral production on one hand, and wind-vigor-based dust sorting on the other. The wind-vigor effect is most evident in Alaska and Siberia because of the shorter distance to dust sources and higher concentrations of magnetic minerals in dust. On the other hand, Alaskan and Siberian soils are often assumed to be seasonally waterlogged, with chemical conditions that promote gleyzation and subsequent reductive dissolution of ultrafine iron oxides (Feng and Khosbayar 2004, Kravchinsky et al. 2008). In this case, magnetic minerals of pedogenic origin are not expected to accumulate, leaving wind-vigor sorting as the sole modulation mechanism for the magnetic properties of loess/paleosols sequences.

An evaluation of the relative importance of pedogenesis and wind-vigor dust sorting requires unmixing the magnetic signatures of lithogenic and pedogenic minerals. The frequency dependent magnetic susceptibility χ_{fd} is a suited parameter for this purpose, because of its highly selective response to magnetic minerals with grain sizes corresponding to the SP/SD boundary (Worm 1998), which happens to be within the grain size distribution of pedogenic magnetite (Liu et al. 2007). Measurements of χ_{fd} indicate that both Alaskan and Siberian soils/paleosols are enriched in ultrafine magnetic minerals with respect to the loess layers (Liu et al. 1999, Kravchinsky et al. 2008). Hydromorphic conditions with subsequent gleyzation might occur locally, mainly driven by local topography and poor drainage (e.g. Grimley et al. 2004), but cannot be considered a systematic characteristic of these soils.

25.2.5 Europe, Russian Steppe, North Africa

Unlike the CLP, European loess deposits are very heterogeneous, reflecting variations in dust provenance, accumulation conditions, post depositional alteration, and presence of tephra (Derbyshire 2001). Loess/paleosols sequences from the Czech Republic (Forster et al. 1996, Oches and Banerjee 1996), southwest Slovakia (Ďurža and Dlapa 2009), along the Danube River (Panaiotu et al. 2001), and in Alsace (France) (Rousseau et al. 1998) are characterized by a magnetic enhancement pattern similar to that of the CLP. On the other hand, the magnetic record of Polish and western Ukrainian sequences is complicated by gley paleosol horizons with susceptibility values lower than those of loess. These horizons indicate that loesses were accumulated in a more humid and cooler climate than the Chinese ones (Nawrocki et al. 1996). East European loess/paleosol sections cannot always be used as a direct climate indicator: spore and pollen analyses suggest that some paleosol horizons correspond to considerable cooling, while traces of warming-up could be revealed within loess horizons (Bolikhovskaya and Molodkov 2006).

A SW-NE climate transect across loess deposits on the Russian Steppe, with MAR ranging from 300 mm/yr (Volgograd) to 500 mm/yr (Stavropol), is characterized by a systematic magnetic enhancement of modern topsoils, with same correlation between susceptibility and MAR as in China (Maher et al. 2003). Present-day dust accumulation in the region is minimal, excluding significant dust flux contributions to magnetic enhancement.

Few data is available for Northern Africa. Loess/paleosols sequences on the Matma Plateau (Tunisia) are characterized by reddened fersillitic paleosols whose susceptibility is a factor 3–9 higher than loess layers. A collection of modern soils from the same area is characterized by a positive correlation with rainfall for MAR values <500 mm/yr, while modern soils formed at >700 mm/yr display the same magnetic enhancement as those with ~250 mm/yr (Dearing et al. 2001).

25.2.6 Non-loessic Soils

Few magnetic studies exist on non-loessic modern soils formed under "aggressive" climates characterized by very high MAR and/or alternation of extremely wet and dry seasons. These soils can provide useful knowledge about magnetic enhancement limits. Torrent et al. (2010) proposed the ratio Hm/χ_{fd} between hematite content and frequency dependence of susceptibility as an indicator of weathering intensity, with values $<5 \times 10^7$ g/m^3 for loessic soils in temperate areas, $\sim 10 \times 10^7$ g/m^3 for red Mediterranean soils, and $>20 \times 10^7$ g/m^3 for well-drained Brazilian ferralsols. The variability of Hm/χ_{fd} could suggest that the total concentration of pedogenic Fe minerals is poorly reflected by magnetic parameters in certain cases. This problem is obvious if magnetic enhancement is assumed to be an intermediate product of ferrihydrite-to-hematite conversion (Torrent et al. 2006). On the other hand, pedogenic magnetite formation *via* redox cycling (e.g. Maher 1998) is not directly linked to pedogenic hematite, and its correlation with climate would not be affected by variations in Hm/χ_{fd}.

Torrent et al. (2010) report geochemical and magnetic parameters of two hematitic soil profiles from the Province of Córdoba (southern Spain). The parent rocks are calcarenite and calcareous orthoquarzite, respectively, containing 0.5–2% Fe_t and <0.3% silicate Fe (Torrent and Cabedo 1986). The present climate is of the warm Mediterranean type, with 600 mm/yr MAR and a long summer drought. A clear magnetic enhancement is observed in the upper horizons, although, on average, to a lesser extent than Chinese loessic soils with same MAR.

We also discuss a series of soils collected along a volcanic mountain transect (Kohala Peninsula, Hawaii) characterized by an altitude-driven climosequence with MAR varying from 160 mm/yr on the coast to 4500 mm/yr at the maximum altitude of 1254 m asl (Chadwick et al. 2003). Age and composition of lavas is homogeneous and rainfall varies to a much greater extent than temperature, making it an ideal site for studying the effect of rainfall on weathering and pedogenesis. Magnetic susceptibility has been measured on soil profiles taken across the climatic gradient, with sampling sites chosen in places with minimal erosion (Singer et al. 1996). Magnetic enhancement is evident in all sites with rainfall <1000 mm/y, while an opposed trend is seen above this limit.

We conclude our review with tropical soils collected across a strong MAR gradient in Ghana (Hendrickx et al. 2005), which are characterized by highly variable magnetic enhancements. These soils formed on a variety of bedrock types including sandstones, phyllites, quartzites, schists, clay shales, and volcanic andesites, schists and amphiboles. MAR values are comprised between 1000 and 2000 mm/yr. Topsoil χ does not correlate significantly with rainfall. One factor responsible for these variations is soil drainage: poorly drained soils on floodplains are one order of magnitude less enhanced than well drained soils with same MAR. On the other hand, strong magnetic enhancement in the uppermost 10 cm is likely produced by burning, which is a prevalent practice in Ghana. This example shows that drainage and human impact must be taken into account when measuring the magnetic properties of modern soils.

25.3 Modeling Soil Moisture and Magnetic Enhancement

One of the most important parameters controlling pedogenic processes is the soil moisture s, which is obviously related to the rainfall R (see Table 25.1 for a definition of all parameters used in this section). Additional factors determining the soil water budget are losses due to leakage L and evapotranspiration E (direct evaporation and transpiration from vegetation). Soil moisture s is defined as the ratio between the actual pore volume filled with water and the total pore volume, so that $0 \leq s \leq 1$. The fraction of soil volume occupied by pores is the soil porosity ϕ. The water balance of a soil slab of thickness H and porosity ϕ is described by the differential equation

$$h\frac{\mathrm{d}s}{\mathrm{d}t} = I(t) - E(s,t) - L(s,t), \qquad (25.2)$$

where I is the rate of water infiltration from rainfall, and $h = \phi H$ is the effective thickness of an equivalent slab with no porosity (Rodriguez-Iturbe et al. 1999). The individual terms of Eq. (25.2) depend in a complex manner on soil properties, vegetation, and climate. Here we use the statistical model of Rodriguez-Iturbe et al. (1999) to estimate the individual terms of Eq. (25.2) and isolate the most important factors controlling s.

25.3.1 Infiltration from Rainfall

Runoff can be neglected on a horizontal soil, and therefore we assume that infiltration equals rainfall: $I = R$. Rainfall is idealized as a series of point events described by a Poisson process with mean rate λ, thereby ignoring the temporal structure within each rain event. Furthermore, the intensity of a rain event is quantified by the *rain depth* Z, defined as the total volume of water per unit surface that reaches the soil during the event. The rain depth is a random variable assumed to have an exponential probability density function

$$p_Z(z) = \frac{1}{\zeta} e^{-z/\zeta}, \quad (25.3)$$

where ζ is the average of Z in a given climate (Rodriguez-Iturbe et al. 1999). The rainfall expected over a fixed period of time (e.g. MAR) is thus given by $\langle R \rangle = \lambda\zeta$. Distinct climates might be characterized by same MAR and different rates and depths of the rain events (Fig. 25.1). For example, rare but strong storms (e.g. λ is small and ζ is large) occur in hot arid climates, while frequent, low-intensity rainfall events are common in cold, humid climates (e.g. λ is large and ζ is small). Both rate and depth of rainfall events display seasonal variations. Typical values for λ and ζ are 0.2–20 events/month and 5–20 mm/event, respectively. In most cases, rainfall depth is positively correlated with temperature (Fig. 25.1).

25.3.2 Evapotranspiration

Evapotranspiration losses are accounted by a simplified but commonly accepted model where E is assumed to be proportional to soil moisture, until a threshold s_E is reached, above which evapotranspiration takes place at a maximum rate $\hat{E}$. This maximum rate is called *potential evapotranspiration* (PET), and depends on several factors such as temperature, wind speed, and vegetation. Evapotranspiration is thus given by:

$$E(s) = \begin{cases} \hat{E}\, s/s_E, & \text{if } 0 < s \le s_E \\ \hat{E}, & \text{if } s_E < s < 1 \end{cases} \quad (25.4)$$

(Fig. 25.2). The threshold s_E depends on soil properties and type of vegetation cover; whereby it is assumed that vegetation is under water deficit stress as long as $s < s_E$. The typical range of s_E is 0.10–0.50, with higher values being characteristic for grasses and desert shrubs (Table 25.3).

PET is a challenging parameter to estimate, and several empirical methods have been developed for this purpose. Here, we use following approximation valid for a short green crop that is completely shading the ground:

$$\hat{E} = 1.6 \times (10\, \tau / J)^a, \quad (25.5)$$

where τ is the mean monthly temperature in °C, a is an appropriate exponent, and J is a so-called heat index given by the sum of the monthly values j (Thornthwaite 1948). The latter two parameters are given by:

$$j = \begin{cases} (\tau/5)^{1.514}, & \text{if } T > 0 \\ 0, & \text{if } T \le 0 \end{cases} .$$
$$a = 0.49 + 1.79 \times 10^{-2} J - 7.71 \times 10^{-5} J^2 + 6.75 \times 10^{-7} J^3 \quad (25.6)$$

Equation (25.5) can be used for any location at which rainfall and daily maximum and minimum temperatures are recorded. It is based on the inherent assumption that a high correlation exists between mean temperature and some of the other pertinent parameters such as radiation, atmospheric moisture, and wind. Accurate PET estimates also take the latitude dependence of daylight time and the variable month length into account. A *daylight-corrected* PET estimate is therefore given by:

$$\hat{E}_c = \hat{E} \times (D/12) \times (n/30), \quad (25.7)$$

where D is the monthly mean daylight duration (photoperiod) in hours, and n is the number of days in a

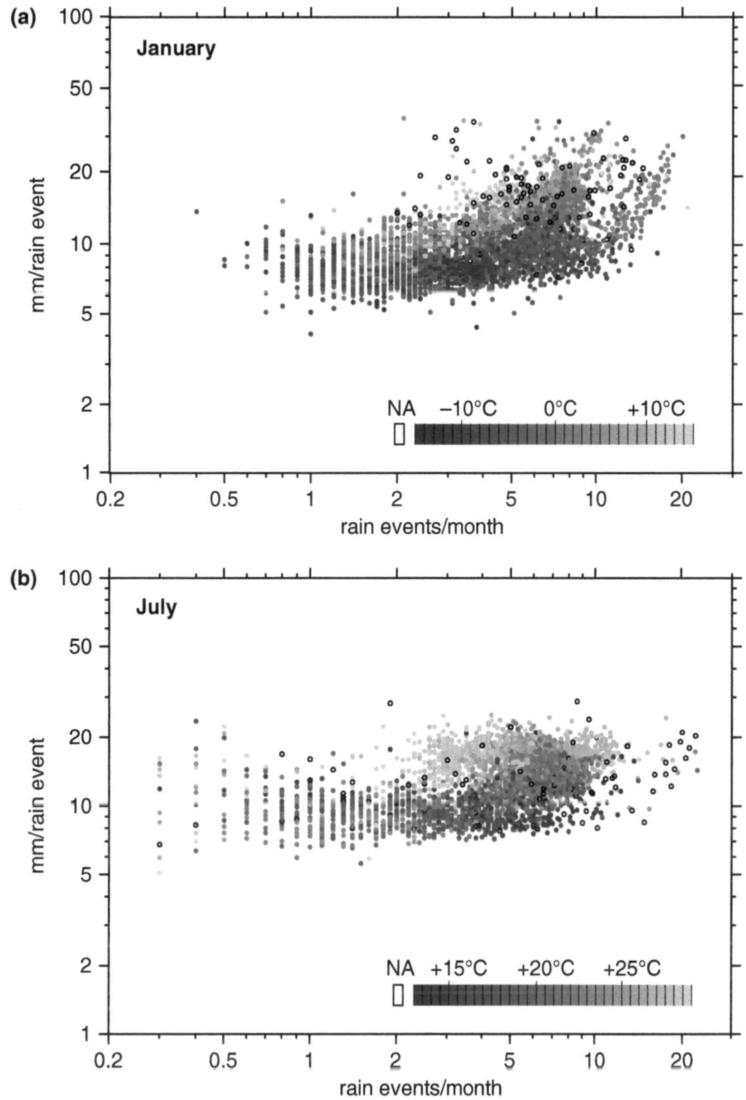

Fig. 25.1 Rainfall frequency λ and rainfall depth ζ at 4460 climate stations in the U.S. for (**a**) January, and (**b**) July, averaged between 1971 and 2000 (source: NOAA). The number of rain events during each month is identified with the number of rainy days with $\zeta > 2.5$ mm. Each station is a point whose gray level is coded according to the mean monthly temperature. Circles are used for stations where temperature data is not available (NA)

month (Dunne and Leopold 1978). Various algorithms have been implemented for the calculation of the interval D between sunrise and sunset, defined as the times of the day when the sun has an altitude Q above (below if $Q < 0$) the horizon (Ligr et al. 1995). One of the simplest solutions is:

$$D = \frac{2}{15} \arccos \left[\frac{\cos(90° + Q) - \sin\varphi \sin\eta \sin L}{\cos\varphi \sqrt{1 - (\sin\eta \sin L)^2}} \right]$$
$$L = M + 1.916° \sin M + 0.02° \sin 2M + 282.565°,$$
$$M = (360°/365.25)\, t - 3.251° \tag{25.8}$$

where φ is the latitude, $\eta = 23.44°$ is the obliquity of the ecliptic, and t is the number of days from the beginning of the year (Keisling 1982). Estimates of D obtained using Eq. (25.8) have a maximum error of 0.33 h for latitudes <60°, which is sufficiently small for practical purposes.

25.3.3 Leakage

Leakage K is modeled as vertical percolation through the lowest active soil layer. Leakage of a saturated soil (i.e. $s = 1$) is equal to its *saturated hydraulic conductivity* $\hat{K}$. For $s < 1$, leakage is given by:

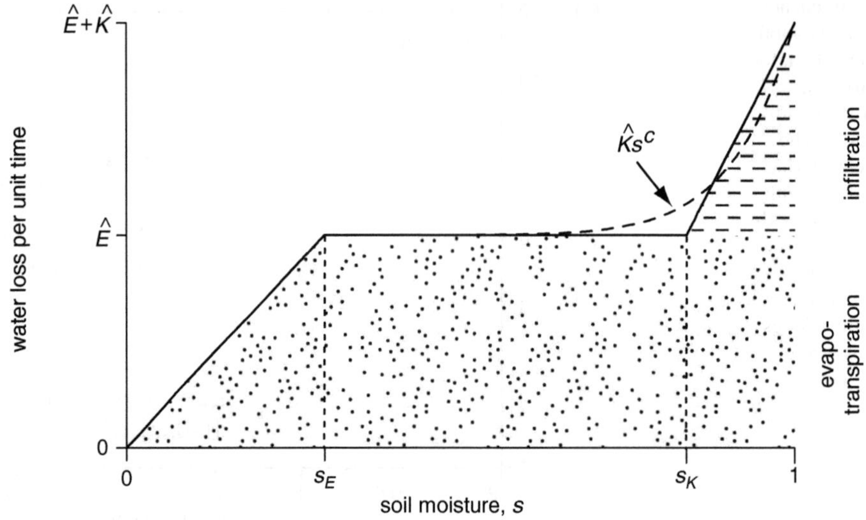

Fig. 25.2 Modeled dependence of the total water loss on soil moisture. The contributions of evapotranspiration and deep infiltration are highlighted (shaded areas). The dashed line represents leakage according to Eq. (25.9)

Table 25.3 Dependence of water balance parameters on soil texture (data from (Laio et al. 2001a, b)). Regular numbers indicate typical average values, while cursive numbers refer to particular cases for specific vegetation types. Cursive abbreviations refer to following species: *Bouteloua gracilis* (short grass, steppe, north-central Colorado), *Prosopis glendulosa* (woody plant, savanna, Texas), *Paspalum setaceum* (grass, savanna, Texas), *Burkea africana* (woody plant, savanna, Nylsvley region, Africa), and *Eragrostis pallens* (grass, savanna, Nylsvley region, Africa)

Soil texture	$\hat{K}$ [mm/day]	ϕ	s_E	s_K	$\hat{E}$ [mm/day]
Sand	>1000	0.35	0.10–0.33	0.85	–
	1030	*0.37*	*0.16 (Bg)*	–	*3.7 (Bg)*
	1098	*0.42*	*0.15 (Ep), 0.11 (Ba)*	–	*6.1 (Ep), 4.7 (Ba)*
Loamy sand	~1000	0.42	0.31	0.86	–
	822	*0.43*	*0.35 (Pg), 0.37 (Ps)*	–	*4.4 (Pg), 0.13 (Ps)*
Sandy loam	~800	0.43	0.46	0.87	–
Loam	~200	0.45	0.57	0.88	–
	330	*0.47*	*0.35 (Bg)*		–
Clay	<200	0.50	0.78	0.93	–
	350	*0.46*	*0.64 (Bg)*	–	–

$$K(s) = \hat{K}s^c, \tag{25.9}$$

where the exponent c depends on soil texture, ranging from $c \approx 11$ for sand to $c \approx 25$ for clay. Since $c \gg 1$, K can be assumed to be effectively zero below a critical moisture threshold s_K, and to vary linearly above it:

$$K(s) \cong \begin{cases} 0, & \text{if } 0 \le s \le s_K \\ \hat{K}\dfrac{s - s_K}{1 - s_K}, & \text{if } s_K < s \le 1 \end{cases} \tag{25.10}$$

(Fig. 25.2). The saturated hydraulic conductivity can vary between 9 and 2000 cm/day, depending on soil texture (Table 25.3) (Clapp and Hornberger 1978). Compaction of deeper soil layers, especially below the root zone, produces an exponential-like decrease of the hydraulic conductivity with depth (Lind and Lundin 1990, Youngs and Goss 1988). Typical s_K values are comprised between 0.84 (sand) and 0.92 (clay).

25.3.4 Solution of the Water Balance Equation

A stochastic solution of Eq. (25.2) for the stationary case can be expressed in terms of a probability density function p_s of the soil moisture (Rodriguez-Iturbe et al. 1999). This function provides an estimate of the average soil moisture $\langle s \rangle$, as well as the probability of drought events, which is particularly important in agricultural science but not relevant here. The analytical expression for p_s is greatly simplified if normalization (obtained by ensuring that the integral of p_s over $0 \le s \le 1$ is equal to 1) is neglected. In this case, p_s is given by:

$$p_s(s) = \begin{cases} \frac{h}{\hat{E}}\left(\frac{s}{s_E}\right)^{\lambda s_E h/\hat{E}-1} e^{-sh/\zeta}, & \text{if } 0 < s \le s_E \\ \frac{h}{\hat{E}} e^{\lambda(s-s_E)h/\hat{E}} e^{-sh/\zeta}, & \text{if } s_E < s \le s_K \\ \frac{h}{\hat{E}}\left[\frac{\hat{K}}{\hat{E}}\frac{s-s_K}{1-s_K}+1\right]^{\lambda(1-s_K)h/\hat{K}-1} e^{\lambda(s_K-s_E)h/\hat{E}-sh/\zeta}, & \text{if } s_K < s \le 1 \end{cases} \quad (25.11)$$

The complex dependence of soil moisture on the parameters described in Sections 25.3.1, 25.3.2, and 25.3.3 is best illustrated by the examples shown in (Rodriguez-Iturbe et al. 1999) for typical climates specified in Table 25.4. Here, we are interested in the average moisture $\langle s \rangle$, which can be calculated from Eq. (25.11) using:

$$\langle s \rangle = \frac{\int_0^1 p_s(u)\, u\, du}{\int_0^1 p_s(u)\, du}. \quad (25.12)$$

In principle, $\langle s \rangle$ depends on seven parameters describing climate, soil properties and vegetation. An important simplification is obtained by grouping the climatic parameters into a single number $W = \lambda\zeta/\hat{E}$, which we call *moisture ratio*, because it is the ratio between MAR $= \lambda\zeta$ and PET. This simplification is possible because different combinations of λ, ζ, and $\hat{E}$ corresponding to the same value of W result in very similar average moistures $\langle s(W) \rangle$ (Fig. 25.3). Analogous ratios have been proposed as soil moisture proxies (Jenny 1941).

The main parameter controlling the shape of the function $\langle s(W) \rangle$ is ζ, which describes the intensity of rain events and therefore discriminates between climates that are more or less "stormy". Figure 25.3 shows the largest climate-driven variations of $\langle s(W) \rangle$ obtained by choosing extreme values of ζ. The error introduced by grouping different parameters into a single number calculable from climatic tables is not larger than the uncertainties involved in estimating λ, ζ, and $\hat{E}$.

If $W < 1$, maximum evapotranspiration exceeds rainfall, and soil moisture remains relatively low on average. As W approaches 1, soil moisture increases until the threshold s_K is reached, above which water starts to percolate. If $W > 1$, soil moisture is maintained constantly near saturation and drainage becomes the main mechanism of water loss. Interestingly, a step-like increase of $\langle s \rangle$ to near-saturation values occurs in a narrow range around $W = 1$ in well-drained soils. Therefore, $W = 1$ is a threshold separating regimes of low/moderate soil moisture from saturation. The transition sharpness increases with soil thickness

Table 25.4 Water balance parameters for typical soils formed in four different climates: (1) deep soil in tropical climate with frequent rainfall of moderate intensity and high maximum evapotranspiration, (2) shallow, sandy soil in a hot arid climate, covered with a mixture of trees, shrubs, and grasses, (3) steppe soil in a cold, arid climate with low maximum evapotranspiration, and (4) forest soil in a humid temperate region (data from Rodriguez-Iturbe et al. 1999)

Climate	λ events/d	ζ mm/event	$\hat{E}$ mm/d	$\hat{K}$ mm/d	h cm	s_E	s_K
Tropical	0.66	15	6.0	900	45	0.30	0.85
Hot arid	0.10	20	6.0	5000	10	0.45	0.80
Cold arid	0.16	10	2.5	200	20	0.30	0.90
Temperate (forest)	0.33	20	4.0	300	30	0.30	0.85

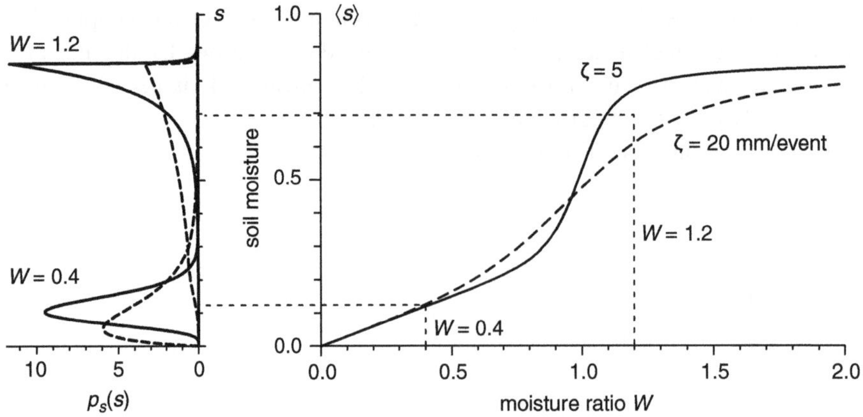

Fig. 25.3 Right plot: dependence of the average soil moisture $\langle s \rangle$ on W for different climates, represented by weak ($\zeta = 5$ mm/event) and intense ($\zeta = 20$ mm/event) rainfall events. Soil properties used for the calculation are: $s_E = 0.3$, $s_K = 0.85$, $h = 30$ cm, $\hat{K} = 0.5$ m/day. Left plot: probability distributions p_s of soil moisture during unsaturated ($W = 0.4$) and saturated ($W = 1.2$) conditions for the same parameters used to calculate $\langle s \rangle$. The effect of potential evapotranspiration on the curves is negligible and is not shown

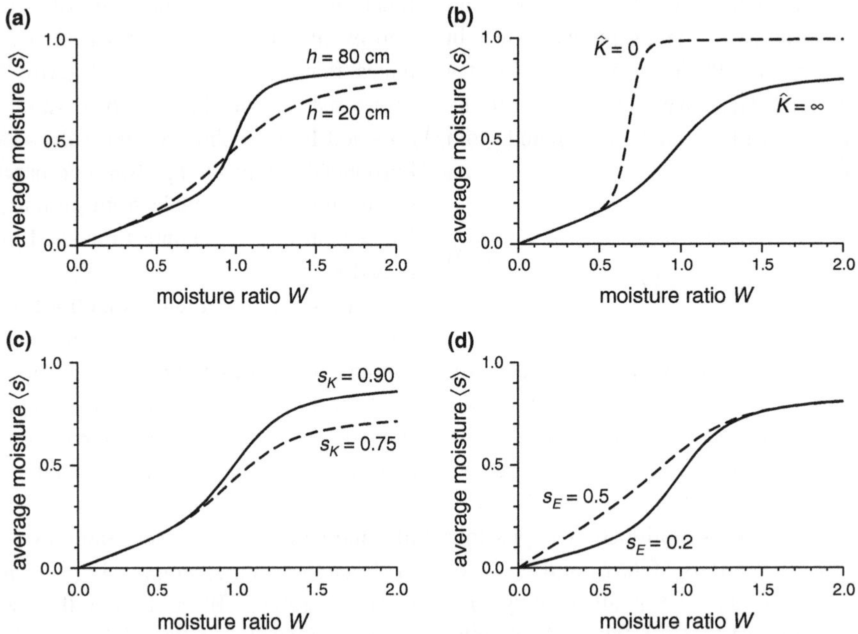

Fig. 25.4 Dependence of $\langle s(W) \rangle$ curves on (**a**) soil thickness h, (**b**) drainage, which is mainly controlled by the saturated hydraulic conductivity $\hat{K}$, and (**c, d**) soil texture (s_K and s_E are small for sand and large for clay). The model parameters correspond to a temperate climate ($\zeta = 15$ mm/event, $\hat{E} = 4$ mm/day), and, unless otherwise specified in the plots, on typical average soil properties ($s_E = 0.3$, $s_K = 0.85$, $h = 30$ cm, $\hat{K} = 0.5$ m/day)

(Fig. 25.4a). Soils characterized by poor drainage (e.g. $\hat{K} \to 0$) experience the transition to a saturation regime when W is as low as 0.5 (Fig. 25.4b). Soil moisture in the saturation regime is mainly controlled by s_K (Fig. 25.4c), which in turn depends on soil texture. The importance of this threshold was recognized by Thornthwaite (1948), who defined a similar parameter, corresponding to $W - 1$, for which zero is the saturation threshold. Alternative parameters, such as the "effective rainfall" PET – MAR, have been defined with the same purpose (Chadwick et al. 2003, Orgeira and Compagnucci 2006).

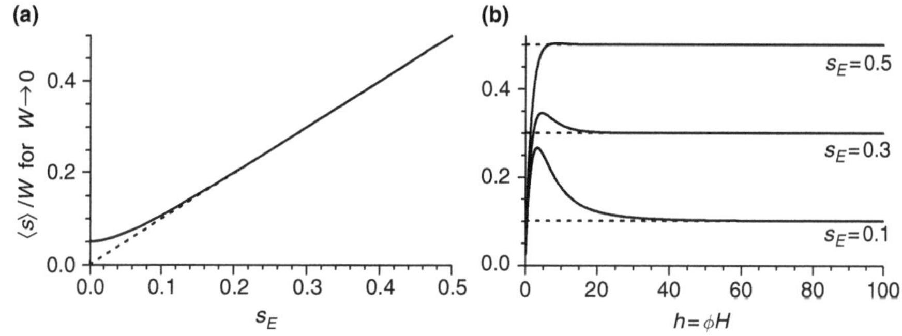

Fig. 25.5 Validity range of Eq. (25.13) tested for (**a**) different values of s_E, and (**b**) different soil thicknesses. Solid lines represent the dependence of the initial slope of $\langle s(W)\rangle$ on s_E and h, compared with the approximation given by Eq. (25.13) (*dotted lines*)

As long as saturation is not reached, a direct proportionality exists between $\langle s\rangle$ and W, and we obtain:

$$\langle s\rangle = s_E\, W \tag{25.13}$$

from the $\lambda \to 0$ limit of Eq. (25.11) (Fig. 25.4d). The proportionality factor s_E is controlled mainly by soil texture and by vegetation cover (Table 25.3). Equation (25.13) is valid over a range of s_E values typical of soils, and for $h > 20$ cm (Fig. 25.5). It can be used to estimate the average moisture of soils in the non-saturated regime (e.g. $W < 1$ in well drained soils and $W < 0.5$ in poorly drained soils) under typical climatic conditions, vegetation cover, and drainage, using data available from climatic tables. Therefore, W can be considered a climatic proxy for soil moisture, and thus an important parameter to take into account for constructing a climofunction.

25.3.5 Enhancement Proxy for Pedogenic Magnetite

Several studies demonstrated that soil magnetic enhancement and climate are correlated (e.g. Maher and Thompson 1992, 1995, Maher et al. 1994, Thompson and Maher 1995, Han et al. 1996, Maher 1998, Porter et al. 2001). Common parameters used to establish such correlation are $\log_{10}\Delta\chi$ and MAR (Eq. (25.1)), as proposed by Maher et al. (1994). Attempts to improve this correlation led to proposing alternative magnetic proxies, such as ARM for loessic soils in the U.S. (Geiss and Zanner 2007), and to the consideration of additional climatic parameters, for example the mean annual temperature (Han et al. 1996). Porter et al. (2001) observed a negative correlation between susceptibility of modern soils on the CLP and the mean annual evapotranspiration, therefore recognizing the role of the latter parameter in the soil water balance. Beside the production of pedogenic minerals, climate-related processes such as weathering of parent magnetic minerals (Liu et al. 1999), syn-pedogenic dust inputs (Heller et al. 1993), and chemical collapse (Anderson and Hallet 1996) are additional factors that influence the magnetic properties of soils. A first model using these factors as a-priori parameters for glacial and interglacial periods was proposed by Anderson and Hallet (1996). The optimal choice of climatic proxies depends critically on understanding how pedogenic magnetite and other magnetic oxides are formed during pedogenesis. For example, the "fermentation process" proposed by Le Borgne (1955) and the ferrihydrite alteration model of Barrón and Torrent (2002) can be expected to have different climatic responses.

In the following, we refine the "fermentation" magnetic enhancement model of Maher (1998) by constructing a quantitative link between pedogenic magnetite and the soil water balance. In simple words, our work consists in replacing MAR with a climatic parameter that accounts for the effects of soil moisture dynamics on the formation of magnetite by redox cycling of Fe sources. The model can be subdivided into distinct processes described as follows (Fig. 25.6):

(1) *Weathering*. Hydrolysis of primary Fe minerals releases Fe^{2+} ions, whose oxidation results in the production of poorly crystallized oxyhydroxides (ferrihydrite). Ferrihydrite is assumed to be the primary source of pedogenic minerals that contribute to the magnetic enhancement. As long

Fig. 25.6 Conceptual model of soil magnetic enhancement

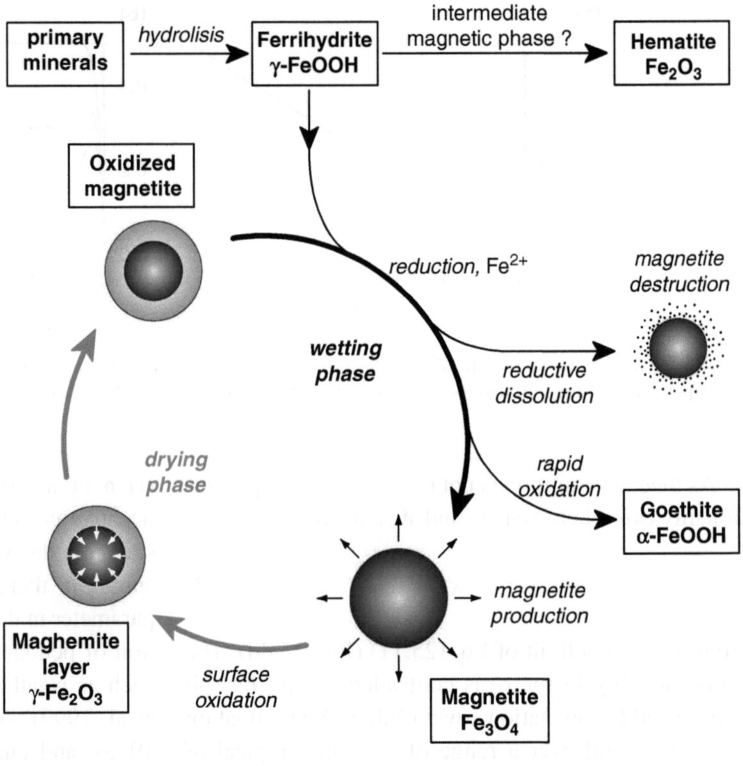

as wheatherable Fe minerals exist or enter into the system, the total concentration of iron available for pedogenic minerals increases. Depending on the composition of primary minerals, weathering can be more or less rapid, leading to a variable time necessary for reaching the mature stage. In our model, along with Maher (1998), we assume that the production of pedogenic minerals responsible for magnetic enhancement is *not limited* by Fe availability, and that a *stationary regime* is reached in mature soils. Lack of Fe limitation in soils formed on most types of parent materials is supported by the fact that the concentration of pedogenic magnetite necessary to explain observed susceptibility enhancements is much smaller than typical Fe_d and Fe_t concentrations. The onset of a stationary regime without significant syn-pedogenic dust inputs implies that iron in pedogenic minerals is recycled.

(2) *Wetting phase 1* (Fig. 25.7). After each rain event, a number of soil pores become saturated with water. In the active soil horizon and in presence of organic matter, oxygen is consumed by microorganisms in wet pores. In some pores, higher initial nutrient concentration, after oxygen depletion,

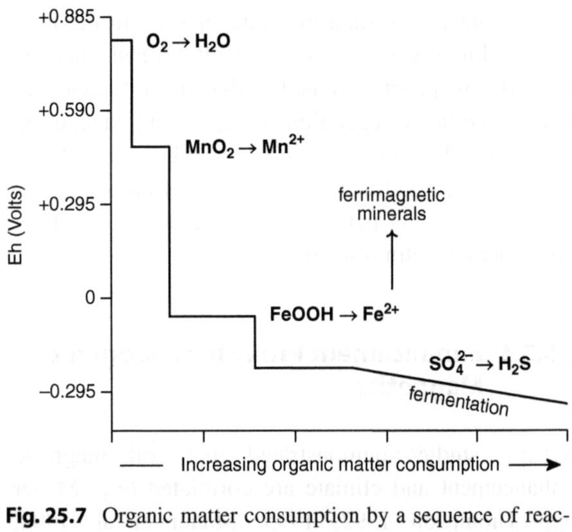

Fig. 25.7 Organic matter consumption by a sequence of reactions involving different electron acceptors, driven by the Eh potential decrease (modified from Chadwick and Chorover 2001). Pedogenic ferrimagnetic minerals are assumed to form by oxidation of Fe^{2+} ions released by iron reduction reactions

might consume all available electron acceptors until Fe^{3+} reduction becomes possible (Deming and Baross 1993, Maher 1998, Chadwick and Chorover 2001). Possible primary sources for Fe reduction are ferrihydrite – which supports much higher reduction rates than other Fe oxides (Roden and Zachara 1996) – and clay minerals (Kostka et al. 2002). The outward diffusion of Fe^{2+} ions in a microscale redox gradient around "reduction spots" is accompanied by re-oxidation and precipitation of a variety of iron oxides and oxyhydroxides, depending on pH and oxidation rate (Dearing et al. 1996, Maher et al. 2003). One of the oxidation products is ultrafine magnetite, which – although not relevant in terms of mass – is the main cause for magnetic enhancement. Because of the order of magnitude difference in concentration, magnetite can be neglected in the mass balance. The invariant grain size distribution of pedogenic magnetite in Chinese soils and paleosols formed at sites with different MAR (Liu et al. 2007) can be interpreted as a circumstantial evidence that pedogenic magnetite is formed under specific, well controlled geochemical conditions, as it is the case with biological mediation by dissimilatory iron reducing bacteria (Maher et al. 2003). Although rapid microscale iron redox cycling is actively sustained by communities of iron oxidizing and reducing bacteria (Sobolev and Roden 2002), the role of dissimilatory metal reducing bacteria in producing a magnetic enhancement could not be proven (Guyodo et al. 2006). Fully inorganic magnetite formation paths are possible, for example by reaction of ferrihydrite with low Fe^{2+} concentrations (Tamaura et al. 1983, Tronc et al. 1992). The invariant grain size distribution of the product might be explained with the specific formation mechanism requiring Fe^{2+} adsorption at pH > 5.0 and subsequent crystal growth by dissolution-recrystallization or solid state reaction. The exact mechanism of magnetite formation is not relevant in our model: we only assume that ultrafine magnetite precipitates during the wetting phase in minute amounts negligible for the Fe mass balance but sufficient produce a magnetic enhancement. Furthermore, we assume that the rate of pedogenic magnetite production depends only on the activity of reduction spots that can support Fe reduction.

(3) *Drying phase*. As water is lost by evapotranspiration, an increasing number of reduction spots dry out and oxidizing conditions are re-established. Magnetite crystals produced during the wetting phase will start to oxidize from the surface towards the interior (Maher 1998). This process can be thought as the growth of a cation-deficient maghemite-like surface layer. The oxidation kinetics is controlled by the diffusion of Fe^{2+} ions in the crystal lattice (Gallagher et al. 1968). For example, magnetite particles with a diameter of 8.7 nm are half-way converted to maghemite after 6.7 days in aqueous solution at 24°C (Tang et al. 2003). Therefore, a significant amount of pedogenic magnetite oxidation can be expected during dry or well oxygenated time intervals (Murad and Schwertmann 1993).

(4) *Wetting phase 2*. Soil pores are subjected to repeated wetting and drying cycles modulated by rain events and successive evapotranspiration (Laio et al. 2001a, b). It is therefore only a matter of time before a pedogenic magnetite particle – after one or more drying and wetting cycles under oxic conditions – experiences a new "reducing event" similar to that during which it was created. If this is the case, reducing conditions promote the dissolution of the oxidized layer, and eventually of the entire particle. Analogous dissolution of surface oxidation layers has been proposed to explain the appearance of the Verwey transition in pelagic sediments below the Fe redox boundary (Smirnov and Tarduno 2000). Simultaneous magnetite production and maghemite dissolution during wetting phases in the *active* soil horizon explains two important observations. The first is that – at least in certain types of soils – magnetic enhancement reaches a dynamic equilibrium with the environment and does not proceed indefinitely with time (Thompson and Maher 1995). The second observation is that maghemite in paleosols is preserved over geological times, implying that dissolution must be promoted by chemical conditions that occur only in active soils. A dynamic equilibrium governing the concentration of pedogenic ferrimagnetic minerals is essential for observing a reliable correlation between magnetic enhancement and climate when comparing soils of different ages.

The conceptual enhancement mechanism described above, which is essentially the same proposed by Maher (1998), is used in the following to implement a semi-quantitative estimate of pedogenic magnetite concentrations at equilibrium with the average climatic parameters described in Sections 25.3.1, 25.3.2, 25.3.3, and 25.3.4. The production rate p is expected to be proportional to the number of reduction spots, which, on a long term, is proportional to the average soil moisture $\langle s \rangle$. Each reduction spot can produce magnetite during a limited amount of time, until it dries out or chemical conditions are no longer favorable. In order to sustain magnetite production over time, reduction spots must be "reset" during dry or oxygenated periods. Therefore, magnetic enhancement is ultimately driven by the alternation of wet and dry phases, whose pace is dictated by the frequency λ of rain events. The last factor to take into account is the onset of a saturation regime at $W \approx 1$, as discussed in Section 25.3.4. During this regime, excess water is drained and evapotranspiration plays a minor rule. Drainage through the active soil layer decreases the chances that pores dry out, limiting the frequency of "reset" events and thus the capability of maintaining appropriate conditions for the production of pedogenic magnetite. Accordingly, we define the *effective soil moisture*, s_{eff}, as the fraction of pores that can sustain magnetite production. This parameter is calculated by multiplying the average fraction $\langle s \rangle$ of wet pores with a function $\theta(W - W_0)$ that equals 1 when W is smaller than a critical threshold $W_0 \approx 1$ marking the onset of a saturated regime, and is 0 if $W > W_0$. Since the onset of saturation occurs between $W = 0.9$ and 1.3 (Fig. 25.3), we assume θ to express a similar, smooth transition (Fig. 25.8).

The instantaneous magnetite production rate is then given by:

$$p = \frac{\mathrm{d}m}{\mathrm{d}t} \propto \lambda s_{\mathrm{eff}}, \tag{25.14}$$

where m is the pedogenic magnetite mass, t is time, and $s_{\mathrm{eff}} = \langle s \rangle \, \theta(W - W_0)$. Since the calculation of $\langle s \rangle$ according to Eqs. (25.11 and 25.12) is elaborated, we approximate $\langle s \rangle$ with Eq. (25.13) and choose a suitable function θ such that s_{eff} has a similar dependence on W as in Fig. 25.8. This is the case when $\theta(x) = [1 - \tanh(5x)]/2$ and $W_0 = 1.0$–1.2. Then, Eq. (25.14) simplifies to:

$$p \propto \lambda s_E \, W \theta(W - W_0). \tag{25.15}$$

The magnetite production rate expressed in Eq. (25.15) is only meaningful when evaluated over a

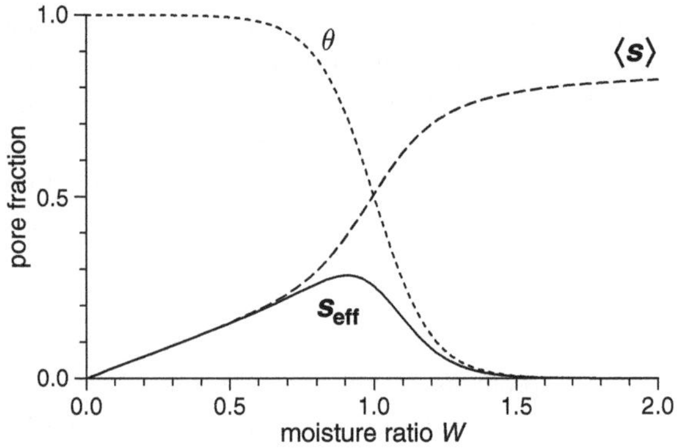

Fig. 25.8 Average fraction $\langle s \rangle$ of wet pores, fraction θ of wet pores that dry by evapotranspiration, and effective fraction s_{eff} of pores where magnetic enhancement occurs, as a function of the moisture ratio W

time interval that is sufficiently long to overcome the statistical nature of λ and W.

On the other hand, low-temperature oxidation of pedogenic magnetite proceeds with a diffusion-controlled kinetics. The fraction δ of Fe^{2+} ions that diffuse out of a magnetite sphere with radius r is given by:

$$\delta = \frac{6}{\sqrt{\pi}} \left(\frac{Dt}{r^2} \right)^{1/2} - 3\frac{Dt}{r^2} \tag{25.16}$$

where D is a temperature-dependent diffusion constant ($D \approx 1.3 \times 10^{-24}$ $\mathrm{m^2 s^{-1}}$ at 24°C), and t is time (Tang et al. 2003). Typical pedogenic magnetite particles have a radius of the order of 10 nm (Maher 1998), so that $Dt/r^2 \ll 1$ over one month: in this case, the second term in Eq. (25.16) can be neglected, and $\delta \propto (Dt)^{1/2} r^{-1}$. The average time interval during which a particle is uninterruptedly exposed to oxygen can be assumed to be inversely proportional to R. Therefore, the mass of the oxidized layer will be proportional to δ, which is in turn proportional to $(D/R)^{1/2} r^{-1}$. This is the amount of pedogenic magnetite that is subsequently removed under reducing conditions. Because the frequency by which reduction spots are activated is λ, the rate q of pedogenic magnetite dissolution is:

$$q \propto \frac{\lambda}{r} \sqrt{\frac{D}{R}} \, m. \tag{25.17}$$

If we assume that the grain size distribution of pedogenic magnetite is climate-independent, the average grain size r in Eq. (25.17) is constant and can be neglected, along with D, which is a material property. The grain size distribution of pedogenic minerals can be estimated using the blocking temperature distribution obtained from the temperature dependence of the out-of-phase susceptibility $\chi_q(T)$ (Shcherbakov and Fabian 2005, Egli 2009). Measurements of $\chi_q(T)$ for a series of CLP paleosols with various magnetic enhancement degrees, supposedly formed under different rainfall regimes, confirm that the blocking temperature distribution does not vary significantly (Liu et al. 2005). Therefore, we will neglect grain size effects in Eq. (25.17).

At equilibrium, the production and dissolution rates given by Eq. (25.15) and Eq. (25.17) must equal each other. This is possible when

$$m \propto s_E R^{1/2} W\theta(W_0 - W). \qquad (25.18)$$

Equation (25.18) describes the equilibrium concentration of pedogenic magnetite in terms of rainfall (R), type of climate (W), and soil properties (s_E). Monthly climatic tables can be used to evaluate Eq. (25.18) over the year. In analogy with the "rainfall effectiveness" introduced by Thornthwaite (1931), we define the *magnetite enhancement proxy*, MEP, as:

$$\mathrm{MEP} = \frac{\hat{E}_0^{1/2}}{s_0} s_E \sum_{k=1}^{12} \frac{R_k^{3/2}}{\hat{E}_k} \theta\left(W_0 - \frac{R_k}{\hat{E}_k}\right), \qquad (25.19)$$

where R_k and $\hat{E}_k$ are monthly values of rainfall and potential evapotranspiration, respectively, and $W_0 = 1.2$. The constant $\hat{E}_0$ is an arbitrarily chosen PET value which ensures that MEP and its monthly values have the same unit as rainfall. Similarly, s_0 is an arbitrarily chosen value of s_E. Normalization of Eq. (25.19) by $\hat{E}_0$ and s_0 eases the comparison between rainfall and MEP, which can be regarded as the "effective rainfall" driving the production of pedogenic magnetite. Given the large number of publications on the magnetic enhancement of soils and paleosols from the CLP, we choose $s_0 = 0.3$ as the typical s_E of those soils, and $\hat{E}_0 = 100$ mm/month, which is comparable with the maximum monthly PET of the region.

In order to understand the proxy defined by Eq. (25.19), we consider a set of similar soils collected from a region with a defined type of climate. We assume that these soils never experienced a water saturation regime, in which case $\theta = 1$. A typical example is the CLP when MAR < 1000 mm/yr. In this case, magnetic enhancement is proportional to $R^{3/2}$, providing a climofunction that fits the modern CLP soil data as well as Eq. (25.1) does. However, MEP also depends on the additional parameters $\hat{E}$ and s_E, which do not necessarily co-vary with MAR. This can explain the relatively large $\Delta\chi$ scatter (about a factor of 2) of modern loessic soils of the same region and similar MAR (e.g. Han et al. 1996, Porter et al. 2001, Maher et al. 2003, Geiss and Zanner 2007). Because age and parent material heterogeneities are negligible in this case, the scatter must originate from "hidden" climatic parameters that are taken into consideration by MEP. Therefore, we expect MEP, which can be calculated from climatic tables, to correlate much better with magnetic enhancement parameters. Furthermore, regional climofunction differences, such as those existing between loessic soils from China and from Midwestern U.S., should disappear if MAR is replaced by MEP.

25.4 Model Verification

In this section we test the magnetite enhancement proxy (MEP) defined in Eq. (25.19), by considering ratios of the form *M*/MEP, where M is a magnetic enhancement parameter (i.e. $\Delta\chi = \chi_B - \chi_C$, χ_{fd}, and χ_{ARM}), and comparing them with *M*/MAR. We expect *M*/MEP to be less scattered than *M*/MAR for any group of soils whose magnetic enhancement occurs as described in Section 25.3, with no regional differences. Since *M*/MEP is a constant in the ideal case, we call this ratio *magnetic enhancement factor*. Persisting regional differences would indicate that the magnetic enhancement mechanism assumed to calculate MEP is not universally valid, or that pedogenic magnetite has not been preserved, for example because of gleyzation. Magnetic enhancement factors of soils from the regions discussed in Section 25.2 are summarized in Tables 25.5 and 25.6 and discussed in the following.

25.4.1 Modern Soils on the CLP

We begin our discussion with the magnetic enhancement threshold at MAR = 900–1100 mm/yr observed for modern soils on the CLP (Han et al. 1996, Porter et al. 2001). Monthly climatic data for two sites

Table 25.5 Magnetic enhancement factors of modern soils discussed in the text

Region	$\Delta\chi$ [10^{-8} m^3/kg] / MEP [mm/yr]	χ_{fd} [10^{-9} m^3/kg] / MEP [mm/yr]	χ_{ARM} [10^{-8} m^3/kg] / MEP [mm/yr]
Chinese Loess Plateau (MAR < 800 mm/yr)	0.18 ± 0.03	0.20 ± 0.03[a]	1.1 ± 0.1[b]
SE China (MAR > 1000 mm/yr)	~0.1	–	–
Russian steppe	0.14–0.18	–	–
Midwestern U.S.	0.10 ± 0.02	–	1.1 ± 0.2
Alaska	–	~0.09	–
Argentina (well drained only)	~0.15	~0.17	–
Montilla (Spain)	0.20 ± 0.08	0.29 ± 0.1	1.05 ± 0.3

[a]Based on the empirical law $\chi_{fd} = 0.112(\chi - \chi_0)$ obtained from data in (Forster et al. 1994, Vidic et al. 2000).
[b]Based on the empirical ratio $\Delta\chi/\chi_{ARM} = 0.165 \pm 0.02$ for well developed Chinese paleosols (Liu et al. 2004).

Table 25.6 Comparison of soil magnetic enhancements (calculated from most reliable magnetic parameters), relative to the CLP

Region	Magnetic enhancement in % of CLP soils with same MAR	Magnetic enhancement in % of CLP soils with same MEP
Chinese Loess Plateau (MAR < 800 mm/yr)	100	100
SE China (MAR > 1000 mm/yr)	~6	~56
Midwestern U.S.	~45	~100
Alaska	~34	~60
Argentina (well drained only)	~27	~85
Montilla (Spain)	~63	~115

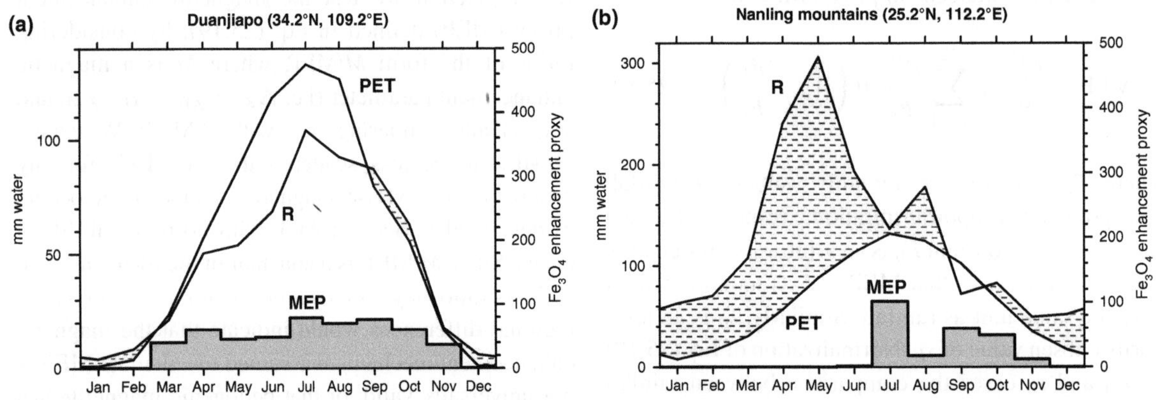

Fig. 25.9 Monthly values of rainfall R, potential evapotranspiration PET, and magnetite enhancement proxy MEP for two sites in China: Duanjiapo (MAR = 578 mm/yr, PET = 704 mm/yr, MEP = 514 mm/yr), and the Nanling mountains (MAR = 1550 mm/yr, PET = 792 mm/yr, MEP = 240 mm/yr). Dashed areas correspond to periods of water excess characterized by $R >$ PET

across this threshold (Duanjiapo, MAR = 578 mm/yr, and Nanling mountains, MAR = 1552 mm/yr) are shown if Fig. 25.9, together with MEP estimates obtained by assuming $s_E = 0.3$ for both soils. In Duanjiapo, monthly PET values are larger than rainfall during spring and summer and about equal in autumn. A different situation occurs in winter, when smaller PET values result from lower temperatures. Elevated W in December and January ensure that the soil is permanently wet, and no magnetite formation is expected during this time. Rainfall is low in December and January as well, meaning that pedogenic magnetite accumulation is not important during wintertime, regardless of PET. Therefore, both rainfall and MEP predict magnetic enhancement to occur mainly between March and November, with maximum values in summer (Fig. 25.9). Using Eq. (25.19) we calculate MEP = 514 mm/yr, which is 90% of MAR.

In southern China, rainfall has a strong monsoonal character, with a maximum monthly value of 300 mm in May. This value corresponds to the total yearly rainfall at the Northern boundary of the CLP. PET is comparable with Duanjiapo; however, rainfall is larger than PET over the entire year, except for September. Maximum values of W are concentrated in winter and spring, and $W < 1.4$ between July and October, which is the only period of the year when pedogenic magnetite production is expected. The resulting MEP is 240 mm/yr, which is only 15% of MAR, and 46% of MEP calculated for Duanjiapo. The susceptibility enhancement expected for this site is $\Delta\chi \approx 250$ mm^3/kg (Han et al. 1996), which is ~1/3 of the susceptibility measured at Duanjiapo (Table 25.2). MEP estimates agree much better with susceptibility measurements than MAR, and correctly predict the non-monotonic dependence of $\Delta\chi$ on MAR. The residual differences in $\Delta\chi$/MEP might reflect additional effects from mountain topography (runoff), or gleyzation, which are not accounted by Eq. (25.19).

25.4.2 Red Mediterranean Soils

Warm Mediterranean climates are characterized by an extended summer drought (June-September) with rainfall concentrated during the winter time, when PET is low and $W > 1$ (Fig. 25.10). Pedogenic magnetite is expected to accumulate only during few rainy months in spring and autumn. The area near Montilla sampled by Torrent and Cabedo (1986) is characterized by MAR = 600 mm/yr and a MEP of only 246 mm/yr, which explains the lower magnetic enhancement with respect to soils with similar MAR in China. Larger MAR values – obtained for example by multiplying the monthly R values with a constant factor >1 – would produce even lower MEP values. The same reasoning can be used to show that the maximum enhancement for this type of climate is expected to occur at MAR = 600 mm/yr.

The magnetic enhancements of the two soil profiles (Montilla and La Ramba) differ by a factor of two (Torrent et al. 2010), which cannot be explained by climatic differences between sites. Geochemistry and magnetic mineralogy of the two soils are similar as well. Interestingly, the product between the enhanced horizon thickness (~60 cm in Montilla and ~120 cm in La Ramba) and magnetic enhancement is constant. Nevertheless, $\Delta\chi$/MEP, χ_{fd}/MEP and χ_{ARM}/MEP values obtained from the *average* of the two profiles are similar to the corresponding factors calculated for Chinese soils. Different magnetic enhancements at Montilla and La Ramba could arise from the extreme sensitivity of MEP to the onset of a saturation regime in this type of climate. A slight shift of the saturation threshold W_0 – caused for example by different values of the hydraulic conductivity – is sufficient to change the duration of magnetite production

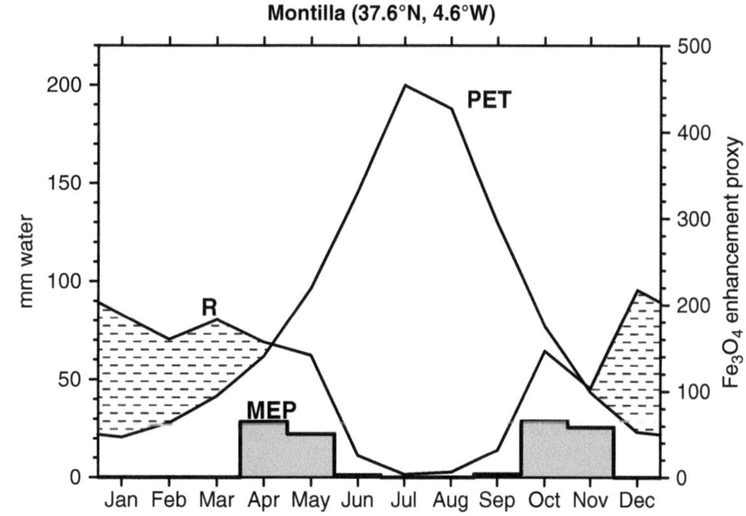

Fig. 25.10 Monthly values of rainfall R, potential evapotranspiration PET, and magnetite enhancement proxy MEP for Montilla (Spain): MAR = 600 mm/yr, PET = 1056 mm/yr, MEP = 246 mm/yr. Dashed areas correspond to periods of water excess characterized by $R >$ PET

and thus MEP. For example, a 15% MEP increase is obtained when $W_0 = 1.2$ in Eq. (25.19) is replaced by $W_0 = 1.4$. Hydraulic conductivity is strongly dependent on soil porosity, which decreases with depth (Lind and Lundin 1990, Youngs and Goss 1988). Therefore, the thinner active soil layer in Montilla is expected to have a larger hydraulic conductivity, a higher saturation threshold W_0, and a larger magnetic enhancement, as indeed observed.

Modern loessic soils on the Matmata Plateau (Tunisia) have formed in a similar climate with rainfall concentrated between December and March, when $W > 1$, and a long summer drought (Dearing et al. 2001). MAR – PET values in excess of 200 mm/yr can be inferred from clay pervection and the development of blocky prismatic structures (Dearing et al. 2001). As for the case depicted in Fig. 25.10, pedogenic magnetite is expected to form during short periods just before and after summer drought. Interestingly, the maximum magnetic enhancement is observed at MAR $\cong$ 500 mm/yr (Dearing et al. 2001), close to 600 mm/yr limit calculated using the climatic data of Fig. 25.10. Magnetic enhancement of soils with similar MAR is highly variable, as seen by typical ratios of ~5 between highest and lowest susceptibility values. The same scatter is observed for χ_{fd}, excluding parent material variability as a possible explanation. As discussed for the two soil profiles in Spain, highly variable magnetic enhancement are expected because of the sensitivity of MEP to small climatic differences and to soil drainage capability.

25.4.3 Volcanic Soils from the Kohala Peninsula, Hawaii

The case study of Hawaiian volcanic soils collected across a strong rainfall gradient provides an interesting test for our enhancement model. Although reliable magnetic enhancement estimates are complicated by the strong magnetic signature of the underlying lava rocks and possible weathering effects on primary magnetic minerals, a clear, non-monotonic dependence of magnetic susceptibility on rainfall can be recognized (Singer et al. 1996). Maximum magnetic enhancement occurs when MAR $\approx$ 1000 mm/yr: above this threshold, χ declines and becomes highly variable. The climatic gradient is primarily determined by altitude, with MAR values increasing from 160 mm/yr on the coast, to 3000 mm/yr at maximum altitude. Evapotranspiration decreases moderately with altitude along with temperature: measured values range from 2200 mm/yr at sea level to 1000 mm/yr at maximum altitude (Shade 1995). Because of opposed MAR and PET trends, the yearly mean soil moisture ratio W increases from ~0.08 at sea level, to ~3 at maximum altitude, with monthly peak values >10. Using the climatic data in (Shade 1995, Chadwick et al. 2003), we calculated MEP values for the sites measured by Singer et al. (1996). Although a direct comparison between χ and MEP should be interpreted with caution because of the strong magnetic signature of the parent material, a common trend can be recognized (Fig. 25.11). Interestingly, MEP estimates become

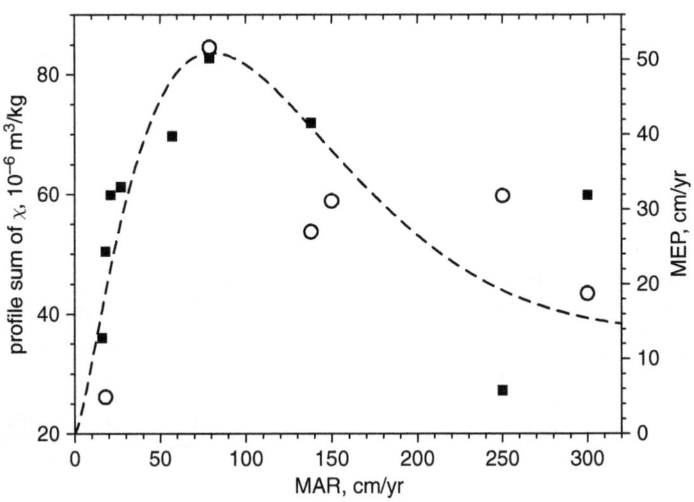

Fig. 25.11 Mean annual rainfall vs. profile sum of magnetic susceptibility (*squares*) for nine soils from the Kohala peninsula, Hawaii (Singer et al. 1996), compared with MEP (*circles*) calculated for sites with available monthly climate data (Shade 1995). The dashed line is a guide for the eye

highly variable at large MAR values, as seen also for χ measurements. The scatter can be explained by the "cutoff" effect occurring near $W = 1$, which makes the number of reduction spots subjected to wetting/drying cycles very sensitive to little changes of the soil moisture. The maximum susceptibility enhancement around MAR = 800 mm/yr coincides with maximum effective cation exchange capacity, and precedes the onset of strong leaching effects at larger MAR values (Chadwick et al. 2003).

The rainfall vs. MEP curve predicted by our model is similar to the rainfall vs. pedogenic magnetite model postulated by Balsam et al. (2004), with a similar peak at MAR $\approx$ 700 mm/y. Poor soil drainage or low PET values can shift the position of this peak to much lower MAR values. It is important to notice that our model does not require reductive dissolution effects to explain the decline of magnetic enhancement for $W > 1$, although gleyzation, if occurring, will strengthen the existing trend.

25.4.4 Comparison Between Loessic Soils from China and North America

In previous sections, we discussed magnetic enhancement examples that were mainly controlled by the decline of pedogenic magnetite production as soils become saturated with water. Most paleorainfall reconstructions, however, are based on loess/paleosol sequences where saturation effects are not sufficient to reverse the positive correlation with MAR. Calibration of susceptibility measurements for paleorainfall reconstruction purposes is based on the study of modern soils collected from worldwide warm temperate regions (Maher 1998, Maher et al. 1994, Maher and Thompson 1995, Maher et al. 2003). The extension of this archive to loessic soils in North America, however, clearly disagrees with the susceptibility-rainfall trends obtained for Asia (Geiss and Zanner 2007). The definition of regional-based climofunctions overcomes this problem, yet the apparent lack of a universal enhancement law in soils with very similar geological settings casts a cloud on the reliability of magnetic enhancement models.

Modern soils on Peoria loess are systematically less enhanced than soils from Asia formed in climates with similar MAR: $\Delta\chi$ and χ_{ARM} are only ~1/3 and ~1/2 of those of Chinese soils, respectively (Fig. 25.12). Pedogenic ARM, defined as the difference between χ_{ARM} of the maximum enhanced horizon and the parent loess, correlates significantly with MAR (r^2 = 0.84) while $\Delta\chi$ does not ($r^2 = 0.3$). A reason for the discrepancy between these two magnetic enhancement

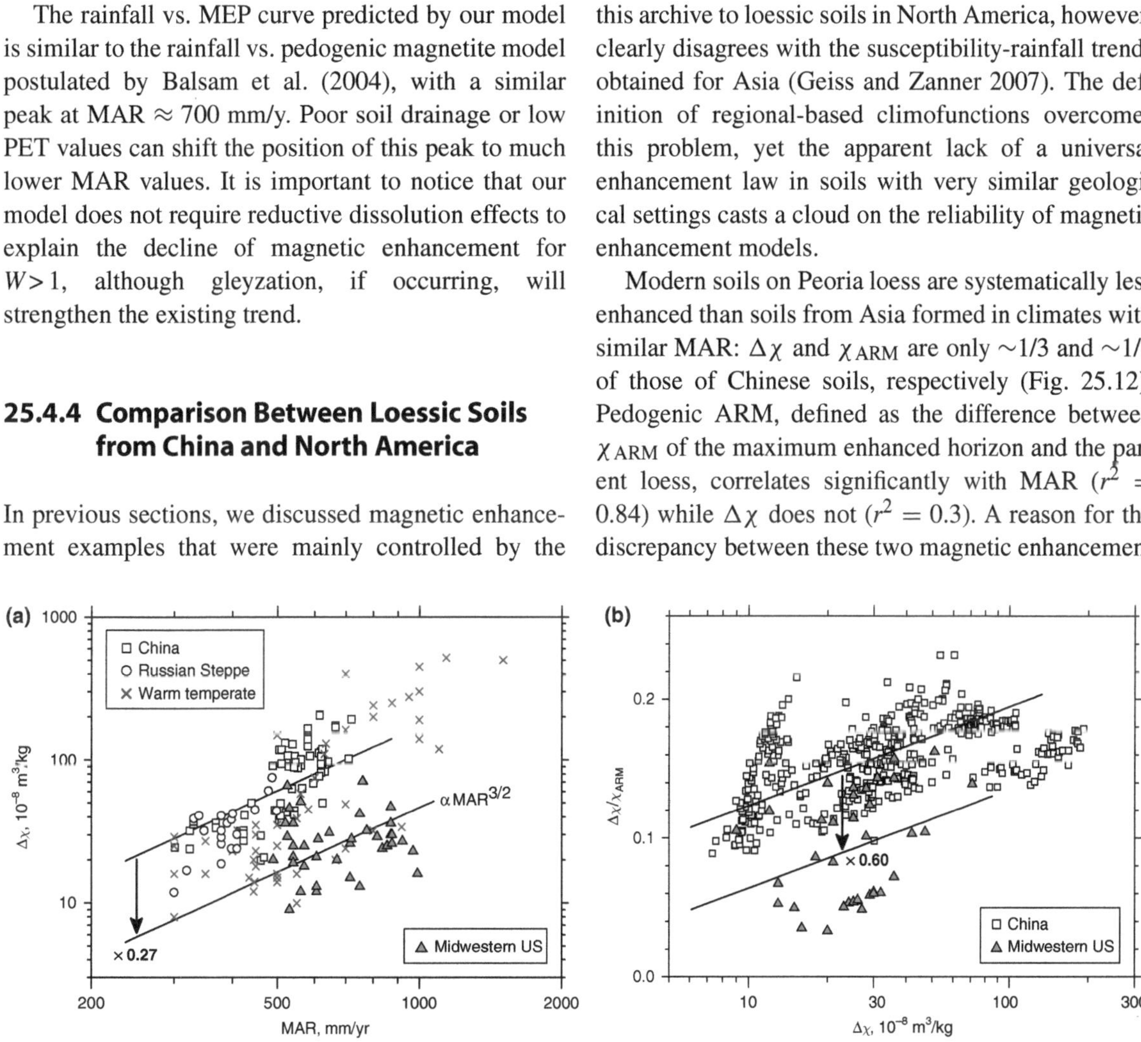

Fig. 25.12 (**a**) Mean annual rainfall vs. susceptibility enhancement $\Delta\chi$ of modern soils in China (Porter et al. 2001), Russian Steppe (Maher et al. 2003), worldwide sites on the northern hemisphere with a warm temperate climate (Maher and Thompson 1995), and Midwestern U.S. (Geiss and Zanner 2007). Lines are least squares fits $\propto$ MAR$^{3/2}$ for China and Russian Steppe, and for Midwestern U.S., respectively. The difference between the two fits defines a ratio of 0.27 between susceptibility enhancements in Asia and Midwestern U.S. (**b**) Susceptibility enhancement $\Delta\chi$ vs. $\Delta\chi/\chi_{ARM}$ for Chinese paleosols (Liu et al. 2004) and modern soils from Midwestern U.S. (Geiss and Zanner 2007). Because of highly magnetic parent material, ARM for the U.S. soils is calculated as the difference between maximum enhanced horizon and the parent loess. Lines are least squares fits for Midwestern U.S. and China, calculated over the susceptibility interval covered by the US soils. The $\Delta\chi/\chi_{ARM}$ ratios of U.S. soils are 60% of those of Chinese paleosols

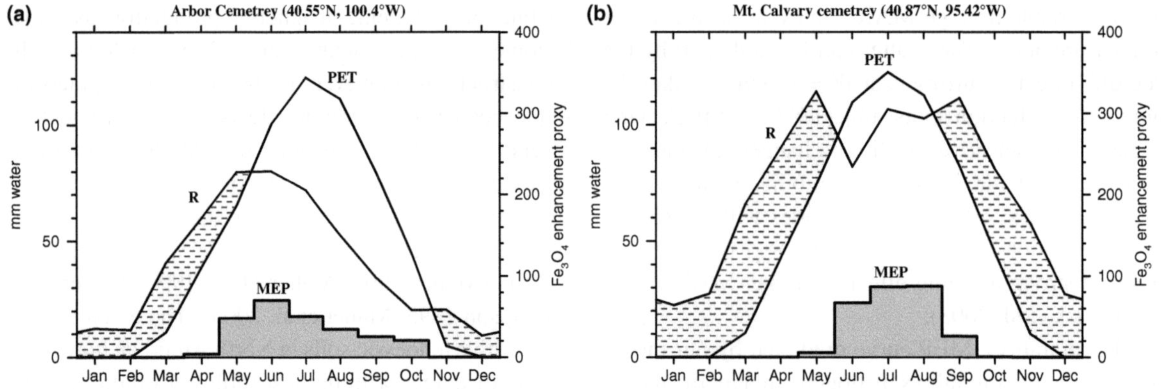

Fig. 25.13 Monthly values of rainfall R, potential evapotranspiration PET, and magnetite enhancement proxy MEP for two sites on Peoria loess (Midwestern US): Arbor cemetery (MAR = 494 mm/yr, PET = 577 mm/yr, MEP = 253 mm/yr), and Mt. Calvary cemetery (MAR = 890 mm/yr, PET = 611 mm/yr, MEP = 273 mm/yr). Dashed areas correspond to periods of water excess characterized by $R >$ PET

parameters could be related to the fact that the assumption of a depth-independent contribution of primary minerals, which underlies the common definition of magnetic enhancement, is not valid. This hypothesis is supported by magnetic unmixing analyses of selected soil profiles, which shows that the magnetization of the coercivity component attributed to non-pedogenic minerals changes with depth (Geiss and Zanner 2006). A depth-dependent magnetic background can result from weathering effects and/or syn-pedogenic dust inputs (Bettis et al. 2003a, b). Because of its strong selectivity to SD particles, ARM is less affected by the parent material and should be considered, at least in this case, a more reliable indicator for the concentration of pedogenic magnetite. Geiss and Zanner (2007) observed that relative enhancement parameters (defined as the ratio between maximum enhancement horizon and parent loess) correlate systematically better with MAR that absolute enhancement parameters, and suggested that magnetic enhancement is limited by the Fe supply from the parent material. This hypothesis is in contradiction with our model, which assumes no Fe limitation.

Monthly climatic data are shown in Fig. 25.13 for two sites: Arbor cemetery, Nebraska (MAR = 494 mm/yr), and Mt. Calvary cemetery, Iowa (MAR = 890 mm/yr). At Arbor cemetery, rainfall is >PET during winter and spring, and pedogenic magnetite is expected to form between May and October. The estimated MEP is 253 mm/yr, which is about 50% of MAR. For comparison, MEP in Duanjiapo is 90% of MAR. Climatic conditions are even less favorable to magnetite pedogenesis at Mt. Calvary cemetery, where rainfall is >PET over the entire year, except for three months in summer. Accordingly, MEP for this site is only 31% of MAR. All sites sampled by Geiss and Zanner (2007) are characterized by climatic parameters that are intermediate between the two examples of Fig. 25.13.

To quantify differences with Chinese soils, we compare magnetic enhancement and MEP of Peoria soils with typical soils on the CLP. A typical Chinese soil is characterized by $\Delta\chi/\text{MEP} \approx 0.18$ and $\chi_{\text{ARM}}/\text{MEP} \approx 1.1$, where $\Delta\chi$ and χ_{ARM} are expressed in 10^{-8} m^3/kg, and MEP in mm/yr (Table 25.5). On the other hand, $\Delta\chi$ and MEP/MAR of Peoria soils are ~27% and ~47% of the values for soils with same MAR on the CLP, respectively, as obtained from Fig. 25.12a and from climatic data of the sites listed in (Geiss and Zanner 2007). Using simple proportions for soils with same MAR, we estimate the magnetite enhancement factor of Midwestern U.S.:

$$\left(\frac{\Delta\chi}{\text{MEP}}\right)_{\text{Peoria}} = \frac{\Delta\chi_{\text{Peoria}}}{\Delta\chi_{\text{CLP}}}\frac{(\text{MEP/MAR})_{\text{CLP}}}{(\text{MEP/MAR})_{\text{Peoria}}}\left(\frac{\Delta\chi}{\text{MEP}}\right)_{\text{CLP}} \approx 0.1 \pm 0.02. \tag{25.20}$$

A similar reasoning can be applied to χ_{ARM}, knowing that $\Delta\chi/\chi_{\text{ARM}} \approx 0.091$ for Peoria soils on average (Fig. 25.12b), which is ~60% of the typical values over the CLP. Then,

$$\left(\frac{\chi_{ARM}}{MEP}\right)_{Peoria} = \left(\frac{\Delta\chi}{MEP}\right)_{Peoria}\left(\frac{\Delta\chi}{\chi_{ARM}}\right)_{Peoria}^{-1} \approx 1.1 \pm 0.2. \tag{25.21}$$

The χ_{ARM}/MEP ratio of CLP and Peoria soils can be considered identical within the standard errors produced by the large scatter of individual sites in Fig. 25.12. A significant difference between the two regions persists if $\Delta\chi$/MEP is considered; however, as discussed before, $\Delta\chi$ is not a reliable enhancement parameter on Peoria loess. Further investigation is needed to obtain unbiased estimates of the pedogenic susceptibility, for example by comparing measurements before and after selective CBD dissolution of pedogenic minerals (Vidic et al. 2000). From the preliminary results obtained with ARM measurements, we can reasonably conclude that a different magnetic enhancement mechanism is not required to explain magnetic data, and that the same model for pedogenic magnetite formation is valid for loessic soils in China and in the US, if climatic differences are taken into account.

25.4.5 Alaska

Loess/paleosol sequences from Alaska provide an interesting test for our model, due to extreme climatic conditions at high latitudes. There is a general agreement that pedogenic enhancement occurred in Alaskan paleosols, although not recognizable from susceptibility measurements, because of the strong opposite trend imposed by the wind-modulated, large concentration of primary magnetic minerals in loess. Nevertheless χ_{fd} is small compared to CLP soils (Table 25.2), suggesting that magnetic enhancement is weak, possibly because of magnetite dissolution due to gleyzation (Liu et al. 1999). The climatic parameters of Halfway House, the site where a loess/paleosol sequence has been characterized by magnetic measurements, are dominated by an extremely pronounced seasonality (Fig. 25.14). Evapotranspiration is absent during the winter months due to temperatures constantly below the freezing point and snow coverage, and rises to moderately large values during summer. Summer evapotranspiration is larger than expected from the relatively cool weather, because of the extended day length at high latitudes (Eq. 25.7). Rainfall is highest in summer, but much lower than evapotranspiration, and obviously occurs in form of snow during the wintertime. Pedogenic magnetite is therefore expected to accumulate during months with mean temperatures >0°C, when evapotranspiration is possible. An exception is given by May, when snow melts. Because a non-zero MEP is predicted for May by Eq. (25.19), while the soil is saturated with melt water, we ignore this month and estimate an annual MEP of 128 mm/yr, which is ~49% of MAR.

Magnetic enhancement estimates for Alaska rely on χ_{fd}, because susceptibility is dominated by the dust signal. Published χ_{fd} values refer to a paleosol from oxygen isotope stage 3 (Liu et al. 1999), which we compare with present day climate. Given these

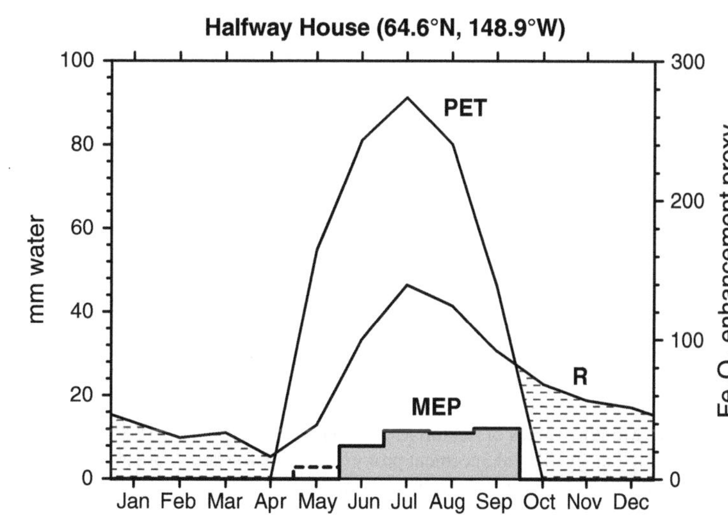

Fig. 25.14 Monthly values of rainfall R, potential evapotranspiration PET, and magnetite enhancement proxy MEP for Halfway House (Alaska): MAR = 264 mm/yr, PET = 354 mm/yr, MEP = 128 mm/yr. Dashed areas correspond to periods of water excess characterized by $R >$ PET. The calculated MEP contribution of May (dashed) was not taken into consideration because of snow melt

uncertainties, we obtain χ_{fd}/MEP $\approx$ 0.094, which is ~60% of the value for China (Table 25.5). The residual discrepancy could arise from a different climate during oxygen isotope stage 3, or from an incipient dissolution of pedogenic magnetite due to gleyzation.

25.4.6 Argentina

The last case we discuss in detail deals with modern soils from two sites on Pampean loess: Verónica (MAR = 885 mm/yr) and Zárate (MAR = 1115 mm/yr). This case is interesting because one site is magnetically enhanced and the other is not, despite similar climates (Orgeira and Compagnucci 2006). Two soil profiles have been measured at each site, one located in the upper watersheds, and the other in a depression with poorer drainage (Orgeira et al. 2008). Besides the orographic differences between profiles of the same site, the Verónica and Zárate soils have distinct textures. Higher clay content in Zárate reduces the saturated hydraulic conductivity (Table 25.3), and thus drainage. Using hydraulic conductivity data of soils with various clay contents (Clapp and Hornberger 1978) and the textural data in (Orgeira et al. 2008), we obtain $K_s \approx$ 100 mm/day for both soils in Verónica, and $K_s \approx$ 380 mm/day for both soils in Zárate. Soils in Verónica are therefore systematically less drained in comparison with Zárate, regardless of the orographic setting. Only the soil profile with best water drainage (upper watershed in Zárate) is magnetically enhanced in the uppermost horizon.

The climatic data of Fig. 25.15 can be used to estimate the expected magnetic enhancement. Monthly rainfall values are constantly larger than PET at both sites, creating a stable saturation regime in the soil. At Verónica, the difference between PET and rainfall is less marked during the summer months, where most of the pedogenic magnetite is expected to form. Only little magnetic enhancement is expected at Zárate, as testified by MEP = 259 mm/yr, which is ~1/4 of MAR. In analogy with the discussion about North America and Alaska, we compare the magnetite production expected from MEP with the actual enhancement of the soil profile in Zárate. Using the data of Table 25.2, we obtain $\Delta\chi$/MEP $\approx$ 0.15 and χ_{fd}/MEP $\approx$ 0.17. These values are similar to our estimates for Chinese soils.

25.4.7 Brief Overview of Other Sites

Having discussed the most relevant soil sites in the previous sections, we conclude with a brief overview of other sites. A rapid assessment of the climatic conditions relevant for pedogenesis is provided by the annually averaged moisture ratio W = MAR/PET, which represents the net balance between water input by rainfall and loss by evapotranspiration. We recall that $W \approx 1$ represent the threshold above which the production of pedogenic magnetite is expected to decline.

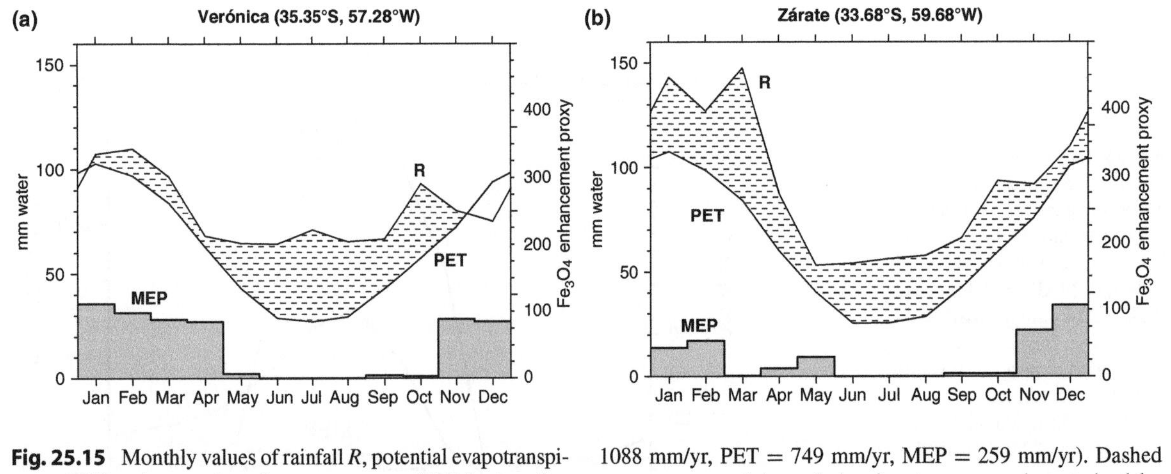

Fig. 25.15 Monthly values of rainfall R, potential evapotranspiration PET, and magnetite enhancement proxy MEP for two sites on Pampean loess (Argentina): Verónica (MAR = 961 mm/yr, PET = 741 mm/yr, MEP = 457 mm/yr), and Zárate (MAR = 1088 mm/yr, PET = 749 mm/yr, MEP = 259 mm/yr). Dashed areas correspond to periods of water excess characterized by $R >$ PET

Yearly averages obscure many of the previously discussed details, and do not provide a sufficient basis for a quantitative magnetic enhancement analysis. This is particularly the case for climates with very strong and opposed seasonal dependences of rainfall and evapotranspiration. For example, the warm Mediterranean climate discussed in Section 25.4.2 is characterized by MAR/PET = 0.568, but a monthly analysis shows that rainfall is larger than evapotranspiration during most part of the rainy season.

Nevertheless, *W* is useful to identify regions of the world where the use of soil magnetic properties as continental climatic proxies might be problematic. For this purpose we compiled maps of *W* for different regions of the World using the University of Delaware's climatic database (Delaware University 2009). This database contains long-term monthly means of air temperature and rainfall obtained from land climatic station and interpolated on a 0.5° latitude/longitude global grid. These maps provide a rough climate classification based on criteria that are similar to those used by Thornthwaite (1931, 1948).

The dominant feature on the map of central Asia (Fig. 25.16) is a strong N-S gradient along which *W* increases from <0.4 on the central CLP to >1.4 in SE China, where the climate is controlled by the eastern summer monsoon and the Indian monsoon. The $W = 1$ isoline is located at ~33°N, and coincides with the MAR = 1100 mm/yr isoline that Han et al. (1996) identified as the maximum enhancement limit. When moving from the central CLP towards Siberia, *W* increases again, this time because of the temperature dependence of PET. The magnetic enhancement of loess/paleosol sequences collected at two localities in Siberia, Lozhok ($W = 0.87$) and Novokuznetsk ($W = 1.11$), follow the trend predicted by *W*, with paleosols in Novokuznetsk being less enhanced than in Lozhok (Table 25.2).

Europe (Fig. 25.17) is characterized by a complex pattern of *W* with overall minimum values towards SE, and *W* slightly >1 in spots centered over Austria and E of the Black Sea. This situation is reflected by the magnetic enhancement of loess/ paleosol profiles in the Czeck Republic and Romania, as well as modern soils from the Russian Steppe (Table 25.2). Parent material heterogeneities and $W \approx 1$ are the main reason for susceptibility-MAR correlations that are not as high as over the CLP.

Peoria loess in the Midwestern U.S. is characterized by a strong EW gradient of *W*, with $W < 1$ in Nebraska, $W \approx 1$ along the Missouri river, and $W > 1$ over Missouri, Iowa, and Minnesota (Fig. 25.18). As discussed in Section 25.4.4, these values are characteristic of a climate that is less favorable to magnetic enhancement. A similar situation is encountered in Alaska, where $W \approx 1$ on average.

Pampean loess extends over regions where *W* is generally >1 with some exceptions in the S (Fig. 25.19).

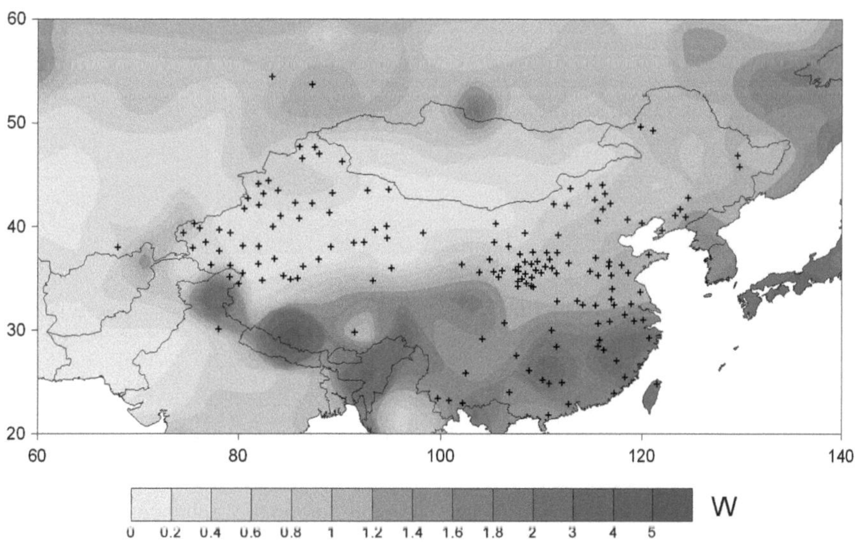

Fig. 25.16 Map of annually averaged $W =$ MAR/PET for Asia. Crosses are soil sample sites on the CLP (Han et al. 1996), and two sites in Siberia (Kravchinsky et al. 2008)

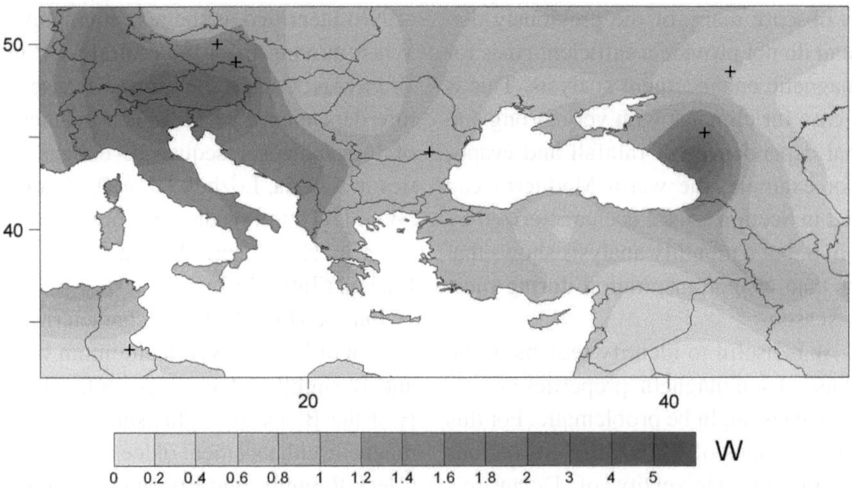

Fig. 25.17 Map of annually averaged W = MAR/PET for Central and Eastern Europe. Crosses are soil sample sites in the Czech Republic (Forster et al. 1996), Romania (Panaiotu et al. 2001), Russian Steppe (Maher and Thompson 1995), and Tunisia (Dearing et al. 2001)

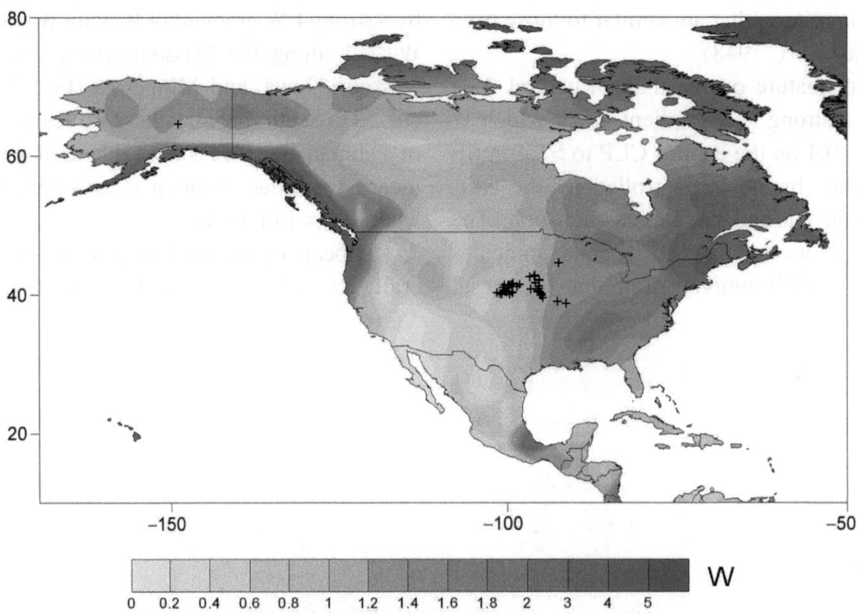

Fig. 25.18 Map of annually averaged W = MAR/PET for North America. Crosses are soil sample sites on Peoria loess (Geiss and Zanner 2007), and Alaska (Liu et al. 1999, Lagroix and Banerjee 2002)

The case of two localities in the Buenos Aires province with $W > 1$ has been discussed in Section 25.4.6. The magnetic enhancement of Pampean soils seems to be more sensitive to water saturation than soils with similar W and MAR values in Asia. For example, a clear enhancement pattern is missing for both soil profiles in Verónica ($W = 1.3$), partially because of poor drainage (Orgeira and Compagnucci 2006). For comparison, magnetic enhancement is observed in soils from a third locality SE of the Buenos Aires Province, where $W \approx 1$ (Bartel et al. 2006), and all soils with $W \approx 1.3$ and MAR $\approx$ 1000 mm/yr in SE China are

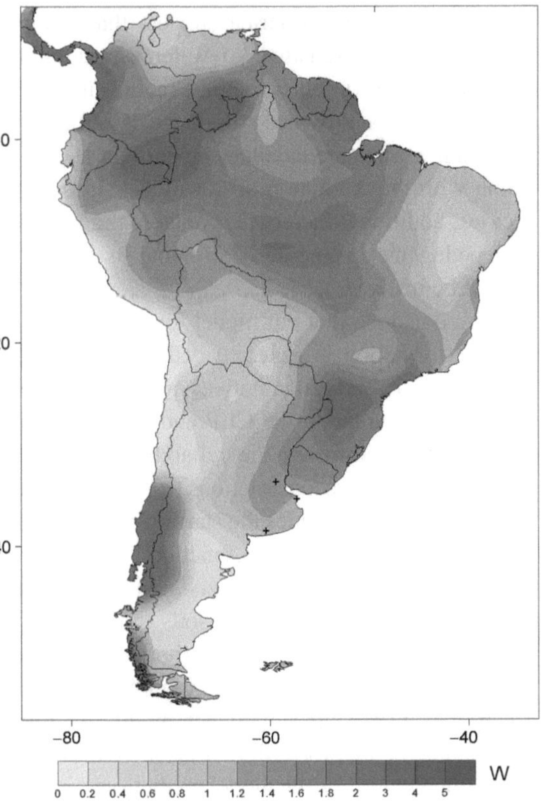

Fig. 25.19 Map of annually averaged W = MAR/PET for South America. Crosses are the soil sample sites near Zárate and Verónica (Orgeira et al. 2008), and SE of the Buenos Aires Province (Bartel et al. 2006)

strongly enhanced (Han et al. 1996). We must conclude from these observations that magnetic enhancement in Argentina is *completely* suppressed above $W = 1.1$–1.3 – yet the magnetite enhancement proxy defined in Eq. (25.19) predicts a gradual decrease and values clearly above zero even for $W = 1.4$ (e.g. North America, Table 25.2).

A possible reason for the elevated sensitivity of Argentinean soils to water saturation regimes is the high concentration of volcanic glass in the Pampean loess. Glass dissolution is expected to release silica in the soil pore water, as seen from the occurrence of trace elements present in the glass, such as As, in groundwater (Nicolli et al. 2004). High silica concentrations can have a twofold effect on magnetic enhancement, by promoting magnetite dissolution (Florindo et al. 2003), and by preventing the crystallization of Fe minerals, as observed in the case of ferrihydrite (Schwertmann and Cornell 2000).

Conclusions

We developed a quantitative description of the climatic modulation of soil magnetic enhancement that integrates Maher's (1998) conceptual model of magnetite formation by pedogenesis with a statistical treatment of the soil water balance (Rodriguez-Iturbe et al. 1999). Our model is based on the following assumptions:

(1) Magnetic enhancement is mainly caused by ultrafine magnetite and maghemite particles.
(2) Pedogenic magnetite forms in "reduction spots" characterized by a microscale redox gradient. Magnetite precipitation is not limited by Fe availability, whose concentration in form of easily reducible minerals is much larger than the magnetite product.
(3) The formation of pedogenic magnetite requires alternating drying and wetting phases. Wetting phases are necessary to support oxygen respiration by microorganisms and the formation of reduction spots. Drying phases are necessary for the precipitation of poorly crystalline Fe(III) minerals, such as ferrihydrite, that serve as a Fe source during the next wetting phase.
(4) The rate of magnetite production is proportional to the frequency of drying/wetting cycles, and to the fraction of wet pores in the soil. The drying/wetting frequency is controlled by rainfall, while the fraction of wet pores, which is related to the soil moisture, is controlled by the balance between water input by rainfall, and water loss by evapotranspiration.
(5) The moisture ratio W, defined as the ratio between rainfall and potential evapotranspiration, controls the average soil moisture. A threshold given by $W_0 \approx 1$ marks the transition from relatively dry conditions to water saturation. The onset of a saturation regime at $W > W_0$ decreases the frequency of drying/wetting cycles and suppresses magnetite production. The saturation threshold is lower ($0.5 < W_0 < 1$) in poorly drained soils.
(6) During the drying phase, surface oxidation of pedogenic magnetite produces a maghemite shell whose growth is controlled by the outward diffusion of Fe^{2+} ions.
(7) If a partially oxidized magnetite particle is exposed to reducing conditions during the wetting phase, the maghemite shell is dissolved.

Maghemitization is thus responsible for the destruction of pedogenic magnetite in an active soil. The destruction rate is proportional to the maghemite shell mass, which depends on the average time interval between rain events, and to the frequency of drying/wetting cycles.

(8) The concentration of pedogenic magnetite at equilibrium is obtained by solving a mass balance equation that includes production and destruction processes. The solution defines a climate-dependent proxy for magnetite concentration in soils, which we called *magnetite enhancement parameter* MEP (Eq. (25.19)). MEP has the same unit as MAR, and is identical to MAR for a reference case chosen to represent typical loessic soils on the CLP. MEP is proportional to the 3/2 power of rainfall, and inversely proportional to evapotranspiration. It also depends on soil properties and vegetation cover. Magnetic parameters used to describe pedogenic magnetite concentration, such as the susceptibility enhancement $\Delta\chi$, the frequency dependence χ_{fd} of susceptibility, and the anhysteretic remanent magnetization ARM, are expected to be proportional to MEP. The proportionality factor is called *magnetic enhancement factor*.

(9) MEP can be calculated from monthly rainfall and PET estimates available from climatic tables. A comparison between calculated MEP values and measured magnetic enhancement provides a test for our model. Site-independent enhancement constants are expected if the model is correct.

We tested our model using magnetic enhancement data for three main loess deposit regions: Asia (China and Siberia), North America (Peoria loess and Alaska), and South America (Pampean loess). In general, our model reduces or eliminates discrepancies observed when comparing magnetic parameters with MAR. It explains the differences between magnetic enhancements in North America and on the CLP in terms of climates characterized by a different ratio between rainfall and evapotranspiration. The estimated magnetic enhancement factors for the two regions agree within the confidence limit given by the variability of individual sites. The model also accounts for the enhancement pattern observed in SE China. On the other hand, results for South America suggest that Pampean soils are more sensitive to water saturation than loessic soils from other regions. In one site (Zárate, Buenos Aires Province), drainage is sufficient to guarantee a magnetic enhancement that is compatible with the prediction of our model. In the other cases, additional factors that prevent the formation of pedogenic magnetite or promote its dissolution must be invoked.

We conclude that our model for pedogenic magnetite formation is generally valid and is capable of accurate and quantitative magnetic enhancement predictions on the CLP and on Peoria loess. This result reinforces the reliability of continental paleoclimate reconstructions from loess/paleosol sequences, and provides the starting point for establishing a universal "climofunction" relating magnetic and climatic parameters. Our model takes some physical properties of soils related to water drainage into account, but it ignores geochemical parameters, such as pH, weathering rate, and cation exchange capacity, which might affect pedogenic magnetite production or promote dissolution. A comparison between model predictions and effective magnetic enhancement in cases where only a qualitative agreement can be obtained might provide the basis for incorporating important soil geochemical parameters, improving our understanding of magnetic mineral pedogenesis.

Acknowledgments We are grateful to Barbara Maher, Christoph Geiss, and an anonymous reviewer for insightful comments that improved our original manuscript significantly.

Appendix: The Susceptibility of Pedogenic Magnetite

Quantitative estimates of pedogenic magnetite concentrations are useful for evaluating the relative abundance of different pedogenic minerals. Because of typical mass concentrations <1%, pedogenic magnetite is not detected by X-ray diffraction and Mössbauer spectrometry on bulk samples. Estimates based on magnetic extracts are generally not reliable, because the extraction efficiency is grain size dependent. On the other hand, the concentration of pedogenic magnetite can be quantified from magnetic measurements if (1) its contribution is separable from that of other minerals, and (2) its properties are known and identical within a large set of soils, for example those from the CLP.

In the following, we assume pedogenic magnetite to be fully oxidized to maghemite, which has a spontaneous magnetization $\mu_s = 73$ Am^2/kg (Dunlop and Özdemir 1997) and contains $n_{Fe} = 73\%$ Fe by weight. Pedogenic magnetite from the CLP is a mixture of SD and SP particles with invariant magnetic properties and a fixed grain size distribution. Magnetic properties of the pedogenic ferrimagnetic phase are defined by following ratios: $\Delta\chi/\chi_{ARM} = 0.165 \pm 0.02$ (Liu et al. 2004), $\chi_{ARM}/M_{rs} = (1.3 \pm 0.2) \times 10^{-3}$ m/A (Egli 2004, Liu et al. 2004), and $M_{rs}/M_s = 0.2 \pm 0.01$ (Liu et al. 2004). Assuming that maghemite is the *only* phase responsible for magnetic enhancement, the susceptibility of pedogenic ferrimagnetic minerals, normalized by their mass, is

$$X_p = \frac{\Delta\chi}{\chi_{ARM}} \frac{\chi_{ARM}}{M_{rs}} \frac{M_{rs}}{M_s} \mu_s = (3.1 \pm 0.6) \times 10^{-3}\ \text{m}^3/\text{kg}. \tag{25.22}$$

If susceptibility is normalized by the Fe mass of pedogenic ferrimagnetic minerals, we obtain $X_{p,Fe} = (4.3 \pm 0.8) \times 10^{-3}$ m^3/kg. The mass concentration C_p in the bulk sample can be estimated using the ratio between the susceptibility enhancement $\Delta\chi$ of the bulk sample and X_p : $C_p = \Delta\chi/X_p$. For example, the strongest susceptibility enhancement on the CLP is $\Delta\chi \approx 2 \times 10^{-6}$ m^3/kg which corresponds to a mass concentration $C_p \approx 6 \times 10^{-4}$ or 0.06%.

References

Ameghino F (1909) Las formaciones sedimentarias de la región litoral de Mar del Plata y Chapalmalán. Anales del Museo Nacional de Buenos Aires 10:343–428

An Z (2000) The history and variability of the East Asian paleomonsoon climate. Quat Sci Rev 19:171–187

Anderson RS, Hallet B (1996) Simulating magnetic susceptibility profiles in loess as an aid in quantifying rates of dust deposition and pedogenic development, Quat Res 45:1–16

Appelo CAJ, Postma D (2005) Geochemistry, groundwater and pollution, 2nd edn. AA Balkema Publishers, Leiden

Balsam W, Ji J, Chen J (2004) Climatic interpretation of the Luochuan and Lingtai loess sections, China, based on changing iron oxide mineralogy and magnetic susceptibility. Earth Planetary Sci Lett 223:335–348

Banerjee SK (2006) Environmental magnetism of nanophase iron minerals: testing the biomineralization pathway. Phys Earth Planetary Inter 154:210–221

Banerjee SK, Hunt CP, Liu XM (1993) Separation of local signals from the regional paleomonsoon record of the Chinese loess plateau: a rock-magnetic approach. Geophys Res Lett 20:843–846

Barrón V, Torrent J (2002) Evidence for a simple pathway to maghemite in Earth and Mars soils. Geochim Cosmochim Acta 66:2801–2806

Bartel A, Bidegain JC, Sinito AM (2006) Señal de incremento magnético en suelos del sur de la región pampeana. X Jornadas pampeanas de Ciencias Naturales, Santa Rosa, La Pampa, Argentina (Actas)

Beer J, Shen C, Heller F et al (1993) ^{10}Be and magnetic susceptibility in Chinese loess. Geophys Res Lett 20: 57–60

Begét JE, Stone DB, Hawkins DB (1990) Paleoclimatic forcing of magnetic susceptibility in Alaskan loess during the late Quaternary. Geology 18:40–43

Bettis III EA, Muhs DR, Roberts HM et al (2003a) Last glacial loess in the conterminous USA. Quat Sci Rev 22:1907–1946

Bettis III EA, Mason JP, Swinehart JB et al (2003b) Cenozoic aeolian sedimentary systems of the USA mid-continent. In: Eastbrook DJ (ed) Quaternary geology of the United State. INQUA 2003, Reno, Nevada, 195–218

Bloemendal J, Liu X (2005) Rock magnetism and geochemistry of two plio-pleistocene Chinese loess-paleosols sequences – implications for quantitative paleoprecipitation reconstruction. Paleogeogr Palaeoclimatol Paleoecol 226:149–166

Bockheim JG (1980) Solution and use of chronofunctions in studying soils development. Geoderma 24:71–85

Bolikhovskaya NS, Molodkov AN (2006) East European loess-paleosol sequences: palynology, stratigraphy and correlation. Quat Int 149:24–36

Brimhall GH, Dietrich WE (1987) Constitutive mass balance relations between chemical composition, volume, density, porosity, and strain in metasomatic hydrochemical systems: results on weathering and pedogenesis. Geochim Cosmochim Acta 51:567–587

Chadwick OA, Chorover J (2001) The chemistry of pedogenic thresholds. Geoderma 100:321–353

Chadwick OA, Gavenda RT, Kelly EF et al (2003) The impact of climate on the biogeochemical functioning of volcanic soils. Chem Geol 202:195–223

Chlachula J, Evans ME, Rutter NW (1998) A magnetic investigation of a Late Quaternary loess/paleosol record in Siberia. Geophys J Int 132:128–132

Clapp RB, Hornberger GM (1978) Empirical equations for some soil hydraulic properties. Water Resources Res 14:601–604

Cornell RM, Schwertmann U (2003) The Iron Oxides, 2nd edn. Wiley-VCH

Crockford RH, Willett IR (1995) Magnetic properties of two soils during reduction, drying, and re-oxidation. Aust J Soil Res 33:597–609

Dearing JA, Hannam JA, Anderson AS et al (2001) Magnetic, geochemical and DNA properties of highly magnetic soils in England. Geophys J Int 144:183–196

Dearing JA, Hay KL, Baban SMJ et al (1996) Magnetic susceptibility of soil: an evaluation of conflicting theories using a national data set. Geophys J Int 127:728–734

Dearing JA, Livingstone IP, Bateman MD et al (2001) Paleoclimate records from OIS 8.0–5.4 recorded in loess-paleosol sequences on the Matmata Plateau, southern Tunisia, based on mineral magnetism and new luminescence dating. Quat Int 76–77:43–56

Delaware University (2009) Monthly air temperature and precipitation long-term means database. http://www.esrl.noaa.gov/psd/data/gridded/data.UDel_AirT_Precip.html

Deming JW, Baross JA (1993) The early diagenesis of organic matter: bacterial activity. In: Engel MH, Macko SA (ed) Organic Geochemistry. Plenum Press, New York, pp 119–144

Deng C, Zhu R, Versoub KL, et al (2000) Paleoclimatic significance of the temperature-dependent susceptibility of Holocene loess along a NW-SE transect in the Chinese loess plateau. Geophys Res Lett 27:3715–3718

Derbyshire E (2001) Characteristics, stratigraphy and chronology of loess and paleosols, and their application to climatic reconstruction: a preface. Quat Int 76–77:1–5

Derbyshire E, Kemp R, Meng X (1995) Variations in loess and paleosol properties as indicators of paleoclimatic gradients across the loess plateau of North China. Quat Sci Rev 14:681–697

Ding ZL, Yang SL, Sun JM et al (2001) Iron geochemistry of loess and red clay deposits in the Chinese Loess Plateau and implications for long term Asian monsoon evolution in the last 7 Ma. Earth Planetary Sci Lett 185:99–109

Dunlop DJ, Özdemir Ö (1997) Rock magnetism: fundamentals and frontiers. Cambridge University Press

Dunne, T, Leopold LB (1978) Water in environmental planning. W.H. Freeman & Co

Ďurža O, Dlapa P (2009) Magnetic susceptibility record of loess/paleosol sequence: case study from south-west Slovakia, Contr Geophys Geod 39:83–94

Egli R (2004) Characterization of individual rock magnetic components by analysis of remanence curves 1: unmixing natural sediments. Stud Geophys Geod 48:391–446

Egli R (2009) Magnetic susceptibility measurements as a function of temperature and frequency 1: inversion theory. Geophys J Int 177:495–420

Evans, ME (1999) Magnetoclimatology: a test of the wind-vigour model using the 1980 Mount St. Helens ash. Earth Planetary Sci Lett 172:255–259

Evans, ME (2001) Magnetoclimatology of aeolian sediments. Geophys J Int 144:495–497

Evans ME, Heller F (1994) Magnetic enhancement and paleoclimate: study of a loess/paleosol couplet across the Loess Plateau of China. Geophys J Int 117:257–264

Evans ME, Heller F (2003) Environmental magnetism. Academic press, Elsevier Science

Evans ME, Rokosh CD (2000) The last interglacial in the Chinese Loess Plateau: a petromagnetic investigation of samples from a north-south transect. Quat Int 68–71:77–82

Evans ME, Rokosh CD, Rutter NW (2002) Magnetoclimatology and paleoprecipitation: evidence from a north-south transect through the Chinese Loess Plateau. Geophys Res Lett 29. doi:10.1029/2001GL013674

Eyre JK, Shaw J (1994) Magnetic enhancement of Chinese loess – the role of γFe_2O_3? Geophys J Int 117:265–271

Fang XM, Ji-Jun L, Banerjee SK et al (1999) Millennial-scale climatic change during the last interglacial period: superparamagnetic sediment proxy from paleosol S1, western Chinese Loess Plateau. Geophys Res Lett 26:2485–2488

Fassbinder JWE, Stanjek H, Vali H (1990) Occurrence of magnetic bacteria in soils. Nature 343:161–163

Feng ZD, Khosbayar P (2004) Paleosubartic Eolian environments along the southern margin of the North American Icesheet and the southern margin of Siberia during the Last Glacial Maximum. Palaeogeogr Palaeoclimatol Palaeoecol 212:265–275

Fine P, Verosub KL, Singer MJ (1995) Pedogenic and lithogenic contributions to the magnetic susceptibility record of the Chinese loess/paleosol sequence. Geophys J Int 122:97–107

Fischer WR (1988) Microbiological reactions of iron in soils. In: Stucki V et al (ed) Iron in Soils and Clay Mineals. Dorotrecht Reidel Publishing Company, 715–748

Fischer H, Luster J, Gehring AU (2008) Magnetite weathering in a Vertisol with seasonal redox-dynamics. Geoderma 143:41–48

Florindo F, Roberts AP, Palmer MR (2003) Magnetite dissolution in siliceous sediments. Geochem Geophys Geosyst 4. doi:10.1029/2003GC000516

Florindo F, Zhu R, Guo B (1999a) Low field susceptibility and paleorainfall estimates: new data along a N-S transect of the Chinese loess plateau. Phys Chem Earth A 24:817–821

Florindo F, Zhu R, Guo B et al (1999b) Magnetic proxy climate results from the Duanjiapo loess section, southernmost extremity of the Chinese loess plateau. J Geophys Res 104:645–659

Forster T, Evans ME, Heller F (1994) The frequency dependence of low field susceptibility in loess sediments. Geophys J Int 118:636–642

Forster T, Heller F (1994) Loess deposits from the Tajik depression (Central Asia): magnetic properties and paleoclimate. Earth Planetary Sci Lett 128:501–512

Forster T, Heller F, Evans ME et al (1996) Loess in Czech republic: magnetic properties and paleoclimate. Stud Geophys Geod 40:243–261

Gallagher KJ, Feitknecht W, Mannweiler U (1968) Mechanism of oxidation of magnetite to γ-Fe_2O_3. Nature 217:1118–1121

Gallet S, Jahn B, Van Vliet Lanoë B et al (1998) Loess geochemistry and its implications for particle origin and composition of the upper continental crust. Earth Planetary Sci Lett 156: 157–172

Geiss CE, Egli R, Zanner CW (2008) Direct estimates of pedogenic magnetite as a tool to reconstruct past climates from buried soils. J Geophys Res 113. doi:10.1029/2008JB005669

Geiss CE, Zanner CW (2006) How abundant is pedogenic magnetite? Abundance and grain size estimates for loessic soils based on rock magnetic analyses. J Geophys Res 111. doi: 10.1029/2006JB004564

Geiss CE, Zanner CW (2007) Sediment magnetic signature of climate in modern loessic soils from the Great Plains. Quat Int 162–163:97–110

Geiss CE, Zanner CW, Banerjee SK et al (2004) Signature of magnetic enhancement in a loessic soil in Nebraska, United States of America. Earth Planetary Sci Lett 228:355–367

Grimley DA, Arruda NK, Bramstedt MW (2004) Using magnetic susceptibility to facilitate more rapid, reproducible, and precise delineation of hydric soils in the Midwestern USA. Catena 58:183–213

Grimley DA, Follmer LR, Hughes RE et al (2003) Modern, Sangamon and Yarmouth soil development in loess of unglaciated southwestern Illinois. Quat Sci Rev 22:225–244

Guo B, Zhu RX, Roberts AP et al (2001) Lack of correlation between paleoprecipitation and magnetic susceptibility of Chinese loess/paleosol sequences. Geophys Res Lett 22:4259–4262

Gupta SK, Sharma P, Juyal N et al (1991) Loess-paleosol sequence in Kashmir: correlation of mineral magnetic stratigraphy with the marine paleoclimatic record. J Quat Sci 6:3–12

Guyodo Y, LaPara TM, Anschutz AJ et al (2006) Rock magnetic, chemical and bacterial community analysis of a modern soil from Nebraska. Earth Planetary Sci Lett 251:168–178

Hallberg GR, Wollenhaupt NC, Miller GA (1978) A century of soil development in spoil derived from loess in Iowa. Soil Sci Soc Am J 42:339–343

Han J, Lu H, Wu N et al (1996) Magnetic susceptibility of modern soils in China and its use for paleoclimate reconstruction. Stud Geophys Geod 40:262–275

Hanesch M, Scholger R (2005) The influence of soil type on the magnetic susceptibility measured throughout soil profiles. Geophys J Int 161:50–56

Hao Q, Guo Z (2005) Spatial variations of magnetic susceptibility of Chinese loess for the last 600 kyr: implications for monsoon evolution. J Geophys Res 110. doi:10.1029/2005JB 003765

Heller F, Liu TS (1984) Magnetism of Chinese loess deposits, Geophys J R Astr Soc 77: 125–141

Heller F, Liu TS (1986) Paleoclimatic and sedimentary history from magnetic susceptibility of loess in China. Geophys Res Lett 13:1169–1172

Heller F, Evans ME (1995) Loess magnetism. Rev Geophys 33:211–240

Heller F, Shen CD, Beer J et al (1993) Quantitative estimates of pedogenic ferromagnetic mineral formation in Chinese loess and paleoclimatic implications. Earth Planetary Sci Lett 114: 385–390

Hendrickx JMH, Harrison JBJ, van Dam RL et al (2005) Magnetic soil properties in Ghana. P Soc Photo-Opt Inst (SPIE) 5794:165–176

Jackson M, Carter-Stiglitz B, Egli R et al (2006) Characterizing the superparamagnetic grain distribution $f(V, H_K)$ by thermal fluctuation tomography. J Geophys Res 111. doi:10.1029/2006JB004514

Jacobs PM, Mason JA (2005) Impact of Holocene dust aggradation on A horizon characteristics and carbon storage in loess-derived Mollisols of the Great Plains, USA. Geoderma 125: 95–106

Jahn B, Gallet S, Han J (2001) Geochemistry of Xining, Xifeng and Jixian sections, Loess Plateau of China: aeolian dust provenance and paleosol evolution during the last 140 ka. Chem Geol 178:71–94

Jenny H (1941) Factors of soil formation. McGraw-Hill

Keisling TC (1982) Calculation of the length of day. Agron J 74:758–759

Kletetschka G, Banerjee SK (1995) Magnetic stratigraphy of Chinese loess as a record of natural fires. Geophys Res Lett 22:1341–1343

Kostka JE, Dalton DD, Skelton H et al (2002) Growth of Iron(III)-reducing bacteria on clay minerals as the sole electron acceptor and comparison of growth yields on a variety of oxidized iron forms. Appl Environ Microbiol 68: 6256–6262

Kukla G, Heller F, Liu XM, et al (1988) Pleistocene climates in China dated by magnetic susceptibility. Geology 16:811–814

Kravchinsky VA, Zykina VS, Zykin VS (2008) Magnetic indicator of global cycles in Siberian loess-paleosol sequences. Earth Planetary Sci Lett 265:498–514

Lagroix F, Banerjee SK (2002) Palaeowind directions from the magnetic fabric of loess in central Alaska. Earth Planetary Sci Lett 195:99–112

Laio F, Porporato A, Ridolfi L et al (2001a) Plants in water-controlled ecosystems: active role in hydrologic processes and response to water stress II: probabilistic soil moisture dynamics. Adv Wat Res 24:707–723

Laio F, Porporato A, Fernandez-Illescas CP et al (2001b) Plants in water-controlled ecosystems: active role in hydrologic processes and response to water stress IV: discussion of real cases. Adv Wat Res 24:745–762

Le Borgne E (1955) Susceptibilité magnétique anormale du sol superficiel. Ann Géophys 11:399–419

Le Borgne E (1960) Influence du feu sur les propriétés magnétiques du sol et sur celles du schiste et du granite. Ann Géophys 16:159–195

Ligr M, Ron C, Nátr L (1995) Calculation of the photoperiod length. Comput Appl Biosci 11:133–139

Lind BB, Lundin L (1990) Saturated hydraulic conductivity of Scandinavian Tills. Nordic Hydrology 21:107–118

Liu Q, Banerjee SK, Jackson MJ et al (2004) Grain sizes of susceptibility and anhysteretic remanent magnetization carriers in Chinese loess/paleosol sequences. J Geophys Res 109. doi:10.1029/2003JB002747

Liu Q, Deng C, Torrent J et al (2007) Review of recent developments in mineral magnetism of the Chinese loess. Quat Sci Rev 26:368–385

Liu XM, Hesse P, Rolf T et al (1999) Properties of magnetic mineralogy of Alaskan loess: evidence for pedogenesis. Quat Int 62:93–102

Liu X, Rolph T, Bloemendal J et al (1995) Quantitative estimates of paleoprecipitation at Xinfeng, in the Loess Plateau of China. Paleogeogr Paleoclimatol Paleoecol 113: 243–248

Liu X, Shaw J, Liu T et al (1992) Magnetic mineralogy of Chinese Loess and its significance. Geophys J Int 108:301–308

Liu Q, Torrent J, Maher BA et al (2005) Quantifying grain size distribution of pedogenic magnetic particles in Chinese loess and its significance for pedogenesis. J Geophys Res 110. doi:10.1029/2005JB 003726

Lovley DR, Stolz JF, Nord GL et al (1987) Anaerobic production of magnetite by a dissimilatory iron reducing microorganism. Nature 330:252–254

Maher BA (1986) Characterization of soils by mineral magnetic measurements. Phys Earth Planetary Inter 42:76–92

Maher BA (1988) Magnetic properties of some synthetic submicron magnetites. Geophys J 94:83–96

Maher BA (1998) Magnetic properties of modern soils and Quaternary loessic paleosols: paleoclimatic implications. Palaeogeogr Palaeoclimatol Palaeoecol 137:25–54

Maher BA (2009) Rain and dust: magnetic records of climate and pollution. Elements 5: 229–234

Maher BA, Hu M (2006) A high-resolution record of Holocene rainfall variations from the western Chinese Loess Plateau: antiphase behavior of the African/Indian and East Asian summer monsoons. The Holocene 16:309–319

Maher BA, Thompson R (1992) Paleoclimatic significance of the mineral magnetic record of the Chinese loess and paleosols. Quat Res 37:155–170

Maher BA, Thompson R (1995) Paleorainfall reconstructions from pedogenic magnetic susceptibility variations in the Chinese Loess and Paleosols. Quat Res 44:383–391

Maher BA, Thompson R (1999) Palaeomonsoons I: the magnetic record of paleoclimate in the terrestrial loess and paleosol sequences. In: Maher BA, Thompson R (ed) Quaternary Climates, Environments and Magnetism. Cambridge University Press, Cambridge, 81–125

Maher BA, Alekseev A, Alekseeva T (2003) Magnetic mineralogy of soils across the Russian Steppe: climatic dependence of pedogenic magnetite formation. Palaeogeogr Palaeoclimatol Palaeocol 201:321–341

Maher BA, Mutch TJ, Cunningham D (2009) Magnetic and geochemical characteristics of Gobi Desert surface sediments: implications for provenance of the Chinese Loess Plateau. Geology 37:279–282

Maher BA, Thompson R, Zhou LP (1994) Spatial and temporal reconstructions of changes in the Asian paleomonsoon: a new mineral magnetic approach. Earth Planetary Sci Lett 125:461–471

Muhs DR, Bettis III EA (2000) Geochemical variations in Peoria loess of western Iowa indicate paleowinds of midcontinental North America during last glaciation. Quat Res 53:49–61

Muhs DR, Zárate M (2001) Late Quaternary aeolian records of the Americas and their paleoclimatic significance. In: Markgraf V (ed) Interhemispheric climate linkages. Academic Press, 183–214

Muhs DR, Ager TA, Bettis III EA et al (2003) Stratigraphy and paleoclimatic significance of Late Quaternary loess-paleosol sequences of last interglacial-glacial cycle in central Alaska. Quat Sci Rev 22:1947–1986

Muhs DR, Bettis III EA, Been J et al (2001) Impact of climate and parent material on chemical weathering in loess-derived soils of the Mississippi river valley. Soil Sci Soc Am J 65:1761–1777

Muhs DR, McGeehin JP, Beann J et al (2004) Holocene loess deposition and soil formation as competing processes, Matanuska valley, southern Alaska. Quat Res 61:265–276

Mullins CE (1977) Magnetic susceptibility of the soil and its significance in soil sequence – a review. J Soil Sci 28:223–246

Murad E, Schwertmann U (1993) Temporal stability of a fine-grained magnetite. Clays Clay Miner 41:111–113

Nawrocki J, Wójcik A, Bogucki A (1996) The magnetic susceptibility record in the Polish and western Ukrainian loess-paleosol sequences conditioned by paleoclimate. Boreas 25: 161–169

Nicolli HB, Tineo A, García JW et al (2004) The role of loess in groundwater pollution at Salí River Basin, Argentina. In: Wanty RB, Seal RR (ed) Water-rock interaction vol. 2. Taylor and Francis, 1591–1595

Nie J, Song Y, King JW et al (2010) Consistent grain size distribution of pedogenic maghemite of surface soils and Miocene loessic soils on the Chinese Loess Plateau. J Quat Sci 25:261–266

Oches EA, Banerjee SK (1996) Rock-magnetic proxies of climate change from loess-paleosols sediments of the Czech Republic. Stud Geophys Geod 40:287–300

Orgeira MJ (1990) Paleomagnetism of late Cenozoic fossiliferous sediments from Barranca de Los Lobos (Buenos Aires Province, Argentina): the magnetic age of the South American land-mammal ages. Phys Earth Planetary Int 64:121–132

Orgeira MJ, Compagnucci R (2006) Correlation between paleosol-soil magnetic signal and climate. Earth Planets Space 58:1373–1380

Orgeira MJ, Pereyra FX, Vásquez C et al (2008) Rock magnetism in modern soils, Buenos Aires province, Argentina. J South Am Earth Sci 26:217–224

Panaiotu CG, Panaiotu EC, Grama A et al (2001) Paleoclimatic record from a loess-paleosol profile in southeastern Romania. Phys Chem Earth 26:893–898

Porter SC, Hallet B, Wu X et al (2001) Dependence of near-surface magnetic susceptibility on dust accumulation rate and precipitation on the Chinese Loess Plateau. Quat Res 55:271–283

Roden EE, Zachara JM (1996) Microbial reduction of crystalline iron(III) oxides: influence of oxide surface area and potential for cell growth. Environ Sci Technol 30:1618–1628

Rodriguez-Iturbe I, Porporato A, Ridolfi L et al (1999) Probabilistic modeling of water balance at a point: the role of climate, soil and vegetation. Proc R Soc Lon A 455:3789–3805

Rousseau DD, Kukla G (1994) Late Pleistocene climate record in the Eustis loess section, Nebraska, based on land snail assemblages and magnetic susceptibility., Quat Res 42:76–187

Rousseau DD, Zöller L, Valet JP (1998) Late Pleistocene climatic variations at Achenheim, France, based on a magnetic susceptibility and TL chronology of loess. Quat Res 49: 255–263

Schwertmann U, Cornell RM (2000) Iron oxides in the laboratory, 2nd ed. Wiley-Vch

Schwertmann U, Murad E, Schulze DG (1982) Is there Holocene reddening (hematite formation) in soils of axeric temperate areas? Geoderma 27:209–223

Shade PJ (1995) Water budget for the Kohala area, island of Hawaii. U.S. Geological Survey Water-resources investigation report 95–4114, Honolulu, Hawaii

Shcherbakov VP, Fabian K (2005) On the determination of magnetic grain-size distribution of superparamagnetic particle ensembles using the frequency dependence of susceptibility at different temperatures. Geophys J Int 162: 736–746

Singer MJ, Fine P, Verosub KL et al (1992) Time dependence of magnetic susceptibility of soil chronosequences on the California coast. Quat Res 37:323–332

Singer MJ, Versoub KL, Fine P et al (1996) A conceptual model for the enhancement of magnetic susceptibility in soils. Quat Int 34–36:243–248

Smirnov AV, Tarduno JA (2000) Low-temperature magnetic properties of pelagic sediments (Ocean Drilling Program site 805C): tracers of maghemitization and magnetic mineral reduction. J Geophys Res 105:16457–16471

Smith J, Vance D, Kemp RA et al (2003) Isotopic constraints on the source of Argentinean loess – with implications for atmospheric circulation and the provenance of Antarctic dust during recent glacial maxima. Earth Planetary Sci Lett 212:181–196

Sobolev D, Roden EE (2002) Evidence for rapid microscale bacterial redox cycling of iron in circumneutral environments. Antoine van Leeuwenhoek 81:587–597

Tamaura Y, Ito K, Katsura T (1983) Transformation of γ-FeO(OH) to Fe_3O_4 by adsorption of iron(II) ion on γ-FeO(OH). J Chem Soc Dalton Trans 2:189–194

Tang J, Myers M, Bosnick KA et al (2003) Magnetite Fe_3O_4 nanocrystals: spectroscopic observation of aqueous oxidation kinetics. J Phys Chem 107:7501–7506

Taylor RM, Maher BA, Self PG (1987) Magnetite in soils: I. The synthesis of single-domain and superparamagnetic magnetite. Clay Minerals 22:411–422

Teruggi ME (1957) The nature and origin of the Argentinean loess. J Sedim Petr 27:322–332

Thompson R, Maher BA (1995) Age models, sediment fluxes, and paleoclimatic reconstructions for the Chinese loess and paleosol sequences. Geophys J Int 123:611–622

Thornthwaite CW (1931) The climates of North America. Geogr Rev 21:633–655

Thornthwaite CW (1948) An approach toward a rational classification of climate. Geogr Rev 38:55–94

Tite MS, Linington RE (1975) Effect of climate on the magnetic susceptibility of soils. Nature 256:565–566

Torrent J, Barrón V, Liu Q (2006) Magnetic enhancement is linked and precedes hematite formation in aerobic soil. Geophys Res Lett 33. doi:10.1029/2005GL024818

Torrent J, Cabedo A (1986) Source of iron oxides in reddish brown soil profiles from calcarenites in southern Spain. Geoderma 37:57–66

Torrent J, Liu QS, Barrón V (2010) Magnetic minerals in calcic Luvisols (Chromic) developed in a warm Mediterranean region of Spain: origin and paleoenvironmental significance. Geoderma 154:465–472

Torrent J, Liu Q, Bloemendal J et al (2007) Magnetic enhancement and iron oxides in the upper Luochuan loess-paleosol sequence, Chinese Loess Plateau. Soil Sci Soc Am J 71: 1570–1578

Tronc E, Belleville P, Jolivet JP et al (1992) Transformation of ferric hydroxide into spinel by Fe^{II} adsorption. Langmuir 8:313–319

Vidic NJ, Singer MJ, Verosub KL (2004) Duration dependence of magnetic susceptibility enhancement in the Chinese loess-paleosols of the past 620 kyr. Palaeogeogr Palaeoclimatol Palaeoecol 211:271–288

Vidic NJ, TenPas JD, Verosub KL et al (2000) Separation of pedogenic and lithogenic components of magnetic susceptibility in the Chinese loess/paleosol sequence as determined by the CBD procedure and a mixing analysis. Geophys J Int 142:551–562

White AF, Peterson ML, Hochella MF (1994) Electrochemistry and dissolution kinetics of magnetite and ilmenite. Geochim Cosmochim Acta 58:1859–1875

Worm HU (1998) On the superparamagnetic–stable single domain transition for magnetite, and frequency dependence of susceptibility. Geophys J Int 133:201–206

Youngs EG, Goss MJ (1988) Hydraulic conductivity profiles of two clay soils. J Soil Sci 39:341–345

Zárate M, Blasi A (1993) Late Pleistocene-Holocene aeolian deposits of the southern Buenos Aires Province, Argentina: a preliminary model. Quat Int 17:15–20

Zhu RX, Matasova G, Kazansky A et al (2003) Rock magnetic record of the last glacial–interglacial cycle from the Kurtak loess section, southern Siberia. Geophys J Int 152: 335–343

26 Palaeoclimatic Significance of Hematite/Goethite Ratio in Bulgarian Loess-Palaeosol Sediments Deduced by DRS and Rock Magnetic Measurements

Diana Jordanova, Tomas Grygar, Neli Jordanova, and Petar Petrov

Abstract

The role of hematite and goethite as palaeoenvironmental indicators in loess-palaeosol sediments is studied by diffuse reflectance spectroscopy (DRS) and rock magnetic methods. Forty five selected samples from four loess-palaeosol profiles in Bulgaria were used to deduce the behaviour of the hematite/goethite ratio in loess and palaeosol units, which were formed under contrasting palaeoclimate conditions – cold (glacial) and warm (interglacial). According to DRS, the pedogenesis is accompanied by preferential formation of hematite over goethite in all sites. At the same time, rock magnetic data prove that this process is concomitant with in-situ formation of a strongly magnetic ferrimagnetic fraction, responsible for the observed magnetic enhancement of palaeosol units. Systematically higher amount of hematite in the loess-palaeosol profile Orsoja, situated at the fifth Danube river terrace and the Durankulak profile at the Black sea coast is supposed to be due to additional coarse-grained hematite in the aeolian dust blown from local dust sources during glacial periods. Component analysis of the curves of stepwise acquisition of isothermal remanence reveals the presence of two components – pedogenic (P) with median coercivity at half-width of the Gaussian function of 31 ± 3.2 mT; detrital unweathered component D_1 in loesses from Lubenovo and Koriten profiles with $B_{1/2}$ of 58 ± 4 mT, and weathred detrital component D_2 ($B_{1/2}$ of 123 ± 46 mT) in all palaeosols and weathered loess units from Orsoja and Durankulak profiles. Mineral phases, carrying these coercivity components are supposed to be pedogenic maghemite (P), aeolian (titano)magnetites (D_1) and hematite (D_2).

26.1 Introduction

The correct reconstruction of past climates is a complex and difficult task in view of the various responses of the natural systems to particular combinations of the number of factors acting, such as temperature, humidity, and seasonality. Thus, an interdisciplinary approach is necessary to retrieve the complete palaeoenvironmental information from climate proxies. Sedimentary sequences, and specifically loess-palaeosol sediments, are thought to represent one of the most complete continental records of palaeoclimate. A number of recent studies report successful use of geochemical, ^{10}Be, quartz grain size and mineral magnetic proxies of palaeoclimate (Gallet et al. 1996,

D. Jordanova (✉)
National Institute of Geophysics, Geodesy and Geography, BAS, 1113 Sofia, Bulgaria; Faculty of Physics, Sofia University "St. Kl. Ohridski", Sofia, Bulgaria
e-mail: vanedi@geophys.bas.bg

E. Petrovský et al. (eds.), *The Earth's Magnetic Interior*, IAGA Special Sopron Book Series 1,
DOI 10.1007/978-94-007-0323-0_26, © Springer Science+Business Media B.V. 2011

Heller et al. 1993, Hunt et al. 1995, Gu et al. 1996, Porter and An 1995, Sun et al. 2000, Verosub et al. 1993, Maher and Thompson 1995, Evans and Heller 2003).

There are many analytical tools used to deduce speciation of Fe in sediments. Among them, both rock magnetic methods and diffuse reflectance spectroscopy (DRS) have an exceptional position because of good sensitivity and selectivity to only certain Fe-bearing minerals. While rock-magnetic methods are best suited for characterization of ferrimagnetic components, DRS is focused on coloured Fe-bearing species, especially hematite (α-Fe_2O_3) and oxyhydroxides (FeOOH). The combination of rock magnetic methods and DRS hence offer a chance to sensitively and selectively characterize wide range of Fe-bearing species even in mineralogically complex mixtures. DRS application in environmental studies is much less common (Ji et al. 2001, Ji et al. 2002, Grygar and van Oorschot 2002, Balsam et al. 2004, Torrent et al. 2007).

The aim of the present study is to evaluate possibilities of using simultaneously the two methods to get deeper view on the palaeoenvironment, deduced from the particular quantitative and grain size distribution of different Fe-oxides in a selected collection of loess-palaeosol samples from four sections on the territory of Bulgaria.

26.2 Samples and Methods

Loess deposits are widely present in North Bulgaria and represent part of the European loess (Haase et al. 2007). Thickness of the units, grain size distribution and other physical properties are mainly controlled by the relative distance to periglacial source area, geomorphology and distance from the Danube river, which is considered as a local source of dust material as well. Secondary pedogenic transformations of the loess material are enhanced towards the south due to reduced thickness of the loess cover and progressively more humid climate conditions (Minkov 1968). Representative samples from four loess-palaeosol profiles (Fig. 26.1) situated at various palaeo- landforms (5th Danube river terrace – Orsoja section; Durankulak section and Koriten borehole – Old Abrasive Accumulative Level (OAAL) and Lubenovo borehole – Pliocene Denudation Surface (PDS) according to Evlogiev (2006), have been used in present investigation. Forty five samples from typical

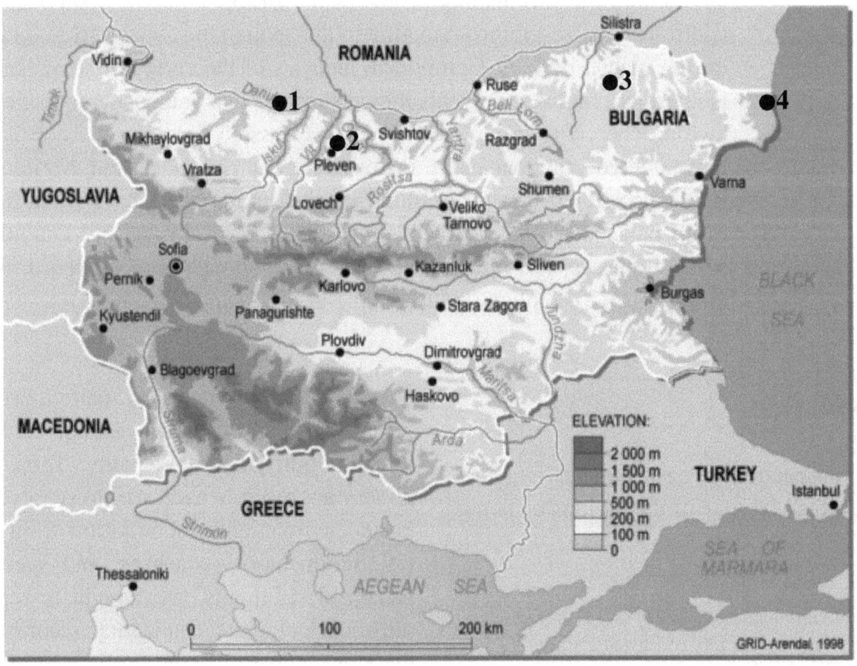

Fig. 26.1 Sketch map of Bulgaria showing the positions of the studied loess-palaeosol profiles: 1 – Orsoja; 2 – Lubenovo; 3 – Koriten; 4 – Durankulak

levels of loess and palaeosol horizons from the four profiles have been chosen, based on the behaviour of low field magnetic susceptibility (χ) along the profiles. Detailed rock magnetic studies on the profiles are given elsewhere (Jordanova and Petersen 1999, Jordanova et al. 2007, Avramov et al. 2006).

26.2.1 Rock Magnetic Measurements

Cubic solid samples were prepared by mixing known amount of powder from loess (palaeosol) material with small amount of gypsum and water. After drying out at ambient conditions, acquisition of saturation isothermal remanence (SIRM) in a pulse magnetizer with maximum field of 2 T was applied. In order to isolate the remanence of high-coercivity phases, samples were demagnetized in an alternating field with 100 mT maximum amplitude. Complementary samples are used for stepwise acquisition of IRM up to field of 2 T (25 steps) using IM 10-30 Impulse Magnetizer (ASC Scientific).

Remanence and susceptibility measurements were done using JR-6A Spinner magnetometer and Kappabridge KLY-2 (Agico, Brno). IRM acquisition curves were fitted by sets of log-Gaussian curves following Heslop et al. (2002) approach and using Heslop's IRMunmix software.

Powder samples were used for measurements of frequency dependent magnetic susceptibility in a Bartington dual-frequency (0.47 and 4.7 kHz) susceptibility meter (Bartington Ltd.). Percent frequency dependent susceptibility after Mullins and Tite (1973) was calculated as:

$$\chi_{FD}\% = 100^{*}(\chi_{LF} - \chi_{HF})/\chi_{LF},$$

where χ_{LF} is susceptibility measured at low frequency (0.47 Hz), and χ_{HF} – susceptibility measured at higher frequency (4.7 kHz).

26.2.2 Diffuse Reflectance Spectroscopy (DRS)

The powder samples were air dried, grind at <2 mm and their diffuse reflectance spectra were obtained by a spectrometer with an integrating sphere (Perkin Elmer Lambda 35 spectrometer equipped with Labsphere). The reflectances were re-calculated to Kubelka-Munk absorbances, plotted against wavenumbers, and deconvoluted to Gaussian components. The fit was performed with 7 components corresponding to the most relevant electron transitions in Fe^{2+} and Fe^{3+} bearing alumosilicates and Fe^{3+} oxides in the range of 1100–300 nm (Hradil et al. 2004, Grygar et al. 2006). The components are: d-d transitions of Fe^{3+} in oxides and alumosilicates, intervalence charge transfer transition Fe^{2+}-O-Fe^{3+}, d-electron pair transition characteristic to α-Fe_2O_3 (hematite) and FeOOH polymorphs (mainly α-FeOOH, goethite in the loess-paleosol specimens studied), and an assemblage of charge transfer transitions in near UV. The d-electron pair transitions were assigned according to Scheinost et al. (1998). To guarantee the stability of the fit by reducing the number of refined parameters, the FWHM of the Gaussian components were fixed and the band positions were refined only within certain physically sound ranges. The areas of the Gaussian components assigned to α-Fe_2O_3 and FeOOH were used as proxies of the concentration of these components assuming that spectral properties (namely reflectivity) of the non-ferric "matrix" of individual strata are the same.

Using the Kubelka-Munk remission function, the band areas are proportional to the concentration of the absorbing species. The ratio of the intensities of the bands E and D is a proxy of the ratio between FeOOH and α-Fe_2O_3. Although that ratio, E/D, cannot be directly interpreted as the ratio of percentages of FeOOH and α-Fe_2O_3, for a given polymorph of FeOOH (goethite, ferrihydrite, and lepidocrocite), given mean crystallinity of α-Fe_2O_3, and a given mineral matrix of the parent loess and related soil, the value of E/D is a good estimate of the trends in the varying composition of the Fe oxide assemblage (Grygar et al. 2003).

Another important proxy of Fe mineral ratios is the ratio of the areas of bands B and C. The band B is due to a pair of neighbouring Fe^{2+} and Fe^{3+} ions in silicates, such as dark micas, amphiboles, pyroxenes, and similar minerals, while band C is only due to Fe^{3+} ions in silicates as well as in oxides. High B/C ratio would hence be typical for unweathered primary minerals, while band C is prevailing in mixtures with Fe mainly as ferric ions. The ratio B/C is hence a proxy of the extent of chemical weathering of a given material. That proxy is useful only if more or less weathered materials of the same origin are to be compared, and that is exactly the case of loess and the soils formed from this parent loess by pedogenesis.

26.3 Experimental Results

26.3.1 Rock Magnetism

Rock magnetic experiments aimed to isolate and characterize properties, relative abundance and origin of the high-coercivity magnetic fraction in the studied samples. At the same time, the relationship between the abundance of strongly magnetic ferrimagnetic component and hematite-goethite content is explored in order to give a clue for possible palaeoclimatic control on the magnetic mineralogy of loess-palaeosol sediments.

The magnetic susceptibility values of palaeosol samples from all four profiles are systematically higher than those of the corresponding parent loess layer. Least enhanced palaeosols from Orsoja section have magnetic susceptibility values 1.3–1.4 times higher that susceptibility of the loess layer L_1. Percent frequency-dependent magnetic susceptibility ($\chi_{FD}\%$) varies from 0 to 11.2% with highest values corresponding to samples with high susceptibility (Fig. 26.2a). Saturation Isothermal Remanent Magnetization ($SIRM_{2T}$) for all profiles shows a similar relationship on low field susceptibility (χ) (Fig. 26.2b). Frequency dependent magnetic susceptibility (χ_{FD}) plotted as a function of isothermal remanence which remains after AF demagnetization with 100 mT peak field ($SIRM_{AFD100mT}$) results in sub-division of data set into two populations – one including samples from Koriten and Lubenovo, and the other including samples from Durankulak and Orsoja (Fig. 26.2c).

IRM component analysis of step-wise acquisition of IRM reveals the presence of one or two components. Samples, which are fitted with one-component model, are the ones from Koriten and Lubenovo profiles. Calculated median values of the three parameters, which describe Gaussian functions – distribution width (dP) at half peak intensity, percent contribution of the corresponding component to the total magnetization, and the mean coercivity at half distribution width of the Gaussian function – ($B_{1/2}$) (Evans and Heller 2003), are listed in Table 26.1. Median coercivity at half width ($B_{1/2}$) shows very consistent values for component 1: 30.9 ± 3.2 mT for palaeosol samples; 31.8 ± 2.5 mT for loess samples fitted with 2-component model (profiles Durankulak

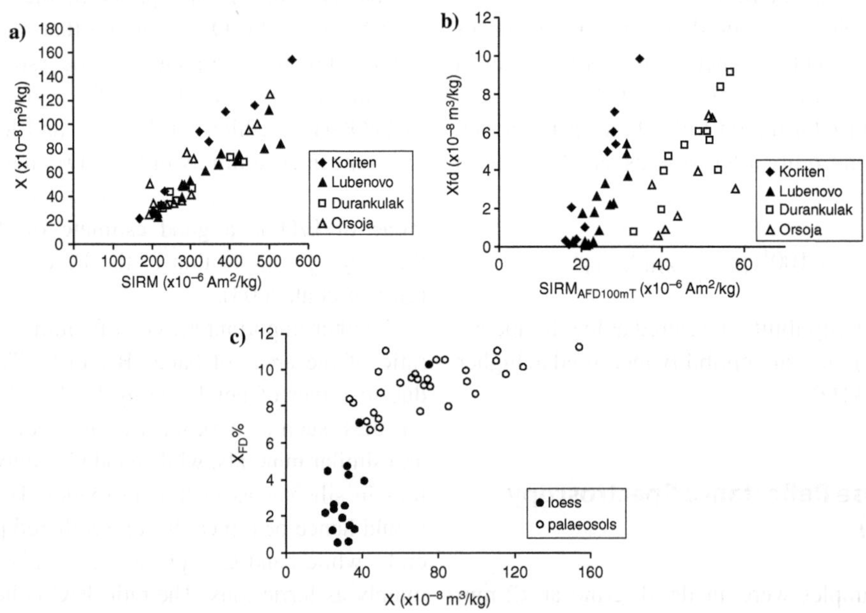

Fig. 26.2 Inter-parametric plots of rock magnetic data: (**a**) low field magnetic susceptibility against percent frequency dependent susceptibility; (**b**) Saturation Isothermal Remanence acquired in a dc field of 2 T versus low field magnetic susceptibility; (**c**) plot of SIRM rest after AF demagnetization at 100 mT peak field ($SIRM_{AFD100mT}$) against frequency dependent magnetic susceptibility

Table 26.1 Median values and standard deviations of the three parameters, describing Gaussian functions fitted to the IRM acquisition curves: dispersion parameter (dP); component contribution to the total SIRM signal (in %); median coercivity $B_{1/2}$ (coercivity at half width of Gaussian function). N is number of samples

Unit		N	Component 1			Component 2		
			dP	Relative contribution (%)	$B_{1/2}$ (mT)	dP	Relative contribution (%)	$B_{1/2}$ (mT)
loess	1-component fit	4	0.443 (0.02)	100 (0.0)	58.2 (4.0)			
	2-components fit	6	0.307 (0.02)	62 (13.9)	31.8 (2.5)	0.416 (0.06)	42 (7.6)	108.8 (15.2)
palaeosols		18	0.322 (0.03)	77.5 (8.9)	30.9 (3.2)	0.533 (0.07)	23.5 (10.4)	122.6 (46.2)

and Orsoja), and 58.2 ± 4 mT for loess samples fitted with 1-component model (Lubenovo and Koriten). Second component is less well defined, having mean coercivity $B_{1/2}$ of 123 ± 47 mT for palaeosol samples and 109 ± 15 mT for loesses. Representative examples as Gradient Acquisition Plots (GAP) with corresponding components fitted are shown in Fig. 26.3. The observed dependency of the relative contribution of two components to the total remanence from mean coercivity at half distribution width ($B_{1/2}$) is depicted in Fig. 26.4. It reveals uniform mean coercivity of the first component in palaeosols and loess samples, fitted by 2-components model. The second component is characterized by increasing mean coercivity as its relative contribution decreases, e.g. from loesses towards palaeosols.

26.3.2 Diffuse Reflectance Spectroscopy (DRS)

Ratio E/D used as a proxy for Goethite/Hematite (G/H) content shows systematically higher values for loess samples compared to the palaeosols (Fig. 26.5). Higher degree of magnetic enhancement of palaeosols for profiles Koriten and Lubenovo, visible through higher susceptibility values is accompanied by more distinct difference in G/H ratio among loess and palaeosol layers. The G/H variations down the profiles are negatively correlated with magnetic susceptibility behaviour (correlation coefficients $R^2 = 0.61$ for Durankulak data; $R^2 = 0.73$ for Orsoja; $R^2 = 0.71$ for Koriten and $R^2 = 0.69$ for Lubenovo).

Weathering degree deduced by the ratio (B/C) as a function of the area of hematite band (Fig. 26.6) exhibits different behaviour for samples from different profiles. Lubenovo and Koriten profiles are characterized by increasing hematite content with increasing degree of weathering (low B/C values). Exceptions are data for recent soil (So) from Koriten and first palaeosol (S_1) from Lubenovo. Data points from Durankulak are subdivided in two groups, the first one following the trend, defined by Koriten and Lubenovo values and the second one – with larger areas of hematite band defining another trend, on which also the exceptions from Lubenovo and Koriten lay. This second group includes samples only from palaeosol units (see Fig. 26.6). Most of the points from Orsoja profile are situated in the middle gap between the two trends. Depth variations of the ratio B/C are shown in Fig. 26.7 and depict regular decrease along depth of Lubenovo and Koriten, while Durankulak and Orsoja do not show such clearly defined trend.

26.3.3 Combined DRS and Rock Magnetic Data

Higher relative hematite content leading to smaller G/H ratio is obtained for palaeosol samples with higher magnetic susceptibilities (Fig. 26.8a) and percent frequency dependent magnetic susceptibility (Fig. 26.8b). Considering the absolute values of frequency-dependent magnetic susceptibility (χ_{FD}) and their relation to the area fit for hematite (Fig. 26.8c), two groups of experimental data can be distinguished based on differences of the DRS estimation of hematite content – one formed by Koriten and Lubenovo and another – by Durankulak and Orsoja. Isothermal remanence which remains after AF demagnetization ($SIRM_{AFD100mT}$) shows well defined linear dependence with DRS bands of both hematite and goethite (Fig. 26.9a, b). Similar relationship is observed between relative contribution of the second component to SIRM from IRM unmixing procedure and the ratio G/H (Fig. 26.9c). Mean coercivity of

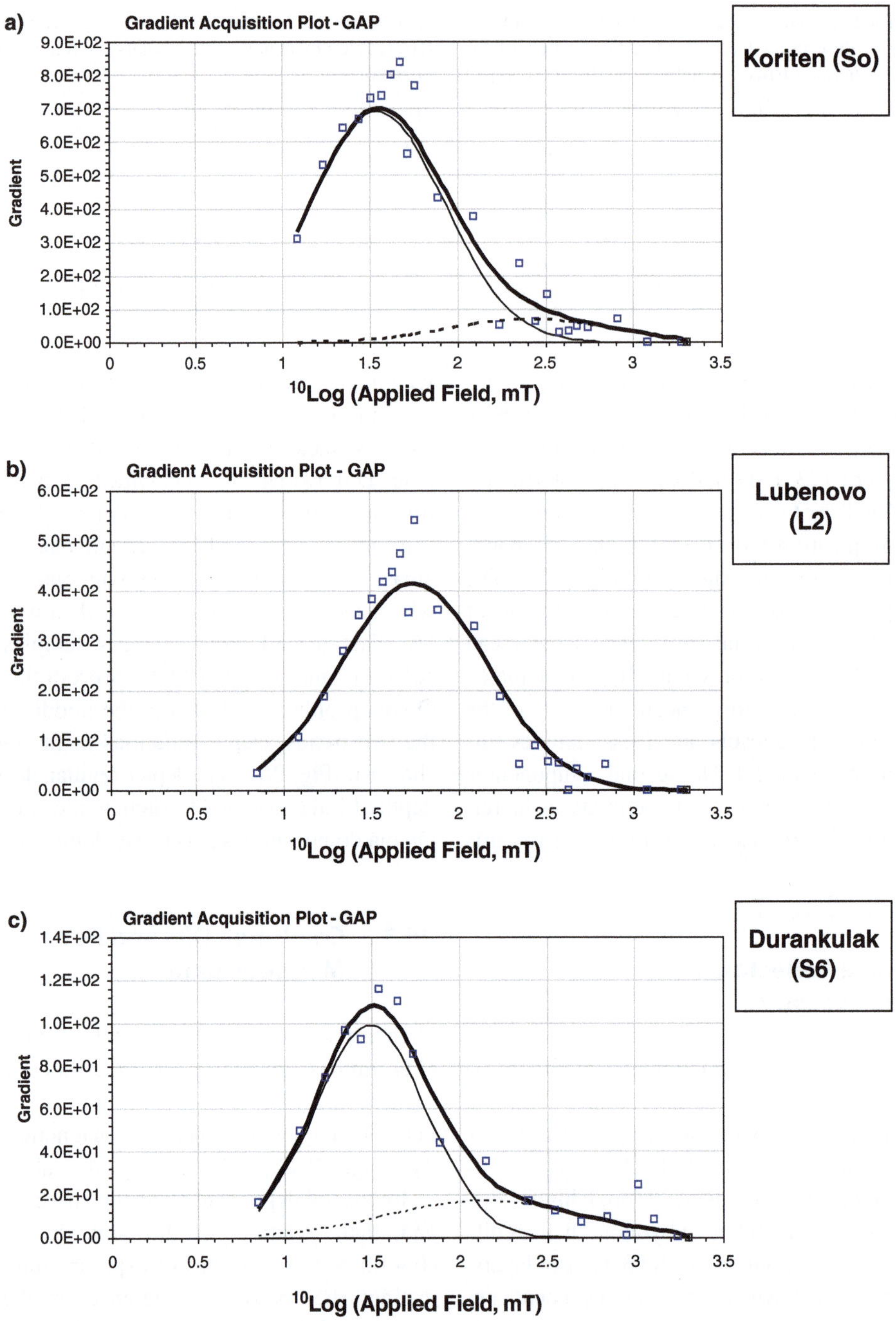

Fig. 26.3 Examples of IRM Gradient Acquisition Plots (GAP) for samples from recent soil (So) (**a**), unweathered loess L2 (**b**) and palaeosol (S6) (**c**). Thick line represents final fit; thin line – first component, dashed line – second components; squares – experimental points

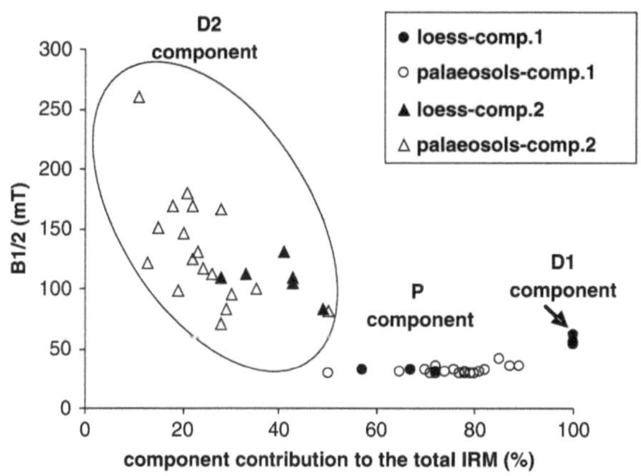

Fig. 26.4 Mean coercivity of the first component deduced by IRM unmixing procedure (after Heslop et al. 2002) as a function of its relative contribution to the total remanence

this second component ($B_{1/2}$) systematically increases when weathering degree, deduced from B/C ratio, also rises (Fig. 26.9d). Exceptions from this trend are samples from recent soils So from Koriten and Durankulak profiles.

26.4 Discussion

26.4.1 Rock Magnetic Data

One of the best preserved terrestrial records of past climate is in loess-palaeosol sequences around the world. Except classical Chinese loess, European sequences are widely used for palaeoenvironmental reconstructions. Low Danube loess deposits are part of the South-eastern European loess cover, linking loess belt in Asia with Central European loess. Its magnetic properties have been extensively studied, putting emphasis on present continental conditions (Jordanova and Petersen 1999, Panaiotu et al. 2001, Jordanova et al. 2007, Buggle et al. 2008, 2009). Use of fine grained pedogenic ferrimagnetic fraction as paleoclimate proxy through magnetic susceptibility variations has been debated due to some inconsistencies between rock magnetic and pedologic criteria about the degree and intensity of soil maturity. A direct palaeoclimate interpretation of susceptibility record in terms of palaeoprecipitation and palaeotemperature is made

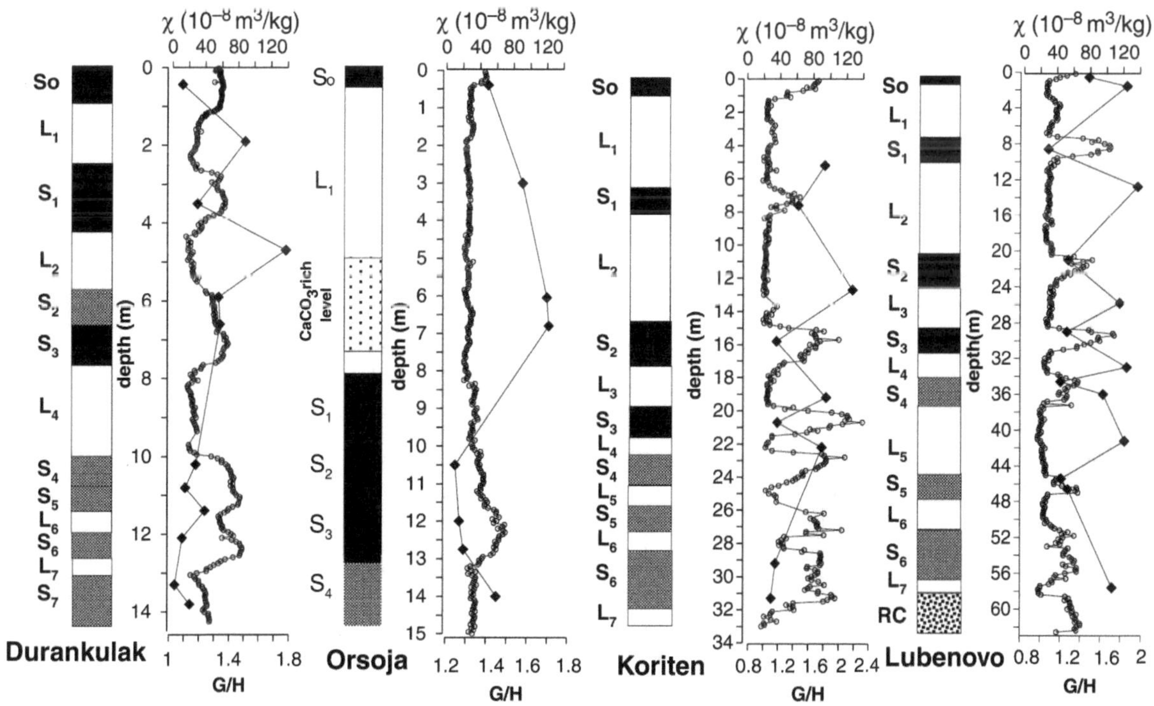

Fig. 26.5 Depth variations of low field magnetic susceptibility along the four profiles studied (*black circles*) and the ratio G/H (*empty squares*). Loess (L) and palaeosol (S) horizons are indicated according to the stratigraphic scheme of the Bulgarian loess-palaeosol complex (Evlogiev 2006). Palaeosols of Chernozem type are presented as black filled columns, reddish palaeosols are represented by grey areas, RC is abbreviation of red clay complex at the bottom of the loess complex at Lubenovo

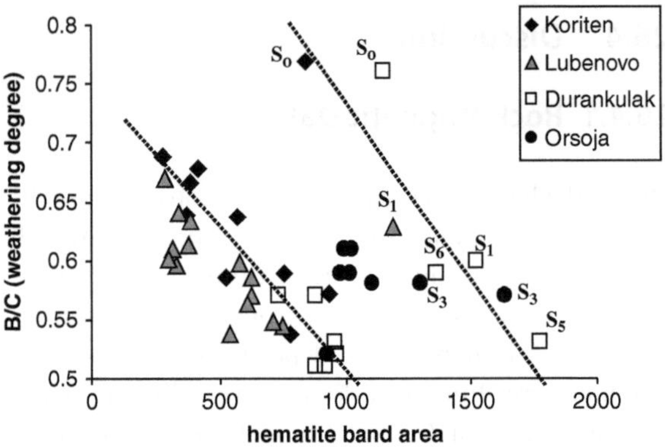

Fig. 26.6 DRS proxy parameter for the degree of weathering (B/C) plotted against fit area of hematite

difficult by the effect of time and duration of pedogenesis (Vidic et al. 2004). As shown by the latter authors, rubification (soil reddening) do not correlate so much with the duration of soil formation. According to the results from previous studies on loess-palaeosol sequences from SE Europe, cited above, magnetic enhancement of palaeosols compared to the signal of parent loess material is typical and widespread phenomena. This is best explained by the model of pedogenic magnetic enhancement, proposed originally for Chinese loess (Heller and Liu 1986, Verosub et al. 1993, Singer et al. 1996). As shown on Fig. 26.2a, paleosol samples, having enhanced magnetic susceptibility as compared to the parent loess, exhibit high percent frequency-dependent magnetic susceptibility (spanning the range 6–12%). This suggests that pedogenesis is accompanied by creation of nm-sized SP ferrimagnetic particles (Dearing et al. 1996, 2007). The extension of grain size distribution of pedogenic component towards sizes of remanence- carrying grains is further elucidated through the observed linear increase of susceptibility and SIRM values (Fig. 26.2b). Distribution of the data points along one trend line suggests that similar mean grain sizes are present in all profiles.

IRM unmixing procedure, aimed to identify and discriminate ferromagnetic components with different coercivities, resulted in defining a 2-component model for most of the loess and palaeosol samples studied. Examples shown on Fig. 26.3 give an overview of the three general cases of IRM acquisition curves, obtained in our study. Recent soils (So) are usually characterized by higher proportion of low-coercivity component if compared to the situation in older palaeosols, for example soil S6 from Durankulak

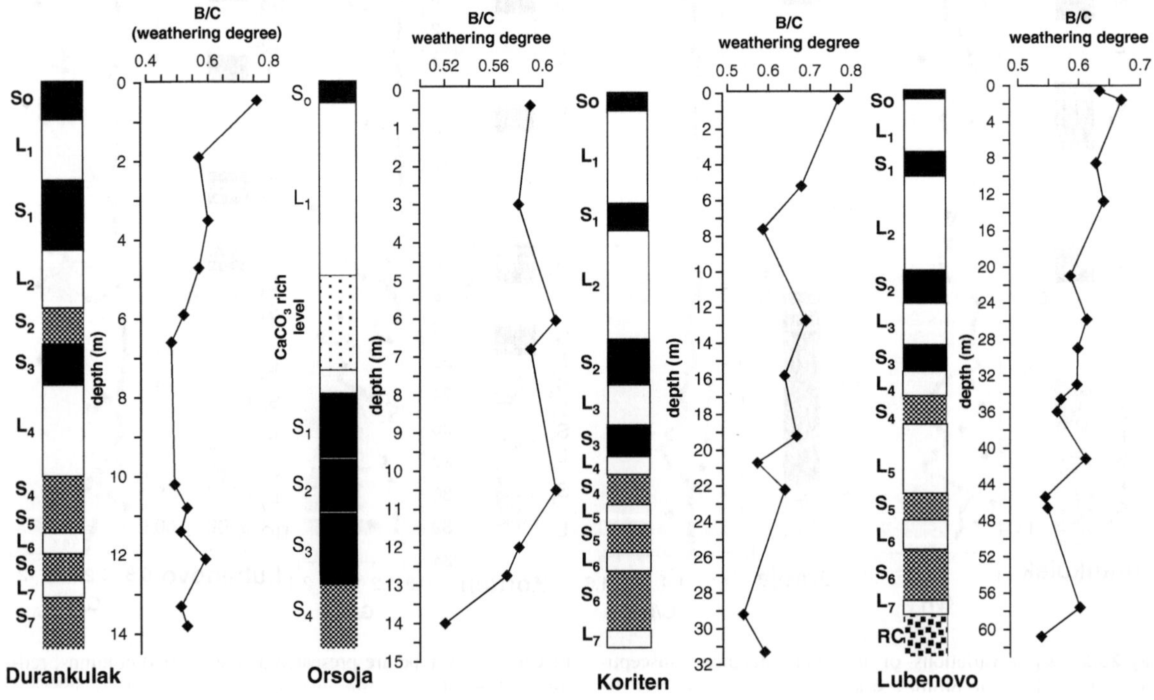

Fig. 26.7 Depth variations of the ratio B/C along the studied profiles

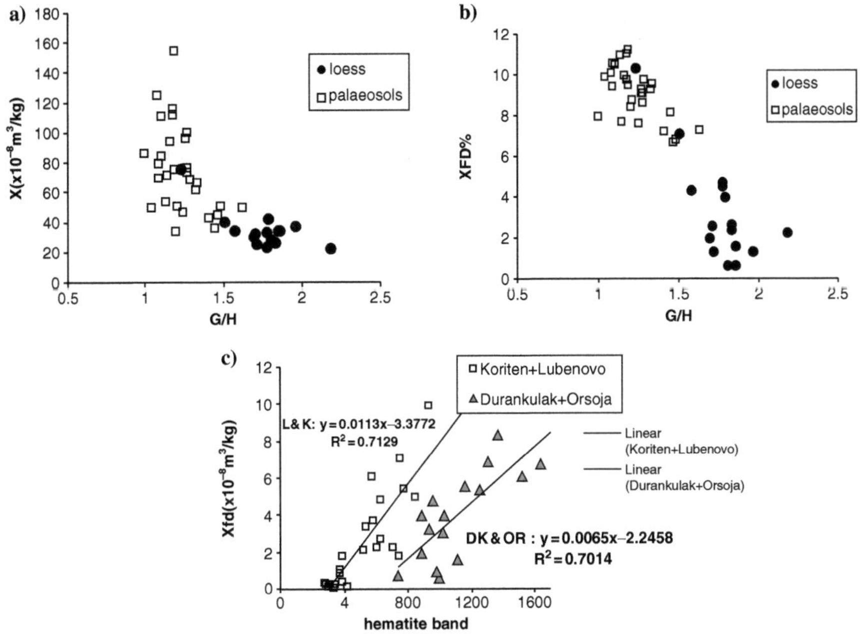

Fig. 26.8 Dependence between rock magnetic and DRS parameters: (**a**) low field magnetic susceptibility versus G/H ratio; (**b**) percent frequency dependent magnetic susceptibility versus G/H ratio; (**c**) area fit for hematite against frequency dependent magnetic susceptibility. Regression equations and the corresponding R^2 for data from Koriten and Lubenovo, and Durankulak and Orsoja are given as well

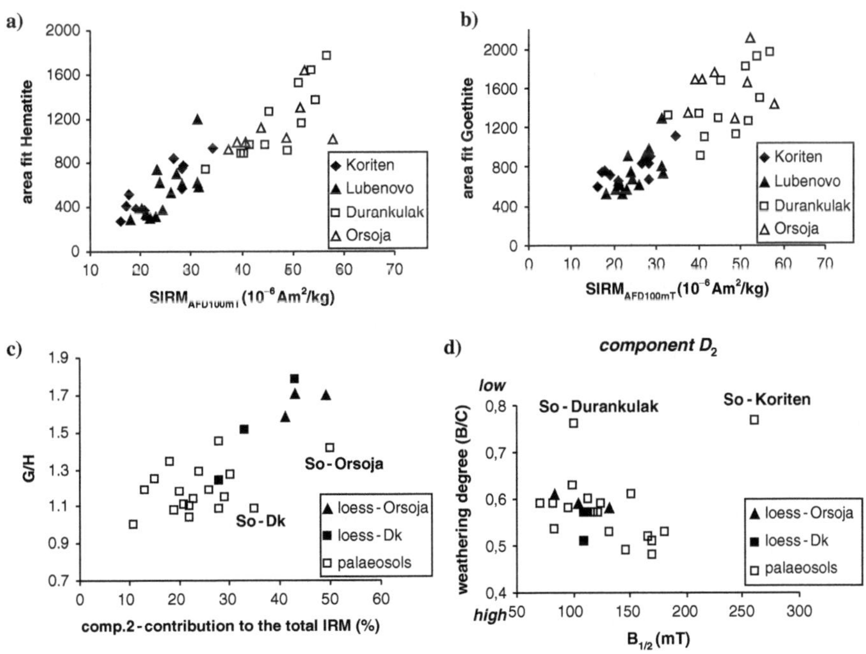

Fig. 26.9 Relationships between magnetic remanence parameters and DRS data: (**a**) $SIRM_{AFD100mT}$ against hematite area and (**b**) against goethite area; (**c**) correlation between the calculated relative contribution of component D_2 from IRM unmixing and the ratio G/H; (**d**) degree of weathering (B/C) vs median coercivity $B_{1/2}$ of component D_2

profile (Fig. 26.3c). IRM acquisition curves for samples from unweathered loess from profiles Koriten and Lubenovo are best fitted with 1 Gaussian function (Fig. 26.3b). Mean coercivity ($B_{1/2}$) of this component is fairly uniform for all loess samples (58 ± 4 mT) (see also Table 26.1) and could be related to the presence of unweathered detrital (titano)magnetite grains – named component D_1. The first component isolated in the rest of the samples has extremely uniform mean coercivity (see Table 26.1), suggesting that it is carried by pedogenically produced phase. Taking into account previously published results on the loess/palaeosol profiles studied here (Jordanova et al. 2007, 2008, Avramov et al. 2006), this low coercivity component is carried by maghemite. Very similar coercivities of pedogenic component are reported in other studies of loess-palaeosol sequences from China and Midwestern United States (Spassov et al. 2003, Geiss et al. 2008). The obtained invariable median coercivity of 31 mT with distribution width dP of 0.3 (Table 26.1) suggests bacterially-mediated pedogenic production of stable single-domain magnetite, subsequently oxidized to maghemite. Varying proportion of this phase in the total IRM, as seen from Fig. 26.4, probably relates to different degree of pedogenesis and particular combination of soil-forming factors. This pedogenic component (P), found in loess samples as well, points to the presence of secondary alterations of loess horizons from Durankulak and Orsoja profiles. Second component (D_2), showing higher and more variable coercivities among samples (Fig. 26.4) has higher contribution in loess samples, accompanied by lower $B_{1/2}$ values. Two possible mineral assignments of this component could be made: oxidized weathered (titano)magnetites or hematite. The latter mineral could have detrital origin, e.g. coarse-grained particles, coming from the dust source area; or it may be present as clusters of nanoparticles precipitated on mineral surfaces after destruction (weathering) of Fe-bearing silicates. Both genetic types of hematite may exist simultaneously in samples from palaeosols (Chen et al. 2010). Separation of data points in Fig. 26.2c of $SIRM_{AFD100mT}$ vs frequency-dependent magnetic susceptibility (χ_{FD}) could be ascribed to such cumulative effect of the presence of two generations of hematite in Durankulak and Orsoja. In an applied DC field of 2T hematite and partly goethite can contribute to the SIRM but with certain limitation of the range of grain size interval. The major part of the high-coercivity fraction of SIRM(2T) is supposed to be due to larger hematite grains, beyond the critical SP-SD threshold size (10–30 μm) (Dunlop and Ozdemir 1997). The ability of antiferromagnetic goethite to carry remanence is ascribed to imperfect spin compensation of the two sublattices (Banerjee 1970). It is the second possible magnetically hard IRM carrier, which has extremely high coercivity and the maximum saturating field used in the present study is only at the lower range of IRM acquisition curves, while saturation is not reached even in fields of 20 T (Dekkers 1989, Rochette and Fillion 1989). Consequently, the contribution of goethite to SIRM(2T) is highly limited to grains with very low coercivities. On the other hand, significant part of red pigmentation of palaeosols is due to fine grained SP hematite, which could acquire remanence of low coercivity at room temperature in case of inter-particle interactions (Hansen et al. 2000).

26.4.2 DRS: Proxies of FeOOH/α-Fe_2O_3 and Chemical Weathering

Variations of the ratio G/H along the profiles studied can be used as a proxy for the degree of soil maturity. Pedogenic formation of hematite and goethite supposes the presence of ferrihydrite as a necessary precursor (Schwertmann 1988). The competing processes of hematite and goethite formation from ferrihydrite strongly depend on the pH, temperature, water activity, and the rate of biomass turnover (Cornell and Schwertmann 1996). Because of the retarding effect of humics on ferrihydrite – goethite transformation (Schwertmann 1988), the pedoenvironment in the upper soil horizons should promote hematite formation over goethite. This phenomenon indicates that the hematite should prevail in interglacial / interstadial conditions with much faster biomas turnover in soils. The observed systematically lower G/H ratio for palaeosols in comparison with loess (Fig. 26.5) is in line with this expectation. The same trend was found by Grygar et al. (2002) in loess-soil profile in a loess-soil section in the Czech Republic, where nanocrystalline hematite was identified in loessic and assumed to be the result of active pedogenesis. On the other hand, Ji et al. (2002) did not find systematic increase in their hematite to goethite ratio in Chinese loess-soil sequence, in spite of visual difference in the colour of soil and loess. The discrepancy could be explained by differences in the analysis of color spectra: we used deconvolution to Gaussian components corresponding to the actual chromophors, Ji et al. (2002) decomposed the spectra to six standard colour bands. The respective Gaussian components of goethite and hematite used in deconvolution (absorption maxima at 481 and 534 nm, respectively) seem better suited to distinguish these two colouring components with partly overlapped absorption bands than separation of the entire spectrum into colour bands of ~50 nm width.

The observed consistent increase of hematite band area with increasing weathering degree (e.g. decreasing B/C ratio) in Fig. 26.6 suggests that main part of hematite fraction is produced as a result of increasing alterations of the primary loess substrate. Palaeosol samples, defining second trend with higher hematite content are probably suffering an additional low-temperature oxidation of detrital magnetite fraction, which leads to extra amount of hematite derived from another formation pathway.

Loess formation results from aeolian dust accumulation and weak post-depositional alterations. Depending on the distance of the source area, grain size and weathering of the primary loess is different (Pye 1987). Higher degree of weathering is typical for loess at great distance from the source. Slowly deposited loess is more prone to post-depositional weathering because the grains stay in surface environment longer time. This effect is combined with the particular climate conditions and geomorphology, which may further enhance the secondary alterations through pedogenic processes, changes in underground water level, etc. DRS data based on the ratio B/C, which indicates almost equal degree of Fe(II)-bearing silicates' weathering both for loess and palaeosol samples from Durankulak and Orsoja (Fig. 26.7) suggests that the loess itself had suffered significant weathering and the subsequent soil formation during warm climatic periods did not result in much more alterations probably because of the completion of the reactions. The other two profiles – Lubenovo and Koriten show more consistent systematic decrease of the ratio B/C with depth. This indicates an increasing degree of weathering with age.

26.4.3 Integrated DRS – Rock Magnetic Approach in Palaeoenvironmental Studies

Magnetic enhancement of palaeosols has been interpreted through various pathways of maghemite (magnetite) formation. Generally two main hypotheses are put forward. Earlier studies on soil magnetites (Maher and Taylor 1988) suggests that abiotic formation of submicron sized Fe_3O_4 takes place during reducing conditions created in the pedo-environment necessary for the presence of FeII in soil solution. Chen et al. (2005) consider bacterially-mediated magnetite formation, which is subsequently oxidized to maghemite. Recently proposed hypothesis considers maghemite in soils as an intermediate product of ferrihydrite transformation to haematite in aerobic soils (Barron and Torrent 2002, Torrent et al. 2006). The latter idea implies existence of close relationship between maghemite and haematite contents in palaeosols. Our DRS and rock magnetic results comply with the reported similar data for loess-palaeosol sections from China (Torrent et al. 2006). The obtained dependence between G/H ratio on the one hand, and low-field susceptibility (Fig. 26.8a) and percent frequency-dependent magnetic susceptibility (Fig. 26.8b) on the other, suggest that both, the fine grained strongly magnetic ferrimagnetic fraction and haematite are mainly formed during pedogenesis. However, there is a separation of data in two populations – Durankulak and Orsoja, and Koriten and Lubenovo (Fig. 26.8c) when the absolute values of χ_{FD} and area of hematite band are compared. Similarity in the range of frequency dependent susceptibility among all profiles, which is accompanied by systematic shift towards larger hematite area fits, indicate that samples from former two profiles also contain at the same time a higher amount of primary lithogenic haematite. Such finding is plausible if the original loess substrate for profiles Durankulak and Orsoja is coming from different source area with higher weathering potential (e.g. warmer climate during cold stadials, favouring hematite formation). This is really the case for both profiles, as far as Orsoja section is situated on fifth Danube river terrace and Durankulak is at the Black sea cost. According to the geological studies in the region (Evlogiev 2006), alluvium of the older river terraces appears as a dust source for loess accumulation on the younger, uppermost laying terrace. Thus, loess material for Orsoja and Durankulak is coming mainly from local sources. Having in mind milder climate conditions during glacial times in this area compared to the source of long distance dust transport (Alpine glaciers), higher hematite content can be readily explained. It most probably results from intense chemical weathering at the source area already before the subsequent pedogenesis in warmer and wetter periods.

Rock magnetic data can be further utilized to get better idea about grain size distribution of hematite (goethite) fractions. Observed good correlation between area fits of hematite (Fig. 26.9a) and goethite (Fig. 26.9b) and the isothermal remanence rest after AF demagnetization ($SIRM_{AFD100mT}$) suggests that at least part of the two fractions is capable to carry remanence, i.e. their mean grain size is larger than the critical SP/SD size for hematite (10–30 μm, Dunlop and Ozdemir 1997). Similar conclusion has been drawn by Balsam et al. (2004) and Ji et al. (2001) in their studies on loess-palaeosol profiles from the Chinese loess, where both goethite and hematite contents were increased in soil layers with respect to loess. Remanence-carrying haematite in the profiles Orsoja and Durankulak is the origin of the second component deduced by unmixing of IRM acquisition curves, clearly indicated in Fig. 26.9c. Some clue on the origin of this hematite gives the observed relationship between the ratio B/C, representing the degree of weathering of primary silicates in loess, and median coercivity of the second IRM component (Fig. 26.9d). The increase of $B_{1/2}$ values with increasing weathering degree for palaeosols suggests that pedogenesis results in higher degree of low-tempearute oxidation and appearance of pedogenic hematite. The outliers in Fig. 26.9d are samples from recent soils, which may be due to anthropogenic contamination.

Conclusions

Integrated rock magnetic and DRS study on samples from four loess-palaeosol profiles in North Bulgaria revealed the potential of such combined approach for obtaining more complete palaeoenvironmental information. The observed increase of hematite content in palaeosol units together with higher percent frequency dependent magnetic susceptibility suggests that both phases are product of pedogenic modification of the loess material. Analysis of coercivity distributions obtained from IRM acquisition curves reveals the presence of pedogenic component (P) with fairly uniform median coercivity of 31 mT and dispersion parameter dP of 0.32 in all palaeosol samples as well as loess samples from profiles Orsoja and Durankulak. This component most probably is carried by maghemite. The best fit of IRM acquisition curves for unweathered loess samples from the other two profiles – Koriten and Lubenovo – is obtained by 1-component model (component D_1) with parameters $B_{1/2} = 58 \pm 4$ mT and $dP = 0.44$. Second component (D_2) identified in the rest of the samples has wider variations of median coercivity – 123 ± 46 mT and dP of 0.53 and is ascribed to the presence of detrital oxidized (titano)magnetites and/or hematite.

Acknowledgments This study is supported by the project "Magnetic properties of soils as a reflection of their ecological status", Grant NZ1510 of the National Science Fund and Grant DO 02-193/2008 "Geophysical investigations of the environmental pollution level and its effect on human health in urban areas".

References

Avramov V, Jordanova D, Hoffmann V, Roesler W (2006) The role of dust source area and pedogenesis of three loess-palaeosol sections from North Bulgaria – a mineral magnetic study. Studia Geophys Geodaet 50:259–282

Balsam W, Ji JF, Chen J (2004) Climatic interpretation of the Luochuan and Lingtai loess sections, China, based on changing iron oxide mineralogy and magnetic susceptibility. Earth Planet Sci Lett 223:335–348

Banerjee S (1970) Origin of thermoremanence in goethite. Earth Planet Sci Lett 8:197–201

Barron V, Torrent J (1986) Use of the Kubelka-Munk theory to study the influence of iron oxides on soil colour. J Soil Sci 37:499–510

Barron V, Torrent J (2002) Evidence for a simple pathway to maghemite in Earth and Mars soils. Geochim Cosmochim Acta 66:2801–2806

Buggle B, Glaser B, Zoeller L, Hambach U, Markovic S, Glaser I, Gerasimenko N (2008) Geochemical characterization and origin of Southeastern and Eastern European Loesses (Serbia, Romania, Ukraine). Quat Sci Rev 27:1058–1075

Buggle B, Hambach U, Glasera B, Gerasimenko N, Markovic S, Glaser I, Zoller L (2009) Stratigraphy, and spatial and temporal paleoclimatic trends in Southeastern/Eastern European loess–paleosol sequences. Quat Int 196:86–106

Chen T, Xu H, Xie Q, Chen J, Ji J, Lu H (2005) Characteristics and genesis of maghemite in Chinese loess and paleosols:

mechanism for magnetic susceptibility enhancement in paleosols. Earth Planet Sci Lett 240:790–802

Chen, T, Xie Q, Hu H, Chen J, Ji J, Lu H, Balsam W (2010) Characteristics and formation mechanism of pedogenic hematite in Quaternary Chinese loess and paleosols. Catena 81:217–225

Cornell RM, Schwertmann U (1996) The iron oxides. Structure, properties, reactions, occurrence and uses. Weinheim, New York, NY

Dearing J, Hay K, Baban S, Huddleston A, Wellington E, Loveland P (1996) Magnetic susceptibility of topsoils: a test of conflicting theories using a national database. Geophys J Int 127:728–734

Dearing JA, Bird PM, Dann RJL, Benjamin SF (2007) Secondary ferrimagnetic minerals in Welsh soils: a comparison of mineral magnetic detection methods and implications for mineral formation. Geophys J Int 130:727–736

Dekkers M (1989) Magnetic properties of natural goethite – I. Grain-size dependence of some low- and high-field related rockmagnetic parameters measured at room temperature. Geophys J 97:323–340

Dunlop D, Ozdemir O (1997) Rock magnetism. In: Edwards D (ed) Fundamentals and frontiers. Cambridge Studies in Magnetism, Cambridge University Press

Evans M, Heller F (2003) Environmental magnetism. Principles and applications of enviromagnetics. Academic, California, MA

Evlogiev J (2006) The pleistocene and holocene in the Danube plain. Doctoral dissertation. Bulg. Acad. Sci. (in Bulgarian)

Gallet S, Jahn BM, Torii M (1996) Geochemical characterization of the LuoChuan loess-palaeosol sequences, China, and palaeoclimatic implication. Chem Geol 133: 67–88

Geiss C, Egli R, Zanner W (2008) Direct estimates of pedogenic magnetite as a tool to reconstruct past climates from buried soils. J Geophys Res 113:B11102. doi:10.1029/2008JB005669

Grygar T, van Oorschot IHM (2002) Voltammetric identification of pedogenic iron oxides in paleosol and loess. Electroanalysis 14:339–344

Grygar T, Dedecek J, Kruiver P, Dekkers MJ, Bezdicka P, Schneeweiss O (2003) Iron oxide mineralogy in late miocene red beds from La Gloria, Spain: Rock-magnetic, voltammetric and vis spectroscopy analyses. Catena 53:115–132

Grygar T, Kadlec J, Pruner P, Swann G, Bezdička P, Hradil D, Lang K, Novotná K, Oberhänsli H (2006) Paleoenvironmental record in Lake Baikal sediments: environmental changes in the last 160 ky. Palaeogeogr Palaeoclimatol Palaeoecol 237:240–254

Gu ZY, Lal D, Liu TS, Southon J, Caffee MW, Guo ZT, Chen MY (1996) Five million year 10Be record in Chinese loess and red clay: climate and weathering relationships. Earth Planet Sci Lett 144:273–287

Haase D, Fink J, Haase G, Ruske R, Pecsi M, Richter H, Altermann M, Jaeger K-D (2007) Loess in Europe – its spatial distribution based on a European Loess Map, scale 1:2,500,000. Quat Sci Rev 26:1301–1312

Hansen M, Koch CB, Morup S (2000) Magnetic dynamics of weakly and strongly interacting hematite nanoparticles. Phys Rev B 62(2):1124–1135

Heller F, Liu TS (1986) Paleoclimatic and sedimentary history from magnetic susceptibility of loess in China. Geophys Res Lett 13:1169–1172

Heller F, Shen CD, Beer J, Liu XM, Liu TS, Bronger A, Suter M, Bonani G (1993) Quantitative estimates of pedogenic ferromagnetic mineral formation in Chinese loess and paleoclimatic implications. Earth Planet Sci Lett 114:385–390

Heslop D, Dekkers M, Kruiver P, van Oorschot H (2002) Analysis of isothermal remanent magnetization acquisition curves using the expectation – maximization algorithm. Geophys J Int 148:58–64

Hradil D, Grygar T, Hrusková M, Bezdicka P, Lang K, Schneeweiss O, Chvatal M (2004) Green earth pigment from Kadan region, Czech Republic: use of rare Fe-rich smectite. Clays Clay Minerals 52:767–778

Hunt, CP, Singer MJ, Kletetschka G, Tenpas J, Hunt CP, Banerjee SK, Han J, Solheid J, Oches E, Sun W, Liu T (1995) Rock-magnetic proxies of climate changes in the loess-palaeosol sequences of the western Loess Plateau of China. Geophys J Int 123:232–244

Ji JF, Balsam W, Chen J (2001) Mineralogic and climatic interpretations of the Luochuan loess section (China) based on diffuse reflectance spectrophotometry. Quaternary Res 56:23–30

Ji JF, Balsam W, Chen J, Liu LW (2002) Rapid and quantitative measurement of hematite and goethite in the Chinese loess-paleosol sequence by diffuse reflectance spectroscopy. Clays Clay Minerals 50:208–216

Jordanova D, Petersen N (1999) Palaeoclimatic record from a loess-soil profile in northeast Bulgaria – II. Correlation with global climatic events during the Pleistocene. Geophys J Int 138(2):533–540

Jordanova D, Hus J, Geeraerts R (2007) Palaeoclimatic implications of the magnetic record from loess/palaeosol sequence Viatovo (NE Bulgaria). Geophys J Int 171:1036–1047

Jordanova D, Hus J, Evlogiev J, Geeraerts R (2008) Palaeomagnetism of the loess/palaeosol sequence in Viatovo (NE Bulgaria) in the Danube basin. Phys Earth Planet Int 167:71–83

Maher B, Thompson R (1995) Paleorainfall reconstructions from pedogenic magnetic susceptibility variations in the Chinese loess and paleosols. Quat Res 44:383–391

Minkov M (1968) Loess in North Bulgaria. Bulgarian Academy of Sciences, Sofia (in Bulgarian)

Mullins C, Tite M (1973) Magnetic viscosity, quadrature susceptibility and frequency dependence of susceptibility in single domain assemblies of magnetite and maghemite. J Geophys Res 78:804–809

Panaiotu C, Panaiotu E, Grama A, Necula C (2001) Paleoclimatic record from a loess-paleosol profile in Southeastern Romania. Phys Chem Earth (A) 26:11–12, 893–898

Porter SC, An ZC (1995) Correlation between climate events in the North Atlantic and China during the last glaciation. Nature 375:305

Pye K (1987) Aeolian dust and dust deposits. Academic, Harcourt Brace Jovanovich Publication, London

Rochette P, Fillion G (1989) Field and temperature dependence of remanence in synthetic goethite: paleomagnetic implications. Geophys Res Lett 16:851–854

Scheinost AC, Chavernas A, Barrón V, Torrent J (1998) Use and limitations of second-derivative diffuse reflectance spectroscopy in the visible to near-infrared range to identify and quantify Fe oxide minerals in soils. Clays Clay Minerals 46:528–536

Schwertmann U (1988) Occurrence and formation of iron oxides in various pedoenvironments. In: Stucki J, Goodman B, Schwertmann U (eds) Iron in soils and clay minerals. NATO ASI series, Series C: mathematical and physical sciences, vol 217. Reidel Publication Company, The Netherlands, pp 267–308

Singer M, Verosub K, Fine P, TenPas J (1996) A conceptual model for the enhancement of magnetic susceptibility in soils. Quat Int 34–36:243–248

Spassov S, Heller F, Kretzschmar R, Evans ME, Yue LP, Nourgaliev DK (2003) Detrital and pedogenic magnetic mineral phases in the loess/paleosol sequence at Lingtai (Central Chinese Loess Plateau). Phys Earth Planet Int 140:255–275

Sun YB, Lu HY, An ZS (2000) Grain size distribution of quartz isolated from Chinese loess/palaeosol. Chin Sci Bull 45:2296–2298

Torrent J, Barron V, Liu QS (2006) Magnetic enhancement is linked to and preceeds hematite formation in aerobic soil. Geophys Res Lett 33. doi:10.1029/2005GL024818

Torrent J, Liu Q, Bloemendal J, Barron V (2007) Magnetic enhancement and iron oxides in the Upper Luochuan Loess-Palaeosol sequence, Chinese Loess Plateau. Soil Sci Soc Am J 71(5):1570–1578

Verosub KL, Fin P, Singer MJ, Ten Pas J (1993) Pedogenesis and palaeoclimate: interpretation of the magnetic susceptibility record of Chinese loess-palaeosol sequences. Geology 21:1011–1014

Vidic N, Singer M, Verosub K (2004) Duration dependence of magnetic susceptibility enhancement in the Chinese loess-palaeosols of the past 620ky. Palaeogeogr Palaeoclimatol Palaeoecol 211:271–288

Magnetic Mapping of Weakly Contaminated Areas

27

Aleš Kapička, Eduard Petrovský, Neli Jordanova, and Vilém Podrázský

Abstract

Soil magnetometry has proved to be a helpful auxiliary method for outlining potentially contaminated areas. Magnetic susceptibility of topsoils may reflect concentration of atmospherically deposited anthropogenic iron oxides, which often coexist with harmful substances, such as heavy metals. Magnetic mapping of topsoils yields unambiguous results in areas with high concentration of pollutants and soils developed on iron-poor geologic basement. In this chapter we review the approach and results of magnetic mapping in a relatively clean area, characterized by rather complex topography. Surface measurements of magnetic susceptibility were carried out after previous examination of its vertical distribution in topsoils and complex laboratory measurements of other magnetic parameters as well as scanning electron microscopy observations. Our results show that over the whole area in concern topsoil magnetic susceptibility is enhanced due to atmospherically deposited anthropogenic iron oxides (prevailing from local sources), and soil magnetometry can thus be used for delineation of anthropogenic effect on soils also in this rather complex and difficult area.

27.1 Introduction

Despite a rather short history, magnetic methods are becoming quite common in studying our environment. The bulk of atmospheric fallouts of industrial origin contain significant portions of anthropogenic ferrimagnets (mostly Fe-oxides), which are deposited on the ground and accumulate in the topsoil layers. Measurements of concentration-dependent magnetic parameters of topsoils can be used to assess and in some cases quantify the amount of atmospherically deposited particles in the given regions (Evans and Heller 2003, Petrovský and Ellwood 1999).

Concentrations of heavy metals in soils often reach environmentally critical values, which exceed the maximum allowed contents of harmful elements, and have negative effect on crop production and human health (Petříková 1990). Therefore, one of the main aims of magnetic studies is to look for the relationship between magnetic parameters and concentrations of heavy metals. The underlying principle for potential correlation is the coexistence of anthropogenic ferrimagnets and heavy metals, from their source through the atmospheric pathways to their deposition. A high degree of correlation between magnetic susceptibility

A. Kapička (✉)
Institute of Geophysics AS CR, v.v.i., Bocni II/1401, 141 31 Prague 4, Czech Republic
e-mail: kapicka@ig.cas.cz

E. Petrovský et al. (eds.), *The Earth's Magnetic Interior*, IAGA Special Sopron Book Series 1,
DOI 10.1007/978-94-007-0323-0_27, © Springer Science+Business Media B.V. 2011

and concentrations of several heavy metals was found in soils around dominant sources of pollution with well-defined emission composition (Petrovský et al. 2000, Scholger 1998) or in urban agglomerations and industrial regions (Hanesch et al. 2003, Spiteri et al. 2005, Heller et al. 1998, Wang and Qin 2005). This relationship is more complicated in regions affected by more than one source of pollution, where the correlation between concentration of magnetic particles and heavy metals cannot be described using linear statistics (Hanesch et al. 2001).

Up to now, magnetic methods have been primarily used in areas with relatively high concentrations of anthropogenic particles in soils. For example, in areas with a high concentration of industries, the annual amount of atmospherically deposited dust reaches several thousands of tons. A single coal-burning power plant can produce hundreds of tons of fly ash per year (e.g., Heller et al. 1998). Since such fly ash contains some 9% of ferrimagnetic particles (Kapička et al. 1999), it is obvious that these particles can significantly influence magnetic properties of soils in the surrounding areas.

However, the reliability of magnetic mapping remains more problematic in relatively unpolluted areas where the contribution of anthropogenic ferrimagnetic particles to total magnetic properties can be relatively low and the contrast with the background values of magnetic parameters is not that pronounced (Magiera and Strzyszcz 2000).

In such cases, detailed laboratory investigations are necessary in order to determine the actual ratio of anthropogenic and natural ferrimagnets in the topsoil layers (Kapička et al. 2003). The concentration distribution of ferrimagnets in a soil column is, in general, determined by (a) the geological basement, (b) pedogenesis, and (c) anthropogenic input and temporal stability of anthropogenic ferrimagnets in real soil environment.

Parent rocks are one of the sources of primary magnetic minerals and in weakly contaminated soils may represent the dominant contribution to the total soil magnetic susceptibility. However, by analyzing vertical soil profiles, the significance of this lithogenic contribution can be well estimated and unsuitable localities can be identified and excluded from further studies or monitoring (Fialová et al. 2006, Magiera et al. 2006). The presence of ferrimagnets of pedogenic origin, which are ultrafine superparamagnetic particles, can be assessed by techniques such as frequency-dependent magnetic susceptibility (Maher and Taylor 1988). Analysis of anthropogenic ferrimagnets under soil conditions (different pH, soil moisture, and time of exposure) showed that their macroscopic magnetic parameters remain practically stable over longer periods of several months (Kapička et al. 2001, Maier et al. 2006). All these results, along with the finding that magnetic mapping of specific areas is reproducible (Boyko et al. 2004), make soil magnetometry useful, provided certain rules are obeyed, also in relatively less polluted areas.

The aim of this study is to demonstrate the application of soil magnetometry on a regional scale in the Giant Mountains National Park, located in northeast Bohemia, which has a low level of immissions relative to the rest of the Czech Republic. There are no major sources of pollution in its near neighborhood. We present the results of detailed magnetic and magnetomineralogic investigations of soil profiles in this region as well as detailed maps of the spatial distribution of topsoil magnetic susceptibility. With respect to the fact that direct atmospheric dust monitoring is carried out on very few sites in this region (Budská 2006), our results represent the first complex information about the distribution of atmospherically deposited dust particles and may serve as a starting point for future monitoring of pollution.

27.2 Description of Studied Area

The Giant Mountains National Park is a natural reserve in the northeast part of Bohemia, Czech Republic, along the Czech–Polish border (Fig. 27.1). The area of the park itself (without the protected band) is 36 ha and its altitude ranges from 400 to 1603 m a.s.l. Geological basement of this territory reflects old geological age of the Giant Mountains Region. Three main formations prevail in this territory (Plamínek 2007), comprising zones of probably W–E orientation. The core zone is formed of Krkonoše-Jizera granitoid pluton of Carbonian Age, metamorphosing at the contact with the neighboring zones of Proterozoic to Paleozoic sediments and volcanites of the Krkonoše-Jizera crystallinicum. Minor part is represented by the Krkonoše-Jizera gneisses. As a result of relatively harsh climate, the soil types are in general poor in nutrient and extremely acidic (Podrázský et al. 2007).

Fig. 27.1 Giant Mountains National Park area with main touristic centers and localities for magnetic and magnetomineralogic investigation of soil profiles (•)

Humic podzols (38% of territory), oligotrophic to podzolized cambisols (26%), and mesotrophic cambisols (10%) prevail on the major areas. Gleys and peats represent some 8% together with waterlogged sites. Leptosols (Rankers and Syrozems), i.e., undeveloped soils, occur in the summit area (6%). Minor sub-types of these soils form the rest of the area of interest. All the soils are strongly acidic (pH between 4 and 5 in the upper soil horizons).

Potential soil contamination can be due to long distance atmospheric transport of fly ash from coal-burning power plants, located in northern Bohemia and southern Poland. In addition, small local sources of pollution (e.g., public boiler houses, local combustion, and traffic) in touristic centers within the protected area and along its border can also play a significant role. The average daily concentration of PM10 (particulate matter with an aerodynamic diameter of $\leq 10\ \mu m$ – one of the major air pollutants consisting of tiny solid or liquid particles of soot, dust, smoke, fumes, and aerosols) in 2006 was 25 $\mu g/m^3$, i.e., about three times less than in industrialized regions (MŽP ČR 2007). Overall, the Giant Mountains National Park belongs to the least polluted areas within the Czech Republic.

27.3 Magnetic and Magnetomineralogic Investigation of Soil Profiles from the Giant Mountains National Park

Detailed magnetic investigations were carried out at a permanent net of 32 open forest soil pits, used for soil monitoring, distributed evenly over the territory of the Giant Mountains National Park (Kapička et al. 2003). The depth of individual pits varies from 40 to 60 cm and in most cases includes a layer of basement rock or soil dominated by the basement rock (C horizon). The pits were available for both in situ measurements and sampling for subsequent laboratory measurements. Our investigations were aimed at magnetic classification of individual soil (sub)horizons and at elucidating the problem of magnetic discrimination of the potentially contaminated surface layer and, eventually, at resolving the depth distribution of the anthropogenic magnetic particles.

Low-field magnetic susceptibility k was measured in situ, using a Bartington MS2 susceptibility meter with MS2F stratigraphic probe, resulting in an almost continuous vertical profile of magnetic susceptibility. Some 160 samples were also collected for further laboratory examination; sampled soil subhorizons were classified and the samples were air-dried and sieved at 2 mm. Laboratory measurements of low-field magnetic susceptibility were performed using an Agico KLY-3 kappabridge. Frequency-dependent magnetic susceptibility, defined as $k_{\mathrm{FD}}(\%) = (k_{\mathrm{lf}} - k_{\mathrm{hf}})/k_{\mathrm{lf}}$, where k_{lf} and k_{hf} represent susceptibility values measured at 0.47 and 4.7 kHz, respectively, was measured using a Bartington MS2 dual-frequency meter. In order to determine magnetic mineral phases, temperature dependence of magnetic susceptibility was measured using an Agico KLY-3 kappabridge and CS-3 furnace. Alternating field demagnetization was performed using a Schonstedt GSD-1 demagnetizer, and isothermal remanent magnetization (IRM) was measured using a JR-5 spinner magnetometer. Magnetic extracts were obtained from soils suspended in isopropanol using a permanent hand magnet. Mineralogy and chemical composition of the extracts were determined by SEM and microprobe analysis, respectively.

27.3.1 Magnetic Mineralogy Results

Changes in low-field magnetic susceptibility of typical soil depth profiles are shown in Fig. 27.2. Both in situ measurements (in most cases carried out on two parallel profiles within one soil pit) and laboratory measurements on samples collected from individual soil horizons are shown. A significant increase of magnetic susceptibility in the uppermost, organic L-F and Ah horizons was commonly observed. Although in some profiles individual soil layers can be less developed or some of them can be missing, increased susceptibility was only observed in depths between 4 and 6 cm below the surface. In deeper layers, in particular in B and C horizons, magnetic susceptibility was considerably lower. Maximum values of volume magnetic susceptibility in the whole region varied from 30 to 60×10^{-5} SI units, while those in deeper layers were between 10 and 20×10^{-5} SI units. A very significant and well-expressed maximum of magnetic susceptibility was observed in several profiles from ombrotrophic peat bogs. A maximum of about 40×10^{-5} SI units was found at a depth of 4 cm, while the deeper layers showed a much lower and practically constant value of $0–3 \times 10^{-5}$ SI units. Only four cambisol profiles exhibited susceptibility behavior, as shown for profile No. 31 in Fig. 27.2, with a comparatively weak maximum in the subsurface layer and a systematically increasing susceptibility with increasing depth.

In order to interpret correctly the magnetic properties of soils, careful evaluation of the contribution due to natural pedogenic/lithogenic processes and anthropogenic input is required. Low magnetic susceptibility values, measured also in the B and C horizons of podzols (Fig. 27.2), are most probably due to advanced acidification and destruction of the primary minerals, products of which are transported outside the soil profile. Cambisols are characterized by slight or moderate weathering of the parent material; therefore, for some of these profiles we obtained increasing magnetic susceptibility with depth, reflecting the influence of the parent material. In contrast, the specific pattern of susceptibility in depth profiles of peat bogs results from negligible lithogenic contribution and these sites are obviously particularly suitable for the purpose of magnetic monitoring of contamination due to deposition of industrial dust.

As a result of pedogenic precipitation of strongly magnetic iron oxides, a mixture of superparamagnetic (SP) and single-domain (SD) magnetite grains (e.g., Oldfield et al. 1985) is produced, showing significantly high values of the frequency-dependent magnetic susceptibility (Thompson and Oldfield 1986, Eyre 1997). Values of k_{FD}, obtained for our samples of organic soil horizons, are relatively low, below 4% (Fig. 27.2). Relatively acid conditions in the soils studied and generally low values of frequency-dependent susceptibility indicate that increased magnetic susceptibility in the surface layers is most probably due to a higher concentration of coarse-grained ferrimagnets, deposited from atmospheric dust. Coarse-grained particles of anthropogenic origin are not subjected to rapid

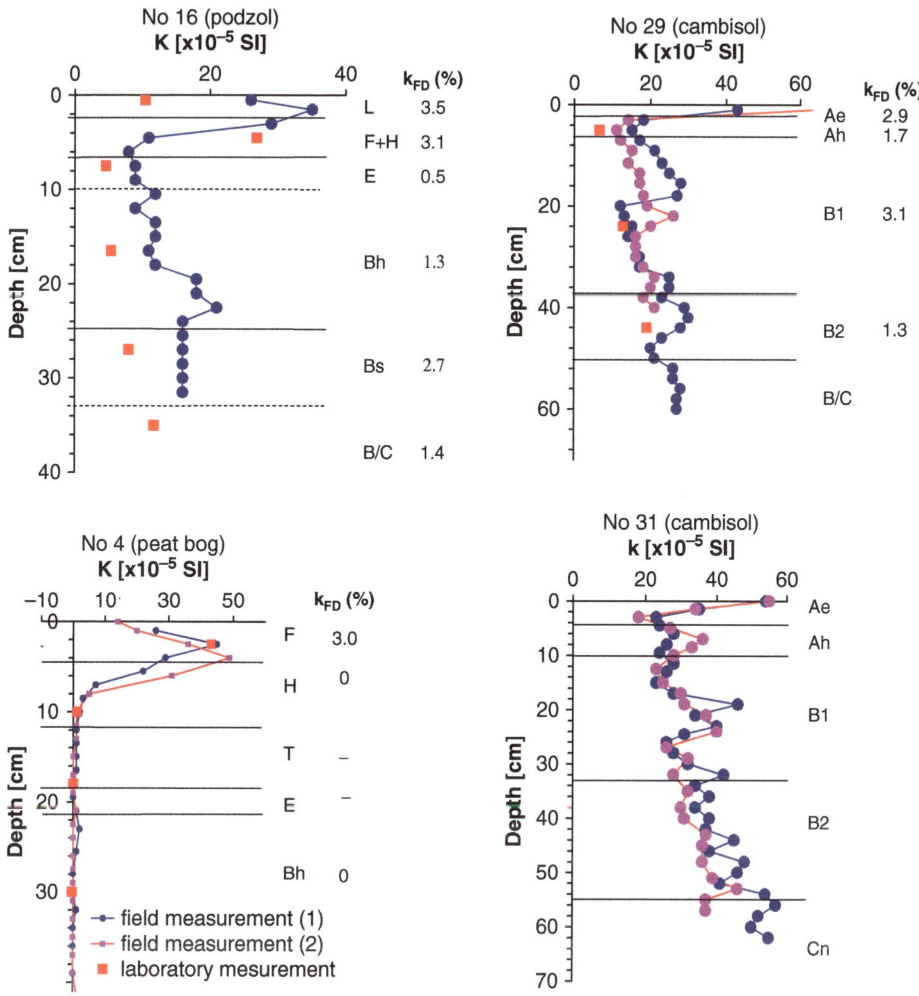

Fig. 27.2 Changes in magnetic susceptibility with depth for typical soil profiles in the Giant Mountains National Park

dissolution because their surface/volume ratio is not as high as for pedogenic ultrafine grains, which are easily dissolved under reducing conditions (Cornell and Schwertmann 1996).

In all soil profiles, we observed the maximum susceptibility in F, H, and Ah soil horizons, but not in an L horizon. This typical distribution of susceptibility can be explained by considering different binding conditions of atmospheric dust in these layers. The litter (L) horizon contains undecayed organic material, so that deposited atmospheric particulates are only physically held on the surface of organic remains (needles and leaves) and could be mechanically washed out and transported down the profile. Deeper F, H, or Ah horizons are characterized by high geochemical activity, so that iron oxide particles are strongly bound to the organic–mineral matter. This could explain the fact that we usually observe maximum magnetic susceptibility at 4–6 cm below the soil surface.

Thermomagnetic measurements were carried out in order to identify and describe in more detail the ferrimagnetic phases present in individual soil layers. Special attention was devoted to the topsoil layer characterized by increased susceptibility values. Temperature dependence of magnetic susceptibility of individual Ah, Ae, B1, and B2 horizons from typical soil pit No. 29 is shown in Fig. 27.3a. The upper soil layer shows the presence of a magnetite-like phase with Curie point T_c of about 580°C. The local maximum of susceptibility below this temperature can be interpreted either as a Hopkinson peak or as the creation of a new magnetite phase during heating.

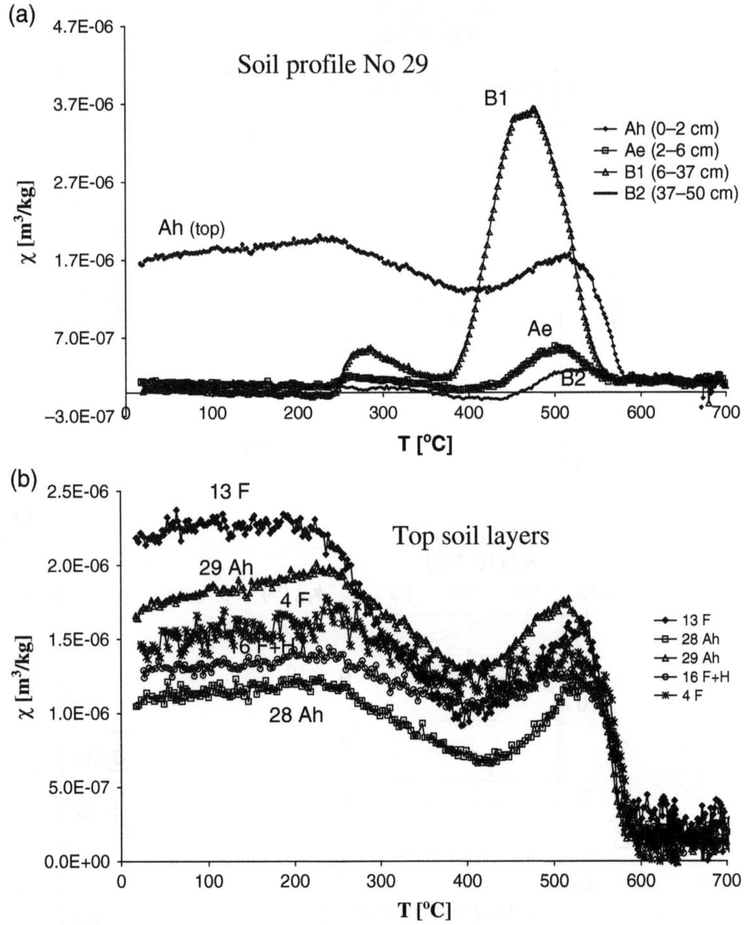

Fig. 27.3 Temperature dependence of magnetic susceptibility for samples from soil profile No. 29 (**a**) and for topsoil layers from different localities in the Giant Mountains National Park (**b**)

Deeper pedogenic horizons (Ae, B1, and B2) show very different and more complex behavior of magnetic susceptibility with increasing temperature. It could be supposed that the high-temperature susceptibility behavior in these subsoil horizons (and especially B horizons, where the secondary transformation products have accumulated) reflects thermal transformations of ferrihydrite, since its preservation in soils is promoted by low pH and the presence of Si and Al cations (Cornell and Schwertmann 1996). The concentration of the initial ferrimagnetic phase in these subsoil horizons is very low and the heating curve is dominated by the creation of secondary ferrimagnets. The maximum above 500°C could reflect neoformation of magnetite as a result of thermal alteration of Fe-rich clay minerals. Moreover, one could suggest creation of another, yet undetermined, magnetic phase at about 260°C in the B horizon. Neoformation of secondary ferrimagnets was also confirmed by cooling curves, ending at room temperature at values much higher than the initial ones. In Fig. 27.3b, heating curves related to potentially contaminated top layers of several soil pits are compared. Despite the fact that these samples are from different podzols and cambisols from the whole territory, their behavior is very similar, suggesting the presence of magnetite with Tc of 580°C and also probably maghemite, taking into account the significant drop in k after ~250°C, usually interpreted in soils as transformation of the maghemite to hematite (e.g., Dunlop and Özdemir 1997).

IRM acquisition curves of samples of individual horizons from typical depth profiles are shown in Fig. 27.4. The curve corresponding to the top layer is clearly distinct, reaching saturation of remanence at a magnetic field of 200 mT, indicating the prevalence of a magnetically soft, magnetite-like phase. On the other hand, samples from deeper horizons contain significant

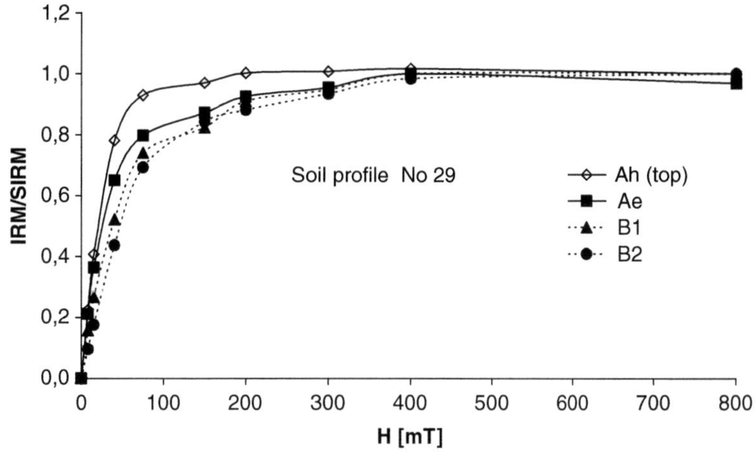

Fig. 27.4 IRM acquisition curves for samples of individual horizons from typical soil profile

portions of a magnetically hard phase and saturate at a much higher field, above 400 mT.

SEM and microprobe analyses were carried out on magnetic extracts from topsoil horizons showing the maximum magnetic susceptibility. Samples with low magnetic susceptibility from deeper soil horizons were also analyzed for comparison. Samples 16F+H, 4F, 12F, 29Ah, 4H, 16Ae, 12H, and 16Bs were investigated. Anthropogenic Fe-oxide particles of typical spherical shape are abundant in the uppermost soil horizons. The concentration of anthropogenic ferrimagnets in the uppermost soil layers is relatively high. They were easily identified in all the samples studied (five individual subsamples were analyzed from each layer) and comprise a dominant part of the ferrimagnetic Fe-oxides (Fig. 27.5, Table 27.1). The situation is different in the deeper soil horizons. Despite the fact that they are immediately below the contaminated uppermost soil horizons, it was practically impossible to identify anthropogenic ferrimagnets by means of SEM observations. In cases where some approximately spherical particles were observed in extracts from these horizons, phase analyses showed that they did not contain Fe-oxides (Table 27.1).

Magnetic and thermomagnetic analyses confirmed that the dominant ferrimagnetic component in the

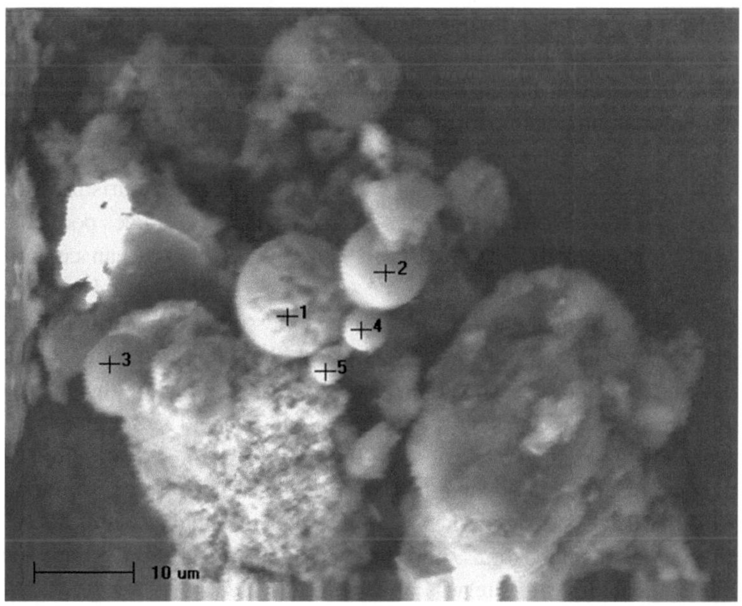

Fig. 27.5 Fe-oxides of anthropogenic origin in topsoil layer (locality No. 4) from the Giant Mountains National Park

Table 27.1 Microprobe analysis of spherical particles in topsoil (F) and subsoil (H) layers

	Topsoil (4F sample)						Subsoil (4H)	
	No. 1	No. 2	No. 3	No. 4	No. 5		No. 1	No. 2
Element	Wt%	Wt%	Wt%	Wt%	Wt%	Oxide	Wt%	Wt%
OK	32.53	34.55	38.9	36.14	42.88	Na_2O	33.99	3.31
MgK	0.19	0	0.61	0.17		MgO	4.82	2.88
AlK	0.79	0	8.23	0.13	0.36	Al_2O_3	3.02	4.16
SiK	0.55	0.44	17.44	0.36	0.63	SiO_2	7.26	9.85
SK	0.2	0.1	0.26	0.13	0.29	SO_3	6.59	3.92
ClK	0.17	0.13	0.11	0.1	0.08	ClO_0	20.06	3.76
KK	0.2	0.16	0.58	0.07	0.1	K_2O	11.94	2.59
CaK	5.5	0.82	1.49	0.35	0.82	CaO	11.31	67.1
TiK	0.27	0.32	0.78	0.14	0.15	TiO_2	0.33	0.82
MnK	0.74	0	0	0.53	0			
FeK	58.87	63.49	31.61	61.89	54.69	Fe_2O_3	0.68	1.61
Total	100	100	100	100	100	Total	100	100

upper organic soil horizons in the Giant Mountains National Park (independently of the particular soil type – podzol, cambisol, and peat bog) is magnetically soft coarse-grained magnetite. Our results are consistent with recently published results by Magiera and Strzyszcz (2000) dealing with weakly contaminated soils in Polish national parks. SEM analysis of magnetic extracts clearly showed that this magnetite is of anthropogenic origin and is responsible for enhancement of magnetic susceptibility in the uppermost organic horizons. Therefore, the surface magnetic susceptibility of forest soils in the Giant Mountains National Park reflects mainly the concentration of anthropogenic ferrimagnets.

27.4 Field Measurements of Topsoil Magnetic Susceptibility

Extensive measurements of topsoil magnetic susceptibility were carried out on 462 sites over the whole park area (Kapička et al. 2008). Magnetic susceptibility was measured using a Bartington MS2D field probe, with declared precision of 10^{-6} SI. Efficient depth of penetration is some 6–8 cm (Lecoanet et al. 1999). All the measured sites were selected in open air grass-covered locations. In order to eliminate local inhomogeneities, the site was always represented by a spot of some 2 × 2 m, where 10–15 readings were taken and averaged. The geographic position of each site was recorded with the precision of ±2 m using a GPS Trimble Pathfinder. In selecting and measuring the sites, we followed the recommendations of Schibler et al. (2002). The obtained data set of site coordinates and topsoil magnetic susceptibilities was used for maps of the susceptibility spatial distribution, compiled using Surfer 8.0 (Golden Software).

27.4.1 Field Mapping Results

The Giant Mountains National Park, with marked touristic centers inside the park (Špindlerův Mlýn, Pec pod Sněžkou, and Harrachov) and important urban centers on the border (Vrchlabí and Svoboda nad Úpou), is shown in Fig. 27.6a along with the distribution of all the measured sites. Although the terrain conditions did not allow regular distribution of the measured sites, their density is high enough and enables compilation of a detailed map of spatial distribution of the measured values.

In situ magnetic susceptibility (k) of topsoils varies from 1.2×10^{-5} SI to 95.5×10^{-5} SI. A histogram of the values, obtained on all the 462 sites, is depicted in Fig. 27.7. The shape of the histogram is very close

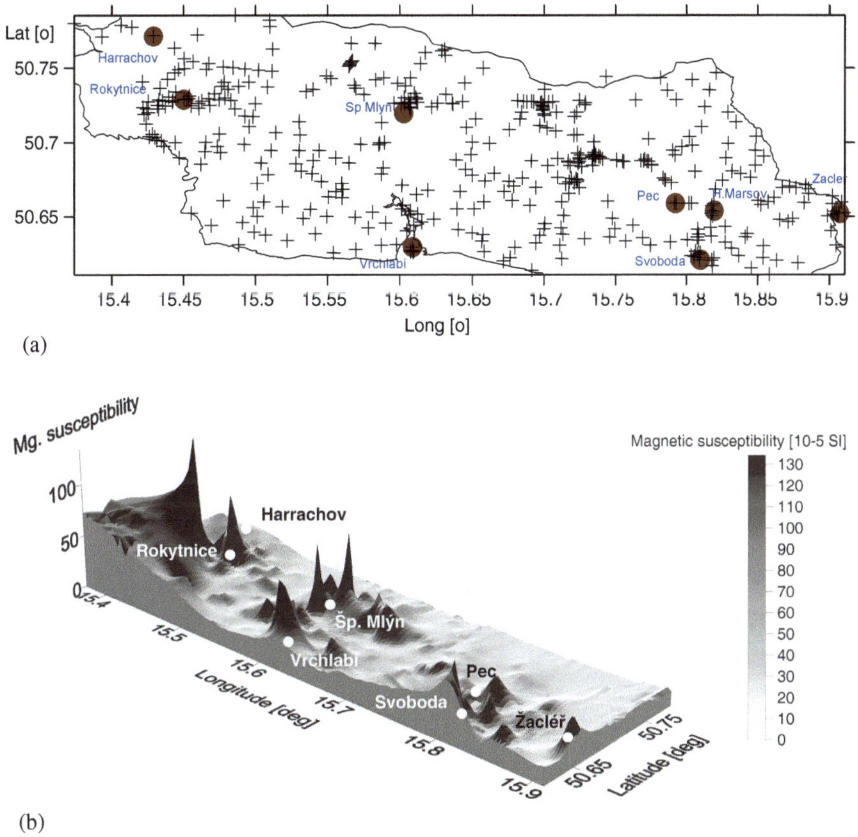

Fig. 27.6 (**a**) Area of Giant Mountains National Park together with main touristic centers (·) and position of measured localities (+). (**b**) Topsoil magnetic susceptibilities in Giant Mountains National Park. *Z*-axis shows values of volume magnetic susceptibility (k) (Kapička et al. (2008), © 2008 Inst. Geophys. AS CR Prague)

to log-normal distribution. The average (k_{mean}) and median (k_{med}) of the measured values are 14.4×10^{-5} SI and 10.7×10^{-5} SI, respectively. However, k_{mean} is more than three times lower than values typical for industrial regions (e.g., Schmidt et al. 2005, Chianese et al. 2006), indicating a much lower concentration of anthropogenic ferrimagnets in these topsoils.

The spatial distribution of topsoil magnetic susceptibility over the whole area of the Giant Mountains National Park is shown in Fig. 27.6b. Evidently, rather low values, smaller than 20–25$\times10^{-5}$ SI, are typical for most of the park area. Susceptibility values are systematically higher in the southwest. Extreme values within the park are associated with the tourist centers of Harrachov, Špindlerův Mlýn, and Pec pod Sněžkou. In these cases, local sources of pollution are responsible for the enhanced topsoil susceptibility. Although the magnetic susceptibility in these areas is very high, these anomalies do not extend far away, are well bordered, and are not interconnected. These results can be interpreted as follows: one of the possible pollution sources, long distance transport of pollutants, affects mainly topsoils in the western part of the region. The geological bedrock is rather uniform in the W-E direction (Plamínek 2007), so no effect of the mineral composition could be expected in this sense. Therefore, variations in the content of pollutants are due to the external sources rather than the bedrock composition. Prevailing wind direction over the park area is west to southwest. Accordingly, atmospherically deposited dust is emitted by sources in an industrial region in North Bohemia and, in general, also in the "Black Triangle" area (ČHMÚ 2003). In addition, contribution from local tourist centers within the National Park is very important.

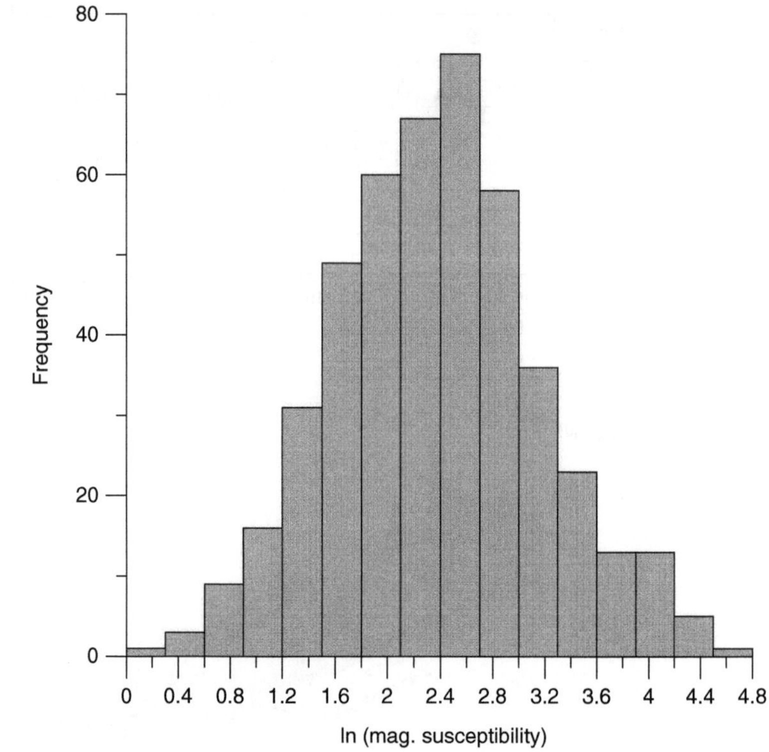

Fig. 27.7 Histogram of the logarithmic values of volume magnetic susceptibility (*k*) for all measured localities in Giant Mountains National Park (Kapička et al. (2008), © 2008 Inst. Geophys. AS CR Prague)

27.5 Heavy Metal Analyses and Correlation with Soil Magnetic Susceptibility

Within this study, heavy metal analyses were carried out on samples from more than 30 vertical soil profiles from the National Park area. These sites were distributed over the whole of the area studied and represent the whole range of immission conditions. Samples from individual soil horizons (F, H, Ah(i), B(i), and Cn) were analyzed, which means that from each site some six to nine specific soil horizons were examined. We focused on heavy metals, which are major contributors to environmental stress (Pb, As, Cd, Co, Zn, and Fe). Their concentration was determined using atomic absorption spectrometry (AAS) on soil solutions obtained by standard leaching with 2M HNO_3, namely 10 g of dried soil material was leached for 17 h in 100 ml of the leaching agent. "Sister" samples from the same pedozones were used for measurements of mass-specific magnetic susceptibility.

27.5.1 Heavy Metal Analyses Results

Concentrations of heavy metals in the region show quite high variability, both within individual sites and between individual pedozones. Significant concentrations, with maximum values exceeding the allowed limits of harmful elements in soils (MŽP ČR 2007), were found for Pb (1.10–342 mg/kg) and Zn (0.8–68.8 mg/kg). High concentrations of these two heavy metals were found in this region and also on the Polish side of the border (Strzyszcz 1999). One of the main tasks in environmental magnetism is to determine those heavy metals that show significant correlation with macroscopic magnetic parameters, such as magnetic susceptibility, in the studied area. From the point of view of the potential of these correlations for future magnetic monitoring of pollution, the main task is to define those elements that are dominantly of anthropogenic origin, while the lithogenic contribution is negligible. For this purpose, we have carried out detailed chemical analyses of individual horizons within vertical soil profiles. The whole set of the studied elements can be divided into two groups. Pb and Zn show typically enhanced values in the topsoil layers. Vertical distribution of the other elements (As and Co)

Table 27.2 Concentrations of heavy metals (HM) in topsoil (F+H) horizon normalized to background mineral (Cn) horizon for representative localities in the Giant Mountains National Park

Localities		HM (anthropogenic)			HM (lithogenic)		
		Zn	Pb	Cd	Fe	Co	As
1	F+H/Cn	3.68	18.84	2.52	0.14	0.07	0.88
	Cn (mg/kg)	5.90	8.60	0.49	9,720.00	1.70	0.30
6	F+H/Cn	1.41	19.50	1.48	0.60	0.03	0.80
	Cn (mg/kg)	26.40	12.60	0.46	2,080.00	8.98	0.19
7	F+H/Cn	1.85	6.32	0.47	0.43	0.05	0.45
	Cn (mg/kg)	15.10	21.60	0.45	2,900.00	7.06	0.75
8	F+H/Cn	1.47	7.94	0.42	0.07	0.04	0.35
	Cn (mg/kg)	17.70	31.40	0.94	25,700.00	6.97	1.09
9	F+H/Cn	1.21	43.95	1.36	0.30	0.03	0.26
	Cn (mg/kg)	22.70	5.10	0.44	4,270.00	5.04	0.64
10	F+H/Cn	1.75	13.41	4.95	0.49	0.98	0.43
	Cn (mg/kg)	14.10	17.00	0.39	3,930.00	0.43	0.97
13	F+H/Cn	2.97	3.56	4.33	0.34	0.11	1.08
	Cn (mg/kg)	11.60	39.80	0.39	6,180.00	6.70	0.57
14	F+H/Cn	1.55	9.06	1.51	0.39	0.14	0.37
	Cn (mg/kg)	20.10	16.10	0.41	4,750.00	2.69	0.89
16	F+H/Cn	5.15	7.86	1.61	0.26	0.92	0.90
	Cn (mg/kg)	5.40	24.20	0.41	5,310.00	0.39	0.36
17	F+H/Cn	1.41	4.88	1.81	0.19	0.16	0.73
	Cn (mg/kg)	20.70	20.30	0.46	4,130.00	6.08	0.93
18	F+H/Cn	3.26	21.58	1.79	0.98	0.35	2.65
	Cn (mg/kg)	5.70	11.20	0.40	1,920.00	0.88	0.16
19	L+H/Cn	2.55	13.92	1.83	0.36	0.04	1.76
	Cn (mg/kg)	9.50	10.40	0.40	8,400.00	8.59	0.23
21	F+H/Cn	2.25	15.42	1.84	0.16	0.02	0.19
	Cn (mg/kg)	13.80	9.00	0.42	6,700.00	12.60	0.89
24	F+H/Cn	1.45	15.25	1.45	0.16	0.07	0.32
	Cn (mg/kg)	11.20	7.30	0.37	8,110.00	6.24	0.96
26	F+H/Cn	5.19	9.96	1.55	0.18	0.37	0.16
	Cn (mg/kg)	6.70	12.30	0.38	5,500.00	1.10	1.01
28	F+H/Cn	4.26	68.21	1.80	0.42	0.14	1.81
	Cn (mg/kg)	9.90	2.80	0.45	2,570.00	3.95	0.17
29	F+H/Cn	3.27	43.53	1.40	0.32	0.12	0.56
	Cn (mg/kg)	12.80	4.80	0.39	10,640.00	4.71	0.42
30	F+H/Cn	9.23	55.36	1.52	0.19	0.09	0.71
	Cn (mg/kg)	7.30	1.10	0.46	3,460.00	5.08	0.13
31	F+H/Cn	5.08	30.96	1.26	0.15	0.05	2.87
	Cn (mg/kg)	9.80	2.80	0.47	8,400.00	5.08	0.05
32	F+H/Cn	3.52	58.44	0.93	0.06	0.03	0.40
	Cn (mg/kg)	14.80	1.60	0.50	11,700.00	4.42	0.34

is completely different, with concentration increasing with increasing depth and reaching maximum values usually in subsoil horizons. In the area of interest, only limited data of heavy metal distribution are available (Podrázský 2001). The microelement contents differ with the bedrock (Phyllite – Metadiabase), altitude, and humus layer transformation. Similar trends in heavy metal and other pollutant distribution were detected by other authors, e.g., Fiala et al. (2008). Distribution of concentration of the elements from the first group correlates well with that of magnetic susceptibility and is, therefore, linked to distribution of anthropogenic ferrimagnets. If we consider the content of heavy metals in Cn mineral horizon (Table 27.2) as the background values for this territory, then the topsoil-related values, normalized with respect to the Cn values, suggest that only the elements of the first group show correlation with magnetic parameters sufficient for magnetic assessment of topsoil contamination by Pb and Zn in the Giant Mountains National Park. Vertical distribution pattern, in particular of the elements of the second group, indicates possible migration of the elements toward the deeper horizons. Fe especially shows such a pattern, and this effect will be the subject of our further investigations. However, distribution of the elements into two groups seems to be useful, because the presented map of spatial distribution of magnetic susceptibility (Fig. 27.6b) may serve for the assessment of actual concentration of specific heavy metals in topsoils within the park area.

Our analyses of topsoils revealed the existence of moderate correlation ($R^2 = 0.44$) between mass-specific magnetic susceptibility (χ) and concentration of Pb. Concentration with Zn, despite remarkably similar patterns of vertical profiles, is only weak, with $R^2 \approx 0.20$ (Kapička et al. 2008). This may be most probably due to the multi-source origin of immissions in the Giant Mountains National Park. Individual sources, including long distance transport, may contain various spectra of heavy metals.

Conclusions

Magnetic mapping has been successfully used in pollution studies primarily in areas with intensive industrial activity and major pollution sources. In contrast, we examine a comparatively unpolluted area of Giant Mountains National Park. Our results confirm that over the whole area in concern the topsoil layer is magnetically enhanced in terms of magnetic susceptibility. This layer is limited to depths of 4–6 cm below the soil surface, usually in F, H, or top of Ah soil horizons. Magnetic properties of the top layers are consistent over the whole territory of the park and magnetomineralogical analysis corroborates that the topsoil horizons are dominated by coarse-grained magnetically soft magnetite of anthropogenic origin. Magnetomineralogy of deeper soil layers is more complex and different from site to site.

Detailed map of topsoil magnetic susceptibility over the area of the Giant Mountains National Park is presented. Within the park and on its margin, local sources of pollution are identified, and the extent of their effect can be delimited. We found that each of these relatively significant local sources of pollution has very limited effect. It is shown that soil contamination in this area has multi-source characteristics (long distance transport and local sources). The mix of sources probably affects the significance of correlation between concentrations of heavy metals of anthropogenic origin and magnetic parameters of topsoils. However, a moderate correlation was found between topsoil magnetic susceptibility and the concentration of Pb in the topsoil. In this specific case, fast and low-cost magnetic mapping can be used as approximate method for monitoring the contents of Pb in topsoils in this area. Our results, based on detailed laboratory investigations of soils from the Giant Mountains National Park, indicate that magnetic mapping of anthropogenic pollution can be used also in areas with relatively lower pollution levels.

Acknowledgments This study was performed with the support of the Grant Agency of the Czech Republic through grant No. 205/07/0941, Grant Agency of the Academy of Sciences of the Czech Republic through project No. IAA300120701, and Ministry of Education, Youth and Sports through project No. LA09015.

References

Boyko T, Scholger R, Stanek H, Magprox Team (2004) Topsoil magnetic susceptibility mapping as a tool for pollution monitoring: repeatability of in situ measurements. J Appl Geophys 55:249–259

Budská E (2006) Monitoring of atmospheric deposition in the area of the Krkonoše Mts. (Giant Mts). Abstracts of

Conference *Geoekologické problémy Krkonoš*. Vrchlabí 3-5.10.2006, KRNAP, p 11
Chianese D, D'Emilio M, Bavusi M, Lapena V, Macchiato M (2006). Magnetic and ground probing radar measurements for soil pollution mapping in the industrial area of Val Basento (Basilicata Region, Southern Italy): a case study. Environ Geol 46:389–404
Cornell R, Schwertmann U (1996) The iron oxides. Structure, properties, reactions, occurrence and uses. VCH Verlagsgesellschaft, Weinheim, pp 338–347
ČHMÚ (2003) Symos 97, verze 02. Systém modelování stacionárních zdrojů. Report CHMU, Praha (in Czech)
Dunlop DJ, Özdemir Ö (1997) Rock magnetism: fundamentals and frontiers, Cambridge University Press, Cambridge
Evans ME, Heller F (2003) Environmental magnetism. Academic Press, San Diego
Eyre JK (1997) Frequency dependence of magnetic susceptibility for populations of single domain grains. Geophys J Int 129:209–211
Fiala P, Reininger D, Samek T (2008) A survey of forest pollution with heavy metals in the Natural Forest Region (NFR) Moravskoslezske Beskydy with particular attention to Jablunkov Pass. J Forest Sci 54:64–72
Fialová H, Maier G, Petrovský E, Kapička A, Boyko T, Scholger R (2006) Magnetic properties of soils from sites with different geological and environmental settings. J Appl Geophys 59:273–283
Hanesch M, Maier G, Scholger R (2003) Mapping heavy metal distribution by measuring the magnetic susceptibility of soils. J Phys IV France 107:605–608
Hanesch M, Scholger R, Dekkers MJ (2001) The application of fuzzy c-means cluster analysis and non- linear mapping to a soil data set for the detection of polluted sites. Phys Chem Earth 26:885–891
Heller F, Strzyszcz Z, Magiera T (1998) Magnetic record of industrial pollution in forest soils of Upper Silesia, Poland. J Geophys Res 103/B8:767–774
Kapička A, Petrovský E, Ustjak S, Macháčková K (1999) Proxy mapping of fly-ash pollution of soils around a coal-burning power plant: a case study in the Czech Republic. J Geochem Explor 66:291–297
Kapička A, Jordanova N, Petrovský E, Podrázský V (2003) Magnetic study of weakly contaminated forest soils. Water Air Soil Pollut 148:31–44
Kapička A, Jordanova N, Petrovský E, Ustjak S (2001) Effect of different soil conditions on magnetic parameters of power-plant fly ashes. J Appl Geophys 48:93–102
Kapička A, Petrovský E, Fialová H, Podrázský V, Dvořák I (2008) High resolution mapping of anthropogenic pollution in the Giant Mountains National Park using soil magnetometry. Stud Geophys Geod 52:271–284
Lecoanet H, Lévêque F, Segura S (1999) Magnetic susceptibility in environmental applications: comparison of field probes. Phys Earth Planet Inter 115:191–204
Magiera T, Strzyszcz Z (2000) Ferrimagnetic minerals of anthropogenic origin in soils of some Polish national parks. Water Air Soil Pollut 124:37–48
Magiera T, Strzyszcz Z, Kapička A, Petrovský E, MAGPROX Team (2006) Discrimination of lithogenic and anthropogenic influences on topsoil magnetic susceptibility in Central Europe. Geoderma 130:299–311
Maher BA, Taylor RM (1988) Formation of ultrafine – grained magnetite in soils. Nature 336:368–370
Maier G, Scholger R, Schon J (2006) The influence of soil moisture on magnetic susceptibility measurements. J Appl Geophys 59:162–175
MŽP R (2007) *Statistická ročenka životního prostředí České republiky 2006 – Statistical Environmental Yearbook of the Czech Republic 2006*. MZP CR, Praha 2007 (in Czech)
Oldfield F, Hunt A, Jones MDH, Chester R, Dearing JA, Olsson L, Prospero JM (1985) Magnetic differentiation of atmospheric dusts. Nature 317:516–518
Petrovský E, Kapička A, Jordanova N, Knab M, Hoffmann V (2000) Low-field magnetic susceptibility: a proxy method of estimating increased pollution of different environmental systems, Environ Geol 39:312–318
Petrovský E, Ellwood BB (1999) Magnetic monitoring of air, land and water pollution. In: Maher BA, Thompson R (eds) Quaternary climates, environments and magnetism. Cambridge University Press, Cambridge
Petříková V (1990) Výskyt imisí v ovzduší a obsah TK v zemědělských plodinách. Rostlinná výroba 36:367–378 (in Czech)
Plamínek J (2007) The geology. In: Flousek J, Hartmanová O, Štursa J, Potocki, J (eds) The giant Mts.- nature, history, life. Baset, Praha, pp 83–102. (in Czech)
Podrázský V (2001) Heavy metals and microelements content of humus forms in different regions of the Czech Republic. In: Borůvka L (ed) Soil Science: Past, Present and Future. Book of Abstracts. Czech University of Agriculture, Praha, p 97
Podrázský V, Vacek S, Mikeska M, Boček M (2007) Status and development of soils in the bilateral biosphere reserve Krkonoše. Opera Concortica 44:129–139
Schibler L, Boyko T, Ferdyn M, Gajda B, Holl S, Jordanova N, Rosler W, Magprox Team (2002) Topsoil magnetic susceptibility mapping: data reproducibility and compatibility, measurement strategy. Stud Geophys Geod 46:43–57
Schmidt A, Yarnold R, Hill M, Ashmore M (2005) Magnetic susceptibility as proxy for heavy metal pollution: a site study. J Geochem Explor 85:109–117
Scholger R (1998) Heavy metal pollution monitoring by magnetic susceptibility measurements applies to sediments of the river Mur (Styria, Austria). Eur J Environ Eng Geophys 3:25–37
Spiteri C, Kalinski V, Rösler W, Hoffmann V, Appel E, MAGPROX team (2005) Magnetic screening of pollution hotspot in the Lausity area, Eastern Germany: correlation analysis between magnetic proxies and heavy metal contamination in soils. Environ Geol 49:1–9
Strzyszcz Z (1999) Heavy metal contamination in mountain soils of Poland as a result of anthropogenic pressure. Biol Bull 26:593–605
Thompson R, Oldfield F (1986) Environmental magnetism. Allen and Unwin, London
Wang XS, Qin Y (2005) Correlation between magnetic susceptibility and heavy metals in urban topsoil: a case study from the city of Xuzhou, China. Environ Geol 49: 10–18

Magnetic Measurements on Maple and Sequoia Trees

28

Gunther Kletetschka

Abstract

Magnetic measurements of soil and tree bark adjacent to a busy highway revealed a significant variation in the concentration of magnetic particles with distance from the highway. Furthermore, forest-facing tree-bark contains significantly more magnetic particles than road-facing tree-bark. Magnetic particles were detected both on the bark of the maple trees and in the first centimeter of the soil cover (O/A horizon). Stability of the Saturation Isothermal Magnetization (SIRM) and the hysteresis parameters of the soil indicates the presence of Single-Domain/Pseudo-Single-Domain (SD/PSD) magnetic carriers. Measurements of the tree bark hysteresis parameters and SIRM detect a significantly lower coercivity component that we interpret to be an indication of more abundant PSD-type magnetic grains. Magnetic measurements around the perimeters of eight tree trunks reveal magnetic carriers whose distribution is antipodal to the source direction (highway). We interpret our observation by adopting an air circulation model, where suspended PSD/SD particles are carried in the air stream. The air stream from the heavy traffic lowers the amount of moisture on the tree trunk surfaces facing the highway and thus reduces an adhesive potential on this side. Therefore, more particles can stay on the moist side of the trunk protected from the direct airflow. A magnetic signature of tree rings was tested as a potential paleo-climatic indicator. We have examined wood from sequoia tree, located in Mountain Home State Forest, California, whose tree ring record spans over the period 600–1700 A.D. We have measured low and high-field magnetic susceptibility, the Natural Remanent Magnetization (NRM), Saturation Isothermal Remanent Magnetization (SIRM), and stability against thermal and Alternating Field (AF) demagnetization. Magnetic investigation of the 200 mm long sequoia material suggests that the magnetic efficiency of natural remanence (=natural remanent magnetization normalized by saturation remanence) may

G. Kletetschka (✉)
NASA's Goddard Space Flight Center, Code 691, Greenbelt, MD, USA; Catholic University of America, Washington, DC, USA; Institute of Geology, Academy of Sciences of the Czech Republic, Prague, Czech Republic
e-mail: gunther.kletetschka@gsfc.nasa.gov

E. Petrovský et al. (eds.), *The Earth's Magnetic Interior*, IAGA Special Sopron Book Series 1,
DOI 10.1007/978-94-007-0323-0_28,

be a sensitive paleoclimate indicator because it is substantially higher (in average $> 0.01 = 1\%$) during the Medieval Warm Epoch (700–1300 A.D.) than during the Little Ice Age (1300–1850 A.D.) where it is $< 0.01 = 1\%$. Diamagnetic behavior has been noted to be prevalent in regions with higher tree ring density. The mineralogical nature of the remanence carrier was not directly detected but maghemite is suggested due to low coercivity and absence of Verwey transition. Tree ring density, along with the wood's magnetic remanence efficiency, records the Little Ice Age (LIA), which is well documented in Europe and elsewhere. Magnetic analysis of the thermal stability reveals the blocking temperatures near 200°C. This phenomenon suggests that the remanent component in this tree may be thermal in origin and was controlled by local thermal conditions.

28.1 Introduction

The provenance distribution and localization of human contamination can be effectively described by utilization of magnetic measurements. We address this issue here by reporting on a curious antipodal effect on tree trunks near busy highways. We invoke an atmospheric circulation model to explain the location of magnetic particles antipodal to the highway facing part of the tree trunk.

Our objective was to confirm that heavy traffic has a measurable magnetic effect on surrounding vegetation surfaces. Contamination by air-borne ash in industrial areas is often reported to be easily detectable by magnetic techniques (Heller et al. 1998, Kapicka et al. 2000, Kapicka et al. 2001, Kapicka et al. 1999, Lecoanet et al. 2001, Magiera and Strzyszcz 2000, Schmidt et al. 2000, Strzyszcz 1999). Apart from pollution of coal-burning power plants we are considering pollution of heavy traffic common in highly industrial areas near large cities. The Washington – Baltimore Parkway is an example of a highway carrying millions of cars (no trucks) each year. Cars wear down at a rate proportional to both the age and the worsening car condition. Highways with high-density traffic crossing the wildlife preserves may accumulate more material from car erosion than local city streets with less traffic.

A cross section from the dead coast redwood tree (Sequoia sempervirens) in Mountain Home State Forest, California, was dendrochronologically cross-dated (tree ring record was compared with living trees that recorded part of the time when this tree was alive) and detected overall age period between 600 and 1700 A.D for this tree cross section. (Adams 1999, Laboratory of tree-ring research, personal communication). Tree ring density may detect climatic variations, however other factors, like fire frequency can also influence tree ring density (Stephens and Libby 2006). Fire paleo-frequency can be detected by variability in formation of pedogenic magnetic particles (Kletetschka and Banerjee 1995) as well as by variability of paleo-climatic recorders (Clark 1988). Uptake of iron via roots requires incorporation of iron-rich solution from the soil and relies on absorption by the root system. Sapwood is the physiologically active part of the xylem (wood). This is the tissue through which water with dissolved iron moves from the roots to the shoots. The heartwood is the older, nonliving central wood of a tree that does not conduct water. Once the sapwood becomes hardwood, it is thermally isolated from the outside environmental changes. Up to three or four annual growth rings of xylem may be active in water transport. Because water movement is related to transpiration, environmental factors such as soil moisture, air temperature, and relative humidity affect the rate of water movement.

Sequoia species are long lived and contain cellular mechanism capable of slowing down or even stop telomere attrition (Flanary and Kletetschka 2005). This is most likely due to cycling in telomerase activity especially within the root cells Flanary and Kletetschka 2006). Such a system should preserve a more or less constant condition of tree ring growth, not related to the tree age, creating ideal condition for climate proxy recorder. The precipitation of iron, therefore, should be related to the change of environment.

Climate change can cause rapid changes in microbacterial communities living within the soil, changing the water acidity, and causing the dissolved iron to precipitate and rather than taking parts in various proteins that manage iron equilibrium, Iron can be stored within the iron oxide particles as it was shown in mammals (Brem et al. 2006). This model creates a convenient test case for magnetic sensing of the tree tissue that may relate to climate changes. In this case, we do not consider atmospheric traffic pollution (Kletetschka et al. 2003) due to the remote location of the redwood specimen and tree age not experiencing the industrial revolution.

28.2 Material and Methods

We chose our sample area along the Washington-Baltimore Parkway, a major highway connecting the two large cities where trucks are prohibited from this road. The collecting area was just north of the exit to the Goddard Space Flight Center, MD at N39° 00.55·, W76° 51.72· (geographical coordinates). Samples of soil and tree bark (Red Maple, *Acer Rubrum*) were collected along a profile starting from the edge of the road (end of the shoulder pavement) and extending about 20 m into the adjacent area covered with deciduous trees (see Fig. 28.1). At first, five soil samples and ten tree-bark samples were collected from the eastern side of the highway. Soil samples were taken from the base of the tree on the roadside direction. Two bark samples were collected from each tree. One sample was from the forest-facing side of the tree and the other from the road-facing side. After detecting the antipodal magnetic signature pattern on the tree-bark samples, we collected an additional two soil samples and four tree-bark samples from the western side of the highway to confirm that the presence of the highway causes the observed pattern. Samples of soil contained the uppermost horizon (not exceeding depth of 1 cm) containing mostly organics (O/A-horizon). The tree bark was scraped with a commercial steel scraper and collected above 30 cm and below 160 cm from the ground. Scraping did not remove more than a 1 mm thick layer of the tree-bark tissue. Samples of soil and tree bark were placed into plastic containers ($\sim$15 cm^3), sealed (to prevent loss of the moisture content) and tested for magnetic hysteresis properties. Hysteresis loops were obtained with Vibrating Sample Magnetometer (VSM) with a LAKE SHORE controller. Maximum available field was $\pm$2 T. The stability of the saturation remanence was tested with alternating field demagnetization up to 0.24 T and measured using a super-conducting rock magnetometer. Magnetic measurements were completed within a week after the samples were collected.

One larger red maple (trunk approximately 30 cm in diameter) was used for measurements of magnetic susceptibility variation along its circumference at a height of 40 cm above the forest floor. These susceptibility measurements were done with a portable magnetic susceptibility meter SM20 (Geofyzika a.s.)

Fig. 28.1 Location map for Washington-Baltimore highway sample collection. North is up. The samples on the eastern side are m6 and m7, closer and further from the road, respectively. Samples on the western side are m1, m2, m3, m4, and m5, where sample number increases with the distance from the road. GPS coordinates are 39°00'38"N, 76°51'42"W and 45 m above the sea level. Purple arrows show traffic direction along with the wind direction

with a noise level of 10^{-6} SI. This handheld instrument has a coil under a case cover about 5 cm in diameter. Measurements are volume normalized. Tree bark and soil samples were measured using Sartorius balance with resolution of 0.1 mg.

A sample of the Sequoia sempervirens (m26 NE3–NE5) was obtained from professor Malcom Hughes on December 17, 1999, in the Laboratory of Tree-ring research, University of Arizona. The tree sample was collected in Mountain Home State Forest, California (SE of Fresno), and dated by Rex K. Adams. The specimen was cross-dated for the time interval between 950 and 1450 years. The rest of the years, estimated based on Fig. 28.8, is not cross-dated and may be associated with some errors (±5 years). In Fig. 28.8, one pinprick (blue dot) indicates the 10th year, two pinpricks in a vertical alignment indicate the 50th year, three pinpricks in a vertical alignment indicate the 100th year, and four pinpricks in a vertical alignment indicate the 1000th year. Magnetic measurements done on sequoia used cubic samples 1 cm in dimension. Number of tree ring of individual samples is shown in Fig. 28.9. Original tree ring analysis in context of absolute age was done by Adams (1999, Laboratory of tree-ring research, personal communication). For temperature dependent measurement (cooling and warming) we used Quantum Designs MPMS2 cryogenic susceptometer with dynamic range 1e-10 to 1e-3 Am^2. The equipment is based in Institute for Rock Magnetism at University of Minnesota, Sheppard Laboratories.

Samples were cut by handheld non-magnetic saw in Pruhonice Paleomagnetic Laboratory (PPL), Czech Republic. Wood was collected in year 1998 and stored for one year in Laboratory of Tree Ring Research, Arizona. Within four months after receiving these samples from Laboratory of Tree Ring Research, samples were cut and measured both at GSFC/NASA and PPL. During this time samples were kept in a dry box at GSFC and in relatively dry storage facility of PPL to avoid moisture exposure. During the process of measurement all parts of the sample holder were cleaned with ethyl alcohol and distilled water to insure absence of magnetic contamination.

The sequoia specimen was cut into a rod, about 600 mm long, with 100 mm^2 cross-section. A sharp scalpel was used to dissect the wooden rod into cubical samples (size ∼1 cm). Tree rings were counted within each cube to obtain tree ring density (Fig. 28.9). We estimated that for every 15 tree rings we may have missed or added an extra ring. This allows obtaining a signal-to-noise ratio value of 15/1. These cubes were used to obtain high-field magnetic susceptibility (Fig. 28.9) at GSFC. Low-field magnetic susceptibilities were obtained from the sister samples sent to PPL along with measurements of Natural Remanent Magnetization (NRM), Saturation Isothermal Remanent Magnetization (SIRM), and thermal/alternating field magnetic stability. High-field magnetic susceptibilities (Fig. 28.9) were obtained at GSFC/NASA from the magnetization change between 1 and 2 T field inside the Vibrating Sample Magnetometer (VSM Model 7300 by Lake Shore, 10 times averaging). The signal-to-noise ratio of this value was estimated to be 12/1 based on repetitive measurements of several samples. Both magnetic slope data and ring density data are approximated using a Stineman function. The output of this function then has a geometric weight applied to the current point and ± 10% of the data range, to arrive at the smoothed curve. Each time the empty holder was measured for the final sample correction. Figure 28.9 contains low-field magnetic susceptibility values from the sister samples measured in PPL using KLY-2 Kappabridge (Jelinek 1973) (frequency 920 Hz, field intensity 300 A/m).

Five samples with SIRM from the older section of the tree (600–1000 years A.D.) were demagnetized by alternating field up to 0.1 T with instrument Schonstedt GSD-1 demagnetizer and subsequently stepwise magnetized by field up to 0.5 T. All magnetic remanence measurements were done such that samples were kept oriented in respect to each other. NRM magnetizations from all samples stayed within ± 30° (see PPL data set, Fig. 28.13).

28.3 Results

28.3.1 Soil Samples

Hysteresis parameters for the soil samples are listed in Table 28.1. Figure 28.2 illustrates the variability of saturation magnetization (J_s) of the soil samples as the distance increases from the edge of the road. We only show the Eastern side for clarity. Hysteresis parameters for Western side are shown in Table 28.1. J_s has the largest value 27.6 e-3 Am^2/kg for the sample closest

Table 28.1 Hysteresis parameter for soils (m*X*s) and tree bark facing forest (m*X*bf) samples. "Side" indicates side of the highway from which the samples were collected. Distance is measured between the base of the tree and start of the pavement perpendicular to the highway. H_c is magnetic coercivity. J_s is saturation magnetization. SIRM is saturation remanence

Sample	Side	Distance [m]	H_c [mT]	J_s [Am^2 kg^{-1}]	SIRM [Am^2 kg^{-1}]
m1s	Eastern	6.2	12.6	27.6×10^{-3}	36.4×10^{-4}
m2s	Eastern	8.3	12.3	21.3×10^{-3}	24.6×10^{-4}
m3s	Eastern	11.2	10.6	16.2×10^{-3}	30.2×10^{-4}
m4s	Eastern	13.3	12.8	15.9×10^{-3}	34.6×10^{-4}
m5s	Eastern	19.3	10.5	16.5×10^{-3}	57.9×10^{-4}
m6s	Western	8.1	11.9	12.6×10^{-3}	25.6×10^{-4}
m7s	Western	9.7	10.6	5.4×10^{-3}	19.6×10^{-4}
m1bf	Eastern	6.2	4.2	0.45×10^{-3}	1.03×10^{-4}
m2bf	Eastern	8.3	9.3	0.75×10^{-3}	1.13×10^{-4}
m3bf	Eastern	11.2	6.5	1.53×10^{-3}	2.03×10^{-4}
m4bf	Eastern	13.3	7.0	1.38×10^{-3}	1.45×10^{-4}
m5bf	Eastern	19.3	2.0	0.27×10^{-3}	3.22×10^{-4}
m6bf	Western	8.1	5.4	1.53×10^{-3}	2.48×10^{-4}
m7bf	Western	9.7	8.5	0.87×10^{-3}	1.49×10^{-4}

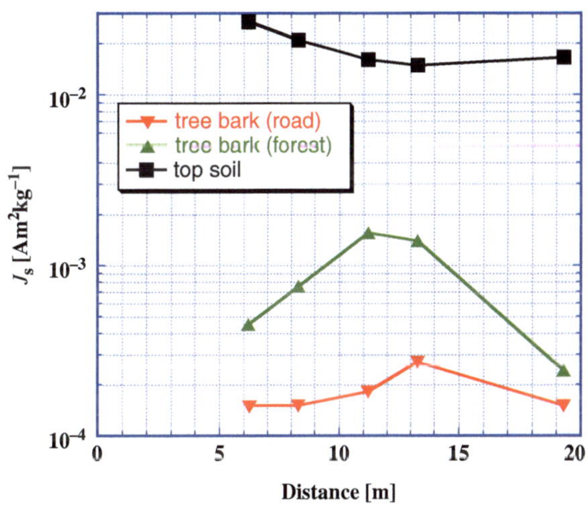

Fig. 28.2 Magnetic distances of the tree bark (red maple) and soil samples from the edge of the road. Upside triangle indicate road-side and upright triangle indicates forest-side of the tree. Note significantly enhanced magnetization of the tree bark facing towards the forest (away from the road)

to the road (6 m). Note that "e-3" indicates 10^{-3} in the rest of this manuscript. As the distance increased, the values of J_s dropped down to about 15.9 e-3 Am^2/kg and 16.5 e-3 Am^2/kg at distances 13.3 m and 19.3 m respectively. Coercivity stayed between 10 and 13 mT (Table 28.1). SIRM values ranged between (25 and 60) e-4 Am^2/kg and were subsequently demagnetized by alternating magnetic field (Figs. 28.3 and 28.4). There was no clear dependence of SIRM as a function of distance from the road. Saturation magnetization was well above the noise limit and maximum demagnetizing alternating field was 250 mT.

28.3.2 Tree Bark Samples

J_s of the forest-facing tree-bark samples increased to its maximum value 1.53 e-3 Am^2/kg at the distance of about 11 m from the road and then dropped to its lowest value 0.27 e-3 Am^2/kg at 19 m distance. J_s of the road-facing tree-bark samples was close to the VSM instrument resolution and thus embodies at least 20% of the measurement error. J_s also increased to its maximum at 13 m distance and stayed at values less than 2 e-4 Am^2/kg for the rest of the samples. Demagnetization of tree-bark samples' SIRM

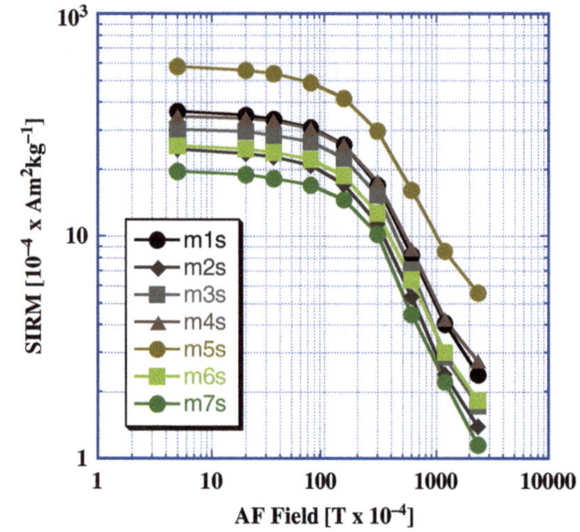

Fig. 28.3 Alternating magnetic field SIRM demagnetization of the soil samples. Samples 1–5 are eastern side of the road and 6 to 7 are samples from the western side of the road (see Fig. 28.1)

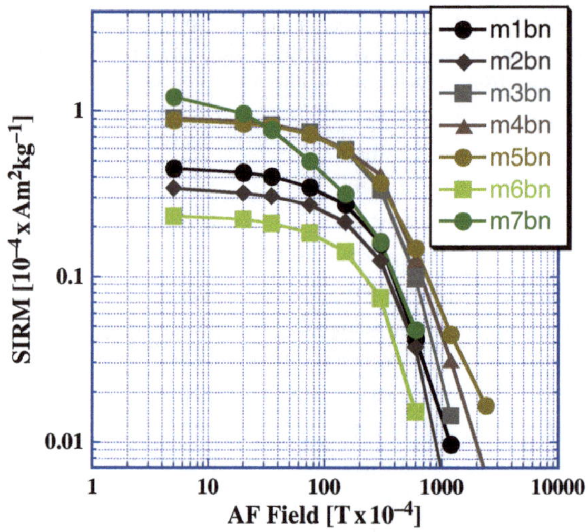

Fig. 28.5 Alternating magnetic field SIRM demagnetization of the highway-facing tree-bark samples. Samples 1 to 5 are eastern side of the road and 6–7 are samples from the western side of the road (see Fig. 28.1)

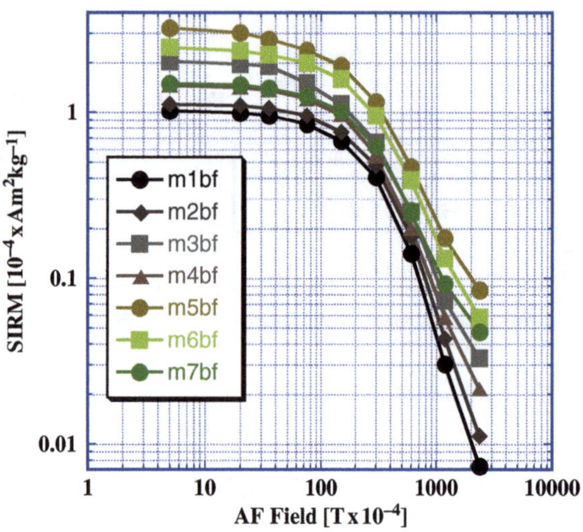

Fig. 28.4 Alternating magnetic field SIRM demagnetization of the forest-facing tree-bark samples. Samples 1 to 5 are eastern side of the road and 6–7 are samples from the western side of the road (see Fig. 28.1)

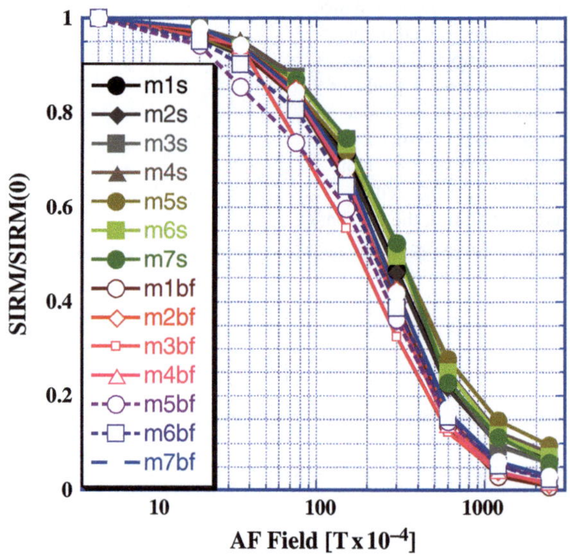

Fig. 28.6 Normalized demagnetization of SIRM of seven soil samples (*solid symbols*) and seven forest-facing tree-bark samples (*empty symbols*). Samples 1–5 are eastern side of the road and 6–7 are samples from the western side of the road (see Fig. 28.1)

is shown in Fig. 28.4 (forest-facing tree-bark samples) and Fig. 28.5 (road-facing tree-bark samples). Figure 28.6 shows normalized demagnetization curves for soil and tree-bark samples.

Magnetic susceptibility measurements around a trunk of the larger red maple tree (~30 cm in diameter) revealed an asymmetric pattern (Fig. 28.7). Magnetic susceptibility on the tree side facing the road was about –3.5 e-6 as oppose to –2.7 e-6 on the tree-side facing the forest. The error of this measurement was around 1 e-6 SI and we took each measurement 3 times to obtain standard deviations ~ 0.5 e-6 SI for each measurement. This tree was located on the western side of the highway at the distance of about 6 m from the road shoulder pavement. We did not collect any samples from this tree.

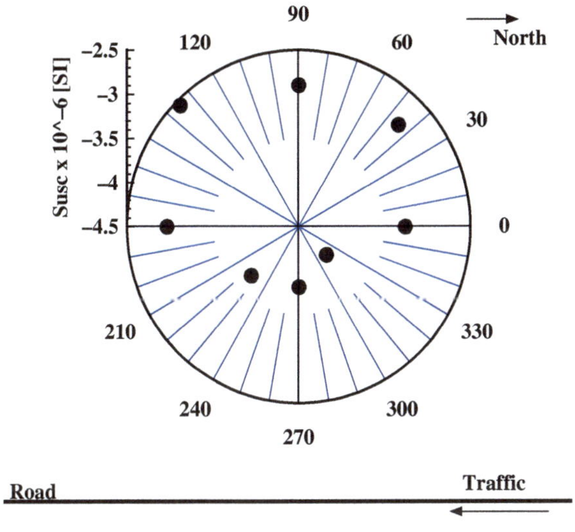

Fig. 28.7 Magnetic susceptibility measurements around the circumference of a tree trunk 30 cm in diameter. Note that the north direction is to the right. Lower line indicates the highway position in respect to susceptibility measurements. The arrow in the *lower right* corner indicates the approaching traffic direction

28.4 Discussion

As we expected the soil samples closest to the road (6 m) contained the largest amount of contamination due to proximity to the traffic. Hysteresis loops of natural materials are sensitive mostly to magnetite content, which is the most common form of oxidized iron on Earth. Even a large content (100 times more hematite than magnetite) of another common mineral, hematite, is essentially invisible in the magnetic hysteresis analysis and/or susceptibility measurements (Kletetschka 1994, Kletetschka et al. 2000) due to the large difference between the J_s of magnetite (90 Am^2/kg) and hematite (0.4 Am^2/kg). Because J_s of pure magnetite is well established (~90 Am^2/kg), the J_s of samples can be used to estimate approximate possible magnetite concentration. In principle some of the particles carried by air could be more reduced if particles are coming from the car's body parts because cars are made of steel. Particles could be also a product of combustion processes, however. Hysteresis measurements indicate sub-micron size magnetite grains and in any case, the large active surfaces of iron would cause quick oxidization to magnetite and/or hematite.

Because we had collected samples from both the eastern and western sides of the road (Fig. 28.1) we have had fairly good control as to whether any magnetic increase is due to local factors within a soil or if it is related to the traffic. Our data in Table 28.1 and Fig. 28.2 show that the J_s and thus the assumed possible magnetite concentration increase with the proximity of the road. The magnetic content increase with distance, from about 6.2–11.2 m, is reflected as an increase in J_s to 11.4 e-3 Am^2/kg (Table 28.1) and thus by an additional 0.1 g of possible magnetite per kg of the uppermost material of the soil (depth <1 cm).

Distributed abundance of this iron rich material in the uppermost organic layer may promote reactions that have various effects on the plants and bacteria that use this soil for cellular growth. Consequently the next step would be to identify the magnetic species. The presence of metallic contamination may interfere with the growth pattern in the vicinity of the highway. The presence of the fine metallic grains both in the soil and on the plants may have significant effects on expression of metal transporter genes and may cause unwanted changes (Assuncao et al. 2001, Badar et al. 2000, Ghaderian et al. 2000, Han et al. 2000, Silver and Phung 1996).

We consider that the steel scraper material may have contaminated the tree-bark samples. The maximum possible extent would be in the case where the road-facing bark has no magnetic carriers and all we see is the steel scraper contamination. In this extreme case the forest-facing samples would also carry about the same amount of steel scraper contamination. Then all of the additional contamination of the forest-facing samples must be due to traffic from the highway.

In order to explain the tree-bark magnetization asymmetry increase with distance we propose an air circulation model driven by the high-speed traffic on the highway to explain the observed magnetic pattern. The ferromagnetic fine particles are entrained in the "high velocity" wind. Encountering the trees enables turbulence behind the trees, lowering the velocity and consequently allowing the magnetic particles to be deposited on the tree bark. Magnetism is lower near the road because air circulation is intense not allowing magnetic particles to settle. With increasing distance from the road (10 m) the air speed is low enough to allow magnetic particles to settle on the tree bark. The tree bark surface facing the highway is exposed to the drying effect of the "high velocity" air stream. Cars on the highway produce an air motion. As the vehicle moves it compress air in front and the compressed air moves along the vehicle and diffuses into sides of the highway. In fact, the motion of the air with the vehicle can be seen on the US bridges over busy highways. Most of the US highways have flags over each direction of the highway. If there is not significant wind due to weather, flags on each side point the traffic direction and presumable are caused by the cars on the highway. As the resulting airstreams diffuses into sides of the highways it produces the drying effect that is confirmed by our measurements.

The backside of the tree is shielded from the diffusing air in the highway and retains more moisture than the front side of the tree, consequently providing a sticking surface for magnetic particles.

SIRM resistance against alternating field demagnetization (Fig. 28.6) indicates that the magnetic carriers are very stable and require AF fields as high as 20 mT to demagnetize to 50% of SIRM. This value is slightly lower than the field that is required to demagnetize the soil particles. This stability indicates mostly SD/PSD magnetic carriers in both materials with grain size close to 0.5 μm (Dunlop and Ozdemir 1997). Particles of this size are easily suspended in airflow and concur with our model hypothesis. However, both hysteresis parameters (Fig. 28.2, Table 28.1) and SIRM demagnetization (Fig. 28.6) indicate lower coercivity for the bark material. This suggests coarser magnetic grain size on the tree barks. These grain sizes would span into the pseudo-single-domain magnetic region based on SIRM/J_s ratios (Table 28.1) (Day et al. 1977). This observation indicates that magnetic grains trapped on the bark undergo less degree of weathering than the particles on the soil surface. Weathering and oxidation decreases the effective magnetic size of the carriers. The soil surface, in general, has a greater amount of moisture than the tree bark surfaces, and thus promotes more intense weathering than on the dryer tree-bark surfaces.

Negative values of the magnetic susceptibility survey of the larger tree trunk detect the diamagnetic signature of the carbon and water in the wood. However, diamagnetism varies around the tree (Fig. 28.7) with the lowest values on a side facing away from the road suggesting a presence of small amount of ferromagnetic particles. The observed asymmetry is consistent with our magnetic observation on smaller trees and indicates an increase of the content of ferromagnetic particles. This observation supports our atmospheric model.

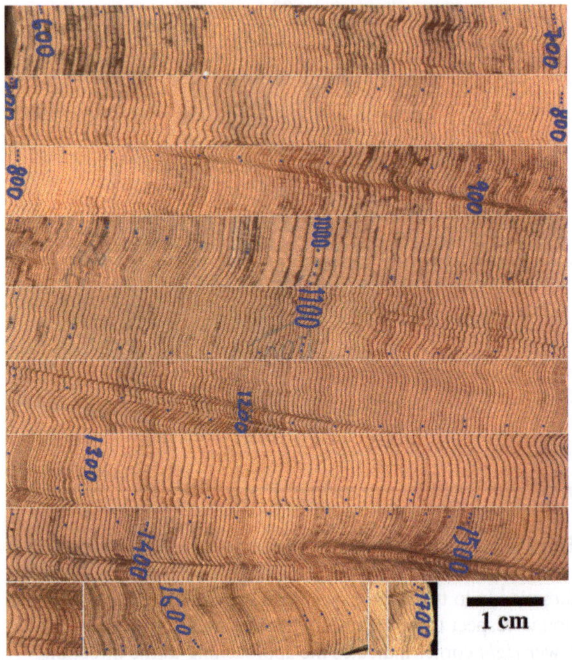

Fig. 28.8 Section of the tree sample that was used for magnetic measurements and tree ring density. Blue numbers indicate a year (AD) when the tree ring was created. Blue dots (pinpricks) help orientation in respect to individual tree ring ages. Each one dot is the 10th year, two vertical dots are the 50th year, three vertical dots are 100th year, and four vertical dots are the 1000th year

Tree Ring Density (TRD) was counted within individual specimens to obtain the tree ring density shown in Figs. 28.8 and 28.9. A more precise variation in tree ring density can be obtained by direct measurements of the tree ring size of the image shown in Fig. 28.8. However, the specific tree ring density should be specifically related to the samples used in the magnetic experiments.

TRD shown in Fig. 28.9 indicates several episodes where little wood material was added, suggesting much slower growth. The first episode is between 900 and 1000 A.D. The second is less pronounced and is between 1200 and 1300 A.D. The most dramatic increase in TRD is in the most recent section of the wood dated between 1400 and 1700 A.D. The relative TRD indicates rapid cellular growth when the climate was likely warmer and wetter. The most pronounced episode based on TRD is between 1300 and 1400 A.D, just near the end of the Medieval Warm Epoch (Bell and Walker 1992) and start of the Little Ice Age in North American Coast Mountains (Grove 2001) (see

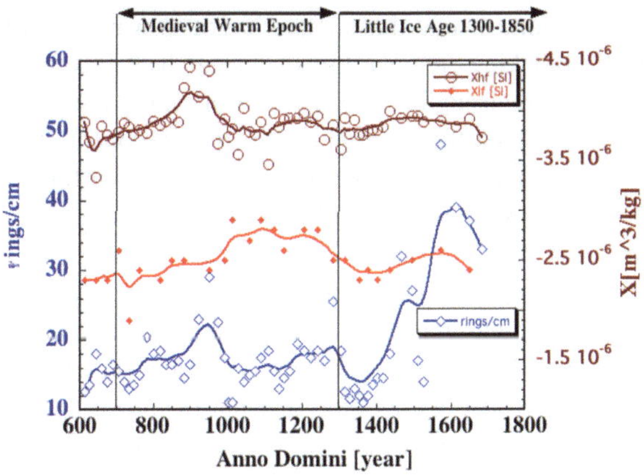

Fig. 28.9 Tree ring density (signal-to-noise ratio ~15/1) and both high- and low-field magnetic susceptibility (susceptibility record has signal-to-noise ratio ~12/1 based on repetitive measurements) are plotted as a function of age. *Top* of the diagram shows intervals for the Medieval Warm Epoch period (Bell and Walker, 1992) and for the Little Ice Age (Grove 2001). The data are approximated with Stineman function. The output of this function then has a geometric weight applied to the current point and ±10% of the data range, to arrive at the smoothed curve. This measurement was done at GSFC/NASA

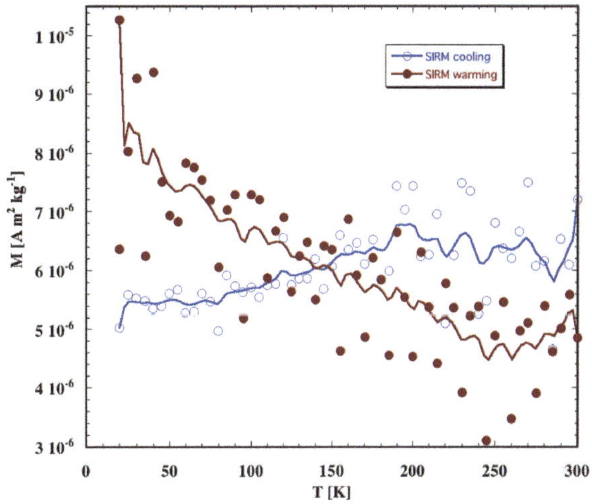

Fig. 28.10 The magnetic remanence of the sequoia is shown as function of temperature during a cryogenic warming (SIRM given at 20 K) and subsequent cryogenic cooling (SIRM given at 300 K). Solid symbols represent warming and empty symbols cooling. Adjacent data variation represents error of the measurement. The signal to noise ratio was 3/1. The data are approximated with the Stineman function. The output of this function then has a geometric weight applied to the current point and ±10% of the data range, to arrive at the smoothed curve. Data were taken at MPMS, Institute for Rock Magnetism, Minnesota

Fig. 28.9). Other periods where the climate favored the cellular proliferation are between 650 and 750 A.D. and between 1000 and 1150 A.D. Interestingly, these outlined episodes of contrasting cellular proliferation weakly correlate with high field diamagnetic susceptibility (the denser the tree rings the more diamagnetic material is present with linear correlation coefficient $R = 0.14$, (see Fig. 28.9). When cellular growth is suppressed the diamagnetic signature is intensified (Alternatively it could be that the amount of iron produced per volume is lowered). The material used for cellular growth may contain more carbon atoms, therefore raising the diamagnetic signature. For the period between 1400 and 1700 A.D., the diamagnetic enhancement is not as dramatic as it is for the tree ring density.

The data suggest that the larger the amount of magnetic carriers the larger the value of high-field diamagnetic slope. Diamagnetic and paramagnetic or superparamagnetic carriers can cause these slope variations. Since a significant part of the magnetic signature appears to be superparamagnetic, our data suggest that less dense wood contains more paramagnetic and superparamagnetic material irrespective of the amount of magnetic remanence carriers (presumably saturated) present.

The cryogenic experiments on MPMS suggest continuous unblocking of the remanence on warming due to the presence of superparamagnetic grains (Fig. 28.10). There is no indication of the Verwey transition. Cooling of Room-temperature SIRM resulted in no significant change in remanence magnetization (Fig. 28.10). The noise level is apparent from the diagram in Fig. 28.10 and amounts to 1–2 e-4 A m^2kg^{-1} and could mask Verwey transition. We did not attempt to image the magnetic carrier as it is likely to be only visible by transmission electron microscopy and we do not have such a facility available currently.

Individual magnetic remanence records (NRM and SIRM) were too noisy to infer any climatic relationship (Fig. 28.11) between the magnetization and tree ring density. Magnetization amplitudes (SIRM) were consistent with the remanence measured at different temperatures with the Quantum Design MPMS instrument, where the SIRM at room temperature corrected for density (500 kg/m^3) is near 3 mA/m (see Fig. 28.10). In summary, NRM and SIRM records shown in Fig. 28.11 indicate that there is no significant correlation between magnetizations and tree ring density.

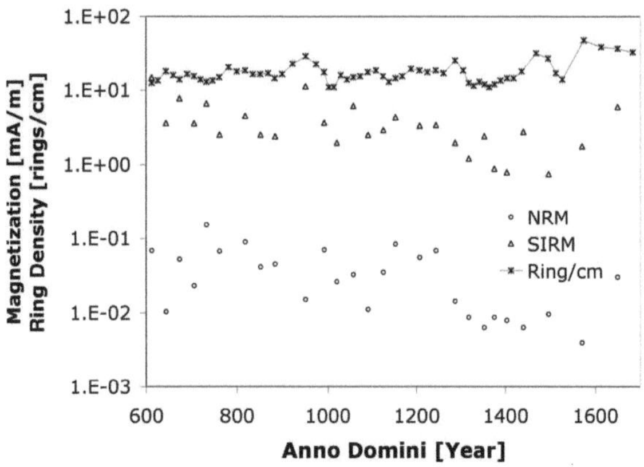

Fig. 28.11 Ring density data are compared with Natural remanent magnetization (NRM, noise limit is 0.0024 mA/m on JR-5A spinner) and Saturation Remanent Magnetization (SIRM) for samples of the wood section sent to Pruhonice laboratory, Prague/Geologic Institute, Czech Republic

28.5 Magnetic Efficiency

The precipitation of the NRM carriers may be completely unrelated to the paleoclimate. This is supported by results in Fig. 28.5. Therefore, we decided to test the wood samples for magnetization efficiency (the NRM/SIRM ratio), which often reveals more detailed magnetic remanence characteristics in terms of the Thermo-Remanent Magnetization (TRM) component. Note that the SIRM is about 75 times larger than the NRM (Fig. 28.5). This is similar to the efficiencies where the NRM remanence of thermal origin (Kletetschka et al. 2004, Kletetschka et al. 2006). TRM is when material is heated above the blocking temperature of the residing magnetic carriers and subsequently cooled down in an ambient magnetic field. Chemical Remanent Magnetization (CRM) has similar physics of remanence acquisition. Magnetic grains grow into the larger volumes during the convenient chemical conditions in ambient temperature. Once the particles' volume reaches the single-domain magnetic state the CRM component is blocked and therefore sample acquires the CRM component.

Another possibility, that we may not be able to exclude, is combination of chemical and a Detridal Remanent Magnetization (DRM). The magnetic particle precipitates from the solution within the tree cell and acquires the CRM. Then it settles within the cellular tissue and acquires the DRM. In the further text we will assume the CRM present.

It is conceivable that magnetic grains warm above their blocking temperatures during heat anomaly events (fire/drought or seasonal heat wave) and cool down to block the thermal remanent component. The relative magnetic signature fluctuation is noisier for the NRM data set than for the SIRM data set (Fig. 28.11) and this possibly relates to a demagnetization event that may have influenced the original remanence after it was acquired in nature. We infer several NRM components. The first component is chemical remanence (CRM) because we assume that the magnetic minerals had to be formed from within the tree tissue below the blocking temperatures of the remanence carriers. Under ambient temperature a significant fraction of the grains is likely to be in the superparamagnetic (SP) domain state with grain size <30 nm. Therefore, the second NRM component can be partial thermoremanent magnetization with blocking temperatures that span across the ambient temperatures. Some evidence supporting this claim is shown in Fig. 28.12, where the SIRM is rapidly declining during the partial thermal demagnetization in ambient air. The rapid decay is due to thermal unblocking of the remanence. Because of the steepness of the slope, it may be conceivable that daily heating due to weather may continuously block and unblock parts of the remanence. When anomalous heating has been recorded it is sealed from the future thermal fluctuations by additional wood growth. We must say, however, that prolonged storage of these samples containing SP grains could also generate a third component of Viscous Remanent Magnetization (VRM) adding to the overall value of the NRM. For example assuming that the subset of these samples had large amount of SP grains, they would be prone to viscous re-magnetization, most likely randomizing the original signature. If this would be the case we would observe significant directional deviation of the NRM within the sample set. In Fig. 28.12, we plot the directions of the Pruhonice data set and show that only few samples (6 out of 27) may have considerable directional change, and indeed, the angular deviation of the largest outliers is associated with the low sample efficiencies in Fig. 28.14.

Samples labeled as AD 952 and AD 613 in Fig. 28.12 show anomalous behavior, both in susceptibility (positive on start) and in shape of the demagnetizing curve (bell-like shape) suggesting a presence of very fine magnetic grains on the surface of these samples that quickly oxidizes and or gets removed

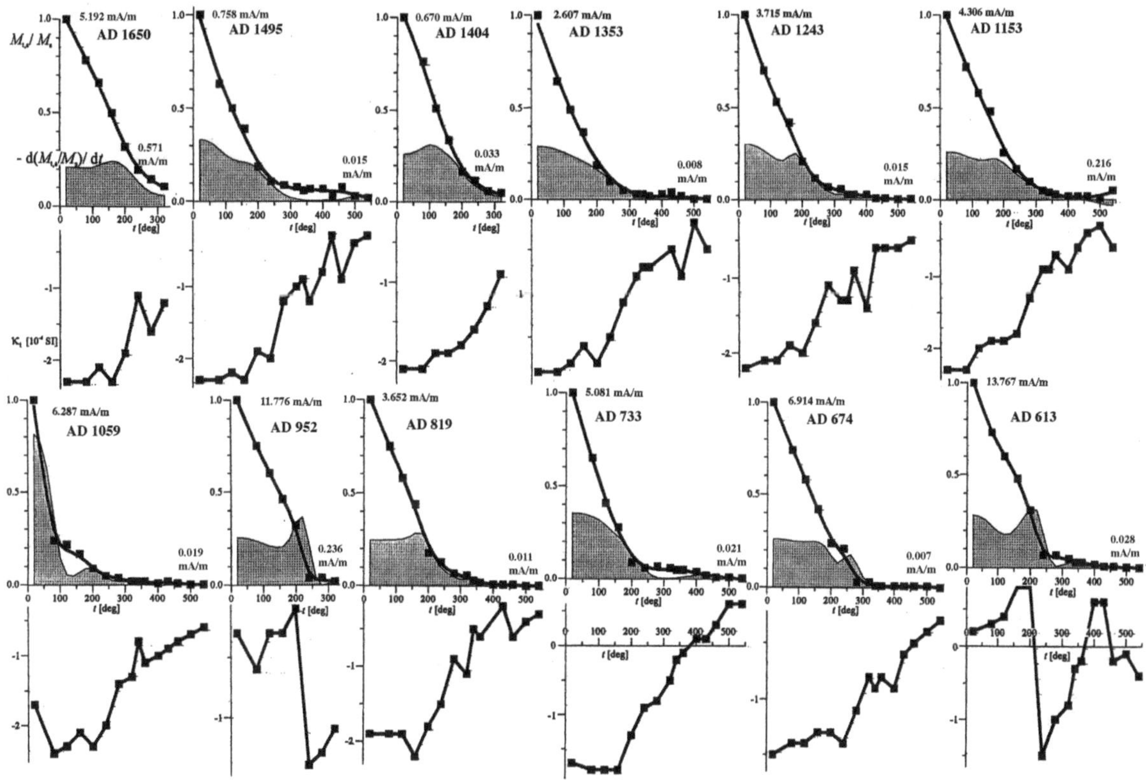

Fig. 28.12 Thermal demagnetization of Saturation Isothermal Remanent Magnetization of selected sequoia samples, indicated by age, normalized by the magnetic remanence at room temperature (Mt,s/Ms). Derivative is based on the smoothed trend of thermal demagnetization. Magnetic susceptibility after each heating step is shown below remanence plots sharing the temperature axis. The remanence data are approximated with the Stineman function. The output of this function then has a geometric weight applied to the current point and ±10% of the data range, to arrive at the smoothed curve. Data are taken at Pruhonice Paleomagnetic Laboratory (noise limit on JR-5A spinner is 0.0024 mA/m)

during the sample handling when heated over 200°C. Note that these two samples have a much higher initial SIRM intensity (12 mA/m and 14 mA/m respectively). Therefore, heating causes a rapid susceptibility swing to negative values (the ferromagnetic part is removed from hysteresis plots) as well as bell-shaped decay curves (samples AD 952 and AD 613 in Fig. 28.12). Most of other samples have smaller initial magnetization and do not show such anomalous behavior. Heating is associated with overall reduction and stabilizing iron-rich complexes into oxide minerals. New ferromagnetic material is evidenced by removal of the diamagnetic component in susceptibility plots (Fig. 28.12). The true nature of the susceptibility and mineralogy of the remanence carriers is speculative, however. We suggest the presence of maghemite due to low thermal magnetic stability along with the absence of the Verwey transition (Figs. 28.3 and 28.5).

Demagnetization by an alternating demagnetizing field along with an IRM acquisition (Fig. 28.15) shows uniform behavior across 5 samples (AD 1059, AD 974, AD 818, AD 691, and AD 613). Sample AD 613 has initial remanence of 14.75 mA/m. Such a high SIRM value may be related to contamination similar to the one shown for samples AD 952 and AD 613 during the thermal heating. This sample showed slightly more positive interaction compared with other samples, which would be consistent with surface contamination where the contaminants are likely clustered together rather then evenly distributed over the surface. The medium demagnetizing field is about 0.01 T suggesting higher coercivity and therefore overall remanent stability. Such behavior indicates that the viscous overprint may be not relevant for these samples.

The rapid decrease of remanence due to slight heating reveals that there may be thermal history recorded

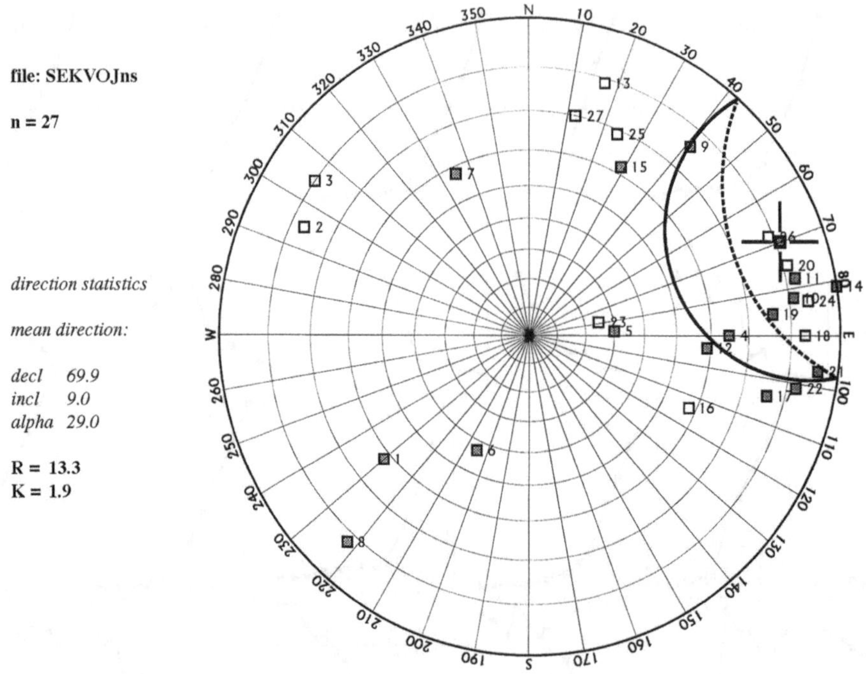

Fig. 28.13 Directional consistency of the sequoia samples measured in the Pruhonice Paleomagnetic Laboratory. Solid symbols indicate positive inclination, empty symbols negative inclination. Numbers denote relate to the following approximate ages: 1 = year 1650, 2 = year 1613, 3 = year 1495, 4 = year 1436, 5 = year 1389, 6 = year 1376, 7 = year 1352, 8 = year 1316, 9 = year 1285, 10 = year 1242, 11 = year 1224, 12 = year 1153, 13 = year 1138, 14 = year 1110, 15 = year 1059, 16 = year 1030, 17 = year 1003, 18 = year 974, 19 = year 899, 20 = year 851, 21 = year 818, 22 = year 782, 23 = year 747, 24 = year 720, 25 = year 691, 26 = year 660, 27 = year 613

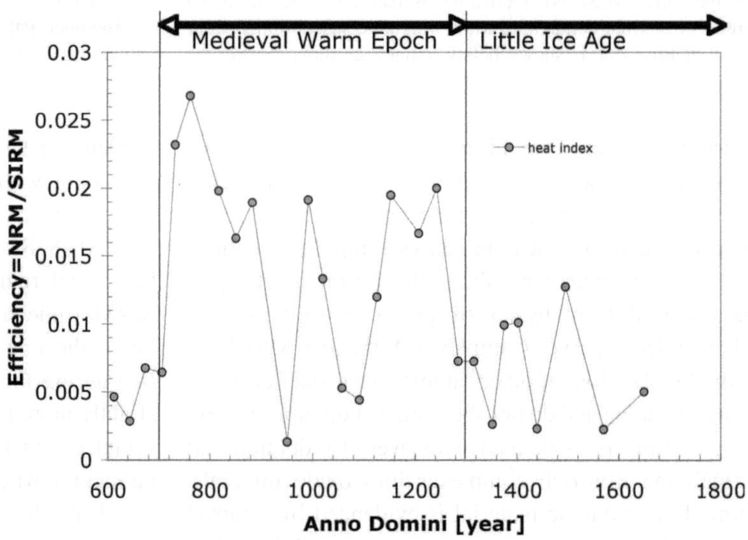

Fig. 28.14 Efficiency of magnetization that approximate the "heat index" is plotted as a function of age for samples of the Sequoia sempervirens. Top of the diagram shows intervals for the Medieval Warm Epoch period (Bell and Walker 1992) and for the Little Ice Age (Grove 2001). Error bars are less than 10% of the data shown

within the wood samples. A remanence record can be characterized by the plotting efficiency (Fig. 28.14) of the Natural Remanent Magnetization (NRM/SIRM). In regular conditions the efficiency of TRM-component of NRM should be very close to 1–2% (Bell and Walker 1992, Kletetschka et al. 2003). However, this

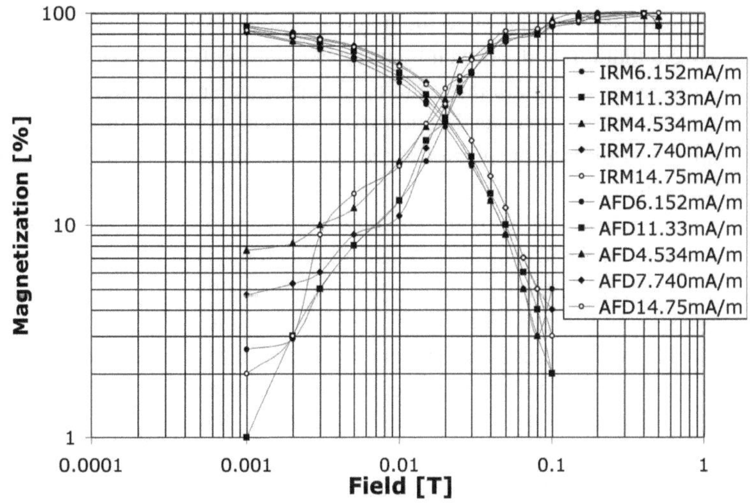

Fig. 28.15 Alternating field demagnetization and Isothermal Remanence acquisition (IRM – volume normalized) measured on JR-5A spinner magnetometer (noise limit 0.0024 mA/m) at the paleomagnetic lab, Pruhonice, Czech Republic. Saturation magnetization is shown in the legend. These 5 samples come from the section with tree rings in the range of 600–1100 years A.D. From the top down they correspond to the following years; 1059, 974, 818, 691, and 613

value often drops well below 0.01 in Fig. 28.14. The efficiency of the CRM component of NRM is likely to be less than efficiency of the TRM component. This is because, if superparamagnetic (SP) grains are present, the saturation magnetizations of the SP grains that do not normally contribute to CRM, cause the SP grains to interact among each other and contribute to the overall saturation remanence. If the SP grains are not saturated (CRM), they do not sense each other and therefore do not contribute to the overall CRM signature. For this reason, the CRM component of efficiency in a sample containing SP grains must be lower.

We propose that there are sections of the wood that have been heated in the past by climate variation (including fire), inducing the partial TRM (pTRM) component into the wood. Plotting the efficiency in Fig. 28.14 depicts wood sections that have been affected by heat and identify these sections as with larger efficiency. Thus the samples with an efficiency exceeding 0.01 would be likely to have undergone some thermal event, strengthening its NRM intensity by mild heating (e.g. forest fire or climate change). This proposed variation is consistent with the rapid unblocking of remanence seen in Fig. 28.12. Our identification of the highly thermally-dependent magnetic signature opens the possibility that trees may contain pTRM record. Since trees are generally only a few thousand years old (e.g. sequoias) the time period may be short enough for a tree to record stable thermal history. Our data outlines this possibility of thermal record preservation and suggests a more general testing of such hypothesis.

We note that the size of the samples is rather large compared to the growth rings and leads to an aliasing of the magnetic record. Using the average tree ring density in Fig. 28.9 as 15 rings per centimeter we estimate that the NRM magnetic signature per one tree ring to be in the range of 10-3 mA/m, which is below the limit of our instruments. However, there are more sensitive instruments being developed by 2G and Quantum Design and paleomagnetic signature of the individual tree rings may not be impossible in near future. In this study the aliasing effect may cause some reduction of significant anomalous climatic events happening on scale smaller than ~15 years. We note, however, that our proposed thermal mechanism for the partial TRM of the tree samples may also cause some degree of aliasing on the tree ring scale due to thermal flow inwards that would be competing with natural cooling capacity of the tree to maintain lower temperature than the ambient temperature.

Conclusions

Magnetic screening of the soil and arboreal vegetation (notable maple trees) in vicinity of a highway with heavy traffic reveals a measurable magnetic contamination likely associated with the vehicular traffic. Contamination of the soil is the heaviest just next to the road and rapidly decreases with distance into the tree area. The concentration change across the measured profile is estimated as 0.1 g per kilogram of the uppermost soil material. Tree bark measurements reveal an interesting antipodal magnetic expression with magnetic enhancement on the opposite side in relation to the magnetic influx. An air-circulation model driven by traffic and humidity variations can explain this pattern. This mechanism must be taken into consideration when studying the spread of disease by aerosol in the agricultural and forest industry.

Wood material from the *Sequoia sempervirens* contains variable tree ring density indicating the environmental changes and health status during the life span of the tree. The tree ring density correlates weakly with the high-field magnetic susceptibility, suggesting accumulation of the diamagnetic material within the zones of high tree ring density. This correlation is less pronounced near the tree perimeter possibly due to proximity of the tree section that was living at the time of the tree death. Correlation with the remanence is absent (Fig. 28.11). Low temperature variations of SIRM suggest continuous unblocking of the remanence during heating and thus lowering the sample intensity in the observed NRM measurements. NRM/SIRM measurements suggest the presence of thermal component of the remanence within the trees. NRM tends to fluctuate to larger amplitudes than the SIRM. This fluctuation may be due to blocking temperature that is very close to room temperature. Therefore, we have attempted to use remanence efficiency to characterize a thermal exposure history of the tree. Such an approach offers a potentially important climate proxy, the record of the peak temperature during the tree ring formations. This assumes that the fluids transporting the nutrients to the tree efficiently cool the interior of the tree and only the very exterior part is exposed to more extreme environmental changes.

The high-field susceptibility variations together with the Tree Ring Density (TRD) is a proxy that documents in detail the cold oscillation between 900 and 1000 A.D. preceding the Medieval Warm Epoch that lasted to the end of the 14th century. The proxies obtained from the Sequoia sempervirens magnetic efficiency (Fig. 28.14) were able to record the steep climate cooling triggered by the Little Ice Age after ca 1400 A.D., which is in agreement with reconstructed Northern Hemispheric temperatures (Moberg et al. 2005, Brazdil 1996).

This work is mostly an exploratory attempt to see if the magnetic analysis of trees may be useful. Our report suggests that there is a signature that may reflect the thermal history and that the tree contains magnetic carriers that recorded ambient field at the time of growth. It may be that in the future, more detailed research of individual tree rings could reveal differences in the magnetic environment at the time of magnetization origin. This could possibly aid in the knowledge of the historical geomagnetic field variation as well as the detailed magnetic field fluctuation on a yearly basis.

Acknowledgements Special thanks to Vojtech Zila, who helped in the collecting of the tree samples, Peter Wasilewski, who provided the magnetic facility, Peter Pruner, Daniela Venhodova, who obtained data in Pruhonice Laboratory, and Jaroslav Kadlec, who helped with the climatic interpretation.

References

Assuncao AGL, Martins PD, De Folter S, Vooijs R, Schat H, Aarts MGM (2001) Elevated expression of metal transporter genes in three accessions of the metal hyperaccumulator Thlaspi caerulescens. Plant Cell Environ 24(2): 217–226

Badar U, Ahmed N, Beswick AJ, Pattanapipitpaisal P, Macaskie LE (2000) Reduction of chromate by microorganisms isolated from metal contaminated sites of Karachi, Pakistan. Biotechnol Lett 22(10):829–836

Bell M, Walker MJ (1992) Late quaternary environmental change. Physical and human perspectives. Pearson Education Ltd., Harlow, Essex, UK

Brazdil R (1996) Reconstruction of past climate from historical sources in the Czech Lands. In: Jones PD, Bradley RS, Jouzel J (eds) Climatic variation and forcing mechanism of the last years. Springer, Berlin, pp 409–431

Brem F, Tiefenauer L, Fink A, Dobson J, Hirt AM (2006) A mixture of ferritin and magnetite nanoparticles mimics the magnetic properties of human brain tissue. Phys Rev B 73, Article no. 224427

Clark JS (1988) Effect of climate change on fire regimes in Northwestern Minnesota. Nature 334:233–235

Day R, Fuller M, Schmidt VA (1977) Hysteresis properties of titanomagnetites: grain-size and compositional dependence. Phys Earth Planetary Inter 13:260–266

Dunlop JD, Ozden Ozdemir (1997) Rock magnetism fundamental and frontiers. Cambridge University Press, Cambridge, UK, p 573

Flanary BE, Kletetschka G (2005) Analysis of telomere length and telomerase activity in tree species of various lifespans, and with age in the bristlecone pine Pinus longaeva. Biogerontology 6:101–111

Flanary BE, Kletetschka G (2006) Analysis of telomere length and telomerase activity in tree species of various lifespans, and with age in the bristlecone pine Pinus longaeva. Rejuvenation Res 9:61–63

Ghaderian YSM, Lyon AJE, Baker AJM (2000) Seedling mortality of metal hyperaccumulator plants resulting from damping off by Pythium spp. New Phytol 146(2):219–224
Grove JM (2001) The onset of the Little Ice Age. In: Jones PD, Ogilvie AEJ, Davies TD, Briffa KR (eds) History and climate memories of the future? Kluwer Academic/Plenum Publishers, New York, NY, pp 153–185
Han FX, Kingery WL, Selim HM, Gerard PD (2000) Accumulation of heavy metals in a long-term poultry waste-amended soil. Soil Sci 165(3):260–268
Heller F, Strzyszcz Z, Magiera T (1998) Magnetic record of industrial pollution in forest soils of Upper Silesia, Poland. J Geophys Res Solid Earth 103(B8):17767–17774
Jelinek V (1973) Precision Ac bridge set for measuring magnetic susceptibility of rocks and its anisotropy. Stud Geophys Geod 17(1):36–48
Kapicka A, Jordanova N, Petrovsky E, Ustjak S (2000) Magnetic stability of power-plant fly ash in different soil solutions. Phys Chem Earth Part A Solid Earth Geodesy 25(5):431–436
Kapicka A, Petrovsky E, Jordanova N, Podrazsky V (2001) Magnetic parameters of forest top soils in Krkonose Mountains, Czech Republic. Phys Chem Earth Part A Solid Earth Geodesy 26(11–12):917–922
Kletetschka G, Acuna MH, Kohout T, Wasilewski PJ, Connerney JEP (2004) An empirical scaling law for acquisition of thermoremanent magnetization. Earth Planet Sci Lett 226:521–528
Kletetschka G, Fuller MD, Kohout T, Wasilewski PJ, Herrero-Bervera E, Ness NF, Acuna MH (2006) TRM in low magnetic fields: a minimum field that can be recorded by large multidomain grains. Phys Earth Planetary Inter 154:290–298
Kapicka A, Petrovsky E, Ustjak S, Machackova K (1999) Proxy mapping of fly-ash pollution of soils around a coal- burning power plant: a case study in the Czech Republic. J Geochem Exploration 66(1–2):291–297
Kletetschka G (1994) A study of the origin and nature of secondary iron oxides in Chinese loess. Master thesis, University of Minnesota, Minneapolis, MN
Kletetschka G, Banerjee SK (1995) Magnetic stratigraphy of Chinese Loess as a record of natural fires. Geophys Res Lett 22:1341–1343
Kletetschka G, Kohout T, Wasilewski PJ (2003) Magnetic remanence in the Murchison meteorite. Meteoritics Planet Sci 38(3):399–405
Kletetschka G, Wasilewski PJ, Taylor PT (2000) Hematite vs. magnetite as the signature for planetary magnetic anomalies? Phys Earth Planet Inter 119(3–4):259–267
Kletetschka G, Zila V, Wasilewski PJ (2003) Magnetic anomalies on the tree trunks. Stud Geophys Geod 47:371–379
Lecoanet H, Leveque F, Arnbrosi JP (2001) Magnetic properties of salt-marsh soils contaminated by iron industry emissions (southeast France). J Appl Geophys 48(2):67–81
Magiera T, Strzyszcz Z (2000) Ferrimagnetic minerals of anthropogenic origin in soils of some Polish national parks. Water Air Soil Pollut 124(1–2):37–48
Moberg A, Sonechkin DM, Holmgren K, Datsenko NM, Karlen W (2005) Highly variable Northern Hemisphere temperatures reconstructed from low- and high-resolution proxy data. Nature 433:613–617
Schmidt MWI, Knicker H, Hatcher PG, Kogel-Knabner I (2000) Airborne contamination of forest soils by carbonaceous particles from industrial coal processing. J Environ Qual 29(3):768–777
Silver S, Phung LT (1996) Bacterial heavy metal resistance: new surprises. Ann Rev Microbiol 50:753–789
Stephens SL, Libby WJ (2006) Anthropogenic fire and bark thickness in coastal and island pine populations from Alta and Baja California. J Biogeogr 33:648–652
Strzyszcz Z (1999) Heavy metal contamination in mountain soils of Poland as a result of anthropogenic pressure. Izvestiya Akademii Nauk Seriya Biologicheskaya (6):722–735

Index

E. Petrovský et al. (eds.), *The Earth's Magnetic Interior*, IAGA Special Sopron Book Series 1,
DOI 10.1007/978-94-007-0323-0, © Springer Science+Business Media B.V. 2011

MIX
Papier aus verantwortungsvollen Quellen
Paper from responsible sources
FSC® C105338
FSC
www.fsc.org

If you have any concerns about our products,
you can contact us on
ProductSafety@springernature.com

In case Publisher is established outside the EU,
the EU authorized representative is:
Springer Nature Customer Service Center GmbH
Europaplatz 3, 69115 Heidelberg, Germany

Printed by Libri Plureos GmbH
in Hamburg, Germany